Inv.-Nr. C-16-7
Informatik
Prof. Dr. A. Zeller

Data Mining

A Knowledge Discovery Approach

Data Mining

A Knowledge Discovery Approach

Krzysztof J. Cios
Witold Pedrycz
Roman W. Swiniarski
Lukasz A. Kurgan

Krzysztof J. Cios
Virginia Commonwealth University
Computer Science Dept
Richmond, VA 23284
& University of Colorado
USA
kcios@vcu.edu

Witold Pedrycz
University of Alberta
Electrical and Computer
Engineering Dept
Edmonton, Alberta T6G 2V4
CANADA
pedrycz@ee.ualberta.ca

Roman W. Swiniarski
San Diego State University
Computer Science Dept
San Diego, CA 92182
USA
& Polish Academy of Sciences
rswiniar@sciences.sdsu.edu

Lukasz A. Kurgan
University of Alberta
Electrical and Computer
Engineering Dept
Edmonton, Alberta T6G 2V4
CANADA
lkurgan@ece.ualberta.ca

Library of Congress Control Number: 2007921581

ISBN-13: 978-0-387-33333-5   e-ISBN-13: 978-0-387-36795-8

Printed on acid-free paper.

© 2007 Springer Science+Business Media, LLC
All rights reserved. This work may not be translated or copied in whole or in part without the written permission of the publisher (Springer Science+Business Media, LLC, 233 Spring Street, New York, NY 10013, USA), except for brief excerpts in connection with reviews or scholarly analysis. Use in connection with any form of information storage and retrieval, electronic adaptation, computer software, or by similar or dissimilar methodology now known or hereafter developed is forbidden.

The use in this publication of trade names, trademarks, service marks, and similar terms, even if they are not identified as such, is not to be taken as an expression of opinion as to whether or not they are subject to proprietary rights.

9 8 7 6 5 4 3 2 1

springer.com

To Konrad Julian – so that you never abandon your inquisitive mind
KJC

To Ewa, Barbara, and Adam
WP

To my beautiful and beloved wife Halinka and daughter Ania
RWS

To the beautiful and extraordinary pianist whom I accompany in life, and to my brother and my parents for their support
LAK

Table of Contents

Foreword xi

Acknowledgement xv

Part 1 Data Mining and Knowledge Discovery Process   **1**

   Chapter 1. Introduction 3
         1. What is Data Mining? 3
         2. How does Data Mining Differ from Other Approaches? 5
         3. Summary and Bibliographical Notes 6
         4. Exercises 7

   Chapter 2. The Knowledge Discovery Process 9
         1. Introduction 9
         2. What is the Knowledge Discovery Process? 10
         3. Knowledge Discovery Process Models 11
         4. Research Issues 19
         5. Summary and Bibliographical Notes 20
         6. Exercises 24

Part 2 Data Understanding   **25**

   Chapter 3. Data 27
         1. Introduction 27
         2. Attributes, Data Sets, and Data Storage 27
         3. Issues Concerning the Amount and Quality of Data 37
         4. Summary and Bibliographical Notes 44
         5. Exercises 46

   Chapter 4. Concepts of Learning, Classification, and Regression 49
         1. Introductory Comments 49
         2. Classification 55
         3. Summary and Bibliographical Notes 65
         4. Exercises 66

   Chapter 5. Knowledge Representation 69
         1. Data Representation and their Categories: General Insights 69
         2. Categories of Knowledge Representation 71
         3. Granularity of Data and Knowledge Representation Schemes 76
         4. Sets and Interval Analysis 77
         5. Fuzzy Sets as Human-Centric Information Granules 78

		6. Shadowed Sets	82
		7. Rough Sets	84
		8. Characterization of Knowledge Representation Schemes	86
		9. Levels of Granularity and Perception Perspectives	87
		10. The Concept of Granularity in Rules	88
		11. Summary and Bibliographical Notes	89
		12. Exercises	90

Part 3 Data Preprocessing — 93

Chapter 6. Databases, Data Warehouses, and OLAP 95
   1. Introduction 95
   2. Database Management Systems and SQL 95
   3. Data Warehouses 106
   4. On-Line Analytical Processing (OLAP) 116
   5. Data Warehouses and OLAP for Data Mining 127
   6. Summary and Bibliographical Notes 128
   7. Exercises 130

Chapter 7. Feature Extraction and Selection Methods 133
   1. Introduction 133
   2. Feature Extraction 133
   3. Feature Selection 207
   4. Summary and Bibliographical Notes 228
   5. Exercises 230

Chapter 8. Discretization Methods 235
   1. Why Discretize Data Attributes? 235
   2. Unsupervised Discretization Algorithms 237
   3. Supervised Discretization Algorithms 237
   4. Summary and Bibliographical Notes 253
   5. Exercises 254

Part 4 Data Mining: Methods for Constructing Data Models — 255

Chapter 9. Unsupervised Learning: Clustering 257
   1. From Data to Information Granules or Clusters 257
   2. Categories of Clustering Algorithms 258
   3. Similarity Measures 258
   4. Hierarchical Clustering 260
   5. Objective Function-Based Clustering 263
   6. Grid - Based Clustering 272
   7. Self-Organizing Feature Maps 274
   8. Clustering and Vector Quantization 279
   9. Cluster Validity 280
   10. Random Sampling and Clustering as a Mechanism
       of Dealing with Large Datasets 284
   11. Summary and Biographical Notes 286
   12. Exercises 287

Chapter 10. Unsupervised Learning: Association Rules 289
   1. Introduction 289
   2. Association Rules and Transactional Data 290
   3. Mining Single Dimensional, Single-Level Boolean Association Rules 295

4. Mining Other Types of Association Rules 301
5. Summary and Bibliographical Notes 304
6. Exercises 305

Chapter 11. Supervised Learning: Statistical Methods 307
1. Bayesian Methods 307
2. Regression 346
3. Summary and Bibliographical Notes 375
4. Exercises 376

Chapter 12. Supervised Learning: Decision Trees, Rule Algorithms, and Their Hybrids ... 381
1. What is Inductive Machine Learning? 381
2. Decision Trees 388
3. Rule Algorithms 393
4. Hybrid Algorithms 399
5. Summary and Bibliographical Notes 416
6. Exercises 416

Chapter 13. Supervised Learning: Neural Networks 419
1. Introduction 419
2. Biological Neurons and their Models 420
3. Learning Rules 428
4. Neural Network Topologies 431
5. Radial Basis Function Neural Networks 431
6. Summary and Bibliographical Notes 449
7. Exercises 450

Chapter 14. Text Mining 453
1. Introduction 453
2. Information Retrieval Systems 454
3. Improving Information Retrieval Systems 462
4. Summary and Bibliographical Notes 464
5. Exercises 465

Part 5 Data Models Assessment   467

Chapter 15. Assessment of Data Models 469
1. Introduction 469
2. Models, their Selection, and their Assessment 470
3. Simple Split and Cross-Validation 473
4. Bootstrap 474
5. Occam's Razor Heuristic 474
6. Minimum Description Length Principle 475
7. Akaike's Information Criterion and Bayesian Information Criterion 476
8. Sensitivity, Specificity, and ROC Analyses 477
9. Interestingness Criteria 484
10. Summary and Bibliographical Notes 485
11. Exercises 486

Part 6 Data Security and Privacy Issues   487

Chapter 16. Data Security, Privacy and Data Mining 489
1. Privacy in Data Mining 489
2. Privacy Versus Levels of Information Granularity 490

  3. Distributed Data Mining 491
  4. Collaborative Clustering 492
  5. The Development of the Horizontal Model of Collaboration 494
  6. Dealing with Different Levels of Granularity
    in the Collaboration Process 498
  7. Summary and Biographical Notes 499
  8. Exercises 501

Part 7 Overview of Key Mathematical Concepts 503

Appendix A. Linear Algebra 505
  1. Vectors 505
  2. Matrices 519
  3. Linear Transformation 540

Appendix B. Probability 547
  1. Basic Concepts 547
  2. Probability Laws 548
  3. Probability Axioms 549
  4. Defining Events With Set–Theoretic Operations 549
  5. Conditional Probability 551
  6. Multiplicative Rule of Probability 552
  7. Random Variables 553
  8. Probability Distribution 555

Appendix C. Lines and Planes in Space 567
  1. Lines on Plane 567
  2. Lines and Planes in a Space 569
  3. Planes 572
  4. Hyperplanes 575

Appendix D. Sets 579
  1. Set Definition and Notations 579
  2. Types of Sets 581
  3. Set Relations 585
  4. Set Operations 587
  5. Set Algebra 590
  6. Cartesian Product of Sets 592
  7. Partition of a Nonempty Set 596

Index 597

Foreword

"If you torture the data long enough, Nature will confess," said 1991 Nobel-winning economist Ronald Coase. The statement is still true. However, achieving this lofty goal is not easy. First, "long enough" may, in practice, be "too long" in many applications and thus unacceptable. Second, to get "confession" from large data sets one needs to use state-of-the-art "torturing" tools. Third, Nature is very stubborn — not yielding easily or unwilling to reveal its secrets at all.

Fortunately, while being aware of the above facts, the reader (a data miner) will find several efficient data mining tools described in this excellent book. The book discusses various issues connecting the whole spectrum of approaches, methods, techniques and algorithms falling under the umbrella of data mining. It starts with data understanding and preprocessing, then goes through a set of methods for supervised and unsupervised learning, and concludes with model assessment, data security and privacy issues. It is this specific approach of using the knowledge discovery process that makes this book a rare one indeed, and thus an indispensable addition to many other books on data mining.

To be more precise, this is a book on knowledge discovery from data. As for the data sets, the easy-to-make statement is that there is no part of modern human activity left untouched by both the need and the desire to collect data. The consequence of such a state of affairs is obvious. We are surrounded by, or perhaps even immersed in, an ocean of all kinds of data (such as measurements, images, patterns, sounds, web pages, tunes, etc.) that are generated by various types of sensors, cameras, microphones, pieces of software and/or other human-made devices. Thus we are in dire need of automatically extracting as much information as possible from the data that we more or less wisely generate. We need to conquer the existing and develop new approaches, algorithms and procedures for knowledge discovery from data. This is exactly what the authors, world-leading experts on data mining in all its various disguises, have done. They present the reader with a large spectrum of data mining methods in a gracious and yet rigorous way.

To facilitate the book's use, I offer the following *roadmap* to help in:

a) reaching certain desired destinations without undesirable wandering, and
b) getting the basic idea of the breadth and depth of the book.

First, an overview: the volume is divided into seven parts (the last one being Appendices covering the basic mathematical concepts of Linear Algebra, Probability Theory, Lines and Planes in Space, and Sets). The main body of the book is as follows: Part 1, Data Mining and Knowledge Discovery Process (two Chapters), Part 2, Data Understanding (three Chapters), Part 3, Data Preprocessing (three Chapters), Part 4, Data Mining: Methods for Constructing Data Models (six Chapters), Part 5, Data Models Assessment (one Chapter), and Part 6, Data Security and Privacy Issues (one Chapter). Both the ordering of the sections and the amount of material devoted to each particular segment tells a lot about the authors' expertise and perfect control of the data mining field. Namely, unlike many other books that mainly focus on the modeling part, this volume discusses all the important — and elsewhere often neglected — parts before and after modeling. This breadth is one of the great characteristics of the book.

A dive into particular sections of the book unveils that Chapter 1 defines what data mining is about and stresses some of its unique features, while Chapter 2 introduces a Knowledge Discovery Process (KDP) as a process that seeks new knowledge about an application domain. Here, it is pointed out that Data Mining (DM) is just one step in the KDP. This Chapter also reminds us that the KDP consists of multiple steps that are executed in a sequence, where the next step is initiated upon successful completion of the previous one. It also stresses the fact that the KDP stretches between the task of understanding of the project domain and data, through data preparation and analysis, to evaluation, understanding and application of the generated knowledge. KDP is both highly iterative (there are many repetitions triggered by revision processes) and interactive. The main reason for introducing the process is to formalize knowledge discovery (KD) projects within a common framework, and emphasize independence of specific applications, tools, and vendors. Five KDP models are introduced and their strong and weak points are discussed. It is acknowledged that the data preparation step is by far the most time-consuming and important part of the KDP.

Chapter 3, which opens Part 2 of the book, tackles the underlying core subject of the book, namely, data and data sets. This includes an introduction of various data storage techniques and of the issues related to both the quality and quantity of data used for data mining purposes. The most important topics discussed in this Chapter are the different data types (numerical, symbolic, discrete, binary, nominal, ordinal and continuous). As for the organization of the data, they are organized into rectangular tables called data sets, where rows represent objects (samples, examples, patterns) and where columns represent features/attributes, i.e., the input dimension that describes the objects. Furthermore, there are sections on data storage using databases and data warehouses. The specialized data types — including transactional data, spatial data, hypertext, multimedia data, temporal data and the World Wide Web — are not forgotten either. Finally, the problems of scalability while faced with a large quantity of data, as well as the dynamic data and data quality problems (including imprecision, incompleteness, redundancy, missing values and noise) are also discussed. At the end of each and every Chapter, the reader can find good bibliographical notes, pointers to other electronic or written sources, and a list of relevant references.

Chapter 4 sets the stage for the core topics covered in the book, and in particular for Part 4, which deals with algorithms and tools for concepts introduced herein. Basic learning methods are introduced here (unsupervised, semi-supervised, supervised, reinforcement) together with the concepts of classification and regression.

Part 2 of the book ends with Chapter 5, which covers knowledge representation and its most commonly encountered schemes such as rules, graphs, networks, and their generalizations. The fundamental issue of abstraction of information captured by information granulation and resulting information granules is discussed in detail. An extended description is devoted to the concepts of fuzzy sets, granularity of data and granular concepts in general, and various other set representations, including shadow and rough sets. The authors show great care in warning the reader that the choice of a certain formalism in knowledge representation depends upon a number of factors and that while faced with an enormous diversity of data the data miner has to make prudent decisions about the underlying schemes of knowledge representation.

Part 3 of the book is devoted to *data preprocessing* and contains three Chapters. Readers interested in Databases (DB), Data Warehouses (DW) and On-Line Analytical Processing (OLAP) will find all the basics in Chapter 6, wherein the elementary concepts are introduced. The most important topics discussed in this Chapter are Relational DBMS (RDBMS), defined as a collection of interrelated data and a set of software programs to access those data; SQL, described as a declarative language for writing queries for a RDBMS; and three types of languages to retrieve and manipulate data: Data Manipulation Language (DML), Data Definition Language (DDL), and Data Control Language (DCL), which are implemented using SQL. DW is introduced as a subject-oriented, integrated, time-variant and non-volatile collection of data in support of management's decision-making process. Three types of DW are distinguished: virtual data warehouse, data mart, and enterprise warehouse. DW is based on a multidimensional data model: the data is visualized using a multidimensional data cube, in contrast to the relational table that is used in the RDBMS. Finally, OLAP is discussed with great care to details. This Chapter is relatively unique, and thus enriching, among various data mining books that typically skip these topics.

If you are like the author of this Foreword, meaning that you love mathematics, your heart will start beating faster while opening Chapter 7 on *feature extraction (FE) and feature selection (FS) methods*. At this point, you can turn on your computer, and start implementing some of the many models nicely introduced and explained here. The titles of the topics covered reveal the depth and breadth of supervised and unsupervised techniques and approaches presented: Principal Component Analysis (PCA), Independent Component Analysis (ICA), Karhunen-Loeve Transformation, Fisher's linear discriminant, SVD, Vector quantization, Learning vector quantization, Fourier transform, Wavelets, Zernike moments, and several feature selection methods. Because FE and FS methods are so important in data preprocessing, this Chapter is quite extensive.

Chapter 8 deals with one of the most important, and often required, preprocessing methods, the overall goal of which is to reduce the complexity of the data for further data mining tasks. It introduces unsupervised and supervised discretization methods of continuous data attributes. It also outlines a dynamic discretization algorithm and includes a comparison between several state of the art algorithms.

Part 4, *Data Mining: Methods for Constructing Data Models*, is comprised of two Chapters on the basic types of unsupervised learning, namely, Clustering and Association Rules; three Chapters on supervised learning, namely Statistical Methods, Decision Trees and Rule Algorithms, and Neural Networks; and a Chapter on Text Mining. Part 4, along with Parts 3 and 6, forms the core algorithmic section of this great data mining volume. You may switch on your computer again and start implementing various data mining tools clearly explained here.

To show the main features of every Chapter in Part 4, let us start with Chapter 9, which covers clustering, a predominant technique used in unsupervised learning. A spectrum of clustering methods is introduced, elaborating on their conceptual properties, computational aspects and scalability. The treatment of huge databases through mechanisms of sampling and distributed clustering is discussed as well. The latter two approaches are essential for dealing with large data sets.

Chapter 10 introduces the other key unsupervised learning technique, namely, association rules. The topics discussed here are association rule mining, storing of items using transactions, the association rules categorization as single-dimensional and multidimensional, Boolean and quantitative, and single-level and multilevel, their measurement by using support, confidence, and correlation, and the association rules generation from frequent item sets (a priori algorithm and its modifications including: hashing, transaction removal, data set partitioning, sampling, and mining frequent item sets without generation of candidate item sets).

Chapter 11 constitutes a gentle encounter with *statistical methods* for *supervised learning*, which are based on exploitation of probabilistic knowledge about data. This becomes particularly visible in the case of Bayesian methods. The statistical classification schemes exploit concepts of conditional probabilities and prior probabilities — all of which encapsulate knowledge about statistical characteristics of the data. The Bayesian classifiers are shown to be optimal given known probabilistic characteristics of the underlying data. The role of effective estimation procedures is emphasized and estimation techniques are discussed in detail. Chapter 11 introduces regression models too, including both linear and nonlinear regression. Some of the most representative generalized regression models and augmented development schemes are covered in detail.

Chapter 12 continues along statistical lines as it describes main types of inductive machine learning algorithms: decision trees, rule algorithms, and their hybrids. Very detailed description of these topics is given and the reader will be able to implement them easily or come up with their extensions and/or improvements. Comparative performances and discussion of the advantages and disadvantages of the methods on several data sets are also presented here.

The classical statistical approaches end here, and neural network models are presented in Chapter 13. This Chapter starts with presentation of biological neuron models: the spiking neuron model and a simple neuron model. This section leads to presentation of learning/plasticity rules used to update the weights between the interconnected neurons, both in networks utilizing the spiking and simple neuron models. Presentation of the most important neuron models and learning rules are unique characteristics of this Chapter. Popular neural network topologies are reviewed, followed by an introduction of a powerful Radial Basis Function (RBF) neural network that has been shown to be very useful in many data mining applications. Several aspects of the RBF are introduced, including its most important characteristic of being similar (almost practically equivalent) to the system of fuzzy rules.

In Chapter 14, concepts and methods related to text mining and information retrieval are presented. The most important topics discussed are information retrieval (IR) systems that concern an organization and retrieval of information from large collections of semi-structured or unstructured text-based databases and the World Wide Web, and how the IR system can be improved by latent semantic indexing and relevance feedback.

Part 5 of the book consists of Chapter 15, which discusses and explains several important and indispensable model selection and model assessment methods. The methods are divided into four broad categories: data re-use, heuristic, formal, and interestingness measures. The Chapter provides justification for why one should use methods from these different categories on the same data. The Akaike's information criterion and Bayesian information criterion methods are also discussed in order to show their relationship to the other methods covered.

The final part of the book, Part 6, and its sole Chapter 16, treats topics that are not usually found in other data mining books but which are very relevant and deserve to be presented to readers. Specifically, several issues of data privacy and security are raised and cast in the setting of data mining. Distinct ways of addressing them include data sanitation, data distortion, and cryptographic methods. In particular, the focus is on the role of information granularity as a vehicle for carrying out collaborative activities (such as clustering) while not releasing detailed numeric data. At this point, the roadmap is completed.

A few additional remarks are still due. The book comes with two important teaching tools that make it an excellent textbook. First, there is an *Exercises* section at the end of each and every Chapter expanding the volume beyond a great research monograph. The exercises are designed to augment the basic theory presented in each Chapter and help the reader to acquire practical skills and understanding of the algorithms and tools. This organization is suitable for both a textbook in a formal course and for self-study. The second teaching tool is a set of PowerPoint presentations, covering the material presented in all sixteen Chapters of the book.

All of the above makes this book a thoroughly enjoyable and solid read. I am sure that no data miner, scientist, engineer and/or interested layperson can afford to miss it.

Vojislav Kecman
University of Auckland
New Zeland

Acknowledgements

The authors gratefully acknowledge the critical remarks of G. William Moore, M.D., Ph.D., and all of the students in their Data Mining courses who commented on drafts of several Chapters. In particular, the help of Joo Heon Shin, Cao Dang Nguyen, Supphachai Thaicharoen, Jim Maginnis, Allison Gehrke and Hun Ki Lim is highly appreciated. The authors also thank Springer editor Melissa Fearon, and Valerie Schofield, her assistant, for support and encouragement.

Part 1

Data Mining and Knowledge Discovery Process

1

Introduction

In this Chapter we define and provide a high-level overview of data mining.

1. What Is Data Mining?

*The aim of data mining is to **make sense** of **large amounts** of **mostly unsupervised data**, in some **domain**.*

The above statement defining the aims of data mining (DM) is intuitive and easy to understand. The users of DM are often domain experts who not only own the data but also collect the data themselves. We assume that data owners have some understanding of the data and the processes that generated the data. Businesses are the largest group of DM users, since they routinely collect massive amounts of data and have a vested interest in making sense of the data. Their goal is to make their companies more competitive and profitable. Data owners desire not only to better understand their data but also to gain new knowledge about the domain (present in their data) for the purpose of solving problems in novel, possibly better ways.

In the above definition, the first key term is **make sense**, which has different meanings depending on the user's experience. In order to make sense we envision that this new knowledge should exhibit a series of essential attributes: it should be *understandable*, *valid*, *novel*, and *useful*. Probably the most important requirement is that the discovered new knowledge needs to be *understandable* to data owners who want to use it to some advantage. The most convenient outcome by far would be knowledge or a model of the data (see Part 4 of this book, which defines a model and describes several model-generating techniques) that can be described in easy-to-understand terms, say, via production rules such as:

IF abnormality (obstruction) in coronary arteries
THEN coronary artery disease

In the example, the input data may be images of the heart and accompanying arteries. If the images are diagnosed by cardiologists as being normal or abnormal (with obstructed arteries), then such data are known as *learning/training data*. Some DM techniques generate models of the data in terms of production rules, and cardiologists may then analyze these and either accept or reject them (in case the rules do not agree with their domain knowledge). Note, however, that cardiologists may not have used, or even known, some of the rules generated by DM techniques, even if the rules are correct (as determined by cardiologists after deeper examination), or as shown by a data miner to be performing well on new unseen data, known as *test data*.

We then come to the second requirement; the generated model needs to be *valid*. Chapter 15 describes methods for assessing the validity of generated models. If, in our example, all the generated rules were already known to cardiologists, these rules would be considered trivial and of no interest, although the generation of the already-known rules validates the generated models and the DM methodology. However, in the latter case, the project results would be considered a failure by the cardiologists (data owners). Thus, we come to the third requirement associated with *making sense*, namely, that the discovered knowledge must be *novel*. Let us suppose that the new knowledge about how to diagnose a patient had been discovered not in terms of production rules but by a different type of data model, say, a neural network. In this case, the new knowledge may or may not be acceptable to the cardiologists, since a neural network is a "black box" model that, in general, cannot be understood by humans. A trained neural network, however, might still be acceptable if it were proven to work well on hundreds of new cases. To illustrate the latter case, assume that the purpose of DM was to automate the analysis (prescreening) of heart images before a cardiologist would see a patient; in that case, a neural network model would be acceptable. We thus associate with the term *making sense* the fourth requirement, by requesting that the discovered knowledge be *useful*. This usefulness must hold true regardless of the type of model used (in our example, it was rules vs. neural networks).

The other key term in the definition is ***large amounts*** of data. DM is not about analyzing small data sets that can be easily dealt with using many standard techniques, or even manually. To give the reader a sense of the scale of data being collected that are good candidates for DM, let us look at the following examples. AT&T handles over 300 million calls daily to serve about 100 million customers and stores the information in a multiterabyte database. Wal-Mart, in all its stores taken together handles about 21 million transactions a day, and stores the information in a database of about a dozen terabytes. NASA generates several gigabytes of data per hour through its Earth Observing System. Oil companies like Mobil Oil store hundreds of terabytes of data about different aspects of oil exploration. The Sloan Digital Sky Survey project will collect observational data of about 40 terabytes. Modern biology creates, in projects like the human genome and proteome, data measured in terabytes and petabytes. Although no data are publicly available, Homeland Security in the U.S.A. is collecting petabytes of data on its own and other countries' citizens.

It is clear that none of the above databases can be analyzed by humans or even by the best algorithms (in terms of speed and memory requirements); these large amounts of data necessarily require the use of DM techniques to reduce the data in terms of both quantity and dimensionality. Part 3 of this book is devoted to this extremely important step in any DM undertaking, namely, data preprocessing techniques.

The third key term in the above definition is ***mostly unsupervised data***. It is much easier, and less expensive, to collect unsupervised data than supervised data. The reason is that with supervised data we must have known inputs corresponding to known outputs, as determined by domain experts. In our example, "input" images correspond to the "output" diagnosis of coronary artery disease (determined by cardiologists – a costly and error-prone process).

So what can be done if only unsupervised data are collected? To deal with the problem, one of the most difficult in DM, we need to use algorithms that are able to find "natural" groupings/clusters, relationships, and associations in the data (see Chapters 9 and 10). For example, if clusters can be found, they can possibly be labeled by domain experts. If we are able to do both, our unsupervised data becomes supervised, resulting in a much easier problem to deal with. Finding natural groupings or relationships in the data, however, is very difficult and remains an open research problem. Clustering is exacerbated by the fact that most clustering algorithms require the user *a priori* to specify (guess) the number of clusters in the data. Similarly, the association-rule mining algorithms require the user to specify parameters that allow the generation of an appropriate number of high-quality associations.

Another scenario exists when the available data are semisupervised, meaning that there are a few known training data pairs along with thousands of unsupervised data points. In our cardiology example, this situation would correspond to having thousands of images without diagnosis (very common in medical practice) and only a few images that have been diagnosed. The question then becomes: Can these few data points help in the process of making sense of the entire data set? Fortunately, there exist techniques of *semi-supervised learning*, that take advantage of these few training data points (see the material in Chapter 4 on partially supervised clustering).

By far the easiest scenario in DM is when all data points are fully supervised, since the majority of existing DM techniques are quite good at dealing with such data, with the possible exception of their scalability. A DM algorithm that works well on both small and large data is called *scalable*, but, unfortunately, few are. In Part 4 of this book, we describe some of the most efficient supervised learning algorithms.

The final key term in the definition is **domain**. The success of DM projects depends heavily on access to domain knowledge, and thus it is crucial for data miners to work very closely with domain experts/data owners. Discovering new knowledge from data is a process that is highly interactive (with domain experts) and iterative (within knowledge discovery; see description of the latter in Chapter 2). We cannot simply take a successful DM system, built for some domain, and apply it to another domain and expect good results.

This book is about making sense of data. Its ultimate goal is to provide readers with the fundamentals of frequently used DM methods and to guide readers in their DM projects, step by step. By now the reader has probably figured out what some of the DM steps are: from understanding the problem and the data, through preprocessing the data, to building models of the data and validating these to putting the newly discovered knowledge to use. In Chapter 2, we describe in detail a knowledge discovery process (KDP) that specifies a series of essential steps to be followed when conducting DM projects. In short, a KDP is a sequence of six steps, one of which is the data mining step concerned with building the data model. We will also follow the steps of the KDP in presenting the material in this book: from understanding of data and preprocessing to deployment of the results. Hence the subtitle: *A Knowledge Discovery Approach*. This approach sets this text apart from other data mining books.

Another important feature of the book is that we focus on the most frequently used DM methods. The reason is that among hundreds of available DM algorithms, such as clustering or machine learning, only small numbers of them are scalable to large data. So instead of covering many algorithms in each category (like neural networks), we focus on a few that have proven to be successful in DM projects. In choosing these, we have been guided by our own experience in performing DM projects, by DM books we have written or edited, and by survey results published at www.kdnuggets.com. This web site is excellent and by far the best source of information about all aspects of DM. By now, the reader should have the "big picture" of DM.

2. How does Data Mining Differ from Other Approaches?

Data mining came into existence in response to technological advances in many diverse disciplines. For instance, over the years computer engineering contributed significantly to the development of more powerful computers in terms of both speed and memory; computer science and mathematics continued to develop more and more efficient database architectures and search algorithms; and the combination of these disciplines helped to develop the World Wide Web (WWW). There have been tremendous improvements in techniques for collecting, storing, and transferring large volumes of data for such applications as image processing, digital signal processing, text processing and the processing of various forms of heterogeneous data. However, along with this dramatic increase in the amount of stored data came demands for better, faster, cheaper ways to deal with those data. In other words, all the data in the world are of no value without mechanisms to efficiently and effectively extract information and knowledge from them. Early pioneers such as U. Fayyad, H. Mannila, G. Piatetsky-Shapiro, G. Djorgovski, W. Frawley, P. Smith, and others recognized this urgent need, and the data mining field was born.

Data mining is not just an "umbrella" term coined for the purpose of making sense of data. The major distinguishing characteristic of DM is that it is *data driven*, as opposed to other methods that are often *model driven*. In statistics, researchers frequently deal with the problem of finding the smallest data size that gives sufficiently confident estimates. In DM, we deal with the opposite problem, namely, data size is large and we are interested in building a data model that is small (not too complex) but still describes the data well.

Finding a good model of the data, which at the same time is easy to understand, is at the heart of DM. We need to keep in mind, however, that none of the generated models will be complete (using all the relevant variables/attributes of the data), and that almost always we will look for a compromise between model completeness and model complexity (see discussion of the bias/variance dilemma in Chapter 15). This approach is in accordance with Occam's razor: simpler models are preferred over more complex ones.

The readers will no doubt notice that in several Chapters we cite our previous monograph on *Data Mining Methods for Knowledge Discovery* (Kluwer, 1998). The reason is that although the present book introduces several new topics not covered in the previous one, at the same time it omits almost entirely topics like rough sets and fuzzy sets that are described in the earlier book. The earlier book also provides the reader with a richer bibliography than this one.

Finally, a word of caution: although many commercial as well as open-source DM tools exist they do not by any means produce automatic results despite the hype of their vendors. The users should understand that the application of even a very good tool (as shown in a vendor's "example" application) to one's data will most often not result in the generation of valuable knowledge for the data owner after simply clicking "run". To learn why the reader is referred to Chapter 2 on the knowledge discovery process.

2.1. How to Use this Book for a Course on Data Mining

We envision that an instructor will cover, in a semester-long course, all the material presented in the book. This goal is achievable because the book is accompanied by instructional support in terms of PowerPoint presentations that address each of the topics covered. These presentations can serve as "templates" for teaching the course or as supporting material. However, the indispensable core elements of the book, which need to be covered in depth, are data preprocessing methods, described in Part 3, model building, described in Part 4 and model assessment, covered in Part 5.

For hands-on data mining experience, students should be given a large real data set at the beginning of the course and asked to follow the knowledge discovery process for performing a DM project. If the instructor of the course does not have his or her own real data to analyze, such project data can be found on the University of California at Irvine website at www.ics.uci.edu/~mlearn/MLRepository.

3. Summary and Bibliographical Notes

In this Chapter, we defined data mining and stressed some of its unique features. Since we wrote our first monograph on data mining [1], one of the first such books on the market, many books have been published on the topic. Some of those that are well worth reading are [2–6].

References

1. Cios, K.J., Pedrycz, W., and Swiniarski, R. 1998. *Data Mining Methods for Knowledge Discovery*, Kluwer
2. Han, J., and Kamber, M. 2006. *Data Mining: Concepts and Techniques*, Morgan Kaufmann
3. Hand, D., Mannila, H., and Smyth, P. 2001. *Principles of Data Mining*, MIT Press

4. Hastie, T., Tibshirani, R., and Friedman, J. 2001. *The Elements of Statistical Learning: Data Mining, Inference and Prediction*, Springer
5. Kecman, V. 2001. *Learning and Soft Computing*, MIT Press
6. Witten, H., and Frank, E. 2005. *Data Mining: Practical Machine Learning Tools and Techniques*, Morgan Kaufmann

4. Exercises

1. What is data mining?
2. How does it differ from other disciplines?
3. What are the key features of data mining?
4. When is a data mining outcome acceptable to the end user?
5. When should not a data mining project be undertaken?

2

The Knowledge Discovery Process

In this Chapter, we describe the knowledge discovery process, present some models, and explain why and how these could be used for a successful data mining project.

1. Introduction

Before one attempts to extract useful knowledge from data, it is important to understand the overall approach. Simply knowing many algorithms used for data analysis is not sufficient for a successful data mining (DM) project. Therefore, this Chapter focuses on describing and explaining the **process** that leads to finding new knowledge. The process defines a sequence of steps (with eventual feedback loops) that should be followed to discover knowledge (e.g., patterns) in data. Each step is usually realized with the help of available commercial or open-source software tools.

To formalize the knowledge discovery processes (KDPs) within a common framework, we introduce the concept of a **process model**. The model helps organizations to better understand the KDP and provides a roadmap to follow while planning and executing the project. This in turn results in cost and time savings, better understanding, and acceptance of the results of such projects. We need to understand that such processes are nontrivial and involve multiple steps, reviews of partial results, possibly several iterations, and interactions with the data owners. There are several reasons to structure a KDP as a **standardized process model**:

1. *The end product must be useful for the user/owner of the data*. A blind, unstructured application of DM techniques to input data, called *data dredging*, frequently produces meaningless results/knowledge, i.e., knowledge that, while interesting, does not contribute to solving the user's problem. This result ultimately leads to the failure of the project. Only through the application of well-defined KDP models will the end product be valid, novel, useful, and understandable.

2. *A well-defined KDP model should have a logical, cohesive, well-thought-out structure and approach that can be presented to decision-makers who may have difficulty understanding the need, value, and mechanics behind a KDP*. Humans often fail to grasp the potential knowledge available in large amounts of untapped and possibly valuable data. They often do not want to devote significant time and resources to the pursuit of formal methods of knowledge extraction from the data, but rather prefer to rely heavily on the skills and experience of others (domain experts) as their source of information. However, because they are typically ultimately responsible for the decision(s) based on that information, they frequently want to understand (be comfortable with) the technology applied to those solution. A process model that is well structured and logical will do much to alleviate any misgivings they may have.

3. *Knowledge discovery projects require a significant project management effort that needs to be grounded in a solid framework.* Most knowledge discovery projects involve teamwork and thus require careful planning and scheduling. For most project management specialists, KDP and DM are not familiar terms. Therefore, these specialists need a definition of what such projects involve and how to carry them out in order to develop a sound project schedule.
4. *Knowledge discovery should follow the example of other engineering disciplines that already have established models.* A good example is the software engineering field, which is a relatively new and dynamic discipline that exhibits many characteristics that are pertinent to knowledge discovery. Software engineering has adopted several development models, including the waterfall and spiral models that have become well-known standards in this area.
5. *There is a widely recognized need for standardization of the KDP.* The challenge for modern data miners is to come up with widely accepted standards that will stimulate major industry growth. Standardization of the KDP model would enable the development of standardized methods and procedures, thereby enabling end users to deploy their projects more easily. It would lead directly to project performance that is faster, cheaper, more reliable, and more manageable. The standards would promote the development and delivery of solutions that use business terminology rather than the traditional language of algorithms, matrices, criterions, complexities, and the like, resulting in greater exposure and acceptability for the knowledge discovery field.

Below we define the KDP and its relevant terminology. We also provide a description of several key KDP models, discuss their applications, and make comparisons. Upon finishing this Chapter, the reader will know how to structure, plan, and execute a (successful) KD project.

2. What is the Knowledge Discovery Process?

Because there is some confusion about the terms data mining, knowledge discovery, and knowledge discovery in databases, we first define them. Note, however, that many researchers and practitioners use DM as a synonym for knowledge discovery; DM is also just one step of the KDP.

Data mining was defined in Chapter 1. Let us just add here that DM is also known under many other names, including *knowledge extraction*, *information discovery*, *information harvesting*, *data archeology*, and *data pattern processing*.

The **knowledge discovery process** (KDP), also called knowledge discovery in databases, seeks new knowledge in some application domain. It is defined as the nontrivial process of identifying valid, novel, potentially useful, and ultimately understandable patterns in data. The process generalizes to nondatabase sources of data, although it emphasizes databases as a primary source of data. It consists of many steps (one of them is DM), each attempting to complete a particular discovery task and each accomplished by the application of a discovery method. Knowledge discovery concerns the entire knowledge extraction process, including how data are stored and accessed, how to use efficient and scalable algorithms to analyze massive datasets, how to interpret and visualize the results, and how to model and support the interaction between human and machine. It also concerns support for learning and analyzing the application domain.

This book defines the term **knowledge extraction** in a narrow sense. While the authors acknowledge that extracting knowledge from data can be accomplished through a variety of methods — some not even requiring the use of a computer — this book uses the term to refer to knowledge obtained from a database or from textual data via the knowledge discovery process. Uses of the term outside this context will be identified as such.

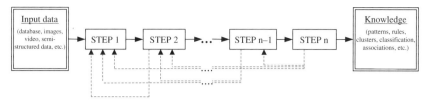

Figure 2.1. Sequential structure of the KDP model.

2.1. Overview of the Knowledge Discovery Process

The KDP model consists of a set of processing steps to be followed by practitioners when executing a knowledge discovery project. The model describes procedures that are performed in each of its steps. It is primarily used to plan, work through, and reduce the cost of any given project.

Since the 1990s, several different KDPs have been developed. The initial efforts were led by academic research but were quickly followed by industry. The first basic structure of the model was proposed by Fayyad et al. and later improved/modified by others. The process consists of multiple steps, that are executed in a sequence. Each subsequent step is initiated upon successful completion of the previous step, and requires the result generated by the previous step as its input. Another common feature of the proposed models is the range of activities covered, which stretches from the task of understanding the project domain and data, through data preparation and analysis, to evaluation, understanding, and application of the generated results. All the proposed models also emphasize the iterative nature of the model, in terms of many feedback loops that are triggered by a revision process. A schematic diagram is shown in Figure 2.1.

The main differences between the models described here lie in the number and scope of their specific steps. A common feature of all models is the definition of inputs and outputs. Typical inputs include data in various formats, such as numerical and nominal data stored in databases or flat files; images; video; semi-structured data, such as XML or HTML; etc. The output is the generated new knowledge — usually described in terms of rules, patterns, classification models, associations, trends, statistical analysis, etc.

3. Knowledge Discovery Process Models

Although the models usually emphasize independence from specific applications and tools, they can be broadly divided into those that take into account industrial issues and those that do not. However, the academic models, which usually are not concerned with industrial issues, can be made applicable relatively easily in the industrial setting and vice versa. We restrict our discussion to those models that have been popularized in the literature and have been used in real knowledge discovery projects.

3.1. Academic Research Models

The efforts to establish a KDP model were initiated in academia. In the mid-1990s, when the DM field was being shaped, researchers started defining multistep procedures to guide users of DM tools in the complex knowledge discovery world. The main emphasis was to provide a sequence of activities that would help to execute a KDP in an arbitrary domain. The two process models developed in 1996 and 1998 are the nine-step model by Fayyad et al. and the eight-step model by Anand and Buchner. Below we introduce the first of these, which is perceived as the leading research model. The second model is summarized in Sect. 2.3.4.

The Fayyad et al. KDP model consists of nine steps, which are outlined as follows:

1. *Developing and understanding the application domain.* This step includes learning the relevant prior knowledge and the goals of the end user of the discovered knowledge.
2. *Creating a target data set.* Here the data miner selects a subset of variables (attributes) and data points (examples) that will be used to perform discovery tasks. This step usually includes querying the existing data to select the desired subset.
3. *Data cleaning and preprocessing.* This step consists of removing outliers, dealing with noise and missing values in the data, and accounting for time sequence information and known changes.
4. *Data reduction and projection.* This step consists of finding useful attributes by applying dimension reduction and transformation methods, and finding invariant representation of the data.
5. *Choosing the data mining task.* Here the data miner matches the goals defined in Step 1 with a particular DM method, such as classification, regression, clustering, etc.
6. *Choosing the data mining algorithm.* The data miner selects methods to search for patterns in the data and decides which models and parameters of the methods used may be appropriate.
7. *Data mining.* This step generates patterns in a particular representational form, such as classification rules, decision trees, regression models, trends, etc.
8. *Interpreting mined patterns.* Here the analyst performs visualization of the extracted patterns and models, and visualization of the data based on the extracted models.
9. *Consolidating discovered knowledge.* The final step consists of incorporating the discovered knowledge into the performance system, and documenting and reporting it to the interested parties. This step may also include checking and resolving potential conflicts with previously believed knowledge.

Notes: This process is iterative. The authors of this model declare that a number of loops between any two steps are usually executed, but they give no specific details. The model provides a detailed technical description with respect to data analysis but lacks a description of business aspects. This model has become a cornerstone of later models.

Major Applications: The nine-step model has been incorporated into a commercial knowledge discovery system called MineSet™ (for details, see Purple Insight Ltd. at http://www.purpleinsight.com). The model has been used in a number of different domains, including engineering, medicine, production, e-business, and software development.

3.2. Industrial Models

Industrial models quickly followed academic efforts. Several different approaches were undertaken, ranging from models proposed by individuals with extensive industrial experience to models proposed by large industrial consortiums. Two representative industrial models are the five-step model by Cabena et al., with support from IBM (see Sect. 2.3.4) and the industrial six-step CRISP-DM model, developed by a large consortium of European companies. The latter has become the leading industrial model, and is described in detail next.

The CRISP-DM (CRoss-Industry Standard Process for Data Mining) was first established in the late 1990s by four companies: Integral Solutions Ltd. (a provider of commercial data mining solutions), NCR (a database provider), DaimlerChrysler (an automobile manufacturer), and OHRA (an insurance company). The last two companies served as data and case study sources.

The development of this process model enjoys strong industrial support. It has also been supported by the ESPRIT program funded by the European Commission. The CRISP-DM Special Interest Group was created with the goal of supporting the developed process model. Currently, it includes over 300 users and tool and service providers.

The CRISP-DM KDP model (see Figure 2.2) consists of six steps, which are summarized below:

1. *Business understanding*. This step focuses on the understanding of objectives and requirements from a business perspective. It also converts these into a DM problem definition, and designs a preliminary project plan to achieve the objectives. It is further broken into several substeps, namely,

   – determination of business objectives,
   – assessment of the situation,
   – determination of DM goals, and
   – generation of a project plan.

2. *Data understanding*. This step starts with initial data collection and familiarization with the data. Specific aims include identification of data quality problems, initial insights into the data, and detection of interesting data subsets. Data understanding is further broken down into

   collection of initial data,
   – description of data,
   – exploration of data, and
   verification of data quality.

3. *Data preparation*. This step covers all activities needed to construct the final dataset, which constitutes the data that will be fed into DM tool(s) in the next step. It includes Table, record, and attribute selection; data cleaning; construction of new attributes; and transformation of data. It is divided into

   – selection of data,
   – cleansing of data,

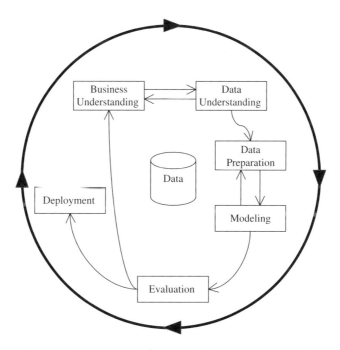

Figure 2.2. The CRISP-DM KD process model (source: http://www.crisp-dm.org/).

- construction of data,
- integration of data, and
- formatting of data substeps.

4. *Modeling.* At this point, various modeling techniques are selected and applied. Modeling usually involves the use of several methods for the same DM problem type and the calibration of their parameters to optimal values. Since some methods may require a specific format for input data, often reiteration into the previous step is necessary. This step is subdivided into

- selection of modeling technique(s),
- generation of test design,
- creation of models, and
- assessment of generated models.

5. *Evaluation.* After one or more models have been built that have high quality from a data analysis perspective, the model is evaluated from a business objective perspective. A review of the steps executed to construct the model is also performed. A key objective is to determine whether any important business issues have not been sufficiently considered. At the end of this phase, a decision about the use of the DM results should be reached. The key substeps in this step include

- evaluation of the results,
- process review, and
- determination of the next step.

6. *Deployment.* Now the discovered knowledge must be organized and presented in a way that the customer can use. Depending on the requirements, this step can be as simple as generating a report or as complex as implementing a repeatable KDP. This step is further divided into

- plan deployment,
- plan monitoring and maintenance,
- generation of final report, and
- review of the process substeps.

Notes: The model is characterized by an easy-to-understand vocabulary and good documentation. It divides all steps into substeps that provide all necessary details. It also acknowledges the strong iterative nature of the process, with loops between several of the steps. In general, it is a very successful and extensively applied model, mainly due to its grounding in practical, industrial, real-world knowledge discovery experience.

Major Applications: The CRISP-DM model has been used in domains such as medicine, engineering, marketing, and sales. It has also been incorporated into a commercial knowledge discovery system called Clementine® (see SPSS Inc. at http://www.spss.com/clementine).

3.3. Hybrid Models

The development of academic and industrial models has led to the development of hybrid models, i.e., models that combine aspects of both. One such model is a six-step KDP model (see Figure 2.3) developed by Cios et al. It was developed based on the CRISP-DM model by adopting it to academic research. The main differences and extensions include

- providing more general, research-oriented description of the steps,
- introducing a data mining step instead of the modeling step,

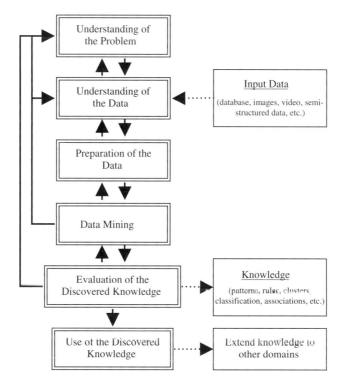

Figure 2.3. The six-step KDP model. Source: Pal, N.R., Jain, L.C., (Eds.) 2005. Advanced Techniques in Knowledge Discovery and Data Mining, Springer Verlag.

- introducing several new explicit feedback mechanisms, (the CRISP-DM model has only three major feedback sources, while the hybrid model has more detailed feedback mechanisms) and
- modification of the last step, since in the hybrid model, the knowledge discovered for a particular domain may be applied in other domains.

A description of the six steps follows

1. *Understanding of the problem domain.* This initial step involves working closely with domain experts to define the problem and determine the project goals, identifying key people, and learning about current solutions to the problem. It also involves learning domain-specific terminology. A description of the problem, including its restrictions, is prepared. Finally, project goals are translated into DM goals, and the initial selection of DM tools to be used later in the process is performed.
2. *Understanding of the data.* This step includes collecting sample data and deciding which data, including format and size, will be needed. Background knowledge can be used to guide these efforts. Data are checked for completeness, redundancy, missing values, plausibility of attribute values, etc. Finally, the step includes verification of the usefulness of the data with respect to the DM goals.
3. *Preparation of the data.* This step concerns deciding which data will be used as input for DM methods in the subsequent step. It involves sampling, running correlation and significance tests, and data cleaning, which includes checking the completeness of data records, removing or correcting for noise and missing values, etc. The cleaned data may be further processed by feature selection and extraction algorithms (to reduce dimensionality), by derivation of new attributes (say, by discretization), and by summarization of data (data granularization). The end results are data that meet the specific input requirements for the DM tools selected in Step 1.

4. *Data mining*. Here the data miner uses various DM methods to derive knowledge from preprocessed data.
5. *Evaluation of the discovered knowledge*. Evaluation includes understanding the results, checking whether the discovered knowledge is novel and interesting, interpretation of the results by domain experts, and checking the impact of the discovered knowledge. Only approved models are retained, and the entire process is revisited to identify which alternative actions could have been taken to improve the results. A list of errors made in the process is prepared.
6. *Use of the discovered knowledge*. This final step consists of planning where and how to use the discovered knowledge. The application area in the current domain may be extended to other domains. A plan to monitor the implementation of the discovered knowledge is created and the entire project documented. Finally, the discovered knowledge is deployed.

Notes: The model emphasizes the iterative aspects of the process, drawing from the experience of users of previous models. It identifies and describes several explicit feedback loops:

- from *understanding of the data* to *understanding of the problem domain*. This loop is based on the need for additional domain knowledge to better understand the data.
- from *preparation of the data* to *understanding of the data*. This loop is caused by the need for additional or more specific information about the data in order to guide the choice of specific data preprocessing algorithms.
- from *data mining* to *understanding of the problem domain*. The reason for this loop could be unsatisfactory results generated by the selected DM methods, requiring modification of the project's goals.
- from *data mining* to *understanding of the data*. The most common reason for this loop is poor understanding of the data, which results in incorrect selection of a DM method and its subsequent failure, e.g., data were misrecognized as continuous and discretized in the *understanding of the data* step.
- from *data mining* to the *preparation of the data*. This loop is caused by the need to improve data preparation, which often results from the specific requirements of the DM method used, since these requirements may not have been known during the *preparation of the data* step.
- from *evaluation of the discovered knowledge* to the *understanding of the problem domain*. The most common cause for this loop is invalidity of the discovered knowledge. Several possible reasons include incorrect understanding or interpretation of the domain and incorrect design or understanding of problem restrictions, requirements, or goals. In these cases, the entire KD process must be repeated.
- from *evaluation of the discovered knowledge* to *data mining*. This loop is executed when the discovered knowledge is not novel, interesting, or useful. The least expensive solution is to choose a different DM tool and repeat the DM step.

Awareness of the above common mistakes may help the user to avoid them by deploying some countermeasures.

Major Applications: The hybrid model has been used in medicine and software development areas. Example applications include development of computerized diagnostic systems for cardiac SPECT images and a grid data mining framework called GridMiner-Core. It has also been applied to analysis of data concerning intensive care, cystic fibrosis, and image-based classification of cells.

3.4. Comparison of the Models

To understand and interpret the KDP models described above, a direct, side-by-side comparison is shown in Table 2.1. It includes information about the domain of origin (academic or industry), the number of steps, a comparison of steps between the models, notes, and application domains.

Table 2.1. Comparison of the five KDP models. The double-lines group the corresponding steps.

Model	Fayyad et al.	Anand & Buchner	Cios et al.	Cabena et al.	CRISP-DM
Domain of origin	Academic	Academic	Hybrid academic/industry	Industry	Industry
# steps	9	8	6	5	6
Steps	1. Developing and Understanding the Application Domain	1. Human Resource Identification 2. Problem Specification	1. Understanding of the Problem Domain	1. Business Objectives Determination	1. Business Understanding
	2. Creating a Target Data Set	3. Data Prospecting 4. Domain Knowledge Elicitation	2. Understanding of the Data	2. Data Preparation	2. Data Understanding
	3. Data Cleaning and Preprocessing 4. Data Reduction and Projection 5. Choosing the Data Mining Task 6. Choosing the Data Mining Algorithm 7. Data Mining 8. Interpreting Mined Patterns	6. Data Preprocessing 5. Methodology Identification 7. Pattern Discovery 8. Knowledge Post-Processing	3. Preparation of the Data 4. Data Mining 5. Evaluation of the Discovered Knowledge	3. Data Mining 4. Analysis of Results	3. Data Preparation 4. Modeling 5. Evaluation

(Continued)

Table 2.1. (Continued)

Model	Fayyad et al.	Anand & Buchner	Cios et al.	Cabena et al.	CRISP-DM
	9. Consolidating Discovered Knowledge		6. Use of the Discovered Knowledge	5. Assimilation of Knowledge	6. Deployment
Notes	The most popular and most cited model; provides detailed technical description with respect to data analysis, but lacks business aspects	Provides detailed breakdown of the initial steps of the process; missing step concerned with application of the discovered knowledge and project documentation	Draws from both academic and industrial models and emphasizes iterative aspects; identifies and describes several explicit feedback loops	Business oriented and easy to comprehend by non-data-mining specialists; the model definition uses non-DM jargon	Uses easy-to-understand vocabulary; has good documentation; divides all steps into substeps that provide all necessary details
Supporting software	Commercial system MineSet™	N/A	N/A	N/A	Commercial system Clementine®
Reported application domains	Medicine, engineering, production, e-business, software	Marketing, sales	Medicine, software	Marketing, sales	Medicine, engineering, marketing, sales

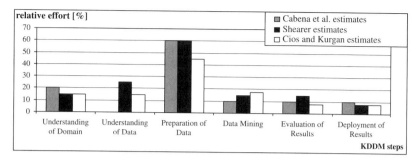

Figure 2.4. Relative effort spent on specific steps of the KD process. Source: Pal. N.R., Jain, L.C., (Eds.) 2005. Advanced Techniques in Knowledge Discovery and Data Mining, Springer Verlag.

Most models follow a similar sequence of steps, while the common steps between the five are domain understanding, data mining, and evaluation of the discovered knowledge. The nine-step model carries out the steps concerning the choice of DM tasks and algorithms late in the process. The other models do so before preprocessing of the data in order to obtain data that are correctly prepared for the DM step without having to repeat some of the earlier steps. In the case of Fayyad's model, the prepared data may not be suitable for the tool of choice, and thus a loop back to the second, third, or fourth step may be required. The five-step model is very similar to the six-step models, except that it omits the data understanding step. The eight-step model gives a very detailed breakdown of steps in the early phases of the KDP, but it does not allow for a step concerned with applying the discovered knowledge. At the same time, it recognizes the important issue of human resource identification. This consideration is very important for any KDP, and we suggest that this step should be performed in all models.

We emphasize that there is no universally "best" KDP model. Each of the models has its strong and weak points based on the application domain and particular objectives. Further reading can be found in the Summary and Bibliographical Notes (Sect. 5).

A very important aspect of the KDP is the relative time spent in completing each of the steps. Evaluation of this effort enables precise scheduling. Several estimates have been proposed by researchers and practitioners alike. Figure 2.4 shows a comparison of these different estimates. We note that the numbers given are only estimates, which are used to quantify relative effort, and their sum may not equal 100%. The specific estimated values depend on many factors, such as existing knowledge about the considered project domain, the skill level of human resources, and the complexity of the problem at hand, to name just a few.

The common theme of all estimates is an acknowledgment that the data preparation step is by far the most time-consuming part of the KDP.

4. Research Issues

The ultimate goal of the KDP model is to achieve overall integration of the entire process through the use of industrial standards. Another important objective is to provide interoperability and compatibility between the different software systems and platforms used throughout the process. Integrated and interoperable models would serve the end user in automating, or more realistically semiautomating, work with knowledge discovery systems.

4.1. Metadata and the Knowledge Discovery Process

Our goal is to enable users to perform a KDP without possessing extensive background knowledge, without manual data manipulation, and without manual procedures to exchange data and knowledge between different DM methods. This outcome requires the ability to store and exchange not only the data but also, most importantly, knowledge that is expressed in terms of data models, and meta-data that describes data and domain knowledge used in the process.

One of the technologies that can be used in achieving these goals is XML (eXtensible Markup Language), a standard proposed by the World Wide Web Consortium. XML allows the user to describe and store structured or semistructured data and to exchange data in a platform- and tool-independent way. From the KD perspective, XML helps to implement and standardize communication between diverse KD and database systems, to build standard data repositories for sharing data between different KD systems that work on different software platforms, and to provide a framework to integrate the entire KD process.

While XML by itself helps to solve some problems, metadata standards based on XML may provide a complete solution. Several such standards, such as PMML (Predictive Model Markup Language), have been identified that allow interoperability among different mining tools and that achieve integration with other applications, including database systems, spreadsheets, and decision support systems.

Both XML and PMML can be easily stored in most current database management systems. PMML, which is an XML-based language designed by the Data Mining Group, is used to describe data models (generated knowledge) and to share them between compliant applications. The Data Mining Group is an independent, vendor-led group that develops data mining standards. Its members include IBM, KXEN, Magnify Inc., Microsoft, MicroStrategy Inc., National Center for DM, Oracle, Prudential Systems Software GmbH, Salford Systems, SAS Inc., SPSS Inc., StatSoft Inc., and other companies (see http://www.dmg.org/). By using such a language, users can generate data models with one application, use another application to analyze these models, still another to evaluate them, and finally yet another to visualize the model. A PMML excerpt is shown in Figure 2.5.

XML and PMML standards can be used to integrate the KDP model in the following way. Information collected during the domain and data understanding steps can be stored as XML documents. These documents can be then used in the steps of data understanding, data preparation, and knowledge evaluation as a source of information that can be accessed automatically, across platforms, and across tools. In addition, knowledge extracted in the DM step and verified in the evaluation step, along with domain knowledge gathered in the domain understanding step, can be stored using PMML documents, which can then be stored and exchanged between different software tools. A sample architecture is shown in Figure 2.6.

5. Summary and Bibliographical Notes

In this Chapter we introduced the knowledge discovery process. The most important topics discussed are the following:

- **Knowledge discovery** is a **process** that seeks new knowledge about an application domain. It consists of many steps, one of which is data mining (DM), each aiming to complete a particular discovery task, and accomplished by the application of a discovery method.
- The KDP consists of **multiple steps** that are executed in a **sequence**. The subsequent step is initiated upon successful completion of the previous step and requires results generated by the previous step as its inputs.
- The KDP ranges from the task of understanding the project domain and data, through data preparation and analysis, to evaluation, understanding and application of the generated knowledge. It is highly **iterative**, and includes many feedback loops and repetitions, which are triggered by revision processes.

- The main reason for introducing **process models** is to formalize knowledge discovery projects within a common framework, a goal that will result in cost and time savings, and will improve understanding, success rates, and acceptance of such projects. The models emphasize **independence** from specific applications, tools, and vendors.
- Five KDP models, including the **nine-step model by Fayyad** et al., the **eight-step model by Anand and Buchner**, the **six-step model by Cios** et al., the **five-step model by Cabena** et al., and the **CRISP-DM model** were introduced. Each model has its strong and weak points, based on its application domain and particular business objectives.
- A very important consideration in the KDP is the relative time spent to complete each step. In general, we acknowledge that the **data preparation step is by far the most time-consuming part of the KDP**.
- The future of KDP models lies in achieving overall **integration** of the entire process through the use of popular industrial standards, such as XML and PMML.

The evolution of knowledge discovery systems has already undergone three distinct phases [16]:

The **first-generation systems** provided only one data mining technique, such as a decision tree algorithm or a clustering algorithm, with very weak support for the overall process framework [11, 15, 18, 20, 21]. They were intended for expert users who already had an understanding of data mining techniques, the underlying data, and the knowledge being sought. Little attention was paid to providing support for the data analyst, and thus the first knowledge discovery systems had very limited commercial success [3]. The general research trend focused on the

```
<?xml version"1.0" encoding="windows-1252"?>
<PMML version="2.0">
<DataDictionary numberOfFields="4">
    <DataField name="PETALLEN" optype="continuous" x-significance="0.89"/>
    <DataField name="PETALWID" optype="continuous" x-significance="0.39"/>
    <DataField name="SEPALWID" optype="continuous" x-significance="0.92"/>
    <DataField name="SPECIES" optype="categorical" x-significance="0.94"/>
    <DataField name="SEPALLEN" optype="continuous"/>
</DataDictionary>
<RegressionModel modelName="..."functionName="regression"
algorithmName="polynomialRegression" modelType="stepwisePolynomialRegression"
targetFieldName="SEPALLEN">
<MiningSchema>
    <MiningField name="PETALLEN" usageType="active"/>
    <MiningField name="PETALWID" usageType="active"/>
    ...
</MiningSchema>
<RegressionTable intercept="−45534.5912666858">
    <NumericPredictor name="PETALLEN" exponent="1" coefficient="8.87" mean="37.58"/>
    <NumericPredictor name="PETALLEN" exponent="2" coefficient="−0.42" mean="1722"/>
    ...
</RegressionTable>
</RegressionModel>
<Extension>
    <X-modelQuality x-rSquared="0.8878700000000001"/>
    ...
</Extension>
</PMML>
```

Figure 2.5. A PMML excerpt that expresses the polynomial regression model for the popular iris dataset generated by the DB2 Intelligent Miner for Data V8.1. Source: http://www.dmg.org/.

22   5. Summary and Bibliographical Notes

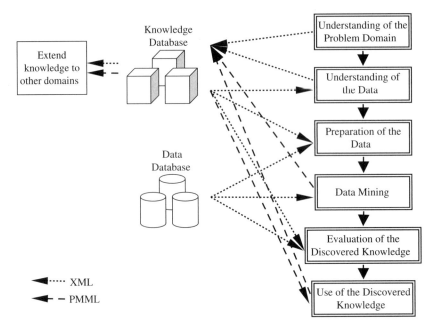

Figure 2.6. Application of PMML and XML standards in the framework of the KDP model.

development of new and improved data mining algorithms rather than on research to support other knowledge discovery activities.
- The **second-generation systems**, called *suites*, were developed in the mid-1990s. They provided multiple types of integrated data analysis methods, as well as support for data cleaning, preprocessing, and visualization. Examples include systems like SPSS's Clementine®, Silicon Graphics's MineSet™, IBM's Intelligent Miner, and SAS Institute's Enterprise Miner.
- The **third-generation systems** were developed in the late 1990s and introduced a vertical approach. These systems addressed specific business problems, such as fraud detection, and provided an interface designed to hide the internal complexity of data mining methods. Some of the suites also introduced knowledge discovery process models to guide the user's work. Examples include MineSet™, which uses the nine-step process model by Fayyad et al., and Clementine®, which uses the CRISP-DM process model.

The KDP model was first discussed during the inaugural workshop on Knowledge Discovery in Databases in 1989 [14]. The main driving factor in defining the model was acknowledgment of the fact that knowledge is the end product of a data-driven discovery process.

In 1996, the foundation for the process model was laid in a book entitled Advances in Knowledge Discovery and Data Mining [7]. The book presented a process model that had resulted from interactions between researchers and industrial data analysts. The model solved problems that were not connected with the details and use of particular data mining techniques but rather with providing support for the highly iterative and complex problem of overall knowledge generation process. The book also emphasized the close involvement of a human analyst in many, if not all, steps of the process [3].

The first KDP model was developed by Fayyad et al. [8–10]. Other KDP models discussed in this Chapter include those by Cabena et al. [4], Anand and Buchner [1, 2], Cios et al. [5, 6, 12], and the CRISP-DM model [17, 19]. A recent survey that includes a comprehensive comparison of several KDPs can be found in [13].

References

1. Anand, S., and Buchner, A. 1998. *Decision Support Using Data Mining*. Financial Times Pitman Publishers, London
2. Anand, S., Hughes, P., and Bell, D. 1998. A data mining methodology for cross-sales. *Knowledge Based Systems Journal*, 10:449–461
3. Brachman, R., and Anand, T. 1996. The process of knowledge discovery in databases: a human-centered approach. In Fayyad, U., Piatetsky-Shapiro, G., Smyth, P., and Uthurusamy, R. (Eds.), *Advances in Knowledge Discovery and Data Mining* 37–58, AAAI Press
4. Cabena, P., Hadjinian, P., Stadler, R., Verhees, J., and Zanasi, A. 1998. *Discovering Data Mining: From Concepts to Implementation*, Prentice Hall Saddle River, New Jersey
5. Cios, K., Teresinska, A., Konieczna, S., Potocka, J., and Sharma, S. 2000. Diagnosing myocardial perfusion from SPECT bull's-eye maps – a knowledge discovery approach. *IEEE Engineering in Medicine and Biology Magazine*, special issue on Medical Data Mining and Knowledge Discovery, 19(4):17–25
6. Cios, K., and Kurgan, L. 2005. Trends in data mining and knowledge discovery. In Pal, N.R., and Jain L.C. (Eds.), Advanced Techniques in Knowledge Discovery and Data Mining, 1–26, Springer Verlag, London.
7. Fayyad, U., Piatesky-Shapiro, G., Smyth, P., and Uthurusamy, R. (Eds.), 1996. *Advances in Knowledge Discovery and Data Mining*, AAAI Press, Cambridge
8. Fayyad, U., Piatetsky-Shapiro, G., and Smyth, P. 1996. From data mining to knowledge discovery: an overview. In Fayyad, U., Piatetsky-Shapiro, G., Smyth, P., and Uthurusamy, R. (Eds.), *Advances in Knowledge Discovery and Data Mining*, 1–34, AAAI Press, Cambridge
9. Fayyad, U., Piatetsky-Shapiro, G., and Smyth, P. 1996. The KDD process for extracting useful knowledge from volumes of data. *Communications of the ACM*, 39(11):27–34
10. Fayyad, U., Piatetsky-Shapiro, G., and Smyth, P. 1996. Knowledge discovery and data mining: towards a unifying framework. *Proceedings of the 2nd International Conference on Knowledge Discovery and Data Mining*, 82–88, Portland, Oregon
11. Klosgen, W. 1992. Problems for knowledge discovery in databases and their treatment in the statistics interpreter explora. *Journal of Intelligent Systems*, 7(7):649–673
12. Kurgan, L., Cios, K., Sontag, M., and Accurso, F. 2005. Mining the Cystic Fibrosis Data. In Zurada, J. and Kantardzic, M. (Eds.), *Next Generation of Data-Mining Applications*, 415–444, IEEE Press Piscataway, NJ
13. Kurgan, L., and Musilek, P. 2006. A survey of knowledge discovery and data mining process models. *Knowledge Engineering Review*, 21(1):1–24
14. Piatetsky-Shapiro, G. 1991. Knowledge discovery in real databases: a report on the IJCAI-89 workshop. *AI Magazine*, 11(5):68–70
15. Piatesky-Shapiro, G., and Matheus, C. 1992. Knowledge discovery workbench for exploring business databases. *International Journal of Intelligent Agents*, 7(7):675–686
16. Piatesky-Shapiro, G. 1999. The data mining industry coming to age. *IEEE Intelligent Systems*, 14(6): 32–33
17. Shearer, C. 2000. The CRISP-DM model: the new blueprint for data mining. *Journal of Data Warehousing*, 5(4):13–19
18. Simoudis, E., Livezey, B., and Kerber, R. 1994. Integrating inductive and deductive reasoning for data mining. *Proceedings of 1994 AAAI Workshop on Knowledge Discovery in Databases*, 37–48, Seattle, Washington, USA
19. Wirth, R., and Hipp, J. 2000. CRISP-DM: towards a standard process model for data mining. *Proceedings of the 4th International Conference on the Practical Applications of Knowledge Discovery and Data Mining*, 29–39, Manchester, UK
20. Ziarko, R., Golan, R., and Edwards, D. 1993. An application of datalogic/R knowledge discovery tool to identify strong predictive rules in stock market data. Working notes from the *Workshop on Knowledge Discovery in Databases*, 89–101, Seattle, Washington
21. Zytow, J., and Baker, J. 1991. Interactive mining of regularities in databases. In Piatesky-Shapiro, G., and Frowley, W. (Eds.), *Knowledge Discovery in Databases*, 31–53, AAAI Press Cambridge

6. Exercises

1. Discuss why we need to standardize knowledge discovery process models.
2. Discuss the difference between terms *data mining* and *knowledge discovery process*. Which of these terms is broader?
3. Imagine that you are a chief data analyst responsible for deploying a knowledge discovery project related to mining data gathered by a major insurance company. The goal is to discover fraud patterns. The customer's data are stored in well-maintained data warehouse, and a team of data analysts who are familiar with the data are at your disposal. The management stresses the importance of analysis, documentation, and deployment of the developed solution(s). Which of the models presented in this Chapter would you choose to carry out the project and why? Also, provide a rationale as to why other models are less suitable in this case.
4. Provide a detailed description of the *Evaluation* and *Deployment* steps in the CRISP-DM process model. Your description should explain the details of the substeps in these two steps.
5. Compare side by side the six-step CRISP-DM and the eight-step model by Anand and Buchner. Discuss the main differences between the two models, and provide an example knowledge discovery application that is best suited for each of the models.
6. Find an industrial application for one of the models discussed in this Chapter. Provide details about the project that used the model, and discuss what benefits were achieved by deploying the model. (hint: see Hirji, K. 2001. Exploring data mining implementation. *Communications of the ACM*, 44(7), 87–93)
7. Provide a one-page summary of the PMML language standard. Your summary must include information about the newest release of the standard and which data mining models are supported by the standard.

Part 2

Data Understanding

3

Data

In this Chapter, we discuss attribute and data types, data storage, and problems of quantity and quality of data.

1. Introduction

The outcome of data mining and knowledge discovery heavily depends on the quality and quantity of available data. Before we discuss data analysis methods, data organization and related issues need to be addressed first. This Chapter focuses on three issues: **data types, data storage techniques**, and **amount and quality of the data**. These topics form the necessary background for subsequent knowledge discovery process steps such as data preprocessing, data mining, representation of generated knowledge, and assessment of generated models. Upon finishing this Chapter, the reader should be able to understand problems associated with available data.

2. Attributes, Data Sets, and Data Storage

Data can have diverse formats and can be stored using a variety of different storage modes. At the most elementary level, a single unit of information is a **value** of a **feature/attribute**, where each feature can take a number of different values. The objects, described by features, are combined to form **data sets**, which in turn are stored as **flat** (rectangular) **files** and in other formats using **databases** and **data warehouses**. The relationships among the above concepts are depicted in Figure 3.1.

In what follows, we provide a detailed explanation of the terminology and concepts introduced above.

2.1. Values, Features, and Objects

There are two key types of values: **numerical** and **symbolic**. Numerical values are expressed by numbers, for instance, real numbers (−1.09, 123.5), integers (1, 44, 125), prime numbers (1, 3, 5), etc. In contrast, symbolic values usually describe qualitative concepts such as colors (white, red) or sizes (small, medium, big).

Features (also known as attributes) are usually described by a set of corresponding values. For instance, height is usually expressed as a set of real numbers. Features described by both numerical and symbolic values can be either **discrete** (categorical) or **continuous**. Discrete features concern a situation in which the total number of values is relatively small (finite), while with continuous features the total number of values is very large (infinite) and covers a specific interval (range).

28   2. Attributes, Data Sets, and Data Storage

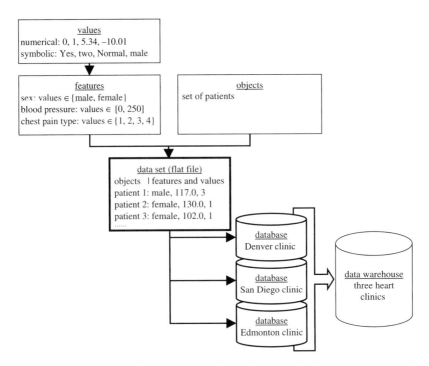

Figure 3.1. Relationships between values, features, objects, data sets, databases and data warehouses.

A special case of a discrete feature is the **binary** (dichotomous) feature, for which there are only two distinct values. A **nominal** (polytomous) feature implies that there is no natural ordering among its values, while an **ordinal** feature implies that some ordering exists. The values for a given feature can be organized as sets, vectors, or arrays. This categorization of data is important for practical reasons. For instance, some preprocessing and data mining methods are only applicable to data described by discrete features. In those cases a process called **discretization** (see Chapter 8) becomes a necessary preprocessing step to transform continuous features into discrete ones, and this step must be completed before the data mining step is performed.

Objects (also known as records, examples, units, cases, individuals, data points) represent entities described by one or more features. The term *multivariate data* refers to situation in which an object is described by many features, while with *univariate data* a single feature describes an object.

Let us consider an example concerning patients at a heart disease clinic. A patient is an object that can be described by a number of features, such as name, sex, age, diagnostic test results such as blood pressure, cholesterol level, and qualitative evaluations like chest pain and its severity type. An example of a "patient" object is shown in Figure 3.2.

An important issue, from the point of view of the knowledge discovery process, is how different types of features and values are manipulated. In particular, any operation on multiple objects, such as the comparison of feature values or the computation of distance, should be carefully analyzed and designed. For instance, the symbolic value "two" usually cannot be compared with the numerical value 2, unless some conversion is performed. Although computation of the distance between two numerical values is straightforward, performing the same computation between two nominal values (such as "white" and "red", or "chest pain of typ 1" and "chest pain of type 4") requires special attention. In other words, how can we measure distance between colors (this could be done using chromaticity diagrams) or distance between chest pain types (this is much more difficult)? In some cases, it might be impossible to calculate such a distance.

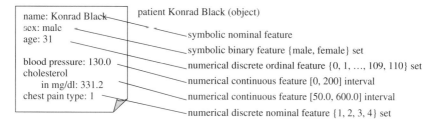

Figure 3.2. Patient record (object).

An important issue concerning data is the limited comprehension of numbers by humans, who are the ultimate users of the (generated) knowledge. For instance, most people will not comprehend a cholesterol value of 331.2, while they can easily understand the meaning when this numerical value is expressed in terms of aggregated information such as a "high" or "low" level of cholesterol. In other words, information is often "granulated" and represented at a higher level of abstraction (aggregation). In a similar manner, operations or relationships between features can be quantified on an aggregated level. In general, **information granulation** means encapsulation of numeric values into single conceptual entities (see Chapter 5). Examples include encapsulation of elements by sets, or encapsulation of intervals by numbers. Understanding of the concept of encapsulation, also referred to as a *discovery window*, is very important in the framework of knowledge discovery. Continuing with our cholesterol example, we may be satisfied with a single numerical value, say 331.2, which expresses the highest level of granularity (Figure 3.3a). Alternatively, we may want to define this value as belonging to an interval [300, 400], the next, lower, granularity level, which captures meaning of the "high" value of cholesterol (Figure 3.3b). Through the refinement of the discovery window, we can change the "crisp" character of the word "high" by using notion of fuzzy sets (Figure 3.3c) or rough sets (Figure 3.3d). In the case of fuzzy sets, we express the cholesterol value as being high to some degree, or being normal to some other degree. The lowest level of granularity (i.e., the highest generality) occurs when the discovery window covers the entire spectrum of values, which implies that the knowledge discovery process focuses on the entire data set.

2.2. Data Sets

Objects described by the same features are grouped to form **data sets**. Many data mining and statistical data analysis tools assume that data sets are organized as **flat files,** in a rectangularly formatted table composed of rows and columns. The rows represent objects and the columns represent features, resulting in a flat file that forms a two-dimensional array. Flat files are used to store data in a simple text file format, and they are often generated from data stored in other, more complex formats, such as spreadsheets or databases.

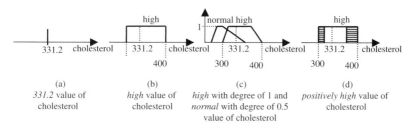

Figure 3.3. Information granularization methods. (a) Numerical; (b) interval based; (c) fuzzy set based; (d) rough set based.

Table 3.1. Flat file data set for heart clinic patients.

Name	Age	Sex	Blood pressure	Blood pressure test date	Cholesterol in mg/dl	Cholesterol test date	Chest pain type	Defect type	Diagnosis
Konrad Black	31	male	130.0	05/05/2005	NULL	NULL	NULL	NULL	NULL
Konrad Black	31	male	130.0	05/05/2005	331.2	05/21/2005	1	normal	absent
Magda Doe	26	female	115.0	01/03/2002	NULL	NULL	4	fixed	present
Magda Doe	26	female	115.0	01/03/2002	407.5	06/22/2005	NULL	NULL	NULL
Anna White	56	female	120.0	12/30/1999	45.0	12/30/1999	2	normal	absent
...	...	...	...	...	...	...	...	...	...

To illustrate, let us design a data set of heart patients, shown in Table 3.1. Each patient (object) is described by the following set of features:

– *name* (symbolic nominal feature)
– *age* (numerical discrete ordinal feature from the {0, 1, ..., 109, 110} set)
– *sex* (symbolic binary feature from the {male, female} set)
– *blood pressure* (numerical continuous feature from the [0, 200] interval)
– *blood pressure test date* (date type feature)
– *cholesterol in mg/dl* (numerical continuous feature from the [50.0, 600.0] interval)
– *cholesterol test date* (date type feature)
– *chest pain type* (numerical discrete nominal feature from the {1, 2, 3, 4} set)
– *defect type* (symbolic nominal feature from the {normal, fixed, reversible} set)
– *diagnosis* (symbolic binary feature from the {present, absent} set)

First, let us note that a new feature type, namely, date, has been introduced. Date can be treated as a numerical continuous feature (after some conversion) but usually is stored in a customary format like mm/dd/yyyy, which is between being numerical and symbolic. A number of other special feature types, such as text, image, video, etc., are briefly described in the next section. Second, a new NULL value has been introduced. This value indicates that the corresponding feature is unknown (not measured or missing) for the object. Third, we observe that it is possible that several different objects are related to the same patient. For instance, Konrad Black first came to get a blood pressure test and later came to do the remaining tests and was diagnosed. Finally, for the object associated with Anna White, the cholesterol value is 45.0, which is outside the interval defined/acceptable for this feature, and thus we may suspect that the value may be incorrect, e.g., it may be that 45.0 was entered instead of the correct value of 450.0. These observations regarding different data types and missing and erroneous data lead to certain difficulties associated with the knowledge discovery process, which are described later in the Chapter.

2.3. Data Storage: Databases and Data Warehouses

Although flat files are a popular format in which to store data, data mining tools used in the knowledge discovery process can be applied to a great variety of other data formats, such as those found in databases, data warehouses, advanced database systems, and the World Wide Web (WWW). Advanced database systems include object-oriented and object-relational databases, as well as data-specific databases, such as transactional, spatial, temporal, text, and multimedia databases.

In the knowledge discovery process we often deal with large data sets that require special data storage and management systems. There are four reasons for using a specialized system:

1. *The corresponding data set may not fit into the memory of a computer used for data mining*, and thus a data management system is used to fetch the data.
2. *The data mining methods may need to work on many different subsets of data*, and therefore a data management system is required to efficiently retrieve the required pieces of data.
3. *The data may need to be dynamically added and updated, sometimes by different people in different locations.* A data management system is required to handle updating (and also offers failure and recovery options).
4. *The flat file may include significant portions of redundant information*, which can be avoided if a single data set is stored in multiple tables (a characteristic feature of data storage and management systems).

A poll (September 2005) performed by KDNuggets (http://www.kdnuggets.com) asked which data formats the users had analyzed in the last 12 months. The results showed that flat files, which are often extracted from relational databases, were used 26% of the time, time series (temporal databases) 13%, text databases 11%, transactional databases 10%, and WWW clickstream data, spatial databases, and WWW content data 5% each. The results indicated that although flat files were still the most often used, other formats gained significant interest and therefore are discussed below.

2.3.1. Databases

The above mentioned reasons result in a need to use a specialized system, a **DataBase Management System** (DBMS). It consists of a **database** that stores the data, and a set of programs for management and fast access to the database. The software provides numerous services, such as the ability to define the structure (**schema**) of the database, to store the data, to access the data in concurrent ways (several users may access the data at the same time) and distributed ways (the data are stored in different locations), and to ensure the security and consistency of the stored data (for instance, to protect against unauthorized access or a system crash). The most common database type is a **relational database**, which consists of a set of tables. Each table is rectangular and can be perceived as being analogous to a single flat file. At the same time, the database terminology is slightly different than that used in data mining. The tables consist of **attributes** (also called columns or fields) and **tuples** (also called records and rows). Most importantly, each table is assigned a unique name, and each tuple in a table is assigned a special attribute, called a **key**, that defines its unique identifiers. Relational databases also include the **entity-relational** (ER) data model, which defines a set of entities (tables, tuples, etc.) and their relationships. Next, we define a relational database for a heart clinic, which will extend our flat file–based data storage and allow better understanding of the above concepts. In order to better utilize the capabilities of a relational database, our original flat file is divided into four relational tables. These include the table *patient*, which stores basic information about patients together with their diagnostic information (the key is the *patient ID* attribute), and two tables *blood_pressure_test* and *cholesterol_test* that store the corresponding test values (see Figure 3.4). The fourth table, *performed_tests*, represents the relationship between multiple other tables. In the case of our relational database, the relationship shows which patients had which tests.

Analysis of the above example shows that the use of multiple tables results in the removal of redundant information included in the flat file. At the same time, the data are divided into smaller blocks and thus are easier to manipulate and fit into the available memory. Another important feature of a DBMS is the availability of a specialized language, called **Structured Query Language** (SQL), that aims to provide fast and convenient access to portions of the entire database. For instance, we may want to extract information about the tests performed between specific dates or obtain a list of all patients who have been diagnosed with a certain diagnosis

patient

Patient ID	Name	Age	Sex	Chest pain type	Defect type	Diagnosis
P1	Konrad Black	31	male	1	normal	absent
P2	Magda Doe	26	female	4	fixed	present
P3	Anna White	56	female	2	normal	absent
...	...	...	...	...	...	...

blood_pressure_test

Blood pressure test ID	Patient ID	Blood pressure	Blood pressure test date
BPT1	P1	130.0	05/05/2005
BPT2	P2	115.0	01/03/2002
BPT3	P3	120.0	12/30/1999
...	...	...	...

cholesterol_test

Cholesterol test ID	Patient ID	Cholesterol in mg/dl	Cholesterol test date
SCT1	P1	331.2	05/21/2005
SCT2	P2	407.5	06/22/2005
SCT3	P3	45.0	12/30/1999
...	...	...	...

performed_tests

Patient ID	Blood pressure test	Cholesterol test
P1	BPT1	SCT1
P2	BPT2	SCT2
P3	BPT3	SCT3
...	...	...

Figure 3.4. Relational database for heart clinic patients

type. This process is relatively simple when a DBMS and an SQL are used. In contrast, with a flat file, the user himself is forced to manipulate the data to select the desired portion – a process that can be tedious and difficult to perform.

SQL allows the user to specify queries that contain a list of relevant attributes and constraints on those attributes. Oftentimes, DBMSs provide a graphical user interface to facilitate query formulation. The user's query is automatically transformed into a set of relational operations, such as join, selection, and projection, optimized for time – and/or resource–efficient processing and executed by the DBMS (see Chapter 6). SQL also provides the ability to aggregate data by computing functions, such as summations, average, count, maximum, and minimum. This ability allows the user to receive answers to more complex and interesting queries. For instance, the user could ask the DBMS to compute how many blood pressure tests were performed in a particular month, or what the average blood pressure is for female patients.

2.3.2. Data Warehouses

Now we consider a more elaborate scenario in which a number of heart clinics in different cities, each having its own database, belong to the same company. Using a database, we are able to analyze (mine) data in individual clinics, but it can be very difficult to perform analysis across all clinics. In this case, a **data warehouse**, which is a repository of data collected in different locations (relational databases) and stored using a unified schema, is used. Data warehouses are usually created by applying a set of processing steps to data coming from multiple databases. The steps usually include data cleaning, data transformation, data integration, data loading, and periodical data update (see Figure 3.5).

While the main purpose of a database is storage of the data, the main purpose of a data warehouse is analysis of the data. Consequently, the data in a data warehouse are organized around a set of subjects of interest to the user. For instance, in case of the heart clinic, these subjects could be patients, clinical test types, and diagnoses. The analysis is performed to provide information

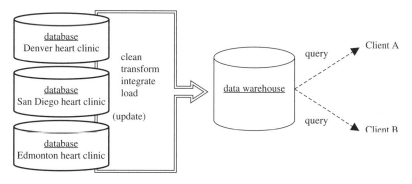

Figure 3.5. Typical architecture of a data warehouse system.

(knowledge) from a historical perspective. For instance, we could ask for a breakdown of the most performed clinical tests in the past five years. Such requests (queries) require the availability of summarized information, and therefore data warehouses do not store the same content as the source databases but rather a summary of these data. For instance, a data warehouse may not store individual blood pressure test values, but rather the number of tests performed by each clinic, the time interval (say a month), and a given set of patient age ranges.

A data warehouse usually uses a multidimensional database structure, where each dimension corresponds to an attribute or a set of attributes selected by the user to be included in the schema. Each cell in the database corresponds to some summarized (aggregated) measure, such as average, count, minimum, etc. The actual implementation of the warehouse can be a relational database or a multidimensional **data cube**. The latter structure provides a three-dimensional view of the data and allows for fast access to the summarized data via precomputation. An example of the data cube for the heart clinic's data warehouse is shown in Figure 3.6. The cube has three dimensions: clinic (Denver, San Diego, Edmonton), time (expressed in months), and age range (0–8, 9–21, 21–45, 45–65, over 65). The values are in thousands and show how many blood pressure tests were performed. Each dimension can be further summarized, e.g., we can collapse months into quarters and can convert age into some numeric intervals (ranges), say, intervals of values within 0–45 and those over 45.

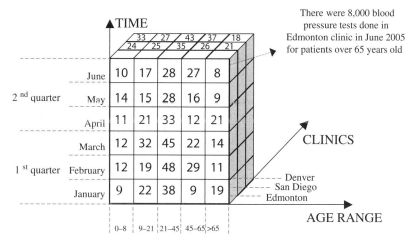

Figure 3.6. A multidimensional data cube. The values are in thousands and show the number of performed blood pressure tests. For readability, only some of the cube cells are shown.

2. Attributes, Data Sets, and Data Storage

The availability of multidimensional data and precomputation gave rise to **On-Line Analytical Processing** (**OLAP**). OLAP makes use of some background knowledge about the domain (which concerns the data in hand) to present data at different levels of abstraction. Two commonly used OLAP operations are roll-up and drill-down. The first operation merges data at one or more dimensions to provide the user with a higher–level summarization, while the latter breaks down data into subranges to present the user with more detailed information. An example for the heart clinics warehouse is shown in Figure 3.7.

A detailed discussion of data warehouses and OLAP is provided in Chapter 6.

2.4. Advanced Data Storage

While relational databases (warehouses) are usually used by businesses like retail stores and banks, other more specialized and advanced database systems have emerged in recent years. The new breed of databases satisfies the needs of more specialized users who must handle more than just numerical and nominal data. The new databases handle **transactional** data; **spatial** data, such as maps; **hypertext**, such as HTML and XML; **multimedia** data, such as combinations of text, image, video and audio; **temporal** data, such as time–related series concerning stock exchange and historical records; and the **WWW**, which is enormously large and distributed (and accessible via the Internet). The special data types require equally specialized databases that utilize efficient data structures and methods for handling operations on such complex data structures. The challenges for databases are to cope with variable-length objects, structured and semi-structured data, unstructured text in various languages, multimedia data formats, and very large amounts of data.

Due to these challenges, numerous specialized databases have been developed. These include object-oriented and object-relational databases, spatial databases, temporal databases, text databases, multimedia databases, and the WWW. Each of these databases not only allows the user to store data in the corresponding format but also provides facilities to efficiently store, update,

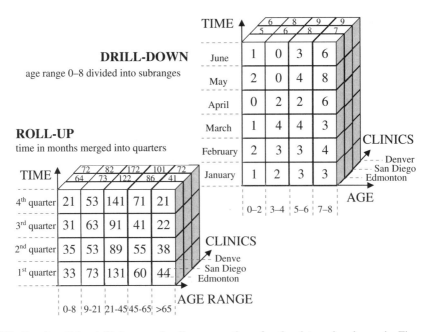

Figure 3.7. Results of the drill-down and roll-up operations for the data cube shown in Figure 3.6. The values are in thousands; for drill-down, we subdivide the age range dimension; for roll-up, we merge the time dimension. For readability, only some of the cube cells are shown.

and retrieve large amount of data. Below we briefly introduce the major types of specialized database systems.

2.4.1. Object-oriented Databases

Object-oriented databases are based on the object-oriented programming paradigm, which treats each stored entity as an object (this object is different than the object or record described earlier in this Chapter). The object encapsulates a set of **variables** (which provide its description), a set of **messages** that the object uses to communicate with other objects, and a set of **methods** that contain code to implement the messages. For instance, in the heart clinic example, the object could be a patient, its variables could be name, address, sex, etc., and methods could implement messages that, for instance, would retrieve particular test values for particular times. The key element of object-oriented databases is the ability to group similar (or identical) objects into classes, which can be organized into hierarchies. For instance, the *patient* class would have the variables name, address, and sex, and its instances would be particular patients. The patient class could have a subclass called *retired patient*, which would inherit all variables of the patient class, but would also have new variables, such as date of home release, that pertain to patients who are not longer under treatment. The inheritance between the classes in a hierarchy allows for better information sharing.

2.4.2. Object-relational Databases

Object-relational databases are based on the object-relational data model. In essence, this is a relational database that has been extended by providing a set of complex data types to handle complex objects. This extension requires availability of a specialized query language to retrieve the complex data from the database. Most common extensions include the ability to handle data types such as trees, graphs, lists, class hierarchies, and inheritance.

2.4.3. Transactional Databases

Transactional databases are stored as flat files and consist of records that represent **transactions**. A transaction includes a unique identifier and a set of **items** that make up the transaction. One example of a transaction is a record of a purchase in a store, which consists of a list of purchased items, the purchase identifier, and some other information about the sale. The additional information is usually stored in a separate file, and may include customer name, cashier name, date, store branch, etc. An example transactional database is shown in Figure 3.8.

The single most important difference between the relational and the transactional databases is that the latter store a set of items (values), rather than a set of values of the related features. A transactional database stores information about the presence/absence of an item, while a relational database stores information about specific feature values that an example (item) possesses. The transactional databases are often used to identify sets of items that frequently coexist in transactions. For instance, given a transactional database for the heart clinic that stores names of tests as items, and where each transaction denotes a single patient's visit to the clinic, we can find out which tests are often performed together. Based on the results, we can modify business practice in a clinic to prescribe these tests together instead of prescribing them separately. Transactional data and corresponding data mining algorithms are described in Chapter 10.

sales

Transation ID	Set of item IDs
TR000001	Item1, Item32, Item52, Item71
TR000002	Item2, Item3, Item4, Item57, Item 92, Item93
TR000003	Item11, Item101
...	...

Figure 3.8. Example transactional database

2.4.4. Spatial Databases

Spatial databases are designed to handle spatially related data. Example data includes geographical maps, satellite images, medical images, and VLSI chip-design data. Spatial data can be represented in two ways: in a **raster format** and in a **vector format**. The raster format concerns using n–dimensional pixel maps, while the vector format requires representing all considered objects as simple geometrical objects, such as lines, triangles, polygons, etc., and using vector-based geometry to compute relations between the objects. Spatial databases allow the user to obtain unique information about the stored data. For instance, for the heart clinic, we can ask whether patients living in a specific neighborhood are more susceptible to certain heart conditions, or whether the clinic accepts more patients from the north side than from the south side of town.

2.4.5. Temporal Databases

Temporal databases (also called time-series databases) are used to store time-related data. Similarly, as in the case of object-relational databases, the temporal databases extend the relational databases to handle time-related features. Such attributes may be defined using timestamps of different semantics such as days and months, hours and minutes, days of the week, etc. The database keeps time-related features by storing sequences of their values that change with time. In contrast, a relational database usually stores the most recent value only. The temporal data allow the user to find unique patterns in the data, which are usually related to trends of changes. In the case of the heart clinic, we can ask whether blood pressure has a tendency to increase or decrease with age for particular patients or patient groups–say, those with high cholesterol–or we could compare rates of decrease or increase.

2.4.6. Text Databases

Text databases include features (attributes) that contain word descriptions for objects. These features are not simple nominal but hold long sentences or paragraphs of text. Examples for the heart clinic would be reports from patient interviews, physician's notes and recommendations, descriptions of how a given drug works for a given patient, etc. The text may be either **unstructured**, like sentences in plain written language (English, Polish, Spanish, etc.); **semistructured**, where some words or parts of the sentence are annotated, such as descriptions of how a drug works, which may use special annotations for the drug's name and the dose; and **structured**, where all the words are annotated, like a physician's recommendation that may be a fixed form listing only specific drugs and doses. Other, easy-to-understand examples include text documents as unstructured text, HTML pages and emails as semistructured text, and library data as structured text. Analysis of such databases requires special tools and integration with text data hierarchies, such as dictionaries, thesauruses, and specialized discipline-oriented term-classification systems, such as those in medicine, engineering, economy, etc. The mining of text databases is described in Chapter 14.

2.4.7. Multimedia Databases

Multimedia databases allow storage, retrieval, and manipulation of image, video, and audio data. The main concern regarding such data sources is their very large size, and therefore, specialized storage and search techniques are required. Additionally, both video and audio data are recorded in real time, and thus the database must include mechanisms that assure a steady and predefined rate of acquisition to avoid gaps, system buffer overflows, etc. An example application of a multimedia database for the heart clinic would be to find the relation between a video of heart motion and a recording of the heart beats.

2.4.8. World Wide Web

The World Wide Web (**WWW**) is an enormous distributed repository of data that is linked together via the use of **hyperlinks**. The hyperlinks link together individual data objects of possibly different types, allowing for interactive access to the information. The most specific characteristic of the WWW is that users seek information traversing between objects via links. The WWW also provides specialized query engines such as Google, Yahoo!, AltaVista, and Prodigy.

Finally, a comment on **heterogeneous databases**. These consist of a set of interconnected databases that communicate between themselves in order to exchange data and provide answers to user queries. The individual databases are connected via intra- and intercomputer networks, and usually they are of different types. The biggest challenge for a heterogeneous database is that objects in the component databases usually differ substantially, resulting in difficulties in developing common semantics to facilitate communication between them. The handling of such databases is beyond the scope of this book.

2.5. Data Storage and Data Mining

We address here several issues concerning data mining from the perspective of the data storage system used. From the point of view of the data warehouse and database users, data mining is often seen as an execution of a set of OLAP commands, but this perception is incorrect. Similarly, the ability to perform data or information retrieval (selection of data from a large repository, such as the WWW) or to find aggregate values should not be confused with data mining but instead should be categorized as using a database or an information retrieval system, respectively. Although the capabilities of the described storage systems may seem advanced and may appear to provide everything a user might need when analyzing data, the field of data mining goes much further. Data mining provides more complex techniques for understanding data and generating new knowledge than the simple summarization-like processing offered by OLAP. It allows for automated discovery of patterns and trends in the data. For instance, based on the information available in the heart clinic database, data mining may allow a user to learn which patients are susceptible to a particular heart condition, to discover that increased blood pressure over a period of time may lead to certain heart defects, or to identify deviations such as patients with unusually high or low blood pressure associated with their given heart condition.

3. Issues Concerning the Amount and Quality of Data

Several fundamental issues related to the data have a significant impact on the quality of the outcome of a knowledge discovery process. These include the following:

- the huge volume of data, and the related problem of scalability of data mining methods
- the dynamic nature of the data, which are constantly being updated/changed
- problems related to data quality, such as imprecision, incompleteness, noise, missing values, and redundancy

The above issues must be addressed before we perform any further work on the data; otherwise, we may encounter serious problems later in the knowledge discovery process. For instance, only some of the data mining methods can be used on data that contain missing values or noise. In this case, we need to make sure that a suitable method is available to analyze the data. When developing a real-time system, we need either to select an appropriate data mining system that can analyze the data in the specified amount of time or to reduce the processing time, for instance, by reducing the size of the data.

In the last decade, a significant amount of work has been devoted to data security and privacy issues. Data miners must be aware of the existing standards for protection against unauthorized access to data and must know how to prepare the data to guarantee confidentiality of the information present therein. The best example is medical data, for which the Health Insurance Portability and Accountability (HIPAA) dictates, in a very stringent way, how to design databases so as to store data securely and to preserve patient confidentiality. These issues are described in Chapter 16.

3.1. High Dimensionality

Although many data mining systems are available, only some can be considered as "truly" mining the data. The distinguishing factor is their ability to handle massive amount of data, which requires the use of **scalable** algorithms and methods. The scalability issue is not related to efficient storage and retrieval of the data (these are handled by using DBMSs) but instead to the algorithm design. Systems that are not capable of handling large quantities of data are known by the terms *machine learning* or *statistical data analysis systems*. For each data mining method under consideration, one needs to evaluate the sensitivity to the size of the data, which translates into the amount of time required for processing such data. Application domains such as the analysis of data from large retail store chains (e.g., Wal-Mart) deal with hundreds of millions of objects, while in the field of bioinformatics, data sets are described by several thousand features (e.g., microarray data analysis). Published results of a survey on the largest and most heavily used commercial databases show that the average size of Unix databases experienced a 6-fold, and of Windows databases a 14-fold increase, between 2001 and 2003. Large commercial databases now average 10 billion objects. Although only a selected portion of these objects is used for data mining purposes, the ability to cope with large quantities of data remains one of the most important data mining issues. Three dimensions are usually considered:

- The number of objects, which may range from a few hundred to a few billion
- The number of features, which may range from a few to several thousand
- The number of values a feature assumes, which may range from two to a few million

The ability of a particular data mining algorithm to cope with highly dimensional inputs is usually described by its asymptotic complexity, i.e., an estimate of an "order of magnitude" of the total number of operations, which translates into the specific duration of run time. The asymptotic complexity describes the growth rate of the algorithm's run time as the size of each dimension increases. The most commonly performed asymptotic complexity analysis describes scalability with respect to the number of objects.

The following example illustrates the importance of the asymptotic complexity of a data mining algorithm. Assume the user wants to generate knowledge in terms of rules from the data using either a decision tree or a rule algorithm (for details, see Chapter 12). Assume we want to use the following algorithms:

- *See5*, which has log-linear complexity, i.e., $O(n^* \log(n))$, n is the number of objects
- *DataSqueezer*, which has linear complexity, i.e., $O(n)$
- *CLIP4*, which has quadratic complexity, i.e., $O(n^2)$
- *C4.5 rules*, which has cubic complexity, i.e., $O(n^3)$

To make the example more compelling, assume that the constant for the linear and log-linear algorithms is 100 times larger than for the quadratic and cubic algorithms–a situation that strongly favors the latter two algorithms. Table 3.2 shows relative difference in the running time in seconds required to compute the rules for the increasing number of objects for the four algorithms.

Table 3.2. Comparison of running time with respect to the "number of objects" for four rule-learning algorithms.

Number of objects in thousands	Running time [sec]			
	DataSqueezer $y = 100^*a^*x$	See5 $y = 100^*a^*x^*\log(x)$	CLIP4 $y = a^*x^2$	C4.5 rules $y = a^*x^3$
100	1000.0	2000.0	1000.0	100000.0
200	2000.0	4602.1	4000.0	800000.0
300	3000.0	7431.4	9000.0	2700000.0
400	4000.0	10408.2	16000.0	6400000.0
500	5000.0	13494.9	25000.0	12500000.0
600	6000.0	16668.9	36000.0	21600000.0
700	7000.0	19915.7	49000.0	34300000.0
800	8000.0	23224.7	64000.0	51200000.0
900	9000.0	26588.2	81000.0	72900000.0
1000	10000.0	30000.0	100000.0	100000000.0
1100	11000.0	33455.3	121000.0	133100000.0
...	...	...	...	...

Thus, for 100,000 objects, the fastest linear algorithm computes the results in 1,000 seconds, while the slowest cubic algorithm needs 100,000 seconds. When the number of objects increases tenfold, the time needed to compute the results increases to 10,000 seconds for the linear algorithm and to 100,000,000 seconds for the cubic algorithm. This example shows that for the linear algorithm, doubling the number of the objects results in doubling the run time, while for the cubic algorithm the run time increases eightfold. Note that the rate at which the run time increases as the number of the objects increases (or any other dimension) is more important than the absolute value of the run time. The functional relationship between the number of objects and the run time is shown in Figure 3.9.

A number of techniques are available to improve the scalability of data mining algorithms. These can be divided into two groups:

– Techniques that **speed up the algorithm**. This outcome can be achieved through use of heuristics, optimization, and parallelization. Heuristics simplify the processing of the data

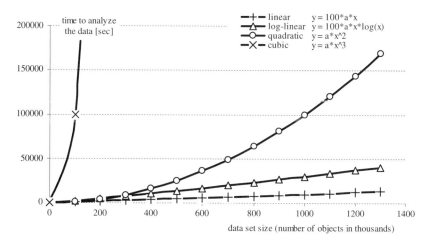

Figure 3.9. Functional relation between data set size and run time for algorithms with linear, log-linear, quadratic, and cubic asymptotic complexity.

performed by the algorithm, e.g., only rules of a certain maximal length are considered, and longer rules are never generated. Optimization of the algorithm is often achieved by using efficient data structures, such as bit vectors, hash tables, and binary search trees, to store and manipulate the data. Finally, parallelization aims to distribute the processing of the data by the algorithm into several processors that work in parallel and thus can speed up the computations.

– Techniques that **partition the data set**. This outcome can be achieved by reducing the dimensionality of the input data set through reduction of the number of objects, features, number of values per feature, and sequential or parallel processing of data divided into subsets. With dimensionality reduction, the data are sampled and only a representative subset of objects and/or features is used (see Chapter 7). This approach is especially viable when dealing with large data sets, in which the objects and/or attributes may be redundant, very similar, or irrelevant, and thus a subset of the objects and/or features may provide representative information about the entire data set. Alternatively, for some algorithms, it is beneficial or necessary to reduce the number of values an attribute assumes by using discretization algorithms (see Chapter 8). Finally, the division of data into subsets and the processing of these subsets sequentially or in parallel is beneficial when the complexity of the data mining algorithms is worse than linear. As an example, let us consider a cubic complexity algorithm from Table 3.2. When used to analyze 1,000,000 objects, the algorithm takes 100,000,000 seconds to compute the result. However, if the data set is partitioned into 10 subsets of 100,000 objects each and the subsets are processed sequentially (processing of each subset takes "only" 100,000 seconds), then all the data can be analyzed in $10*100,000 = 1,000,000$ seconds – a whopping 99,000,000 seconds savings. Note that the division of the data into subsets should be performed only when the results generated for each of the subsets can be combined into a result that spans the entire data set.

3.2. Dynamic Data

Data sets are often dynamic in nature, i.e., new objects and/or features may be added and some objects and/or features may be removed or replaced by new ones. In this case, data mining algorithms should also evolve with time, which means that the knowledge derived so far should be also incrementally updated. The difference between **incremental** and **nonincremental** DM algorithms is shown in Figure 3.10.

The main challenge in incremental data mining methods is merging the newly generated knowledge from new data with the existing, previous knowledge. The merger may be as simple as adding the new knowledge to the existing knowledge, but most often it requires modifying the existing knowledge to preserve consistency.

3.3. Imprecise, Incomplete, and Redundant Data

Real data often include **imprecise data** objects. For instance, we may not know the exact value of a given medical test, but instead we know whether the value is high, average, or low. In such cases, fuzzy and/or rough sets can be used to process such imprecise information (see Chapter ??).

The term **incomplete data** refers to the situation in which the available data do not contain enough information to discover new (desired) knowledge. Incompleteness may be a result of an insufficient feature description of the objects, insufficient number of objects, or missing values for a given feature. For instance, when analyzing heart patient data, if one wanted to distinguish between sick and healthy patients but only demographic information was available, the task could be impossible to complete. In dealing with incomplete data, we first need to identify the problem and then take measures to remove it. To detect incompleteness, the user must analyze the existing data set and determine whether the existing features and objects give a sufficiently rich representation of the problem at hand. A common sign of incompleteness is when generated

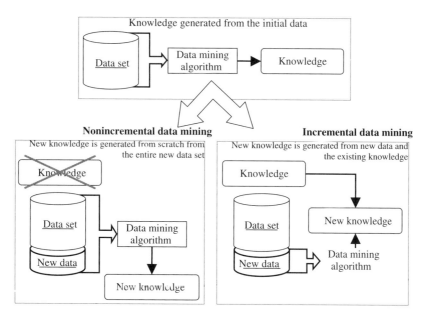

Figure 3.10. Incremental vs. nonincremental data mining.

(new) knowledge is of low quality and cannot be improved by using other data mining methods. The most obvious way to improve the incomplete data is to gather and record additional data, such as new features and/or new objects. A discussion of missing values can be found in the next section.

The term **redundant data** refers to a problem when there are two or more identical objects, or two or more features that are strongly correlated. Redundant data (objects, one of the correlated features) are thus removed to speed up the processing time. In some cases, however, redundant data may reveal useful information, i.e., frequency of identical objects may provide useful information. A special case of redundant data is **irrelevant data**, where some objects and/or features are insignificant with respect to data analysis. For instance, we can expect that a patient's name is irrelevant with respect to his/her heart condition and thus can be removed. Redundant data can be identified by feature selection and extraction algorithms (see Chapter 7).

3.4. Missing Values

Many data sets are plagued by the problem of **missing values**. These can result from incomplete manual data entry, incorrect measurements, equipment errors, etc. Such values are usually denoted by "NULL", "*" and "?" values. In some domains (as in medicine), it is common to encounter data with a large percentage of missing values, even over 50% of all values. The methods for dealing with these missing values can be divided into two categories:

– **Removal of missing data**. In this case the objects and/or features with missing values are simply discarded. This approach is effective only when the removed features are not crucial to the analysis, since the removal would than result in decreasing the information content of the data. It is practical only when the data contain small amounts of missing values and when analysis of the remaining complete data will not be biased by the removal. For example, in distinguishing between sick and healthy heart patients, the removal of the blood pressure feature will result in biasing the discovered knowledge towards other features, although obviously the blood pressure feature is important.

- **Imputation (filling in) of missing data.** Imputation is performed using a number of different algorithms, which can be subdivided into single and multiple imputation methods. In single-imputation methods a missing value is imputed by a single value, while with multiple-imputation methods, several likelihood-ordered choices for imputing the missing value are computed and one "best" value is selected.

Two missing data imputation methods, the mean and the hot deck, are described next. In **mean imputation**, the mean of the values of a feature that contains missing data is used to fill in the missing values. For a symbolic (categorical) feature, a mode, which is the most frequent value, is used instead of the mean. The algorithm imputes missing values for each attribute separately. Mean imputation can be conditional or unconditional. The conditional mean method imputes a mean value that depends on the values of the complete features for the incomplete object. In **hot deck imputation**, for each object that contains missing values, the most similar object is found, and the missing values are imputed from that object. If the most similar record also contains missing values for the same features as in the original object, then the similar record is discarded and another closest object is found. The procedure is repeated until all the missing values are successfully imputed or the entire data set is searched. When no similar object with the required values is found, the closest object with the minimum number of missing values is chosen to impute the missing values. The similarity between objects is usually measured using **distance**. One of the commonly used measures for both numerical and categorical values assumes a distance of 0 between two corresponding features if both have the same value; otherwise, the distance is 1. A distance of 1 is also assumed for a feature for which at least one object has a missing value. The data set for heart clinic patients given in Table 3.1 can be used to illustrate the working of the missing data methods (see Figure 3.11). An example conditional-mean imputation rule could state that the mean value for the *chest pain type* feature is imputed only if the *diagnosis* feature has the value *absent*; otherwise, a 2*mean value is imputed. Another popular conditional-mean imputation rule constrains the computation of the mean value to a certain subset of the objects: for instance, the imputed mean values of features for a given object might be computed by using only objects with the same value of the *diagnosis* feature, i.e., *absent* or *present*.

Several other more complex missing-data imputation methods can also be used. Some compute the most probable value to replace the missing value based on information from the complete portion of the data. An example is a regression imputation, which is performed by regression of the missing values using complete values for a given object, e.g., missing value for the *cholesterol* feature of object 3 would be computed by using a linear function of values for the remaining numerical features, such as *age*, *blood pressure*, and *chest pain type*. Several regression models can be used, including linear, logistic, polytomous, etc. Usually, the logistic regression model is applied for binary features, linear regression for numerical continuous features, and polytomous regression for discrete features (see Chapter 11 for information about regression methods).

3.5. Noise

Noise in the data is defined as a value that is a random error or variance in a measured feature. Depending on the amount in the data, noise it can be a substantial problem that can jeopardize the knowledge discovery process. The influence of noise on the data can be prevented by imposing constraints on features in order to detect anomalies when the data are entered. For instance, DBMS usually provides means to define customized constrains for individual attributes. When noise is already present, it can be removed by using one of the following methods: manual inspection with the use of predefined constraints on feature values, binning, and clustering.

Original data set with missing values

Object ID	Name	Age	Sex	Blood pressure	Cholesterol in mg/dl	Chest pain type	Defect type	Diagnosis
1	Konrad Black	31	male	130.0	NULL	NULL	NULL	NULL
2	Konrad Black	31	male	130.0	331.2	1	normal	absent
3	Magda Doe	26	female	115.0	NULL	4	fixed	present
4	Magda Doe	26	female	115.0	407.5	NULL	NULL	NULL
5	Anna White	56	female	120.0	45.0	2	normal	absent

Data set with removed objects with missing values

Object ID	Name	Age	Sex	Blood pressure	Cholesterol in mg/dl	Chest pain type	Defect type	Diagnosis
2	Konrad Black	31	male	130.0	331.2	1	normal	absent
5	Anna White	56	female	120.0	45.0	2	normal	absent

Data set with removed features with missing values

Object ID	Name	Age	Sex	Blood pressure
1	Konrad Black	31	male	130.0
2	Konrad Black	31	male	130.0
3	Magda Doe	26	female	115.0
4	Magda Doe	26	female	115.0
5	Anna White	56	female	120.0

Data set with missing values imputed using the mean imputation method

Object ID	Name	Age	Sex	Blood pressure	Cholesterol in mg/dl	Chest pain type	Defect type	Diagnosis
1	Konrad Black	31	male	130.0	261.2	2	normal	absent
2	Konrad Black	31	male	130.0	331.2	1	normal	absent
3	Magda Doe	26	female	115.0	261.2	4	fixed	present
4	Magda Doe	26	female	115.0	407.5	2	normal	absent
5	Anna White	56	female	120.0	45.0	2	normal	absent

The imputed values for the "cholesterol" feature equal (331.2+407.5+45)/3.

The imputed values for the "chest pain type" equal (1+4+2)/3 rounded to the nearest integer.

The imputed values for the "defect type" equal the most frequent value "normal".

Data set with missing values imputed using the hot deck imputation

Object ID	Name	Age	Sex	Blood pressure	Cholesterol in mg/dl	Chest pain type	Defect type	Diagnosis
1	Konrad Black	31	male	130.0	331.2	1	normal	absent
2	Konrad Black	31	male	130.0	331.2	1	normal	absent
3	Magda Doe	26	female	115.0	407.5	4	fixed	present
4	Magda Doe	26	female	115.0	407.5	4	fixed	present
5	Anna White	56	female	120.0	45.0	2	normal	absent

The distance between objects 1 and 2 (object ID is not considered as a feature) equals 0+0+0+0+1+1+1+1=4.

The distance between objects 1 and 3 equals 1+1+1+1+1+1+1+1=8.

The object most similar to object 3 that has a complete value for "cholesterol" is object 4 (distance 4).

Figure 3.11. Results of using missing-data handling methods for a data set describing heart clinic patients.

In **manual inspection**, the user checks feature values against predefined constraints and manually removes (or changes into missing values) all values that do not satisfy these constraints. For example, for object 5 in Figure 3.11, the *cholesterol* value is 45.0, which is outside the predefined acceptable interval for this feature, namely, within [50.0, 600.0].

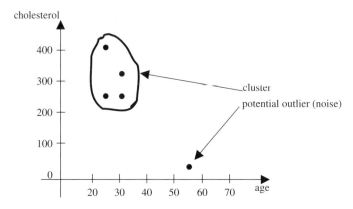

Figure 3.12. Noise detection with use of clustering.

Binning first orders the values of the noisy feature and then replaces the values with a mean or median value for the predefined bins. As an example, let us consider the *cholesterol* feature, with its values 45.0, 261.2, 331.2, and 407.5. If the bin size equals two, two bins are created: bin_1 with 45.0 and 261.2, and bin_2 with 331.2 and 407.5. For bin_1 the mean value is 153.1, and for bin_2 it is 369.4. Therefore the values 45.0 and 261.2 would be replaced with 153.1 and the values 331.2 and 407.5 with 369.4. Note that the two new values are within the acceptable interval.

Clustering finds groups of similar objects and simply removes (or changes into missing values) all values that fall outside the clusters. For instance, clustering that uses the *age* and *cholesterol* features from Figure 3.11 (in the data set with missing values imputed using the mean imputation) is shown in Figure 3.12. See Chapter 9 for more information on clustering.

4. Summary and Bibliographical Notes

In this Chapter, we defined concepts related to **data** and **data sets**. We also covered various data storage techniques and issues related to quality and quantity of data that are used for data mining purposes. The most important topics discussed in this Chapter are the following:

– different data types, including **numerical** and **symbolic** values, which can be subdivided into **categorical** (discrete), **binary** (dichotomous), **nominal** (polytomous), **ordinal** and **continuous**
– data are organized into rectangularly formatted tables called **data sets**, where rows represent **objects** and columns represent **features/attributes**, which describe the objects
– data sets are stored as **flat** (rectangular) **files** and in other formats using **databases** and **data warehouses**
– specialized data types, including **transactional** data, **spatial** data, **hypertext**, **multimedia** data, **temporal** data, and the **WWW**
– important issues related to the data used in data mining including

  – **large quantities** of data and the related problem of **scalability** of data mining methods
  – **dynamic data**, which require the use of **incremental** data mining methods
  – data quality problems, which include **imprecision, incompleteness, redundancy missing values**, and **noise**

A tutorial-style article that introduces data types, their organization into data sets, and their relation to DM can be found in [3]. Another good source for fundamental concepts concerning data types, data sets and data quality and quantity issues is provided in [4], in particular Chapters 4, 6, and 14.

There are three main **repositories** of flat file data sets from different domains, donated by data mining practitioners:

- http://www.ics.uci.edu/~mlearn/ (**Machine Learning repository**)
- http://kdd.ics.uci.edu/ (**Knowledge Discovery in Databases archive**)
- http://lib.stat.cmu.edu/ (**StatLib repository**)

These sites provide free access to numerous data sets and are used mainly for benchmarking and comparative purposes. Data sets are posted by the original authors along with results concerning the data. One prominent example of data mining in specialized data is a prototype multimedia data mining system called MultiMediaMiner [16].

The issues related to the quantity and quality of data used in data mining have been addressed in numerous publications. A survey of the largest commercial databases was published in [15]. The availability of huge data sets that can be extracted from these databases calls for the development of **scalable algorithms** for data mining. A good overview of scalable algorithms is given in [2]. Data-quality research can be divided into two streams: data quality assessment and methods for improving data quality. A method that combines subjective and objective **assessments of data quality** and proposes practical data quality matrices was introduced in [7]. A framework for the identification and analysis of data quality issues, covering management responsibilities, operation and assurance costs, research and development, production, distribution, personnel management, and legal function, is described in [14]. The second approach addresses **data cleaning** methods. A methodology for data cleaning for relational data is reported in [5]. An important branch of data cleaning is related to the handling of missing values. Traditional imputation methods for missing values are based on statistical analysis and include simple algorithms such as mean and hot deck imputation, as well as more complex methods including regression and the Expectation-Maximization (EM) algorithm. The **multiple imputations** method was first described by Rubin [8], while subsequent methods [6, 10] used Bayesian algorithms that perform multiple imputations using posterior predictive distribution of the missing data based on the complete data. Another method imputes each incomplete attribute by cubic spline regression [1]. A detailed description of multiple imputation algorithms can be found in [9, 12, 17], and a primer can be found in [13].

There is significant industrial and academic interest in data cleaning research and development. Several groups exist that aim to develop data cleaning solutions, including the following:

- The *Data Cleaning* group at Microsoft:
  http://research.microsoft.com/dmx/DataCleaning
- The *Data Quality* group at AT&T:
  http://www.dataquality-research.com
- The *Total Data Quality Management* group at MIT:
  http://web.mit.edu/tdqm/www/ index.shtml

Finally, data security and privacy issues with a focus on HIPAA rules in the United States, Canada and Europe are discussed in [11].

References

1. Alzola, C., and Harrell, F. 1999. An introduction of S-Plus and the Hmisc and design libraries (download from http://www.med.virginia.edu/medicine/clinical/hes)
2. Ganti, V., Gehrke, J., and Ramakrishnan, R. Aug. 1999. Mining very large databases. *IEEE Computer*, 32(8):38–45
3. Holsheimer, M., and Siebes, A. 1994. *Data Mining: The Search for Knowledge in Databases*. Report CS-R9406, ISSN 0169–118X, CWI: Dutch National Research Center, Amersterdam, Netherlands

4. Klosgen, W., and Zytkow, J. (Eds.). 2002. *Handbook of Data Mining and Knowledge Discovery*, Oxford University Press New York, USA
5. Lee, M-L., Ling, T.W., Lu, H., and Ko, Y.T. 1999. Cleansing data for mining and warehousing, *Proceedings of the International Conference on Database and Expert Systems Applications* (DEXA), Florence, Italy, *Lecture Notes in Computer Science*, 1677:751–760
6. Li, K-H. 1988. Imputation using markov chains. *Journal of Statistical Computation and Simulation*, 30:57–79
7. Pipino, L., Lee, Y., and Wang, R. 2002. Data quality assessment. *Communications of the ACM*, 45(4):211–218
8. Rubin, D.B. 1977. Formalizing subjective notions about the effect of nonrespondents in sample surveys. *Journal of American Statistical Association*, 72:538–543
9. Rubin, D.B. 1987. *Multiple Imputations for Nonresponse in Surveys*, John Wiley and Sons: New York
10. Rubin, D.B., and Schafer, J.L. 1990. Efficiently creating multiple imputation for incomplete multivariate normal data. *Proceedings of the Statistical Computing Section*, Alexandria: ASA, 83–8
11. Saul, J.M. 2000. Legal policy and security issues in the handling of medical data. In Cios, K.J. (Ed.), *Medical Data Mining and Knowledge Discovery*, Springer Verlag, 21–40
12. Shafer, J.L. 1997. *Analysis of Incomplete Multivariate Data*, Chapman and Hall Heidelberg, Germany
13. Shafer, J.L. 1999. Multiple imputations: a primer. *Statistical Methods in Medical Research*, 8:3–15
14. Wang, R., Storey, V., and Firth C. 1995. A framework for analysis of data quality research. *IEEE Transactions on Knowledge and Data Engineering*, 7(4):623–640
15. Winter, R., and Auerbach, K. May 2004. Contents under pressure. *Intelligent Enterprise*, available online at http://www.intelligententerprise.com/showArticle.jhtml?articleID=18902161
16. Zaïane, O., Han, J., Li, Z.N., and Hou, J. 1998. Mining multimedia data, *Proceeding of the CASCON'98: Meeting of Minds*, 83–96, Toronto, Canada
17. Zhang, P. 2003. Multiple imputation: theory and method (with discussion). *International Statistical Review*, 71(3):581–592

5. Exercises

1. Discuss feature types and organization of data in the *Adult*, *Mushrooms*, and *SPECT* data sets from the ML repository (http://www.ics.uci.edu/~mlearn/). Using this repository, find at least three data sets that consist of only symbolic features, and another three that consists only of numerical features. Finally, find the data set that includes the largest number of features.
2. Convert the data set from Table 3.1 into the transactional format (an example transactional database is shown in Figure 3.8). Briefly describe how such a conversion could be performed. Remember to use the appropriate vocabulary.
3. Considering the data given in Table 3.2, calculate the best partitioning of a data set containing 1,000,000 objects with respect to minimizing the total run time when applying the See5 algorithm to these subsets sequentially. The data set can be divided into two or more subsets, where each subset size is limited to those listed in Table 3.2, i.e. 100,000, 200,000, ..., 900,000, that is, the data set can be partitioned into ten subsets of 100,000, one of 800,000 and one of 200,000, or one of 800,000 and two 100,000. You must use the estimated run time values from Table 3.2 and add an overhead of 1,000 seconds for the processing of each subset. For example, division of the data set into ten 100,000 subsets gives a total run time of $10*2,000 + 10*1,000 = 30,000$ seconds compared with $30,000 + 1,000 = 31,000$ seconds if the entire data set is used. Compute how many seconds can be saved by such sequential processing.
4. Write a program that performs imputation of missing values using both the mean and hot deck methods for the *Water Treatment Plant* data set from the ML repository (http://www.ics.uci.edu/~mlearn/). After performing the imputation, discuss which imputation method, in your opinion, gave better results based on a comparison between the two resulting complete data sets.

5. Investigate other methods (except manual, binning, and clustering) for dealing with noise in the data sets. Find and briefly summarize two alternative methods. Write a one-page report that includes links and references to the source of the information that you have used.
6. Briefly define and describe the data *inconsistency* problem. This data quality issue was not described in this Chapter but may be important with respect to some data mining projects. You report should be similar to the description of the data redundancy problem from this Chapter and should include an example that is based on the heart clinic data set.

4

Concepts of Learning, Classification, and Regression

In this Chapter, we introduce the main concepts and types of learning, classification, and regression, as well as elaborate on generic properties of classifiers and regression models (regressors) along with their architectures, learning, and assessment (performance evaluation) mechanisms.

1. Introductory Comments

In data mining, we encounter a diversity of concepts that support the creation of models of data. Here we elaborate in detail on **learning**, **classification** and **regression** as being the most dominant categories of developments of a variety of models.

1.1. Main Modes of Learning from Data-Problem Formulation

To deal with huge databases, we first describe several fundamental ways to complete their analyses, build underlying models, and deliver major findings. We present the fundamental approaches to such analyses by distinguishing between supervised and unsupervised learning. We stress that such a dichotomy is not the only taxonomy available since a number of interesting and useful alternatives lie somewhere in-between the two thus forming a continuum of options that could be utilized. The relevance and usefulness of such alternatives are described below.

1.2. Unsupervised Learning

The paradigm of unsupervised learning, quite often referred to as clustering involves a process that automatically reveals (discovers) structure in data and does not involve any supervision. Given an N-dimensional dataset $\mathbf{X} = \{x_1, x_2, \ldots, x_N\}$, where each x_k is characterized by a set of attributes, we want to determine the structure of $\mathbf{X}$, i.e., identify and describe groups (clusters) present within it. To illustrate the essence of the problem and build some conceptual prerequisites, let us consider the examples of two-dimensional data shown in Figure 4.1. What can we say about the first one, shown in Figure 4.1(a)? Without any hesitation, we can distinguish three well-separated spherical groups of points. These are the clusters expressing the structure in the data.

The clusters could also exhibit a very different geometry. For instance, in Figure 4.1(b), we see two elongated structures and one ellipsoidal cluster. All the clusters are well separated and clearly visible. In Figure 4.1(c), the structure is much less apparent; the clusters overlap significantly. Perhaps the two that are close to each other could be considered to form a single cluster. In

50   1. Introductory Comments

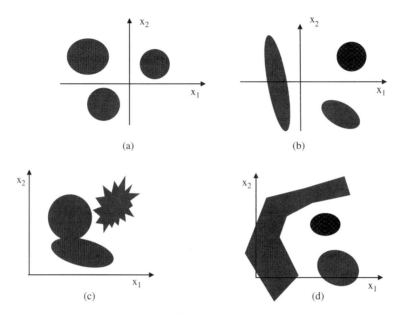

Figure 4.1. Examples of two-dimensional data (all in the two-dimensional feature space $x_1 - x_2$) and the search for their structures: geometry of clusters and their distribution in the two-dimensional data space.

Figure 4.1(d), the shapes are even more complicated: a horseshoe exhibits far higher geometric complexity than the two remaining spherical clusters.

The practicality of clustering is enormous. In essence, we perform clustering almost everywhere. Clusters form aggregates or, to put it differently, build an abstraction of the dataset. Rather than dealing with millions of data points, we focus on a few clusters and doing so is evidently very convenient. Note, however, that clusters do not have a numeric character; instead we perceive "clouds" of data and afterwards operate on such structures. Hopefully, each cluster comes with well-defined semantics that capture some dominant and distinguishable parts of the data. Consider, for instance, a collection of transactions in a supermarket; the data sets generated daily are enormous. What could we learn from them? Who are the customers? Are there any clusters – segments of the market we should learn about and study for the purpose of an advertising campaign? Discovering these segments takes place through clustering. A concise description of the clusters leads to an understanding of structure within the collection of customers. Obviously, these data are highly dimensional and involve a significant number of features that describe the customers. A seamless visual inspection of data (as we have already seen in the case of examples in Figure 4.1) is not feasible. We need a powerful "computer eye" that will help us to explore the structure in any space even highly-dimensional one. Clustering delivers an algorithmic solution to this problem.

In spite of the evident diversity of **clustering mechanisms** and their algorithmic underpinnings, the underlying principle of grouping is evident and quite intuitive. We look for the closest data points and put them together. The clusters start to grow as we expand them by bringing more points together. This stepwise formation of clusters is the crux of **hierarchical clustering**. We could look at some centers (prototypes) and request that the data points be split so that the given distance function assumes its lowest value (i.e., their similarity or dissimilarity is highest).

Here we note that all strategies rely heavily on the concept of **similarity** or **distance** between the data. Data that are close to each other are likely to be assigned to the same cluster. Distance impacts clustering in the sense that it predefines a character of the "computer eye" we use when searching for structure. Let us briefly recall that for any elements – patterns x, y, z (treated formally as vectors) the distance function (metric) is a function producing a nonnegative numeric value $d(x,y)$ such that the following straightforward conditions are met:

(a) $d(x,x) = 0$
(b) $d(x,y) = d(y,x)$ symmetry
(c) $d(x,z) + d(z,y) \geq d(x,y)$ triangle inequality

While the above requirements are very general, there are a number of commonly encountered ways in which distances are described. One frequently used class of distances is known as the **Minkowski distance** (which includes Euclidean, Manhattan, and Tchebyshev as special cases). Examples of selected distances are given below with and their pertinent plots shown in Figure 4.2. We should note that while the distances here concern vectors of numeric values, they could be defined for nonnumeric descriptors of the patterns as well.

$$\text{Hamming distance}: \quad d(\mathbf{x},\mathbf{y}) = \sum_{i=1}^{n} |x_i - y_i| \quad (1)$$

$$\text{Euclidean distance}: \quad d(\mathbf{x},\mathbf{y}) = \sqrt{\sum_{i=1}^{n} (x_i - y_i)^2} \quad (2)$$

$$\text{Tchebyschev distance}: \quad d(\mathbf{x},\mathbf{y}) = \max_i(|x_i - y_i|) \quad (3)$$

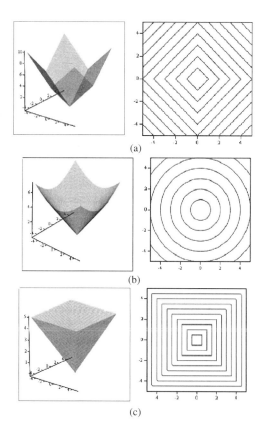

(a)

(b)

(c)

Figure 4.2. Graphic visualization of distances: equidistant regions: (a) Hamming, (b) Euclidean, and (c) Tchebyshev. Shown are a three-dimensional plot and the two-dimensional contour plots with the distances computed between **x** and the origin, $d(\mathbf{x},\mathbf{0})$.

52   1. Introductory Comments

The graphic illustration of distances is quite revealing and deserves attention. Each distance comes with its own geometry. The Euclidean distance imposes spherical shapes of equidistant regions. The Hamming distance imposes diamond – like geometry, while the Tchebyschev distance forms hyper – squares. The weighted versions affect the original shape. For the weighted Euclidean, we arrive at ellipsoidal shapes. The Tchebyschev distance leads to hyperboxes whose sides are of different length. It is easy to see that the use of a certain distance function in expressing the closeness of the patterns implies which patterns will be treated as closest neighbors and candidates for belonging to the same cluster. If we are concerned with, say, the Euclidean distance, the search for the structure leads to the discovery of spherical shapes. This means that the method becomes predisposed towards searching for such geometric shapes in the structure.

Clustering methods are described in Chapter 9, while another class of unsupervised learning methods, namely, association rules, is described in Chapter 10.

1.3. Supervised Learning

Supervised learning is at the other end of the spectrum from unsupervised learning in the existing diversity of learning schemes. In unsupervised learning, we are provided with data and requested to discover its structure.

In supervised learning, the situation is very different. We are given a collection of data (patterns) and their characterization, which can be expressed in the form of some discrete labels (in which case we have a classification problem) or some values of auxiliary continuous variables. In which case we are faced with a regression problem or an approximation problem. In classification problems, each data point x_k comes with a certain **class label**, say ω_k, where the values of ω_k come from a small set of integers $\omega_k \in \{1, 2, \ldots, c\}$, where "c" stands for a number of classes. Some examples of two-dimensional classification data are shown in Figure 4.3. The objective here is to build a classifier that is a construct of a function Φ, called a classifier, that generates a class label as its output, $\Phi(x_k) = \omega_k$.

The geometry of the classification problem depends on the distribution of classes. Depending upon the distribution, we can design linear or nonlinear classifiers. Several examples of such

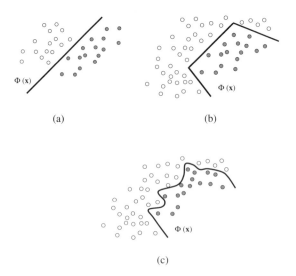

Figure 4.3. Examples of classification problems in a two-dimensional space: (a) linear classifier, (b) piecewise linear classifier, (c) nonlinear classifier. Two classes of patterns are denoted by black and white dots, respectively.

classifiers are illustrated in Figure 4.3. The geometry of the classifiers is reflective of the distribution of the data. Here we emphasize the fact that the nonlinearity of the classifier depends upon the geometry of the data. Likewise, the patterns belonging to the same class could be distributed in several disjoint regions of **X**.

1.4. Reinforcement Learning

Reinforcement learning is a learning paradigm positioned between unsupervised and supervised learning. In unsupervised learning, there is no guidance as to the assignment of patterns to classes. In supervised learning, class assignment is known. In reinforcement learning, we are offered less detailed information (the supervision mechanism) than that encountered in supervised learning. This information (guidance) comes in the form of some reinforcement (the reinforcement signal). For instance, given "c" classes, the reinforcement signal $r(w)$ could be binary in its nature:

$$r(\mathbf{w}) = \begin{cases} 1, & \text{if class label is even}(\omega_2, \omega_4, \ldots) \\ -1, & \text{otherwise} \end{cases} \quad (4)$$

See also Figure 4.4. When used to supervise development of the classifier, reinforcement offers a fairly limited level of supervision. For instance, we do not "tell" the classifier to which class the pattern belongs but only distinguish between the two super – categories that are composed of odd and even numbers of class labels. Nevertheless, this information provides more guidance than no labeling at all. In the continuous problem (regression), reinforcement results from the discretization of the original continuous target value. Another situation arises when detailed supervision over time is replaced by the mean values regarded as its aggregates – a certain reinforcement signal, as shown in Figure 4.4(c).

In a nutshell, reinforcement learning is guided by signals that could be sought as an aggregation (generalization) of the more detailed supervision signals used in "standard" supervised learning. The emergence of this type of learning could be motivated by scarcity of available supervision, which in turn that could have been dictated by economical factors (less supervision effort).

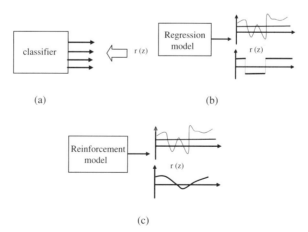

Figure 4.4. Examples of reinforcement signals: (a) classification provides only partial guidance by combining several classes together as exemplified by (4), and (b) regression offers a threshold version of the continuous target signal, (c) reinforcement supervision involves aggregates (mean value) over time.

1.5. Learning with Knowledge Hints and Semi-supervised Learning

Supervised and unsupervised learning are two extremes and reinforcement learning is an aggregate positioned in-between them. However, a number of other interesting options exist that fall under the umbrella of what could generally be called **learning with knowledge-based hints**. These options reflect practice: rarely are we provided with complete knowledge and rarely do we approach a problem with no domain knowledge. Both these cases are quite impractical. Noting this point, we now discuss several possibilities in which domain knowledge comes into the picture.

In a large dataset **X**, we have a small portion of labeled patterns that lead to the notion of clustering with **partial supervision**, see Figure 4.5.

These labeled patterns in an ocean of data form some "anchor" points that help us navigate the process of determining (discovering) clusters. The search space and the number of viable structures in the data are thus reduced, simplifying and focusing the overall search process.

The format in which partial supervision directly encapsulates available knowledge could be present in numerous situations in data mining. Imagine a huge database of handwritten characters (e.g., digits and letters used in postal codes). Typically there will be millions of characters (digits and letters). A very small fraction of these are labeled by an expert, who chooses some handwritten characters (maybe those that are difficult to decipher) and labels them. In this way, we produce a small labeled dataset.

Knowledge hints come in different formats. Envision a huge collection of digital pictures. In this dataset we have some pairs of data (patterns) whose proximity has been quantified by an expert or user (see Figure 4.6). These **proximity hints**, are a useful element of supervision during data clustering or in this case when organizing a digital photo album.

We note that knowledge hints of this nature are very different from those we had in partial supervision. The striking difference is this: in the case of clustering with partial supervision, we assume that the number of clusters is equal to the number of classes and that this number is known. This assumption could be true in many instances; for example when dealing with handwritten characters. In proximity-based clustering, however we do not specify the number of clusters, and in this sense the format of the hints is far more general. In a photo album, the number of clusters is obviously unknown. Thus the use of proximity-based hints under these circumstances is fully justifiable.

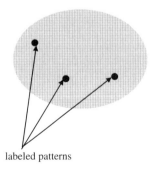

labeled patterns

Figure 4.5. Clustering with partial supervision–the highlighted patterns come with class labeling.

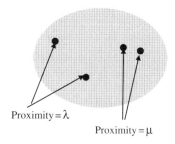

Proximity = λ
Proximity = μ

Figure 4.6. Clustering with proximity hints-selected pairs of data are assigned some degree of proximity (closeness).

2. Classification

2.1. Problem Formulation

In the previous section, we introduced the concept of classification. **Classifiers** are constructs (algorithms) that discriminate between classes of patterns. Depending upon the number of classes in the problem, we may encounter **two-class** or **many-class** classifiers. The design of a classifier depends upon the character of the data, the number of classes, the learning algorithm and the validation procedures. Let us recall that the development of the classifier gives rise to the mapping (Φ)

$$\Phi : \mathbf{X} \to \{\omega_1, \omega_2, \ldots, \omega_c\} \tag{5}$$

that maps any pattern x in $\mathbf{X}$ to one of the labels (classes). In practice, both linear and nonlinear mapping require careful quality assessment. Building classifiers requires prudent use of data so that we reach a sound balance between accuracy and the generalization abilities of the constructed classifier. This goal calls for the arrangement of data into training and testing subsets and for the running training procedures in some mode. The remainder of this section covers these topics.

2.2. Two- and Many-class Classification Problems

Classification tasks involve two or more classes. This taxonomy is quite instructive as it reveals several ways to form classifier architectures and to organize them into a certain topology. In the simplest case, let us consider two classes of data (patterns). Here the classifier (denoted by Φ) generates a single output whose value depends on the class to which the given pattern is assigned as shown in Figure 4.7.

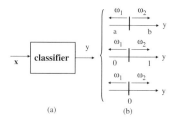

Figure 4.7. (a) a two-class classifier; (b) some alternative ways to code its single output (y) to represent two classes in the problem.

Since there are two classes, their coding could be realized in several ways. As shown, a given range of real numbers could be split into two intervals, Figure 4.7(b). In particular, if we use the range [0,1], the coding could assume the following form:

$$[0, \tfrac{1}{2}] \text{ if the pattern belongs to class } \omega_1$$
$$[\tfrac{1}{2}, 1] \text{ if the pattern belongs to class } \omega_2 \qquad (6)$$

In another typical coding alternative, we use the entire real space, and the coding assumes the form:

$$\Phi(x) < 0 \text{ if } x \text{ belongs to } \omega_1$$
$$\Phi(x) \geq 0 \text{ if } x \text{ belongs to } \omega_2 \qquad (7)$$

We will come back to this classification rule when we discuss the main classifier architectures.

The multiclass problem can be handled in two different ways. An intuitive way to build a classifier is to consider all classes together and to create a classifier with "c" outputs, where the class to which x belongs to is assigned by identifying the output for which it attains the maximal value. We express this option in the form:

$$i_0 = \arg\max\{y_1, y_2, \ldots, y_c\}$$

where $y_1, y_2, \ldots, y_c$ are the outputs of the classifier (see Figure 4.8).

The other general approach is to split the c-class problem into a subset of two-class problems. In each, we consider one class, say ω_1, with the other class composed of all the patterns that do not belong to ω_1. In this case, we come up with a dichotomy:

$$\Phi_1(x) \geq 0 \text{ if } x \text{ belongs to } \omega_1$$
$$\Phi_1(x) < 0 \text{ if } x \text{ does not belong to } \omega_1 \qquad (8)$$

(here the index in the classifier formula pertains to the class label).

In the same way, we can design a two-class classifier for $\omega_2, \omega_3, \ldots, \omega_c$. When used, these classifiers are invoked by some pattern **x** and return decisions about class assignment (see Figure 4.9).

If only one classifier generates a nonnegative value, the assignment of the class is obvious. There are two other possible outcomes, however: (a) several classifiers identify the pattern as belonging to a specific class, in which case we have a conflicting situation that needs to be resolved, or (b) no classifier issued a classification decision, in which case the class assignment of the pattern becomes undefined. An example of the resulting geometry of the two-class linear classifiers is shown in Figure 4.9.

Below, we discuss the main categories of classifiers i.e., elaborate on various forms of the classifier (viz. function Φ) being used to realize discrimination.

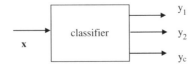

Figure 4.8. A single-level c-output classifier $\Phi(x)$.

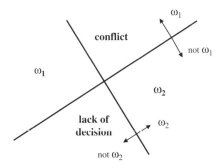

Figure 4.9. Two-class linear classifiers and their classification decisions. Note the regions of conflict and lack of decision.

2.3. Classification and Regression: A General Taxonomy

As we have noted, the essence of classification is to assign a new pattern to one of the classes when the number of classes is usually quite low. We usually make a clear distinction between classification and regression. In **regression**, we encounter a continuous output variable and our objective is to build a model (regressor) so that a certain approximation error is minimized. More formally, consider a data set formed by some pairs of input-output data (x_k, y_k), $k = 1, 2, \ldots, N$, where now $y_k \in \mathbf{R}$. The regression model (regressor) comes in the form of some mapping $F(x)$ such that for any x_k we obtain $F(x_k) \approx y_k$. As illustrated in Figure 4.10, the regression model attempts to "capture" most of the data by passing through the area of the highest density of data. The quality of the model depends on the nature of the data (including their dispersion) and its functional form. Several cases are illustrated in Figure 4.10.

Regression becomes a standard model when revealing dependencies between input and output variables. For instance, one might be interested in finding a meaningful relationship between the spending of customers and their income, marital status, job, etc.

One could treat classification as a discrete version of the regression problem. Simply discretize the continuous output variable existing in the regression problem and treat each discrete value as a class label.

As we will see in the ensuing discussion, the arguments, architectures, and algorithms pertaining to classification models are also relevant when discussing and solving regression problems.

2.4. Main Categories of Classifiers and Selected Architectures

The diversity of classifiers is amazing. It is motivated by the geometry of the classification problems, the ensuing approaches to design, and complexity of the problem at hand. The reader may have been exposed to names like linear classifier, decision trees, neural networks, k-nearest neighbor, polynomial classifier, and the alike. The taxonomy of classifiers could be presented in several ways depending on which development facet we decide to focus. In the ensuing discussion, we concentrate on two different viewpoints.

2.4.1. Explicit and Implicit Characterization of the Classifier

The distinction made here is concerned with the form in which the classifier arises. It could be described as some function, say, $\Phi(x)$, where Φ could be quite complex yet described in an explicit manner. In other words, Φ could be a linear function, a quadratic relationship, some high order polynomial, etc. Implicit characterization of a classifier takes place when we do not have a formula but rather the classifier is described in some graphic form, such as a decision tree, a nearest neighbor classifier, or a cognitive map. To illustrate this point, let us concentrate on

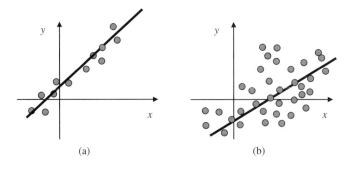

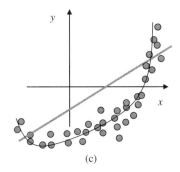

Figure 4.10. Examples of regression models with single input and single output and their relationships with data: (a) linearly distributed data with a very limited dispersion – a an excellent fit offered by some linear model; (b) linearly distributed data with significant dispersion; no model could result in acceptable fit; (c) nonlinearly distributed data well captured by the corresponding nonlinear model yet leading to poor performance of any linear model.

the first two of these. The **nearest neighbor** classifier is perhaps one of the simplest (yet quite powerful) classification mechanisms. Given a collection of labeled data (we could refer to these as reference patterns), the new data x is classified on a basis of the distance between it and other labeled data. We are tempted to classify x in the same class as its closest labeled neighbor (see Figure 4.11). Because of its evident simplicity and intuitive appeal, this classification principle is very attractive.

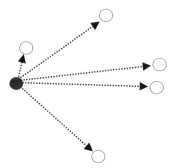

Figure 4.11. The principle of the nearest neighbor (NN) classifier; the dark point to be classified on the basis of distances computed from some labeled patterns (empty points).

Formally, we write the classification rule in the following form:

$$L = \arg\min_k \|x - x_k\| \quad \text{class of } x \text{ is the same as the class to which } x_L \text{ belongs}$$

where the minimum is taken over all data; here k ranges across the set of labeled data. There are also some variations to this generic rule in which we consider several (K) closest neighbors of x. In this case we talk about a K-nearest neighbor classification rule. We consider K to be a small odd number, say, 3, 5, and 7. Choosing more than one neighbor helps to increase our confidence in the classification (labeling) of the new data point. What has to be stressed, though, is that the distance in the above classification rule plays a pivotal role since it quantifies the concept of the nearest neighbor. Depending on our choice of distance, we encounter different classification boundaries for each classifier (viz. the boundaries of the region where each point is closer to the given reference point in comparison with the others); in essence the region consists of all elements of the space where the classification rule for the x_L th pattern applies. The concept is illustrated in Figure 4.12. Consider three labeled data points in a two-dimensional space. The regions of "attraction" (where the points are found to be the closest to the given labeled data) determine the classification (decision) boundaries. As shown in the Figure, they depend on the distance being accepted in the determination of the closest neighbor.

The concept of **decision trees** reflects a sequential process of forming piecewise decision boundaries. The tree involves a series of decision blocks that offer a series of splits of the variables. By directly traversing the tree, we come up with the classification decision. Several illustrative examples of decision trees concerning two variables with several split points are provided in Figure 4.13.

The classifiers given in an explicit functional form give rise to a general taxonomy that reflect the character of the decision boundary being captured by the function.

2.4.2. Linear and Nonlinear Classifiers

This very general view describes classifiers in terms of the properties of the underlying mapping. The **linear classifier** is governed by the following expression:

$$\Phi(x) = w_0 + w_1 x_1 + w_2 x_2 + \ldots + w_n x_n \tag{9}$$

where $w_0, w_1, w_2, \ldots, w_n$ are the parameters of the classifier. Since $\Phi(x)$ is a linear relationship of the inputs, we call this construct a linear classifier. The classification boundary is determined by the values of its parameters. Given the linear dependency captured by (9), the classification boundary is a line, plane or a hyperplane (which one depends upon the dimensionality of the

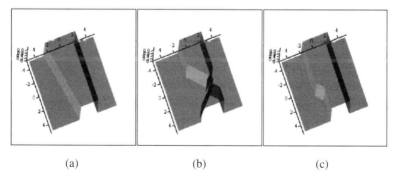

(a)   (b)   (c)

Figure 4.12. Classification boundaries implied by different distance functions used in the nearest neighbor rule : (a) Hamming; (b) Euclidean; (c) Tchebyschev.

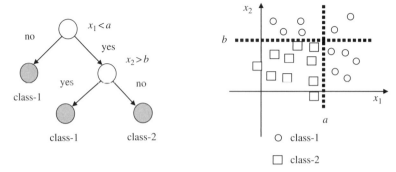

Figure 4.13. Example of (a) a decision tree and (b) its ensuing decision boundaries. Note the series of guillotine cuts (dashed line) parallel to the coordinates. There are two classes of patterns.

problem). The plot of the characteristics of the classifier in the two-dimensional case is shown in Figure 4.14. It becomes apparent that such classifiers produce good results when data (classes) are linearly separable. Otherwise, as shown in Figure 4.14, we always end up with a nonzero classification error.

The linear classifier can be conveniently described in a compact form by introducing the following vector notation:

$$\Phi(x) = w^T x^\sim \tag{10}$$

where $w = [w_0\, w_1 \ldots w_n]^T$ and $x^\sim = [1\, x_1\, x_2 \ldots x_n]$.

Note that $x^\sim$ is defined in an extended input space that is $x^\sim = [1\, x]^T$. To emphasize the fact that the classifier comes with some parameters, we use them in the characterization of the classifier by writing it in the form $\Phi(x; w)$. To cope with any nonlinear character of the classification boundary, we introduce nonlinear classifiers. Here the number of alternative approaches is very high. One interesting category of such classifiers is formed by polynomial functions. We can then talk about quadratic, cubic, and higher-order polynomials. For instance, the quadratic form reads as follows:

$$\begin{aligned}\Phi(x) =\, & w_0 + w_1 x_1 + w_2 x_2 + \ldots + w_n x_n + \\ & + w_{n+1} x_1^2 + w_{n+2} x_2^2 + \ldots + w_{2n} x_n^2 + \\ & + w_{2n+1} x_1 x_2 + \ldots .\end{aligned} \tag{11}$$

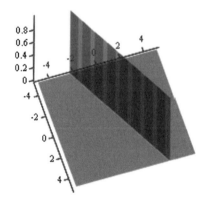

Figure 4.14. Classification boundary of the linear classifier $\Phi(x_1, x_2) = 0.7 + 1.3 x_1 - 2.5 x_2$.

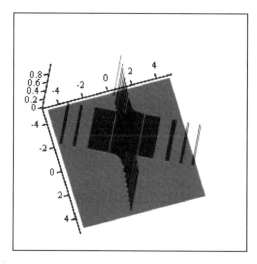

Figure 4.15. An example of nonlinear boundaries produced by the quadratic classifier; $\Phi(x_1, x_2) = 0.1 + 0.6x_1^2 - 2*x_2^2 + 4x_1 x_2 + 1.5x_1 - 1.6x_2$.

Thus, in addition to the linear part, we have quadratic terms and a series of bilinear components in which we have products of all combinations of the variables. Here the number of parameters is far higher than in the case of a linear classifier. This increased complexity is the price we pay for enhanced functionality of the classifier. The higher-order polynomials help us deal with the nonlinear classification boundaries, but at significant expense to the architectural complexity, Figure 4.15.

2.5. Learning Algorithms

As stressed above, the role of a classifier is to discriminate between classes of data. The discrimination happens through the use of a certain **classification boundary**, realized by the classifier. The form of the classifier depends not only on the nature of the classifier (linear or nonlinear) but also on the values of its parameters. We observed quite vividly that by changing the parameters of a linear classifier, we affect the position of the resulting hyperplane that forms the classification boundary. The changes in parameters have translated into the rotation and translation of the hyperplane. By changing the location of the decision boundary, we affect the quality of the classifier and minimize the resulting **classification error**. The minimization of this error through optimization of the parameters of the classifier is achieved through **learning**. In other words, we could view learning as a process of optimizing the classifier. While we will discuss the nature of performance and its evaluation later on, it is instructive to look into the linear classifier and emphasize how the optimization is defined and what this process entails. Alluding to the fundamental requirement of the "ideal" (viz. zero-error) classifier, the process translates into the satisfaction of the following system of inequalities:

$$\Phi(\mathbf{x}_k) \geq 0 \text{ for } \mathbf{x}_k \in \omega_1 \text{ and } \Phi(\mathbf{x}_k) < 0 \text{ for } \mathbf{x}_k \in \omega_2 \qquad (12)$$

which is equivalent to the system

$$\Phi(\mathbf{x}_k) \geq 0 \text{ for } \mathbf{x}_k \in \omega_1 \text{ and } -\Phi(\mathbf{x}_k) > 0 \text{ for } \mathbf{x}_k \in \omega_2 \qquad (13)$$

(note that we have multiplied the second set of the constraints by -1). We also observe that the optimization boils down to the determination of the line parameters for which the above system of inequalities is satisfied.

2. Classification

Likewise, we can think of a general formulation in which we consider a certain performance of the classifier, say Q, whose value depends upon the values of the parameters of the classifier. Depending upon the form of the classifier, the optimization could be carried out in a closed-loop format or in an iterative fashion. This choice very much depends upon the formulas of the classifier. If the classifier is a linear function of its parameters and Q is differentiable, we could arrive at a closed-type formula. A solution of this nature may lead to the global minimum of the performance index. Note, that even polynomial classifiers (such as those described by (11)) are linear with respect to their parameters and as such could be optimized through the use of some closed-type formula. If the problem is indeed nonlinear, some optimization techniques are essential, including gradient-based techniques, evolutionary optimization, and others.

2.5.1. Performance Assessment: Basic Classification Rates and Loss Functions

The performance index of the classifier that we used in the previous section was treated in a somewhat general manner. We switch now to its detailed specification and show different ways of forming it depending upon the nature of the problem at hand. In one possible scenario, the performance of the classifier can be expressed (estimated) by counting the number of misclassified patterns. For any pattern, when we encounter class ω_i while the classifier issues, a decision about some other class ω_j where $i \neq j$, we regard this to be an error and increment some error counter. Checking this condition for each pattern, we finally obtain a total count which determines performance of the classifier. The higher the value of this count, the higher the error and the lower the performance of the classifier. Ideally, we can have zero value of this counter. In this case, the way in which we counted the error was very simple, yet it did not take into consideration the specificity of the problem. We viewed each error to be the same – an assumption that could be an obvious oversimplification of the problem. It could well be (and is usually the case) that misclassifying patterns produces different consequences that could be associated with different costs. Say we have two classes in some inspection problem, ω_1 and ω_2. ω_1 stands for good parts, while ω_2 denotes faulty ones. If we misclassify the pattern that actually belongs to ω_1 and classify it as ω_2, then the consequences of this misclassification could be quite different. In the first case, in which $L(\omega_1, \omega_2)$ denotes losses encountered when we classify as a faulty part (ω_2) as the good one (ω_1), the costs could be quite high (testing and replacement costs when the part get used in the systems). However the losses we are faced with when classifying as a good part (ω_1) as a faulty one (ω_2) might be far lower (we discard the part but do not face troubleshooting and replacement afterwards); let us denote the losses by $L(\omega_2, \omega_1)$. Obviously, the losses when the classification decision is correct are equal zero. We note an evident asymmetry: $L(\omega_1, \omega_2) \neq L(\omega_2, \omega_1)$. All of four these situations are visualized in Figure 4.16.

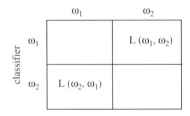

Figure 4.16. Classification results. Four possible situations are shown, of which two lead to the classification error.

The overall performance index we use to measure the quality of the classifier can be expressed in the following form for the k-th data

$$e(x_k) = \begin{cases} L(\omega_1, \omega_2) \text{ if } \mathbf{x}_k \text{ was misclassified as belonging to } \omega_1 \\ L(\omega_2, \omega_1) \text{ if } \mathbf{x}_k \text{ was misclassified as belonging to } \omega_2 \\ 0, \text{ otherwise} \end{cases} \quad (14)$$

We sum up the above expressions over all data to compute a cumulative error

$$Q = \sum_{k=1} e(\mathbf{x}_k) \quad (15)$$

A detailed discussion of performance assessment is given in Chapter 15.

2.6. Generalization Abilities of Classification and Regression Models

2.6.1. Generalization Abilities

So far we have focused on architectures and the learning of the classifiers, and we have shown how to assess classifier performance. Performance, like classification error, is computed with the same data that have been used for its design. Consequently, the results could be overly optimistic. A classifier is intended to be used on new, unseen, data. It is crucial to gain some sense of its anticipated classification rate in such cases. Performance is linked with the ability of the classifier to generalize. We refer to this important property of the classifier as its **generalization** ability. High generalization stipulates that the classifier can perform well in a broad range of situations (new data). This feature is highly desirable and practically relevant given the future use of the classifier. This means to achieve high generalization abilities of the classifier relates directly to ways of using the available data in the overall design process.

2.6.2. Dealing with Experimental Data: Training, Validation, and Testing Sets

A classifier that is being trained and evaluated on one dataset produces overly optimistic results that could lead to catastrophic behavior of the classifier in practice. To achieve some confidence, the development of the classifier must be based on prudent utilization of the data. This approach helps quantify expectations about the future performance of the classifier on unseen data. Typically, the data are split into three disjoint subsets:

- training data
- validation data, and
- testing data

Each of these subsets plays a different role in the design process. The **training data** set is essentially used to complete the training (learning) of the classifier. In particular, all optimization activities are guided by the performance index and its changes as reported on the basis of the training data. The **validation set** is particularly relevant when we are concerned with the structural optimization of the classifier. Suppose we can adjust the order of the classifier polynomial. The noticeable tendency is that with increased order, the accuracy (approximation abilities) of the classifier increases. Since we want to minimize error, we could make the classifier overly complex. The validation set is essential in avoiding this pitfall. Even though performance on the training set could be steadily improving, the classification accuracy reported from the validation set could indicate a different tendency and point to a structure that is not excessive. Likewise, we can use the validation set to monitor the learning process, particularly if this process relies on some form

64 2. Classification

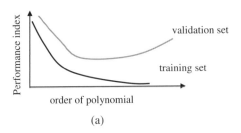

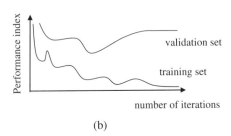

Figure 4.17. Example performance of a polynomial classifier on the training and validation sets; (a) monitoring structural complexity- identification of the preferred order of the polynomial; (b) monitoring the learning process – avoidance of overtraining.

of gradient-based optimization that usually involves a number of iterations. Finally, the **testing set** is used to evaluate the classifier design in the same manner. The plots shown in Figure 4.17 show typical examples of performance as reported for the training and validation sets.

The division of the data into three subsets could be done in many different ways. There could be different protocols for building the classifier, using which we could quantify its future performance on unseen patterns. A detailed discussion is given in Chapter 15.

2.6.3. Approximation, Generalization, and Memorization

In general, we note an interesting tradeoff in the performance of the same classifier reported from training and testing data. Excellent performance of the classifier on the training set quite often results in relative poor performance on the testing set. This effect is referred to as **memorization**.

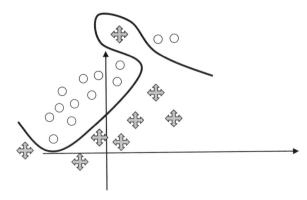

Figure 4.18. Approximation and generalization abilities of the classifier at the learning phase, the nonlinear classifier produced a zero classification error through building a more complex boundary; however, its generalization abilities deteriorated.

The classifier in this case has high approximation abilities but low generalization power. It cannot easily and efficiently move beyond the data it was trained on and fails on the unseen (testing) data. We are thus faced with the **approximation-generalization dilemma** which is no trivial and requires careful attention during the development phase. During its training, the classifier has almost memorized all data but has failed to acquire any generalization abilities. Noisy data, if exist, gets incorporated into the classifier and afterwards may create a highly detrimental effect on future classifier efficiency and contribute to a low classification rate. To illustrate this effect, let us refer to Figure 4.18. Note that the two classes are almost linearly separable, with the exception of three patterns that seem to be more like outliers. If these are to be accommodated, the resulting classifier needs to be nonlinear, as displayed in the Figure. The generalization abilities could be quite detrimental to the formation of classification boundaries. We say that the classifier has memorized all data but lost its generalization capabilities.

3. Summary and Bibliographical Notes

The fundamental concepts of **classification**, the mechanisms of classification and **regression** and general architectures form the backbone of the paradigm of **pattern recognition.** This framework is essential in casting a broad category of problems in a general framework with a well-supported methodology and algorithms. The distinction between several modes of learning is crucial from the practical standpoint: it helps address the needs of the problem and effectively utilizes all domain knowledge available within the context of the task at hand. It is also worth stressing that a significant number of learning modes are spread between purely **supervised** and **unsupervised** learning.

Supervised and unsupervised learning has been intensively studied in the framework of pattern recognition; the reader may refer here to [3, 4, 5, 14] (**supervised learning**) and [1, 2, 8, 11] (**unsupervised learning**). Classic references concerning statistical pattern recognition include [7, 9, 14]. Various parameter estimation problems can be discussed and studied in the context of system identification [12, 13] and signal processing [10].

References

1. Anderberg, M.R. 1973. *Cluster Analysis for Applications*, Academic Press
2. Bezdek, J.C. 1981. *Pattern Recognition with Fuzzy Objective Function Algorithms*, Plenum Press
3. Bezdek, J.C., Keller, J., Krishnampuram, R., and Pal, N.R. 1999. *Fuzzy Models and Algorithms for Pattern Recognition and Image Processing*, Kluwer Academic Publishers
4. Bishop, C.M. 1995. *Neural Networks for Pattern Recognition*, Oxford University Press
5. Devijver, P.A., and Kittler, J (Eds.). 1987. *Pattern Recognition Theory and Applications*, Springer Verlag
6. Duda, R.O., Hart, P.E. and Stork D.G. *2001 Pattern Classification*, 2nd edition, John Wiley
7. Fukunaga, K. 1990. *Introduction to Statistical Pattern Recognition*, 2nd edition, Academic Press
8. Jain, A.K, Murthy M.N. and Flynn, P.J. 1999. Data clustering: A review, *ACM Computing Survey*, 31(3): 264–323
9. Jain, A.K., Duin, R.P.W., and Mao, J. 2000. Statistical Pattern recognition: a review, *IEEE Transactions on Pattern Analysis and Machine Intelligence*, 22(1): 4–37
10. Kalouptsidis, N. 1997. *Signal Processing Systems*, John Wiley
11. Kaufmann, L., and Rousseeuw, P.J. 1990. *Finding Groups in Data: An Introduction to Cluster Analysis*, John Wiley
12. Norton, J. 1986. *Introduction to System Identification*, John Wiley
13. Soderstrom, T., and Stoica, P. 1986. *System Identification*, John Wiley
14. Webb, A. 2002. *Statistical Pattern Recognition*, 2nd edition, John Wiley

4. Exercises

1. The Hamming distance is used to design robust classifiers-that is, classifiers whose construction is not affected by outliers. Explain why.
   Hint: Plot a one-dimensional distance d(x,0) for the Hamming distance and compare it with the plot of the Euclidean distance. Compare these two curves and elaborate on how they change with respect to changes in x. Which one changes more rapidly for larger values of x?

2. Plot the classification boundary of the classifier for two classes, where the discriminant function is described in the form

$$\Phi(x_1, x_2) = 6.5 + 0.5x_1 - 0.9x_2 + 0.1x_1 x_2$$

   Do the patterns (1.2, 7.5) and (10.4, −15.0) belong to the same class? Justify your answer.

3. What would you identify as the main drawback of the nearest neighbor classifier?

4. Think of some possible way to introduce mechanisms of supervision in the following problems.(a) records of patients, (b) collection of integrated circuits, (c) collection of movies. Justify your choice of the pertinent mechanism.

5. It is known that the number of clusters is crucial in unsupervised learning. Consider that the number of "true" clusters is equal to "c". Discuss what happens if you go for the smaller number of clusters and the larger number of clusters.

6. The Minkowski distance d(x,y) is defined in the form

$$d(\mathbf{x}, \mathbf{y}) = \sqrt[m]{\sum_{i=1}^{n} |x_i - y_i|^m}$$

   $m > 0$. Plot this distance in the two-dimensional case for several values of m. In particular, select values of m equal to 1, 2, and 20. What do you notice about the geometry of the equidistant curves?

7. A two-class problem comes with the following distribution of data. What type of classifier would you suggest here? Justify your choice.

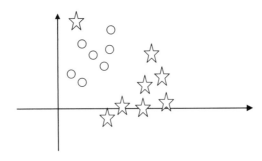

8. Give some everyday examples for which you could use clustering. What would be the objective of its use?

9. It is evident that regression can be realized in the form of linear and nonlinear relationships. Pick one everyday example and elaborate on the choice of the form of the model. What would be your preference? Justify your choice.

10. The decision tree for a two-class problem is shown below. Write down the expressions describing its boundaries and sketch them. The variables x_1 and x_2 assume nonnegative values.

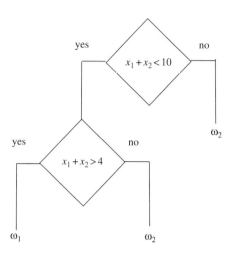

5

Knowledge Representation

In this Chapter, we discuss several categories of the fundamental models of knowledge representation. We also describe some generic schemes of data representation and their essential features.

Organizing available domain knowledge, as well as dealing with the knowledge acquired through data mining, could be realized in many different ways. The organization of knowledge is fundamental to all pursuits of data mining and impacts their effectiveness, accuracy of findings, and interpretability of results.

1. Data Representation and their Categories: General Insights

The data available for data mining purposes are highly diversified. It is instructive to elaborate on their main categories since these directly affect the character of the generic schemes of knowledge representation used in data mining. There are several categories of data types (data variables) as shown in Figure 5.1 and described below.

Continuous quantitative data. These data concern continuous variables, such as pressure, temperature, height, etc., Figure 5.1(a). They occur very often in relationship to physical phenomena as these data naturally generate such continuous variables.

Qualitative data. They typically assume some limited number of values. For which either a linear organization which could be established (**ordinal qualitative** data) or no such order could be formed (**nominal qualitative** data). For instance, when quantifying academic qualifications we can enumerate successive levels of education starting from junior high and high school and finishing with graduate school as shown in Figure 5.1 (b). Qualitative data could also be the result of a transformation of original continuous quantitative data, subject to some generalization. As an example, consider temperature. This continuous quantitative variable can be granulated (abstracted) by defining several labels such as low, medium, and high. These could ordered by noting that low ⊲ medium ⊲ high where ⊲ denotes a certain intuitively appealing ordering of the labels. We will return to this issue later on in the context of information granulation and information granules. Some qualitative data are nominal, for which no specific order could be established (that makes sense). An example is the enumeration of various types of employment, refer to Figure 5.1(c).

Structured data. These data become more complex as we build a **hierarchy** of specialized concepts; an example shown in Figure 5.2 illustrates the case where the general concept of *vehicles* forms a hierarchy of more specialized and sometimes nested subconcepts. This process gives rise to tree structures whose nodes represent the values encountered in the problem while the links (edges) indicate relationships between them. Another example, structure of colors, as

70   1. Data Representation and their Categories: General Insights

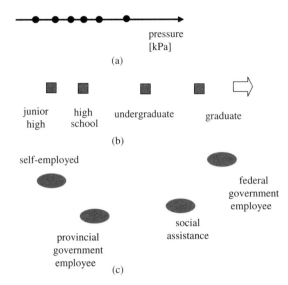

Figure 5.1. Examples of quantitative and qualitative data as described in the text: (a) continuous quantitative; (b) ordinal qualitative; (c) nominal qualitative.

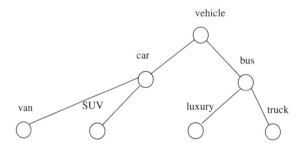

Figure 5.2. Examples of structured data showing their hierarchical organization.

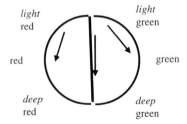

Figure 5.3. An example of a lattice of colors as perceived by a human observer.

perceived by a human observer, forms a lattice structure. This means that we encounter some extreme elements (viz. maximal and minimal); here those will be white and black. However, not all colors can be ordered in a linear fashion. To be more specific, along each edge of the structure shown in Figure 5.3, the colors could be organized with respect to their intensity (light... dark) and their position in the spectrum (yellow, green, red, etc.). Those located on different edges of the graph are not comparable from the standpoint of our intuitive perception of colors (see Figure 5.3).

2. Categories of Knowledge Representation

We now consider the main categories of **knowledge representation schemes** such as rules, graphs, and networks.

2.1. Rules

In their most generic format, rules are conditional statements of the form

$$\text{IF condition THEN conclusion (action)} \qquad (1)$$

where the **condition** and **conclusion** are descriptors of pieces of knowledge about the domain, while the rule itself expresses the relationship between these descriptors. namely, For instance, the rule "IF the temperature is *high* THEN the electricity demand is *high*" captures a piece of domain knowledge that is essential to planning the activities of an electric company. Notably, rules of this sort are quite qualitative yet highly expressive. We are perhaps not so much concerned with detailed numeric quantification of the descriptors occurring in such rules since we appreciate that the rule exhibits an interesting relationship that is pertinent to the problem. We note that both the condition and the conclusion are formed as information granules-conceptual entities that are semantically sound abstractions. The operational context within which information granules are formalized and used can be established by considering any of the available formal frameworks such as sets, fuzzy sets, and rough sets.

In practice, domain knowledge is typically structured into a family of rules, with each of these assuming the same format or a similar format, e.g.,

$$\text{IF condition is } A_i \quad \text{THEN conclusion is } B_i \qquad (2)$$

where A_i and B_i are **information granules**. The rules articulate a collection of meaningful relationships existing within the problem.

We may envision more complex rules whose left-hand side may include several conditions of the form

$$\text{IF condition}_1 \text{ and condition}_2 \text{ and } \ldots \text{ and condition}_n \text{ THEN conclusion} \qquad (3)$$

with multidimensional input space composed as a Cartesian product of the input variables. Note that the individual conditions are aggregated together by the *and* logic connective. Rules give rise to the highly modular form of a granular model, and expansion of the model requires addition of some new rules while the existing ones are left intact.

There are two important points regarding the development of this form of knowledge representation. First, information granules can be expressed within various formal frameworks. Second, rules exhibit various architecture and a richness of possible extensions.

2. Categories of Knowledge Representation

2.1.1. Gradual Rules

In gradual (graded) rules, rather than expressing an association between condition and conclusion, we capture the trend within information granules and hence the condition and conclusion parts will each contain a term referring to that trend. For instance, the gradual rules rules of the form:

$$\text{"the higher the values of the condition, the higher the values of the conclusion"} \quad (4)$$

or

$$\text{"the lower the values of the condition, the higher the values of the conclusion"} \quad (5)$$

represent knowledge about the relationships between changes in the condition and the conclusion, as in "the higher the income, the higher the taxes". Using the previous notation, we can rewrite such rules in the form

$$\text{"IF } \tau(A_i) \text{ THEN } \mu(B_i)\text{"} \quad (6)$$

where τ and μ are modifiers acting upon the values of A_i and B_i and causing their shift in a certain direction (hence capturing the existing increasing or decreasing trends in their values).

2.1.2. Quantified Rules

It is quite common to quantify the relevance (confidence) of a specific rule. For instance, a rule of the form

$$\text{the likelihood that } high \text{ fluctuations in real estate prices lead to a}$$
$$significant\ migration \text{ of population within the province is } quite\ moderate \quad (7)$$

is an example of a quantified rule whose likelihood of satisfaction is quite moderate. Again, we may have several schemes for the realization of the quantification effect itself. In some expert systems, these take the form of confidence factors, that is numeric values in the range of $[-1, 1]$. In other cases, we may use linguistic characters of quantification captured in the language of fuzzy sets.

2.1.3. Analogical Rules

Here the rules focus on levels of similarity (closeness, resemblance, etc.) between pairs of items standing in the condition and the conclusion. The analogical rules assume the form

$$\text{IF similarity } (A_i, A_j) \quad \text{THEN similarity } (B_i, B_j) \quad (8)$$

Present in this form, the rules constitute a framework for analogical reasoning. In a nutshell, we are provided with an information granule A in the condition space and wish to infer a corresponding conclusion B in the conclusion space.

2.1.4. Rules with Regression Local Models

In such rules the conclusion comes in the form of a "local" regression models whose scope is narrowed down to the condition portion of the rule, that is,

$$\text{IF condition is } A_i \quad \text{THEN } y = f_i(\mathbf{x}, \mathbf{a}_i) \quad (9)$$

The regression model (f_i) could be linear or nonlinear and applies only to the inputs ($\mathbf{x} \in \mathbf{R}^n$) that belong to the information granule represented by A_i.

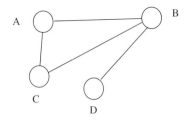

Figure 5.4. An example graph representing dependencies between concepts.

Let us reiterate two general and highly essential observations that hold despite the significant diversity of formats of the rules themselves and of their applications. First, the rules give rise to highly modular architectures, and this an organization becomes crucial to the efficient realization of the mechanisms of model formation, reasoning, and maintenance. Secondly, rules are always formulated in the language of information granules; hence, they constitute an abstract reflection of the problem or of problem solving. Since we never reason in terms of numbers, the explicit use of numbers in the rules is highly unjustifiable.

2.2. Graphs and Directed Graphs

Graphs are fundamental constructs representing relationships between concepts. Concepts are represented as nodes of a graph. The linkages (associations, dependencies, etc.) between the concepts are represented as the edges of the graph. An example of a graph is shown in Figure 5.4; here we have four nodes (A, B, C, and D) with several links between them (A-B, A-C, B-C, and C-D).

Graphs are helpful in visualizing a collection of concepts and presenting key relationships (dependencies) between them. They are highly appealing to designers as well as to users of developed systems. When properly displayed and augmented by colors, graphs help provide a solid insight into the behavior of the phenomenon under discussion.

There are many variations and augmentations of generic graphs. First, graphs can be directed, that is, their links can indicate directional relationships between the nodes. For instance, an occurrence of concept (node) A triggers occurrence of concept (node) B. We can represent a chaining effect (A-B-C) or looping (C-D-E -C) as illustrated in Figure 5.5.

Second, graphs can also show numeric quantification of the links that indicate the strength of dependency between the concepts (nodes). This quantification is shown in Figure 5.5 as well. Here, the strength of the relationship between E and C is far lower than that between B and C.

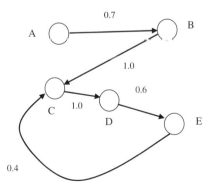

Figure 5.5. A directed graph showing a chaining effect (A-B-C) and a loop (C-D-E-C) in the graph.

74   2. Categories of Knowledge Representation

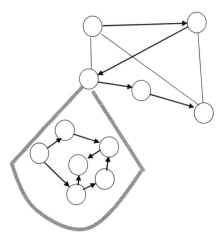

Figure 5.6. A hierarchy of graphs; note that the lower graph (presenting more details) expands a node of the structure at the higher level of abstraction.

To deal with a significant number of nodes, a graph can be structured as a hierarchy of subgraphs where the nodes on the upper level are expanded to a collection of nodes in the graph on the lower level. An example of a hierarchy of graphs is illustrated in Figure 5.6.

2.3. Trees

Trees are a special category of graphs in which there is a single root node and a collection of terminal nodes. There are no loops in trees. An example tree is shown in Figure 5.7.

As in the case of the graphs themselves, each node represents a certain concept or attribute while the edges of the graph present the relationships between the concepts.

Decision trees are one of the most commonly used tree structures. Here each node of the tree represents an attribute that assumes a finite number of discrete values while the edges originating from each node are labeled with the corresponding values. An example of a decision tree is shown in Figure 5.8.

We note that decision trees can be easily translated into a collection of rules. For each rule, we traverse the tree starting from its root and moving down to one of the terminal nodes. Along the way, we collect the attributes and the values that they assume. For instance, the rules obtained from the tree shown in Figure 5.8 read as follows:

$$
\begin{aligned}
&\text{IF } A \text{ is } c \text{ and } B \text{ is } w \quad \text{THEN } \phi \\
&\text{IF } A \text{ is } c \text{ and } B \text{ is } z \quad \text{THEN } \omega \\
&\text{IF } A \text{ is } a \text{ and } C \text{ is } k \quad \text{THEN } \omega
\end{aligned}
$$

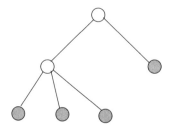

Figure 5.7. An example tree. Terminal nodes are shaded.

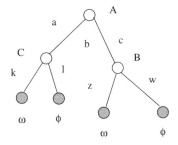

Figure 5.8. A decision tree showing nodes with the edges labeled by the discrete values (a, b, c...) of the attributes (A, B, C). The terminal nodes are shaded.

$$\text{IF A is a and C is l} \quad \text{THEN } \phi \tag{10}$$

Note that the length of each rule is different since it is the result of the tree traversal and each path being traveled could be of a different length. It is important to observe that the traversal process sets up the order of the conditions (attributes) in the rule. This organization of conditions is not captured when the rules are formed on a basis of available domain knowledge.

2.4. Networks

Networks can be regarded as generalized graphs in the sense that at each node of the graph we encounter some local processing capability. In other words, the network not only represents the knowledge itself but also contains the underlying processing being realized at the local level. An example network, illustrated in Figure 5.9 shows the nodes at which we compute the logic expressions of the inputs. In essence, these are logic functions operating on the input variables through the use of the logic operations of conjunction, disjunction, and negation.

Networks can be built in a hierarchical structure where a node at a higher conceptual level unfolds into a collection of nodes at a lower level; the nodes existing there may be endowed with some limited processing capabilities or they may contain no processing at all. The hierarchical organization of the network is helpful in representation of structure in data.

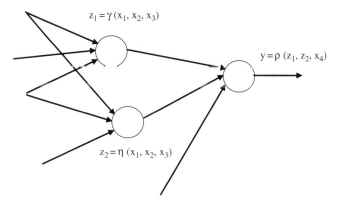

Figure 5.9. A network whose nodes provide logic computing (logic formulas $\gamma, \eta, \rho, \ldots$) over the variables associated with the incoming links.

3. Granularity of Data and Knowledge Representation Schemes

We now proceed with fundamentals of **information granules** that support human-centric computing. In a broad sense, by **human-centricity** we mean all faculties of a computing systems that facilitate interaction with human users; either by improving and enhancing the quality of communication of findings or by accepting user inputs that can be realized in a more flexible and friendly manner (say, in some linguistic form). We discuss key concepts and show how they translate into detailed algorithmic constructs.

Information granules tend to permeate human endeavors. No matter what a given task is, we usually cast it in a certain conceptual framework of basic entities that we regard to be of relevance in the given formulation. Using this framework, we formulate generic concepts adhering to some level of abstraction, carry out processing, and communicate the results to the external environment. Consider image processing: in spite of continuous progress in the field a human being assumes a dominant position with reference to image understanding and interpretation. Humans do not focus on individual pixels and process them afterwards, but instead group them together into semantically meaningful constructs – objects we deal with in everyday life. These involve regions that consist of pixels or groups of pixels drawn together because of their proximity in the image, their similar texture, their color, etc. This remarkable ability of humans, so far unchallenged by computer systems, is based on our ability to construct information granules and manipulate them. As another example, consider a collection of time series. From our perspective we can describe these in a semiqualitative manner by pointing to specific regions of the signals. Specialists easily interpret ECG signals by distinguishing some segments and their combinations. Again, the individual samples of the signal are not the focus of the analysis and signal interpretation. We granulate all phenomena (no matter whether they are originally discrete or analog). Time is a variable that is subject to granulation. We use seconds, minutes, days, months, years, and so on. Depending on a specific problem and who the user is, the size of information granules (time intervals) could vary dramatically. To high-level management, a time interval of quarters or a few years could be a meaningful temporal information granule on the basis of which management may develop models. For a designer of high-speed digital systems the temporal information granules are nanoseconds and microseconds. These commonly encountered examples are convincing enough to lead us believe that (a) information granules are key components of knowledge representation and processing, (b) the level of granularity of information granules (their size) becomes crucial to the problem description, and (c) there is no universal level of granularity of information; the size of granules is problem specific and user dependent.

What has been said so far touches on the qualitative aspect of the problem. The challenge is to develop a computing framework within which all these representation and processing endeavors could be formally realized. The common platform emerging within this context comes under the name **granular computing**, an emerging paradigm of information processing. While we have already described a number of important conceptual and computational constructs developed in the domain of system modeling, machine learning, image processing, pattern recognition, and data compression in which various abstractions (information granules) came into existence,

Granular Computing is innovative in several ways:

– It identifies essential commonalities between surprisingly diversified problems and technologies, which can be cast into a unified framework that we usually refer to as a granular world. Fully operational processing entity interacts with the external world (which could be another granular or numeric world) by collecting necessary granular information and returning the outcomes of granular computing.

- With the emergence of the unified framework of granular processing we get a better grasp on the role of interaction between various formalisms and can visualize the way in which they communicate.
- It brings together the existing formalisms of set theory (interval analysis), fuzzy sets, rough sets, etc. under the same roof by clearly showing that in spite of their visibly distinct underpinnings (and ensuing processing), they also possess fundamental commonalities. In this sense, Granular Computing provides an environment of synergy between the individual approaches.
- By building upon the commonalities of the existing formal approaches, Granular Computing helps build heterogeneous and multifaceted models of processing of information granules by recognizing the orthogonal nature of some existing and well probability density functions, and fuzzy sets with their membership functions). The orthogonality implies that some of these technologies supplement each other rather than taking some form of competition.
- Granular Computing fully uses a notion of variable granularity, whose range may cover detailed numeric entities and very abstract and general information granules. It looks at the aspects of compatibility of information granules and communication mechanisms between the granular worlds.

To cast Granular Computing in a historic perspective, it is worth acknowledging that a number of fundamental mechanisms of granulation have come from interval analysis, fuzzy sets, uncertain variables, and rough sets.

4. Sets and Interval Analysis

The two-valued world of sets and interval analysis ultimately depends upon a collection of intervals in the line of reals, $[a, b]$, $[c, d]$,..., etc. Conceptually, sets are rooted in a two-valued logic using the fundamental predicate of membership "belongs to" or "element of" ($\in$). An important isomorphism holds between the structure of two-valued logic endowed with truth values (false-true) and the set theory with sets described by characteristic functions. Sets are fundamental manifestations of abstraction whose role is to organize knowledge about the world and to form the most suitable perspective from which that knowledge could be described. The most essential manipulation of sets is carried out through basic operations such as union, intersection, complement, exclusive or, and the like. While their definitions are straightforward, very often we consider the use of characteristic functions that assume values of 0 and 1, depending on whether a certain element is included in a set or excluded. The list of essential operations along with their operational facets as expressed in the language of characteristic functions is shown in Table 5.1.

Interval analysis is a cornerstone of reliable computing that is ultimately associated with digital computing, in which any variable comes with a finite accuracy (implied by the fixed number of

Table 5.1. Sets, their description and their generic set-based operations (sets A and B are defined in the same space $\mathbf{X}$). The description of sets is provided in terms of their characteristic functions.

Concept	Description
Characteristic function of a set, $A(x)$	Offers a complete description of a set $$A(x) = \begin{cases} 1 & \text{if } x \in A \\ 0, & \text{otherwise} \end{cases}$$
Union of A and B	$(A \cup B)(x) = \max(A(x), B(x))$
Intersection of A and B	$(A \cap B)(x) = \min(A(x), B(x))$
Complement of A	$\bar{A}(x) = 1 - A(x)$

Table 5.2. Algebraic processing in interval analysis. Two numeric intervals, $[a, b]$ and $[c, d]$ are considered.

Operation	Result
Addition $[a, b] + [c, d]$	$[a+c, b+d]$
Multiplication $[a, b]^*[c, d]$	$[\min(ac, ad, bc, bd), \max(ac, ad, bc, bd)]$
Division $[a, b] / [c, d]$	$[\min(a/d, a/c, b/c, b/d), \max(a/d, a/c, b/c, b/d)]$

bits used to represent numbers). This limited accuracy gives rise to a pattern of propagation of computing error. Interval analysis offers detailed calculations that produce results that are very much on the "conservative" side. In practice, the results of propagation of the outcomes of interval computing lead to a rapid growth of the ranges of the results, especially if such results are the outcome of successive processing of intervals. For instance, the addition of two intervals $[a, b]$ and $[c, d]$ leads to some broader interval of the form $[a+c, b+d]$. Detailed formulas governing algebraic processing of numeric intervals are shown in Table 5.2. When intervals degenerate into a single numeric quantity the results are the same as obtained for "standard" numeric computing.

The accumulation of uncertainty (or decreased granularity of the result) depends upon the specific algebraic operation realized for the given intervals. Interestingly, intervals formed uniformly in a certain space realize an analog-to-digital conversion; the higher the number of bits, the finer the intervals and the higher their number. The well-known fundamental relationship states that with "n" bits we can build a collection of 2^n intervals of width $(b-a)/2^n$ for the original range of $[a, b]$. Traditionally, given a universe of discourse **X**, a family of intervals (sets) defined therein is denoted by $P(\mathbf{X})$.

5. Fuzzy Sets as Human-Centric Information Granules

The main difference between sets (intervals) and **fuzzy sets** is that in fuzzy sets we allow a concept of partial membership so that we can discriminate between elements that are "typical" to the concept and those of borderline character. Information granules such as *high* speed, warm weather, *fast* car are examples of information granules falling into this category. We cannot specify a single, well-defined element that forms a solid border between full belongingness and full exclusion. Fuzzy sets, with their *soft* or gradual transition boundaries, are an ideal vehicle to capture the notion of partial membership. An example of a fuzzy set of *safe* highway speed is shown in Figure 5.10.

A fuzzy set A defined in **X** is characterized by its membership function:

$$A : \mathbf{X} \rightarrow [0, 1] \tag{11}$$

where $A(x)$ denotes a degree of membership of x in A.

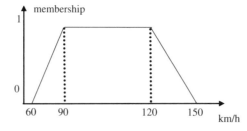

Figure 5.10. A membership function of *safe* speed.

Table 5.3. Generic descriptors used in the characterization of fuzzy set A.

Notion and definition	Description
α-cut $A_\alpha = \{x \mid A(x) \geq \alpha\}$	Set induced by some threshold consisting of elements belonging to A to an extent not lower than α. By choosing a certain threshold, we convert A into the corresponding set representative. α-cuts provide important links between fuzzy sets and sets.
Height of A, $\text{hgt}(A) = \sup_x A(x)$	Supremum of the membership grades; A is normal if $\text{hgt}(A) = 1$. Core of A is formed by all elements of the universe for which $A(x)$ attains 1.
Support of A, $\text{supp}(A) = \{x \mid A(x) > 0\}$	Set induced by all elements of A belonging to it with nonzero membership grades
Cardinality of A, $\text{card}(A) = \int_X A(x)dx$ (assumed that the integral does exist)	Counts the number of elements belonging to A; characterizes the granularity of A. Higher $\text{card}(A)$ implies higher granularity(specificity) or, equivalently, lower generality

A family of fuzzy sets defined in $\mathbf{X}$ is denoted as $F(\mathbf{X})$. As the semantics of A is far richer than the one encountered in sets, fuzzy sets come with several important characterizations which are summarized in Table 5.3.

Fuzzy sets are provided in the form of **membership functions** either in their continuous or discrete format. In the first case, we have an analytical expression for the membership function, say:

$$A(x) = \begin{cases} \exp(-0.7x), & \text{if } x \geq 0 \\ 0, & \text{otherwise} \end{cases} \quad (12)$$

In the second case, the membership function is defined in discrete elements of the universe of discourse and could be expressed by some formula (say, $A(x_i) = i/100$, $i = 1, 2, \ldots, 100$) or could come in a tabular format:

x_i	1	2	3	4	5	6
$A(x_i)$	0.6	1.0	0.0	0.5	0.3	0.9

Fuzzy sets defined in the line of real numbers ($\mathbf{R}$) whose membership functions satisfy several intuitively appealing properties such as (a) unimodality, (b) continuity, and (c) convexity, are referred to as **fuzzy numbers**. They generalize the concept of a single numeric quantity by providing an envelope of possible values they can assume.

The calculus of fuzzy numbers generalizes the idea of operations on intervals and follows the **extension principle**. Given two fuzzy numbers A and B, their function $C = f(A, B)$ returns a fuzzy number with the membership function

$$C(z) = \sup_{x, y: z = f(x, y)} [\min(A(x), B(y))] \quad (13)$$

5.1. Operations on Fuzzy Sets

Fuzzy sets defined in the same space are combined logically through **logic operators** of intersection, union, and complement, originally defined by Polish logician Lukasiewicz in the 1920s. The realization of these operations is completed via t- and s-norms (t-conorms) commonly used as models of the logic operators *and* and *or*, respectively. This description gives rise to the following expressions

$$A \cap B: \quad (A \cap B)(x) = A(x) \text{ t } B(x)$$
$$A \cup B: \quad (A \cup B)(x) = A(x) \text{ s } B(x)$$
$$\bar{A}: \quad \bar{A}(x) = 1 - A(x) \tag{14}$$

The negation operation, denoted by an overbar symbol, is usually taken by subtracting the membership function from 1, that is, $\bar{A}(x) = 1 - A(x)$. Let us recall that by a t-norm we mean a two-argument function:

$$t: [0, 1]^2 \to [0, 1] \tag{15}$$

that satisfies the following collection of conditions:

(a) $x \text{ t } y = y \text{ t } x$ commutativity
(b) $x \text{ t } (y \text{ t } z) = (x \text{ t } y) \text{ t } z$ associativity
(c) if $x \leq y$ and $w \leq z$ then $x \text{ t } w \leq y \text{ t } z$ monotonicity
(d) $0 \text{ t } x = 0, 1 \text{ t } x = x$ boundary conditions

Several examples of commonly used t-norms are shown in Table 5.4. The most commonly used t-norms are the minimum (min), product, and Lukasiewicz's *and*-operator. Given any t-norm, we can generate a dual t-conorm (s-norm) through the following expression:

$$a \text{ s } b = 1 - (1-a) \text{ t } (1-b) \tag{16}$$

$a, b \in [0, 1]$ which is the Morgan law (recall that $\overline{A \cup B} = \bar{A} \cap \bar{B}$). In other words, we do not need any separate table for s-norms as these could be easily generated. Again, the list of s-norms (t-conorms) being in common use involves the operation of the maximum (max), the probabilistic sum, and Lukasiewicz's *or*-operator.

5.2. Fuzzy Relations

Fuzzy sets are defined in a given space. **Fuzzy relations** are defined in Cartesian products of some spaces and represent composite concepts. For instance, the notion "*high* price and *fast* car" can be represented as a fuzzy relation R defined in the Cartesian product of price and speed.

Table 5.4. Examples of t-norms.

t-norm	Comments
$x \text{ t } y = \min(x, y)$	The first model of the *and* operation used in fuzzy sets (as proposed by Zadeh in his seminal 1965 paper). Note that this model is noninteractive, meaning that the result depends on only a single argument. More specifically, $\min(x, x+\varepsilon) = x$ regardless of the value of ε.
$x \text{ t } y = \max(0, (1+p)(x+y-1) - pxy), p \geq -1$	For $p = 0$, this yields the Lukasiewicz *and* operator
$x \text{ t } y = xy$	The product operator is commonly encountered in applications; the operator is interactive.
$x \text{ t } y = 1 - \min(1, \sqrt[p]{(1-x)^p + (1-y)^p}), p > 0$	The parametric flexibility is assured by the choice of values of "p"
$x \text{ t } y = \dfrac{xy}{p + (1-p)(x+y-xy)}, p \geq 0$	As above, this family of t-norms is indexed by the auxiliary parameter, whose value could be adjusted.
$x \text{ t } y = \dfrac{xy}{\max(x, y, p)}, p \in [0, 1]$	See the note above.
$x \text{ t } y \begin{cases} x \text{ if } y = 1 \\ y \text{ if } x = 1 \\ 0, \text{ otherwise} \end{cases}$	Drastic product – exhibits a "drastic" behavior that is it returns a nonzero argument when one of them is equal to 1, otherwise it returns 0

Note that R could be formally treated as a two-dimensional fuzzy set, $R : \mathbf{X} \times \mathbf{Y} \to [0, 1]$, with $\mathbf{X}$ and $\mathbf{Y}$ being the corresponding spaces of price and speed.

Fuzzy partition matrices generated by fuzzy clustering provide a discrete characterization of fuzzy relations. Given "c" clusters, the partition matrix consists of "c" fuzzy relations A_1, $A_2, \ldots, A_c$ whose membership grades appear as individual rows of this matrix. In other words, U can be written down in the form

$$U = \begin{bmatrix} A_1 \\ A_2 \\ A_c \end{bmatrix} \tag{17}$$

The partition matrix is a manifestation of granulation of original numeric data.

5.3. Second-Type Fuzzy Sets

Fuzzy sets are constructs with membership grades in the unit interval. There are several interesting generalizations justified by some specific applications. Two among these are of particular interest.

Second type fuzzy sets are fuzzy sets whose membership grades are fuzzy sets defined in [0,1]. Thus we depart from individual numeric membership grades and acknowledge that the degrees of membership themselves could be given in an approximate way. This implies a model of fuzzy sets as membership valuation. In particular, we can admit some ranges of membership, thus arriving at interval-valued fuzzy sets.

The generalization along this line is of particular interest when we are dealing with situations where the granularity in quantification of membership cannot be ignored and has to be incorporated into further processing. An example of second-type fuzzy sets is shown in Figure 5.11; here we illustrate both an interval-valued fuzzy set as well as a second order fuzzy set, with fuzzy sets regarded as their memberships.

5.4. Fuzzy Sets of Order-2

Fuzzy sets of order 2 are another conceptual extension of fuzzy sets, where we define a certain fuzzy set over a universe of several reference fuzzy sets. For instance, the term *comfortable* temperature can be defined by a collection of generic terms such as *cold* temperature, *around* zero,

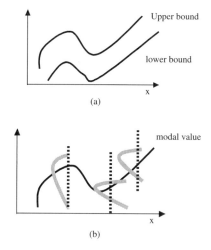

Figure 5.11. Examples of second type (type-2) fuzzy sets. (a) Interval valued; and (b) membership functions as grades of belongingness.

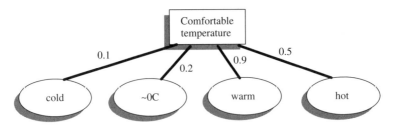

Figure 5.12. Fuzzy set of order-2 of *comfortable* temperature. Note the collection of referential fuzzy sets over which the order-2 fuzzy set is formed.

warm, *hot*, etc. Fuzzy sets of order-2 are more abstract than fuzzy sets of order-1 (the "standard" fuzzy sets). We portray the fuzzy set of order-2 of *comfortable* temperature in Figure 5.12.

Some further generalizations and extensions are possible as well e.g., we can consider fuzzy sets of type 2 of order 2.

With all these formalisms of representation of information granules, it is worth viewing them vis-à-vis the main agenda of data mining. Indisputably, the users are various domain experts who own the data and who often also collect the data themselves. For our purposes, we assume that the data owner has deep understanding of the data and of the processes that generated the data. Businesses are the largest group of DM users because they routinely collect massive amounts of data and have a vested interest in making sense of them. They are interested in making their businesses more competitive and profitable. Domain experts desire not only to better understand their data but also to gain new knowledge about the domain (represented by the data) for the purpose of solving problems in novel, possibly better ways. The various formalisms of information granulation and information granules are one of the possibilities arise with new and efficient ways of data mining. The interpretation of the results, taking full advantage of such formalisms, is another essential asset that has to be taken into consideration.

6. Shadowed Sets

Fuzzy sets help describe and quantify concepts of continuous boundaries. By introducing an α-cut, we can convert a fuzzy set into a set. By choosing a threshold level (α) that is high enough, we admit elements whose membership grades are meaningful (as viewed from the standpoint of the used threshold). The fact that an α-cut transforms a fuzzy set into a set, could create the impression that any fuzzy set can be made equivalent to some set. This point of view is highly deceptive. In essence, by building any α-cut, we elevate some membership grades to 1 (full membership) and eliminate other with lower membership grades (total exclusion). Surprisingly, this process does not take into account the distribution of elements with partial membership so this effect cannot be quantified in the resulting construct. The idea of **shadowed sets** aims at alleviating this problem by constructing regions of complete ignorance about membership grades. In essence, a shadowed set $A^\sim$ induced by a given fuzzy set A defined in $\mathbf{X}$ is an interval-valued set in $\mathbf{X}$ which maps elements of $\mathbf{X}$ into 0, 1, and the entire unit interval, that is [0,1]. Formally, $A^\sim$ is a mapping:

$$A^\sim : \mathbf{X} \to \{0, 1, [0, 1]\} \qquad (18)$$

Given $A^\sim(x)$, the two numeric values (0 and 1) take on a standard interpretation: 0 denotes complete exclusion from $A^\sim$, while 1 stands for complete inclusion in A. $A^\sim(x)$ equal to [0,1] represents a complete ignorance – nothing is known about the membership of x in $A^\sim$: we *neither* confirm its belongingness to $A^\sim$ *nor* commit to its exclusion. In this sense, such as "x" is the most "questionable" point and should be treated as such (e.g., this outcome could trigger some action to analyze this element in more detail, exclude it from further analysis, etc.). The name *shadowed set* is a descriptive reflection of a set that comes with "shadows" positioned around the edges of the characteristic function, as illustrated in Figure 5.13.

Shadowed sets are isomorphic with a **three-valued logic**. Operations on shadowed sets are the same as in this logic. The underlying principle is to retain the vagueness of the arguments (shadows of the shadowed sets being used in the aggregation). The following tables capture the description of the operators on shadowed sets:

Union

$$\begin{matrix} & 0 & 1 & [0,1] \\ 0 & 0 & 1 & [0,1] \\ 1 & 1 & 1 & 1 \\ [0,1] & [0,1] & 1 & [0,1] \end{matrix}$$

Intersection

$$\begin{matrix} & 0 & 1 & [0,1] \\ 0 & 0 & 0 & 0 \\ 1 & 0 & 1 & [0,1] \\ [0,1] & 0 & [0,1] & [0,1] \end{matrix}$$

Complement

$$\begin{matrix} 0 & 1 \\ 1 & 0 \\ [0,1] & [0,1] \end{matrix}$$

From the design point of view, shadowed sets are induced by fuzzy sets, and in this setting their role is to help interpret results given in the form of fuzzy sets and to reduce computational overhead. Since shadowed sets do not focus on detailed membership grades and process only 0, 1, and 1/2 (considering that the numeric value of 1/2 is used to code the shadow), all processing is very simple and computationally appealing.

The development of shadowed sets starts from a given fuzzy set. The underlying criterion governing this transformation is straightforward: maintain a balance of uncertainty in the sense that, while reducing low membership grades to zero and bringing high membership grades to 1, maintain the overall balance of change in membership grades. The changes of membership grades to 0 and 1 are compensated for by the construction of the shadow that "absorbs" the previous elimination of partial membership at low and high ranges of membership. This design principle for a unimodal fuzzy set is illustrated in Figure 5.14. The transformation is guided by the value of threshold β; more specifically, we are concerned with two individual thresholds, namely, β and 1-β.

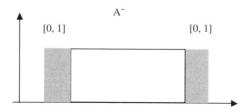

Figure 5.13. A shadowed set $A^\sim$. Note "shadows" produced at the edges of the characteristic function.

84   7. Rough Sets

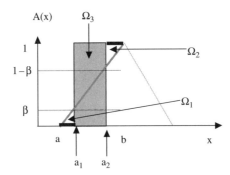

Figure 5.14. Induced shadowed set. The elimination of regions of partial membership is counterbalanced by the formation of shadows "absorbing" the reduction realized in the region of partial membership grades.

The retention of balance translates into the following dependency:

$$\Omega_1 + \Omega_2 = \Omega_3$$

where the corresponding regions are illustrated in Figure 5.14. Note that we are dealing with the increasing and decreasing portions of the membership functions separately. The integral form of the above relationship is

$$\int_a^{a_1} A(x)\mathrm{d}x + \int_{a_2}^b (1 - A(x))\mathrm{d}x = \int_{a_1}^{a_2} \mathrm{d}x \qquad (19)$$

For its detailed interpretation, refer again to Figure 5.14. A certain threshold value of $\beta, \beta \in [0, 1/2)$ that satisfies this expression is treated as a solution to the problem. Based on this result, we form a shadow of the shadowed set. In the case of commonly encountered membership functions, the optimal value of β can be determined in an analytical manner. For the triangular membership function, we consider each segment (the increasing and decreasing portion of the membership function) separately and focus on the linearly increasing portion of the membership function governed by an expression of the form $(x - a)/(b - a)$. Simple calculations reveal that the cutoff points a_1 and a_2 are equal to $a + \beta(b - a)$ and $a + (1 - \beta)(b - a)$. Subsequently, the resulting optimal value of β is equal to $\frac{2^{3/2}-2}{2} = 0.4142$. Similarly, when dealing with a square root type of membership function, that is $A(x) = \sqrt{\frac{x-a}{b-a}}$ in $x \in [a, b]$ and zero outside this interval we get $a_1 = a + \beta^2(b - a)$ and $a_2 = a + (1 - \beta)^2(b - a)$. The only root that satisfies the requirements imposed on the threshold values is equal to 0.405.

7. Rough Sets

The fundamental concept represented and quantified by **rough sets** concerns the description of a given concept in the language of a collection (vocabulary) of generic terms. Depending upon this collection of elements of the vocabulary relative to the concept, we encounter situations where a concept cannot be fully and uniquely described with the aid of the elements that form a certain vocabulary. The result is an approximate, or "rough" description of the given concept. To illustrate this, let us consider a concept of temperature in the range [a, b] that we intend to describe using a vocabulary of terms that are uniformly distributed intervals, as shown in Figure 5.15. Clearly the concept (shown as a solid thick line) to be described fully "covers" (includes) one interval, which is I_3. There are also intervals that have at least some limited overlap with the

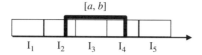

Figure 5.15. Concept (set) $[a, b]$ represented in the language of uniformly distributed intervals. Note the upper and lower bounds forming their representation.

concept, namely I_2, I_3, and I_4. In other words, we say that the concept, when characterized by the predefined vocabulary of intervals, does not lend itself to a unique description and therefore the best characterization we can produce comes in the form of some bounds. The tighter the bounds are, the better the description is.

Rough sets are directly related to the families of predefined information granules whose definition has been provided above. Using Figure 5.16, let us consider that a certain algorithm operating on the data set **X** forms a collection of information granules – sets. Their number is equal to $c[1]$. When using some other algorithm on the same data, we end up with $c[2]$ information granules. In general, there is no need to consider these two numbers to be equal to each other, $c[1] \neq= c[2]$.

Let us denote the information granules produced by the first algorithm by $A_1, A_2, \ldots, A_{c[1]}$. The other algorithm gives rise to the information granules denoted as $B_1, B_2, \ldots, B_{c[2]}$. Now each information granule B_i is characterized in the language of information granules formed by the first algorithm. The result of such description is a rough set. More specifically, we identify those A_j's that are fully "covered" by B_is (that is, they are fully included):

$$B_{i*} = \{A_j | A_j \subset B_i\} \tag{20}$$

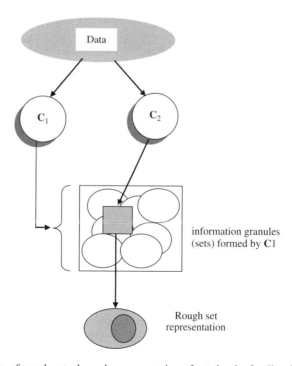

Figure 5.16. Emergence of rough sets through representation of sets by the family of sets constructed by C_1.

Similarly, we list all A_j's such that their intersection with B_i is nonempty, that is,

$$B_i^* = \{A_j | A_j \cap B_i \neq \emptyset\} \tag{21}$$

For the given family $\{A_i\}$, the resulting rough set is fully characterized by its upper and lower bound that are described in the form $< B_{i*}, B_i^* >$.

Example: Let us consider $c[1] = 4$ where the corresponding sets are as follows:

$$A_1 = [1\ 0\ 1\ 1\ 0\ 1\ 1]$$
$$A_2 = [0\ 1\ 0\ 0\ 1\ 0\ 0]$$
$$A_3 = [0\ 0\ 0\ 0\ 0\ 1\ 0]$$
$$A_4 = [0\ 0\ 0\ 0\ 0\ 0\ 1]$$

Let B_1 be given by the following characteristic function:

$$B_1 = [0\ 0\ 0\ 0\ 0\ 1\ 1]$$

Then B_1 expressed in the language of the first family of sets is described with the lower and upper bound, where:

$$B_{1*} = \{A_3, A_4\} \quad B_1^* = \{A_1, A_3, A_4\} \tag{22}$$

In this manner we have arrived at the rough set representation of B_1.

8. Characterization of Knowledge Representation Schemes

When we consider and select one of the categories of models of knowledge representation a number of aspects need to be taken into account. The most essential facets worth considering are the following:

- expressive power of the model
- computational complexity and associated tradeoffs vs. flexibility of the knowledge representation scheme as well as its scalability
- designer and user familiarity with the specific knowledge representation scheme
- effectiveness of creating such a model on the basis of the existing domain knowledge and experimental data

– character of information granulation and the level of specificity of the information granules to be used in the constructs of knowledge representation.

Interestingly, this list embraces various aspects that are human-centric (user and designer oriented) and includes issues that are driven by explicit problem requirements and algorithmic needs.

We are not concerned here with purely symbolic representations of information granules, as those might give rise to a great deal of ambiguity. To clarify, let us consider temperature as one of the variables in a problem under consideration. One could easily arrive at a collection of linguistic terms to describe temperature such as cold, warm, and hot. Let us denote them by C, W, and H, respectively. Conceptually, we can look at these entities as being symbols and manipulate them using some syntactic rules (as we use them in grammars). If we treat these terms as sets, fuzzy sets, or rough sets, in essence we endow them with some underlying numeric information that could be further processed according to the nature of the specific information granules.

9. Levels of Granularity and Perception Perspectives

Information granules play a pivotal role in knowledge representation when regarded as building blocks that capture the essence of the problem and facilitate its handling. A given problem can also be looked at from different standpoints. In particular, the level of detail (specificity) is particularly relevant and is directly associated with the level of information granularity at which we represent knowledge about the problem. By delivering a certain mechanism of **abstraction**, information granules can be a flexible means to deal with the required level of **specificity/generality** of the problem description. So far, we have not discussed granularity in a formal fashion, even though the essence of this concept seems to be self-explanatory. Intuitively, we associate the notion of granularity with the number of elements or concepts that are included in the form of some information granule. The larger the number of components embraced in this way, the higher the level of abstraction (generality); in other words, the level of granularity gets lower. On the other hand, if only a few elements are contained in a certain information granule, we say that this granule is very specific. Hence, the terms *granularity* and *specificity* are reciprocal: an increase in one of them leads to a decrease in the other.

Given that granularity is reflective of the number of elements within the information granule, in quantifying this concept we need to engage in some sort of counting of the number of elements. In the case of sets, the process is straightforward, since the corresponding characteristic functions indicate the belongingness of the given element to the set. In the case of other formalisms of information granulation, such as fuzzy sets or rough sets, the choice is not that evident.

Table 5.5 shows some selected ways to compute the level of granularity of information granules.

Table 5.5. Computing the granularity of information granules expressed as fuzzy sets and rough sets.

Information granules	Granularity
Fuzzy sets	σ-count, for given fuzzy set A, its σ-count is determined as $\int_x A(x)dx$ (we assume that the integral does exist). In the case of a finite space **X**, the integral is replaced by a sum of the membership grades
Rough sets	cardinality of the lower and upper bound, card (A_+), card(A_-); the difference between these two describes roughness of A

All the measures of granularity shown in Table 5.5 could be also presented in their normalized version by dividing their values by the cardinality of the overall space over which they are defined.

10. The Concept of Granularity in Rules

Rules are generic constructs of knowledge representation whose components are information granules. By virtue of their very nature, they tend to capture general relationships between variables (attributes). Furthermore, to make such dependencies meaningful, they have to be expressed at the level of abstract entities – information granules. For instance, a rule of the form "if high temperature then a substantial sale of cold drinks" is quite obvious. The relationship "if temperature of 28C then sale of 12,566 bottles of cold drinks" does not tell us too much and might not be a meaningful rule. The bottom line is that rules come hand in hand with information granules.

To emphasize the relevance of the use of **granular information in rule-based computing**, let us consider a rule of the form "If A then B" where A and B are represented as numeric intervals in the space of real numbers. In this context, the idea of information granularity comes with a clearly articulated practical relevance. A low level of granularity of the condition associated with a high level of granularity of the conclusion describes a rule of high relevance: it applies to a wide range of situations (as its condition is not very detailed) while offering a very specific conclusion. On the other hand, if we encounter a rule containing a very specific (detailed) condition with quite limited applicability while the conclusion is quite general, we may view a rule's relevance to be quite limited. In general, increasing granularity (high specificity) of the condition and decreasing granularity of the conclusion decrease the quality of the rule. We can offer some qualitative assessment of rules by distinguishing between those rules that are still acceptable and those whose usefulness (given the specificity of conditions and conclusions) could be questioned. A hypothetical boundary between these two categories of rules is illustrated in Figure 5.17. Obviously the detailed shape of the boundary could be different; our primary intent was to illustrate the main character of such a relationship.

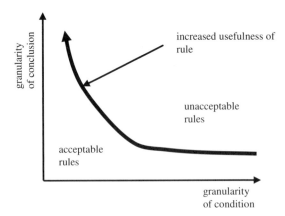

Figure 5.17. Identifying relevant rules with reference to the granularity of their conditions and conclusions. The boundary is intended to reflect the main tendency by showing the reduced usefulness of the rule when associated with an increased level of granularity of the condition and a decreased level of granularity of the conclusion.

11. Summary and Bibliographical Notes

In this Chapter, we covered the essentials of **knowledge representation** by presenting the most commonly encountered schemes such as **rules**, **graphs**, **networks**, and their generalizations. The fundamental issue of abstraction of information captured by **information granulation** and the resulting information granules was discussed in detail. We stressed that the formalisms existing within the realm of **granular computing** offer a wealth of possibilities with reference it comes to representation of generic entities of information. Similarly, by choosing a certain level of granularity (specificity), one can easily cater the results to the needs of the user. In this way, **information granules** offer the important feature of customization of data mining activities. We showed that the choice of a certain formalism of information granulation depends upon a number of essential factors spelled out in Sec. 5.8. Given that in data mining we are faced with an enormous diversity of data one has to make prudent decision about the underlying schemes of knowledge representation.

An excellent coverage of generic knowledge-based structures including rules is offered in [4, 6, 9, 21]. **Graph** structures are discussed in depth in [5, 14]. A general overview of the fundamentals and realization of various schemes of **knowledge representation** is included in [23]. **Granular computing** regarded as a unified paradigm of forming and processing information granules, is discussed in [1, 19, 24]. The role of information granulation is stressed in [26]. Set-theoretic structures (in particular, interval analysis) are presented in [8, 11, 24] Fuzzy sets introduced by Zadeh [25] are studied in depth in [16, 27–29, 31]. An in-depth coverage of rough sets developed by Pawlak is provided in [13, 20, 22]. Shadowed sets are presented in [15, 17, 18]. Some generalizations of fuzzy sets are discussed in [10, 12].

References

1. Bargiela, A., and Pedrycz, W. 2003. *Granular Computing: An Introduction*, Kluwer Academic Publishers
2. Bubnicki, Z. 2002. *Uncertain Logics, Variables and Systems. Lecture Notes in Control and Information Sciences*, no. 276, Springer Verlag
3. Butnariu, D., and Klement, E.P. 1983. *Triangular Norm Based Measures and Games with Fuzzy Coalitions*, Kluwer Academic Publishers
4. Cowell, R.G., David A.P., Lauritzen, S.L., and Spiegelhalter, D.J. 1999. *Probabilistic Networks and Expert Systems*, Springer-Verlag
5. Edwards, D. 1995. *Introduction to Graphical Modelling*, Springer-Verlag
6. Giarratano, J., and Riley, G. 1994. *Expert Systems: Principles and Programming*, 2nd edition, PWS Publishing
7. Hansen, E. 1975. A generalized interval arithmetic, *Lecture Notes in Computer Science*, Springer Verlag, 29: 7–18
8. Jaulin, L., Kieffer, M., Didrit, O., and Walter, E. 2001. *Applied Interval Analysis*, Springer Verlag
9. Lenat, D.B., and Guha, R.V. 1990. *Building Large Knowledge-Based Systems*, Addison Wesley
10. Mendel J.M., and John RIB. 2002. Type-2 fuzzy sets made simple, *IEEE Transactions. on Fuzzy Systems*, 10 (2002): 117–127
11. Moore R. 1966. *Interval Analysis*, Prentice Hall
12. Pal, S.K., and Skowron, A. (Eds.). 1999. *Rough Fuzzy Hybridization. A New Trend in Decision-Making*, Springer Verlag
13. Pawlak, Z. 1991. *Rough Sets. Theoretical Aspects of Reasoning About Data*, Kluwer Academic Publishers
14. Pearl, J. 1988. *Probabilistic Reasoning in Intelligent Systems*, Morgan Kaufmann
15. Pedrycz, W. 1998. Shadowed sets: representing and processing fuzzy sets, *IEEE Transactions. on Systems, Man, and Cybernetics, part B*, 28: 103–109

16. Pedrycz, W., and Gomide, F. 1998. *An Introduction to Fuzzy Sets; Analysis and Design*. MIT Press
17. Pedrycz, W. 1999. Shadowed sets: bridging fuzzy and rough sets, In: *Rough Fuzzy Hybridization. A New Trend in Decision-Making*, Pal, S.K., and Skowron, A. (Eds), Springer Verlag, 179–199
18. Pedrycz, W., and Vukovich, G. 2000. Investigating a relevance of fuzzy mappings, *IEEE Transactions. on Systems Man and Cybernetics*, 30: 249–262
19. Pedrycz, W. (Ed.). 2001. *Granular Computing: An Emerging Paradigm*, Physica Verlag
20. Polkowski, L., and Skowron, A. (Eds.). 1998. *Rough Sets in Knowledge Discovery*, Physica Verlag
21. Russell, S., and Nonig, P. 1995. *Artificial Intelligence: A Modern Approach*, Prentice-Hall
22. Skowron, A. 1989. Rough decision problems in information systems, *Bulletin de l'Academie Polonaise des Sciences (Tech)*, 37: 59–66
23. Sowa, J. 2000. *Knowledge Representation*, Brooks/Cole
24. Warmus, M. 1956. Calculus of approximations, *Bulletin de l'Academie Polonaise des Sciences*, 4(5): 253–259
25. Zadeh, L.A. 1965. Fuzzy sets, *Information & Control*, 8: 338–353
26. Zadeh, L.A. 1979. Fuzzy sets and information granularity, In: Gupta, M.M., Ragade, R.K., and Yager, R. R. (Eds.), *Advances in Fuzzy Set Theory and Applications*, North Holland, 3–18
27. Zadeh, L.A. 1996. Fuzzy logic = Computing with words, *IEEE Transactions on Fuzzy Systems*, 4: 103–111
28. Zadeh, L.A. 1997. Toward a theory of fuzzy information granulation and its centrality in human reasoning and fuzzy logic, *Fuzzy Sets and Systems*, 90: 111–117
29. Zadeh, L.A. 1999. From computing with numbers to computing with words-from manipulation of measurements to manipulation of perceptions, *IEEE Transactions. on Circuits and Systems*, 45: 105–119
30. Zadeh, L.A., and Kacprzyk, J. (Eds.). 1999. *Computing with Words in Information/Intelligent Systems*, Physica-Verlag
31. Zimmermann, H.J., 2001. *Fuzzy Set Theory and Its Applications*, 4th edition, Kluwer Academic Publishers

12. Exercises

1. Offer some examples of quantitative and qualitative variables.
2. What would be a shadowed set induced by the membership function $A(x) = \cos(x)$ defined over $[0, \pi/2]$?
3. What rules would you suggest to describe the following input-output relationship?

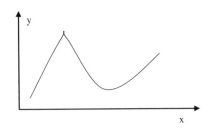

4. Suggest a rule-based description of the problem of buying a car. What attributes (variables) would you consider? Give a granular description of these attributes. Think of a possible quantification of relevance of the rules.
5. Derive a collection of rules from the decision tree shown below; order the rules with respect to their length (number of conditions).

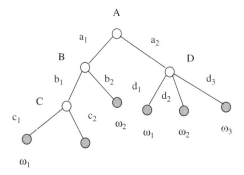

6. The "curse of dimensionality" is present in rule based systems. Consider that we are given "n" variables and each of them assumes "p" values. What is the number of rules in this case? To get a better sense as to the growth of this number, take $p = 5$ and vary n from 5 to 20. Plot your findings, treating the number of rules as a function of n. How could you avoid this curse?

7. Obtain the rules from the following network:

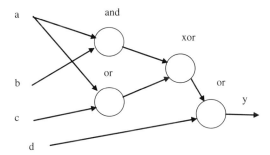

8. Suggest a membership function for the concept of *fast* speed. Discuss its semantics. Specify conditions under which such a fuzzy set could be effectively used.

9. Given are two fuzzy sets A and B with the following membership functions:

$$A = [0.7\ 0.6\ 0.2\ 0.0\ 0.9\ 1.0]\ B = [0.9\ 0.7\ 0.5\ 0.2\ 0.1\ 0.0].$$

Compute their union, intersection and the expression $C = A \cap \bar{B}$.

10. You are given a two-dimensional grid in the x-y space where the size of the grid in each coordinate is one unit. How could you describe the concept of a circle $(x-10)^2 + (y-5)^2 = 4$ in terms of the components of this grid?

Part 3

Data Preprocessing

6

Databases, Data Warehouses, and OLAP

In this Chapter, we explain issues related to databases and data warehouses and show their relationships to data mining. Specific topics covered include relational database management systems, data warehouses, SQL, and OLAP.

1. Introduction

Databases and data warehouses provide efficient **data retrieval** and summarization capabilities, which are necessary to prepare and select data for the subsequent steps of the knowledge discovery process. Therefore, prior to presenting data mining methods, we provide an overview of data storage and retrieval technologies.

This Chapter introduces and describes the following topics: **database management systems**, **SQL**, **data warehouses**, and **OLAP**. It explains the architecture and basic functionality, from a data mining point of view, of the most commonly used **relational database management systems**. Next, it discusses the differences between these systems and data warehouses. Finally, it describes different data warehouse architectures and their functionality and outlines the most important operations.

A diagram that visualizes relationship between databases/data warehouses and data mining is shown in Figure 6.1. Both databases and data warehouses are used to select data that are relevant to a data mining task from the available data. The main difference between a database and a data warehouse is that the latter provides integrated, subject-centered, organization-wide, historical data, while the former provides only a snapshot of the organization's data. Upon finishing this Chapter, the reader will understand basic concepts and terminology, architectures, and implementation techniques required for performing data mining using databases and data warehouses as the source of data.

2. Database Management Systems and SQL

Database management systems (DBMSs) were initially introduced in Chapter 3. A DBMS is defined as a collection of interrelated data and a set of programs to access the data. Two primary goals for a DBMS are

– to provide an environment that is both convenient and efficient to use in retrieving and storing data
– to provide design, update, and maintenance capabilities

2. Database Management Systems and SQL

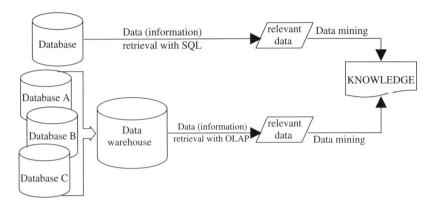

Figure 6.1. Databases and data warehouses and their relation to knowledge generated using data mining.

From the perspective of data mining and knowledge discovery, the primary objective of a DBMS system is to provide efficient data retrieval capabilities, while the other functionalities are of lesser importance. Therefore, this Chapter describes only the basic principles of the DBMS and focuses on data retrieval.

2.1. Architecture of Database Management Systems

A DBMS has a three-level structure, as shown in Figure 6.2:

– the view level, which is defined as the part of the database that is interesting to the user, and that consists of a selected portion of the data stored in the DBMS
– the logical level, which describes what data is stored in the database and what relationships are defined among these data
– the physical level, which describes how the data are stored.

This structure aims to shield the user from the internal physical details of how data are stored and organized. The user interfaces with the DBMS through the views and is able to define and update the logical level, but is not entangled in the details of the physical level.

The most common database type is a **relational database management system** (RDBMS), which consists of a set of rectangular tables. Each table is assigned a unique name and consists of attributes (also called columns or fields) and tuples (also called records and rows). Each table includes a special attribute called **key** that is used to uniquely identify tuples stored in the table. All activities performed with a RDBMS, including data retrieval, data storage, design, update, and maintenance, are performed based on a basic working unit called a **transaction**, which consists of a sequence of queries. A **query** is defined as an instance of access to a database. A transaction may include multiple queries and is characterized by the following (ACID) properties:

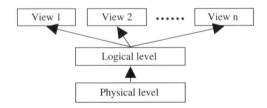

Figure 6.2. The three-layered structure of an RDBMS.

- **A**tomicity — a transaction should be run in its entirety or not at all ("all-or-nothing")
- **C**onsistency — a transaction always transforms an RDBMS from a consistent state to another consistent state
- **I**solation — any transient state of the RDBMS caused by a specific transaction is invisible to other transactions until it is finalized (committed)
- **D**urability — once a transaction has been committed, its results are permanent

The RDBMS architecture consists of three major components:

- a **query processor**, which handles the translation of queries into physical read/write requests. Queries are written in a language that hides the details of the data storage representation. The query processor performs query optimization, which aims to come up with the best (most efficient) strategy for extracting the data needed to satisfy a particular query.
- a **storage manager**, which handles the physical aspects of the RDBMS, such as disk space allocation, read/write operations, buffer and cache management, etc.
- a **transaction manager**, which handles issues related to concurrent multiuser access and issues related to system failures and recovery.

A schematic RDBMS architecture and the relation among its components is shown in Figure 6.3. The user sends queries, which are packed into transactions, to the RDBMS through the query processor, which communicates with both the transaction and storage managers to provide the required data. To speed up the location of the required data, the storage manager uses metadata indices, which allow for more efficiency in finding the physical location of the data.

We will explain and demonstrate concepts related to the data retrieval capabilities of an RDBMS using an example database, which stores information about personal banking accounts. A database is defined by a **schema**, which provides information about individual tables, attributes, and relations among attributes. The example schema consists of three tables, namely, *Own, Borrow*, and *Account*, and is shown in Figure 6.4. The key attribute for all three tables is the *account number*. The *Account* table stores information about checking, savings, and loan accounts for the clients. These accounts are divided into those in which the clients own their money (table *Own*) and those in which the clients borrow (table *Borrow*) the money.

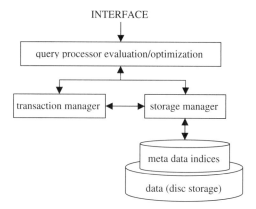

Figure 6.3. Architecture of an RDBMS system.

Own		Borrow		Account		
CustomerName	AccountNumber	CustomerName	AccountNumber	AccountNumber	AccountType	Balance
Will Smith	1000001	Will Smith	1000005	1000001	checking	1605
Joe Dalton	1000002	Will Smith	1000006	1000002	saving	1000
Joe Dalton	1000004	Joe Dalton	1000003	1000003	loan	5000
				1000004	checking	1216
				1000005	loan	205
				1000006	loan	1300

Figure 6.4. Example RDBMS schema.

2.2. Introduction to SQL

Several programming languages support data retrieval, but the **structured query language** (SQL)[1] is the standard for writing queries for an RDBMS. SQL is a declarative language (in contrast to procedural languages). A **procedural language** requires the user to specify explicit sequences of steps to follow in order to produce a result, while a **declarative language** is based on describing the relations among attributes (variables) and the language executor (often called the interpreter, compiler, or query processor in the case of SQL), who applies an algorithm to these relations to produce a result. In other words, with SQL, the user specifies only what he/she wants, and the RDBMS determines the best way to obtain it. The advantage of a procedural language is speed, since it specifies explicitly and in advance what and how things must be done. These instructions are compiled into machine language to speed up the execution. At the same time, a procedural language lacks flexibility in terms of its lack of adaptation to changes. A declarative language references items by name rather than by physical location and thus requires additional processing, but it provides flexibility for dealing with changes. Examples of declarative languages include Prolog and SQL, while examples of procedural languages are Visual Basic, C, and C++.

Being a declarative language, SQL does not have control flow constructs, such as if-then-else or do-while, and function definitions. Instead, it offers a fixed set of data types–that is, the user cannot create his/her own data types, as is common with other languages–and consists of only about 30 core commands. SQL is a relatively simple language and is used in all major RDBMSs, such as IBM's Informix, Microsoft SQL Server, Microsoft Access, Oracle, and Sybase.

An RDBMS uses the following types of languages to retrieve and manipulate data:

– **Data Manipulation Language** (DML), which is used to retrieve and modify the data
– **Data Definition Language** (DDL), which is used to define the structure of the data (database schema), i.e., statements that create, alter, or remove tables, attributes, and relations between attributes
– **Data Control Language** (DCL), which is used to define the privileges of database users

DDL and DCL are used only by a DBA (database administrator) and by users authorized by the DBA, while DML is used by "regular" RDBMS users. These three languages are expressed with SQL.

The main applications of SQL are as follows:

– **defining schema** for each relation using SQL DDL, which includes creating and managing tables, keys, and indices. **Indices** are used to speed up the process of finding the physical location of the data. Corresponding SQL commands include CREATE TABLE, ALTER TABLE, DROP TABLE, CREATE INDEX, and DROP INDEX.

---

[1] SQL is pronounced "sequel" or "es cue el," although the former pronunciation is more commonly used

- **defining privileges** for users using SQL DCL, which includes creating of objects related to user access and privileges and assigning/revoking permissions to read and alter data. Corresponding SQL commands include ALTER PASSWORD, GRANT, REVOKE, and CREATE SYNONYM.
- **populating the database**, which includes initial loading of the tuples into the tables created in the database. Corresponding SQL commands include SELECT and INSERT.
- **writing queries**, which are used to perform various operations on the existing database, such as selecting and modifying existing tuples, creating views, updating privileges, etc. Corresponding SQL commands include SELECT, INSERT, UPDATE, DELETE, and VIEW.

Once the database has been created and initially populated, SQL statements are prepared and executed to retrieve and update the data. The latter process happens online, i.e., while the RDBMS is running. In contrast, some of the commands included in the first three application types have to be executed offline, i.e., when the RDBMS is not being accessed by the users.

2.3. Data Retrieval with SQL

This section introduces Data Manipulation Language (DML), which provides functionality for data retrieval, the core functionality of an RDBMS for data mining. The most popular SQL DML commands are

- SELECT, which is used to scan the content of tables
- VIEW, which is used to create a new database view (the view can be seen as a filter applied to one or more tables, that does not constitute a new physical table, and is often used to design more complex queries)
- INSERT, which is used to insert new data into a table
- UPDATE, which is used to modify existing data in a table, but not to remove or add new tuples
- DELETE, which is used to remove a tuple from a table

The most commonly used SQL commands (queries) consist of three main clauses: **select**, **from**, and **where**. To write a query, the user needs to specify

- attributes that will be retrieved inside the SELECT clause
- all tables (relations) that are affected or used in the query inside the FROM clause
- conditions that constrain the desired operations in the WHERE clause

Before defining specific commands, we describe several common problems and mistakes related to writing SQL queries. First, the same attributes may appear in different tables under different names, and it is up to the user to provide the correct names in the correct tables. Second, although SQL itself is case insensitive, users should be cautious when retrieving the contents of a field (attribute value), since the stored data may be case sensitive. Finally, every SQL statement must be terminated by a semicolon, even if it is long and extended over many lines.

Below we describe the syntax and provide examples for each of the above data retrieval commands. The examples are based on the database defined in Figure 6.4. The text inside square brackets [ ] denotes optional conditions; | denotes or; '' (single quotation marks) are used to express a string of characters; and SQL keywords are given in bold.

2.3.1. SELECT Command

    SELECT   [*| **all** | **distinct**] column1, column2, ...
    FROM   table1 [, table2, ...]
    [**WHERE**   condition1 | expression1]
    [**AND**   condition2 | expression2]
    [**GROUP BY** column1, column2, ...]
    [**HAVING**   conditions1 | expression1]
    [**ORDER BY** column1 | integer1, column2 | integer2, ... [**ASC** | **DESC**]]

The SELECT command must contain a SELECT list, i.e., a list of attributes (columns) or expressions to be retrieved, and a FROM clause, i.e., the table(s) from which to retrieve the data. The *distinct* keyword is used to prevent duplicate rows being returned, *all* keyword assumes that a query concerns all columns from a given table(s), and the WHERE clause is used to filter out records (tuples) of interest. Finally, GROUP BY allows computing values for a set of tuples (see the next section for details).

Example: Find all account numbers (and their balances) with loan balances bigger than 1000.

    SELECT   AccountNumber, Balance
    FROM   Account
    WHERE   Balance > 1000
    AND   AccountType = 'loan'
    ORDER BY Balance DESC;

The query will return

AccountNumber	Balance
1000003	5000
1000006	1300

Example: (a join between two tables): Find all customers who have both a loan and some other account type.

    SELECT   distinct CustomerName
    FROM   Own, Borrow
    WHERE   Own.CustomerName = Borrow.CustomerName
    ORDER BY CustomerName;

The query will return

CustomerName
Joe Dalton
Will Smith

This query, as well as virtually all other queries, can be written in several ways, e.g.,

    SELECT   distinct CustomerName
    FROM   Own, Borrow

WHERE	Own.CustomerName = Borrow.CustomerName
ORDER BY	CustomerName;
SELECT	distinct CustomerName
FROM	Borrow
WHERE	CustomerName IN (SELECT CustomerName FROM Own)
ORDER BY	CustomerName;
SELECT	distinct CustomerName
FROM	Borrow
WHERE	EXISTS (SELECT CustomerName FROM Own WHERE
	Own.CustomerName = Borrow.CustomerName)
ORDER BY	CustomerName;

The two latter alternatives use **nested queries**. The nested query imbeds another query (or queries) to compute its own result. The ability to express a given query in alternative ways in SQL is necessary since not all commercial products support all SQL features. It also provides flexibility in designing complex queries.

Example: (a join with aliases among three tables): Find all customers (and their account types) who have both a loan and another type of account. Rename the corresponding columns (attributes) as Name and Type.

SELECT	distinct O.CustomerName Name, A.AccountType Type
FROM	Account A, Borrow B, Own O
WHERE	O.CustomerName = B.CustomerName
AND	(O.AccountNumber = A.AccountNumber OR
	B.AccountNumber = A.AccountNumber)
ORDER BY	CustomerName;

The query will return

AccountNumber	Balance
Joe Dalton	saving
Joe Dalton	checking
Joe Dalton	loan
Will Smith	checking
Will Smith	loan

2.3.2. Aggregate Functions

SQL provides several **aggregate functions**, which map a collection of values into a single value. These functions allow the user to compute simple statistics from the data based on the SELECT command. The aggregate functions are

- avg(x) — average of a collection of numbers x
- sum(x) — sum of a collection of numbers x
- max(x) — max value among a collection of numbers or nonnumeric data x
- min(x) — min value among a collection of numbers or nonnumeric data x
- count(x) — cardinality of a collections of numbers or nonnumeric data x

Example: (using aggregate functions): Find an average balance and number of all loans.

```
SELECT   avg(Balance) average loan, count(Balance) number of loans
FROM     Account
WHERE    AccountType = 'loan';
```

The query will return

average loan	number of loans
2168.3	3

Example: (using aggregate functions with GROUP BY): Find all account types, and their maximum balances, but only if their average balance is more than 1000.

```
SELECT     AccountType, max(Balance)
FROM       Account
GROUP BY   AccountType
HAVING     avg(Balance) >1000
```

The query will return

AccountType	max(Balance)
checking	1605
loan	5000

2.3.3. VIEW Command

CREATE VIEW view [(column_name_list)]
AS SELECT   query

The *view* is the name of a view to be created, *column_name_list* is an optional list of names to be used for columns in the view (these names override the column names that would be deduced from the SQL query), and *query* is an SQL query, which is usually given as a SELECT statement, that provides the columns and rows (tuples) of the view.

Example: Design a view that lists all customers that have a non-loan account together with their account types.

```
CREATE VIEW   CustomerAccounts (Name, Type)
AS            SELECT CustomerName, AccountType FROM Own, Account
              WHERE Own.AccountNumber = Account.AccountNumber;
```

The query will return
CustomerAccounts

Name	Type
Will Smith	checking
Joe Dalton	saving
Joe Dalton	checking

Once created, the view can be used to formulate more complex queries that are based on the data from the view. Often a single view can be used as a basis to formulate multiple different complex queries.

2.3.4. INSERT Command

INSERT INTO   table_name [('column1', 'column2')]
VALUES   ('values1', 'value2', [**NULL**]);

The SELECT command can be used with the INSERT command to insert data into the table based on the results of a query from another table.

INSERT INTO   table_name [('column1', 'column2')]
SELECT   [*| ('column1', 'column2')]
FROM   table_name
[**WHERE**   condition(s)];

Example: Add a new savings account for Will Smith with balance of $10,000. This requires the following two queries:

INSERT INTO   Own (AccountNumber, CustomerName)
VALUES   (1000007,'Will Smith');
INSERT INTO   Account
VALUES   (1000007,'saving','10000');

The two queries will update the *Account* and *Own* tables to read as follows:

Own

CustomerName	AccountNumber
Will Smith	1000001
Joe Dalton	1000002
Joe Dalton	1000004
Will Smith	1000007

Account

AccountNumber	AccountType	Balance
1000001	checking	1605
1000002	saving	1000
1000003	loan	5000
1000004	checking	1216
1000005	loan	205
1000006	loan	1300
1000007	saving	10000

2.3.5. UPDATE Command

UPDATE   table_name
SET   column1 = 'value',
            [column2 = 'value',]
            [column3 = 'value']
[**WHERE**   condition];

The UPDATE command usually uses the WHERE clause to specify tuple(s) that should be updated. In the case when this clause is omitted, all tuples in the table for the specified column(s) are updated.

Example: The just-established savings account for Will Smith should have a balance of $1,000 instead of $10,000.

```
UPDATE   Account
SET      Balance = 1000
WHERE    AccountNumber = 1000007;
```

The query will update the *Account* table to

Account

AccountNumber	AccountType	Balance
1000001	checking	1605
1000002	saving	1000
1000003	loan	5000
1000004	checking	1216
1000005	loan	205
1000006	loan	1300
1000007	saving	1000

2.3.6. DELETE Command

DELETE FROM table_name
[**WHERE** condition];

This command removes a single, or several, tuples from the specified table. Similarly to the UPDATE command, the DELETE is usually used with the WHERE clause that selects the tuples to be deleted. In the case when the WHERE clause is omitted, all tuples in the table are deleted, but an empty table is kept.

Example: Will Smith has closed his checking account with a balance of $1,605, and thus this account should be removed. Doing so requires the following two queries:

```
DELETE FROM   Account
WHERE         Account Number =
              (SELECT Account.AccountNumber FROM Own,
              Account
              WHERE Own.AccountNumber =
              Account.AccountNumber
              AND Account.Balance = 1605
              AND Own.CustomerName = 'Will Smith');
DELETE FROM   Own
WHERE         CustomerName = 'Will Smith' AND AccountName
              = 1000001;
```

The query will update the *Account* and *Own* tables to

Own

CustomerName	AccountNumber
Joe Dalton	1000002
Joe Dalton	1000004
Will Smith	1000007

Account

AccountNumber	AccountType	Balance
1000002	saving	1000
1000003	loan	5000
1000004	checking	1216
1000005	loan	205
1000006	loan	1300
1000007	saving	1000

2.3.7. Finalizing the Changes to the Database

When using SQL DML commands, such as INSERT, UPDATE, and DELETE, the changes are finalized, i.e., the data are physically updated on the hard drive, by using the following three commands:

– COMMIT, which makes the changes permanent
– ROLLBACK, which undoes the current transaction (a transaction is understood to be the last block of SQL queries)
– SAVEPOINT, which marks and names the current point in processing a transaction and thus allows the undoing of part of a transaction instead of the whole transaction.

Example: Committing a transaction.

```
DELETE FROM   Account
WHERE         AccountNumber = 1000002;
COMMIT;
```

As a result, one tuple in the *Account* table is removed and the database is physically updated.

2.3.8. Query Optimization

A very important issue from the perspective of data mining is how fast the data can be retrieved and updated. Given a user-specified query, the RDBMS interprets it and plans a strategy for carrying it out. This means that the RDBMS, not the user, is responsible for **query optimization**, i.e., the execution of a given query in the most efficient way. All but the simplest queries can be executed in many ways. Each execution can have a different total processing cost, which can vary even by several orders of magnitude.

To process and optimize an SQL query, the RDBMS goes through five steps:

1. **Parsing**. The query is broke up into individual words called tokens, and the query processor makes sure that the query contains valid verb and legal clauses, i.e., syntax errors and misspellings are detected.
2. **Validation**. The query processor checks the query against the schema to verify that all tables named in the query exist in the database, all columns exist and their names are unambiguous, and the user has the required privileges to execute the query.

3. **Optimization**. The query processor explores various ways to carry out the query. For instance, it may choose between first applying a condition to table A and then merging it with table B, or first merging the two tables and then applying the condition. In general, optimization aims to use predefined indices to speed up the search for data and to avoid sequential searches through entire tables by first reducing them though applying conditions. After the alternatives have been explored, the optimal sequence of actions is chosen.
4. **Execution plan preparation**. The query processor generates an execution plan for the query. This includes the generation of an "executable code" that translates the query into a sequence of low-level operations, such as read/write.
5. **Execution**. The query is executed according to the prepared execution plan.

The parsing of an SQL query does not require access to the database and typically can be done very quickly. Optimization is CPU intensive and requires access to the database schema. For a complex query that uses multiple tables, the optimization often explores more than a dozen different alternatives. At the same time, the cost of doing so in a nonoptimized way is usually much higher than the combined cost of optimization and execution of the query in the optimized way.

3. Data Warehouses

While databases offer basic data storage and retrieval solution, businesses require more sophisticated systems in order to systematically use, organize, and understand data for data mining and, ultimately, for decision making. Many large companies have spent millions of dollars to develop enterprise-wide data warehouses. In the current fast-paced and highly competitive marketplace, data warehousing is one of the must-have support tools.

A **data warehouse** (DW) is a database that is maintained separately from the organization's operational database for the purpose of decision support. **Data warehousing** is a process of constructing and using a DW. A DW provides integrated, enterprise-wide, historical data and focuses on providing support for decision makers with respect to data modeling and analysis. A DW is a collection of data specific to the entire organization, not only to a certain group of users. Finally, a DW, in contrast to an RDBMS, is not used for daily operations and transaction processing.

DWs provide online analytical processing tools for fast and interactive analysis of multidimensional data at different levels of granularity. This process boils down to providing tools to visually represent aggregated information and to enable interactive manipulation of these views. The corresponding DW tools are integrated with certain data mining techniques, such as classification, prediction, clustering, and association mining, resulting in an all-in-one solution. Finally, a DW is characterized by very high data retrieval performance.

3.1. Data Warehouses vs. RDBMS

W. Inmon, the "father of the data warehouse," defined the DW as "a subject-oriented, integrated, time-variant and non-volatile collection of data in support of management's decision making process." Following Inmon let us explain each term used in the above definition:

- **Subject oriented**. A DW provides a simple and concise view around a particular subject, such as customer, product, or sales, instead of the general organization's ongoing operations. This is done by excluding data that are not useful with respect to the subject and including all data needed by the users to understand the subject.
- **Integrated**. A DW integrates multiple, heterogeneous data sources, such as RDBMS, flat files, and online transaction records. Doing so requires performing data cleaning and integration during data warehousing to ensure consistency in naming conventions, attribute types, etc., among different data sources.

- **Time-variant**. The DW has a much longer time horizon than the operational RDBMS. The latter keeps only current-value data (a current data snapshot), while the DW provides information from a historical perspective, e.g., the past 10 years of data. Every key attribute in the DW contains a time-defining element.
- **Nonvolatile**. The DW is a physically separate data storage, which is transformed from the source operational RDBMS. The operational updates of data do not occur in the DW, i.e., update, insert, and delete operations are not performed. In fact, in the DW only two data processing operations are performed: (1) initial loading of data, and (2) read. Therefore, the DW does not require transaction processing, recovery, and concurrency capabilities, which allows for substantial speedup of data retrieval.

RDBMSs use **on-line transaction processing** (OLTP) to process queries grouped in transactions. The transactions are used to insert, read, and update data for day-to-day operations. RDBMSs are tuned for OLTP, which provides functionalities such as data accessing, indexing, concurrency control, recovery, etc. On the other hand, DWs use **on-line analytical processing** (OLAP) to perform data analysis based on a static copy of data and using exclusively read-only operations. DWs are tuned for OLAP, which is designed for efficient execution of complex read-based queries, and for providing multidimensional views involving the GROUP BY and aggregative functions.

The DW includes historical data that are not maintained by the RDBMS; uses integrated and reconciled (and thus consistent) data representations, codes, and formats; and provides a basis for analysis and exploration, which are used to identify useful trends and to create data summaries. A side-by-side comparison between OLTP and OLAP is given in Table 6.1.

3.2. Virtual Data Warehouses, Data Marts, and Enterprise Data Warehouses

A DW is usually established through an incremental process. This evolution involves these three steps:

- Creation of a **virtual data warehouse**. The virtual DW is created based on a set of views defined for an operational RDBMS. This warehouse type is relatively easy to build but requires excess computational capacity of the underlying operational database system. The users directly access operational data via middleware tools. This architecture is feasible only if queries are posed

Table 6.1. Comparison between OLTP (RDBMS) and OLAP (data warehouse).

Dimension	Specific feature	OLTP	OLAP
users	target users	clerks, IT professionals	decision support specialists
	# concurrent users	thousands	several (up to hundreds)
data	type of data	current, detailed, flat relational, and isolated	historical, multidimensional, integrated, and summarized
	size	MB to GB	MB to TB
	# accessed records / work unit	tens	up to millions
goals	target of the analysis	customer oriented	market oriented
	main use	day-to-day operations	decision support
	designed to provide	application-oriented solution	subject-oriented solution
main features	type of access	read and update	read-only
	queries	less complex	very complex
	unit of work	transaction	complex query
	data accessing pattern	frequently	ad hoc
	type of underlying DB design	ER diagrams	star model

infrequently, and usually is used as a temporary solution until a permanent data warehouse is developed.
- Creation of a **data mart**. The data mart contains a subset of the organization-wide data that is of value to a small group of users, e.g., marketing or customer service. This is usually a precursor (and/or a successor) of the actual data warehouse, which differs with respect to the scope that is confined to a specific group of users.
- Creation of an **enterprise warehouse**. This warehouse type holds all information about subjects spanning the entire organization. For a medium- to a large-size company, usually several years are needed to design and build the enterprise warehouse.

The differences between the virtual and the enterprise DWs are shown in Figure 6.5.

Data marts can also be created as successors of an enterprise data warehouse. In this case, the DW consists of an enterprise warehouse and (several) data marts (see Figure 6.6).

3.3. Architecture of Data Warehouses

The DWs have a three-level (tier) architecture that includes

- a bottom tier that consists of the DW server, which may include several specialized data marts and a metadata repository
- a middle tier that consists of an OLAP server (described later) for fast querying of the data warehouse
- a top tier that includes front-end tools for displaying results provided by OLAP, as well as additional tools for data mining of the OLAP-generated data

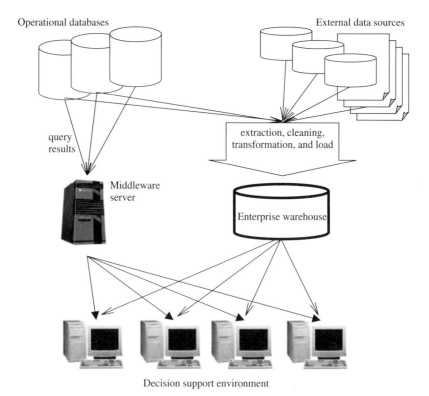

Figure 6.5. A virtual data warehouse (left) and an enterprise data warehouse (right).

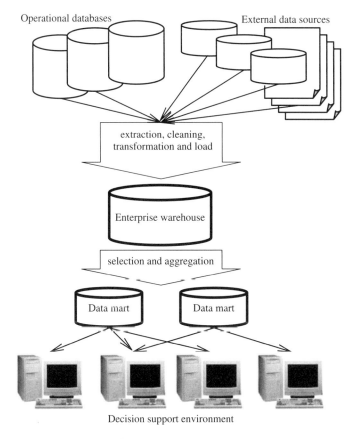

Figure 6.6. Data warehouse consisting of an enterprise warehouse and data marts.

The overall DW architecture is shown in Figure 6.7.

The **metadata repository** stores information that defines DW objects. It includes the following parameters and information for the middle and the top tier applications:

- a description of the DW structure, including the warehouse schema, dimensions, hierarchies, data mart locations and contents, etc.
- operational meta-data, which usually describe the currency level of the stored data, i.e., active, archived or purged, and warehouse monitoring information, i.e., usage statistics, error reports, audit trails, etc.
- system performance data, which includes indices used to improve data access and retrieval performance
- information about mapping from operational databases, which includes source RDBMSs and their contents, cleaning and transformation rules, etc.
- summarization algorithms, predefined queries and reports
- business data, which include business terms and definitions, ownership information, etc.

Similarly to the RDBMS, the internal structure of a DW is defined using a warehouse **schema**. There are three major types of warehouse schemas:

- the **star schema**, where a so-called **fact table**, which is connected to a set of **dimension tables**, is in the middle

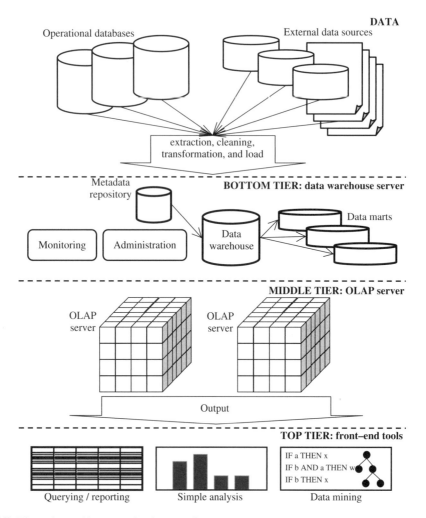

Figure 6.7. Three-tier architecture of a data warehouse.

- the **snowflake schema**, which is a refinement of the star schema, in which some dimensional tables are normalized into a set of smaller tables, forming a shape similar to a snowflake
- the **galaxy schema**, in which there are multiple fact tables that share dimension tables (this collection of star schemas is also called a **fact constellation**).

Each schema type is illustrated using the example of a computer hardware reseller company. The company sells various computer hardware (CPUs, printers, monitors, etc.), has multiple locations (Edmonton, Denver, San Diego), and sells products from different vendors (CompuBus, CyberMax, MiniComp). The subject of this DW is "the sells," e.g., the number of sold units and the related costs and profits.

3.3.1. Star Schema

An example star schema for the hardware reseller company is shown in Figure 6.8.

The *item, time, location*, and *producer* are dimension tables, associated with the fact table *sales* through key attributes, such as *item_code*, *time_code*, etc. These tables are implemented as regular

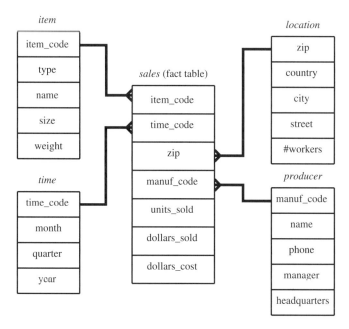

Figure 6.8. Example star schema.

relational tables. The *fact* table stores time-ordered data that concern the predefined DW subject, while the dimension tables store supplementary data that allow the user to organize, dissect, and summarize the subject-related data.

The **star schema**-based DW contains a single fact table that includes nonredundant data. Each tuple in the fact table can be identified using a composite key attribute that usually consists of key attributes from the dimension tables. Each dimension also consists of a single table, i.e., items, time, location, and producer are described using a single table. These tables may be denormalized, i.e., they may contain redundant data. Using the star schema, the data are retrieved by performing a **join operation** between the fact table and one or more dimension tables followed by a **projection operation** and a **selection operation**. The join operation selects common data between two or more tables, the projection operation selects a set of particular columns, and the selection operation selects a set of particular tuples. The main benefits of the star schema include ease of understanding and a reduction in the number of joins needed to retrieve data. This translates into higher efficiency when compared with the other two schemas. On the other hand, the star schema does not provide support for concept (attribute) hierarchies, which are explained later in the Chapter. The star schema for our example is shown in Figure 6.9.

3.3.2. Snowflake and Galaxy Schemas

The **snowflake schema** is a refinement of the star schema. Similarly to the star schema, it has only one fact table, but dimensional tables are normalized into a set of smaller tables, forming a shape similar to snowflake. The normalization results in tables that do not contain redundant data. The normalized dimensions improve the ease of maintaining the dimension tables and also save storage space. However, the space savings are, in most cases, negligible in comparison with the magnitude of the size of the fact table. The snowflake schema is suitable for concept hierarchies, but it requires the execution of a much larger number of join operations to provide answers to most of the queries and thereby has a strong negative impact on the data retrieval performance. Finally, the **galaxy schema** is a collection of several snowflake schemas, in which there are

112   3. Data Warehouses

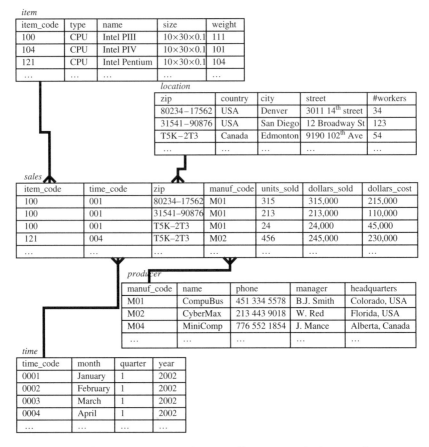

Figure 6.9. Example data for the computer hardware reseller company in the star schema.

multiple fact tables that may share some of their dimension tables. Example snowflake and galaxy schemas for our example are shown in Figure 6.10.

3.3.3. Concept Hierarchy

A **concept hierarchy** defines a sequence of mappings from a set of very specific, low-level concepts to more general, higher-level concepts. In a data warehouse, it is usually used to express different levels of granularity of an attribute from one of the dimension tables. To illustrate, we use the concept of location, in which each street address is mapped into a corresponding city, which is mapped into the state or province, which is finally mapped into the corresponding country. The location concept hierarchy is shown in Figure 6.11.

Concept hierarchies are crucial for the formulation of useful OLAP queries. The hierarchy allows the user to summarize the data at various levels. For instance, using the location hierarchy, the user can retrieve data that summarize sales for each individual location, for all locations in a given city, a given state, or even a given country without the necessity of reorganizing the data.

3.4. Multidimensional Data Models and Data Cubes

One of the most important characteristics of a data warehouse is that it is based on a **multidimensional data model**. The data are visualized not as a table, as is the case for the RDBMS, but rather as a multidimensional **data cube**. Each dimension of the cube represents different

Chapter 6 Databases, Data Warehouses, and OLAP   113

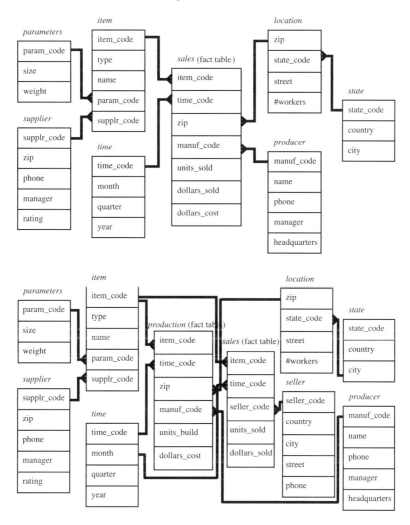

Figure 6.10. Example snowflake (top) and galaxy (bottom) schemas for the computer hardware reseller company.

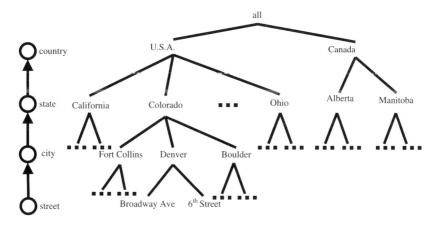

Figure 6.11. "Location" concept hierarchy.

114   3. Data Warehouses

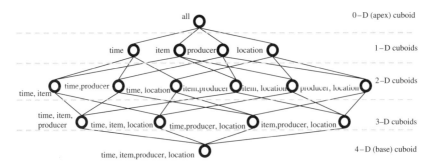

Figure 6.12. Example lattice of cuboids.

Relational (two-dimensional) table

location = "Edmonton", *producer* = "MiniComp"

month	CPU_Intel	CPU_AMD	Prnt_HP	Prnt_Lexmark	Prnt_Canon
January 2002	442	401	201	302	187
February 2002	224	289	134	89	121
March 2002	211	271	75	76	312
April 2002	254	208	143	108	112
...	...	...	...	...	...

Three-dimensional data cube
location, time, and *item (producer = "MiniComp")*

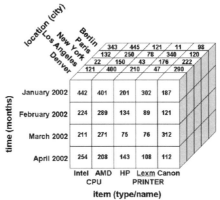

Four-dimensional data cube
producer, location, time and *item*

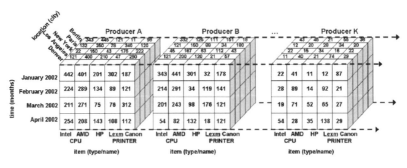

Figure 6.13. Example multidimensional data models.

information and can contain multiple levels of granularity (abstraction) defined by the corresponding concept hierarchies. For our computer hardware company, the example dimensions are *item* (name and type), *producer* (name), *location* (cities), and *time* (day, week, month, quarter and year). Figure 6.12 shows a **lattice of cuboids** that is defined based on combinations of the four dimensions. The apex cuboid is the topmost 0-D cuboid, which is at the highest level of summarization, while the *n*-D base cuboid is at the bottom.

The difference between the two-dimensional, table-based data model of the RDBMS and the multidimensional data model of the data warehouse is shown in Figure 6.13. It shows the number of sold units when using the two-dimensional relational table, and the three- and four-dimensional cuboids.

The concept hierarchies for the *time, item*, and *location* dimensions of the three-dimensional cuboid that represents sold units for the computer hardware company are shown in Figure 6.14.

The data cubes can be interactively manipulated by the user through a point-and-click interface, making the data retrieval and analysis process relatively intuitive and user-friendly (at least when compared with relational tables). For instance, a data cube can be rotated and the user can focus on a particular dimension, summarization, and cross-section. These manipulations are implemented using OLAP, which is discussed next.

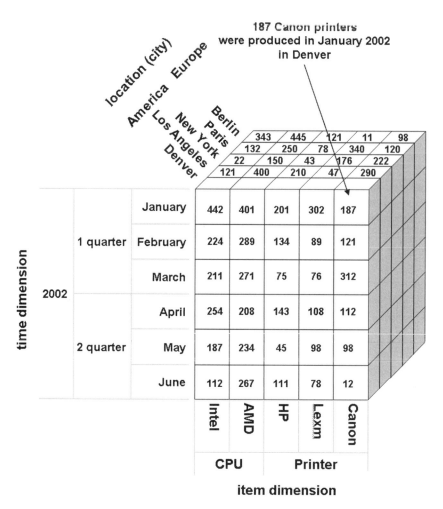

Figure 6.14. Example concept hierarchies for a three-dimensional cuboid.

4. On-Line Analytical Processing (OLAP)

As described above, the DWs use **on-line analytical processing** (OLAP) to formulate and execute user queries. OLAP is an SLQ-based methodology that provides aggregate data (measurements) along a set of dimensions, in which

- each dimension is described by a set of attributes, i.e., each dimension table includes a set of attributes
- each measure depends on a set of dimensions that provide context for the measure, e.g. for the reseller company DW, the measure is the number of sold units, which are described by the corresponding location, time, and item type
- all dimensions are assumed to uniquely determine the measure, e.g., for the reseller company, the location, time, producer, and item type provide all necessary information to determine context of a particular number of sold units.

Below, we define and illustrate basic OLAP commands that are used to perform data retrieval from a DW.

4.1. Data Retrieval with OLAP

There are five basic **OLAP commands**:

- ROLL UP, which is used to navigate to lower levels of details for a given data cube. This command takes the current data cube (object) and performs a GROUP BY on one of the dimensions, e.g., given the total number of sold units by month, it can provide sales summarized by quarter.
- DRILL DOWN, which is used to navigate to higher levels of detail. This command is the opposite of ROLL UP, e.g., given the total number of units sold for an entire continent, it can provide sales in the U.S.A.
- SLICE, which provides a cut through a given data cube. This command enables users to focus on some specific slice of data inside the cube, e.g., the user may want to look at the data concerning unit sales only in Denver.
- DICE, which provides just one cell from the cube (the smallest slice), e.g. it can provide data concerning the number of sold Canon printers in May 2002 in Denver.
- PIVOT, which rotates the cube to change the perspective, e.g., the "time item" perspective may be changed into "time location."

These commands, in terms of their specification and execution, are usually carried out using a point-and-click interface, and therefore we do not describe their syntax. Instead, we give examples for each of the above OLAP commands.

4.1.1. ROLL UP Command

The ROLL UP allows the user to summarize data into a more general level in hierarchy. For instance, if the user currently analyzes the number of sold CPU units for each month in the first half of 2002, this command will allows him/her to aggregate this information into the first two quarters. Using ROLL UP, the view

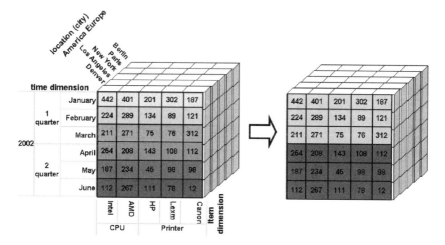

Figure 6.15. Example ROLL UP command.

# sold units		2002					
		January	February	March	April	May	June
CPU	Intel	442	224	211	254	187	112
	AMD	401	289	271	208	234	267

is transformed into

# sold units		2002	
		Quarter 1	Quarter 2
CPU	Intel	877	553
	AMD	961	709

From the perspective of a three-dimensional cuboid, the time (y) axis is transformed from months to quarters; see the shaded cells in Figure 6.15.

4.1.2. DRILL DOWN Command

The DRILL DOWN command provides a more detailed breakdown of information from lower in the hierarchy. For instance, if the user currently analyzes the number of sold CPU and Printer units in Europe and U.S.A., it will allows him/her to find details of sales in specific cities in the U.S.A., i.e., the view

# sold units		CPU		Printer		
		Intel	AMD	HP	Lexm	Canon
All	USA	2231	2134	1801	1560	1129
	Europe	1981	2001	1432	1431	1876

is transformed into

# sold units		CPU		Printer		
		Intel	AMD	HP	Lexm	Canon
All	Denver	877	961	410	467	620
	LA	833	574	621	443	213
	NY	521	599	770	650	296

Again, using a data cube representation, the location (z) axis is transformed from summarization by continents to sales for individual cities; see the shaded cells in Figure 6.16.

4.1.3. SLICE and DICE Command
These commands perform selection and projection of the data cube onto one or more user-specified dimensions. The specific possibilities are shown in Figure 6.17.

SLICE allows the user to focus the analysis of the data on a particular perspective from one or more dimensions. For instance, if the user analyzes the number of sold CPU and Printer units in all combined locations in the first two quarters of 2002, he/she can ask to see the units in the same time frame in a particular city, say in Los Angeles. The view

# sold units		CPU		Printer		
		Intel	AMD	HP	Lexm	Canon
2002	1 quarter	2231	2001	2390	1780	1560
	2 quarter	2321	2341	2403	1851	1621

is transformed into the L.A. table

# sold units		CPU		Printer		
		Intel	AMD	HP	Lexm	Canon
2002	1 quarter	666	601	766	187	730
	2 quarter	1053	759	323	693	501

The DICE command, in contrast to SLICE, requires the user to impose restrictions on all dimensions in a given data cube. An example SLICE command, which provides data about sales only in L.A., and DICE command, which provides data about sales of Canon printers in May 2002 in L.A., are shown in Figure 6.18.

4.1.4. PIVOT Command
PIVOT is used to rotate a given data cube to select a different view. Given that the user currently analyzes the sales for particular products in the first quarter of 2002, he/she can shift the focus to see sales in the same quarter, but for different continents instead of for products, i.e., the view

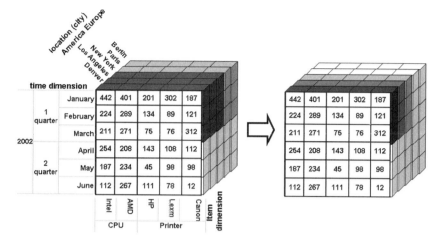

Figure 6.16. Example DRILL DOWN command.

# sold units		CPU		Printer		
		Intel	AMD	HP	Lexm	Canon
1 quarter	January	442	401	201	302	187
	February	224	289	134	89	121
	March	211	271	75	76	312

is transformed into

# sold units		America			Europe	
		Denver	LA	NY	Paris	Berlin
1 quarter	January	556	321	432	432	341
	February	453	564	654	213	231
	March	123	234	345	112	232

Again, in the data cube representation, the location-time view is pivoted to the product-time view (see Figure 6.19).

4.1.5. SQL Realization of OLAP Commands

Although the user is shielded from writing OLAP statements in a command-line style, as in the case of SQL, we have decided to explain how the OLAP commands are implemented behind the scenes. As mentioned before, the OLAP commands are translated and executed using SQL. A three-dimensional cuboid shown in Figure 6.20, which concerns the computer reseller example, is used to show SQL-based implementations of several OLAP commands. The example concerns three dimension tables, i.e., time, location, and item, and a fact table that stores the number of produced units for each combination of location, time, and item codes.

4. On-Line Analytical Processing (OLAP)

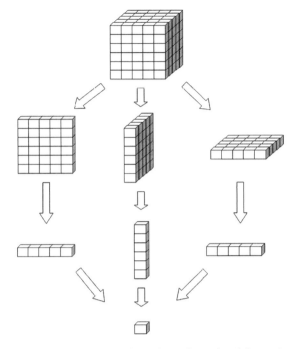

Figure 6.17. Possible projections and selections for a three-dimensional data cube.

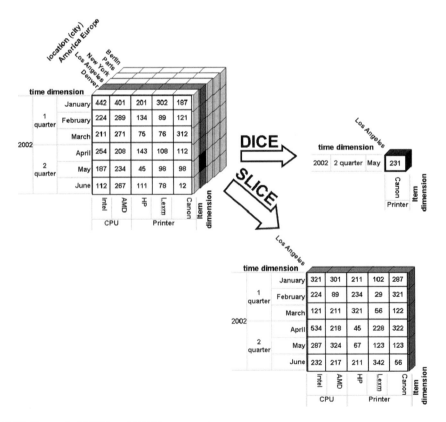

Figure 6.18. Example SLICE and DICE commands.

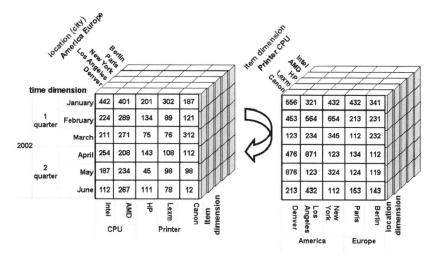

Figure 6.19. Example PIVOT command.

The DICE command shown in Figure 6.20 is implemented using the following SQL statement:

Example: (DICE command):
SELECT   units_produced
FROM   location L, time T, item I, facts F
WHERE   F.location_code = L.location_code
         AND F.time_code = T.time_code
         AND F.item_code = I.item_code
         AND L.city = 'Denver'
         AND T.month = 'January'
         AND I.brand = 'Canon';

The statement consists of two distinct parts. The first five lines define the cuboid that should be used, and are independent of the executed OLAP command. The last three lines, which are shown

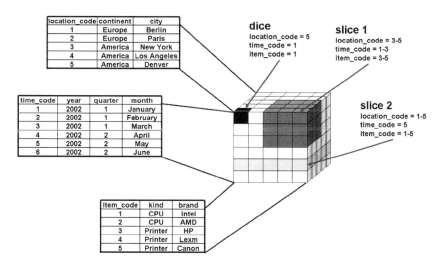

Figure 6.20. A three-dimensional cuboid for the computer hardware reseller example.

122   4. On-Line Analytical Processing (OLAP)

in italics, are specific to the DICE command and select a single cell in the cube. The command will return number of sold Canon units in January 2002 in Denver.

There are two different SLICE commands shown in Figure 6.20. In the case of slice 1, a subset of the original cuboid, which is narrowed down to sales of printers in America in the first quarter of 2002, will be returned to the user. The following SLQ statement will be used.

Example: (SLICE command):
```
SELECT   units_produced
FROM     location L, time T, item I, facts F
WHERE    F.location_code = L.location_code
         AND F.time_code = T.time_code
         AND F.item_code = I.item_code
         AND L.continent = 'America'
         AND T.quarter = '1'
         AND I.kind = 'Printer';
```

In the case of slice 2, the user narrows down the cube with respect to only one dimension, i.e., time, while the other two dimensions are left untouched. Again, the corresponding SQL statement is shown below.

Example: (SLICE command):
```
SELECT   units_produced
FROM     location L, time T, item I, facts F
WHERE    F.location_code = L.location_code
         AND F.time_code = T.time_code
         AND F.item_code = I.item_code
         AND T.month = 'May';
```

Since the above three examples are executed on the same cuboid, they share the first five lines. Next, we focus on implementation of the ROLL UP and DRILL DOWN commands. The following SQL statement concerns ROLL UP on the time dimension, i.e., instead of focusing on the individual months, the command moves two levels up in the hierarchy and summarizes sales by years.

Example: (ROLL UP command):
```
SELECT   SUM(units_produced)
FROM     location L, time T, item I, facts F
WHERE    F.location_code = L.location_code
         AND F.time_code = T.time_code
         AND F.item_code = I.item_code
         GROUP BY T.year;
```

Again, lines 2 through 5 are the same as in the previous examples, since the same cuboid is at stake. At the same time, summarization of the data requires the execution of SUM on the data in the original cube (first line) and the specification that it should be performed based on the year "level" of the time-dimension hierarchy through the application of the GROUP BY command (last line). Once the new cuboid that uses years for the time dimension is created, DRILL DOWN can be executed to change the focus of this dimension to one level down in the hierarchy, i.e., the quarters.

The only difference between this and the previous SQL implementations is the last line, which specifies the level in the hierarchy on which the data should be summarized.

Example: (DRILL DOWN command):

```
SELECT   SUM(units_produced)
FROM     location L, time T, item I, facts F
WHERE    F.location_code = L.location_code
         AND F.time_code = T.time_code
         AND F.item_code = I.item_code
         GROUP BY T.quarter;
```

4.2. OLAP Server Architectures

We describe here the physical implementation of an OLAP server in a DW. There are three different possible designs:

– **Relational OLAP** (ROLAP)
– **Multidimensional OLAP** (MOLAP)
– **Hybrid OLAP** (HOLAP)

ROLAP stores the data based on the already familiar relational DBMS technology. In this case, data and the related aggregations are stored in RDBMS, and OLAP middleware is used to implement handling and exploration of data cubes. This architecture focuses on the optimization of the RDBMS back end and provides additional tools and services such as data cube navigation logic. Due to the use of the RDBMS back end, the main advantage of ROLAP is scalability in handling large data volumes. Example ROLAP engines include the commercial IBM Informix Metacube (www.ibm.com) and the Microstrategy DSS server (www.microstrategy.com), as well as the open-source product Mondrian (mondrian.sourceforge.net).

In contrast to ROLAP, which uses tuples as the data storage unit, the MOLAP uses a dedicated n-dimensional array storage engine and OLAP middleware to manage data. Therefore, OLAP queries are realized through a direct addressing to the related multidimensional views (data cubes). Additionally, this architecture focuses on pre-calculation of the transactional data into the aggregations, which results in fast query execution performance. More specifically, MOLAP precalculates and stores aggregated measures at every hierarchy level at load time, and stores and indexes these values for immediate retrieval. The full precalculation requires a substantial amount of overhead, both in processing time and in storage space. For sparse data, MOLAP uses sparse matrix compression algorithms to improve storage utilization, and thus in general is characterized by smaller on-disk size of data in comparison with data stored in RDBMS. Example MOLAP products are the commercial Hyperion Ebasse (www.hyperion.com) and the Applix TM1 (www.applix.com), as well as Palo (www.opensourceolap.org), which is an open-source product.

To achieve a tradeoff between ROLAP's scalability and MOLAP's query performance, many commercial OLAP servers are based on the HOLAP approach. In this case, the user decides which portion of the data to store in the MOLAP and which in the ROLAP. For instance, often the low-level data are stored using a relational database, while higher-level data, such as aggregations, are stored in a separate MOLAP. An example product that supports all three architectures is Microsoft's OLAP Services (www.microsoft.com/), which is part of the company's SQL Server.

4.3. Efficiency of OLAP

Since a DW is built to store and retrieve/manipulate huge volumes of data, it employs methodologies for efficient computation and indexing of multidimensional data cubes. These methods are especially valuable for data mining applications, since they are aimed at improving data retrieval speed. Two approaches are considered to improve performance of the OLAP queries:

4. On-Line Analytical Processing (OLAP)

- **Materialization of cuboids.** Here the most frequently accessed cuboids are materialized, i.e., stored on the hard drive.
- **Data indexing.** Two popular indexing methods are **bitmap indexing** and **join indexing**. The former is used to perform searches through cuboids, while the latter is used to perform joins between the fact table and the dimension tables.

4.3.1. Materialization of Cuboids

The optimization of the OLAP performance with respect to materialization of cuboids can be done using the following methods:

- **Full cube materialization**, in which case the entire data cube (including all cuboids) is physically materialized. This results in the fastest query response but requires heavy precomputing and very large storage space. For large DWs, it is often unrealistic to precompute and materialize all the cuboids that can be generated for a given data cube. For instance, for our example cube that includes just four dimensions (*item*, *time*, *location*, and *producer*), sixteen cuboids can be created (see Figure 6.12).
- **No cube materialization**, in which case nothing is materialized. This architecture gives the slowest query response and always requires dynamic query evaluation, but at the same time it requires no overhead and needs significantly smaller amount of storage space. For a DW, which requires execution of complex queries, this architecture results in a very slow response time.
- **Partial cube materialization**, in which case selected parts of a data cube, i.e., selected cuboids, are materialized. This hybrid architecture implements a balance between the response time and the required storage space and overhead. It requires identification of a subset of cuboids that will be materialized as well as efficient updating of the materialized cuboids during each load and refresh of the data. The selection of cuboids for materialization can be performed using system performance data stored in the **metadata repository**.

4.3.2. Data Indexing

DWs also use data indexing techniques to improve response time. The **bitmap index** is defined for chosen attributes (columns) in the dimension and fact tables. In this case, each attribute value is represented by a bit vector of a length equal to the number of values, i.e., a bit in the vector is set to 1 if the corresponding row in the table has the corresponding value for the indexed attribute. Using bitmap indexing, the join and aggregation operators executed by OLAP (and SQL) are reduced to bit arithmetic, which is very fast, even faster than hash and tree indexing. This index works best for attributes with low cardinality, i.e., a low number of values. In case of high cardinality attributes, a compression technique may be used to implement the index.

For example, for the *item* dimension table

Item

item_code	type	name
1	DVD drive	HP
2	DVD drive	Intel
3	HDD	HP
4	HDD	Seagate
5	DVD drive	Samsung
6	HDD	Intel
7	HDD	Seagate

the corresponding bitmap indices for the *type* and *name* attributes are

index on type

record_code	DVD drive	HDD
1	1	0
2	1	0
3	0	1
4	0	1
5	1	0
6	0	1
7	0	1

index on name

record_code	HP	Intel	Seagate	Samsung
1	1	0	0	0
2	0	1	0	0
3	1	0	0	0
4	0	0	1	0
5	0	0	0	1
6	0	1	0	0
7	0	0	1	0

The *name* index can be used to find all rows in the *item* table where the name is either HP or Intel, which translates into binary 1000 OR 0100. Therefore, using the mask of 1100 the rows 1, 2, 3, and 6 are selected.

The **join index** addresses computationally expensive join operations. While traditional indices map the values of an attribute to a list of record IDs, e.g., a bitmap index, join indices are used to register the joinable rows between two or more tables. In the case of a DW, join indices speed up the process of finding related values (rows) between a dimension table and a fact table, and between multiple dimension tables and a fact table. In the latter case, they are called **composite join indices** and are used to select interesting cuboids. An example definition of a join index for subsets of the schema related to the computer hardware reseller company is shown in Figure 6.21. The join indices for the *type*, which is a part of the *item* dimension, and the fact table, between *continent*, which is part of the *location* dimension, and the fact table, and finally the composite join index between the *type*, the *continent*, and the fact table, are defined based on sample data from the corresponding three tables.

4.4. Fast Analysis of Shared Multidimensional Information (FASMI) Test

Dozens of relatively new, OLAP tools were developed by smaller vendors, in contrast to the modern DBMSs that were developed by well-established companies like IBM, Microsoft, Oracle, and Sybase. Below we explain how the user can decide whether a particular tool actually provides the desired OLAP functionalities and what these functionalities are.

Although many vendors claim to have "OLAP compliant" products, the user should be aware of the vendors' own descriptions. The techniques used to achieve OLAP functionality include different flavors of client/server architectures, time series analysis, object-orientation, optimized (and often proprietary) data storage, multithreading, and various patented ideas of which vendors are very proud. Although these techniques are important in order to implement the OLAP functionalities, different tools should be compared independently of the underlying technology. Therefore,

4. On-Line Analytical Processing (OLAP)

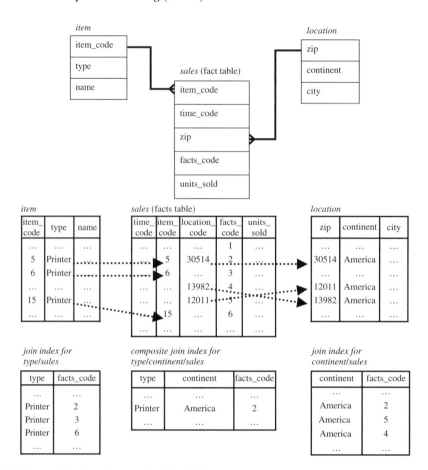

Figure 6.21. Example join and composite join indices.

a standard test called **Fast Analysis of Shared Multidimensional Information** (FASMI) was developed to evaluate the OLAP products. The FASMI test was first used in the mid-1990s and has been widely adopted ever since. The test summarizes the OLAP tools using five dimensions that were used to derive its acronym:

– Fast; An OLAP system must deliver most responses to users within about five seconds, with the simplest analyses taking no more than one second and very few taking more than 20 seconds. Slow query response is consistently the most often-cited technical problem with OLAP products.
– Analysis; An OLAP system must be able to cope with any business logic and statistical analysis that is relevant for the user of the system and application, and to keep the analysis easy enough for the target user. The system must permit the definition of new ad hoc calculations and report on the data in any desired way, without needing low-level programming.
– Shared; An OLAP system must implement the security mechanisms necessary to provide confidentiality (possibly down to a single-cell level) and concurrent update locking capabilities (if multiple write access is needed).
– Multidimensional; Multidimensionality is a key requirement for any OLAP tool. An OLAP system must provide a multidimensional conceptual view of the data, which includes full support for hierarchies and multiple hierarchies.
– Information; Information is defined as all the data and derived information stored in the underlying DW, wherever it is and however much of it is relevant for a particular application. An

OLAP tool is evaluated in terms of how much data it can handle, not how many Gb it takes to store the data. The "largest" OLAP products can hold at least three orders of magnitude more data than the "smallest" tools.

The FASMI web site can be found at www.olapreport.com/fasmi.htm. It currently stores evaluations of about 25 OLAP products. A significant downside of this test is that the actual evaluations have to be purchased by the user.

4.5. Example Commercial OLAP Tools

A recent report by the OLAP Report (www.olapreport.com) describes the top ten commercial OLAP products together with their market shares: Microsoft (28.0%), Hyperion (19.3%), Cognos (14.0%), Business Objects (7.4%), MicroStrategy (7.3%), SAP (5.9%), Cartesis (3.8%), Systems Union/MIS AG (3.4%), Oracle (3.4%), and Applix (3.2%). About 25 currently supported and over 20 discontinued OLAP products are also listed.

Specific commercial OLAP products include Microsoft SQL Server 2000 and 2005 Analysis Services, Hyperion Essbase 7X, Cognos PowerPlay 7.3, BusinessObjects XI, MicroStrategy 7i, SAP BW 3.1, Cartesis Magnitude 7.4, Oracle Express and the OLAP Option 6.4, and Applix TM1 8.3. Also, a number of open-source OLAP products, including Mondrian and Palo, have been developed.

5. Data Warehouses and OLAP for Data Mining

A DW integrated with an OLAP server can be used to perform these important tasks:

- **information processing**, which is implemented through querying, providing basic statistical analysis, and reporting using tables, charts, and graphs
- **analytical processing**, in which multidimensional analysis of the data by using basic OLAP operations, such as slice and dice, drilling, pivoting, etc. is performed
- **data mining**, in which new and interesting hidden patterns in the data are found. Some of the DW are furnished with, or can be integrated with, data mining tools that support discovery of data associations, construction of analytical models, performance of classification and prediction, and presentation of the results using visualization tools.

The above three functionalities are aligned with the typical evolution process of a DW. Initially, they are usually used for the generation of reports and for providing answers to predefined user queries. As the amount of stored data accumulates, the DWs are used to analyze summarized data, and the results are often presented in the form of charts and reports. Mature DWs are usually used for decision making purposes through the performance of multidimensional data analysis and complex OLAP operations. Finally, the ultimate application is to perform strategic decision making through the use of knowledge discovery and data mining tools that are integrated or interfaced with OLAP.

As mentioned in Chapter 3, although OLAP functionalities may seem to be advanced and as all that a user may need when analyzing the data, data mining provides additional and different services. It offers more complex techniques for understanding the data and generating new knowledge rather than simple user-directed data summarizations and comparisons provided by OLAP. More specifically, data mining allows the user to perform association discovery, classification, prediction, clustering, time-series analysis, etc., which are described in subsequent Chapters.

There are many reasons why data mining systems are integrated or interfaced with DWs. DWs contain high-quality data, which are already integrated, cleaned, and consistent. Therefore, the amount of work necessary to prepare data for a data mining project is substantially reduced (as

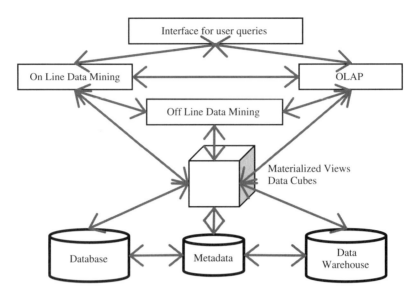

Figure 6.22. Integrated architecture for OLAP and data mining in a DW environment.

we know from Chapter 2, preparation of the data is one of the most time-consuming tasks in the knowledge discovery process). DWs also provide an information-processing infrastructure that includes the following:

- Open Database Connectivity (ODBC), a widely accepted application programming interface (API) for database access
- Object Linking and Embedding for Databases (OLEDB), which is a COM-based data access object that provides access to data in a DBMS
- OLAP tools
- reporting capabilities
- web-based interface

Finally, DWs provide OLAP-based exploratory data analysis, which allows for efficient selection and processing of relevant data by means of drilling, dicing, pivoting, and other operators.

The overall architecture that integrates DWs, OLAP, and data mining is shown in Figure 6.22.

The integration is performed though concurrent use of on-line data mining tools and the OLAP technology. As some of the data mining techniques require substantial computational resources and time, they are used to analyze the data off-line, and only the results of their analysis are used during the on-line interaction with the user. The data are extracted from the underlying DW (or a DBMS) and presented to the data mining algorithms through the use of data cubes.

6. Summary and Bibliographical Notes

In this Chapter, we introduced **databases**, **DataBase Management Systems** (DBMSs), **Structured Query Language** (SQL), **Data Warehouses** (DWs), and **On-Line Analytical Processing** (OLAP). The most important topics discussed in this Chapter are the following:

- **Relational DBMS** (RDBMS) is defined as a collection of interrelated data and a set of software programs to access those data. It consists of three major components: a **query processor**, a **storage manager**, and a **transaction manager**. All activities performed with an RDBMS, including data retrieval, data storage, design, update, and maintenance, are performed based on a basic working unit called **transaction**, which consists of a sequence of queries. A **query** is defined as an instance of access to a database.

- **SQL** is a declarative language for writing queries for an RDBMS. Three types of languages to retrieve and manipulate data, which include **Data Manipulation Language** (DML), **Data Definition Language** (DDL), and **Data Control Language** (DCL), are implemented using SQL. DML provides functionality for data retrieval, which is the core functionality of an RDBMS with respect to data mining. The most popular SQL DML commands are SELECT, VIEW, INSERT, UPDATE, and DELETE. Given a user-specified SQL query, the RDBMS interprets it and plans a strategy for carrying it out in the most efficient way, i.e., the RDBMS is responsible for **query optimization**.
- A **DW** is a subject-oriented, integrated, time-variant and nonvolatile collection of data in support of management's decision-making process. Three types of DW can be distinguished: the **virtual data warehouse**, the **data mart**, and the **enterprise warehouse**. The internal structure of a DW is defined using a warehouse schema, which can be defined using one of the three main types: **star schema**, **snowflake schema**, and **galaxy schema**.
- A **concept hierarchy** is a sequence of mappings from a set of very specific, low-level concepts to more general, higher-level concepts in a DW. Concept hierarchies allow the formulation of data summarizations and thus are crucial to the writing of useful **OLAP queries** (described below).
- A DW is based on a **multidimensional data model**. The data are shown using a multidimensional **data cube**, in contrast to the relational table used in the case of the RDBMS. Each dimension of the cube represents different information and can contain multiple levels of granularity (abstraction) defined by the corresponding concept hierarchies.
- RDBMSs use **on-line transaction processing** (OLTP) to process queries grouped in transactions and to provide functionalities such as data accessing, indexing, concurrency control, recovery, etc. In contrast, DWs use **OLAP** to perform data analysis based on a static copy of data and using exclusively read-only operations.
- **OLAP** provides aggregate data (measurements) along a set of predefined dimensions. There are five basic **OLAP commands**: ROLL UP, DRILL DOWN, SLICE, DICE, and PIVOT.
- Physical implementation of an OLAP server in a DW can be realized through **relational OLAP** (ROLAP), **multidimensional OLAP** (MOLAP), and **hybrid OLAP** (HOLAP).
- Two approaches are taken to improve the performance of OLAP queries: **materialization of cuboids** and **data indexing**.
- The **FASMI test** was developed to evaluate different OLAP products.
- A DW integrated with an OLAP server can be used to perform three main tasks: **information processing**, **analytical processing**, and **data mining**.

Some of the introductory-level textbooks on **data warehousing** and **OLAP** are [3, 8, 9, 12, 18]. A comparison between statistical databases and OLAP can be found in [16]. Good articles that provides an overview and describe the relation between data warehousing and OLAP are [5, 13], while the **integration of OLAP and data mining** is addressed in [10].

The history of data warehousing and OLAP goes back to the early 1990s, when the term OLAP was coined [6]. The **data cube** was proposed as a relational aggregation operator in [7]. A greedy algorithm for the **partial materialization** of cuboids was proposed in [1]. Several methods for fast computation of data cubes can be found in [2, 4, 15]. The use of **join indices** and the **bitmap join index** method to speed up OLAP query processing were proposed in [14, 19]. Operations for modeling multidimensional databases were proposed in [1] and methods for **selection of materialized cuboids** for efficient OLAP query processing in [5, 11, 17]. More information about **data warehousing** and an **OLAP bibliography** can be found at www.daniel-lemire.com/OLAP.

References

1. Agrawal, R., Gupta, A., and Sarawagi, S. 1997. Modeling multidimensional databases. *Proceedings of the 1997 International Conference on Data Engineering*, Birmingham, U.K., 232–243
2. Agarwal, S., Agrawal, R., Deshpande, P., Gupta, A., Naughton, J., Ramakrishnan, R., and Sarawagi, S. 1996. On the computation of multidimensional aggregates. *Proceedings of the 1996 International Conference on Very Large Databases*, Bombay, India, 506–521
3. Berson, A., and Smith, S.J. 1997. *Data Warehousing, Data Mining and OLAP*, McGraw-Hill New York, USA
4. Beyer, K., and Ramakrishnan, R. 1999. Bottom-up computation of sparse and iceberg cubes. *Proceedings of the 1999 ACM-SIGMOD Conference on Management of Data*, Philadelphia, PA, USA, 359–370
5. Chaudhuri, S., and Dayal, U. 1997. An overview of data warehousing and OLAP technology. *ACM SIGMOD Record*, 26(1):65–74
6. Codd, E., Codd, S., and Salley, C. 1993. Beyond decision support. *Computer World*, 27(30)
7. Gray, J., Chaudhuri, S., Bosworth, A., Layman, A., Reichart, D., Venkatrao, M., Pellow, F., and Pirahesh, H. 1997. Data cube: a relational aggregation operator generalizing group-by, cross-tab and sub-totals. *Data Mining and Knowledge Discovery*, 1:29–54
8. Inmon, W. 2005. *Building a Data Warehouse*, 4th edition, John Wiley and Sons Indianapolis, IN, USA
9. Jarke, M., Lenzerini, M., Vassiliou, Y., and Vassiliadis, P. 2003. *Fundamentals of Data Warehouses*, Springer-Verlag New York, USA
10. Han, J. 1997. OLAP Mining: An integration of OLAP with data mining. *Proceedings of the 1997 IFIP Conference on Data Semantics*, Leysin, Switzerland, 1–11
11. Harinarayan, V., Rajaraman, A., and Ullman, J. 1996. Implementing data cubes efficiently. *Proceedings of the 1996 ACM-SIGMOD International Conference on Data*, Montreal, Canada, 205–216
12. Kimball, R., and Ross, M. 2002. *The Data Warehouse Toolkit — The Complete Guide to Dimensional Modeling*, 2nd edition, John Wiley and Sons
13. Ma, C., Chou, D., and Yen, D. 2000. Data warehousing, technology assessment and management. *Industrial Management and Data Systems*, 100(3):125–134
14. O'Neil, P., and Graefe, G. 1995. Multi-table joins through bitmapped join indices. *SIGMOD Record*, 24:8–11
15. Ross, K., and Srivastava, D. 1997. Fast computation of sparse data cubes. *Proceedings of the 1997 International Conference on Very Large Databases*, Athens, Greece, 116–125
16. Shoshani, A. 1997. OLAP and statistical databases: similarities and differences. *Proceedings of the 16th ACM Symposium on Principles of Database Systems*, Tucson, AZ, USA, 185–196
17. Sristava, D., Dar, S., Jagadish, H., and Levy, A. 1996. Answering queries with aggregation using views. *Proceedings of the 1996 International Conference on Very Large Databases*, Bombay, India, 318–329
18. Thomsen, E. 1997. *OLAP Solutions: Building Multidimensional Information Systems*, John Wiley and Sons New York, USA
19. Valduriez, P. 1987. Join indices. *ACM Transactions on Database Systems*, 12:218–246

7. Exercises

1. Discuss which SLQ commands can be executed online and which are executed only offline.
2. Explain what does it means that the data warehouses are nonvolatile. Give three example SQL commands that would not be executed in an online data warehouse.
3. Contrast a normalized relational schema and a star schema. You may use an example to demonstrate the differences and commonalities.
4. Design a data warehouse schema for a global weather center. In your design you should assume that the center gathers data from 10,000 weather stations, which record and transmit basic weather information including temperature, pressure, precipitation, wind speed, and wind direction every hour. The stations are located in different cities around the globe, which can be divided into different continents, countries, and states (provinces); those that are located

inland, on the coast, and on the ocean; and finally those that are located at different altitudes. The center has already accumulated five years of historical data and now requires a tool that allows for efficient querying and deriving of general weather patterns in a multidimensional space.

5. Consider an example data warehouse that consists of five dimensions, namely, *students, courses, professors, departments, semesters*, and two measures, namely, *GPA* and *number_of_credits*. At the lowest granularity level, GPA stores a student's grade in a course taken from a specific professor, in a specific department and semester, while *number_of_credits* stores how many credits the course is worth. Given this description:

   a. list three types of data warehouse schema that can be used to model this data
   b. draw a schema diagram for the schema that uses multiple fact tables
   c. considering the four-dimensional data cube, identify what specific OLAP commands should be used to derive the GPA of students from the Electrical and Computer Engineering department from the fourth-year courses in 2003
   d. write down the SQL implementation of the command from (c)
   e. assuming that each dimension has four levels, which includes all, e.g., course < course year < course college < all, determine how many cuboids can be generated for the above data cube (you should include both base and apex cuboids)

6. Find at least four commercial OLAP tools that utilize the ROLAP architecture. For each of these tools, list the corresponding URL or publication.

7. Explain the main differences between the three types of data warehouse usage, i.e., information processing, analytical processing, and data mining. List at least five specific names of data mining algorithms that can be used to perform classification.

7

Feature Extraction and Selection Methods

This Chapter provides background and algorithms for feature extraction and feature selection from numerical data. Both methods are performed to reduce the dimensionality of the original data. Feature extraction methods do it by generating new transformed features and selecting the informative ones while feature selection methods choose a subset of original features.

1. Introduction

Nowadays, we deal with large datasets that include up to billions of objects (examples, patterns) and up to several thousands of features. This Chapter provides an introduction to data preprocessing methods, which are concerned with the extraction and selection of features to reduce the dimensionality and improve the data for subsequent data mining analysis. **Feature selection** selects a subset of features among the set of all features from the original dataset. On the other hand, **feature extraction** generates new features based on the original dataset.

This Chapter describes both **supervised** and **unsupervised** feature extraction methods. These include dimensionality reduction and feature extraction via unsupervised **Principal Component Analysis**, unsupervised **Independent Component Analysis**, and **supervised Fisher's linear discriminant analysis**. The first two methods are linear transformations that optimally reduce dimensionality, in terms of the number of features, of the original unsupervised dataset. The Fisher's method also implements a linear transformation that optimally converts supervised datasets into a new space that includes fewer features, which are more suitable for classification. While the above methods are mainly used with numerical (time-independent) data, we also describe two groups of methods for preprocessing of **time-series** data. These include **Fourier transform** and **Wavelets** and their two-dimensional versions. We also discuss **Zernike moments** and **Singular Value Decomposition**.

The second part of the Chapter describes a wide variety of feature selection methods. The design of these methods is based on two components, namely, **selection criteria** and **search methods**.

2. Feature Extraction

Data preprocessing may include transformation (projection) of the original **patterns** (also called **examples** or **objects**) into the transformed pattern space, frequently along with **reduction of**

2. Feature Extraction

dimensionality of a pattern by extraction of only the most informative features. The transformation and reduction of pattern dimensionality may improve the recognition process through a consideration of only the most important data representation, possibly with uncorrelated-pattern elements retaining maximum information about the original data. These approaches may also lead to better generalization abilities of a subsequently designed model, such as a classifier.

Reduction of the original pattern dimensionality refers to a transformation of original n-dimensional patterns into other m-dimensional feature patterns ($m \leq n$). The pattern transformation and dimensionality reduction can be considered as a nonlinear transformation (mapping)

$$\mathbf{y} = \mathbf{F}(\mathbf{x}) \tag{1}$$

of n-dimensional original patterns $\mathbf{x}$ (vectors in the n-dimensional pattern space) into m-dimensional transformed patterns $\mathbf{y}$ (vectors in the m-dimensional transformed pattern space). The m-dimensional transforming function $\mathbf{F}(\mathbf{x})$ may be designed based on the available knowledge about a domain and data statistics. Elements y_i ($i = 1, 2, \cdots, m$) of the transformed patterns $\mathbf{y}$ are called **features** and the m-dimensional transformed patterns $\mathbf{y}$ are called **feature vectors**. Feature vectors represent data objects in the **feature space**. However, the general name pattern is also adequate in this context.

The projection and reduction of a pattern space may depend on the goal of processing. The purpose of transformation is to obtain a pattern representing data in the best form for a given processing goal. For example, one can choose features in order to characterize (model) a natural phenomenon that generates patterns. Another goal may be finding the best features for the classification (recognition) of objects.

Below we present an optimal linear transformation that guarantees the preservation of maximum information by the extracted feature vector.

The reasons for performing data transformation and dimensionality reduction of patterns are as follows:

- Removing redundancy in data
- Compression of data sets
- Obtaining transformed and reduced patterns containing only a relevant set of features that help to design classifiers with better generalization capabilities
- Discovering the intrinsic variables of data that help design a data model, and improving understanding of phenomena that generate patterns
- Projecting high-dimensional data (preserving intrinsic data topology) onto low-dimensional space in order to visually discover clusters and other relationships in data

2.1. Principal Component Analysis

Probably the most popular statistical method of linear pattern transformation and feature extraction is **Principal Component Analysis** (PCA). This linear transformation is based on the statistical characteristics of a given data set represented by the **covariance matrix** of data patterns, its **eigenvalues**, and the corresponding **eigenvectors**.

Principal Component Analysis (PCA) is a technique, developed in a biological context, to represent a linear regression analysis as fitting planes to data in the sense of least-squares error.

PCA determines an optimal linear transformation

$$\mathbf{y} = \mathbf{W}\mathbf{x} \tag{2}$$

of a real-valued n-dimensional random data pattern $\mathbf{x} \in \mathbb{R}^n$ into another m-dimensional ($m \leq n$) transformed vector $\mathbf{y} \in \mathbb{R}^m$. The $m \times n$ linear transformation matrix $\mathbf{W} \in \mathbb{R}^{m \times n}$ is optimal from the point of view of obtaining the maximal information retention. PCA is realized through exploring statistical correlations among elements of the original patterns and finding (possibly reduced) data representation that retains the maximum nonredundant and uncorrelated intrinsic information of the original data. Exploration of the original data set, represented by the original n-dimensional patterns $\mathbf{x}^i$, is based on computing and analyzing a data covariance matrix, its eigenvalues, and the corresponding eigenvectors arranged in descending order. The arrangement of subsequent rows of a transformation matrix $\mathbf{W}$ as the normalized eigenvectors, corresponding to the subsequent largest eigenvalues of the data covariance matrix, will result in an optimal linear transformation matrix $\hat{\mathbf{W}}$. The elements of the m-dimensional transformed feature vector $\mathbf{y}$ will be uncorrelated and arranged in decreasing order according to decreasing information content. This allows for a straightforward reduction of dimensionality (and thus data compression) by discarding trailing feature elements with the lowest information content. Depending on the nature of an original data pattern, one can obtain a substantial reduction of feature vector dimensionality $m << n$ compared with the dimensionality of original data patterns. First, having determined the optimal transformation matrix $\hat{\mathbf{W}}$, one can reduce the decorrelated feature vector dimension and use reduced feature vectors for classification. Second, all original n-dimensional data patterns can be optimally transformed to data patterns in the feature space with lower dimensionality. This means that the original data will be compressed with the minimal information loss when the data are reconstructed (preserving the maximal information content of the original data).

A PCA-based linear transformation of an original data pattern can also be interpreted as a projection of original patterns into m-dimensional feature space with orthonormal bases (guaranteeing that one obtains decorrelation of feature vector elements).

We can think of a PCA as an **unsupervised learning** from data. Indeed, PCA does not use knowledge about a class associated with a pattern, but only discovers correlation among patterns and their elements, as well as ordered intrinsic directions where the data patterns change most (with maximum variance), as shown in Figure 7.1.

Despite the fact that PCA is an unsupervised method, it can also be used in classifier design for the projection and reduction of feature patterns. Here, PCA is applied solely to patterns in order to determine an optimal transformation of original patterns into a principal component space and possibly to reduce the dimensionality of the projected pattern. Once an optimal transformation

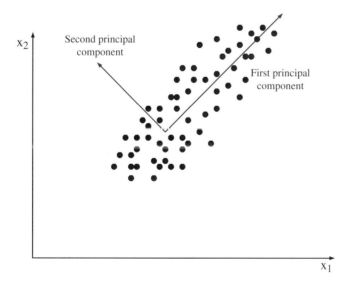

Figure 7.1. Principal components.

has been completed, the projected patterns will have the same class assignments as those in the original data set.

2.1.1. Statistical Characteristics of Data Required by PCA

Let us consider data objects characterized by n-dimensional column patterns $\mathbf{x} \in \mathbb{R}^n$ in n-dimensional pattern space whose elements take real values $x_i \in \mathbb{R}$. We assume that our knowledge about a domain is represented as a limited size sample (from a certain domain) of N random patterns $\mathbf{x}^i$ gathered as an unlabeled training data set T_{tra}:

$$T_{tra} = \{\mathbf{x}^1, \mathbf{x}^2, \cdots, \mathbf{x}^N\} \qquad (3)$$

The entire training set data will be represented as an $N \times n$ data pattern matrix:

$$\mathbf{X} = \begin{bmatrix} (\mathbf{x}^1)^T \\ (\mathbf{x}^2)^T \\ \vdots \\ (\mathbf{x}^N)^T \end{bmatrix} \qquad (4)$$

One row of the data matrix contains one transposed pattern. If a data set contains patterns labeled by classes, for unsupervised PCA analysis we need to extract only patterns from this data set.

The data can be characterized by second-order statistics, namely, by the n-dimensional **mean** vector

$$\boldsymbol{\mu} = E[\mathbf{x}] = \big[E[x_1], E[x_2], \cdots, E[x_n]\big]^T \qquad (5)$$

and the square $n \times n$-dimensional **covariance** matrix

$$\mathbf{R}_{xx} = \boldsymbol{\Sigma} = E\big[(\mathbf{x} - \boldsymbol{\mu})(\mathbf{x} - \boldsymbol{\mu})^T\big] \qquad (6)$$

where $E[\cdot]$ denotes the expectation operator and $\boldsymbol{\mu}$ is the mean vector of a pattern vector $\mathbf{x}$. The square, semipositive definite, symmetric ($r_{ij} = r_{ji}$), real-valued **covariance matrix** $\mathbf{R}_{xx}$ describes correlations between elements of pattern vectors (treated as random variables). The PCA technique assumes that the original data patterns are zero-mean random vectors

$$\boldsymbol{\mu} = E[\mathbf{x}] = \mathbf{0} \qquad (7)$$

If this condition is not satisfied, one can convert an original pattern $\mathbf{x}$ to the zero mean representation by the operation $\mathbf{x} - \boldsymbol{\mu}$. For zero mean patterns, the covariance matrix (equal to the correlation matrix) is defined as

$$\mathbf{R}_{xx} = \boldsymbol{\Sigma} = E[\mathbf{x}\mathbf{x}^T] \qquad (8)$$

The true values $\boldsymbol{\mu}$ and $\mathbf{R}_{xx}$ for the mean vectors and the covariance matrix are in practice not available, since we usually do not know the exact probabilistic characteristics of patterns generated by nature. Our knowledge about a pattern-generation mechanism is included in a given data set T_{tra} containing a finite number of N patterns $\{\mathbf{x}^1, \mathbf{x}^2, \cdots, \mathbf{x}^N\}$. Under these circumstances, we find estimates for the mean

$$\hat{\boldsymbol{\mu}} = \frac{1}{N} \sum_{i=1}^{N} \mathbf{x}^i \qquad (9)$$

and the covariance matrix (unbiased estimate)

$$\hat{\mathbf{R}}_{xx} = \frac{1}{N-1} \sum_{i=1}^{N} (\mathbf{x}^i - \boldsymbol{\mu})(\mathbf{x}^i - \boldsymbol{\mu})^T \qquad (10)$$

based on a given limited sample. For zero-mean data, the covariance estimate becomes

$$\hat{\mathbf{R}}_{xx} = \frac{1}{N-1} \sum_{i}^{N} \mathbf{x}^i (\mathbf{x}^i)^T = \frac{1}{N-1} \mathbf{X}^T \mathbf{X} \qquad (11)$$

where $\mathbf{X}$ is a whole $N \times n$ original data pattern matrix (data set).

The intrinsic characteristic of given data $\mathbf{X}$ can be found as a set of n eigenvalues λ_i and the corresponding eigenvectors $\mathbf{e}^i$ by solving the **eigenvalue problem**

$$\mathbf{R}_{xx} \mathbf{e}^i = \lambda_i \mathbf{e}^i, \quad i = 1, 2, \cdots, n \qquad (12)$$

We consider the orthonormal eigenvectors, which is a legitimate approach since a covariance matrix $\mathbf{R}_{xx}$ is symmetric and real valued. This means that the eigenvectors are orthogonal $(\mathbf{e}^i)^T \mathbf{e}^j = 0 (i, j = 1, 2, \cdots, n, \ i \neq j)$ with unit length $\|\mathbf{e}^i\| = \sqrt{(\mathbf{e}^i)^T \mathbf{e}^i} = 1 (i = 1, 2, \cdots n)$.

In the PCA analysis, it is essential that the eigenvalues of the matrix $\mathbf{R}_{xx}$ are arranged in the decreasing order

$$\lambda_1 \geq \lambda_2 \geq \cdots \lambda_n \geq 0 \qquad (13)$$

with $\lambda_1 = \lambda_{\max}$. The corresponding orthonormal eigenvectors $\mathbf{e}^i$ will be composed as the square $n \times n$ matrix

$$\mathbf{E} = [\mathbf{e}^1, \mathbf{e}^2, \cdots, \mathbf{e}^n] \qquad (14)$$

with the ith column representing one eigenvector $\mathbf{e}^i$ corresponding to the eigenvalue λ_i. The most dominant first eigenvector $\mathbf{e}^1$ in the first column of the matrix $\mathbf{E}$ corresponds to the first most dominant eigenvalue λ_1 of the covariance matrix $\mathbf{R}_{xx}$. The second most dominant eigenvector $\mathbf{e}^2$ in the second column corresponds to the second most dominant eigenvalue λ_2, etc.

The arrangement of eigenvalues and corresponding eigenvectors in descending order is essential for data dimensionality reduction. Only the first m principal components of projected feature vectors (those carrying the most information) and corresponding to the first m dominant eigenvalues should be considered.

The eigenvalue problem equation can be written in the matrix form

$$\mathbf{R}_{xx} \mathbf{E} = \mathbf{E} \boldsymbol{\Lambda} \qquad (15)$$

where $\boldsymbol{\Lambda} = diag[\lambda_1, \lambda_1, \cdots, \lambda_n]$.

We can observe that for the orthonormal matrix $\mathbf{E}$ we have

$$\mathbf{E}^T \mathbf{E} = \mathbf{I} \qquad (16)$$

where $\mathbf{I}$ is the $n \times n$ unit matrix. Consequently, we have

$$\mathbf{E}^{-1} = \mathbf{E}^T \qquad (17)$$

In light of the above equality, we can write formulas for the so-called **orthogonal similarity transformation**

$$\mathbf{E}^{-1} \mathbf{R}_{xx} \mathbf{E} = \mathbf{E}^T \mathbf{R}_{xx} \mathbf{E} = \boldsymbol{\Lambda} \qquad (18)$$

and consequently the **spectral factorization** of the covariance matrix $\mathbf{R}_{xx}$,

$$\mathbf{R}_{xx} = \mathbf{E} \boldsymbol{\Lambda} \mathbf{E}^T \qquad (19)$$

2.1.2. The Optimization Criterion of PCA

The goal of PCA is to find the optimal linear transformation $\mathbf{y} = \mathbf{Wx}$ of the original n-dimensional data patterns $\mathbf{x}$ into m-dimensional feature vectors $\mathbf{y}$, possibly with lower dimensionality ($m < n$). More formally, PCA can be considered as a static optimization problem, with a specifically defined optimization criterion, which will guarantee obtaining (through optimal projection) a transformed feature vector possessing the desired characteristics. In PCA, it is required that:

- optimal transformation is orthogonal (with orthonormal basis)
- elements of the transformed feature vector $\mathbf{y}$ are uncorrelated
- orthonormal basis of the linear projections shows, in decreasing order, the orthogonal intrinsic directions in data along which the data changes (variances) are maximal
- pattern reconstruction error will be minimal in the least-squares sense

For the orthonormal linear transformation $\mathbf{y} = \mathbf{Wx}$, with the $m \times n$-dimensional orthonormal tranformation matrix $\mathbf{W}$, an estimate of the reconstructed pattern is $\hat{\mathbf{x}} = \mathbf{W}^{-1}\mathbf{y}$. Since for the orthonormal matrices we have $\mathbf{W}^{-1} = \mathbf{W}^T$, thus

$$\hat{\mathbf{x}} = \mathbf{W}^{-1}\mathbf{y} = \mathbf{W}^T\mathbf{y} = \mathbf{W}^T\mathbf{Wx} \tag{20}$$

The criterion for the optimal, PCA-based, linear transformation is selected in order to guarantee obtaining a minimum of the reconstruction error metric. The reconstruction error-based criterion has the form

$$J_{\text{lse}}(\mathbf{W}) = E\left[\|\mathbf{x} - \hat{\mathbf{x}}\|^2\right] \tag{21}$$

where $\|\mathbf{x} - \hat{\mathbf{x}}\|^2 = \|\mathbf{x} - \hat{\mathbf{x}}\|_2^2$ denotes the square of the Euclidean distance, denoted by the subscript *lse* indicating that the criterion used in optimization is based on the mean least squares error criterion. For practical computations, one can use the criterion

$$J_{\text{lse}}(\mathbf{W}) = \frac{1}{2}\sum_i^N \|\mathbf{x}^i - \hat{\mathbf{x}}^i\|^2 = \frac{1}{2}\sum_i^N \sum_j^n (x_j^i - \hat{x}_j^i)^2 \tag{22}$$

PCA seeks the optimal transformation matrix $\mathbf{W}$ that guarantees minimization of the mean squares reconstruction error (criterion $J_{\text{lse}}(\mathbf{W})$) for a given data set T_{tra}.

In order to better understand the goal of PCA, we provide a more detailed interpretation of the PCA criterion:

$$\begin{aligned} J(\mathbf{W}) &= E\left[\|\mathbf{x} - \hat{\mathbf{x}}\|^2\right] \\ &= E\left\{\text{trace}\left[(\mathbf{x} - \hat{\mathbf{x}})(\mathbf{x} - \hat{\mathbf{x}})^T\right]\right\} \\ &= \text{trace}(\mathbf{R}_{xx}) - \text{trace}(\mathbf{W}\mathbf{R}_{xx}\mathbf{W}^T) \end{aligned} \tag{23}$$

In the above equations, we used the facts that $\text{trace}(\mathbf{W}) = \text{trace}(\mathbf{W}^T)$ and $\mathbf{W}\mathbf{W}^T = \mathbf{W}\mathbf{W}^{-1} = \mathbf{I}$, along with the following equalities:

$$\begin{aligned} \|\mathbf{x} - \hat{\mathbf{x}}\|^2 &= \text{trace}\{(\mathbf{x} - \hat{\mathbf{x}})(\mathbf{x} - \hat{\mathbf{x}})^T\} \\ \text{trace}\{E[\mathbf{W}^T\mathbf{W}\mathbf{xx}^T\mathbf{W}^T\mathbf{W}]\} &= \text{trace}\{\mathbf{W}\mathbf{W}^T\mathbf{W}\mathbf{R}_{xx}\mathbf{W}^T\} \\ &= \text{trace}(\mathbf{W}\mathbf{R}_{xx}\mathbf{W}^T) \end{aligned} \tag{24}$$

We can observe that in the PCA criterion given in Equation (23), the second term $\text{trace}(\mathbf{W}\mathbf{R}_{xx}\mathbf{W}^T) = J_{\text{variance}}(\mathbf{W})$ and is equal to the variance of the projected feature vector $\mathbf{y}$, or consequently, equal to to the variance of the reconstructed pattern vector $\hat{\mathbf{x}}$:

$$J_{\text{variance}}(\mathbf{W}) = \text{trace}(\mathbf{W}\mathbf{R}_{xx}\mathbf{W}^T) = E[\text{trace}(\mathbf{y}\mathbf{y}^T)] = \sum_{i=1}^{m} y_i^2 \qquad (25)$$

$$J_{\text{lse}}(\mathbf{W}) = \text{trace}(\mathbf{W}^T\mathbf{W}\mathbf{R}_{xx}\mathbf{W}^T\mathbf{W}) = E[\text{trace}(\hat{\mathbf{x}}\hat{\mathbf{x}}^T)] = \sum_{i=1}^{n} \hat{x}_i^2$$

The conclusion from the above criterion analysis is that minimization of the mean square error criterion $J_{\text{lse}}(\mathbf{W})$ is in fact equivalent to maximization of the projected feature vector $\mathbf{y}$ variance (with the criterion J_{variance}). One can interpret PCA as minimization of the reconstruction error in the mean least squares sense, or equivalently as maximization of the resulting projection (feature vector) variance. We see that the optimal PCA-based linear transformation will result in a projection of the original patterns into the feature vectors, with elements located in the feature space in the directions with maximal variances (maximal variabilities).

2.1.3. PCA Theorem
For a given data set (a training set), let $T_{\text{tra}} = \{\mathbf{x}^1, \mathbf{x}^2, \cdots, \mathbf{x}^N\}$, containing N n-dimensional zero-mean randomly generated patterns $\mathbf{x} \in \mathbb{R}^n$ with real-valued elements and the symmetric, real valued $n \times n$ covariance matrix $\mathbf{R}_{xx} \in \mathbb{R}^{n \times n}$. Let the eigenvalues of the covariance matrix $\mathbf{R}_{xx}$ be arranged in decreasing order $\lambda_1 \geq \lambda_2 \geq \cdots \lambda_n \geq 0$ (with $\lambda_1 = \lambda_{\max}$). Assume that the corresponding orthonormal eigenvectors (orthogonal with unit length $\|\mathbf{e}\| = 1$) $\mathbf{e}^1, \mathbf{e}^2, \cdots, \mathbf{e}^n$ compose the $n \times n$ orthonormal matrix

$$\mathbf{E} = [\mathbf{e}^1, \mathbf{e}^2, \cdots, \mathbf{e}^n] \qquad (26)$$

with columns being orthonormal eigenvectors. Then the optimal linear transformation

$$\hat{\mathbf{y}} = \hat{\mathbf{W}}\mathbf{x} \qquad (27)$$

transforms the original n-dimensional patterns $\mathbf{x}$ into m-dimensional ($m \leq n$) feature patterns, minimizing the mean least squares reconstruction error criterion $J_{\text{lse}}(\mathbf{W})$ given by Equation (25) (or maximizing the variance of projected patterns) and provides for the $m \times n$ optimal transformation matrix $\hat{\mathbf{W}}$, denoted also by $\mathbf{W}_{KL}$ (under the constraints $\mathbf{W}\mathbf{W}^T = \mathbf{I}$), as

$$\hat{\mathbf{W}} = \begin{bmatrix} (\mathbf{e}^1)^T \\ (\mathbf{e}^2)^T \\ \vdots \\ (\mathbf{e}^m)^T \end{bmatrix} \qquad (28)$$

composed with m rows that are the first m orthonormal eigenvectors of the original data covariance matrix $\mathbf{R}_{xx}$.

The resulting optimal linear transformation $\mathbf{y} = \hat{\mathbf{W}}\mathbf{x}$, with the optimal transformation matrix $\hat{\mathbf{W}}$, is called the **Karhunen-Loéve** (KLT) or **Hotelling** transformation.

2.1.4. Properties of the Karhunen-Loéve Transformation
The optimal KLT transformation guarantees the minimum reconstruction error in the least squares sense, with the minimal value

$$\min J_{\text{lse}}(\mathbf{W}) = \sum_{i=m+1}^{n} \lambda_i \qquad (29)$$

The minimum value of the reconstruction error is equal to the sum of the trailing $n-m$ eigenvalues $\lambda_{m+1}, \lambda_{m+2}, \cdots, \lambda_n$ (from the ordered eigenvalues) of the covariance matrix $\mathbf{R}_{xx}$, where m is the possibly reduced length of the projected feature vector $\mathbf{y}$ ($m \leq n$).

Simultaneously, the transformation guarantees the maximum of the projected feature vector variance, with the maximum value

$$\max J_{\text{variance}}(\mathbf{W}) = \sum_{i=1}^{m} \lambda_i \tag{30}$$

equal to the sum of the first m eigenvalues of $\mathbf{R}_{xx}$. The orthonormal eigenvectors $\mathbf{e}^1, \mathbf{e}^2, \cdots, \mathbf{e}^n$ (rows of the optimal transformation matrix $\hat{\mathbf{W}}$), corresponding to the descending-order eigenvalues $\lambda_1, \lambda_2, \cdots, \lambda_n$ of the data covariance matrix $\mathbf{R}_{xx}$, are called the **principal eigenvectors**. They show orthogonal directions (in descending order, corresponding to the principal eigenvectors and eigenvalues) in the pattern space where data change maximally (with maximal variance). The m principal eigenvectors (arranged as rows) compose the optimal transformation matrix $\hat{\mathbf{W}}$.

For a given n-dimensional random pattern $\mathbf{x}$, the optimal transformation $\mathbf{y} = \hat{\mathbf{W}}\mathbf{x}$ will produce the optimally projected m-dimensional feature vector $\mathbf{y} = [y_1, y_2, \cdots, y_m]^T$. The elements $y_1, y_2, \cdots, y_m$ of the feature vector are called the **principal components** of a pattern $\mathbf{x}$. The principal components are statistically uncorrelated with covariances

$$E[y_i y_j] = \mathbf{e}_i^T \mathbf{R}_{xx} \mathbf{e}_j = 0 \tag{31}$$

and with variances equal to the corresponding eigenvalues

$$E[y_i y_i] = E[y_i^2] = (\mathbf{e}^i)^T \mathbf{R}_{xx} \mathbf{e}^i = \lambda_i \tag{32}$$

The covariance matrix $\mathbf{R}_{yy} = E[\mathbf{y}\mathbf{y}^T]$ of the projected feature vectors $\mathbf{y}$ is diagonal, with eigenvalues of the original pattern covariance matrix $\mathbf{R}_{xx}$ on the main diagonal given in descending value order. The variances are arranged in descending variance value order $E[y_1^2] \leq E[y_2^2] \leq \cdots \leq E[y_m^2]$.

The ith principal component y_i of the original pattern $\mathbf{x}$, corresponding to the ith largest eigenvalue λ_i of the covariance matrix $\mathbf{R}_{xx}$, is obtained as an inner product

$$y_i = (\mathbf{e}^i)^T \mathbf{x} = e_1^i x_1 + e_2^i x_2 + \cdots + e_n^i x_n \tag{33}$$

of the ith orthonormal eigenvector $\mathbf{e}^i$ (i^{th} row of $\hat{\mathbf{W}}$) and a given pattern $\mathbf{x}$. It is just a linear combination of the elements of pattern vector $\mathbf{x}$.

The first principal component $y_1 = (\mathbf{e}^1)^T \mathbf{x}$, corresponding to the first most dominant eigenvalue λ_1 (and first eigenvector $\mathbf{e}^1$) of the covariance matrix, is such that its variance

$$\text{variance}(y_1) = E[y_1^2] = E[(\mathbf{e}^1 \mathbf{x})^2] = \mathbf{e}^1 E[\mathbf{x}\mathbf{x}^T](\mathbf{e}^1)^T = \mathbf{e}^1 \mathbf{R}_{xx}(\mathbf{e}^1)^T = \lambda_1 \tag{34}$$

is maximal. The first most dominant principal component y_1 is along the first eigenvector direction $\mathbf{e}^1$, with maximum variance equal to the most dominant eigenvalue λ_1 of the covariance matrix.

Subsequently, the second principal component $y_2 = (\mathbf{e}^2)^T \mathbf{x}$, corresponding to the second dominant eigenvalue λ_2 (and the second eigenvector $\mathbf{e}^2$), is such that its variance

$$\text{variance}(y_2) = E[y_2^2] = \lambda_2 \tag{35}$$

is maximal. The second dominant principal component y_2 is along the second eigenvector direction $\mathbf{e}^2$, with the second maximum variance equal to the second dominant eigenvalue λ_2 of the covariance matrix. The direction of the second principal component is perpendicular to the direction of the first most dominant principal component. Similarly, the third dominant principal component $y_3 = (\mathbf{e}^3)^T\mathbf{x}$ is along the third eigenvector direction $\mathbf{e}^3$, with the third maximum variance equal to the third dominant eigenvalue λ_3 of the covariance matrix. The direction of the third principal component is perpendicular (orthogonal) to the direction of the first and second dominant principal component. Generally, the ith principal component will be in the direction orthogonal to all prior principal components $y_1, y_2, \cdots, y_{i-1}$, with the maximal value of variance in this direction equal to λ_i. The m principal components form principal component space into which patterns $\mathbf{x}$ are optimally projected. Most information is contained along the first principal component.

2.1.5. Optimal KLT Transformation of the Original Patterns

Once we have defined the optimal KLT transformation matrix $\hat{\mathbf{W}}$, the optimal transformation of a given n-dimensional original pattern $\mathbf{x}$ into the m-dimensional optimal feature pattern $\mathbf{y}$ is given by $\mathbf{y} = \hat{\mathbf{W}}\mathbf{x}$. The inverse transformation can be obtained from

$$\hat{\mathbf{x}} = \hat{\mathbf{W}}^{-1}\mathbf{y} = \hat{\mathbf{W}}^T\mathbf{y} = \sum_{i=1}^{m} y_i \mathbf{e}^i \quad (36)$$

The optimal transformation of the entire original data set $\mathbf{X}$ is given by the formula

$$\mathbf{Y} = (\hat{\mathbf{W}}\mathbf{X}^T)^T = \mathbf{X}\hat{\mathbf{W}}^T \quad (37)$$

The $m \times m$ covariance matrix $\mathbf{R}_{yy} = E[\mathbf{y}\mathbf{y}^T]$ for the projected patterns $\mathbf{y}$ can be estimated as

$$\mathbf{R}_y = \frac{1}{N-1}\mathbf{Y}^T\mathbf{Y} = \hat{\mathbf{W}}\mathbf{R}_{xx}\hat{\mathbf{W}}^T = \mathrm{diag}[\lambda_i] \quad (38)$$

The entire reconstructed pattern set is given by

$$\hat{\mathbf{X}} = \mathbf{Y}\hat{\mathbf{W}} \quad (39)$$

2.1.6. Dimensionality Reduction

PCA can be effectively used for feature extraction and dimensionality reduction. Instead of the entire n-dimensional original data pattern $\mathbf{x}$, one can form the m-dimensional ($m \leq n$) feature vector $\mathbf{y} = [y_1, y_2, \cdots, y_m]^T$ containing only the first m most dominant principal components of $\mathbf{x}$, corresponding to the first m most dominant eigenvalues $\lambda_1, \lambda_2, \cdots, \lambda_m$ of the original data pattern covariance matrix. PCA is the best technique for linear feature extraction from the original set of patterns, in the sense of minimization of the reconstruction error. Projection of patterns from highly dimensional pattern space ($\mathbf{x} \in \mathbb{R}^n$) onto reduced principal component space (with dimension $m << n$) will result in the square of the approximation error

$$\mathbf{e}^2 = \sum_{i=m+1}^{n} \lambda_i \quad (40)$$

which is equal to the sum of $n - m$ discarded eigenvalues. We also can say that the m principal components y_i are the **most expressive** features of a data set. PCA provides a way to reduce (compress) data representation by choosing the feature vector with lower dimensionality. The reduced-size feature vectors represent data in a new feature space, where the feature vector elements are uncorrelated and placed along the orthogonal directions of principal components with maximal variances. This characteristic might be desirable in classifier design for some types of data. However, PCA does not consider the improvement of classification in principal component space, as a transformation criterion. This mean that most expressive features obtained through principal components are well suited for data representation (the model) and compression. However, they might not be good for classification. The data in the feature space will be an approximation (a model) of the original data patterns.

The approximation error vector **e** is orthogonal to the reconstructed data pattern $\hat{\mathbf{x}}$. The least squares error between **x** and $\hat{\mathbf{x}}$ is

$$||\mathbf{e}|| = ||\mathbf{x} - \hat{\mathbf{x}}|| = \left[\sum_{i=1}^{n}(x_i - \hat{x}_i)^2\right]^{1/2} \tag{41}$$

The least mean squares error is equal to

$$E[||\mathbf{x} - \hat{\mathbf{x}}||^2] = \sum_{i=m+1}^{n} \lambda_i \tag{42}$$

and it is also equal to the sum of $n - m$ eigenvalues not used.

One method (criterion) for (data representation oriented) selection of a dimension of a reduced feature vector **y** is to choose a minimal number of the first m most dominant principal components $y_1, y_2, \cdots, y_m$ of **x** for which the mean square reconstruction error is less than the heuristically set error threshold ϵ. Another, more practical method may be to select the minimal number of the first m most dominant principal components for which a percentage V of a sum of unused eigenvalues of a sum of all eigenvalues

$$V = \frac{\sum_{i=m+1}^{n} \lambda_i}{\sum_{i=1}^{n} \lambda_i} \, 100\% \tag{43}$$

and is less than a defined threshold ζ: $P < \zeta$.

One can try to use principal components for classification. Selection of the best principal components for classification purposes is outside the scope of this Chapter. In such a case, Fisher's Linear Discriminant Analysis (described in Section 2.2) should be used.

Example: Let us consider a data set (shown in Table 7.1) containing 10 two-feature patterns $\mathbf{x} \in \mathbb{R}^2$ from two classes $c_1 = 0$ and $c_2 = 1$ (five patterns in each class) drawn according to the Gaussian normal density distribution.

The patterns $\mathbf{x} \in \mathbb{R}^2$ from both classes of the original data set can be composed as a 10×2 ($n=2$, $N=10$) data pattern matrix $\mathbf{X}_{\text{orig}}$ (Table 7.2). The mean vector for the original patterns **x**

Table 7.1. Two-class data set.

x_1	x_2	class
1	2	1
2	2	1
2	3	1
3	1	1
3	2	1
6	8	2
7	8	2
8	7	2
8	8	2
7	9	2

Table 7.2. Data pattern matrix

$$\mathbf{X}_{\text{orig}} = \begin{bmatrix} 1 & 2 \\ 2 & 2 \\ 2 & 3 \\ 3 & 1 \\ 3 & 2 \\ 6 & 8 \\ 7 & 8 \\ 8 & 7 \\ 8 & 8 \\ 7 & 9 \end{bmatrix}$$

is $\boldsymbol{\mu} = [4.7,\ 5.0]^T$, and the zero mean data pattern can be obtained by extraction $\mathbf{x} - \boldsymbol{\mu}$, yielding the zero mean data pattern matrix $\mathbf{X}$:

$$\mathbf{X} = \begin{bmatrix} -3.7 & -3.0 \\ -2.7 & -3.0 \\ -2.7 & -2.0 \\ -1.7 & -4.0 \\ -1.7 & -3.0 \\ 1.3 & 3.0 \\ 2.3 & 3.0 \\ 3.3 & 2.0 \\ 3.3 & 3.0 \\ 2.3 & 4.0 \end{bmatrix}$$

The 2×2 covariance matrix $\mathbf{R}_{xx}$ of the data patterns $\mathbf{X}$ is

$$\mathbf{R}_{xx} = \begin{bmatrix} 7.5667 & 8.1111 \\ 8.1111 & 10.4444 \end{bmatrix}$$

$\lambda_1 = 0.7678$, $\lambda_2 = 17.2433$, and $[0.6424\ 0.7624]^T$. Arranged in decreasing order, the eigenvalues and corresponding orthogonal eigenvectors of matrix $\mathbf{R}_{xx}$ are $\lambda_1 = 17.2433$, $\lambda_2 = 0.7678$, and $\mathbf{e}^1 = [0.6424\ 0.7664]^T$, $\mathbf{e}^2 = [-0.7664\ 0.6424]^T$. Matrix $\mathbf{E}$, composed with eigenvectors $\mathbf{e}^1$ and $\mathbf{e}^2$ as columns, is

$$\mathbf{E} = \begin{bmatrix} 0.6424 & -0.7664 \\ 0.7664 & 0.6424 \end{bmatrix}$$

We can easily see that the eigenvectors are orthonormal: orthogonal $((\mathbf{e})^1)^T \mathbf{e}^2 = 0$ with unit length $\|\mathbf{e}^i\| = 1$. Finally, the optimal KLT transformation matrix $\hat{\mathbf{W}}$, with rows of eigenvectors corresponding to the decreasing-order eigenvalues of a covariance matrix, is

$$\hat{\mathbf{W}} = \begin{bmatrix} 0.6424 & 0.7664 \\ -0.7664 & 0.6424 \end{bmatrix}$$

The projected first vector $\mathbf{x}^1 = [-3.7,\ -3.0]^T$ from $\mathbf{X}$ gives, as a feature vector in the principal component space, $\mathbf{y}^1 = [-4.6760,\ 0.9084]^T$. Here, $y_1^1 = -4.6760$ is the first principal component of the pattern $\mathbf{x}^1$ along the direction of the first eigenvector $\mathbf{e}^1$. The data have maximal variance $\lambda_1 = 17.2433$ in this direction. Consequently, $y_2^1 = 0.9084$ is the second principal component of the pattern $\mathbf{x}^1$ along the direction of the second eigenvector $\mathbf{e}^2$ orthogonal to $\mathbf{e}^1$. The projection of all vectors from $\mathbf{X}$ onto principal components are $\mathbf{Y} = \mathbf{X}\hat{\mathbf{W}}^T$:

$$\mathbf{Y} = \begin{bmatrix} -4.6760 & 0.9084 \\ -4.0336 & 0.1421 \\ -3.2672 & 0.7844 \\ -4.1576 & -1.2667 \\ -3.3912 & -0.6243 \\ 3.1342 & 0.9309 \\ 3.7766 & 0.1645 \\ 3.6526 & -1.2443 \\ 4.4190 & -0.6019 \\ 4.5430 & 0.8069 \end{bmatrix}$$

The covariance matrix of $\mathbf{Y}$ is diagonal, with elements equal to λ_1 and λ_2:

$$\begin{bmatrix} 17.2433 & 0.0000 \\ 0.0000 & 0.7678 \end{bmatrix}$$

The first original pattern vector $\mathbf{x}^1 = [-3.7, -3.0]^T$ can be reconstructed, by the inverse operation $\hat{\mathbf{x}} = \hat{\mathbf{W}}^T \mathbf{y}$, from the principal component space vectors $\mathbf{y}^1 = [-4.6760, 0.9084]^T$:

$$\hat{\mathbf{x}}^1 = \begin{bmatrix} 0.6424 & -0.7664 \\ 0.7664 & 0.6424 \end{bmatrix} \begin{bmatrix} -4.6760 \\ 0.9084 \end{bmatrix} = \begin{bmatrix} -3.7 \\ -3.0 \end{bmatrix}$$

The reconstruction is exact, since the dimension of the original pattern and the feature patterns are equal ($m = n$), and consequently we used all eigenvectors in the optimal transformation matrix $\hat{\mathbf{W}}$. Reconstruction of all the original pattern space vectors $\hat{\mathbf{X}}$, by the inverse operation $\hat{\mathbf{X}} = \mathbf{Y}\mathbf{W}$, gives

$$\hat{\mathbf{X}} = \begin{bmatrix} -3.7 & -3.0 \\ -2.7 & -3.0 \\ -2.7 & -2.0 \\ -1.7 & -4.0 \\ -1.7 & -3.0 \\ 1.3 & 3.0 \\ 2.3 & 3.0 \\ 3.3 & 2.0 \\ 3.3 & 3.0 \\ 2.3 & 4.0 \end{bmatrix}$$

The fully reconstructed original pattern vectors, with the mean vector $\hat{\mathbf{x}}_{\text{orig}} = \hat{\mathbf{x}} + \boldsymbol{\mu}$ added, are equal to the original patterns $\mathbf{X}_{\text{orig}}$:

$$\hat{\mathbf{X}}_{\text{orig}} = \begin{bmatrix} 1 & 2 \\ 2 & 2 \\ 2 & 3 \\ 3 & 1 \\ 3 & 2 \\ 6 & 8 \\ 7 & 8 \\ 8 & 7 \\ 8 & 8 \\ 7 & 9 \end{bmatrix}$$

Let us consider now how we can model considered zero mean data patterns $\mathbf{X}$ by only one latent variable $\mathbf{y} = [y_1] \in \mathbb{R}(m = 1)$ being the first principal component of two dimensional patterns $\mathbf{x}$. Here, we form the 1×2 optimal KLT transformation matrix $\hat{\mathbf{W}}$ by choosing as its sole row the first eigenvector $\mathbf{e}^1$ of the covariance matrix $\mathbf{R}_{xx}$, corresponding to the first most dominant eigenvalue λ_1:

$$\hat{\mathbf{W}} = [\mathbf{e}^1] = [0.6424 \quad 0.7664]$$

The projected first vector $\mathbf{x}^1 = [-3.7, \ -3.0]^T$ from $\mathbf{X}$ gives, as the feature vector in the principal component space, $\mathbf{y}^1 = [-4.6760]$. Here, $y_1^1 = -4.6760$ is the first principal component of the pattern $\mathbf{x}^1$ along the direction of the first eigenvector $\mathbf{e}^1$. The data have maximal variance λ_1 in this direction. The projection of all vectors from $\mathbf{X}$ onto the first principal component gives ($\mathbf{Y} = \mathbf{X}\hat{\mathbf{W}}^T$)

$$\mathbf{Y} = \begin{bmatrix} -4.6760 \\ -4.0336 \\ -3.2672 \\ -4.1576 \\ -3.3912 \\ 3.1342 \\ 3.7766 \\ 3.6526 \\ 4.4190 \\ 4.5430 \end{bmatrix}$$

The covariance of $\mathbf{Y} = [17.2433]$ for λ_1. The first original pattern vector, $\mathbf{x}^1 = [-3.7, \ -3.0]^T$, can be reconstructed, by the inverse operation $\hat{\mathbf{x}} = \hat{\mathbf{W}}^T \mathbf{y}$, from the principal component space vectors $\mathbf{y}^1 = [-4.6760]$:

$$\hat{\mathbf{x}}^1 = \begin{bmatrix} 0.6424 \\ 0.7664 \end{bmatrix} [-4.6760] = \begin{bmatrix} -3.0038 \\ -3.5836 \end{bmatrix}$$

The reconstructed error vector is $\mathbf{x}^1 - \hat{\mathbf{x}}^1 = [-0.6962, \ 0.5836]^T$. Reconstruction of all original pattern space vectors $\hat{\mathbf{X}}$, by the inverse operation $\hat{\mathbf{X}} = \mathbf{YW}$, gives

$$\hat{\mathbf{X}} = \begin{bmatrix} -3.0038 & -3.5836 \\ -2.5911 & -3.0913 \\ -2.0988 & -2.5039 \\ -2.6708 & -3.1863 \\ -2.1785 & -2.5989 \\ 2.0134 & 2.4020 \\ 2.4261 & 2.8943 \\ 2.3464 & 2.7993 \\ 2.8387 & 3.3866 \\ 2.9184 & 3.4817 \end{bmatrix}$$

The reconstructed error vectors are

$$\mathbf{X}-\hat{\mathbf{X}} = \begin{bmatrix} 0.6962 & 0.5836 \\ -0.1089 & 0.0913 \\ -0.6012 & 0.5039 \\ 0.9708 & -0.8137 \\ 0.4785 & -0.4011 \\ -0.7134 & 0.5980 \\ -0.1261 & 0.1057 \\ 0.9536 & -0.7993 \\ 0.4613 & -0.3866 \\ -0.6184 & 0.5183 \end{bmatrix}$$

The least mean squares error is equal to $\lambda_2 = 0.7678$ (its numerical estimate as the mean of the pattern error equals 0.7478).

2.2. Supervised Feature Extraction Based on Fisher's Linear Discriminant Analysis

Fisher's linear discriminant method, with a linear transformation of the original patterns, is the classic method of real-valued feature extraction and pattern **dimensionality reduction**. This linear transformation is obtained based on statistical characteristics extracted from a given data set represented by data pattern **scatter matrices** (proportional to covariance matrices). Fisher's linear transformation is constructed based on a given limited-size data set T_{tra}, containing N examples. Each example $(\mathbf{x}^i, c^i_{\text{target}})$ ($i = 1, 2, \cdots, N$), representing one object of recognition, is constituted with an n-dimensional real-valued pattern $\mathbf{x} \in \mathbb{R}^n$ with corresponding target class c^i_{target}. We assume that a data set T_{tra} contains N_i ($\sum_i^l N_i = N$) examples from each categorical class c_i, with the total number of classes denoted by l.

Fisher's linear discriminant analysis is a method of **supervised learning** from data, since it considers patterns labeled by target classes and reveals and uses measures of pattern scatter through the total data set, as well as within and between classes. A linear transformation of an original data pattern can also be interpreted as a projection of original patterns into a reduced m-dimensional feature space. Here, feature space reduction and feature extraction via transformation are included in one transformation.

Fisher's linear discriminant analysis, based on statistical data-analysis techniques, determines an optimal linear transformation

$$\mathbf{y} = \mathbf{W}\mathbf{x} \tag{44}$$

of a real-valued n-dimensional data pattern $\mathbf{x}$ into another m-dimensional ($m \leq n$) transformed pattern $\mathbf{y}$. The $m \times n$ transformation matrix $\mathbf{W}$ is designed optimally from the point of view of maximal interclass separability of projected patterns. This design allows us to find a reduced compact data representation, in lower-dimensionality pattern space, with maximal separability between classes, thereby allowing us to design a better classifier.

Designing Fisher's linear transformation as a static optimization problem with specifically defined optimization criterion leads to obtaining an optimal transformation. The criterion J_{Fisher} is selected for the optimal Fisher linear transformation in order to ensure that we obtain a minimum interclass separability of transformed patterns. The evaluation of interclass separability is based on scatter matrices (proportional to covariance matrices) estimated for a given data set for each class and between classes. It is constructed to choose new features, by the transformation of original features to fewer new features, which will ensure a small within-class scatter and a large between-class scatter.

2.2.1. Two-class Data and Fisher's Projection onto a Line

First, we assume a two-class data set and analyze a linear transformation of n-dimensional patterns $\mathbf{x} \in \mathbb{R}^n$ into a **one-dimensional** feature space, with patterns $y \in \mathbb{R}$ containing one feature. This transformation has the linear form

$$y = \mathbf{w}^T \mathbf{x} \qquad (45)$$

where $\mathbf{w} = [w_1, w_2, \cdots, x_n]^T$ is an n-dimensional transformation vector (containing n adjustable coefficients). We see that this is a linear projection of multidimensional patterns in the n-dimensional pattern space onto a line in the one-dimensional reduced feature space y. Here, y is an extracted and reduced data feature. Even for well-clustered and separated patterns from two classes in the original n-dimensional pattern space, such linear projection onto a line may result in a large loss of information and a confusing overlap of patterns from both classes. However, by rotating a line of projection (by changing a transformation coefficient w_i), we can find a line position that will give maximal separability of projected patterns from the two classes. This observation sets the foundation for Fisher's linear transformation (projection), which can be generalized also for multiclass data sets.

We will discuss Fisher's linear transformation as constructed based on a given limited-size data set T_{tra}. The set T_{tra} contains N examples (containing patterns labeled by classes). Each example $(\mathbf{x}^i, c^i_{target})(i = 1, 2, \cdots, N)$ is constituted with an n-dimensional real-valued pattern $\mathbf{x} \in \mathbb{R}^n$ with corresponding target class c^i_{target} ($i = 1, 2$). We assume that a data set T_{tra} (a training set) contains N_i ($N_1 + N_2 = N$) examples from each of two categorical classes c_1, c_2, with the total number of classes denoted by $l = 2$.

2.2.2. Statistical Characteristics of Patterns from an Original n-feature Data Set T_{tra}

First, we should provide measures of the statistical characteristics of patterns from an original n-feature data set T_{tra}. The mean for each class, which shows a separation of patterns from each class, can be estimated by

$$\boldsymbol{\mu}_i = \frac{1}{N_i} \sum_{\mathbf{x} \in xc_i} \mathbf{x}, \quad (i = 1, 2) \qquad (46)$$

where $\mathbf{x} \in xc_i$ denotes a set of patterns from the class c_i. In order to calculate the scatter of patterns within one class c_i around this class mean, the $n \times n$-dimensional **within-class c_i scatter matrix** is defined as

$$\mathbf{S}_i = \sum_{\mathbf{x} \in xc_i} (\mathbf{x} - \boldsymbol{\mu}_i)(\mathbf{x} - \boldsymbol{\mu}_i)^T, \quad (i = 1, 2) \qquad (47)$$

This matrix is proportional to the covariance matrix; it is symmetric and positive semidefinite (and, for $N_i > n$, usually nonsingular). The summarizing measure for the scatter of patterns around means for all l classes, the so-called **within-class scatter matrix**, can be defined as

$$\mathbf{S}_w = \mathbf{S}_1 + \mathbf{S}_2 \qquad (48)$$

The **total scatter matrix** (for all N_{all} patterns from T_{tra})

$$\mathbf{S}_t = \sum_{j=1}^{N} (\mathbf{x}^j - \boldsymbol{\mu})(\mathbf{x}^j - \boldsymbol{\mu})^T \qquad (49)$$

illustrates a between-class scatter. The **total data mean** can be estimated by

$$\boldsymbol{\mu} = \frac{1}{N} \sum_{j=1}^{N} \mathbf{x}^j = \frac{1}{N}(N_1 \boldsymbol{\mu}_1 + N_2 \boldsymbol{\mu}_2) \qquad (50)$$

2. Feature Extraction

We find that the total scatter matrix can be decomposed into two matrices

$$\mathbf{S}_t = \mathbf{S}_w + \mathbf{S}_b \tag{51}$$

where $\mathbf{S}_w$ is the within-class scatter matrix, and the $n \times n$ square matrix $\mathbf{S}_b$ is the **between-class scatter matrix**, defined as

$$\mathbf{S}_b = N_1(\boldsymbol{\mu}_1 - \boldsymbol{\mu})(\boldsymbol{\mu}_1 - \boldsymbol{\mu})^T + N_2(\boldsymbol{\mu}_2 - \boldsymbol{\mu})(\boldsymbol{\mu}_2 - \boldsymbol{\mu})^T \tag{52}$$

2.2.3. Statistical Characteristics of the Projected Original Data Set Patterns onto a Single Feature y Pattern Space

Similar statistical characteristics can be provided for projected original data set patterns onto a single feature y pattern space (with linear transformation $y = \mathbf{w}^T \mathbf{x}$). All projected patterns y form a projected data set $T_{\text{tra, proj}}$ with N cases $(y^i, c^i_{\text{target}})$ containing labeled single feature patterns $y \in \mathbb{R}$. The mean of projected patterns y in $T_{\text{tra, proj}}$ for each class c_1 and c_2 can be estimated from

$$\mu_{i,p} = \frac{1}{N_i} \sum_{y \in yc_i} y = \frac{1}{N_i} \sum_{x \in xc_i} \mathbf{w}^T \mathbf{x} = \mathbf{w}^T \boldsymbol{\mu}_i \quad (i = 1, 2) \tag{53}$$

and the scatter of a feature y (from projected original data) for each class is

$$s^2_{i,p} = \frac{1}{N_i} \sum_{y \in yc_i} (y - \mu_{i,p})^2, \quad (i = 1, 2) \tag{54}$$

Note that the estimate of variance of the projected patterns for each class is $(1/N_i) s^2_{i,p}$. The total within-class scatter of y in an entire data set $T_{\text{tra, proj}}$ is defined by

$$s^2_{t,p} = s^2_{1,p} + s^2_{2,p} \tag{55}$$

We can see that the distance between projected means can be found to be $|\mu_{1,p} - \mu_{2,p}| = |\mathbf{w}^T(\boldsymbol{\mu}_1 - \boldsymbol{\mu}_2)|$.

The design criterion for optimal interclass separability

$$J_{\text{Fisher}}(\mathbf{W}) = \frac{|(\mu_{1,p} - \mu_{2,p})^2|}{s^2_{1,p} + s^2_{2,p}} \tag{56}$$

guarantees a large value either for larger between-class scatter or for smaller within-class scatter.

The above criterion for a single feature, based on interclass separability, is called the Fisher **F-ratio**. It guarantees finding an optimal linear transformation $y = \mathbf{w}^T \mathbf{x}$ of n-feature patterns into one feature pattern y, maximizing the between-class variance while simultaneously minimizing the within-class variance.

Let us sketch a solution for an optimal linear transformation. We try to find an optimal transformation vector $\hat{\mathbf{w}}$ providing maximization of a criterion $J_{\text{Fisher}}(\mathbf{W})$. Derivations show that $(\mu_{1,p} - \mu_{2,p})^2 = \mathbf{w}^T \mathbf{S}_b \mathbf{w}$, with

$$\mathbf{S}_b = (\boldsymbol{\mu}_1 - \boldsymbol{\mu}_2)(\boldsymbol{\mu}_1 - \boldsymbol{\mu}_2)^T \tag{57}$$

and $s^2_{t,p} = s^2_{1,p} + s^2_{2,p} = \mathbf{w}^T \mathbf{S}_w \mathbf{w}$. Thus, the criterion J for interclass separability of projected patterns y as a function of $\mathbf{w}$ is given by

$$J_{\text{Fisher}}(\mathbf{w}) = \frac{\mathbf{w}^T \mathbf{S}_b \mathbf{w}}{\mathbf{w}^T \mathbf{S}_w \mathbf{w}} \tag{58}$$

It is known that the optimal transform vector **w** that minimizes the criterion functional $J(\mathbf{w})$ must be a solution of the generalized eigenvalue problem

$$\mathbf{S}_b \mathbf{w} = \lambda \mathbf{S}_w \mathbf{w} \tag{59}$$

or, written in another form (after multiplying both sides of the criterion by $\mathbf{S}^{-1}$),

$$\mathbf{S}_w^{-1} \mathbf{S}_b \mathbf{w} = \lambda \mathbf{w} \tag{60}$$

For a nonsingular $\mathbf{S}_w$, we have a final solution for the optimal value of Fisher's linear transformation vector $\hat{\mathbf{w}}$ (Fukunaga, 1990):

$$\hat{\mathbf{w}} = \mathbf{S}_w^{-1}(\boldsymbol{\mu}_1 - \boldsymbol{\mu}_2) \tag{61}$$

2.2.4. Fisher Linear Discriminant – Classification

The main purpose of Fisher's linear transformation is to optimally convert patterns into a pattern space with a reduced dimensionality of projected pattern more suitable for the classification. However, we can design a classifier based on Fisher's linear transformation by using an optimal pattern projection formula that acts like a discriminant:

$$y = d(\mathbf{x}) = \hat{\mathbf{w}}^T \mathbf{x} \tag{62}$$

If we choose a threshold value $y_{\text{threshold}}$ for the projected pattern y, a new pattern **x** could be classified as belonging to class $c_i = 1$ if $y = d(\mathbf{x}) = \hat{\mathbf{w}}^T \mathbf{x}) \geq y_{\text{threshold}}$, and otherwise as belonging to class $c_i = 2$.

Let us consider the original l-class data set T_{tra} with n-dimensional class labeled patterns. We assume that for the class data set T_{tra} with n dimensions, the dimensionality of the original patterns is smaller than a number of classes $n > l$. We will discuss how to transform n-dimensional patterns from an original data set T_{tra} into reduced-size m-dimensional feature patterns (with $m > 1$), thereby ensuring maximal interclass separability in the projected space. Later we will show that $m > l$.

Here we have m linear transformations (discriminants)

$$y_i = \mathbf{w}_i^T \mathbf{x}, \quad (i = 1, 2, \cdots, m) \tag{63}$$

for all features y_i of projected patterns, with $\mathbf{w}_i$ being the transformation column vectors (for y_i discriminant). All m features of projected patterns form an m-dimensional projected feature pattern $\mathbf{y} = [y_1, y_2, \cdots, y_m]^T \in \mathbb{R}^m$. The linear transformation has the form

$$\mathbf{y} = \mathbf{W}\mathbf{x} \tag{64}$$

where **W** is an $m \times n$-dimensional transformation matrix, with each row $\mathbf{w}_i$ being a transformation vector for a corresponding feature y_i ($l = 1, 2, \cdots, m$).

One can generalize definitions for the statistical characteristics for both data sets, the original T_{tra} and the projected $T_{\text{tra, proj}}$, needed to find an optimal transform matrix $\hat{\mathbf{W}}$.

The $n \times n$ **within-class c_i scatter matrix** is defined as

$$\mathbf{S}_i = \sum_{\mathbf{x} \in xc_i} (\mathbf{x} - \boldsymbol{\mu}_i)(\mathbf{x} - \boldsymbol{\mu}_i)^T, \quad (i = 1, 2, \cdots, l) \tag{65}$$

where $\boldsymbol{\mu}_i$ is the mean

$$\boldsymbol{\mu}_i = \frac{1}{N_i} \sum_{\mathbf{x} \in xc_i} \mathbf{x}, \quad i = 1, 2, \cdots, l \tag{66}$$

for patterns within each class. The **within-class scatter matrix** is defined as

$$\mathbf{S}_w = \sum_{i=1}^{l} \mathbf{S}_i \tag{67}$$

The **total scatter matrix** for all patterns is defined as

$$\mathbf{S}_t = \sum_{j=1}^{N} (\mathbf{x}^j - \boldsymbol{\mu})(\mathbf{x}^j - \boldsymbol{\mu})^T \tag{68}$$

where $\boldsymbol{\mu}$ is the estimate of the **total data mean**

$$\boldsymbol{\mu} = \frac{1}{N} \sum_{j=1}^{N} \mathbf{x}^j = \frac{1}{N} \sum_{i=1}^{l} N_i \boldsymbol{\mu}_i \tag{69}$$

We can find the decomposition

$$\mathbf{S}_t = \mathbf{S}_w + \mathbf{S}_b \tag{70}$$

where $\mathbf{S}_w$ is the within-class scatter matrix, and the $n \times n$ **between-class scatter matrix** $\mathbf{S}_b$ is defined as

$$\mathbf{S}_b = \sum_{i=1}^{l} N_i (\boldsymbol{\mu}_i - \boldsymbol{\mu})(\boldsymbol{\mu}_i - \boldsymbol{\mu})^T \tag{71}$$

Now, the projection of an n-dimensional pattern space into an $(l-1)$-dimensional discriminant space is given by $(l-1)$ functions

$$y_i = \mathbf{w}_i^T \mathbf{x}, \; i = 1, \cdots, l-1 \tag{72}$$

We can also write the matrix version of above equation, assuming that $\mathbf{y} \in \mathbb{R}^{(l-1)}$ and that $\mathbf{W}$ is the $n \times (l-1)$ matrix containing, as rows, the transpose of the transformation weigh vectors $\mathbf{w}_i$:

$$\mathbf{y} = \mathbf{W}^T \mathbf{x} \tag{73}$$

Projections of all the original patterns $\{\mathbf{x}^1, \cdots, \mathbf{x}^N\}$ will result in the set of corresponding projected patterns $\{\mathbf{y}^1, \cdots, \mathbf{y}^N\}$. One can find similar statistical characteristics (denoted by p) for the projected data set $T_{\text{tra, proj}}$ with m-dimensional projected patterns $\mathbf{y}$:

$$\boldsymbol{\mu}_{i,p} = \frac{1}{N_i} \sum_{\mathbf{y} \in yc_i} \mathbf{y}, \; (i = 1, 2, \cdots, l) \tag{74}$$

$$\boldsymbol{\mu}_p = \frac{1}{N} \sum_{j=1}^{N} \mathbf{y}^j = \frac{1}{N} \sum_{i=1}^{l} N_i \boldsymbol{\mu}_{i,p} \tag{75}$$

$$\mathbf{S}_{i,p} = \sum_{\mathbf{y} \in yc_i} (\mathbf{y} - \boldsymbol{\mu}_{i,p})(\mathbf{y} - \boldsymbol{\mu}_{i,p})^T, \; (i = 1, 2, \cdots, l) \tag{76}$$

$$\mathbf{S}_{w,p} = \sum_{i=1}^{l} \mathbf{S}_{i,p}, \; \mathbf{S}_{t,p} = \sum_{j=1}^{N} (\mathbf{y}^j - \boldsymbol{\mu}_p)(\mathbf{y}^j - \boldsymbol{\mu}_p)^T \tag{77}$$

It can be seen that

$$\mathbf{S}_{w,p} = \mathbf{W}^T \mathbf{S}_w \mathbf{W} \tag{78}$$

and

$$\mathbf{S}_{b,p} = \mathbf{W}^T \mathbf{S}_b \mathbf{W} \tag{79}$$

For multiple classes and multifeature patterns, the following scalar between-class separability criterion may be defined:

$$J(\mathbf{W}) = \frac{|\mathbf{S}_{b,p}|}{|\mathbf{S}_{w,p}|} = \frac{|\mathbf{W}^T \mathbf{S}_b \mathbf{W}|}{|\mathbf{W}^T \mathbf{S}_w \mathbf{W}|} \tag{80}$$

Here, $|\mathbf{S}_{b,p}|$ denotes a scalar representation of the between-class scatter matrix, and similarly, $|\mathbf{S}_{w,p}|$ denotes a scalar representation of the within-class scatter matrix for projected patterns.

An optimal transformation matrix $\hat{\mathbf{W}}$ can be found as a solution to the general eigenvalue problem

$$\mathbf{S}_b \mathbf{w}_i^T = \lambda_i \mathbf{S}_w \mathbf{w}_i^T, \quad (i = 1, 2, \cdots, n) \tag{81}$$

Multiplying both sides by $\mathbf{S}_w^{-1}$ gives

$$\mathbf{S}_w^{-1} \mathbf{S}_b \mathbf{w}_i^T = \lambda_i \mathbf{w}_i^T, \quad (i = 1, 2, \cdots, n) \tag{82}$$

The solution of this problem is based on computing, and rearranging in decreased order, the eigenvalues and corresponding eigenvectors for the matrix $\mathbf{S}_w^{-1}\mathbf{S}_b$. Let us assume that the eigenvalues of the matrix $\mathbf{S}_w^{-1}\mathbf{S}_b$ are arranged in decreasing order $\lambda_1 \geq \lambda_2 \geq \cdots \lambda_n \geq 0$ (with $\lambda_1 = \lambda_{\max}$). Consequently, assume that the corresponding eigenvectors $\mathbf{e}^1, \mathbf{e}^2, \cdots, \mathbf{e}^n$ compose the $n \times n$ matrix

$$\mathbf{E} = [\mathbf{e}^1, \mathbf{e}^2, \cdots, \mathbf{e}^n] \tag{83}$$

with columns being eigenvectors. Then the optimal linear transformation matrix $\hat{\mathbf{W}}$ is composed with rows being the first m columns of matrix $\mathbf{E}$:

$$\hat{\mathbf{W}} = \mathbf{E}^T = \begin{bmatrix} (\mathbf{e}^1)^T \\ (\mathbf{e}^2)^T \\ \vdots \\ (\mathbf{e}^m)^T \end{bmatrix} \tag{84}$$

Thus, an optimal Fisher transformation matrix $\hat{\mathbf{W}}$ (denoted also by $\mathbf{W}_F$) is composed with m rows being the first m eigenvectors of the matrix $\mathbf{S}_w^{-1}\mathbf{S}_b$.

The linear Fisher transformation faces problems when the within-class scatter matrix S_w becomes degenerate (noninvertible). This can happen when the number of cases (objects) is smaller than the dimension of a pattern.

2.3. Sequence of PCA and Fisher's Linear Discriminant Projection

The PCA, with its resulting linear Karhunen-Loéve projection, provides feature extraction and reduction that are optimal from the point of view of minimizing the reconstruction error. However, PCA does not guarantee that selecting (reduced) principal components as a feature vector will be adequate for classification (will have discriminatory power). Nevertheless, the projection of high-dimensional patterns into lower-dimensional orthogonal principal-component feature vectors might ensure better classification for some data types.

On the other hand, the linear Fisher transformation is designed optimally from the point of view of maximal interclass separability of projected patterns. This transformation is a linear projection of the original patterns into a reduced **discriminative** feature space (suitable for classification).

However, we recall that the linear Fisher transformation faces problems when the within-class scatter matrix S_w becomes degenerate (noninvertible). This can happen when the number of cases (objects) in a data set is smaller than the pattern dimension. One solution to this problem is first to transform a data set with high-dimensional patterns into a lower dimensional feature space, for example by using a KLT transform, and then to apply a discriminative linear Fisher projection to the lower-dimensional patterns. This approach may lead to feature vectors that are more suitable for classification (and with better discriminatory power than the original patterns).

The entire projection procedure of the original high dimensional (n-dimensional) patterns $\mathbf{x}$ into the lower-dimensional (c-dimensional), more discriminative feature vectors can be decomposed into two subsequent linear transformations:

– PCA with the resulting Karhunen-Loéve projection into the m-dimensional principal component feature vectors $\mathbf{y}$
– A linear Fisher projection of m-dimensional principal component vectors $\mathbf{y}$ into c-dimensional Fisher discriminative feature vectors $\mathbf{z}$

Let us consider a given limited size data set T containing N cases labeled by associated classes. Each case $(\mathbf{x}^i, c^i_{\text{target}})(i = 1, 2, \cdots, N)$ is composed of an n-dimensional real-valued pattern $\mathbf{x} \in \mathbb{R}^n$ and a corresponding target class c^i_{target}. The total number of classes is l.

The selection of the projection dimensions m and c can be done in the following way. One can choose m such that for a number N of cases in a data set we have $m + l \leq N$. We know that it is impossible for m to be greater than $N - 1$ (since there is a maximum of $N - 1$ nonzero eigenvalues in the Karhunen-Loéve projection). We further constrain the final dimension of the reduced PCA feature vector to be less than the rank of the within-class scatter matrix S_w in order to make S_w nondegenerate (invertible). However, m cannot be smaller than the number of classes l. Since there are at most $l - 1$ nonzero eigenvalues of the matrix $S_w^{-1} S_b$ (where S_b is the between-class scatter matrix), one can choose $c \leq l - 1$ as a resulting dimension in the discriminative Fisher feature space. This will be the final dimension of a sequence of two projections. The relations between the dimensions of each projection are $c + 1 \leq l \leq m \leq N - l$.

The algorithm for the Karhunen-Loéve-Fisher transformations is as follows.

Given: A data set T, containing N cases $(\mathbf{x}^i, c^i_{\text{target}})$ labeled by associated categorical classes (with a total of l classes)

1. Extract from data set T only the pattern portion represented by the $n \times N$ matrix $\mathbf{X}$ with n-dimensional patterns $\mathbf{x}$ as rows.
2. Select a dimension m for the feature vectors containing principal components (already projected by the Karhunen-Loéve transformation), satisfying the inequalities $l \leq m \leq N - l$.
3. Compute, for the data in matrix $\mathbf{X}$, the $m \times n$-dimensional optimal linear Karhunen-Loéve transformation matrix $\mathbf{W}_{KL}$.
4. Transform (project) each original pattern $\mathbf{x}$ onto a reduced-size m-dimensional pattern $\mathbf{y}$ by using the formula $\mathbf{y} = \mathbf{W}_{KL}\mathbf{x}$ (or, for all projected patterns, by using the formula $\mathbf{Y} = \mathbf{X}\mathbf{W}_{KL}^T$).
5. Select a dimension c for the final reduced feature vectors $\mathbf{z}$ (projected by Fisher's transformation), satisfying the inequality $c + 1 \leq l$.
6. Compute, for the projected data in matrix $\mathbf{Y}$, the $c \times m$-dimensional optimal linear Fisher transformation matrix $\mathbf{W}_F$.
7. Transform (project) each projected pattern $\mathbf{y}$ from $\mathbf{Y}$ into the reduced-size c-dimensional pattern $\mathbf{z}$ by using the formula $\mathbf{z} = \mathbf{W}_F \mathbf{y}$ (or, for all projected patterns, by the formula $\mathbf{Z} = \mathbf{Y}\mathbf{W}_F^T$).

Result: The $m \times n$-dimensional optimal linear Karhunen-Loéve transformation matrix is $\mathbf{W}_{KL}$. The $c \times m$ dimensional optimal linear Fisher's transformation matrix is $\mathbf{W}_F$. The projected pattern matrix is $\mathbf{Z}$.

2.4. Singular Value Decomposition as a Method of Feature Extraction

One of the most important matrix decompositions is the **Singular Value Decomposition** (SVD). SVD can be used both as a powerful method of extracting features from images and as a method of image compression. We know from linear algebra theory that any symmetric matrix can be transformed into a diagonal matrix by means of orthogonal transformation. Similarly, any rectangular $n \times m$ real image represented by an $n \times m$ matrix **A**, where $m \leq n$, can be transformed into a diagonal matrix by singular value decomposition. SVD decomposes a rectangular matrix $\mathbf{A} \in \mathbb{R}^{m \times n}$ into two orthogonal matrices $\mathbf{\Psi}_r$ of dimension $n \times n$ and $\mathbf{\Phi}_r$ of dimension $m \times m$, a pseudodiagonal matrix $\mathbf{\Lambda}$ of dimension $r \times r$, and a pseudodiagonal matrix containing singular values of the transposed matrix. Here, $\mathbf{\Lambda} = \mathrm{diag}\{\sigma_1, \sigma_2, \ldots, \sigma_p\}$, where $p = \min(m, n)$. The real nonnegative numbers $\sigma_1 \geq \sigma_2 \geq \cdots \geq \sigma_p$ are called the singular values of matrix **A**.

The following equalities hold:

$$\mathbf{\psi}_r^T \mathbf{\psi}_r = \mathbf{\psi}_r \mathbf{\psi}_r^T = \mathbf{I} \quad \text{and} \quad \mathbf{\phi}_r^T \mathbf{\phi}_r = \mathbf{\phi}_r \mathbf{\phi}_r^T = \mathbf{I} \tag{85}$$

Now, assume that the rank of matrix **A** is $r \leq m$. The matrices $\mathbf{A}\mathbf{A}^T$ and $\mathbf{A}^T\mathbf{A}$ are nonnegative and symmetric and have identical eigenvalues λ_i. For $m \leq n$, there are at most $r \leq m$ nonzero eigenvalues. The SVD transform decomposes matrix **A** into the product of two orthogonal matrices $\mathbf{\psi}$ of dimension $n \times r$, and $\mathbf{\phi}$ of dimension $m \times r$ and a diagonal matrix $\mathbf{\Lambda}^{1/2}$ of dimension $r \times r$. The SVD of a matrix **A** (for example, representing an image) is given by

$$\mathbf{A} = \mathbf{\psi} \mathbf{\Lambda}^{1/2} \mathbf{\phi}^T = \sum_{i=1}^{r} \sqrt{\lambda_i} \psi_i \phi_i^T. \tag{86}$$

where the matrices $\mathbf{\psi}$ and $\mathbf{\phi}$ have r orthogonal columns $\psi_i \in \mathbb{R}^n$, $\phi_i \in \mathbb{R}^m$ ($i = 1, \cdots, r$), respectively (representing the orthogonal eigenvectors of $\mathbf{A}\mathbf{A}^T$ and $\mathbf{A}^T\mathbf{A}$, respectively). The square matrix $\mathbf{\Lambda}^{1/2}$ has diagonal entries defined by

$$\mathbf{\Lambda}^{1/2} = \mathrm{diag}(\sqrt{\lambda_1}, \sqrt{\lambda_2}, \cdots, \sqrt{\lambda_r}) \tag{87}$$

where $\sigma_i = \sqrt{\lambda_i}(i = 1, 2, \cdots, r)$ are the singular values of the matrix **A**. Each λ_i, $(i = 1, 2, \cdots, r)$ is the nonzero eigenvalue of $\mathbf{A}\mathbf{A}^T$ (and of $\mathbf{A}^T\mathbf{A}$). Since the columns of $\mathbf{\psi}$ and $\mathbf{\phi}$ are the eigenvectors of $\mathbf{A}\mathbf{A}^T$ and $\mathbf{A}^T\mathbf{A}$, respectively, the following equations must therefore be satisfied:

$$(\mathbf{A}\mathbf{A}^T - \lambda_i \mathbf{I}_{n \times n})\psi_i = \mathbf{0}, \quad i = 1, 2, \cdots, n \tag{88}$$

$$(\mathbf{A}^T\mathbf{A} - \lambda_i \mathbf{I}_{m \times m})\phi_i = \mathbf{0}, \quad i = 1, 2, \cdots, m \tag{89}$$

Note that in both cases, we assume $\lambda_i = 0$ if $i > r$. We also include additional ψ_i such that $\mathbf{A}^T \psi_i = \mathbf{0}$, for $i = r+1, \cdots, n$, and additional ϕ_i such that $\mathbf{A}\phi_i = \mathbf{0}$, for $i = r+1, \cdots, m$. The above equations are also defined for $i = 1, 2, \cdots, r$. In this case the SVD transform is unitary, with unitary matrices $\mathbf{\psi}$ and $\mathbf{\phi}$.

Having decomposed matrix **A** (an image) as $\mathbf{A} = \mathbf{\psi}\mathbf{\Lambda}^{1/2}\mathbf{\phi}^T$, and since $\mathbf{\psi}$ and $\mathbf{\phi}$ have orthogonal columns, the **singular value decomposition transform** (SVD transform) of the image **A** is defined as

$$\mathbf{\Lambda}^{1/2} = \mathbf{\psi}^T \mathbf{A} \mathbf{\phi} \tag{90}$$

The r singular values $\sqrt{\lambda_i}(i = 1, 2, \cdots, r)$ from the main diagonal of the matrix $\mathbf{\Lambda}^{1/2}$ represent in condensed form the matrix **A**. If the matrix **A** represents an $n \times m$ image, then r singular values can be considered as extracted features from an image.

Unlike the PCA, the SVD is purely a matrix processing technique and not a statistical technique. However, SVD may relate to statistics if we consider processing a matrix related to statistical observations. If the singular values λ_i are arranged in decreasing order of magnitude, i.e., $\lambda_1 > \lambda_2 > \cdots > \lambda_r$, the error in approximation of the image is minimized in the least squares error sense. The nonzero diagonal singular values are unique for a given image. They can be used as features for textural image modeling, compression, and possibly classification purposes. The SVD features have many excellent characteristics, such as stability and rotational and translation invariances.

The SVD transform provides a means of extraction of the most expressive features for minimization of reconstruction error. Let us assume that, for the $n \times m (m \leq n)$ image represented by the matrix $\mathbf{A}$ of rank $r \leq m$, we find the SVD decomposition $\mathbf{A} = \mathbf{\psi} \mathbf{\Lambda}^{1/2} \mathbf{\phi}^T$, corresponding to r eigenvalues λ_i of $\mathbf{AA}^T$. Now, assume that we will represent the original image $\mathbf{A}$ by the reduced number $k \leq r$ of features $\sqrt{\lambda_i}(i = 1, 2, \cdots, k)$. Assume that eigenvalues $\lambda_i (i = 1, 2, \cdots, r)$ are arranged in decreasing order. The reconstructed $n \times m$ image $\mathbf{A}_k$ based on k eigenvalues of $\mathbf{AA}^T$ can be obtained by

$$\mathbf{A}_k = \sum_{i=1}^{k} \sqrt{\lambda_i} \mathbf{\psi}_i \mathbf{\phi}_i^T, \quad k \leq r \tag{91}$$

The reconstruction matrix $\mathbf{A}_k$ of rank $k \leq r$ is the best approximation, in the least squares error sense, of the original matrix $\mathbf{A}$. The reconstruction least squares error is computed by

$$e_k^2 = \sum_{i=1}^{n} \sum_{j=1}^{m} |a_{ij} - a_{k,ij}|^2 \tag{92}$$

or equivalently by

$$e_k^2 = \sum_{i=k+1}^{r} \lambda_i \tag{93}$$

We see that the least squares reconstruction error is equal to the sum of $r - k$ trailing eigenvalues of $\mathbf{AA}^T$.

Despite the expressive power of the SVD transform image features, it is difficult to say arbitrarily how powerful the SVD features could be for the classification of images.

2.5. Independent Component Analysis

Independent component analysis (ICA) is a computational and statistical method for discovering intrinsic independent factors in the data (sets of random variables, measurements, or signals). ICA is an unsupervised data processing method that exploits higher-order statistical dependencies among data. It discovers a generative model for the observed multidimensional data (given as a database). In the ICA model, the observed data variables x_i are assumed to be linear mixtures of some unknown independent sources s_i (latent variables, intrinsic variables) $x_i = h_1 s_1 + \cdots + h_m s_m$. A mixing system is assumed to be unknown.

Independent variables are assumed to be nongaussian and mutually statistically independent. Latent variables are called the independent components or sources of the observed data. ICA tries to estimate unknown **mixing matrix** and **independent components** representing processed data.

2.5.1. The Cocktail-party Problem

One interesting example of independent component analysis is the blind source separation problem known as "the cocktail party." Let us consider a room where two people are speaking simultaneously. Two microphones in different locations record speech signals $x_1(t)$ and $x_2(t)$. Each recorded signal is a weighted sum of speech signals generated by the two speakers $s_1(t)$ and $s_2(t)$. The mixing of speech signals can be expressed as a linear equation:

$$x_1(t) = a_{11}s_1(t) + a_{12}s_2(t)$$
$$x_2(t) = a_{21}s_1(t) + a_{22}s_2(t) \qquad (94)$$

where a_{11}, a_{12}, a_{21}, and s_{22} are parameters. The estimation of two original speech signals $s_1(t)$ and $s_2(t)$ using only the recorded signals $x_1(t)$ and $x_2(t)$ is called the **cocktail-party problem**. For known parameters a_{ij}, one could solve the linear mixing equation using classical linear algebra methods. However, the cocktail-party problem is more difficult, since the parameters a_{ij} are unknown. Independent component analysis can be used to estimate the mixing parameters a_{ij} based on information about their independence. This, in turn, allows us to separate the two original source signals $s_1(t)$ and $s_2(t)$ from their recorded mixtures $x_1(t)$ and $x_2(t)$.

Figure 7.2 shows an example of two original source signals and their linear mixtures.

2.5.2. ICA: Data and Model

ICA can be considered as an extension of Principal Component Analysis (PCA). We know that PCA finds the linear transformation of data patterns such that transformed patterns will have uncorrelated elements. The transformation matrix is formed with the ordered **eigenvectors** (corresponding to ordered **eigenvalues** in descending order) of the patterns covariance matrix. Discovered uncorrelated orthogonal principal components are optimal in the sense of minimizing mean-squares reconstruction error, and they show the directions in which data change the most. PCA decorrelates patterns but does not assure that uncorrelated patterns will be statistically independent.

ICA provides data representation (through the linear transformation) based on discovered statistically independent **latent variables** (independent components). The observed data signal can

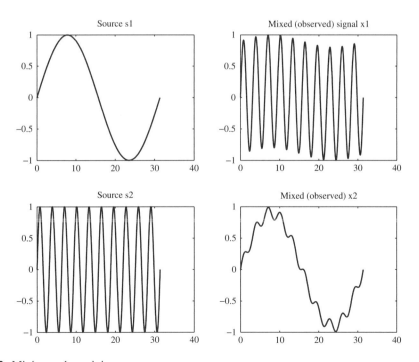

Figure 7.2. Mixing and unmixing.

be expressed as a linear mixture of statistically independent components (with highly nongaussian distributions). ICA is optimal in the sense of maximal statistical independence of sources.

The ICA model assumes that the n observed sensory signals x_i are given as the n-dimensional column pattern vectors $\mathbf{x} = [x_1, x_2, \cdots, x_n]^T \in \mathbb{R}^n$. ICA is applied based on a sample from the given domain of observed patterns. This sample is given as a set T of N pattern vectors $T = \{\mathbf{x}_1, \mathbf{x}_2, \cdots, \mathbf{x}_N\}$, which can be represented as an $n \times N$ **data matrix X** of measured N data patterns $\mathbf{x}_i (i = 1, \cdots, N)$. Columns of matrix $\mathbf{X}$ are composed with patterns $\mathbf{x}_i$. Generally, data pattern elements are considered to be random variables.

The **ICA model** for the x_i is given as linear mixtures of m source-independent variables s_j,

$$x_i = \sum_{j=1}^{m} h_{i,j} s_j = h_{i1} s_1 + h_{i2} s_2 + \cdots + h_{im} s_m, \quad i = 1, 2, \cdots, n \tag{95}$$

where x_i is the observed variable, s_j is the jth independent component (source signal), and $h_{i,j}$ are mixing coefficients. The source variable constitutes the m-dimensional column source vector (source pattern, independent component pattern) $\mathbf{s} = [s_1, s_2, \cdots, s_m]^T \in \mathbb{R}^m$. Without loss of generality, we can assume that both the observable variables and the independent components have zero mean. The observed variables x_i can always be centered by subtracting the sample mean.

The ICA model can be presented in the matrix form

$$\mathbf{x} = \mathbf{Hs} \tag{96}$$

where $\mathbf{H} \in \mathbb{R}^{n \times m}$ is the $n \times m$ unknown mixing matrix whose row vector $[h_{i,1}, h_{i,2}, \cdots, h_{i,m}]$ represents the mixing coefficients for observed signal x_i. We can also write

$$\mathbf{x} = \sum_{i=1}^{m} \mathbf{h}_i s_i \tag{97}$$

where $\mathbf{h}_i$ denotes the ith column of the mixing matrix $\mathbf{H}$.

In statistical independent component analysis, a data model is a generative type of model. This model expresses how the observed data are generated by a process of mixing of the independent components s_i. The independent components (sources) are latent variables that are not directly observable. The mixing matrix $\mathbf{H}$ is also assumed to be unknown.

The task of ICA is to estimate both the mixing matrix $\mathbf{H}$ and the sources $\mathbf{s}$ (independent components) based on the set of observed pattern vectors $\mathbf{x}$ (see Figure 7.3).

The ICA model can be discovered based on the $n \times N$ data matrix $\mathbf{X}$. The ICA model for the set of patterns from matrix $\mathbf{X}$ can be written as

$$\mathbf{X} = \mathbf{H}\,\mathbf{S} \tag{98}$$

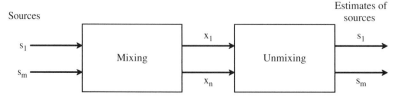

Figure 7.3. Mixing and unmixing.

where $\mathbf{S}$ is the $m \times N$ matrix whose columns contain m-dimensional independent component vectors $\mathbf{s}_i = [s_{i,1}, s_{i,2}, \cdots, s_{i,m}]^T$ discovered from the observation vectors. We can also write

$$\mathbf{X}^T = \mathbf{S}^T \mathbf{H}^T \tag{99}$$

ICA seems to be an underdetermined system. However, despite some constraints and ambiguities, one can discover independent components based on the principles of statistical independence of sources.

The following ambiguities are associated with ICA:

- Energies (variances) of independent components cannot be discovered because a scalar multiplier in s_i results in the same vector $\mathbf{x}$ as a scaling of the ith column in matrix $\mathbf{H}$.
- It is not possible to determine the order of independent components because exchanging two sources results in the same $\mathbf{x}$ as swapping the corresponding columns in $\mathbf{H}$.

These ambiguities arise from the fact that both $\mathbf{s}$ and $\mathbf{H}$ are unknown. The ICA assumes that the components s_i are statistically independent and have nongaussian distributions. Once the $n \times m$ mixing matrix $\mathbf{H}$ has been estimated, we can compute its $m \times n$ inverse matrix (demixing, **separation matrix**) $\mathbf{B} = \mathbf{H}^{-1}$ (for $m = n$), or pseudo-inverse $\mathbf{B} = \mathbf{H}^+$ (for $m \le n$), then the independent component vector for the observation vector $\mathbf{x}$ can be computed by

$$\mathbf{s} = \mathbf{B}\mathbf{x} \tag{100}$$

The inverse ICA model for the set of patterns T (matrix $\mathbf{X}$) can be written as

$$\mathbf{S} = \mathbf{B} \mathbf{X} \tag{101}$$

where $\mathbf{S}$ is the $m \times N$ matrix in which columns contain m-dimensional independent component vectors discovered from the observation vectors. We can also write

$$\mathbf{S}^T = \mathbf{X}^T \mathbf{B}^T \tag{102}$$

The extracted independent components s_i are as independent as possible, but can be evaluated by an information-theoretic cost criterion such as minimum Kulback-Leibler divergence.

ICA closely relates to the technique of **blind source separation** (BSS) or blind signal separation. In BSS, source signals (i.e., unknown independent components) are extracted from the observed patterns with very little knowledge about the mixing matrix or the statistics of the source signals.

2.5.3. Preprocessing

ICA can be preceded by necessary preprocessing, including centering and whitening, in order to simplify operations and make them better conditioned.

Centering of $\mathbf{x}$ is the process of subtracting its mean vector $\boldsymbol{\mu} = E\{\mathbf{x}\}$:

$$\mathbf{x} = \mathbf{x} - E\{\mathbf{x}\} \tag{103}$$

This process makes $\mathbf{x}$ (and consequently also $\mathbf{s}$) a zero-mean variable.

Once the mixing matrix $\mathbf{H}$ has been estimated for the centered data, then we can compute the final estimation by adding the mean vector of $\mathbf{s}$ back to the centered estimates of $\mathbf{s}$. The mean vector of $\mathbf{s}$ is given by $\mathbf{H}^{-1}\boldsymbol{\mu}$, where $\boldsymbol{\mu}$ is the mean that has been computed in preprocessing.

The second frequent preprocessing step in ICA (provided for centered signals) is **whitening**. The whitening is a linear transformation of measured patterns $\mathbf{x}$ that produces decorrelated white patterns $\mathbf{y}$, possibly with reduced dimensionality compared with n. Speaking more formally, whitening of the sensor signal vector $\mathbf{x}$ is a linear transformation

$$\mathbf{y} = \mathbf{Wx} \quad \text{so} \quad E\{\mathbf{yy}^T\} = \mathbf{I}_l \tag{104}$$

where $\mathbf{y} \in \mathbb{R}^l$ is the l-dimensional ($l \leq n$) whitened vector, $\mathbf{W}$ is the $l \times n$ **whitening matrix** and $\mathbf{I}_l$ is the $l \times l$ identity matrix.

First, let us assume that $l = n$ (the dimension of whitened signal equals dimension of the original pattern $\mathbf{x}$).

Generally, the observed sensor signals x_i are mutually correlated, and their covariance matrices $\mathbf{R}_{xx} = E\{\mathbf{xx}^T\}$ are of full rank (not diagonal). The purpose of whitening is to transform the observed vector $\mathbf{x}$ linearly to a new vector $\mathbf{y}$ (which is white) whose elements are uncorrelated and whose variances equal unity. This mean that the covariance matrix of $\mathbf{y}$ is equal to the identity matrix:

$$\mathbf{R}_{yy} = E\{\mathbf{yy}^T\} = E\{\mathbf{Wxx}^T\mathbf{W}^T\} = \mathbf{WR}_{xx}\mathbf{W}^T = \mathbf{I}_l \tag{105}$$

This transformation (which is always possible) is called **sphering**. We can notice that the matrix $\mathbf{W} \in \mathbb{R}^{l \times n}$ is not unique. Whitening also allows dimensionality reduction because it considers only the $l \leq n$ largest eigenvalues and the corresponding l eigenvectors of the covariance matrix of $\mathbf{x}$. Since the covariance matrix of observed vectors $\mathbf{x}$ is usually symmetric positive definite, whitening can therefore be realized using, for example, the eigenvalue decomposition of the covariance matrix $E\{\mathbf{xx}^T\} = \mathbf{R}_{xx} \in \mathbb{R}^{n \times n}$ of the observed vector $\mathbf{x}$:

$$\mathbf{R}_{xx} = E\{\mathbf{xx}^T\} = \mathbf{E}_x \mathbf{\Lambda}_x \mathbf{E}_x^T \tag{106}$$

Here, $\mathbf{E}_x \in \mathbb{R}^{n \times n}$ is the orthogonal matrix of eigenvectors of $\mathbf{R}_{xx} = E\{\mathbf{xx}^T\}$ and $\mathbf{\Lambda}$ is the $n \times n$ diagonal matrix of the corresponding eigenvalues

$$\mathbf{\Lambda}_x = \text{diag}(\lambda_1, \lambda_2, \cdots, \lambda_n) \tag{107}$$

Note that $E\{\mathbf{xx}^T\}$ can be estimated from the available sample $\mathbf{x}_1, \cdots, \mathbf{x}_N$ of the observed vector $\mathbf{x}$. For $l = n$, we consider all $l = n$ eigenvalues $\lambda_1, \lambda_2, \cdots, \lambda_n$ (and all $l = n$ corresponding eigenvectors of the covariance matrix $\mathbf{R}_{xx}$).

The whitening matrix can be computed as

$$\mathbf{W} = \mathbf{E}_x \mathbf{\Lambda}_x^{-1/2} \mathbf{E}_x^T \tag{108}$$

Thus, the whitening operation can be realized using the formula

$$\mathbf{y} = \mathbf{E}_x \mathbf{\Lambda}_x^{-1/2} \mathbf{E}_x^T \mathbf{x} = \mathbf{Wx} \tag{109}$$

here the matrix $\mathbf{\Lambda}^{-1/2}$ is computed as $\mathbf{\Lambda}^{-1/2} = \text{diag}(\lambda_1^{-1/2}, \cdots, \lambda_n^{-1/2})$. One can now find that for whitened vectors we have $E\{\mathbf{yy}^T\} = \mathbf{I}_l$.

Recalling that $\mathbf{y} = \mathbf{Wx}$ and $\mathbf{x} = \mathbf{H\,s}$, we can find from the above equation that

$$\mathbf{y} = \mathbf{E}_x \mathbf{\Lambda}_x^{-1/2} \mathbf{E}_x^T \mathbf{H\,s} = \mathbf{H}_w \mathbf{s} \tag{110}$$

We can see that whitening transforms the original mixing matrix $\mathbf{H}$ into a new one, $\mathbf{H}_w$:

$$\mathbf{H}_w = \mathbf{E}_x \mathbf{\Lambda}_x^{-1/2} \mathbf{E}_x \mathbf{H} \tag{111}$$

An important feature of whitening is producing the new mixing matrix $\mathbf{H}_w$, which is orthogonal ($\mathbf{H}_w^{-1} = \mathbf{H}_w^T$):

$$E\{\mathbf{yy}^T\} = \mathbf{H}_w E\{\mathbf{ss}^T\}\mathbf{H}_w^T = \mathbf{H}_w \mathbf{H}_w^T = \mathbf{I}_l \qquad (112)$$

If during the whitening process $l = n$ we consider all $l = n$ eigenvalues $\lambda_1, \lambda_2, \cdots, \lambda_n$ (and all $l = n$ corresponding eigenvectors of the covariance matrix $\mathbf{R}_{xx}$), the dimension of the matrix $\mathbf{W}$ is $n \times n$, and there is no reduction of the size of the observed transformed vector $\mathbf{y}$. For the fully dimensional matrix $\mathbf{W} \in \mathbb{R}^{n \times n}$, whitening provides only decorrelation of observed vectors and orthogonalization of the mixing matrix. We can see that whitening for $l = n$ can reduce the number of parameters to be estimated by ICA processing. Instead of having to estimate the n^2 parameters that are the elements of the original mixing matrix $\mathbf{H}$, we only need to estimate the new, orthogonal mixing matrix $\mathbf{H}_w$. In this orthogonal matrix, we have to estimate only $n(n-1)/2$ parameters.

Whitening allows us to reduce the dimensionality of the whitened vector by considering only the l ($l \leq n$) largest eigenvalues and the corresponding l eigenvectors of the covariance matrix $\mathbf{E}_{yy} = \mathbf{R}_{xx}$. This will result in obtaining a reduced whitening matrix $\mathbf{W}$ of dimension $l \times n$ and a reduced l-dimensional whitened vector $\mathbf{y} \in \mathbb{R}^l$. Consequently, the dimension of the new mixing matrix $\mathbf{H}_w$ will be reduced to $l \times n$. This is similar to the dimensionality reduction of patterns in PCA. Then the resulting dimension of the matrix $\mathbf{W}$ is $l \times n$, and there is a reduction of the size of the observed transformed vector $\mathbf{y}$ from n to l.

For the set of N original patterns arranged in matrix $\mathbf{X}$, whose columns are patterns $\mathbf{x}$, the whitening is given by

$$\mathbf{Y} = \mathbf{W}\mathbf{X} \qquad (113)$$

where $\mathbf{Y}$ is the $N \times l$ matrix, whose columns are constituted by prewhitened vectors $\mathbf{y}$. The dewhitening operation is given by

$$\tilde{\mathbf{X}} = \mathbf{D}\mathbf{Y} \qquad (114)$$

where the dewhitening matrix is defined as $\mathbf{D} = \mathbf{W}^{-1}$ for $l = n$ and as $\mathbf{D} = \mathbf{W}^+$ for $l = n$ (where $\mathbf{W}^+$ denotes the pseudoinverse of the rectangular matrix).

The output vector of the whitening process $\mathbf{y}$ can be considered as an input to the ICA algorithm: an input to the unmixing operation (separation, finding an independent components vector).

We can see that instead of estimating the mixing matrix $\mathbf{H}$, we can just estimate the orthogonal matrix $\mathbf{W}_w$, which simplifies the computation.

Whitening results in a new mixing matrix $\mathbf{H}_w$ (and in the ICA model $\mathbf{y} = \mathbf{H}_w\mathbf{s}$) being found, as well as a set of N whitened patterns $\mathbf{Y}$. Assume that the mixing matrix $\mathbf{H}_w$ and $\mathbf{Y}$ have been found. Recall that $\mathbf{y} = \mathbf{H}_w\mathbf{s}$; therefore, the unmixing operation for one vector $\mathbf{y}$ is given by

$$\mathbf{s} = \mathbf{B}_w\mathbf{y} \qquad (115)$$

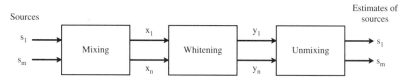

Figure 7.4. Whitening and unmixing.

where the demixing matrix is computed as inverse $\mathbf{B}_w = \mathbf{H}_w^{-1}$ (for $l = n$) or as pseudo-inverse $\mathbf{B} = \mathbf{H}_w^+$ (for $l \leq n$) of $\mathbf{H}_w$.

For a set of N whitened patterns $\mathbf{Y}$, we can compute a set of N corresponding independent component vectors arranged as the $m \times N$ matrix $\mathbf{S}$ whose columns are constituted by m-dimensional independent component vectors $\mathbf{s}_i = [s_{i,1}, s_{i,2}, \cdots, s_{i,m}]^T$:

$$\mathbf{S} = \mathbf{B}_w \mathbf{Y} \tag{116}$$

Since $\mathbf{Y} = \mathbf{WX}$, therefore we can also write

$$\mathbf{S} = \mathbf{B}_w \mathbf{WX} \tag{117}$$

Discovered independent components $\mathbf{s}$ are only estimates. Thus, inverse operations will also be just estimates. Once the mixing matrix $\mathbf{H}_w$ has been estimated, one can estimate the original mixing matrix $\mathbf{H}$, denoted by $\mathbf{H}_{esty}$, by applying the inverse (for $l = n$) (or pseudo-inverse) of the whitening operation on the new mixing matrix ($\mathbf{H}_w = \mathbf{W}\mathbf{H}$):

$$\mathbf{H}_{esty} = \mathbf{W}^{-1} \mathbf{H}_w \tag{118}$$

An approximation (**reconstruction**) of the original observed vector $\mathbf{x}$ from a given $\mathbf{s}$ can be computed as

$$\tilde{\mathbf{x}} = \mathbf{H}_{esty} \mathbf{s} \tag{119}$$

For the set of N independent component vectors arranged as columns of the $m \times N$ matrix $\mathbf{S}$, we can provide an estimation of the ICA model:

$$\tilde{\mathbf{X}} = \mathbf{H}_{esty} \mathbf{S} \tag{120}$$

Whitening can be realized using the PCA transformation. Here, the eigenvalues of the pattern covariance matrix $\mathbf{R}_{xx}$ are arranged in decreasing order $\lambda_1 \geq \lambda_2 \geq \cdots \geq \lambda_n$, and the corresponding eigenvectors, arranged in decreasing order, constitute rows of the whitening transformation matrix $\mathbf{W}$.

The dimensionality reduction of whitened patterns can be realized by considering only the first l eigenvectors in constructing the whitening matrix.

The ICA estimates of the **demixing** (separation) **matrix B** and **independent component vectors S** are based on a set of patterns $\mathbf{X}(\mathbf{S} = \mathbf{BX})$. Estimations can be realized using different criteria and algorithms. In a computationally efficient algorithm, the following maximization criterion has been used:

$$J(\tilde{\mathbf{s}}) = \sum_{i=1}^{m} \left| E\{\tilde{s}_i^4\} - 3[E\{\tilde{s}_i^2\}]^2 \right| \tag{121}$$

where m is number of independent and m-dimensional principal component vectors. The above equation corresponds to Fourth-order cumulant kurtosis. Based on the gradient operation of J, the independent component separation matrix guarantees that the independency characteristic will be preserved:

$$p(s_1, s_2, \cdots, s_m) = \Pi_{i=1}^{m} p_i[s_i] \tag{122}$$

where $p[]$ is the **probability density function**.

Let us consider two scalar-valued random variables y_1 and y_2. The variables y_1 and y_2 are **independent** if information about the value of y_1 does not provide any information about the value of y_2 (and vice versa). Practically, independence can be explained by probability densities. Let us denote by $p(y_1, y_2)$ the joint probability density function (pdf) of y_1 and y_2. Additionally, let us denote by $p_1(y_1)$ the pdf of y_1 when it is considered alone:

$$p_1(y_1) = \int p(y_1, y_2) dy_2 \tag{123}$$

and similarly for y_2. Now, one can state that y_1 and y_2 are independent if and only if the joint pdf can be expressed as

$$p(y_1, y_2) = p_1(y_1) p_2(y_2) \tag{124}$$

This definition can be generalized for n random variables (the joint probability density must be a product of n pdfs).

By virtue of that definition, for two functions, h_1 and h_2, we have

$$E\{h_1(y_1) h_2(y_2)\} = E\{h_1(y_1)\} E\{h_2(y_2)\} \tag{125}$$

We can write generally for n independent random variables that

$$p(y_1, y_2, \cdots, y_n) = p_1(y_1) p_2(y_2) \ldots p_n(y_n) \tag{126}$$

Consequently, we can write, for any functions f_i,

$$E\{f_1(y_1), \cdots, f_n(y_n)\} = E\{f_1(y_1)\} \cdots E\{f_n(y_n)\} \tag{127}$$

It is known that uncorrelated variables are only partly independent. A weaker definition of variable independence is **uncorrelatedness**. Two random variables v_1 and v_2 are uncorrelated, if their covariance is zero:

$$E\{y_1 y_2\} - E\{y_1\} E\{y_2\} = 0 \tag{128}$$

Independent variables are uncorrelated (independency implies uncorrelatedness); however, uncorrelated variables do not have to be independent. The majority of ICA methods provide uncorrelated estimates of the independent components.

ICA assumes that independent components must be nongaussian. The basis for estimating the ICA model is this assumption about nongaussianity.

The most natural measure of nongaussianity is kurtosis or the fourth-order cumulant. The **kurtosis** of random variable y defined as

$$\text{kurt}(y) = E\{y^4\} - 3(E\{y^2\})^2 \tag{129}$$

Assuming that y has unit variance, the above equation simplifies to $\text{kurt}(y) = E\{y^4\} - 3$. We see that kurtosis is a normalized version of the fourth moment $E\{y^4\}$. Since for a Gaussian variable y, the fourth moment is $3(E\{y^2\})$, therefore the kurtosis value is zero for a Gaussian random variable. For most nongaussian random variables, kurtosis is nonzero. Random variables that have a negative kurtosis are called **subgaussian**, and those with positive kurtosis are called **supergaussian**.

Usually nongaussianity is measured by the absolute (or square) value of kurtosis. This measure is equal to zero for a Gaussian variable, and greater than zero for most nongaussian random variables.

Kurtosis has been frequently used as a measure of nongaussianity in ICA and related fields. In practical computations, kurtosis can be estimated by using the fourth moment of the sample data.

The linearity property helps in the simplification of kurtosis:

$$\text{kurt}(x_1 + x_2) = \text{kurt}(x_1) + \text{kurt}(x_2)$$
$$\text{kurt}(\alpha x_1) = \alpha^4 \text{kurt}(x_1) \tag{130}$$

where x_1 and x_2 are two independent random variables, and α is a coefficient. Kurtosis values can be very sensitive to outliers, and thus they may be not robust measures of nongaussianity. As a remedy for this weakness, negentropy can be used for robust approximations of kurtosis.

A **negentropy** (which is based on the information-theoretic quantity of differential entropy) can be considered as a robust measure of nongaussianity. Thus, it can be used as a performance criterion in the optimal solution (estimation) of the ICA model.

We recall that entropy of a random variable can be considered as the degree of information given by observation of the variable. For more unpredicted, "random" (unstructured, disarrayed) variables, the entropy is larger. Formally, entropy is expressed as the coding length of the random variable. **Entropy** H_{entr} for a discrete random variable Y, with e_i as the possible values of Y, is defined as

$$E_{\text{entr}}(Y) = -\sum_i P(Y = e_i) \log P(Y = e_i) \tag{131}$$

For a continuous-valued random variable $\mathbf{y}$ with known probability density $f(\mathbf{y})$, the entropy (called **differential entropy**) H_{dentr} is defined as

$$H_{\text{dentr}}(\mathbf{y}) = -\int f(\mathbf{y}) \log f(\mathbf{y}) \, d\mathbf{y} \tag{132}$$

Information theory shows that a Gaussian variable has the largest entropy among all random variables of equal variance. This property predisposes entropy to be a good measure of nongaussianity. The Gaussian distribution is the "most random" (the least structured) among all distributions.

A good measure of nongaussianity is differential entropy, called **negentropy**:

$$J(\mathbf{y}) = H_{\text{entr}}(\mathbf{y}_{\text{gauss}}) - H_{\text{dentr}}(\mathbf{y}) \tag{133}$$

where $\mathbf{y}_{\text{gauss}}$ is a Gaussian random variable with the same covariance matrix as $\mathbf{y}$. Negentropy has nonnegative values, and it is zero for the Gaussian distribution of y. Furthermore, negentropy is invariant for invertible linear transformations. Computation of negentropy requires an estimation of the probability density of y, which makes the use of pure definition impractical.

For the random variable y (zero mean with unit variance), a computationally feasible approximation of negentropy using higher-order moments can be expressed as

$$J(y) \approx \frac{1}{12} E\{y^3\}^2 + \frac{1}{48} \text{kurt}(y)^2 \tag{134}$$

The applicability of this approximation may be limited due to the nonrobustness associated with kurtosis. Better approximations of negentropy are given by

$$J(y) \approx \sum_{i=1}^{r} a_i \Big[E\{G_i(y)\} - E\{G_i(\nu)\} \Big]^2 \tag{135}$$

where $G_i (i = 1, \cdots, r)$ are chosen r nonquadratic functions. Here a_i are positive constants and ν is a zero mean and unit variance Gaussian variable. Variable y is assumed to be of zero mean and unit variance. The above measure can be used to test nongaussianity. It is always nonnegative and is equal to zero if y has a Gaussian distribution.

If we use only one nonquadratic function G, the approximation results in

$$J(y) \approx \left[E\{G(y)\} - E\{G(v)\} \right]^2 \qquad (136)$$

for practically any nonquadratic function G. For $G(y) = y^4$, one can obtain exactly a kurtosis-based approximation. The following choices of G are recommended:

$$G_1(u) = \frac{1}{a_1} \log(\cosh\, a_1 u), \quad G_2(u) = -\exp(-u^2/2) \qquad (137)$$

where $1 \leq a_1 \leq 2$ is a constant.

Measures of nongaussianity can be used as performance criteria for ICA estimation.

Two fundamental applications of ICA are blind source separation and feature extraction.

In blind source separation, the observed values of $\mathbf{x}$ correspond to the realization of an m-dimensional discrete-time signal $\mathbf{x}(t)$, ($t = 1, 2, \ldots$). The components $s_i(t)$ are called source signals, which are usually original, uncorrupted signals or noise sources. Frequently such sources are statistically independent from each other, and thus the signals can be recovered from linear mixtures x_t by finding a transformation in which the transformed signals are as independent as possible, as in ICA.

In feature extraction based on ICA, the ith s_i independent component represents the ith feature in the observed data pattern vector $\mathbf{x}$.

2.5.4. Feature Extraction using ICA

In feature extraction, which is based on independent component analysis, one can consider an ith independent component s_i as the ith feature of the recognized object represented by the observed pattern $\mathbf{x}$. The feature pattern can be formed from m independent components of the observed data pattern. A procedure of pattern formation using ICA follows:

1. Extract n-element original feature patterns $\mathbf{x}$ from the recognized objects. Compose the original data set T containing N cases $\{\mathbf{x}_i^T, c_i\}$ that contain patterns and the corresponding class c_i. The original patterns are represented as the $n \times N$ pattern matrix $\mathbf{X}$ (composed with patterns as columns) and the corresponding categorical classes (represented as column $\mathbf{c}$).
2. Subtract a mean vector from each pattern.
3. Perform linear feature extraction by projection (and reduction) of the original patterns $\mathbf{x}$ from matrix $\mathbf{X}$ through ICA.

    (a) Whiten the data set $\mathbf{X}$, including dimensionality reduction. Obtain l-element decorrelated whitened patterns $\mathbf{x}_w$ gathered in an $l \times N$ matrix $\mathbf{X}_w$.
    (b) Estimate (for whitened patterns) the $m \times n$ unmixing matrix $\mathbf{B}$.
    (c) Estimate m independent components forming an m-element independent component pattern $\mathbf{x}_s$ for each pattern $\mathbf{x}_w$. This result can be realized through projections of patterns $\mathbf{x}_w$ into an m-element pattern $\mathbf{x}_s$ (in independent component space). This projection can be realized using the discovered unmixing matrix $\mathbf{B}$. The independent component patterns will have decorrelated and independent elements.
    (d) Form the final $m \times N$ pattern set $\mathbf{X}_s$ with N ICA patterns $\mathbf{x}_s$ as columns.

4. Construct the final class-labeled data set as the matrix $\mathbf{X}_{sf}$ which is formed by adding class column vector $\mathbf{c}$ to matrix $\mathbf{X}_s$.

ICA does not guarantee that the independent components selected first, as a feature vector, will be the most relevant for classification. In contrast to PCA, ICA does not provide an intrinsic order for the representation features of a recognized object (for example, an image). Thus, one cannot reduce an ICA pattern just by removing its trailing elements (which is possible for PCA patterns). Selecting features from independent components is possible through the application of **rough set** theory (see Chapter 5). Specifically, defined in rough sets, the computation of a reduct can be used to select some independent components-based features (attributes) constituting a reduct. These reduct-based independent component features will describe all concepts in a data set. The rough set method is used for finding reducts from discretized ICA patterns. The final pattern is formed from ICA patterns based on the selected reduct.

2.6. Vector Quantization and Compression

Data compression is particularly important for images, time series and speech, where the amount of data to be handled and stored is very large. When data sets contain a lot of redundant information, or when specific applications do not require high precision of data representation, then **lossy compression** might be a viable option among techniques of data storing or transmitting. Vector quantization is a powerful technique of lossy data representation (approximation) that is widely used in data compression and in classification.

Vector quantization(VQ) is a technique of representing (encoding) input data patterns by using a smaller finite set of **codevectors (template vectors, reference vectors)** that is a good approximation of the input pattern space. When an input data pattern is processed, it can be encoded (approximated) by the nearest codevector.

The objective of vector quantization for a given training data set T is to design (discover) the optimal **codebook**, containing a predetermined number of reference code vectors, which guarantee minimization of the chosen distortion metric for all encoded patterns from the data set. Each code vector in the codebook has an associated integer index used for referencing.

Once the optimal set of codevectors in the codebook has been computed, then it can be used for the encoding and decoding of input patterns (for example in data transmission). This process may result in substantial data compression.

Vector quantization belongs to **unsupervised learning** techniques. It uses an unlabeled training set in the design phase. The fundamental role played by unsupervised quantization involves **distortion measures**. Vector quantization approximates data and naturally will cause some information loss. One distortion metric is used to measure a distance (similarity) between an input pattern $\mathbf{x}$ and the codevector $\mathbf{w}$. It can also be used for the selection, for a given $\mathbf{x}$, of the nearest codevector $\mathbf{w}_j$ from a codebook. The other metric is used to determine the average distortion measure for all training patterns and a given codebook. This metric measures the quality of the entire codebook with respect to a given data set T, and it is instrumental in the design of an optimal ("best") codebook. The optimal codebook depends on the distortion metrics used. Examples of distortion metrics are discussed in the next sections.

The design process for vector quantization comprises the discovery of an optimal set of codevectors (a codebook) that best approximates a given training set of patterns. An optimal vector quantizer yields the minimum of the average distortion for a given data set. The processing of an input pattern using VQ relates to finding the reference vector that is closest to a pattern, according to the selected similarity metric.

2.6.1. Vector Quantization as a Clustering Process

Vector quantization, which is based on the optimal design of a codebook, can be seen as **clustering** (see Chapter 9). It corresponds to a partition of the input pattern space into encoding regions/clusters. Here, for each discovered cluster (encoding region), one can find the vector that is a representation of the cluster. For example, the mean vector of a cluster (a cluster **centroid**) can be selected as a representation of the entire cluster. Such a centroid vector can be considered as the codevector associated with a given cluster. This codevector can be added as one entry (codevector) in the codebook. A set of all indexed cluster centroids is treated as the codebook of codevectors. A codebook represents, in an approximate way, a given training set. Based on a given codebook, a training set of patterns, and selected distortion measures, one can find the optimal partition of the input pattern space into a number of encoding regions (clusters) with codevectors representing entire regions. For an optimal partition and an optimal codebook, the average distortion measure as an optimization criterion will be minimal. This means that a distortion measure between any input pattern **x** and its quantized representation as codevector **w** is minimal.

The procedure of deciding to which cluster an input pattern belongs can be realized by using selected distortion measures between a pattern and codevector. Here, a specific role is played by the Euclidean distance (or the squared Euclidean distance)

$$d(\mathbf{x}, \mathbf{y}) = \| \mathbf{x} - \mathbf{y} \|_2 = \sqrt{\sum_{i=1}^{n}(x_i - y_i)^2} \tag{138}$$

The optimal partition, which uses the Euclidean distance as a similarity measure (a distortion measure between **x** and **w**) in order to find the closest codevector to the input pattern, produces a special vector quantizer called the **Voronoi quantizer**. The Voronoi quantizer partitions the input pattern space into Voronoi cells (clusters, encoding regions). One Voronoi cell is represented by one corresponding codevector in a codebook. A cell contains all those input patterns **x** that are closer to the codevector $\mathbf{w}_j$ (in the sense of Euclidean distance) than to any other codevector. The data set patterns that fall inside a Voronoi cell will be assigned to one corresponding codevector.

This partition of the input pattern space is called **Voronoi tessellation**. The "rigid" boundaries between Voronoi cells are perpendicular bisector planes of lines that join codevectors in neighboring cells. Topologically, the Voronoi partition resembles a honeycomb.

A codebook contains all the codevectors representing a data set. The set of all encoding regions is called the partition of the input pattern space.

2.6.2. Vector Quantization – Design and Processing

Design of an optimal codebook is a minimization process. Let us consider a given data set T (training set) containing N n-dimensional continuous-valued input patterns $\mathbf{x} \in \mathbb{R}^n$. Vector quantization is the process of categorization of input patterns **x** into M distinct **encoding regions** (**cells, clusters**) $C_1, C_2, \cdots, C_M$. Each encoding region C_i is represented by one n-dimensional real-valued codevector $\mathbf{w}_i \in \mathbb{R}^n (i = 1, 2, \cdots, M)$. A set of codevectors will be denoted by $W = \{\mathbf{w}_1, \mathbf{w}_2, \cdots, \mathbf{w}_M\}$ or, in $M \times n$ matrix form, as

$$\mathbf{W} = \begin{bmatrix} \mathbf{w}_1^T \\ \mathbf{w}_2^T \\ \vdots \\ \mathbf{w}_M^T \end{bmatrix} \tag{139}$$

The indexed set of M codevectors $\mathbf{w}_i$ ($i = 1, 2, \cdots, M$) is called the **codebook** $W_c = \{(i, \mathbf{w}_i)\}_1^M$ of the quantization. Table 7.3 shows the form of the codebook.

The vector quantization for a given data set contains two steps:

1. Designing the optimal (i.e., having minimum average distortion) codebook $W_c = \{(i, \mathbf{w}_i)\}_{i=1}^M$, which is an indexed set of M reference codevectors representing M entirely distinct cells in the input pattern space. A codebook of codevectors is designed for a given training set T of n-dimensional patterns $\mathbf{x} \in \mathbb{R}^n$.

Table 7.3. The code book.

Index	Reference codevector w
1	$\mathbf{w}_1$
2	$\mathbf{w}_2$
⋮	⋮
M	$\mathbf{w}_M$

2. Processing input patterns: encoding/decoding.

Vector quantization can also be seen as a combination of two functions: vector **encoding** and vector **decoding**. During the encoding process, every input pattern vector $\mathbf{x}$ is compared with each of the M codevectors in a codebook, according to some distortion measure $d(\mathbf{x}, \mathbf{w}_i)$ ($i=1, 2, \ldots, M$). The codevector $\mathbf{w}_j$ that yields the minimum distortion is selected as a quantized representation of the input pattern, and the index j associated with this codevector is generated for transmission or storage. In other words, the representative codevector for the pattern $\mathbf{x}$ (an encoding vector) is determined to be the closest codevector $\mathbf{w}_j$ in the sense of a given pattern distortion metric (for example, the Euclidean distance). This codevector is a unique representation of all input patterns from the encoding region represented by the considered codevector.

If indexed codevectors (a codebook) are known in advance, then instead of representing the pattern $\mathbf{x}$ by the whole corresponding codevector $\mathbf{w}_j$, one can provide only the the codevector index j (the encoding region number j into which the pattern falls). This encoding region number can then be stored or transmitted instead of the pattern (as the compressed representation of the pattern).

In the decoding phase, a decoder has a copy of the codebook and uses the received index to find the corresponding codeword. The decoder then outputs the codevector $\mathbf{w}_j$ as a substitute for the original vector $\mathbf{x}$. Figure 7.5 shows the basic quantization model.

In other words, determining an encoding region for a given input pattern $\mathbf{x}$ is provided by choosing the "winning" codebook vector $\mathbf{w}_j$, representing the jth encoding region C_j, which has the smallest value of the distortion metric used to determine the distance of the input pattern $\mathbf{x}$ to the codevector $\mathbf{w}$:

$$\text{winning } j^{\text{th}} \text{ codevector}: \mathbf{w}_j \min_{i=1,2,\cdots,M} d(\mathbf{x}, \mathbf{w}_i) \qquad (140)$$

The pattern $\mathbf{x}$ thus belongs to the same region C_j as the "winning" closest codebook vector $\mathbf{w}_j$.

One of the most important applications of competitive learning is data compression through vector quantization. The idea of applying competitive learning for vector quantization, which is based on a winner selection determined by the distortion (distance) measure, is quite natural and straightforward.

Vector quantization is understood as the process of dividing the original input pattern space into a number of distinct regions (Voronoi cells). To each region, a codevector $\mathbf{w}_i$ ($i = 1, 2, \cdots, M$))

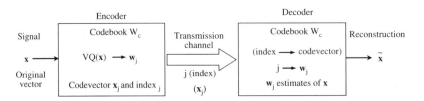

Figure 7.5. Encoding and decoding.

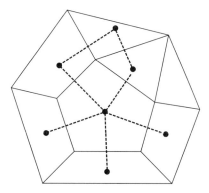

Figure 7.6. Voronoi cells and Voronoi tesselation.

is assigned. This codevector represents all the patterns from the entire region (cell), as indicated in Figure 7.6.

In other words, a vector quantization maps N n-dimensional patterns $\mathbf{x}$ from the input pattern space $\mathbb{R}^n$ into a finite set of M codevectors $\mathbf{w}_i \in \mathbb{R}^n$ belonging to a codebook. The nearest-neighbors region (**Voronoi cell**) is associated with each $\mathbf{w}_i$ codevector. A region is defined as

$$C_i = \{\mathbf{x} \in \mathbb{R}^n : d(\mathbf{x}, \mathbf{w}_i) \leq d(\mathbf{x}, \mathbf{w}_j), \quad \text{for all} \quad i \neq j\} \tag{141}$$

The set of Voronoi regions partitions the entire space $\mathbb{R}^n$ such that

$$\bigcup_i C_i = \mathbf{R}^n \quad \text{and} \quad C_i \cap C_j = \emptyset \quad \text{for} \quad i \neq j \tag{142}$$

Vector quantization is a technique that partitions the input pattern space into M distinct regions (cells), each with a codevector representing the entire region. Vector quantization maps any input pattern vector $\mathbf{x} \in \mathbb{R}^n$ from an n-dimensional input pattern space $\mathbb{R}^n$ into a finite set of M n-dimensional codevectors $\mathbf{w}_i \in \mathbb{R}^n$ $(i = 1, 2, \cdots, M)$:

$$VQ : \mathbf{x} \in \mathbb{R}^n \rightarrow W \tag{143}$$

where $W = \{\mathbf{w}_i\}_1^M$ is the set of M codevectors. The set of indexed codevectors $\mathbf{w}_c = \{i, \mathbf{w}_i\}_1^M$ constitutes a **codebook** of vector quantization.

Since the codebook has M entries, the quantity R, called the **rate** of the quantizer, is defined as

$$R = \log_2 M \tag{144}$$

measured in bits per input pattern vector. Since the input pattern has n components, it follows that

$$r = \frac{R}{n} \tag{145}$$

is the rate in bits per vector component.

In the topological interpretation, the knowledge of the mapping function VQ determines a partition of input patterns into M subsets (cells):

$$C_i = \{\mathbf{x} \in \mathbb{R}^n : VQ(x) = \mathbf{w}_i\}, \quad (i = 1, \cdots, M) \tag{146}$$

A cell C_i in $\mathbb{R}^n$ is associated with each codevector $\mathbf{w}_i$ of the vector quantizer, that is, $C_i = \{\mathbf{x} \in \mathbb{R}^n : VQ(x) = \mathbf{w}_i\}$. It is sometime called the inverse image of a codevector $\mathbf{w}_i$ under mapping VQ and is denoted by $C_i = VQ^{-1}(\mathbf{w}_i)$.

From the definition of a cell, it follows that

$$\bigcup_i C_i = \mathbb{R}^n \text{ and } C_i \cap C_j = \emptyset \text{ for } i \neq j \tag{147}$$

which indicates that the cells form a partition of $\mathbb{R}^n$.

When the input pattern $\mathbf{x}$ is quantized and represented as a codevector $\mathbf{w}$ ($\mathbf{w} = VQ(\mathbf{x})$), a quantization error results and a distortion measure can be defined between $\mathbf{x}$ and $\mathbf{w}$. As with the measurement of a distance between two vectors (see Appendix A), several distortion measures have been proposed for vector quantization algorithms, including **Euclidean distance**, the **Minkowski norm**, the **weighted-squares distortion**

$$d(\mathbf{x}, \mathbf{y}) = \sum_{i=1}^{n} a_i |x_i - y_i|^2, \quad a_i \geq 0 \tag{148}$$

the **Mahalanobis distortion**, and the **general quadratic distortion**, defined as

$$d(\mathbf{x}, \mathbf{y}) = (\mathbf{x} - \mathbf{y}) \mathbf{B} (\mathbf{x} - \mathbf{y})^T = \sum_{i=1}^{n} \sum_{j=1}^{n} b_{ij} (x_i - y_i)(x_j - y_j) \tag{149}$$

where $\mathbf{B}$ is an $n \times n$ positive definite symmetric matrix.

These distortion measures are computed for two vectors, and they depend on the error vector $(\mathbf{x} - \mathbf{y})$. They are called **difference distortion measures**.

It is also possible to define the following average distortion measure for a given set $T = \{\mathbf{x}_i\}_{i=1}^{N}$ of N input pattern vectors, a given set of M codevectors $W = \{\mathbf{w}_1, \mathbf{w}_2, \cdots, \mathbf{w}_M\}$, and a given partition $P(W)$ of the input vector space on cells:

$$d_a = d_a(W, P(W)) = \frac{1}{N} \sum_{j=1}^{N} \min_{\mathbf{w} \in W} d(\mathbf{x}_j, \mathbf{w}) \tag{150}$$

where $d(\mathbf{x}_j, \mathbf{w})$ is the distortion measure for two vectors.

One can also define an average distortion measure as

$$D_a = \frac{1}{N} \sum_{i=1}^{N} d(\mathbf{x}_i, VQ(\mathbf{x}_i)) = \frac{1}{N} \sum_{i=1}^{N} \| \mathbf{x}_i - VQ(\mathbf{x}_i) \|_2^2 \tag{151}$$

or as

$$D_a = \frac{1}{M} D_i \tag{152}$$

where D_i is the total distortion for cell C_i:

$$D_i = \sum_{\mathbf{x} \in C_i} d(\mathbf{x}, \mathbf{w}_i) \tag{153}$$

where $VQ(\mathbf{x}_i)$ denotes a mapping of the input pattern $\mathbf{x}_i$ into a codevector $\mathbf{w} = VQ(\mathbf{x}_i)$.

The goal of designing an optimal quantizer is to find a codebook of M codevectors such that the average distortion measure in encoding all patterns from the training set is minimal. To design an M-level codebook, the n-dimensional input pattern space is partitioned into M distinct regions or cells $\{C_i : i = 1, 2, \cdots, M\}$ and a codevector $\mathbf{w}_i$ is associated with each cell C_i. Figure 7.7 shows the partition.

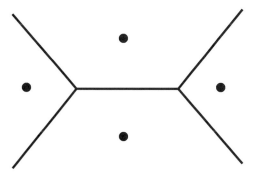

Figure 7.7. Partition of a pattern space into encoding regions (cells).

The quantizer then assigns the $\mathbf{w}_i$ if an input vector $\mathbf{x}$ belongs to C_i. Two necessary conditions (criteria) must be fulfilled in order to obtain an optimal codebook: the nearest neighbor condition and the centroid condition.

The first condition is that the partition of the input pattern space should be obtained by using a minimum-distortion or **nearest-neighbor selection rule**:

$$VQ(\mathbf{x}) = \mathbf{w}_i \quad \text{iff} \quad d(\mathbf{x}, \mathbf{w}_i) \leq d(\mathbf{x}, \mathbf{w}_j), \; j \neq i, \; 1 \leq j \leq M \tag{154}$$

Here, the quantizer selects for $\mathbf{x}$ the codevector $\mathbf{w}_j$ that results in the minimum distortion with respect to $\mathbf{x}$. In other words, a cell C_i should consist of all patterns that are closer to $\mathbf{w}_i$ than any of the other codevectors:

$$\{\mathbf{x} \in T : \| \mathbf{x} - \mathbf{w}_i \|_2^2 \leq \| \mathbf{x} - \mathbf{w}_j \|_2^2 \}, \quad j = 1, \cdots M \tag{155}$$

The nearest neighbor conditions permits the realization of an optimal partition.

The second necessary condition to achieve an optimal codebook is that each codevector $\mathbf{w}_i$ is chosen to minimize the average distortion in cell C_i. That is, for each cell there exists a minimum distortion codevector for which

$$E[d(\mathbf{x}, \mathbf{w}) \mid \mathbf{x} \in C_i] = \min_j \left(E[d(\mathbf{x}, \mathbf{w}_j)] \mid \mathbf{x} \in C_i \right) \tag{156}$$

where E denotes the expectation with respect to the underlying probability distribution

$$E[d(\mathbf{x}, \mathbf{w}) \mid \mathbf{x} \in C_i] = \int_{\mathbf{x} \in C_i} d(\mathbf{x}, \mathbf{w}) \, p(\mathbf{x}) \, d\mathbf{x} \tag{157}$$

The codevector $\mathbf{w}_i$ is called the **centroid** of the cell C_i:

$$\mathbf{w}_i = \text{cent}(C_i) \tag{158}$$

One can also say that codevector $\mathbf{w}_i$ should be the average of all those training patterns that are in the encoding region (a cell):

$$\mathbf{w}_i = \frac{\sum_{\mathbf{x} \in C_i} \mathbf{x}}{\sum_{\mathbf{x} \in C_i} 1}, \quad i = 1, \cdots, M \tag{159}$$

Computing the centroid for a particular region or cell depends mainly on the definition of the distortion measure. The cells are known as **nearest neighbor cells** or Voronoi cells. Generally, the design of an optimal codebook requires knowledge of the probability distribution of the

2. Feature Extraction

source pattern vectors. However, in most cases the distribution is not known and the codebook determination usually involves training from examples. Usually, a set of N training vectors $T = \{x_j\}_{j=1}^{N}$ is given that is representative of the data the quantizer is more likely to encounter in practice.

We briefly describe two vector quantization (clustering) algorithms:

- the generalized Lloyd algorithm (LBG centroid algorithm), and
- the iterative cluster splitting algorithm.

The idea of the **LBG centroid algorithm** is as follows. The algorithm starts with an initial codebook selected. For the given codebook (with its set of codevectors) the minimum distortion is iteratively improved until a local minimum is reached. In each iteration, a partition is realized, and then for a given partition the optimal set of codevectors is found. This can be done in two steps. In the first step, for each iteration, each pattern is mapped to the nearest codevector (according to the defined distortion measure between the two patterns) in the current codebook. In the second step, all the codevectors of the code book are recalculated as the centroids of new encoding regions of the new partitions. The algorithm continues as long as improvement is achieved. For a given set of codevectors and a data set of patterns, the partition of the input pattern space that minimizes the average distortion measure is achieved by mapping each pattern $\mathbf{x}_i$ ($i = 1, \cdots, N$) to the codevector $\mathbf{w}_j$ ($j = 1, \cdots, M$), for which the distortion $d(\mathbf{x}_i, \mathbf{w}_j)$ is the minimum over all $\mathbf{w}_j$. The centroid vector quantization clustering algorithm (which uses a squared Euclidean distortion metric) can be described as follows:

Algorithm:
Given: A training set $T = \{\mathbf{x}_i\}_1^N$ of N unlabeled pattern vectors, given number M of codebook reference vectors (quantization levels), and distortion measure threshold $\epsilon \geq 0$.

1. Select randomly an initial setting of the M-level codebook containing a set of M codevectors $\hat{W}^0 = \{\hat{\mathbf{w}}_i^0\}_{i=1}^{M}$.
2. Select a distortion measure threshold $\epsilon \geq 0$.
3. Set the previous average distortion measure $d_a^0 = 0$.
4. Set the iteration step number $k = 1$.
5. Given $\hat{W}^k = \{\hat{\mathbf{w}}_i^k\}_{i=1}^{M}$, find the minimum distortion partition $P(\hat{W}^k) = \{C_i\}_{i=1}^{M}$ of the training set patterns (based on the nearest-neighbor condition)

$$\mathbf{x}_j \in C_i \quad \text{if} \quad d(\mathbf{x}_j, \mathbf{w}_i) \leq d(\mathbf{x}_j, \mathbf{w}_l) \quad \text{for all} \quad l = 1, 2, \cdots, M \tag{160}$$

where $d(\mathbf{x}, \mathbf{w})$ is considered to be the distortion measure. If the distortion measures are the same for a few codevectors, then the assigning of patterns to cells is random.

6. Compute the average distortion for a given codebook of codevectors $\hat{W}^k$ and partition $P(W^k)$ at step k:

$$d_a^k = d_a^k[\hat{W}^k, P(\hat{W}^k)] = \frac{1}{N}\sum_{i=1}^{N} d(\mathbf{x}_i, VQ(\mathbf{x}_i)) = \frac{1}{N}\sum_{i=1}^{N} \|\mathbf{x}_i - VQ(\mathbf{x}_i)\|_2^2 \tag{161}$$

7. If the average distortion measure d_a^k at iteration step k relative to d_a^{k-1} is below a certain threshold

$$\frac{|d_a^{k-1} - d_a^k|}{d_a^k} \leq \epsilon \tag{162}$$

then stop the iteration, with $\hat{W}^k$ being the final codebook of the codevector set. Otherwise, proceed to the next step.

8. Find the optimal codebook (codevectors) (based on the centroid condition) $\hat{W}^{k+1}(P(\hat{W}^k)) = \{\hat{\mathbf{w}}^{k+1}(C_i) : i = 1, 2, \cdots, M\}$ for $P(\hat{W}^k)$. For the squared-error distortion measure, $\hat{\mathbf{w}}^{k+1}(C_i)$ is the Euclidean center of gravity (or centroid) for a cell C_i and given by the equation

$$\hat{\mathbf{w}}^{k+1}(C_i) = \frac{1}{\|C_i\|} \sum_{\mathbf{x}_j \in C_i} \mathbf{x}_j \qquad (163)$$

Here $\hat{\mathbf{w}}^{k+1}(C_i)$ is averaged in a componentwise manner, where $\|C_i\|$ denotes the number of training vectors in the cell C_i. If $\|C_i\| = 0$, set $\hat{\mathbf{w}}^{k+1}(C_i) = \mathbf{w}_i^k$ (the previous codevector).

Set $k = k + 1$. Continue the iteration from step 5.

In the above algorithm, the initial codebook vector set $\hat{W}^0$ must be guessed. Frequently, the M vectors randomly chosen from the training set T are selected as initial values for the codevectors.

Alternatively, an initial codebook vector set can be obtained by computing the centroid of the training set vectors and dividing this vector into two vectors. Each of these two vectors is then split into two new codevectors, and this process continues until the initial M level codebook vector set has been created. Each vector $\mathbf{w}_i$ is split by adding a fixed perturbation vector ϵ, thereby producing two new codevectors $\mathbf{w}_i + \epsilon$ and $\mathbf{w}_i - \epsilon$.

The top-down **cluster (region) splitting algorithm** starts with a single cluster including all patterns of the training set T, with one codevector $\mathbf{w}_1$ computed as a mean of all patterns belonging to the first cluster. Then new clusters are created (one at a time) by splitting the existing clusters. This can be done by "splitting" the codevector $\mathbf{w}_1$ into two close codevectors $\mathbf{w}_1 + \epsilon$ and $\mathbf{w}_1 - \epsilon$, where ϵ is an n-dimensional vector with small values ϵ of all elements. Such new codevectors are considered in the following iterations. For a given codevectors, a two-step procedure is realized: finding the minimum average distortion partitions, and finding the optimal set of code vectors (codebook) for a given partition. The cluster splitting process continues until the required number of clusters is reached.

The performance of VQ is frequently measured as the **signal-to-distortion ratio** (SDR):

$$SDR = 10 \, \log_{10} \frac{\sigma^2}{D_{\text{average}}} \quad [db] \qquad (164)$$

computed in dB. Here σ^2 is the source variance and D_{average} is the average squared-error distortion. Larger values correspond to better performance.

Example: We provided vector quantization of the black and white version of the Halinka image (Figure 7.8(a)) composed with $I \times J = 720 \times 680$ pixels of 256 gray levels. All gray levels are represented by an integer value included in $[0, 255]$.

The image shown in Figure 7.8(a) was divided into $N = I/(n_b) \times J/(n_b) = 180 \times 170 = 30600$ blocks, each of $n_b \times n_b = 4 \times 4$ pixels, where $n_b = 4$ is the number of rows in a block (equal to the number of columns). Each block has been unfolded (row by row) into the resulting $n = n_b \times n_b = 4 \times 4 = 16$-element block vector $\mathbf{x}_b$ (a column vector). A sequence of n-element ($n = 16$) block vectors, constituted with a sequence of considered image blocks (from left to right, row by row) forms the unsupervised training set T_{tra} containing $N = 30600$ block vectors. The number of codevectors in the codebook was set to $M = 64$.

For the training set T_{tra} and a codebook size equal to $M = 64$, the optimal codebook was found using the generalized Lloyd algorithm. Figure 7.8(b) shows the reconstructed black and white version of the Halinka image.

(a)   (b)

Figure 7.8. Original (a) and reconstructed (b) Halinka images.

In image compression, the so-called **peak signal-to-noise ratio** (PSNR) is frequently used to evaluate the quality of the resulting images after the quantization process. The *PSNR* is defined as

$$PSNR = 10 \, \log_{10} \frac{\text{MAX}^2}{\frac{1}{IJ} \sum_{i=0}^{i=I-1} \sum_{j=0}^{i=J-1} \left(f(i,j) - \tilde{f}(i,j) \right)^2} \qquad (165)$$

where $f(i,j)$ and $\tilde{f}(i,j)$ are, respectively, the gray level of the original image and of the reconstructed one and I and J are the number of rows and columns of an image. Here, MAX denotes the maximum value of an image intensity. For example, for an 8-bit image, $\text{MAX} = 2^8 - 1 = 255$.

2.6.3. Complexity Issues

The computational and storage costs impose a very real and practical limitation on the applicability of vector quantization. The computational cost refers to the amount of computation needed per unit of time, and the storage cost refers to the memory size required to store the codebook for a particular application.

With a basic vector quantizer designed, each n-dimensional input vector **x** can be encoded by computing the distortion measure between the vector **x** and each of the codebook codevectors, keeping track of the one with minimum distortion and continuing until every codevector has been tested. For a quantizer with a codebook of size M, the number of distortion computations is M, and for a training set containing N patterns, each distortion computation requires N multiply-add operations for the squared-error distortion (other distortion measures can have higher computational demands). Therefore, the computational cost κ for encoding each input vector **x** is

$$\kappa = M\,N \qquad (166)$$

If we encode each codevector into $R = \log_2 M$ bits, since $R = rn$ (where $r = \frac{R}{n}$ is a rate in bits per vector component) then

$$\kappa = N 2^{rn} \qquad (167)$$

The computational cost grows exponentially with the number of pattern dimensions and the number of bits per dimension. The storage cost can be measured assuming one storage location per vector component

$$\mu = n 2^{rn} \qquad (168)$$

Again, the storage cost grows exponentially in the number of dimensions and the number of bits per dimension.

2.7. Learning Vector Quantization

Kohonen proposed the so-called **Learning Vector quantization (LVQ)** algorithm (and its neural network implementation), which combines **supervised learning** and vector quantization. The algorithm guarantees the performance of vector quantization, but in addition uses the idea of **competitive learning** to select the winning neuron, whose weights will be adjusted and punish-reward learning utilized for the direction of weight adjustment. In a nutshell, assuming that Euclidean distance is used for pattern similarity measures, LVQ provides fine tuning (moving) of the rigid boundaries of the Voronoi tessellation regions. This technique utilizes additional information about regions, which comes from the class labels associated with data set patterns. The learning vector quantization algorithm assumes that the training set $T = \{(\mathbf{x}_i, C_i)\}_{i=1}^{N}$ of N n-element input pattern vectors $\mathbf{x} \in \mathbb{R}^n$ labeled by corresponding l categorical classes C_i is available for supervised learning. Generally, it is also assumed that the probability of classes is known as well as the probability density of patterns $p(\mathbf{x})$.

From this perspective, the LVQ algorithm can be seen as a classifier design based on a given supervised training set. However, the LVQ algorithm with Kohonen learning also provides vector quantization of input patterns from the population represented by a given, representative, labeled training set. This means that the LVQ algorithm guarantees the mapping of input patterns from the input pattern space $\mathbb{R}^n$ into one of the reference code vectors from the limited-size codebook $\left(W_c = \{(i, \mathbf{w}_i)\}_{i=1}^{M}\right)$. This is an approximation of input pattern vectors by their quantized values.

In the LVQ algorithm, for the purpose of precise vector quantization approximation, usually several reference code vectors of the codebook are assigned to represent the labeling patterns of each class C_i from the training set

$$W_{C_i} = \{\mathbf{w}_j\} \text{ for all } j, \; \mathbf{w}_j \text{ representing class } C_i \tag{169}$$

The pattern $\mathbf{x}$ is considered to belong to the same class to which the nearest reference code vector from the codebook $\mathbf{w}_j$ belongs. For vector $\mathbf{x}$ and for the Euclidean distance metric, we can write

$$j^{\text{th}} \text{ nearest reference vector } \mathbf{w}_j \; \min_{i=1,2,\cdots,M} ||\mathbf{x} - \mathbf{w}_i|| \tag{170}$$

Kohonen proposed a competitive supervised learning algorithm that approximately minimizes misclassification errors of vector quantization, stated as the **nearest-neighbors classification**.

During proposed supervised learning with a punish-reward idea of weights adjustment, the optimal reference vectors $\mathbf{w}_i$ ($i = 1, 2, \cdots, M$) of the codebook may be found as the asymptotic values of the following competitive learning technique.

First, for a given input pattern $\mathbf{x}$ belonging to the class C_l and a previous value $\{\mathbf{w}_j^k\}_{i=1}^{M}$ of the codebook reference code vectors, the code vector that is nearest to the input pattern $\mathbf{x}$ is selected as the winner of the competition:

$$j^{\text{th}} \text{ nearest code vector } \mathbf{w}_j \; \min_{i=1,2,\cdots,M} ||\mathbf{x} - \mathbf{w}_i|| \tag{171}$$

This reference code vector belongs to a certain class C_r. Then only that j^{th} code vector $\mathbf{w}_j$ nearest to $\mathbf{x}$ will be adjusted in the following way:

$$\begin{aligned}
\mathbf{w}_j^{k+1} &= \mathbf{w}_j^k + \alpha(k)[\mathbf{x} - \mathbf{w}_j] \quad \text{if} \quad C_l = C_r \\
\mathbf{w}_j^{k+1} &= \mathbf{w}_j^k - \alpha(k)[\mathbf{x} - \mathbf{w}_j] \quad \text{if} \quad C_l \neq C_r \\
\mathbf{w}_i^{k+1} &= \mathbf{w}_i^k \quad \text{if} \quad i \neq j
\end{aligned} \tag{172}$$

where $0 < \alpha(k) < 1$ is the learning rate. The above weights adjustment is based on the "winner-take-all" and the punish-reward idea. Only the reference code vector $\mathbf{w}_j$ nearest to the pattern $\mathbf{x}$

is adjusted. If the class C_r of the reference code vector $\mathbf{w}_j$ is the same as the class C_l of the input pattern $\mathbf{x}$, then this winning reference code vector is rewarded by a positive adjustment in the direction of improving the match with vector $\mathbf{x}$:

$$\mathbf{w}_j^{k+1} = \mathbf{w}_j^k + \alpha(k)[\mathbf{x} - \mathbf{w}_j] \qquad (173)$$

If the class of the winning reference code vector and the class of the input pattern are different, than the reference code vector is punished:

$$\mathbf{w}_j^{k+1} = \mathbf{w}_j^k - \alpha(k)[\mathbf{x} - \mathbf{w}_j] \qquad (174)$$

Reference code vectors other than the jth reference code vectors nearest to the pattern $\mathbf{x}$ remain unchanged. This learning scheme for vector quantization, which adjusts only the reference vector nearest to $\mathbf{x}$, is called **LVQ1**.

The main reason for the punish-reward learning scheme is to minimize the number of misclassifications. This scheme also ensures that the reference code vectors will be pulled away from the pattern overlap areas when misclassifications persist.

Vector quantization through supervised learning allows more fine tuning of the decision surfaces between classes (the borders between Voronoi cells represented by reference vectors). This outcome is understandable in the light of the idea of pattern classification, where the decision surface between pattern classes is most important and not the precision of the pattern distribution within classes.

In the above learning rule, an $\alpha(k)$ is a learning rate. The learning rate values are not critical for learning (as long as the number of learning steps is sufficiently large). The value may be set as a small constant or as a decreasing function with learning time (with a starting value for examples smaller than 0.1). Frequently a linear decreasing function of learning time is chosen $-\alpha(k) = \alpha_{\max}\left(1 - \frac{k}{N_l}\right) -$ where N_l is the maximal number of learning steps.

2.7.1. Learning Vector Quantization Neural Network

The feedforward static competitive learning vector quantization neural network LVQ1 consists of an input and an output layer fully connected through weights (see Figure 7.9). The weightless input layer is connected via weights with the output layer. The weightless n neurons of the input layer receive only the n-dimensional input pattern $\mathbf{x} \in \mathbb{R}^n$. The number of input layer neurons equals the size of the input pattern $\mathbf{x} \in \mathbb{R}^n$. The subsequent output layer with M neurons is fully connected with the input layer neurons via weights. The number of output layer neurons is equal

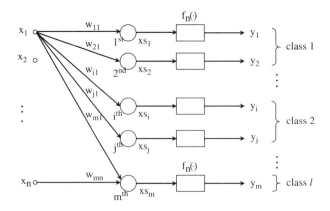

Figure 7.9. LVQ neural network.

to the size (or number of reference code vectors) of codebook M. The connection weights are described by the $n \times M$ weights matrix

$$\mathbf{W} = [\mathbf{w}_i], \quad i = 1, 2, \cdots, M \tag{175}$$

where $\mathbf{w}_i = [w_{i1}, w_{i2}, \cdots, w_{iM}]$ gives the vector weights of the ith output neuron. The weights vectors $\mathbf{w}_i$ of each ith output neuron represent the reference code vector of the codebook. Each output neuron represents a certain class of the training set.

The type of output activation function of the output neurons is not critically important and may be assumed to be an identity function $y = f_h(xs) = xs$, where xs is the intermediate value of an output neuron. Each ith $(i = 1, 2, \cdots, M)$ output neuron, with value y_i, has a known class $C_{y_i} \in \{C_1, C_2, \cdots, C_l\}$ that represents it.

Several output neurons $i, j, \cdots, s$ representing the distinct reference vectors $\{\mathbf{w}_i, \mathbf{w}_j, \cdots, \mathbf{w}_s\}$ may be assigned to the same class C_g. Generally, we have $M \geq M$ in order to provide a better approximation of the input pattern space by reference vectors.

The processing model for immediate outputs $\mathbf{xs}$ and outputs $\mathbf{y}$ of the output neurons can be written as

$$\mathbf{xs} = \mathbf{Wx}$$
$$\mathbf{y} = F(\mathbf{xs}) \tag{176}$$

or written in the scalar form

$$xs_i = \mathbf{w}_i \mathbf{x} = \sum_{l=1}^{n} w_{il} x_l, \quad i = 1, 2, \ldots, M$$
$$y_i = f_h(xs_i) = xs_i \ i = 1, 2, \cdots, M \tag{177}$$

where $\mathbf{W}$ is the $M \times n$ weights matrix.

2.7.2. Learning Algorithm for LVQ1 Neural Network

Vector quantization by the LVQ1 neural network requires

– Supervised learning from the class-labeled patterns of training set T. This learning is in the design of the optimal codebook $W_c = \{(i, \mathbf{w})\}_{i=1}^{M}$ composed of reference code vectors indexed by i, that is, $W = \{\mathbf{w}_i\}_{i=1}^{M}$, which are the asymptotic values of the learned vector weights of the output neurons.
– Processing of the input patterns by the already-designed neural network. For a given input pattern $\mathbf{x}$, the most responsive output neurons will be declared as the winners and will denote the reference vector, which will represent input pattern.

Given: The training set $T = \{\mathbf{x}_i, C_{x_i}\}_{i=1}^{L}$ containing N class-labeled patterns $\mathbf{x} \in \mathbb{R}^n$, the number of learning steps N_l, and the learning rate function $\alpha(k)$ (for example, $\alpha(k) = \alpha_{\max}\left(1 - \frac{k}{N_l}\right)$).

1. Select randomly M input patterns from the training set T and assign them as the starting values of the codebook reference vectors

$$\mathbf{w}_i^0 = \mathbf{x}_i, \quad i = 1, 2, \cdots, M \tag{178}$$

The remaining patterns of the training sets will be used for learning.
2. Present a randomly selected input pattern $\mathbf{x}$ to the network. This pattern will represent a known class C_l.

2. Feature Extraction

3. Adjust the learning rate $\alpha(k)$ for the kth learning steps.
4. Select the jth winning neuron. The winner is the neuron whose weights vector (representing the current codebook reference vector $\mathbf{w}_j = \mathbf{x}_{c,j}$) best matches (is nearest) the input vector $\mathbf{x}_i$ according to the minimal Euclidean distance:

$$j^{th} \text{ nearest reference vector } \mathbf{w}_j \quad \min_{i=1,2,\ldots,M} ||\mathbf{x} - \mathbf{w}_i|| \tag{179}$$

This reference vector $\mathbf{w}_j$ belongs to a certain class C_r.

5. Adjust only the jth winning neuron's weights $\mathbf{w}_j$, related to the jth reference vector of the codebook, by an amount depending on whether the class C_l of input pattern $\mathbf{x}$ is in the same as the class C_r of the jth winning neuron:

(a)
$$\mathbf{w}_j^{k+1} = \mathbf{w}_j^k + \alpha(k)[\mathbf{x} - \mathbf{w}_j] \quad \text{if} \quad C_l = C_r$$
$$\mathbf{w}_j^{k+1} = \mathbf{w}_j^k - \alpha(k)[\mathbf{x} - \mathbf{w}_j] \quad \text{if} \quad C_l \neq C_r$$

(b)
$$\mathbf{w}_i^{k+1} = \mathbf{w}_i^k \quad \text{if} \quad i \neq j \tag{180}$$

where $0 < \alpha(k) < 1$ is the learning rate. The above weights adjustments is based on the "Winner-Take-All" and punish-reward ideas.

6. Test stopping condition. This condition may be specified by reaching a fixed maximal number of iterations $k \leq N_l$ or by the learning rate reaching a minimal value $\alpha_{\min}$. If the stopping condition is not met, then continue iterations from step 2. If the stopping condition is satisfied, stop the iterations.

Result: A weight vector $\mathbf{w}_i (i = 1, 2, \cdots, M)$ that constitutes the reference vectors of the codebook.

Modification of the learning rate is frequently realized after all the patterns from the training sets have been learned (after learning of one epoch).

During the learning computation of the network output neurons, intermediate outputs xs_i ($i = 1, 2, \cdots, M$) are not required.

The inputs are processed using the following algorithm.

Given: The network architecture with an n-neuron input layer fully connected via weights with M neurons of the output layer. The weights vectors of the output neurons represent M stored reference code vectors of the codebook. The learned weights matrix $\mathbf{W} = [\mathbf{w}_i](i = 1, 2, \cdots, M)$ whose vectors $\mathbf{w}_i$ represent the reference vectors of a codebook is also given.

For the given input pattern $\mathbf{x}$, the LVQ1 neural network provides the following forward computations.

1. The input-layer neurons receive the input pattern $\mathbf{x}$.
2. The intermediate values xs of the output neurons are computed by the matrix equations

$$\mathbf{xs} = \mathbf{Wx} \tag{181}$$

where

$$xs_i = \sum_{j=1}^{n} w_{ij} x_j, \quad i = 1, 2, \cdots, M \tag{182}$$

3. The final values of the output neurons are computed through the output activation functions f_h

$$y_i = f_h(xs_i) \qquad (183)$$

or in the vector form

$$\mathbf{y} = \mathbf{F}_h(\mathbf{xs}) \qquad (184)$$

where $\mathbf{F}_h(\cdot) = [f_h(xs_1), f_h(xs_2), \cdots, f_h(xs_M)]^T$ is the M-dimensional output activation vector function.

As a result of processing, many output neurons may have nonzero values, depending on the distance from the input pattern to the neuron's weight vector. However, an output neuron may have the dominating value. This neuron can be considered as indicating the input vector class (the cluster to which the input vector belong). To specify the reference vector best representing an input vector, one can identify the neuron with the most responsive output value:

$$j^{\text{th}} \text{ code vector } y_j = \max_{l=1,2,\ldots,M} y_i, \qquad (185)$$

2.7.3. Practical Design Issues for LVQ Vector Quantizers

There is no existing rigorous proven design method for optimal LVQ vector quantizers. Good design still relies on heuristics and simulation-based findings. The goal of LVQ-type vector quantization is not to best approximate the probability density function of the class pattern samples but to directly define between-class borders based on the nearest-neighbor technique. The performance and accuracy of an LVQ quantizer can be evaluated based on its ability to perform accurate and generalized classification or to quantize input patterns for reduction purposes with minimal distortion. The accuracy of an LVQ may depend in the design phase on the following:

– An optimal number of reference vectors assigned to each class defined by training set
– Initial values of the codebook reference vectors
– Concrete, detailed implementation of a learning algorithm: an appropriate learning rate and its decreasing rule, an appropriate stopping criterion, and an order of selecting the input pattern from the training set, as well as a mix of different LVQ types of algorithms in the multistep hybrid sequence of learning
– A stopping rule and generalization through crossvalidation

2.8. The Fourier Transform

Fourier analysis is one of the most important data processing methods. This section focuses on the role of **Fourier transform** in feature extraction and pattern formation in timeseries.

An appropriate transform of data or functions into another space may result in better understanding and representation, with simpler and computationally more efficient processing algorithms. The Fourier transform maps a function, or a signal, into the **frequency domain**, providing information about the periodic frequency components of a function or of a signal (generated by a function).

The Fourier transform decomposes a function into a **spectrum** of its frequency components, and the inverse transform synthesizes a function from its spectrum of frequency components. In experimental domains, the transform of a signal can be thought of as that signal in the "frequency domain." In other words, the Fourier transform decomposes a function or a signal into sine waves with different frequencies. The Fourier transform localizes a function or a signal in frequency space. However, it does not localize such frequencies in time. The results of the Fourier transform are Fourier coefficients $F(u)$ (a spectrum) related to different frequencies. Multiplication of **spectral coefficients** by a sinusoid of corresponding frequency results in reconstruction (an inverse transform) of the sinusoidal component of the original function or signal.

The Fourier transform allows us to form patterns of timeseries or of images in frequency domain. Fourier transform processing comes from the observation that a periodic continuous function $f(t)$ with period 2π can be represented by the infinite sum of sine and cosine functions

$$a_0 + \sum_{k=1}^{\infty}(a_k \cos(kt) + b_k \sin(kt)) \qquad (186)$$

where parameters a_0, a_k, b_k are defined as

$$a_0 = \frac{1}{2\pi}\int_0^{2\pi} f(t)\,dt$$

$$a_k = \frac{1}{\pi}\int_0^{2\pi} f(kt)\sin(kt)\,dt$$

$$b_k = \frac{1}{\pi}\int_0^{2\pi} f(kt)\cos(kt)\,dt \qquad (187)$$

Fourier transform is a linear mapping of functions to other functions (in a frequency domain). In other words, the Fourier transform decomposes a function into a continuous spectrum of its frequency components. The inverse Fourier transform reconstructs (synthesizes) a function from its spectrum of frequency components back into the original domain. In empirical domains, the Fourier transform provides representation of a signal in the time domain into the "frequency domain."

For a continuous function $f(t)$ of a real variable t, the continuous **Fourier transform** $FT(f(t)) = F(u)$ is defined as

$$FT(f(t)) = F(u) = \int_{-\infty}^{\infty} f(t)e^{-j2\pi ut}\,dt \qquad (188)$$

where $j = \sqrt{-1}$ is the imaginary unit and u is the **frequency variable**. Given $F(u)$, we can obtain an original function $f(t)$ by using the **inverse Fourier transform**:

$$f(t) = \int_{-\infty}^{\infty} F(u)e^{j2\pi ut}\,du \qquad (189)$$

We can also write

$$F(j\omega) = \int_{-\infty}^{\infty} f(t)e^{-j\omega t}\,dt \qquad (190)$$

where $\omega = 2\pi u$ is the angular frequency (in radians) and u is the oscillatory frequency (in HZ). Here, the inverse Fourier transform is given by

$$f(t) = \frac{1}{2\pi}\int_{-\infty}^{\infty} F(j\omega)e^{j\omega t}\,d\omega \qquad (191)$$

The term frequency variable comes from the fact that, using Euler's formula, the exponential term $e^{-j2\pi ut}$ may be written in the form

$$e^{-j2\pi ut} = \cos 2\pi ut - j\sin 2\pi ut \qquad (192)$$

$F(u)$ is composed of an infinite sum of sine and cosine terms, and each value of u determines the frequency of its corresponding sine-cosine pairs.

We can also write

$$F(u) = \int_{-\infty}^{\infty} f(t)(\cos 2\pi ut - j \sin 2\pi ut) dt \qquad (193)$$

The Fourier transform (even for a real function) is generally complex, that is

$$F(u) = R(u) + j I(u) \qquad (194)$$

where $R(u)$ and $I(u)$ are, respectively, the real and imaginary components of $F(u)$. The above equation is often more convenient to express in the exponential form

$$F(u) = |F(u)| \, e^{j\phi(u)} \qquad (195)$$

where

$$|F(u)| = [R^2(u) + I^2(u)]^{1/2} \qquad (196)$$

The $|F(u)|$ (magnitude) is called the **Fourier spectrum** (frequency spectrum) of function $f(t)$. The term

$$\phi(u) = \tan^{-1}\left[\frac{I(u)}{R(u)}\right] \qquad (197)$$

is called the **phase angle** of the Fourier transform. The square of Fourier spectrum values

$$P(u) = |F(u)|^2 = R^2(u) + I^2(u) \qquad (198)$$

is called the **power spectrum** (spectral density) of function $f(t)$.

2.8.1. Basic Properties of the Fourier Transform
Time – Frequency Duality.

$$F(t) \Leftrightarrow f(-u) \qquad (199)$$

Linearity.

$$F(af(t) + bg(t)) = aF(f(t)) + bF(g(t)) \qquad (200)$$

Symmetry. The Fourier transform is symmetric, since $F(u) = FT(f(t))$ implies $F(-u) = FT(f(-t))$:

$$\text{if } f(x) \text{ is real, then } F(-u) = F(u)^*$$
$$\text{if } f(x) \text{ is imaginary, then } F(-u) = -F(u)^*$$
$$\text{if } f(x) \text{ is even, then } F(-u) = F(u)$$
$$\text{if } f(x) \text{ is odd, then } -F(-u) = -F(u) \qquad (201)$$

where $*$ indicates a complex conjugate operation.

180   2. Feature Extraction

Orthogonality. Functions $\frac{1}{\sqrt{2\pi}}e^{j\omega t}$ form an orthogonal basis

$$\int_{-\infty}^{\infty}\left(\frac{1}{\sqrt{2\pi}}e^{jat}\right)\left(\frac{1}{\sqrt{2\pi}}e^{-jbt}\right)dt = \delta(a-b) \qquad (202)$$

where δ is Kronecker delta. The Fourier transform can be considered as a transformation of coordinate bases in this space.

Scaling.

$$f(at) \Leftrightarrow \frac{1}{|a|}F\left(\frac{u}{a}\right) \qquad (203)$$

Time Shift.

$$f(t-t_0) \Leftrightarrow e^{-j2\pi u t_0}F(u) \qquad (204)$$

Convolution.

$$f(t) \otimes g(t) \Leftrightarrow F_f(u)F_g(u) \qquad (205)$$

where $\otimes$ denotes the convolution operation. Here, we have

$$f(t) \otimes g(t) = \int_{-\infty}^{\infty} f(\tau)g(t-\tau)\,d\tau \qquad (206)$$

Releigh Property.

$$\int_{-\infty}^{+\infty} |f(t)|^2 dt = \int_{-\infty}^{+\infty} |F(u)|^2 du \qquad (207)$$

2.8.2. Discrete Data and the Discrete Fourier Transform

In real life processing we deal mostly with discrete data and we apply Discrete Fourier Transform.

The **sampling rate** or sampling **frequency** determines the number of samples per second taken from a continuous signal in order to create a discrete signal. The inverse of the sampling frequency is the **sampling period** or **sampling time**, which is the time between samples.

The sampling frequency is measured in Hertz. The Hertz (denoted by Hz) is the unit of frequency. One Hz denotes one sample of signal per second, 50 Hz means 50 samples per second, etc. The unit may be applied to any periodic event.

2.8.3. The Discrete Fourier Transform

Frequently a continuous function $f(t)$ is given as a finite discrete set of N uniformly spaced samples of the function values

$$\{f(t_0), f(t_0+\Delta t), f(t_0+2\Delta t), \cdots, f(t_0+(N-1)\Delta t)\} \qquad (208)$$

denoted also as

$$\{f(k), k=0, 1, \cdots, N-1\}, \quad \{f(k)\}_{k=0}^{N-1} \qquad (209)$$

in N sampling points of variable t in the range of values $[t_0, t_0+(N-1)\Delta t]$. Here, Δt is the time increment between samples.

For the discrete sampled function $f(k)$, the **discrete Fourier transform** (DTF) is defined as

$$F(u) = \sum_{k=0}^{N-1} f(k)e^{-j2\pi uk/N}, \quad \text{for } u = 0, 1, 2, \cdots, N-1 \tag{210}$$

and the **inverse of the discrete Fourier transform** is given by

$$f(k) = \frac{1}{N}\sum_{u=0}^{N-1} F(u)e^{j2\pi uk/N}, \quad \text{for } k = 0, 1, 2, \cdots, N-1 \tag{211}$$

The discrete values of samples $u = 0, 1, \cdots, N-1$ in the discrete Fourier transform correspond to the samples of the continuous Fourier transform given for the values $0, \Delta u, 2\Delta u, \cdots, (N-1)\Delta u$. Hence, $F(u)$ represents $F(u\,\Delta u)$. The relation between Δt and Δu is given by

$$\Delta u = \frac{1}{N\,\Delta t} = \frac{1}{T} \tag{212}$$

We can see that the following equalities hold in the time and frequency domains.

Time Domain.

- T : total sampling time (sampling interval, total sampling length of a time signal)
- N : total number of discrete samples taken (sample size)
- Δt : time increment between samples (time spacing) $\Delta t = \frac{T}{N}$
- length : $T = (N-1)\Delta t$
- period : $T = N\Delta t$

Frequency Domain.

- N : number of components in the spectrum
- f_s : the sampling frequency (sampling rate) $f_s = \frac{1}{\Delta t} = \frac{N}{T}$
- $\Delta u = \Delta f$: the **frequency increment** (**frequency resolution**) $\Delta u = \frac{1}{T} = f_s/N = \frac{1}{N\Delta t}$
- f_p : the frequency period (spectrum period) $f_p = Nf_s = \frac{N}{\Delta t}$ in Hz.
- f_{max} : the maximum frequency $f_{max} = \frac{1}{2}f_s$

The time domain values are measured in seconds, and the frequency domain values in Hz. Both FT and DFT deal with discrete functions in time and frequency. The DFT transforms a series of N values in time $f(t_i)$, $(i = 0, 1, \cdots, N-1)$, into a series of N components $F(f_i)$, $(i = 0, 1, \cdots, N-1)$, in the frequency space, where f_i is a discrete frequency in the spectrum. For samples equally spaced in time, with total sampling time T [seconds] and time increment between samples Δt (signal sampled every Δt seconds for $(N-1)\Delta t$ seconds), the N discrete samples taken create a time signal of length

$$T = (N-1)\Delta t \tag{213}$$

which is equal to the total sampling period T in seconds. For a signal with N samples, the frequency spectrum has N components. The numbers of samples N in time is usually taken to be a power of $2(N = 2^r)$.

The **sampling frequency** is

$$f_s = \frac{1}{\Delta t} = \frac{N}{T} \tag{214}$$

2. Feature Extraction

The spacing interval for frequencies (the frequency resolution)

$$\Delta f = \frac{1}{N\Delta t} = \frac{1}{T} \tag{215}$$

is determined for given N and Δt. The highest frequency is determined as

$$f_{max} = \frac{N}{2}\Delta f = \frac{1}{2}f_s \,(Hz) \tag{216}$$

In DFT, a computing aliasing situation might happen when false spectral lines appear in a spectrum from frequencies above the measured bandwidth. Application of the **Nyquist frequency** criterion helps to prevent such situations. The Nyquist frequency (critical frequency) is half the sampling frequency for a signal. In principle, a Nyquist frequency equal to the signal bandwidth is sufficient to allow perfect reconstruction of the signal from the samples. The critical $1/2f_s$ frequency point is known as the Nyquist frequency. The Nyquist criterion states that for signal sampling at some sampling frequency f_s, it is possible to obtain proper (without aliasing) frequency information only for frequencies less than $f_s/2$. In other words, only half of the N outputs of DFT can be used.

Example: Assume that we sample a signal for $T = 2$ seconds, at a sampling rate of $f_s = 100\,Hz$. The number of samples taken in 2 seconds is $N = Tf_s = 2 \times 100 = 200$. The time interval between samples is $\Delta t = 1/f_s = 1/100 = 0.01$ seconds. The number of useful samples that has been obtained is $N/2 = 100$. The frequency resolution is $\Delta f = \frac{1}{T} = 0.5\,Hz$. The maximum frequency for which the DFT output is reliable is $f_{max} = f_s/2 = (100\,Hz)/2 = 50\,Hz$ (or $f_{max} = \frac{N}{2}\Delta f = 50\,Hz$).

The real part of the transform is even whenever the signal is real. Hence, the values for $k < N/2$ are negative frequency results. Negative frequency values are mirror images of the positive values around the middle frequency $f_p/2$ called the **folding frequency**. For N points in time space, there are N points in the frequency domain. For real signals, there are N complex values in the transform with N real and complex parts. Around the folding frequency, the real part is even and the imaginary part is odd. Including the component for frequency $f = 0$, there are in fact only $N/2 + 1$ points in the frequency domain.

For the signal $f(t)$, with N samples, with sampling time Δt and corresponding spacing interval for frequencies $\Delta u = \frac{1}{N\Delta t}$, the frequency components of the spectrum in HZ are

$$f = -\frac{N\Delta u}{2}, \cdots, 0, \Delta u, , \cdots, \left(\frac{N}{2} - 1\right)\Delta u \tag{217}$$

The Fourier spectrum of the signal $f(t)$ is periodic, and the spectrum period is $f_p = N/\Delta t = Nf_s$ in Hz. Since the signal spectrum is symmetric in pattern formation, one can only consider spectral values for $N/2 + 1$ frequencies:

$$f = 0, \Delta u, 2\Delta u, \cdots, \left(\frac{N}{2} - 1\right)\Delta u \tag{218}$$

2.8.4. Fast Fourier Transform

A **fast Fourier transform** (FFT) is a computationally efficient algorithm to compute the discrete Fourier transform (DFT) and its inverse. Due to reduced computation costs FFT plays an important role in digital signal processing and its applications.

For a sequence of discrete complex values $x_0, \cdots, x_{N-1}$, a DFT is defined by the formula

$$F(u) = \sum_{k=0}^{N-1} x_k e^{\frac{-j2\pi}{N}ku} \quad u = 0, \cdots, N-1 \tag{219}$$

A direct evaluation of the Fourier transform for this sequence would requires $O(N^2)$ arithmetical operations. An FFT algorithm computes the same result in only $O(n \log n)$ operations (where log is the base-2 logarithm). The FFT algorithm can also be easily adapted for computation of an inverse Fourier transform.

2.8.5. Properties of the Discrete Fourier Transform

The basic properties of the DFT are similar to that for the continuous Fourier transform. We can list some specific properties of the DFT:

- **Orthogonality**. Vectors $e^{j2\pi ku/N}$ form an orthogonal basis over the set of N-dimensional complex vectors:

$$\sum_{k=0}^{N-1} e^{j2\pi ku/N} e^{-j2\pi k'u/N} = \begin{cases} N & \text{if } k = k' \\ 0 & \text{otherwise} \end{cases} \quad (220)$$

- **Autocorrelation** from the Fourier transform. It can be shown that the Fourier transform of the power spectrum is the autocorrelation function. In applications where the full autocorrelation function is needed, it may be faster to use this method than the direct computation.

2.8.6. Short Fourier Transform

The **discrete short-term Fourier transform** (SDFT) is the DFT performed for the short N-elements frame (usually windowed) of discrete complex (or just real) values $x(k)_{k=0}^{N-1}$. The result is an N-element array of complex values that is the discrete frequency spectrum of a signal spaced at interval $f_s = 1/(N\Delta t)$, where Δt is the spacing between time samples.

The SFT is widely used in speech processing, where a speech signal is divided into overlapping short sliding frames (containing 64, 128, 256, or 512 samples). It is believed that the parameters of the spectral model of the speech frame are stationary within a frame, and a sequence of frames represents variability and the natural sequential character of the speech signal. The SFT spectral components are extracted from windowed frame signals $x_w(k)$:

$$x_w(k) = x(k) \cdot w(k) \quad (221)$$

The popular Hamming window is defined as $w(k) = 0.54 - 0.46 \cos(2\pi k/(n_w - 1))$, where n_w is the window size.

2.8.7. Spectrogram

In order to grasp the sequential character of longer signals (for example, speech signals), the subsequent windows from a sliding frame along the speech signal are arranged in a so-called **spectrogram**. A spectrogram shows the sequential character of a speech signal in the time-frequency coordinate system on a plane. Time runs along the horizontal axis, while frequency runs along the vertical axis. The amount of energy (the power spectrum of the window's short Fourier transform) in any region in the time-frequency plane is depicted by the darkness of shading. This spectrogram image is composed with subsequent power spectrum vectors (as subsequent columns) obtained by taking discrete short Fourier transforms of short subsequent windows of the original signal in the time domain. A spectrogram is a signal representation in the time domain that captures the sequential nature of processed signals. A spectrogram might be further processed, for example, by using image processing transforms. We can also compose a pattern vector representing a sequence of frames as a concatenation of spectrogram columns.

184   2. Feature Extraction

2.8.8. Pattern Forming for a Spectral DFT Signal

Since spectrum of signal is symmetric in pattern forming, one can only consider spectral values for $n_{dft} = N/2 + 1$ frequencies

$$f = \{0, f_s, 2f_s, \cdots, \left(\frac{N}{2} - 1\right)f_s\} \tag{222}$$

which corresponds to the following indices of DFT values

$$n_{dft}, n_{dft} + 1, \cdots, n_{dft} + N/2 + 1 \tag{223}$$

1. The **spectral patterns** $\mathbf{x}_{sdft}$.
   The spectral DFT pattern can be formed as

$$\mathbf{x}_{sdft} = \left[|F(n_{dft})|, |F(n_{dft}+1)|, \cdots, |F(n_{dft}+N/2+1)|\right]^T \tag{224}$$

   where $|F(i)|$ is an amplitude of DFT transform value for frequency related to index i of DFT.

2. The **power spectrum patterns** $\mathbf{x}_{pdft}$.
   The power spectrum pattern can be formed as

$$\mathbf{x}_{pdft} = \left[PF(n_{dft}), PF(n_{dft}+1), \cdots, PF(n_{dft}+N/2+1)\right]^T \tag{225}$$

   where $PF(i)$ is the power spectrum value for frequency related to index i of DFT.

2.9. Two-dimensional Fourier Transform and 2D Spectral Features

One of the most powerful feature extraction methods from images or sub-windows of images is based on the two-dimensional Fourier transform. The Fourier transform allows us to represent and interpret an image in the frequency domain. Let us begin with the general continuous two-dimensional Fourier transform.

2.9.1. The Continuous 2D Fourier Transform

The **two-dimensional continuous Fourier transform** (2DFT) for the two-dimensional continuous function $f(x, y)$ is defined as

$$F(u, v) = \int_{-\infty}^{\infty} \int_{-\infty}^{\infty} f(x, y) e^{-j2\pi(ux+vy)} \, dx \, dy \tag{226}$$

and the **inverse** of the Fourier transform is defined as

$$f(x, y) = \int_{-\infty}^{\infty} \int_{-\infty}^{\infty} F(u, v) e^{j2\pi(ux+vy)} \, du \, dv \tag{227}$$

The Fourier transform $F(u, v)$ (even for a real function) is generally complex. From Euler's formula, we have

$$e^{-j(ux+vy)} = \cos(ux+vy) - j\sin(ux+vy)$$

Here, x and y are spatial coordinates, u and v are the spatial frequencies, and $F(u, v)$ is the **frequency spectrum**.

For the continuous transform, (x, y) and (u, v) take on a real continuum of values. The spectrum $F(u, v)$ is complex and periodic.

2.9.2. The Discrete 2D Fourier Transform

For the two-dimensional continuous function sampled in the 2D grid of $M \times N$ points, with divisions of width Δx and Δy for the x- and y-axis, respectively, we can define the **two-dimensional discrete Fourier transform**. Here, the discrete function $f(x, y)$ represents discrete samples of the function $f(x_0 + x\Delta x, y_0 + y\Delta y)$ for $x = 0, 1, \cdots, M-1$ and $y = 0, 1, \cdots, N-1$. Similarly, the discrete function $F(u, v)$ represents samples of the function $F(u\Delta u, v\Delta v)$ at $u = 0, 1, \cdots, M-1$ and $v = 0, 1, \cdots, N-1$. The sampling increments in the frequency domain are given by

$$\Delta u = \frac{1}{M\Delta x}$$
$$\Delta v = \frac{1}{N\Delta y} \tag{228}$$

The two-dimensional discrete Fourier transform is given by the formula

$$F(u, v) = \frac{1}{MN} \sum_{x=0}^{M-1} \sum_{y=0}^{N-1} f(x, y) e^{-j2\pi(\frac{ux}{M} + \frac{vy}{N})}$$

$$= \frac{1}{M} \sum_{x=0}^{M-1} \left[\frac{1}{N} \sum_{y=0}^{N-1} f(x, y) e^{-j2\pi(\frac{vy}{N})} \right] e^{-j2\pi(\frac{ux}{M})};$$

$$u = 0, 1, \cdots, M-1, \quad v = 0, 1, \cdots, N-1 \tag{229}$$

The term in square brackets [] is the one-dimensional discrete Fourier transform of the xth line (row) and can be computed using standard Fourier transform procedures (usually assuming $N = 2k$). Each line is replaced by its Fourier transform, and the one-dimensional discrete Fourier transform of each column is computed.

The **inverse 2D discrete Fourier transform** is given by the equation

$$f(x, y) = \sum_{u=0}^{M-1} \sum_{v=0}^{N-1} F(u, v) e^{j2\pi(\frac{ux}{M} + \frac{vy}{N})};$$

$$x = 0, 1, \cdots, M-1, \quad y = 0, 1, \cdots, N-1 \tag{230}$$

The kernel function for the 2D discrete Fourier transform is

$$e^{-j2\pi\left(\frac{ux}{M} + \frac{vy}{N}\right)} \tag{231}$$

The Fourier transform $F(u, v)$ (even for a real function) is generally complex and consists of real and imaginary parts. Using Euler's formula, it can be expressed as

$$F(u, v) = Re(u, v) + j\,Im(u, v), \quad u = 0, \cdots, M-1, \quad v = 0, \cdots, N-1 \tag{232}$$

where

$$Re(u, v) = \frac{1}{MN} \sum_{x=0}^{M-1} \left[\sum_{y=0}^{N-1} f(x, y) \cos\left(2\pi \left(\frac{ux}{M} + \frac{vy}{N}\right)\right) \right] \tag{233}$$

is the real part, and

$$Im(u, v) = \frac{1}{MN} \sum_{x=0}^{M-1} \left[\sum_{y=0}^{N-1} -f(x, y) \sin\left(2\pi \left(\frac{ux}{M} + \frac{vy}{N}\right)\right) \right] \tag{234}$$

is the imaginary part of the transform. We can also express $F(u, v)$ in the exponential form

$$F(u, v) = |F(u, v)|e^{i\phi(u,v)} \qquad (235)$$

where the norm of magnitude (amplitude) $|F(u, v)|$

$$|F(u, v)| = \sqrt{Re^2(u, u) + Im^2(u, v)} \qquad (236)$$

is called the **Fourier spectrum** (frequency spectrum) of $f(x, y)$ and the term

$$\phi(u, v) = \tan^{-1}\left[\frac{Im(u, v)}{Re(u, v)}\right] \qquad (237)$$

is the **phase spectrum** (phase angle). The square of the amplitude

$$P(u, v) = |F(u, v)|^2 = Re^2(u, v) + Im^2(u, v) \qquad (238)$$

is called the **power spectrum** $P(u, v)$ of $f(x, y)$. The power spectrum $P(u, v)$ is also called the **spectral density**.

Let us consider the Fourier transform for images, which are defined on a finite support. In computing the Fourier transform of an image, we will consider an image as an $M \times N$ matrix, where M is a number of rows and N is a number of columns:

$$\begin{bmatrix} f(0,0) & f(0,1) & \cdots & f(0, N-1) \\ f(1,0) & f(1,1) & \cdots & f(1, N-1) \\ \vdots & \vdots & \ddots & \vdots \\ f(M-1, 0) & f(M-1, 0) & \cdots & f(M-1, N-1) \end{bmatrix} \qquad (239)$$

where $f(x, y)$ denotes pixel brightness at the integer coordinates (x, y) of an image. If an image has width N and height M with the origin in a center, then

$$F(u, v) = \sum_{-M/2}^{M/2} \sum_{-N/2}^{N/2} f(x, y) e^{-j2\pi(ux+vy)} \qquad (240)$$

Here, we assume that $f(x, y)$ is extended, with $f(x, y) = 0$ outside the image frame.

The 2D discrete Fourier transform is an important image processing tool that is used to decompose an image into its sine and cosine components. The input to 2DDFT is an image in the real domain, whereas output of the transformation represents the image in the Fourier or frequency space. In the Fourier space image, each point represents a specific frequency contained in the real domain image.

The Fourier transform is used in a wide range of applications, such as image analysis, filtering, recognition, compression, and image reconstruction.

2.9.3. Basic Properties of 2DDFT

There are several properties associated with the two-dimensional Fourier transform and the 2D inverse Fourier transform. Generally, the properties of 2DDFT are the same as those for one-dimensional DFT. These properties have an interesting interpretation when 2DDFT is applied to images. We can list some of the most important properties of 2DDT as applied to digital image processing. The Fourier transform is, in general, a complex function of real frequency variables. As such, the transform can be written in terms of its magnitude and phase.

The Fourier transform is **linear**. For 2DDFT of images, this means, that

– adding two images together results in adding the two Fourier transforms together
– multiplying an image by a constant multiplies the image's Fourier + transform by the same constant

Separability means that the Fourier transform of a two-dimensional function is the Fourier transform in one dimension of the Fourier transform in the other direction. This means that we can compute the two-dimensional Fourier transform by providing a one-dimensional Fourier transform of the rows and then taking a one-dimensional Fourier transform of the columns of the result.

Rotational Invariance. Rotation of an image results in the rotation of its Fourier transform.

Translation and Phase. Translation of an image does not change the magnitude of the Fourier transform but does change its phase.

Scaling. Changing the spatial unit of distance changes the Fourier transform. If the 2D signal $f(x, y)$ is scaled $(M_x x, M_y y)$ in its spatial coordinates (x, y), then $F(u, v)$ becomes $F(u/M_x, v/M_y)/|M_x M_y|$.

Periodicity and Conjugate Symmetry. The Fourier transform in discrete space is periodic in both space and in frequency. The periodicity of the Fourier transform can be explained by

$$F(u, -v) = F(u, N_p - v) \; ; \; F(-u, v) = F(M_p - u, v)$$
$$F(aM_p + u, bN_p + v) = F(u, v) \; ; \; F(-x, y) = F(M_p - x, y)$$
$$f(x, -y) = f(x, N_p - y) \; ; \; f(aM_p + x, bN_p + y) = f(x, y) \qquad (241)$$

where N_p and M_p are periods. If a 2D signal $f(x, y)$ is real, then the Fourier transform possess certain symmetries:

$$F(u, v) = {}^*F(-u, -v) \qquad (242)$$

The symbol $(*)$ indicates complex conjugation of $F(u, v)$. For real signals,

$$|F(u, v)| = |F(-u, -v)| \qquad (243)$$

If a 2D signal is real and even, then the Fourier transform is real and even.

Energy. According to Parseval's theorem, the energy in a 2D signal can be computed either in the spatial domain or in the frequency domain. For a continuous 2D signal with finite energy,

$$E = \int_{-\infty}^{+\infty} \int_{-\infty}^{+\infty} |f(x, y)|^2 \, dx \, dy = \frac{1}{4\pi^2} \int_{-\infty}^{+\infty} \int_{-\infty}^{+\infty} |F(u, v)|^2 \, du \, dv \qquad (244)$$

For a discrete 2D signal with finite energy,

$$E = \sum_{-\infty}^{+\infty} \sum_{-\infty}^{+\infty} |f(x, y)|^2 = \frac{1}{4\pi^2} \int_{-\pi}^{+\pi} \int_{-\pi}^{+\pi} |F(u, v)|^2 \, du \, dv \qquad (245)$$

Convolution. For three given two-dimensional signals a, b, and c and their Fourier transforms F_a, F_b, and F_c,

$$c = a \otimes b \rightarrow F_a \cdot F_b \tag{246}$$

$$c = a \cdot b \rightarrow F_c = \frac{1}{4\pi^2} F_a \otimes F_b \tag{247}$$

where $\otimes$ denotes convolution operation. Convolution in the spatial domain is equivalent to multiplication in the Fourier (frequency) domain and vice versa. This property provides a method for the implementation of a convolution.

2.9.4. 2DDFT Patterns

A power spectrum of 2DDFT of an image can be used to form an image **spectral pattern**. In the first phase of feature extraction from an image, the 2DDFT can be used in order to convert the gray-scale image pixels into the corresponding spatial frequency representation. The 2DDFT complex features are extracted from an $M \times N$ pixel image by formulas:

$$F(u, v) = Re(u, v) + j\, Im(u, v), \quad u = 0, \cdots, M-1, \quad v = 1, \cdots, N-1 \tag{248}$$

where

$$Re(u, v) = \frac{1}{MN} \sum_{x=-\frac{M}{2}}^{\frac{M}{2}-1} \left[\sum_{y=-\frac{N}{2}}^{\frac{N}{2}-1} f(x, y) \cos\left(2\pi \left(\frac{ux}{M} + \frac{vy}{N}\right)\right) \right]$$

$$Im(u, v) = \frac{1}{MN} \sum_{x=-\frac{M}{2}}^{\frac{M}{2}-1} \left[\sum_{y=-\frac{N}{2}}^{\frac{N}{2}-1} -f(x, y) \sin\left(2\pi \left(\frac{ux}{M} + \frac{vy}{N}\right)\right) \right] \tag{249}$$

and (x, y) and (u, v) denote the image pixel integer-valued coordinates.

In the above equation, $Re(u, v)$ denotes the real component and $Im(u, v)$ the imaginary component of the discrete Fourier transform of an image. For each pixel $f(u, v)$ of an original image, we can compute the real and imaginary part of $Re(u, v)$ and $Im(u, v)$ and then the real-valued power spectrum

$$P(u, v) = |F(u, v)|^2 = Re^2(u, v) + Im^2(u, v) \tag{250}$$

This power spectrum can be represented as an $m \times n$ image (a power spectrum map). In the most general situation, a two-dimensional transform takes a complex array. The most common application is for image processing, where each value in the array represents a pixel; the real value is the pixel value, and the imaginary value is 0. Two-dimensional Fourier transforms simply involve a number of one-dimensional Fourier transforms. More precisely, a two-dimensional transform is achieved by first transforming each row, replacing each row with its transform, and then transforming each column and replacing each column with its transform. Thus a 2D transform of a 1000×1000 image requires 2000 1D transforms. This conclusion follows directly from the definition of the Fourier transform of a continuous variable or the discrete Fourier transform of a discrete system.

2.9.5. Constructed Spatial 2DDFT Power Spectrum Features and Patterns

One of the natural ways of constructing a pattern from 2DDFT of 2D data (for example an image) is forming a column vector containing concatenated subsequent column of the power spectrum image (map) of the 2DDFT of 2D data. For $m \times n$ 2D data the resulting pattern will have $m \times m$ elements.

In order to capture the spatial relation (a shape) of the components of the "power spectrum image" (map), a number of spatial features can be computed from the power spectrum array treated as an image.

We have assumed that the numerical characteristics (measures) of the shape of the spatial frequency spectrum, such as location, size, and orientation of peaks and entropy of the normalized spectrum in regions of spatial frequency, can be used as object features (pattern elements) suitable for recognition of an image. Some spatial features of the normalized "power spectrum" image require the computation for such an image of the covariance matrix and its eigenvalues and eigenvectors as well as principal components. Numerous subsequent spatial spectral features (computed in the frequency domain) can be extracted or created from the normalized "power spectrum image" $P(u, v)$:

$$P(u, v) = \frac{P(u, v)}{\sum_{u,v \neq 0} P(u, v)} \quad (251)$$

Energy of the major peak

$$f_1 = p(u_1, v_1) \times 100 \quad (252)$$

where u_1, v_1 are the frequency coordinates of the maximum peak of the normalized power spectrum. Here, f is a percentage of the total energy.

Laplacian of the major peak

$$\begin{aligned} f_2 &= \nabla^2 P(u_1, v_1) \\ &= P(u_1 + 1, v_1) + P(u_1 - 1, v_1) + P(u_1, v_1 + 1) + P(u_1, v_1 - 1) - 4P(u_1, v_1) \end{aligned} \quad (253)$$

Laplacian of the secondary peak

$$f_3 = \nabla^2 P(u_2, v_2) \quad (254)$$

where u_2, v_2 are the coordinates of the second largest peak in the $P(u, v)$ map.

Spread of the major peak
f_4 is the number of adjacent neighbors of u_1, v_1 with

$$P(u, v) \geq \frac{1}{2} k P(u_1, v_1) \quad (255)$$

where the neighbors are $u_1 \pm 1$, v_1, and u_1, $v_1 \pm 1$.

Squared frequency of the major peak in $P(u, v)$

$$f_5 = u_1^2 + v_1^2 \quad (256)$$

Relative orientation of the major and secondary peaks

$$f_6 = \left| \tan^{-1} \frac{v_1}{u_1} - \tan^{-1} \frac{v_2}{u_2} \right| \quad (257)$$

Isotropy of the normalized power spectrum $P(u, v)$

$$f_7 = \frac{|\delta_u - \delta_v|}{\left[(\delta_u - \delta_v)^2 - 4\delta_{uv}^2\right]^{\frac{1}{2}}} \tag{258}$$

where

$$\delta_u = \sum_u \sum_v u^2 p(u, v)$$
$$\delta_v = \sum_u \sum_v v^2 p(u, v)$$
$$\delta_{uv} = \sum_u \sum_v uv p(u, v)$$

Here, f_7 measures the elongation of the normalized power spectrum and is maximum for parallel line faces.

Circularity of the normalized power spectrum

$$f_8 = \frac{A_D}{A_C} \tag{259}$$

where

A_D = number of nonzero frequency components within a circle of radius $\sqrt{\lambda_1}$
A_C = number of distinct frequency components within a circle of radius $\sqrt{\lambda_1}$
λ_1 = maximum eigenvalue of the covariance matrix of $p(u, v)$

Major peak horizontal frequency

$$f_9 = u_1 \tag{260}$$

Major peak vertical frequency

$$f_{10} = v_1 \tag{261}$$

Secondary peak horizontal frequency

$$f_{11} = u_2 \tag{262}$$

Secondary peak vertical frequency

$$f_{12} = v_2 \tag{263}$$

Squared distance between the major and secondary peak

$$f_{13} = (u_1 - u_2)^2 + (v_1 - v_2)^2 \tag{264}$$

Principal component magnitude (squared)

$$f_{14} = \lambda_1 \tag{265}$$

Principal component direction

$$f_{15} = \cos^{-1}(\phi_1) \tag{266}$$

where $\boldsymbol{\phi} = \begin{bmatrix} \phi_1 \\ \phi_2 \end{bmatrix}$ is a normalized eigenvector for eigenvalue λ_1

Ratio of the minor to major principal axis

$$f_{16} = \left(\frac{\lambda_2}{\lambda_1}\right)^{\frac{1}{2}} \tag{267}$$

where λ_2 is the minimum eigenvalue of the covariance matrix of $p(u, v)$

Moment of inertia, quadrant I

$$f_{17} = \sum_{u>0}\sum_{v>0} (u^2 - v^2)^{\frac{1}{2}} p(u, v) \tag{268}$$

Moment of inertia, quadrant II

$$f_{18} = \sum_{u<0}\sum_{v>0} (u^2 \ v^2)^{\frac{1}{2}} p(u, v) \tag{269}$$

Here, in f_{17} and f_{18}, the power spectrum is normalized within quadrants I and II, respectively.

Moment ratio

$$f_{19} = \frac{f_{18}}{f_{17}} \tag{270}$$

Percentage energy, quadrant I

$$f_{20} = \sum_{u>0}\sum_{v>0} p(u, v) \tag{271}$$

Percentage energy, quadrant II

$$f_{21} = \sum_{u<0}\sum_{v>0} p(u, v) \tag{272}$$

Ratio of nonzero components

$$f_{22} = \frac{n_1}{n_2} \tag{273}$$

where n_i denotes a number of nonzero frequency components in quadrant i

Laplacian of the major peak phase

$$f_{23} = \nabla^2 \phi(u_1, v_1) \tag{274}$$

Laplacian of the secondary peak phase

$$f_{24} = \nabla^2 \phi(u_2, v_2) \tag{275}$$

2. Feature Extraction

Relative entropy of the normalized power spectrum (R_1)

$$f_{25} = \frac{\left[-\sum_{u,v \in R_1} P_1(u,v) \log P_1(u,v)\right]}{\log K_1} \tag{276}$$

where

$P_1(u,v) = \frac{P(u,v)}{\sum_{u,v \in R_i} P(u,v)}$
K_i = number of distinct frequencies in R_i

$$R_i = \left\{ u, v : \frac{i-1}{4} u_m < |u| < \frac{i}{4} u_m \quad \text{and} \quad \frac{i-1}{4} v_m < |v| < \frac{i}{4} v_m \right\}$$

where u_m, v_m are the maximum frequency components for the local spectrum

Relative entropy (R_2)

$$f_{26} = \frac{\left[-\sum_{u,v \in R_2} P_2(u,v) \log P_2(u,v)\right]}{\log K_2} \tag{277}$$

Relative entropy (R_3)

$$f_{27} = \frac{\left[-\sum_{u,v \in R_3} P_3(u,v) \log P_3(u,v)\right]}{\log K_3} \tag{278}$$

Relative entropy (R_4)

$$f_{28} = \frac{\left[-\sum_{u,v \in R_4} P_4(u,v) \log P_4(u,v)\right]}{\log K_4} \tag{279}$$

Histogram subpattern x_h:

$$f_{29}, \cdots, f_{29} + n_h \times N.$$

For an $M \times N$ "image" of normalized $P(u,v)$, the following histogram features can be extracted and presented as the $n_h \times N$-element vector $\mathbf{x}_h$:

(a) For each column of an image (treated as a matrix), the column elements are binned into n_h equally spaced containers and the number of elements in each container is computed. The result forms a histogram matrix $\mathbf{H}_p$ of dimension $n_h \times N$.
(b) Now we can consider the histogram subpattern as an $n_h \times N$-element vector $\mathbf{x}_h$, obtained through columnwise concatenation of the matrix $\mathbf{H}_p$.

Magnitude of complex Zernike moments (defined later in the Chapter) of order from $(1, 1)$ through (p, q) for normalized $P(u, v)$.

The 2DDFT provides a powerful spectral representation of images in the frequency domain, with significant predispositions toward features of image patterns used in image recognition. A spectral pattern of an image can be formed as concatenated columns of a "normalized power spectrum image." One can also form a spectral pattern as any subset of the spatial spectral features defined above.

2.10. Wavelets

Wavelets are another powerful technique for processing timeseries and images. The foundation of wavelets comes from orthogonality, function decomposition, and multiresolution approximation. **Wavelets** (wavelet analysis, the wavelet transform) provide representation (approximation) of a function (or a signal) by a fast-decaying oscillating waveform (known as the **mother wavelet**). This waveform can be scaled and translated in order to best match the function or input signal.

A **wavelet** is a special kind of oscillating waveform of substantially limited duration with an average value equal to zero. Wavelets can be considered to be one step ahead of the Fourier transform.

The Fourier transform decomposes a function or signal into sine waves with different frequencies (or, in other words, is the sum over all time of the signal $f(t)$ multiplied by a complex exponential). However, it does not localize these frequencies in time. The results of the Fourier transform are Fourier coefficients $F(\omega)$:

$$FT\ f(t) = F(\omega) = (2\pi)^{-\frac{1}{2}} \int_{-\infty}^{+\infty} f(t) e^{-j\omega t}\ dt \qquad (280)$$

(a spectrum) related to different frequencies. Multiplication of spectral coefficients by a sinusoid of corresponding frequency results in the reconstruction (an inverse transform) of the sinusoidal component of the original function or signal.

Wavelets break up a function or a signal into a shifted and scaled instance of the mother wavelet. The **continuous wavelet transform** is the sum over all time of a function or a signal multiplied by the shifted, scaled instance of the wavelet function. The results of the wavelet transform are coefficients C:

$$C(\text{position, scale}) = \int_{-\infty}^{+\infty} f(t) \Psi(\text{position, scale}, t)\ dt \qquad (281)$$

which are a function of position and scale. In the **inverse wavelet transform**, multiplication of each coefficient by the corresponding shifted and scaled wavelet results in constituent wavelets of the original function or signal. Wavelets provide localization both in frequency and in time (or in space). A function $f(t)$ (or signal $x(k)$) can be more easily analyzed or described when expressed as a linear decomposition

$$f(t) = \sum_r a_r \psi_r(t) \qquad (282)$$

where r is an integer index (for a finite or infinite sum). The $a_r \in \mathbb{R}$ are **expansion coefficients**, and the $\psi_r(t)$ are real-valued functions of t (the expansion set). For a unique expansion, the set of functions $\psi_r(t)$ is called a basis. Especially important expansions of a function can be obtained for the orthogonal basis when we have

$$<\psi_r(t), \psi_l(t)> = \int \psi_r(t)\psi_l(t)\ dt = 0,\ r \neq l \qquad (283)$$

$$a_r = <f(t), \psi_r(t)> = \int f(t)\psi_r(t)\ dt \qquad (284)$$

For the best-known Fourier transform, the orthogonal basis functions $\psi_k(t)$ are $\sin(k\omega t)$ and $\cos(k\omega t)$ function of $k\omega t$. This transformation maps a one-dimensional function of the continuous variable into one-dimensional sequence of coefficients.

The main difference between wavelets and the Fourier transform is that wavelets are localized in both time and frequency, whereas the Fourier transform is localized in frequency space. The Short-time Fourier Transform (STFT) could be called a prelude to wavelets, since it is also time and frequency localized. However, wavelets provide better tools for muliresolution in frequency and in time. For two-parameter wavelet transformation, with the wavelet expansion function $\psi_{j,k}(t)$ forming the orthogonal basis, we have

$$f(t) = D_\psi^{-1} \int \int \frac{1}{j^2} <f(t), \psi_{j,k}> \psi_{j,k} dj\, dk \tag{285}$$

and the corresponding discrete version

$$f(t) = \sum_k \sum_j a_{j,k} \psi_{j,k}(t) \tag{286}$$

where j and k are integer indices and D_ψ^{-j} is the scaling factor. The set of coefficients $a_{j,k}$ ($j, k \in \mathbb{Z}$) is the discrete wavelet transform $DW f(t)$ of a function $f(t)$.

The wavelet expansion maps a one-dimensional function into a two-dimensional array of coefficients (discrete wavelets transform), allowing localization of the signal in both time and frequency simultaneously.

Wavelets are defined by the wavelet function $\psi(t)$ (wavelet) and the scaling function $\phi(t)$ (also called the **father wavelet**) in the time domain. The wavelet function acts as a band-pass filter, and scaling it for each level halves its bandwidth, whereas scaling the function filters the lowest level of the transform and ensures that the entire spectrum is covered.

2.10.1. A Wavelet Function

The wavelet transform is a method of approximating a given function $f(t) \in L^2(\mathbb{R})$ or a signal (as a sequence of values) using the functions $\psi(t)$ and $\phi(t)$. The function $\psi(t)$ is a scalable approximation curve localized on a definite time (or space) interval. The function $\psi(t)$ is called a **mother function** (**wavelet function**, or **generating wavelet**). The mother wavelet must satisfy the following admissibility conditions:

$$c_\psi = \int_0^\infty \frac{|\Psi(\omega)|^2}{|\omega|} d\omega < \infty$$

or

$$\int_{-\infty}^\infty \psi(t) dt = 0 \tag{287}$$

where $\Psi(\omega) = (2\pi)^{-\frac{1}{2}} \int_{-\infty}^\infty \psi(t) e^{-j\omega t} dt$ is the Fourier transform of $\psi(t)$.

The second derivative of the Gaussian function is an example of the wavelet function:

$$\psi(t) = (1 - t^2) e^{-\frac{t^2}{2}} \tag{288}$$

Two-dimensional parametrization, with a dilation parameter a and a translation parameter b, yields the possibility of scaling and shifting the wavelet function over a certain time (or space) domain. Wavelets constitute a family of functions designed from **dilations** and **translations** of a single function – the mother wavelet. For signal expansion, the mother wavelet is a band-pass filter. Incorporating continuous variation of the dilation parameter a and the translation parameter b, we can find a family of continuous wavelets

$$\psi_{a,b}(t) = |a|^{-\frac{1}{2}} \psi\left(\frac{t-b}{a}\right), \quad a, b \in \mathbb{R}, \ a \geq 0 \tag{289}$$

where a and b may vary over $\mathbb{R}$. This is a band-pass filter with two parameters: the dilation and translation parameters a and b. The dilation parameter a determinates the frequency of information (interpreted as changing the bandwidth of the filter). Varying the dilation parameter a generates different spectra of $\psi_{a,b}(t)$. For a smaller dilation parameter a, the wavelet is narrowed; for increased a, the wavelet stretches in time. The translation parameter b determinates the time information (location in time) or space information (location in space). It localizes the wavelet curve on a specific time interval with center at $t = b$. For this reason, wavelets are time (or space) and frequency localized. These properties are essential for wavelets. Wavelets functions have finite energy, and they are locally concentrated.

2.10.2. Continous Wavelet Transform

For a given function $f(t)$, the continuous wavelet transform is defined as

$$WT\, f(a, b) = |a|^{-\frac{1}{2}} \int_{-\infty}^{\infty} f(t) \overline{\psi\left(\frac{t-b}{a}\right)} dt \qquad (290)$$

where a and b are dilation and translation parameters, and $\psi(\cdot)$ is a wavelet function. One can see that the wavelet transform is the scalar product of two functions, namely, $f(t)$ and $\psi_{a,b}$ in $L^2(\mathbb{R})$:

$$WT\, f(a, b) = <f(t), \psi_{a,b}(t)> \qquad (291)$$

A function can be characterized by its wavelet coefficients $<f(t), \psi_{a,b}(t)>$. A function $f(t)$ can be reconstructed from its wavelet transform $WT\, f(a, b)$ by the inverse transform

$$f(t) = \frac{1}{D_\psi} \int_{-\infty}^{\infty} \int_{-\infty}^{\infty} \frac{1}{a^2} WT\, f(a, b) \psi_{a,b}(t)\, da\, db \qquad (292)$$

where D_ψ is a scaling factor representing an average energy of the wavelet function

$$D_\psi = 2\pi \int_{-\infty}^{\infty} \frac{|\psi(\omega)|^2}{|\omega|} d\omega \qquad (293)$$

2.10.3. Discrete Wavelet Transform

In the discrete case, parameters a and b take only discrete values. The **discrete wavelets** are obtained, after sampling parameters a and b as $a = a_0^j$ and $b = kab_0 = ka_0^j b_0$ (where $j, k \in \mathbb{Z}$, $i, j = \overline{+1, +2, \cdots}$)), as

$$\psi_{j,k}(t) = |a_0|^{-\frac{j}{2}} \psi(a_0^{-j} t - kb_0), \quad j, k \in \mathbb{Z} \qquad (294)$$

where $\psi_{j,k}(t)$ constitutes basis for $L^2(\mathbb{R})$. The selection of a_0 and b_0 depends on an application.
The discrete wavelet transform is defined for sampled parameters by the equation

$$DWT\, f(j, k) = |a_0|^{-\frac{j}{2}} \int_{-\infty}^{\infty} f(t) \psi(a_0^{-j} t - kb_0)\, dt \qquad (295)$$

For $a_0 = 2$ ($a = 2^j$) and $b_0 = 1$ ($b = 2^j k$), functions $\psi_{j,k}(t)$ form an orthogonal wavelet base for $L^2(\mathbb{R})$.

$$\psi_{j,k}(t) = 2^{-\frac{j}{2}} \psi(2^{-j} t - k) \qquad (296)$$

For a function $f(t) \in L^2(\mathbb{R})$, the discrete wavelet expansion of $f(t)$ is represented by

$$f(t) = \sum_j \sum_k b_{j,k} \psi_{j,k}(t) \qquad (297)$$

where the expansion coefficients $b_{j,k}$ are the inner product of $f(t)$ and $\psi_{j,k}(t)$, i.e.,

$$b_{j,k} = <\psi_{j,k}(t), f(t)> \qquad (298)$$

2.10.4. Multiresolution Analysis of a Function

A continuous wavelet transform is redundant. This redundancy can be avoided in a discrete transform by using the fast wavelet transform. The idea of multiresolution analysis of a function $f(t)$ is to construct a ladder of close subspaces of $\mathbb{Z}$, $\{V_n : n \in \mathbb{Z}\}$ (nested subspaces), for representing functions with successive resolution. This result can be realized with a basis **scaling function** $\phi(t)$ in $L^2(\mathbb{R})$ (low pass or smoothing function):

$$\phi_k(t) = \phi(t-k), \quad k \in \mathbb{Z} \tag{299}$$

with a spanned subspace V_0 of $L^2(\mathbb{R})$ for this function for all integers from $-\infty$ to ∞, and with

$$f(t) = \sum_k a_k \phi_k(t) \text{ for any } f(t) \in V_0 \tag{300}$$

Subspaces in $L^2(\mathbb{R})$ are

$$\{0\} \subset \cdots \subset V_{-2} \subset V_{-1} \subset V_0 \subset V_1 \subset V_2 \subset \cdots \subset L^2(\mathbb{R}) \tag{301}$$

These representations satisfy the following conditions:

1. A subspace V_n is contained in V_{n+1}:

$$V_n \subset V_{n+1}, \quad n \in \mathbb{Z} \tag{302}$$

A space containing high resolution signals will also contain lower resolution signals.

2. An intersection of all subspaces V_n for all $n \in \mathbb{Z}$ is null $\bigcap V_n = \{0\}$, and the union of subspaces is $\bigcup V_n$ is dense in $L^2(\mathbb{R})$.
3. A subspace V_n is invariant under integral translations

$$f(t) \in V_n \Leftrightarrow f(t-k) \in V_n, \quad k \in \mathbb{Z} \tag{303}$$

4. There exists a scaling function $\phi(t) \in V_0$ that, with its translated version $\phi(t-k)$, forms an orthonormal basis in V_0 so that following conditions hold:

$$\int \phi(t)\,dt = 1, \quad \text{normalization}$$

$$\int \phi(t)\overline{\phi(t-k)}\,dt = \delta(k), \quad \text{orthogonality} \tag{304}$$

The subspaces satisfy the natural scaling condition

$$f(t) \in V_n \Leftrightarrow f(2t) \in V_{n+1} \tag{305}$$

which means that elements in a space V_{n+1} are scaled version of elements in the next subspace. The scaling function can also be dilated and translated (similarly to the wavelet function):

$$\phi_{j,k} = |a|^{-\frac{1}{2}} \phi(a^{-1} - k), \quad a = 2 \tag{306}$$

However, the scaling function is not orthogonal to its dilation. Nesting conditions show that $\phi(t)$ can be expressed in terms of the weighted sum of shifted $\phi(\cdot)$:

$$\phi(t) = \sum_k h_k \phi_{-1,k}(t) \tag{307}$$

where h_k ($k = 0, 1, \cdots, K-1$) is a set of scaling coefficients $h_k = <\phi(t), \phi_{-1,k}(t)>$, and $\sum_{k \in \mathbb{Z}} |h_k|^2 = 1$:

$$\phi_{-1,k} = 2^{-\frac{1}{2}} \phi(2^{-1}t - k) \tag{308}$$

2.10.5. Construction of the Wavelet Function from the Scaling Function

Based on an idea of multiresolution and the property of the orthonormal complement W_i of subspaces V_i (contained in V_{i-1}), we find that wavelet function $\psi(t)$ can be expressed as a linear combination of the basis scaling functions $\phi_{-1,k}(t)$:

$$\psi(t) = \sum_k g_k \phi_{-1,k}(t) \tag{309}$$

where $g_k = (-1)^k h_{-k+1}$ or, in matrix notation,

$$\psi(t) = \boldsymbol{\phi}_{-1}\mathbf{g} \tag{310}$$

with $\mathbf{g} = [(-1)^{K-1}h_{-K+2}, \cdots, h_{-1}, h_0, h_1]^T$.

The coefficients h_k satisfy the orthonormality condition $\sum_{k=0}^{K-1} h_k \overline{h_{k+2n}} = \delta(n)$ and

$$\sum_{k=0}^{K-1} h_k = \sqrt{2}, \quad \sum_k (-1)^k h_k = 0 \tag{311}$$

A function $f(t) \in L^2(\mathbb{R})$ can be analyzed using multiresolution idea with scaling and wavelet functions $\phi_{j,k}(t)$ and $\psi_{j,k}(t)$ ($j, k = -1, 0, 1, 2, \cdots$), respectively. These functions constitute the orthonormal bases of the approximation spaces W_j and V_j. Based on functions $\phi_{-1,k}(k)$ and $\psi_{-1,k}$, one can decompose a space V_{-1} into two subspaces V_0 and W_0. Similarly, the subspace V_0 can be decomposed into V_1 and W_1, and so forth. In general, we have the following decomposition:

$$\begin{aligned} \phi_{j,k}(t) & \quad V_j \to V_{j+1} \\ \psi_{j,k}(t) & \quad V_j \to W_{j+1} \end{aligned} \tag{312}$$

Each expanded subspace has a different resolution specified by index j. Thus, we may have the multiresolution expansion of a given space, with the resulting wavelet expansion of a function $f(t)$ in this space.

For signals, scaling functions encode the low spatial (or time) frequency information, whereas wavelets encode signals in different frequency bands to a certain level of frequency.

Let us consider a function uniquely determined by N discrete samples

$$\{f(1), f(2), \cdots, f(N)\}$$

It can be shown that it is possible to expand this function as a series of N orthogonal basis functions.

2.10.6. Discrete Wavelet Transform: Wavelet and Scaling Functions

For discrete parameters a and b, a **discrete wavelets transformation** decomposes a function into an expansion (using dilations and translations) of two functions: a scaling function $\phi(t)$ and a wavelet function $\psi(t)$. The basis sets for a scaling function (nonnormalized) are

$$\phi_{L,k}(t) = \phi(2^L t - k), \quad k = 1, 2, \cdots, K_L, \quad K_L = N 2^{-L} \tag{313}$$

where L is an expansion level, and for the wavelet function

$$\psi_{j,k}(t) = \psi(2^j t - k), \quad j = 1, 2, \cdots, L; \quad k = 1, 2, \cdots, K; \quad K = N 2^{-j} \tag{314}$$

where the level of expansion L satisfies $0 < L \leq \log_2(N)$.

An L-level discrete wavelet transform of function $f(t)$ described by N samples contains:

1. a set of parameters $\{a_{L,k}\}$ defined by the inner products of $f(t)$ with $N2^{-j}$ translations of the scaling function $\phi(t)$ at L different widths

$$\{a_{L,k}(t)\} = \{<f(t), \phi_{L,k}(t)>;\ k=1,2,\cdots,K_L,\ K_L = N2^{-L}\} \qquad (315)$$

2. a set of parameters $\{b_{j,k}\}$ defined by the inner products of $f(t)$ with $N2^{-L}$ translations of the wavelet function $\psi_{j,k}(t)$ at a single width

$$\{b_{j,k}(t)\} = \{<f(t), \psi_{j,k}(2^j t - k)>;\ j=1,2,\cdots,L,$$
$$k=1,2,\cdots,K;\ K = N2^{-j}\}$$

The reconstruction of a function $f(t)$ based on wavelet transform coefficients can be obtained by the inverse transform

$$f(t) = \sum_{k}^{K_L} a_{L,k} \phi_{L,k}(t) + \sum_{j}^{L} \sum_{k}^{K} b_{j,k} \psi_{j,k}(t) \qquad (316)$$

The number of parameters of L-level wavelet transform is equal to

$$\sum_{j}^{L} 2^{-j} + N2^{-L} = N(1 - 2^L + 2^L) = N \qquad (317)$$

which is the same as the corresponding number of coefficients of a Fourier transform.

2.10.7. Haar Wavelets

One of the simplest orthogonal wavelets is generated from the Haar scaling function and wavelet. The **Haar transform** uses square pulses to approximate the original function. The basis functions for Haar wavelets at some level all look like a unit pulse, shifted along the x-axis. Haar scales are all of unit pulses.

The Haar wavelet is defined as follows

$$\psi(t) = \begin{cases} 1, & \text{if } t \in [0, 0.5) \\ -1, & \text{if } t \in [0.5, 1) \\ 0, & \text{otherwise} \end{cases} \qquad (318)$$

The dilations and translations of the Haar wavelet function form an orthogonal wavelet base for $L^2(\mathbb{R})$. The **mother Haar wavelets** are defined as

$$\psi_{j,k}(t) = \psi(2^j t - k),\ j, k \in \mathbb{Z} \qquad (319)$$

The Haar scaling function $\phi(t)$ is the unit-width function $\phi(t)$

$$\phi(t) = \begin{cases} 1, & \text{if } 0 \leq t \leq 1, \\ 0, & \text{otherwise} \end{cases} \qquad (320)$$

Figure 7.10 shows the Haar scaling and wavelet functions.

We can easily see that the Haar scaling function $\phi(t)$ can be constructed using $\phi(2t)$:

$$\phi(t) = \phi(2t) - \phi(2t - 1) \qquad (321)$$

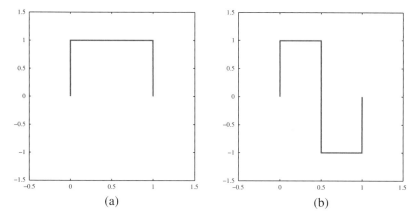

Figure 7.10. Haar scaling (a) and Haar wavelet (b) functions.

The Haar decomposition of a function $f(t) \in L^2(\mathbb{R})$ or finite-dimensional vector for a signal in the time domain can be expressed as

$$f(t) = a_{0,0}\phi_{0,0}(t) + \sum_{j=1}^{d} \sum_{k=0}^{2^{j-1}} b_{j,k}\psi_{j,k}(t) \tag{322}$$

where $f(t) \in L^2[0, 1]$, $a_{0,0}$ is the parameter, $\phi_{0,0}(t)$ is a scale function on the interval [0,1), and $\psi_{j,k}(t)$ is a set of wavelets with different resolutions. Due to the property of local analysis, the time interval of a function is not [0,1); the desired interval can be shifted to [0,1) to get the expected results. For the function $f(t) \in L^2(\mathbb{R})$, the discrete wavelet expansion of $f(t)$ is represented as

$$f(t) = a_{0,0}\phi_{0,0}(t) + \sum_{j=1}^{d} \sum_{k=0}^{2^{j-1}} b_{j,k}\psi_{j,k}(t) = a_{0,0}\phi_{0,0}(t) + \mathbf{b}\boldsymbol{\psi}_{(j)}(t) \tag{323}$$

where $\phi_{0,0}(t)$ is a scale function on interval [0,1), $\psi_{j,k}(t)$ is the set of wavelets with different resolution, $\mathbf{b}$ is the Haar wavelet coefficient vector, and $\boldsymbol{\psi}_j$ is a set of Haar wavelets, $\psi_{j,k}(t)$, which is chosen according to necessity. The two vectors $\mathbf{b}$ and $\boldsymbol{\psi}_j$ are defined by

$$\mathbf{b} = [b_0, b_1, \cdots, b_{j-1}]^T$$
$$\boldsymbol{\psi}_j = [\psi_0, \psi_1, \cdots, \psi_{j-1}]^T \tag{324}$$

where ψ_i, $(i = 0, 1, \cdots, j-1)$, is some $\psi_{j,k}(t)$.

The Haar integral operational matrix $\mathbf{P}$ is given as

$$\int_{t_0}^{t} \boldsymbol{\psi}_{(j)}(t)\,dt = \mathbf{P}\boldsymbol{\psi}_{(j)}(t) \tag{325}$$

The coefficients $p(i, j)$ of $\mathbf{P}$ can be found numerically as the discrete wavelet expansion

$$\mathbf{P}(i, j) = <\int_{t_0}^{t} \psi_i dt, \ \psi_j> \tag{326}$$

2.10.8. Two-dimensional Wavelets

In general, the **two-dimensional wavelet transform** can be realized by successively applying the one-dimensional wavelet transform to data in every dimension. First, the rows are transformed by using a one-dimensional transform, and a similar transformation is provided for all columns of the intermediate results.

For a two-dimensional wavelet expansion of the two-dimensional function $f(t_1, t_2)$, we have to define a wavelet function of two variables t_1 and t_2 $\psi(t_1, t_2)$. This function should satisfy the following condition:

$$C_\psi = \int_{-\infty}^{\infty} \frac{|\Psi(vt_1, vt_2)|^2}{|v|} dv < \infty \tag{327}$$

where Ψ denotes the Fourier transform of ψ.

The two-variable wavelet function can be defined as a product of two one-dimensional mother (generating) wavelets $\psi(t_1)$ and $\psi(t_2)$:

$$\psi(t_1, t_2) = \psi(t_1)\psi(t_2) \tag{328}$$

With dilation and translation parameters, the two-dimensional wavelet function is defined as

$$\psi_{(a_1,a_2),(b_1,b_2)}(t_1, t_2) = \psi_{(a_1,b_1)}(t_1)\psi_{(a_2,b_2)}(t_2) \tag{329}$$

where $\psi_{a,b}(t) = |a|^{-\frac{1}{2}}\psi(\frac{t-b}{a})$. We can also write

$$\psi_{(a_1,a_2),(b_1,b_2)}(t_1, t_2) = \frac{1}{\sqrt{|a_1 a_2|}}\psi_{(a_1,b_1)}\left(\frac{t_1-b_1}{a_1}\right)\psi_{(a_2,b_2)}\left(\frac{t_2-b_2}{a_2}\right) \tag{330}$$

Assuming that $a_1 = a_2$ the **two-dimensional continuous wavelet expansion** of the two-variable function $f(t_1, t_2)$ can be expressed as

$$TWT f((a), (b_1, b_2)) = |a|^{-1} \int_{-\infty}^{\infty}\int_{-\infty}^{\infty} f(t_1, t_2)\psi_{(a),(b_1,b_2)}(t_1, t_2) dt_1 dt_2 \tag{331}$$

The **inverse of the continuous wavelet transform** is defined as

$$f(t_1, t_2) = \frac{1}{C_\psi} \int_{-\infty}^{\infty}\int_{-\infty}^{\infty}\int_{-\infty}^{\infty} \frac{1}{a^4} TWT f((a), (b_1, b_2))\psi_{(a),(b_1,b_2)}(t_1, t_2) da\, db_1\, db_2 \tag{332}$$

For the discretized parameters a, b_1, and b_2,

$$a = a_0^j, \quad b_1 = k_1 ab_{1,0}, \quad b_2 = k_2 ab_{2,0} \tag{333}$$

the two-dimensional discrete wavelet expansion can be written as

$$f(t_1, t_2) = \sum_{k_1}\sum_{k_2} a_{L,k_1,k_2}\phi_{L,k_1,k_2}(t_1, t_2) + \sum_{i=H,D,V}\sum_{j=1}^{L}\sum_{k_1}\sum_{k_2} b_{j,k_1,k_2}^i \psi_{j,k_1,k_2}^i(t_1, t_2) \tag{334}$$

For a two-dimensional grid of $2^n \times 2^n$ values (for example, image pixels) and for discrete parameters $a_1 = 2^{n-j_1}$, $a_2 = 2^{n-j_2}$, $b_1 = 2^{n-j_1}k_1$, $b_2 = 2^{n-j_2}k_2$, with integer values for j_1, j_2, k_1, and k_2, the 2D discrete wavelet function can be defined as

$$\psi_{j_1,j_2,k_1,k_2}(t_1, t_2) = 2^{\frac{(j_1+j_2)}{2}-n}\psi(2^{j_1-n}t_1 - k_1)\psi(2^{j_2-n}t_2 - k_2) \tag{335}$$

where j_1, j_2, k_1, and k_2 are the dilation and the translation coefficients for each variable, satisfying the conditions

$$0 \leq j_1, \ j_2 \leq n-1, \ 0 \leq k_1 \leq 2^{j_1}-1, \ 0 \leq k_2 \leq 2^{j_2}-1 \tag{336}$$

The resolution level is $j = \frac{j_1+k_1}{2}$ and corresponds to 2^{n-j}.

Additionally, defining the scaling function

$$\phi_{j_1,j_2,k_1,k_2}(t_1, t_2) = 2^{\frac{j_1+j_2}{2}-n} \phi(2^{j_1-n}t_1 - k_1) \phi(2^{j_2-n}t_2 - k_2) \tag{337}$$

allows us to define a complete basis to reconstruct a discrete function $f(t_1, t_2)$ (for example, a discrete image):

$$\phi_{0,0,0,0}(t_1, t_2) = 2^{-n} \phi(2^{-n}t_1) \phi(2^{-n}t_2)$$

$$\gamma^H_{0,j_2,0,k_2}(t_1, t_2) = 2^{\frac{j_2}{2}-n} \phi(2^{-n}t_1) \psi(2^{j_2-n}t_2 - k_2)$$

$$\gamma^V_{j_1,0,k_1,0}(t_1, t_2) = 2^{\frac{j_1}{2}-n} \psi(2^{j_1-n}t_1 - k_1) \phi(2^{-n}t_2) \tag{338}$$

The discrete bases satisfy orthonormality conditions

$$< \lambda_{j_1,j_2,k_1,k_2}, \lambda_{j'_1,j'_2,k'_1,k'_2} > = \delta_{j_1,j'_1} \delta_{j_2,j'_2} \delta_{k_1,k'_1} \delta_{k_2,k'_2} \tag{339}$$

where $\lambda_{\ldots}$ denotes any of the previous orthonormal bases. The 2D discrete wavelet coefficients are defined as

$$2DWT \ w_{j_1,j_2,k_1,k_2} = \int \int f(t_1, t_2) \lambda_{j_1,j_2,k_1,k_2} \, dt_1 \, dt_2 \tag{340}$$

where $\lambda_{j_1,j_2,k_1,k_2}$ denotes any of the previously defined orthonormal bases. These coefficients can be formed as the Haar wavelet coefficient matrix **P**.

An **inverse discrete wavelet reconstruction** (an image reconstruction) can be described by the following expression:

$$f(t_1, t_2) = \sum_{j_1} \sum_{j_2} \sum_{k_1} \sum_{k_2} w_{j_1,j_2,k_1,k_2} \lambda_{j_1,j_2,k_1,k_2}(t_1, t_2) \tag{341}$$

Wavelet Patterns. Patterns, representing recognition objects (time-series or images), can be formed based on the coefficient matrices of certain level of wavelets transform. One can constitute a pattern through concatenation of subsequent parameter matrices rows as one pattern.

2.11. Zernike Moments

A robust pattern recognition system must be able to recognize an image (or an object within an image) regardless of its orientation, size, or position. In other words, **rotation-**, **scale-**, and **translation-invariance** are desired properties for extracted features. For example, as shown in Figure 7.11, all the images should be recognized as "8." In this section, we will introduce moments and complex Zernike moments for robust feature extraction from images. In statistics, the concept of **moments** is used extensively. Moments were first introduced for two-dimensional pattern recognition in the early 1960s. However, the recovery of an image from these moments is quite difficult and computationally expensive.

Major extension of moments has been provided through introduction of **Zernike moments** by using the idea of orthogonal moment invariants and the theory of orthogonal polynomials. This

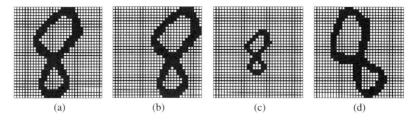

Figure 7.11. The image of character "8" (a) and its translated (b), scaled (c), and rotated (d) versions.

approach allows moment invariants to be constructed to an arbitrarily high order. In addition, Zernike moments are also rotation – invariant. Rotating the image does not change the magnitudes of the moments. Another property of Zernike moments is the simplicity of image reconstruction.

Studies have shown that orthogonal moments including Zernike moments are better than other types of moments in terms of information redundancy and image representation. Since Zernike moments are only rotation invariant, to obtain scale and translation invariance, an image must be normalized via **image normalization**. In order to understand the image normalization process, in the next section we briefly describe three basic image processing operations: **translation**, **scaling**, and **rotation**.

2.11.1. An Image Description

A computer image is a collection of **pixels** in 2D coordinate space, with the horizontal axis usually labeled x and the vertical axis usually labeled y. A gray-scale spatial domain image can be defined as

$$\{f(x, y) \in \{0, 1, \cdots, 255\} : x = 0, 1, \cdots, M-1; \; y = 0, 1, \cdots, N-1\} \tag{342}$$

and a binary spatial domain image as

$$\{f(x, y) \in \{0, 1\} : x = 0, 1, \cdots, M-1; \; y = 0, 1, \cdots, N-1\} \tag{343}$$

where x is the column index, y is the row index, M is the number of columns, N is the number of rows, and $f(x, y)$ is the pixel value at location (x, y).

2.11.2. Basic Image Transformations

Let f denote the original image and f^t the transformed image. **Translation**,

$$f^t(x + x', y + y') = f(x, y) \tag{344}$$

is a transformation that allows an image to be changed in position the along x-axis by x' and along the y-axis by y'.

The operation of **scaling**,

$$f^s(x, y) = f(x/a_x, y/a_y) \tag{345}$$

where a_x and a_y are scaling factors and $a_x, a_y > 0$, allows an image to be changed in size. If $a_x/a_y < 1$, the image is shrunk along the x-axis / y-axis. Similarly, if $a_x/a_y > 1$, the image is enlarged along x-axis / y-axis.

The **rotation** operation,

$$f^r(x, y) = f^r(\rho, \theta) = f(\rho, \theta - \alpha) \tag{346}$$

allows an image to be rotated about its center point through any arbitrarily specified angle. The angle of rotation is counterclockwise. Let f^r denote a rotated image and f the original image. The relationship between the rotated and original image can be explained as follows: ρ is the length of the vector from the origin to the (x, y) pixel, θ is the angle between ρ and the x-axis in the counterclockwise direction, and α is the angle of rotation in the counterclockwise direction.

2.11.3. Image Normalization

Zernike moments are only rotationally invariant, but the images might have scale and translation differences. Therefore, prior to the extraction of Zernike moments, the images should be normalized with respect to scaling and translation.

Translation invariance can be achieved by moving the origin to the center (centroid) of an image. To obtain the centroid location of an image, **general moments** (or **regular moments**) can be used. General moments are defined as

$$m_{pq} = \int_{-\infty}^{\infty} \int_{-\infty}^{\infty} x^p y^q f(x, y)\, dx dy \qquad (347)$$

where m_{pq} is the $(p+q)^{th}$ order moment of the continuous image function $f(x, y)$. For digital images, the integrals can be replaced by summations. Given a two-dimensional $M \times N$ image, the moment m_{pq} is given by

$$m_{pq} = \sum_{x=0}^{M-1} \sum_{y=0}^{N-1} x^p y^q f(x, y) \qquad (348)$$

To keep the dynamic range of m_{pq} consistent for any size of image, the $M \times N$ image plane should be first mapped onto a square defined by $x \in [-1, +1]$, $y \in [-1, +1]$. This mapping implies that grid locations will no longer be integers but will have real values in the $[-1, +1]$ range. This changes the definition of m_{pq} to

$$m_{pq} = \sum_{x=-1}^{+1} \sum_{y=-1}^{+1} x^p y^q f(x, y) \qquad (349)$$

Hence, we can find the centroid location of an image by the general moment. According to Zernike, the coordinates of the image centroid $(\bar{x}, \bar{y})$ are

$$\bar{x} = \frac{m_{10}}{m_{00}}; \quad \bar{y} = \frac{m_{01}}{m_{00}} \qquad (350)$$

To achieve translation invariance, one can transform the image into a new one whose first-order moments, m_{01} and m_{10}, are both equal to zero. This can be done by transforming the original image into the $f(x+\bar{x}, y+\bar{y})$ image, where $\bar{x}$ and $\bar{y}$ are centroid locations of an original image computed as in Equation (350). In other words, we need to move the origin of the coordinates to the image centroid. Let $g(x, y)$ represent the translated image; then the new image function becomes

$$g(x, y) = f(x+\bar{x}, y+\bar{y}) \qquad (351)$$

Scale invariance is accomplished by enlarging or reducing each image such that its zero-order moment, m_{00}, is set equal to a predetermined value β. That is, we can achieve this outcome by transforming the original image function $f(x, y)$ into a new function $f(x/a, y/a)$, with scaling factor a, where

$$a = \sqrt{\frac{\beta}{m_{00}}} \qquad (352)$$

Note that in the case of binary images, m_{00} is equal to the total number of object pixels in the image. β is chosen based on the size of the image and the object in the image. For example, one can choose $\beta = 800$ for 64×64 binary images of characters "A" to "Z"; consequently, the lower case "a" to "z" might need a lower value of β than that of upper case "A" to "Z". Another choice – for example, for 32×32 images of digits "0" to "9" – could be $\beta = 256$.

Let $g(x, y)$ be the scaled image. After scale normalization, we will obtain

$$g(x, y) = f\left(\frac{x}{a}, \frac{y}{a}\right) \tag{353}$$

2.11.4. Translation and Scale Normalization

In summary, an image function can be normalized with respect to scale and translation by transforming it into $g(x, y)$, where

$$g(x, y) = f\left(\frac{x}{a} + \bar{x}, \frac{y}{a} + \bar{y}\right) \tag{354}$$

with $(\bar{x}, \bar{y})$ being the centroid of $f(x, y)$ and $a = \sqrt{\frac{\beta}{m_{00}}}$, with β a predetermined value. However, $(x/a + \bar{x}, y/a + \bar{y})$ might not correspond to a grid location. To solve this problem, an interpolation method known as **nearest neighborhood approximation** can be used. In this technique, four nearest pixels are used and the fractional address of a pixel is truncated to the nearest integer pixel address.

2.11.5. Zernike Moments

As a result of image normalization, an image has obtained translation and scale invariance. To achieve rotational invariance, complex **Zernike moments** are used.

Zernike introduced a set of complex polynomials that form a complete orthogonal set over the interior of a unit circle, i.e., $x^2 + y^2 = 1$. Let the set of these polynomials be denoted by $V_{nl}(x, y)$. The form of these polynomials is

$$V_{nl}(x, y) = V_{nl}(\rho \sin \theta, \rho \cos \theta) = V_{nl}(\rho, \theta) = R_{nl}(\rho) \exp(il\theta), \tag{355}$$

where n is a positive integer or zero; l is a positive or negative integer, subject to the constraints $n - |l| =$ even; $|l| \leq n$, i is the complex number $i = \sqrt{-1}$; ρ is the length of the vector from the origin to the (x, y) pixel; and θ is the angle between vector ρ and the x-axis in the counterclockwise direction.

In the following, we assume that the notations $[V_{nl}(x, y)]^*$ and $V_{nl}^*(x, y)$ are equivalent where the symbol * denotes the complex conjugate. Radial polynomials $R_{nl}(\rho)$ are defined as

$$R_{nl}(\rho) = \sum_{s=0}^{\frac{n-|l|}{2}} \frac{(-1)^s [(n-s)!] \rho^{n-2s}}{s!(\frac{n+|l|}{2} - s)!(\frac{n-|l|}{2} - s)!} \tag{356}$$

We note that $R_{n,-l}(\rho) = R_{nm}(\rho)$. These polynomials are orthogonal and satisfy the equality

$$\iint_{x^2+y^2\leq 1} [V_{nl}(x,y)]^* V_{jk}(x,y)\, dxdy = \frac{\pi}{n+1}\delta_{nj}\delta_{lk}$$

where the meaning of Kronecker delta is as follows:

$$\delta_{ab} = \begin{cases} 1 & a=b \\ 0 & \text{otherwise} \end{cases} \tag{357}$$

where the symbol $*$ denotes the complex conjugate. Zernike moments are projection of the image function onto these orthogonal basis functions. The Zernike moments of order n, with repetition l for a continuous image function $f(x,y)$, are

$$A_{nl} = \frac{n+1}{\pi}\iint_{x^2+y^2\leq 1} f(x,y)[V_{nl}(\rho,\theta)]^*\, dxdy = (A_{n,-l})^* \tag{358}$$

For a digital image,

$$A_{nl} = \frac{n+1}{\pi}\sum_x\sum_y f(x,y) V_{nl}^*(\rho,\theta); \quad x^2+y^2 \leq 1 \tag{359}$$

The real Zernike moments for $l \neq 0$ are

$$\begin{bmatrix} C_{nl} \\ S_{nl} \end{bmatrix} = \frac{2n+2}{\pi}\iint_{x^2+y^2\leq 1} f(x,y) R_{nl}(\rho) \begin{bmatrix} \cos l\theta \\ -\sin l\theta \end{bmatrix} dxdy \tag{360}$$

or

$$C_{nl} = 2\,Re\,(A_{nl})$$
$$= \frac{2n+2}{\pi}\iint_{x^2+y^2\leq 1} f(x,y) R_{nl}(\rho) \cos l\theta\, dxdy \tag{361}$$

$$S_{nl} = -2\,Im\,(A_{nl})$$
$$= \frac{-2n-2}{\pi}\iint_{x^2+y^2\leq 1} f(x,y) R_{nl}(\rho) \sin l\theta\, dxdy \tag{362}$$

and for $l = 0$

$$C_{n0} = A_{n0} = \frac{1}{\pi}\iint_{x^2+y^2\leq 1} f(x,y) R_{n0}(\rho)\, dxdy$$
$$S_{n0} = 0 \tag{363}$$

For a digital image, when $l \neq 0$,

$$\begin{bmatrix} C_{nl} \\ S_{nl} \end{bmatrix} = \frac{2n+2}{\pi}\sum_{x=-1}^{+1}\sum_{y=-1}^{+1} f(x,y) R_{nl}(\rho) \begin{bmatrix} \cos l\theta \\ -\sin l\theta \end{bmatrix} \tag{364}$$

and when $l = 0$,

$$C_{n0} = A_{n0} = \frac{1}{\pi}\sum_{x=-1}^{+1}\sum_{x=-1}^{+1} f(x,y) R_{n0}(\rho)$$
$$S_{n0} = 0 \tag{365}$$

The connection between real and complex Zernike moments is ($l > 0$)

- $C_{nl} = 2Re(A_{nl})$
- $S_{nl} = -2Im(A_{nl})$
- $A_{nl} = \frac{(C_{nl} - iS_{nl})}{2} = (A_{n,-l})^*$

To compute the Zernike moments of a given image, the center of the image is taken as the origin, and pixel coordinates are mapped to the range of the unit circle, i.e., $x^2 + y^2 \leq 1$. Those pixels that fall outside the unit circle are not used in the computation.

The image **translation** and **scale normalization** processes affect two of the Zernike features; namely, $|A_{00}|$ and $|A_{11}|$, the magnitude of Zernike moments A_{00} and A_{11}.

1. $|A_{00}|$ is going to be the same for all images.

$$C_{00} = \frac{2}{\pi} \int\int_{x^2+y^2 \leq 1} g(x,y) R_{00}(\rho) \, dxdy = \frac{2}{\pi} m_{00}; \quad S_{00} = 0 \tag{366}$$

Since $m_{00} = \beta$,

$$|A_{00}| = \left| \left(\frac{C_{00}}{2} \right) - i \left(\frac{S_{00}}{2} \right) \right| = \frac{\beta}{\pi} \tag{367}$$

2. $|A_{11}|$ is equal to zero.

$$C_{11} = \frac{4}{\pi} \int\int_{x^2+y^2 \leq 1} g(x,y) R_{11}(\rho) \cos\theta \, dxdy$$

$$= \frac{4}{\pi} \int\int_{x^2+y^2 \leq 1} g(x,y) \, \rho \cos\theta \, dxdy$$

$$= \frac{4}{\pi} \int\int_{x^2+y^2 \leq 1} g(x,y) \, x \, dxdy$$

$$= \frac{4}{\pi} m_{10} \tag{368}$$

and

$$S_{11} = \frac{4}{\pi} \int\int_{x^2+y^2 \leq 1} g(x,y) R_{11}(\rho) \sin\theta \, dxdy$$

$$= \frac{4}{\pi} \int\int_{x^2+y^2 \leq 1} g(x,y) \, \rho \sin\theta \, dxdy$$

$$= \frac{4}{\pi} \int\int_{x^2+y^2 \leq 1} g(x,y) \, y \, dxdy$$

$$= \frac{4}{\pi} m_{01}, \tag{369}$$

Since $m_{10} = m_{01} = 0$ for all **normalized** images, then

$$|A_{11}| = \left| \left(\frac{C_{11}}{2} \right) - i \left(\frac{S_{11}}{2} \right) \right| = 0 \tag{370}$$

Therefore, $|A_{00}|$ and $|A_{11}|$ are not taken as features utilized in the classification.

Image reconstruction (**inverse transform**) from Zernike moments can be done in a simple way. Suppose we know all moments A_{nl} (C_{nl} and S_{nl}) of an image $f(x,y)$ up to a given order n_{max}. We can reconstruct an image $\hat{f}$ by

$$\hat{f}(x,y) = \sum_{n=0}^{n_{max}} \sum_l A_{nl} V_{nl}(\rho,\theta) \quad (371)$$

where $n - |l| =$ even and $|l| \leq n$.

Since it is easier to work with real-valued functions, we can expand Equation (371) to

$$\hat{f}(x,y) = \sum_{n=0}^{n_{max}} \sum_l (C_{nl} \cos l\theta + S_{nl} \sin l\theta) R_{nl}(\rho) \quad (372)$$

where $n - |l| =$ even and $|l| \leq n$.

The reconstructed image can be generated by mapping $f(x,y)$ to the $[0, 255]$ range. To generate a binary image, we can use a threshold of 128. One can choose the values of β and order as 256 and 12, respectively. After image normalization, the input image would turn out to be close to the original.

2.11.6. Pattern Formation from Zernike Moments

Let us assume that $n_f = (p+1)(q+1)$ subsequent Zernike moments, of orders from m_{00} to m_{pq}, have been extracted from a given normalized image. One can form the Zernike moment-based pattern representing an image. Each Zernike moment of a given order is a complex number with real part C and imaginary part S. In the pattern-forming phase, we represent a given i^{th} moment by its real-valued magnitude $\sqrt{C^2 + S^2}$, and we set this value as the ith element of the Zernike pattern $\mathbf{x}_{zer}$.

The **translation** and **scale normalization** processes affect two of the Zernike features, namely, $|A_{00}|$ and $|A_{11}|$. The magnitude of the Zernike moments A_{00} is the same for all images, and the moment $|A_{11}|$ is equal to zero. Therefore, $|A_{00}|$ and $|A_{11}|$ are not considered as pattern features utilized in the classification. Consequently, the length of the Zernike pattern is equal $n_z = n_f - 2$.

Values of m_{pq} (and the resulting value n_f) are found heuristically.

Example: Zernike moments have been successfully applied to handwritten character recognition. Here, Zernike moments from (2,0) through (12,12) are extracted from binarized, thinned, and normalized 32×32 pixel images of characters. From the initial number $169 - 2 = 167$ of Zernike moments, the final (reduced) 6 element patterns are selected by the rough sets methods and used in recognition. The back-propagation neural network-based classifier yielded 92% accuracy in this application.

3. Feature Selection

Pattern **dimensionality reduction** (and thus data set compression), via feature extraction and feature selection, belongs to the most fundamental steps in data processing. Feature selection can be an inherent part of feature extraction (for example, using principal component analysis) or even a processing algorithm design (as in decision tree design). However, feature selection is often isolated as a separate step in processing sequence.

We can define **feature selection** as a process of finding a subset of features, from the original set of features forming patterns in a given data set, according to the defined criterion of feature selection (a **feature goodness criterion**). Here, we consider feature selection as a process of finding the best **feature subset** X_{opt} from the original set of pattern features, according to the defined feature goodness criterion $J_{\text{feature}}(X_{\text{feature\_subset}})$, without additional feature transformation or construction. Feature selection should be stated in terms of the optimal solution of the selection problem (according to the defined goal and criterion) and with the resulting algorithm of such optimal selection.

3.1. Optimal Feature Selection

Assume that a limited-size data set T_{all} is given (consisting of N_{all} cases), constituted with n-feature patterns **x** (labeled or unlabeled by target values), sometimes accompanied by a priori knowledge about domain. Let all n features of the pattern (the pattern vector elements x_i ($i = 1, 2, \cdots, n$)) form the entire original feature set $X_{\text{all}} = \{x_1, x_2, \cdots, x_n\}$. The **optimal feature selection** is the process of finding, for a given type of predictor, a subset $X_{\text{opt}} = \{x_{1,\text{opt}}, x_{2,\text{opt}}, \cdots, x_{m,\text{opt}}\}$ containing $m \leq n$ features from the set of all original features $X_{\text{opt}} \subseteq X_{\text{all}}$ that guarantee accomplishment of a processing goal while minimizing a defined feature selection criterion (a feature goodness criterion) $J_{\text{feature}}(X_{\text{feature\_subset}})$. The optimal feature set will depend on the type of predictor designed. More precisely, optimal feature selection will depend on the overall processing goal and its performance evaluation criterion, type of predictor designed, existing data set, a priori domain knowledge, original set of pattern features, overall processing algorithm applied, and defined criterion of feature subset goodness J_{feature} (feature selection criterion). A solution for the optimal feature selection may not be unique. Different subsets of original features may result in the same performance.

The goals of data processing, roles of features, and performance evaluation criteria of processing algorithms may be different. Feature selection algorithms, based on defined feature selection (feature goodness) criteria and the resulting optimal features, will depend on these conditions. For example, for the same data set from the same domain, optimal features found for the classification task, with minimum average classification error probability criterion, might be different that those found for the data compression task with the minimum sum squares error criterion. Similarly, an optimal feature set found for a Bayesian quadratic discriminant-based classifier could be different that that found for a back-propagation neural network classifier.

Generally, the criterion J of performance evaluation for an overall processing algorithm and the criterion J_{feature} of feature goodness are different, although one can design optimal feature selection where these criteria can be the same.

3.1.1. Paradigms of Optimal Feature Selection

We will shortly discuss two paradigms in optimal feature selection: **minimal representation** and **maximal class separability**.

The general goal of feature selection typically includes the key ability of the processing algorithm (using an optimal feature subset) to best process novel instances of domain data that were not seen or used during the design (generalization problem).

Since the processing algorithm (for example, a classifier) is a data model, and since optimal feature selection influences processing algorithm complexity, we can state that optimal feature selection should have much in common with finding an optimal data model. This similarity implies that optimal feature selection might possibly be supported by some general paradigms of data model building. Even though these paradigms have mostly theoretical value, experience shows that they may have also practical implications. The most prominent paradigms in data model building, and potentially in optimal feature selection, are the so-called **minimum construction paradigms**: **Occam's razor**, **minimum description length**, and **minimum message length** (see Chapter 15).

In the light of minimum construction, a straightforward technique of best feature selection could be to choose a minimal feature subset that fully describes all concepts (for example, classes in prediction-classification) in a given data set. However, this approach, while applicable for a given (possibly limited) data set, may not be useful for processing unseen patterns, since these methods might not provide good generalization.

Methods based on minimum construction paradigms for limited-size data sets should take into account generalization problem. This kind of approach relates to the general solution of the **bias-variance dilemma** in data processing design based on limited-size data sets.

Informally, let us transform these somewhat overlapping minimum construction paradigms into the terms of optimal feature selection. The paradigms deal with generalization versus the complexity of a processing algorithm (a data model complexity) and are influenced by the size of the feature set. They indicate that in order to obtain the best generalization, we should find the processing algorithm that has minimal complexity guaranteed by the minimal feature set and that well represents the available data set. Such a paradigm sheds some light on the design of the generalizing algorithm for optimal feature selection. However, it does not provide a rigorous design procedure.

Designers of algorithms based on optimal feature selection face the **bias/variance dilemma** (see Chapter 15). This dilemma underlines the controversy related to the selection of a processing algorithm and optimal feature set complexity: namely, the need to find the best process for a given data set and simultaneously to provide the best generalization for future patterns. Given this dilemma, designers divide the generalization error criteria into the sum of two parts: squared **bias** and **variance**. If the designer too precisely fits the complex processing algorithm to given data in a large feature set, then the algorithm's ability to generalize for unseen patterns may deteriorate. By increasing the complexity of the processing algorithm and feature set, we can reduce the bias and increase the variance. On the other hand, a processing algorithm with a small feature set may not be able to process a given data set satisfactorily. A processing algorithm that is too simple and thus inflexible (with too small a number of parameters), influenced by its small feature set, may have too big a bias and too small a variance. The robust processing algorithm, with its associated set of features (reflecting complexity), implements a tradeoff between its best ability to process a given data set and its generalization capability. This problem is also called the **bias/variance tradeoff**. Despite the theoretical power of design paradigms with minimal structures (for example, with a minimal set of features), in practice, for limited size data sets, the optimal solution is not always a minimal one.

The second general paradigm of optimal feature selection, mainly used in classifier design, relates to selecting the feature subset that guarantees maximal between-class separability for a reduced data set and thus helps design a better predictor-classifier. This paradigm relates to discriminatory power of features, i.e., their ability to distinguish patterns from different classes.

Selection of the best feature subset for a given prediction task corresponds to **feature relevancy**. The relevance of a feature can be understood as its ability to contribute to improving the predictor's performance. For a predictor-classifier, relevance would mean the ability to improve classification accuracy.

A few attempts (both deterministic and probabilistic) have been made in machine learning to define feature relevancy. Let us assume a labeled data set T with N cases (**x**, **target**), containing n-feature patterns **x** and associated **targets**. For classification, a **target** is a categorical class target c_{target} (a concept c) with values from the set of l discrete classes $\{c_1, c_2, \cdots, c_l\}$. For regression, a target is the desired output (scalar or vector) of a real valued predictor (see Chapter 4).

The following definition of deterministic relevancy was proposed for Boolean features in noise-free data sets for the classification task.

Definition 1. A feature x_i is **relevant to a class** c (a concept c) if x_i appears in every Boolean formula that represents c, and is **irrelevant** otherwise.

Definition 2. A feature x_i is **relevant** if there exists some value of that feature a_{x_i} and a predictor output **y** value $\mathbf{a_y}$ (generally a vector) for which $P(x_i = a_{x_i}) > 0$ such that

$$P(\mathbf{y} = \mathbf{a_y} | x_i = a_{x_i}) \neq P(\mathbf{y} = \mathbf{a_y}) \qquad (373)$$

According to this definition, a feature x_i is relevant if knowledge of its value can change the estimates of **y**, or in another words, if an output vector **y** is conditionally dependent on x_i. Since the above definitions do not deal with the relevance of features in the parity concept, a modification was proposed. Let us denote a vector of features $\mathbf{v}_i = (x_1, x_2, \cdots, x_{i-1}, x_{i+1}, \cdots, x_n)^T$ (with its values denoted by $\mathbf{a_{v_i}}$) obtained from an original feature vector **x** by removing the x_i feature.

Definition 3. A feature x_i is **relevant** if there exists some value of that feature a_{x_i} and a predictor output **y** value $\mathbf{a_y}$ (generally a vector) for which $P(x_i = a_{x_i}) > 0$ such that

$$P(\mathbf{y} = \mathbf{a_y}, \mathbf{v}_i = \mathbf{a_{v_i}} | x_i = a_{x_i}) \neq P(\mathbf{y} = \mathbf{a_y}, \mathbf{v}_i = \mathbf{a_{v_i}}) \qquad (374)$$

According to this definition, a feature x_i is relevant if the probability of a target (given all features) can change if we remove knowledge about a value of that feature. Since the above definitions are quite general and may provide unexpected relevancy judgments for a specific data set (for example, one with the nominal features numerically encoded by indicators), more precise definitions of so-called strong and weak relevance were introduced.

Definition 4. A feature x_i is **strongly relevant** if there exists some value of that feature a_{x_i}, a predictor output **y** value $\mathbf{a_y}$, and a value $\mathbf{a_{v_i}}$ of a vector $\mathbf{v}_i$ for which $P(x_i = a_{x_i}, \mathbf{v}_i = \mathbf{a_{v_i}}) > 0$ such that

$$P(\mathbf{y} = \mathbf{a_y} | \mathbf{v}_i = \mathbf{a_{v_i}}, x_i = a_{x_i}) \neq P(\mathbf{y} = \mathbf{a_y} | \mathbf{v}_i = \mathbf{a_{v_i}}) \qquad (375)$$

Strong relevance indicates that a feature is indispensable, which means that its removal from a feature vector will decrease **prediction accuracy**.

Definition 5. A feature x_i is **weakly relevant** if it is not strongly relevant, and there exists some subset of features (forming a vector $\mathbf{z}_i$) from a set of features forming patterns $\mathbf{v}_i$ for which there exist some value of that feature a_{x_i}, a predictor output value $\mathbf{a_y}$, and a value $\mathbf{a_{z_i}}$ of a vector $\mathbf{z}_i$, for which $P(x_i = a_{x_i}, \mathbf{z}_i = \mathbf{a_{z_i}}) > 0$ such that

$$P(\mathbf{y} = \mathbf{a_y} | \mathbf{z}_i = \mathbf{a_{z_i}}, x_i = a_{x_i}) \neq P(\mathbf{y} = \mathbf{a_y} | \mathbf{z}_i = \mathbf{a_{z_i}}) \qquad (376)$$

Weak relevance indicates that a feature might be dispensable but sometimes (in the company of some other features) may improve prediction accuracy.

In the light of the above definitions, a feature is **relevant** if it is either **strongly relevant** or **weakly relevant**; otherwise, it is **irrelevant**. By definition an irrelevant feature will never contribute to prediction accuracy and thus can be removed.

The theory of rough sets defines deterministic strong and weak relevance for discrete features and discrete targets. For a given data set, a set of all strongly relevant features forms a **core**. A minimal set of features satisfactory to describe concepts in a given data set, including a core and possibly some weakly relevant features, form a **reduct**. A core is an intersection of reducts.

It has been shown that, for some predictor designs, feature relevancy (even strong relevancy) does not imply that the feature must be in an optimal feature subset. Relevancy, although helpful in feature assessment, does not necessarily contribute to optimal predictor design with generalization ability.

Since an optimal feature set depends on the type of predictor used, definitions of absolute irrelevant, conditionally irrelevant, and conditionally relevant are suggested. **Absolute irrelevant** features, equivalent to irrelevant features as defined above, are those that cannot contribute to prediction performance, and thus can be removed. The remaining features are either conditionally irrelevant or relevant, depending on the designed predictor type. For a given type of predictor, **conditionally irrelevant** features are these not included in an optimal set of features (for which a predictor achieves maximal performance). **Conditionally irrelevant** features are not included in the optimal set for a given predictor and thus can be removed. However, conditionally irrelevant features for one type of a predictor could be conditionally relevant for another types. The conditional relevance depends not only on the type of predictor used but also on the applied feature optimality criterion.

3.1.2. Feature Selection Methods and Algorithms

Although optimal feature selection, related to data model discovery and processing algorithm design, is a rather general problem, so far the statistical, machine learning, and automatic control communities have developed slightly different methods of solution. The existing feature selection methods, depending on the feature selection criterion used, include two main streams:

– open loop methods (filter, preset bias, front end)
– closed loop methods (wrapper, classifier feedback)

Open loop methods, also called **filter**, **preset bias**, or the **front end** methods (Figure 7.12), are based mostly on selecting features through the use of between-class separability criteria. These methods do not consider the effect of selected features on the performance of an entire processing algorithm (for example, a classifier), since the feature selection criterion does not involve predictor evaluation for reduced data sets containing patterns with selected feature subsets only. Instead, these methods select, for example, those features for which the resulting reduced data set has maximal between-class separability, usually defined based on between-class and between-class covariances (or scatter matrices) and their combinations. The ignoring of the effect of a selected feature subset on the performance of the predictor (lack of feedback from predictor performance) is a weak side of open-loop methods. However, these methods are computationally less expensive.

Closed loop methods, also called **wrapper**, **performance bias**, or **classifier feedback** methods (Figure 7.13), are based on feature selection using predictor performance (and thus providing processing feedback) as a criterion of feature subset selection. The goodness of a selected feature subset is evaluated using as a criterion $J_{feature} = J_{predictor}$, where $J_{predictor}$ is the performance evaluation of a whole prediction algorithm for a reduced data set containing patterns with the selected features as pattern elements. Here, the selection algorithm is a "**wrapper**" around the prediction algorithm.

Closed loop methods generally provide better selection of a feature subset, since they fulfill the ultimate goal and criterion of optimal feature selection, i.e., they provide best prediction. The

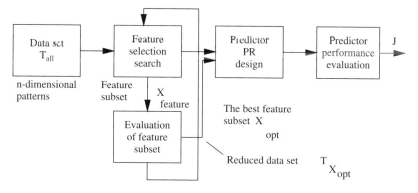

Figure 7.12. An open loop feature selection method.

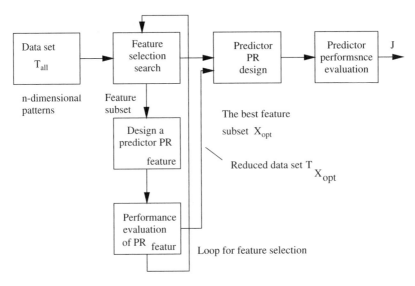

Figure 7.13. A closed loop feature selection method.

prediction algorithm used in closed-loop feature selection may be the final predictor *PR* (which provides best selection, $J_{feature} = J$), or a simpler predictor $PR_{feature}$ may be used for reasons of computational feasibility.

A procedure for optimal feature selection contains

- **feature selection criterion** $J_{feature}$ that allows us to judge whether one subset of features is better than another (evaluation method)
- systematic **search procedure** that allows us to search through candidate subsets of features and includes the initial state of the search and stopping criteria

A search procedure selects a feature subset from among possible subsets of features, and the goodness of this subset is evaluated using the feature selection (optimality judgement) criterion.

Some feature selection criteria do not obey the monotonicity property, which limits the application of the dynamic programming type of search. Ideally, a feature selection criterion should be the same as the criterion for evaluating an entire predictor algorithm (a prediction quality evaluation). This is the ideal approach for closed-loop type (wrapper) feature selection. In practice, simplified selection criteria could be used with simpler predictors that are used exclusively for feature selection.

An ideal search procedure would implement an exhaustive search through all possible subsets of features. This approach is, in fact, the only method that ensures finding an optimal solution. In practice, for large number of features, an exhaustive search is not feasible. Thus, in order to reduce computational complexity, simplified nonexhaustive search methods are used. However, these methods usually provide only a suboptimal solution for feature subset selection.

3.1.3. Feature Selection Criteria

Depending on the criterion used, feature selection, like other optimization problems, can be described in two ways: either as the maximization of a criterion or as the minimization of the reversed criterion. Here, we will consider criteria based on maximization, where a better subset of features always gives a bigger value of a criterion, and the optimal feature subset gives the maximum value of the criterion.

A feature selection algorithm is based on defined criteria for feature selection (goodness), which ideally should be the same as the criteria for the design of a generalizing prediction.

In other words, the general goal of feature selection typically takes into account the ability of the processing algorithm to best process new data patterns. This constitutes a tradeoff between predictor generalization capability and the dimension and type of features used for pattern formation.

In the light of the generalization goal, we may expect specific behavior of the feature selection criteria during an optimal search process. Reduction of feature dimensionality initially improves the generalization ability of an entire predictor (with increasing values of the feature goodness criterion). However, when a particular reduction of pattern dimensionality is reached, the generalization ability starts to degrade. This change may correspond to the point when the best generalizing feature subset is obtained, when the selection criterion reaches its maximum. Surprisingly, however, many feature selection criteria do not behave in this way. The **monotonicity** property is defined as

$$J_{\text{feature}}(X_{\text{feature}}^+) \geq J_{\text{feature}}(X_{\text{feature}}) \tag{377}$$

where X_{feature} denotes a feature subset, and X_{feature}^+ denotes a larger feature subset that contains X_{feature} as a subset. This definition means that adding a feature to a given feature set results in a criterion value that stays the same or increases:

$$J_{\text{feature}}(\{x_1\}) \leq J_{\text{feature}}(\{x_1, x_2\}) \leq J_{\text{feature}}(\{x_1, x_2, x_3\}) \leq$$
$$\leq J_{\text{feature}}(\{x_1, x_2, \cdots, x_n\}) \tag{378}$$

On the other hand, deleting a feature does not improve performance. Several criteria, such as the class separability (based on covariance computations), the Bayes average probability of error, and some distance measures such as the Mahalanobis or Bhattacharyya distance satisfy the monotonicity condition. However, some criteria used in machine learning (such as inconsistency rate) do not obey the monotonicity condition.

Criteria that include monotonicity properties cannot be used to compare the goodness of different size feature subsets when a large subset contains a smaller one. However, such criteria can still be used to compare feature subsets of equal size. In practice, for a limited-size data set and performance estimation based on that data set, removing a feature may improve performance. Thus, by using estimates of ideal performance as criteria, we still can seek an optimal reduced subset of features.

Ideally, the feature selection criterion should be the same as the criterion for evaluating an entire predictor algorithm (a prediction quality evaluation), i.e., $J_{\text{feature}} = J$. For example, for prediction-regression the criterion could be a sum of squares error, and for prediction-classification a classification error rate. In an ideal closed-loop type (wrapper) feature selection approach, the entire predictor is used in order to evaluate the goodness of a feature subset, and this predictor's performance criterion should be the same as the feature selection criterion. In practice, simplified closed-loop feature selection criteria could be used, based on a simpler predictor PR_{feature} and only for feature evaluation, with a performance evaluation criterion equal to the feature extraction criterion, i.e., $J_{\text{feature}} = J_{PR_{\text{feature}}}$. For example, instead of evaluating feature goodness by training and testing a complex neural network-type predictor, a simpler predictor, such as k-nearest neighbors, along with its performance evaluation, for example the error rate, can be used for feature evaluation.

Open-loop feature selection criteria are frequently designed differently for real-valued features and prediction-regression and for discrete features for prediction-classification, whereas closed loop criteria are usually adequate for both real-valued and discrete features. Some criteria may include penalty terms that favor lower dimensionality of optimal feature subsets.

3.1.4. Open Loop Feature Selection Criteria

Open loop feature selection criteria are usually based on information (such as interclass separability) contained in the data set alone. They do not consider the direct influence of the selected feature subset on the performance of an entire predictor. They do not provide feedback from the predictor quality assessment to the feature selection process.

3.1.5. Criteria Based on Minimum Concept Description

Feature selection criteria based on the **minimum concept description** paradigm have been studied in machine learning and in statistics for discrete features of noise-free data sets. One technique for best feature selection is to choose a minimal feature subset that fully describes all the concepts in a given data set. Here, a criterion of feature selection could be defined as a Boolean function $J_{\text{feature}}(X_{\text{feature}})$ with value 1 if the feature subset X_{feature} is satisfactory in describing all concepts in a data set, and otherwise having a value 0. The final selection would be based on choosing a minimal subset for which the criterion gives value 1.

3.1.6. Criteria Based on Mutual Information

Based on information theory analysis, the **mutual information** (MI) measure of data sets (based on entropy) can be used as a criterion for feature selection. For two variables, mutual information can be considered to provide a reduction of uncertainty about one variable given the other one. Let us consider mutual information for classification for a given data set T containing n-dimensional pattern vectors $\mathbf{x}$ labeled by l classes $c_{\text{target}} \in \{c_1, c_2, \cdots, c_l\}$ in feature pattern vector $\mathbf{x}$. The entire set of original features is a collection of pattern vector elements: $X = \{x_1, x_2, \cdots, x_l\}$. The mutual information for the classification problem is the reduction of uncertainty about classification given a subset of features X forming a feature pattern $\mathbf{x}$. It can be understand as the suitability of the feature subset X for classification. If we consider initially only probabilistic knowledge about classes, the uncertainty is measured by entropy as

$$E(c) = -\sum_{i=1}^{l} P(c_i) \log_2 P(c_i) \tag{379}$$

where $P(c_i)$ is the a priori probability of a class c_i occurrence (which may be estimated based on the data set). **Entropy** $E(c)$ is the expected amount of information needed for class prediction. The entropy is maximal when a priori probabilities $P(c_i)$ are equal. The uncertainty about class prediction can be reduced by knowledge abut feature patterns $\mathbf{x}$ formed with features from a subset X, characterizing recognized objects and their class membership. The conditional entropy $E(c|\mathbf{x})$ (a measure of uncertainty), given pattern $\mathbf{x}$, is defined as

$$E(c|\mathbf{x}) = \sum_{i=1}^{l} P(c_i|\mathbf{x}) \log_2 P(c_i|\mathbf{x}) \tag{380}$$

The **conditional entropy**, given the subset of features X, is defined for discrete features as

$$E(c|X) = -\sum_{\text{all } \mathbf{x}} P(\mathbf{x}) \left(\sum_{i=1}^{l} P(c_i|\mathbf{x}) \log_2 P(c_i|\mathbf{x}) \right) \tag{381}$$

The outer sum considers all feature vectors $\mathbf{x}$ in a feature space. Using equality $P(c_i|\mathbf{x}) = \frac{P(c_i, \mathbf{x})}{P(\mathbf{x})}$, we can obtain

$$E(c|X) = -\sum_{\text{all } \mathbf{x}} P(\mathbf{x}) \left(\sum_{i=1}^{l} \frac{P(c_i, \mathbf{x})}{P(\mathbf{x})} \log_2 \frac{P(c_i, \mathbf{x})}{P(\mathbf{x})} \right)$$

$$= -\sum_{\text{all } \mathbf{x}} \sum_{i=1}^{l} P(c_i, \mathbf{x}) \log_2 \frac{P(c_i, \mathbf{x})}{P(\mathbf{x})} \tag{382}$$

For patterns with continuous features, the outer sum should be replaced by an integral and the probabilities $P(\mathbf{x})$ by the probability density function $p(\mathbf{x})$:

$$E(c|X) = -\int_{\text{all } \mathbf{x}} p(\mathbf{x}) \left(\sum_{i=1}^{l} P(c_i|\mathbf{x}) \log_2 P(c_i|\mathbf{x}) \right) \tag{383}$$

Using Bayes's rule,

$$P(c_i|\mathbf{x}) = \frac{p(\mathbf{x}|c_i) P(c_i)}{p(\mathbf{x})} \tag{384}$$

The probabilities $P(c|\mathbf{x})$ that are difficult to estimate can be replaced by $p(\mathbf{x})$ and $P(\mathbf{x}|c_i)$. The initial uncertainty (based on a priori probabilities $P(c_i)$ only), might decrease given knowledge about feature pattern $\mathbf{x}$. The **mutual information** $MI(c, X)$ between the classification and the feature subset X is measured by a decrease in uncertainty about the prediction of classes, given knowledge about patterns $\mathbf{x}$ formed from features X:

$$J_{\text{feature}}(X) = MI(c, X) = E(c) - E(c|X) \tag{385}$$

Since for discrete features we can derive the equation

$$J_{\text{feature}}(X) = MI(c, X) = \sum_{\text{all } \mathbf{x}} \sum_{i=1}^{l} P(c_i, \mathbf{x}) \log_2 \frac{P(c_i, \mathbf{x})}{P(\mathbf{x}) P(c_i)} \tag{386}$$

the mutual information is a function of c and $\mathbf{x}$; if they are independent, the mutual information is equal to zero (knowledge of $\mathbf{x}$ does not improve class prediction).

Mutual information is the unbiased information about the ability of feature subset X to predict classes. The criterion $MI(c, X)$ is a theoretical limit for feature goodness (similarly, a classifier design based on the Bayes optimal decision is a theoretical limit of accuracy for predictors). The informative power of a feature subset X is never larger than its mutual information with a predicted class. Features from a subset X are absolutely irrelevant for the classification task if their mutual information is equal to zero.

The mutual information criterion is difficult to use in practice due to the difficulties and inaccuracy of estimating conditional probabilities for limited-size data sets. These problems surface when the dimensionality of feature patterns is high and the number of cases small. For low-dimensional data patterns, application of the mutual information criterion (with probability density estimations) can be used to choose the best feature subset from all possible feature subsets. In the simplified application of the mutual information criterion for feature selection, a greedy algorithm adds one most-informative feature at a time. The added feature is chosen as that which has the maximal mutual information with a class and minimal mutual information with already selected features. This method does not solve the redundancy problem between groups of features.

3.1.7. Criteria Based on Inconsistency Count

Another criterion for feature subset evaluation for discrete feature data sets is the **inconsistency measure**. Let us consider a given feature subset X_{feature} and a reduced data set $T_{X_{\text{feature}}}$, with all N_{all} cases $(\mathbf{x}_f, c_{\text{target}})$. Each case contains a pattern $\mathbf{x}_f$ constituted with m features from a subset X_{feature} and labeled by classes c_{target}. The inconsistency criterion $J_{\text{inc}}(T_{X_{\text{feature}}})$ for a data set $T_{X_{\text{feature}}}$ can be defined as the ratio of all inconsistency counts divided by the number of cases. Two cases $(\mathbf{x}_f^j, c_{\text{target}}^j)$ and $(\mathbf{x}_f^k, c_{\text{target}}^k)$ are inconsistent if both have the same patterns $\mathbf{x}_f^j = \mathbf{x}_f^k$ but different associated classes $c_{\text{target}}^j \neq c_{\text{target}}^k$. We can find the inconsistency count of a given set of the same patterns $\mathbf{x}_f^i$ for which cases are inconsistent. Here, for the same matching patterns $\mathbf{x}_f^i$ we compute the inconsistency count as a number $n_{\text{inc},i}$ of all inconsistent cases for the matching pattern minus the largest number of cases in one of the classes from this set of inconsistent cases. For example, let us assume that we have, in a reduced data set $T_{X_{\text{feature}}}$, v inconsistent cases for a pattern $\mathbf{x}_f^i$, with q_1 cases from a class c_1, q_2 cases from a class c_2, and q_3 cases from a class $c_3 (v = q_1 + q_2 + q_3)$. If q_2 is the largest number among the three, then the inconsistency count for matching pattern $\mathbf{x}_f^i$ is $I_i = n - q_2$. The **inconsistency rate** criterion is defined for a reduced data set $T_{X_{\text{feature}}}$ as a ratio of sum of all inconsistency counts and a number of all cases N_{all} in the data set

$$J_{\text{inc}}(T_{X_{\text{feature}}}) = \frac{\sum_{\text{all inconsistent patterns}} I_i}{N_{\text{all}}} \qquad (387)$$

3.1.8. Criteria Based on Rough Sets' Quality of Classification

For prediction-classification, feature subset goodness can be measured by its ability to classify concepts (family of classes) in a given data set. This idea comes from rough sets theory. Let us consider a data set represented by the information system S with the close universe U (object sets with card $U = N$) and a full set of attributes X. Consider a subset of attributes $A \subseteq X$. Let $\Upsilon = \{Y_1, Y_2, \ldots, Y_m\}$ for every $Y_i \subseteq U (1 \leq i \leq m)$ be a classification (a partition, a family of subsets) of U. The family of sets $\Upsilon = \{Y_1, Y_2, \cdots, Y_m\}$ is a classification in U in S, if $Y_i \cap Y_j = \emptyset$ for every $i, j \leq m, i \neq j$ and $\bigcup_{i=1}^m X_i = U$. X_i are called classes of Υ. Here, we will consider a classification based on a subset of attributes A. The **quality of classification** Υ by A, imposed by the set of attributes A, is defined as follows:

$$\rho_A(\Upsilon) = \frac{\sum_{i=1}^m \text{card}\,(\underline{A}Y_i)}{\text{card}\,(U)} \qquad (388)$$

which represents the ratio of all A correctly classified objects to all objects in the information system S. The term $\underline{A}Y_i$ denotes the lower approximation of class Y_i by the set of attributes A.

This measure can be considered as a feature selection criterion if we assume $X_{\text{feature}} = A$:

$$J_{\text{feature}}(X_{\text{feature}}) = \rho_A(\Upsilon) = \rho_{X_{\text{feature}}} = \frac{\sum_{i=1}^m \text{card}\,(\underline{X}_{\text{feature}} Y_i)}{\text{card}\,(U)} \qquad (389)$$

The above criterion is an open-closed loop type from the point of view of a given data set (the training set), since it considers selection of a feature subset (a reduct) that guarantees correct classification for a given data set. However, this criterion provides no assurances about how well this feature subset will perform for new unseen cases.

3.1.9. Criteria Based on Interclass Separability

Prediction-classification open loop criteria for feature selection are frequently based on **interclass separability**, computed based on covariances (or scatter matrices) estimated for given data sets for each class and between classes. They are constructed based on an evaluation paradigm such that a good feature (with high discernibility power) will cause a small within-class scatter and a large between-class scatter.

Let us first study a given original data set T_{all} containing N_{all} cases $(\mathbf{x}^i, c_{\text{target}}^i)$ with patterns $\mathbf{x}$ constituted with n features and labeled by one target class c_{target}^i from all possible l classes. For a data set T_{all}, we denote a number of cases in each class $c_i (i = 1, 2, \cdots, l)$ by $N_i (\sum_{i=1}^l N_i = N_{\text{total}})$. Recalling Fisher's analysis, one can estimate the expected value (a mean) for patterns within each class by

$$\boldsymbol{\mu}_i = \frac{1}{N_i} \sum_{j=1, \mathbf{x}^j \text{ in } c_i}^{N_i} \mathbf{x}^j, \quad (i = 1, 2, \cdots, l) \qquad (390)$$

In order to present a scatter of patterns around a mean and within patterns of one class c_i, one ideally would consider an unbiased estimate of a squared covariance matrix of class $n \times n$,

$$\Sigma_i = \frac{1}{N_i - 1} \sum_{j=1, \mathbf{x}^j \text{ in } c_i}^{N_i} (\mathbf{x}^j - \boldsymbol{\mu}_i)(\mathbf{x}^j - \boldsymbol{\mu}_i)^T, \quad (i = 1, 2, \cdots, l) \tag{391}$$

However, instead of a covariance matrix, a squared $n \times n$ **within-class** c_i **scatter matrix** is considered instead:

$$\mathbf{S}_i = \sum_{j=1, \mathbf{x}^j \text{ in } c_i}^{N_i} (\mathbf{x}^j - \boldsymbol{\mu}_i)(\mathbf{x}^j - \boldsymbol{\mu}_i)^T, \quad (i = 1, 2, \cdots, l) \tag{392}$$

This matrix is proportional to the covariance matrix, symmetric, and positive semidefinite (and, for $N_{\text{all}} > n$, usually nonsingular). In order to provide a summarizing measure for scatter patterns around means for all l classes, the so-called **within-class scatter matrix** is defined:

$$\mathbf{S}_w = \sum_{i=1}^{l} \mathbf{S}_i \tag{393}$$

To illustrate between-class scatter, first we define estimates of the total data mean and the total scatter matrix (for all N_{all} patterns from T_{all}) as a proportional representation of a covariance estimate. The **total data mean** can be estimated by

$$\boldsymbol{\mu} = \frac{1}{N} \sum_{j=1}^{N} \mathbf{x}^j = \frac{1}{N} \sum_{i=1}^{l} N_i \boldsymbol{\mu}_i \tag{394}$$

and the **total scatter matrix** for all patterns as

$$\mathbf{S}_t = \sum_{j=1}^{N_{\text{all}}} (\mathbf{x}^j - \boldsymbol{\mu})(\mathbf{x}^j - \boldsymbol{\mu})^T \tag{395}$$

We find that the total scatter matrix can be decomposed into two matrices

$$\mathbf{S}_t = \mathbf{S}_w + \mathbf{S}_b \tag{396}$$

where $\mathbf{S}_w$ is the within-class scatter matrix and the $n \times n$ square matrix $\mathbf{S}_b$ is the so-called **between-class scatter matrix**, defined as

$$\mathbf{S}_b = \sum_{i=1}^{l} N_i (\boldsymbol{\mu}_i - \boldsymbol{\mu})(\boldsymbol{\mu}_i - \boldsymbol{\mu})^T \tag{397}$$

To form a final scalar feature selection criterion involving interclass separability, we need to define a function that gives a larger value when within-class scatter is smaller or between-class scatter is larger. Generally, this function gives a larger value when interclass separability is larger. For one feature, one can say that a feature is "good" (has a large discriminatory or predictive power) if its within-class variance is small and its between-class variance is large. For multiple classes and multifeature patterns, the following feature selection criteria, based on interclass separability, are defined:

3. Feature Selection

1. Ratio of determinants for between-class and within-class scatter matrices:

$$J_{\text{feature}} = \frac{|\mathbf{S}_b|}{|\mathbf{S}_w|} = \frac{\det(\mathbf{S}_b)}{\det(\mathbf{S}_w)} \qquad (398)$$

where the determinant $|\mathbf{S}_b|$ denotes a scalar representation of the between-class scatter matrix, and similarly the determinant $|\mathbf{S}_w|$ denotes a scalar representation of the within-class scatter matrix.

2. Ratio of determinants for between-class and total scatter matrices:

$$J_{\text{feature}} = \frac{|\mathbf{S}_b|}{|\mathbf{S}_t|} = \frac{|\mathbf{S}_b|}{|\mathbf{S}_b + \mathbf{S}_w|} \qquad (399)$$

which in older literature is referred to as **Wilks' lambda**.

3. Trace of $\mathbf{S}_w^{-1}\mathbf{S}_b$:

$$J_{\text{feature}} = \text{trace}(\mathbf{S}_w^{-1}\mathbf{S}_b) \qquad (400)$$

where **trace** denotes a matrix trace.

4. Logarithm of $\mathbf{S}_w^{-1}\mathbf{S}_b$:

$$J_{\text{feature}} = \ln\left(\mathbf{S}_w^{-1}\mathbf{S}_b\right) \qquad (401)$$

For single feature patterns x, and for two-class classification, the following version of the interclass separation criterion can be used:

$$J_{\text{feature}} = F_{\text{Fisher}} = \frac{(\mu_1 - \mu_2)^2}{s_1^2 + s_2^2} \qquad (402)$$

Here, data patterns mean, for each class,

$$\mu_i = \frac{1}{N_i} \sum_{j=1, x^j \text{ in } c_i}^{N_i} x^j, \quad (i = 1, 2) \qquad (403)$$

and the scatter of a feature x for each class is

$$s_i^2 = \frac{1}{N_i} \sum_{j=1, x^j \text{ in } c_i}^{N_i} (\mu_i - x^j)^2, \quad (i = 1, 2) \qquad (404)$$

Finally, the total scatter of feature x in an entire data set is

$$s_t^2 = s_1^2 + s_2^2 \qquad (405)$$

The above single feature selection criterion, which is based on the Fisher linear discriminant analysis and on interclass separability, is called the **Fisher F-ratio**. It allows us to find a feature guaranteeing maximization of the between-class variance while simultaneously minimizing the within-class variance.

3.1.10. Closed Loop Feature Selection Criteria

The closed-loop-type feature selection criteria and their estimations are similar both for classification and for regression type predictors.

Let us consider a feature selection criterion for a prediction task based on the original data set T_{all} that includes N_{all} cases (**x, target**) consisting of n-dimensional input patterns **x** (whose elements represent all features X) and a **target** value. Let us assume that the m-feature subset $X_{feature} \subseteq X$ ought to be evaluated based on the closed-loop-type criterion. First, a reduced data set $T_{feature}$, with patterns containing only m features from the subset $X_{feature}$, should be constructed. Then a type of predictor $PR_{feature}$ (for example, k-nearest neighbors or a neural network), used for feature goodness evaluation, should be chosen. This predictor ideally should be the same as the final predictor PR for an entire design; however, in a simplified suboptimal solution, a computationally less expensive predictor can be used only for feature selection purposes. After a reduced data set $X_{feature}$ has been constructed and a predictor algorithm $PR_{feature}$ has been decided for the considered feature subset $X_{feature}$, then evaluation of feature goodness, equivalent to the predictor evaluation criterion, can be performed. Doing so will require defining a performance criterion $J_{PR_{feature}}$, of a predictor $PR_{feature}$, and an error counting method showing how to estimate a performance and how to grasp its statistical character through averaging of results. Consider as an example a holdout error-counting method for predictor performance evaluation. In order to evaluate the performance of a predictor $PR_{feature}$, an extracted feature data set $T_{feature}$ is split into a N_{tra}-case training set $T_{feature,tra}$, and a N_{test}-case test set $T_{feature,test}$ (the holdout for testing). Each case ($\mathbf{x}_f^i$, **target**^{i}) of both sets contains a feature pattern $\mathbf{x}_f$ labeled by a target.

The criteria for evaluating the performance of predictor $PR_{feature}$ should be considered separately for regression and classification.

In prediction-regression, the predicted output variables are continuous. Prediction-regression performance criteria are based on the counting error between target and guessed real values. Let us consider defining a feature selection criterion for a prediction-regression task, with a q-dimensional output vector $\mathbf{y} \in \mathbb{R}^q$ whose elements take real values. Here, the design is based on a reduced feature data set $T_{feature}$ (with N_{all} cases). This data set's case ($\mathbf{x}_f^i$, $\mathbf{y}_{target}^i$) includes m-dimensional feature input patterns $\mathbf{x}_f$ and associated real-valued $\mathbf{y}_{target} \in \mathbb{R}^q$ target vectors for outputs. Elements of patterns $\mathbf{x}_f$ are features from the subset $X_{feature}$. One example of the predictor performance criterion $J_{PR_{feature}}$, being here equivalent to the feature selection criterion $J_{feature} = J_{PR_{feature}}$, could be

$$J_{feature} = J_{PR_{feature}} = \hat{J}_{squared} = \sum_{i=1}^{N_{all}} (\mathbf{y}_{target}^i - \mathbf{y}^i)^T (\mathbf{y}_{target}^i - \mathbf{y}^i)$$

$$= \sum_{i=1}^{N_{all}} \sum_{j=1}^{m} (y_{j,target}^i - y_j^i)^2 \qquad (406)$$

which boils down to the sum of squares errors. This criterion is evaluated based on a limited-size test set $T_{feature,test}$ obtained from the example by splitting the entire set $T_{feature}$ into subsets $T_{feature,tra}$ for training (design) and $T_{feature,test}$ for testing.

In prediction-classification, cases in the feature subset $T_{feature}$ are pairs ($\mathbf{x}_f$, c_{target}) that include the feature input pattern $\mathbf{x}_f$ and a categorical-type target c_{target} being one of the possible l classes c_i. The quality of classifier $PR_{feature}$, computed, for example (for holdout error counting), based on the limited-size test set $T_{feature,test}$ with N_{test} patterns, can be measured using the performance criterion $J_{PR_{feature}}$ below (here equal to the feature selection criteria $J_{feature}$), which estimate the probabilities of errors (expressed in percent) by relative frequencies of errors:

$$J_{PR_{feature}} = \hat{J}_{all\_miscl} = \frac{n_{all\_miscl}}{N_{test}} \cdot 100\% \qquad (407)$$

where $n_\text{all miscl}$ is the number of all misclassified patterns, and N_test is the number of all tested patterns. This measure is an estimate of the probability of error $P(\textbf{assigned class different than target class})$, expressed in percent (percentage of all misclassified patterns, an error rate, or the relative frequency of errors).

3.1.11. Computing Feature Selection Criteria

Let us consider a given m-feature subset X_feature, created from a n-feature set of original features $X = \{x_1, x_2, \cdots, x_n\}$. Computing a feature selection criterion $J_\text{feature}(X_\text{feature})$, for a given m-feature subset X_feature, first requires the creation of a reduced data set T_feature extracted from the original total data set T_all. This reduced data set T_feature contains N_all cases with m-dimensional feature pattern vectors $\mathbf{x}_f$, containing, as elements, features from the considered subset X_feature, and the same targets as in the original entire data set. Other $n - m$ "columns" of the entire data set are discarded. For this reduced data set T_feature, a feature selection criterion could be computed.

The computing of closed-loop-type feature selection criteria, which is based on the evaluation of feature goodness by testing the performance of an entire predictor, is computationally expensive. It involves design (training) of a predictor PR_feature and its performance evaluation, both based on a reduced data set T_feature. First, for a given m-feature subset X_feature, a reduced m-feature pattern data set T_feature is constructed (with N_all cases). Then, based on this reduced data set, a chosen predictor PR_feature is designed. Finally, the performance of this designed predictor is evaluated, according to its defined evaluation criterion J_{PR_feature}, which is equal to feature selection criterion J_feature. Design and performance evaluation of a predictor used for feature selection can be realized using one of the methods of predictor design-evaluation. Such a method splits the reduced data set T_feature into training and test sets and then uses a statistical error counting method. For example, the average holdout, leave-one-out, or leave-k-out method of predictor design/performance evaluation could be used.

For open-loop-type feature selection criteria, which are based on interclass separability, first the within-class $\mathbf{S}_w$ and between-class $\mathbf{S}_b$ scatter matrices are computed for a reduced data set X_feature, and then the final value of criterion J_feature is computed as a function of these scatter matrices. No predictor performance evaluation is considered for the tested features.

3.1.12. Search Methods

Given the large number of features constituting a pattern, the number of possible feature subsets evaluated by using exhaustive search-based feature selection could be too high to be computationally feasible. For n features, a total of 2^n subsets (including an empty subset) can be formed. For $n = 12$ features, the number of possible subsets is 4096; however, for $n = 100$, the number of possible subset is larger than 10^{30}, which makes exhaustive search unrealizable. If, for some design reason, we are searching for a feature subset containing exactly m features, then for n feature patterns, the total number of possible m-feature subsets is

$$\binom{n}{m} = \frac{n!}{(n-m)!m!} \qquad (408)$$

This number (which can be much smaller than 2^n) can be, from a computational perspective, too high.

Generally, feature selection is an NP-hard problem, and for highly dimensional patterns, an exhaustive search can be impractical and suboptimal selection methods should be used. Next, we discuss an exhaustive search technique and the optimal branch and bound method. Then we consider two suboptimal greedy search methods: forward and backward search. We also discuss random search, and finally we give pointers suboptimal search methods such as simulated annealing and genetic programming.

For a small number of pattern features, the **exhaustive search** could be acceptable and could guarantee an optimal solution.

Algorithm: Feature selection based on exhaustive search

Given: A data set T_{all} with N_{all} labeled patterns constituted with n features $X = \{x_1, x_2, \cdots, x_n\}$. A feature selection criterion $J_{feature}$ with a defined computation procedure based on a limited-size data set $T_{X_{feature}}$.

1. Set $j = 1$ (a counter of the feature subset number).
2. Select a distinct subset of features $X^j \subseteq X$ (with the number of elements $1 \leq N_{X^j} \leq n$).
3. For a selected feature subset X^j, compute a feature selection criterion $J_{feature}(X^j)$.
4. If $j \leq 2^n$, continue from step 2; otherwise, go to the next step.
5. Chose an optimal subset $\hat{X}_{opt}$ with a maximal value of the selection criterion

$$J_{feature}(\hat{X}_{opt}) \geq J_{feature}(\hat{X}^j), \quad j = 1, 2, \cdots, 2^n \tag{409}$$

A sequence of generated distinct feature subsets X^j is not important for the above algorithm. For example, for the three feature patterns $X = \{x_1, x_2, x_3\}$, one can generate the following exhaustive collection of $2^3 = 8$ feature subsets (including an empty subset):

$$\{\,\}, \{x_1\}, \{x_2\}, \{x_3\}, \{x_1, x_2\}, \{x_1, x_3\}, \{x_2, x_3\}, \{x_1, x_2, x_3\} \tag{410}$$

The algorithm guarantees finding an optimal feature subset. Depending on the feature selection criterion (and the procedure for its estimation), either an open loop or a closed loop feature selection scheme is consequently employed.

3.1.13. Branch and Bound Method for Optimal Feature Selection

In combinatorial optimization, in order to avoid a costly exhaustive search, **branch and bound** methods were developed and adapted to optimal feature selection. The branch and bound search can be used for feature selection, assuming that a feature selection criterion satisfies the monotonicity relation. This method allows us to find an optimal set of features without needing to test all possible feature subsets. Using a tree representation for the exhaustive subset search, the branch and bound methods with monotonic performance criterion are based on the idea of discarding some subtrees from the exhaustive search. The removed subtrees contain subsets of feature that would not improve performance in the search procedure. To explain the method, we consider an attempt to construct a fully exhausted search tree (containing all possible subsets of features) designed in a way that allows reduction of a search by employing the idea of monotonicity of a feature selection criterion.

We start with pattern feature of all n-elements, represented in this algorithm as a sequence $x_1, x_2, \cdots, x_n$, and search for the best subset containing a known number of m features. Let us denote the indices of the $d = n - m$ discarded features (from all sets of n features X) by $z_1, z_2, \cdots, z_d$. Here, each variable z_i can take an integer number (the index number of the feature from the sequence of features taken from X) from the set of indices $\{1, 2, \cdots, n\}$. We note that each variable z_i has a distinct index number, since each feature can be discarded only one time. The order of variables z_i is not important, since any permutation of the sequence $z_1, z_2, \cdots, z_d$ gives an identical value for the feature selection criteria. In order to better design a search tree, we consider sequences of variables z_i that satisfy the relation

$$z_1 < z_2 < \cdots < z_d \tag{411}$$

and that determine a convenient way to design a search tree. By definition, a feature selection criterion $J_{feature}(X_{feature})$ is a function of the $m = n - d$ feature subset $X_{feature}$ remaining after discarding d features from the set of all features. For convenience, we will also use the notation

$J_{\text{feature}}(z_1, z_2, \cdots, z_d)$ for the same criterion. The goal of the optimal search is to find the best m-feature subset X_{opt}, with discarded $d = n - m$ features (from the original sequence of n features) indicated by an "optimal" sequence of indices $\hat{z}_1, \hat{z}_2, \cdots, \hat{z}_d$. Searching for an optimal feature subset is then equivalent to searching for the "optimal" set of discarded features:

$$J_{\text{feature}}(\hat{z}_1, \hat{z}_2, \cdots, \hat{z}_d) = \max_{z_1, z_2, \cdots, z_d} J_{\text{feature}}(z_1, z_2, \cdots, z_d) \qquad (412)$$

Using the defined convention, we can now design a specific search tree starting from all sets of n features at the root (level 0). At each subsequent level, we attempt to generate a limited number of subtrees by deleting, from the ancestor node's pool of features, one specific feature at a time (limited by convention). In the search tree, each node at the jth level is labeled by a value of the variable z_j, which is equal to the index of a feature discarded from the sequence of features in the ancestor node at the previous level $j - 1$. Each node at the jth level can be identified by a sequence of already-discarded j features starting from the root. Here, according to defined convention, at each jth level the largest value of a variable z_i must be $m - j$. This method allows the design of a search tree containing all possible $\frac{n!}{(n-m)!m!}$ subsets with m features out of all n. However, in the branch and bound search, not all subtrees need to be searched.

Assume that a feature selection criterion satisfies monotonicity:

$$J_{\text{feature}}(z_1) \geq J_{\text{feature}}(z_1, z_2) \geq J_{\text{feature}}(z_1, z_2, z_3) \geq \cdots \geq J_{\text{feature}}(z_1, z_2, \cdots, z_d) \qquad (413)$$

Let as assume that at a certain level of the search, the best feature set identified so far has been found by deleting d features indexed by a sequence $z_1, z_2, \cdots, z_d$, with maximal performance criterion value $J_{\text{feature}}(z_1, z_2, \cdots, z_d) = \beta$ (set as a current threshold). Then, for a new feature subset obtained by deleting r features ($r < d$) indexed by $z_1, z_2, \cdots, z_r$, if

$$J_{\text{feature}}(z_1, z_2, \cdots, z_r) \leq \beta \qquad (414)$$

then the monotonicity property yields

$$J_{\text{feature}}(z_1, z_2, \cdots, z_r, z_{r+1}, \cdots z_d) \leq J_{\text{feature}}(z_1, z_2, \cdots, z_r) \leq \beta \qquad (415)$$

for all possible sequences $z_r, z_{r+1}, \cdots z_d$. This new feature subset, obtained by deleting r features ($r < d$), cannot be optimal, nor can its successors in the search tree. If the described technique of designing a search tree has been applied, then the above observation reveals the main idea of the branch and bound search, namely, if the value of a selection criterion evaluated at any node of a search tree (for corresponding feature subset) is smaller than the current value of a threshold β (corresponding to the best subset found so far by using best evaluation criterion value), then all nodes in the tree, including successors of that node, have a selection criterion value less than the threshold β. Consequently, this node cannot be optimal (nor can all its successors), and consequently this subtree can be removed from the search. This is why, in branch and bound feature selection, we obtain an optimal solution without needing to evaluate all possible feature subsets.

Algorithm: Feature selection by branch and bound search.

Given: A data set T_{all} with N_{all} labeled patterns constituted with n features $X = \{x_1, x_2, \cdots, x_n\}$. A number m of features in the resulting subset of best features. A feature subset selection criterion J_{feature} (satisfying the monotonicity property) with a defined procedure for its computation based on a limited-size data set $T_{X_{\text{feature}}}$.

1. Set a level number $j = 0$, $z_0 = 0$ (the notation of a node at level j), and an initial value of the threshold $\beta = -\infty$.
2. Create successors by generating a list S_j,

$$S_j = \{z_{j-1}+1, z_{j-1}+2, \cdots, m+j\}, \quad (j = 1, 2, \cdots, m) \tag{416}$$

of all possible values that z_j at level j can take (assuming given values from previous levels $z_1, z_2, \cdots, z_{j-1}$), with a maximal index $m+j$. The successor nodes contains feature subsets with one feature deleted from the list of the previous level.
3. Select a new node. If a list S_j is empty, then go to step 5. Otherwise, find a value k (with maximal value of the criterion) for which

$$J_{\text{feature}}(z_1, z_2, \cdots, z_{j-1}, k) = \max_{i \in S_j} J_{\text{feature}}(z_1, z_2, \cdots, z_{j-1}, i) \tag{417}$$

Set $z_j = k$, and delete k from the list S_j.
4. Test a bound. If $J_{\text{feature}}(z_1, z_2, \cdots, z_j) < \beta$, then go to the step 5. If the last level has been reached, go to step 6; otherwise, advance to a new level by setting $j = j+1$ and continuing from step 2.
5. Return (backtrack) to a lower level. For $j = 0$, terminate; otherwise, continue from step 3.
6. The last level: Set $\beta = J_{\text{feature}}(z_1, z_2, \cdots, z_d)$ and $\hat{z}_1, \hat{z}_2, \cdots, \hat{z}_d = z_1, z_2, \cdots, z_d$. Continue from step 5.

Result: An optimal subset of features with the largest value of the criterion.

Figure 7.14 shows an example of the branch and bound algorithm for selecting two features ($m = 2$) from the total number of $n = 5$ features.

The black nodes show examples of subtrees that do not need to be searched for optimal subsets because they will not provide better solutions. The number next to each node shows the index of the feature deleted from a list of ancestor features. A node at level j is labeled with the value z_j. The set of all possible subsets of two features out of five is represented by nodes of the last level (each node represents one feature subset obtained by deleting corresponding features). According to the defined convention, the maximal value of z_j is $m+j$; thus at the first level it equals 3, at the next level it equals 4, etc. To show how the algorithm work, we assume that so far the node marked by A has given the best value of the criterion set as a current threshold β. Then, according to criterion monotonicity, if at any step of the algorithm an intermediate node is considered (such as that marked by B) for which the criterion value is less

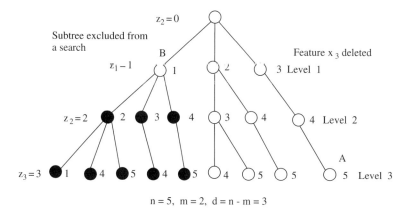

Figure 7.14. Illustration of the branch and bound algorithm.

than the current threshold, then all subtrees starting from that node can be excluded from the search. If for any node in the final level a selection criterion has a larger value than the current threshold, this criterion value becomes a new threshold. The algorithm terminates when each final level node has been evaluated or excluded based on the monotonicity property.

3.1.14. Feature Selection with Individual Feature Ranking

One of the simplest feature selection procedures is based on first evaluating the individual predictive power of each feature alone, then ranking such evaluated features, and eventually choosing the best first m features. The criterion for the individual feature could be of either open loop or closed loop type. This algorithm assumes that features are independent and that the final selection criterion can be obtained as a sum or product of criteria evaluated for each feature independently. Since these conditions are rarely satisfied, the algorithm does not guarantee an optimal selection. A single feature alone may have very low predictive power. However, this feature in combination with another feature may provide substantial predictive power. A decision concerning how many best m-ranked features should be chosen for the final feature set could be made based on experience from using another search procedure. Here, one could select the minimal number $\hat{m}$ of best-ranked features that guarantee a performance better than or equal to a predefined threshold according to a defined criterion $J_{\text{feature,ranked}}$.

Algorithm: Feature selection with individual feature ranking

Given: A data set T_{all} with N_{all} labeled patterns consisting of n features $X = \{x_1, x_2, \cdots, x_n\}$; a feature evaluation criterion $J_{\text{feature,single}}$ with a defined procedure for its computation based on a limited-size data set $T_{X_{\text{feature}}}$; and an evaluation criterion $J_{\text{feature,ranked}}$ for a final collection of m ranked features.

1. Set $j = 1$, and choose a feature x_j.
2. Evaluate the predictive power of a single feature x_j alone by computing the criterion $J_{\text{feature,single}}(x_j)$.
3. If $j \leq n$, continue from step 1; otherwise, go to the next step.
4. Rank all n features according to the value of the computed criterion $J_{\text{feature,single}}$:

$$x_a, x_b, \cdots, x_m, \cdots, x_r, \quad J_{\text{feature,single}}(x_a) \geq J_{\text{feature,single}}(x_b), \text{ etc.} \tag{418}$$

5. Find the minimal number of first-ranked $\hat{m}$ features according to the criterion $J_{\text{feature,ranked}}$.
6. Select the first $\hat{m}$ best-ranked features as a final subset of selected features.

Result: An optimal subset of features.

3.1.15. Sequential Suboptimal Forward and Backward Feature Selection

In order to reduce the computational burden associated with an exhaustive search, several suboptimal feature selection methods have been proposed. Let us present two suboptimal sequential (stepwise) methods of feature selection: **forward** and **backward feature selection**.

Let us consider selecting the best m-feature subset from $n(m < n)$ features constituting an original pattern. Figure 7.15 shows an example of finding an $m = 3$ feature subset from an $n = 4$ feature pattern.

A **forward selection** search starts with individual evaluation of each feature. For each feature, a feature selection criterion, J_{feature}, is evaluated, and the feature with the best score (maximal value of the performance criterion) is selected for the next step of the search (a "winner" – an ancestor of the subtree). Then, in the second step, one additional feature is added to the selected "winner" feature (having the best value of the criterion) from previous step, forming all possible

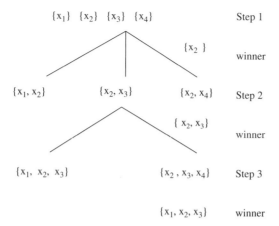

Figure 7.15. Sequential forward feature selection search.

two-feature subsets containing a "winner." Each subset with its pair of features is evaluated, and those presenting the maximal increase of the performance criterion are selected as a winner and successor of the next step. The procedure continues until the best m-feature subset (the "winner" of the mth step) has been processed.

Algorithm: Feature selection by stepwise forward search.

Given: A data set T_{all} with N_{all} labeled patterns consisting of n features $X = \{x_1, x_2, \cdots, x_n\}$; a number m of features in the resulting subset of best features; and a feature subset evaluation criterion J_{feature} with a defined procedure for its computation based on a limited-size data set $T_{X_{\text{feature}}}$.

1. Set an initial "winner" feature subset as an empty set $X_{\text{winner},0} = \{\ \}$.
2. Set a step number $j = 1$.
3. Form all possible $n - j + 1$ subsets, with a total of j features, that contain a winning $j - 1$ feature subset $X_{\text{winner},j-1}$ from the previous step, with one new feature added.
4. Evaluate the feature selection criterion for each feature subset formed in step j. Select as a winner a subset $X_{\text{winner},j}$ with a larger increase Δ of the performance criterion J_{feature} as compared with the maximal criterion value (for the winner subset $X_{\text{winner},j-1}$) from the previous step.
5. If $j = m$, then stop. The winner $X_{\text{winner},j}$ subset in step j is the final selected subset of m features. Otherwise, set $j = j + 1$ and continue from step 3.

The forward selection algorithm provides a suboptimal solution, since it does not examine all possible subsets of features.

The basic forward selection procedure assumes that the number of features m in a resulting subset is known. This procedure will require exactly m steps. In some cases, the proper number of features m has to be found. This situation defines another search process with stopping criterion $J_{\text{feature,length}}$. Here, a possible stopping criterion for finding the proper number m of features in a final selected feature subset could be, for example, a defined threshold ϵ_{length} of maximal performance increase for two consecutive steps. In other words, the stopping point is reached when the increase in the feature selection criterion for the jth-step winning feature subset $X_{\text{winner},j}$, as compared with the corresponding performance for a winner feature subset from the previous step $j - 1$, is less than the defined threshold ϵ_{length}:

$$J_{\text{feature,length}} = J_{\text{feature}}(X_{\text{winner},j}) - J_{\text{feature}}(X_{\text{winner},j-1}) < \epsilon_{\text{length}} \tag{419}$$

3. Feature Selection

Backward selection is similar to forward selection, but it applies a reversed procedure of feature selection, starting with the entire feature set and eliminating features one at a time. In backward selection, assuming a known number m of final features, the search starts with the evaluation of the entire set of n features. For the entire feature set, a selection criterion $J_{feature}$ is evaluated. Then, in the next step, all possible subsets containing features from the previous step with one feature discarded are formed and their performance criteria are evaluated. At each step, one feature, which gives the smallest decrease in the value the feature selection criterion included in the previous step, is discarded. The procedure continues until the best m-feature subset is found. Figure 7.16 depicts an example of finding an $m = 2$ optimal feature subset from an $n = 4$ feature pattern.

Algorithm: Feature selection by stepwise backward search.

Given: A data set T_{all} with N_{all} labeled patterns consisting of n features $X = \{x_1, x_2, \cdots, x_n\}$; a number m of features in the resulting subset of best features; and a feature subset evaluation criterion $J_{feature}$ with a defined procedure for its computation based on a limited-size data set $T_{X_{feature}}$.

1. Evaluate a feature selection criterion $J_{feature}(X)$ for a set X of all n features.
2. Set a step number $j = 1$ with a list X of all n features.
3. Form all $n - j + 1$ possible subsets with $n - j$ features by discarding one feature at a time from the list of features of the previous step.
4. Evaluate a feature selection criterion for each feature subset formed in step j. Select as a "winner" a subset $X_{winner,j}$ with the smallest decrease of a performance criterion $J_{feature}(X_{winner,j})$ as compared with the criterion value from the previous step (which corresponds to its biggest value for this step from a pool of all subsets). The discarded feature from the previous step, which caused the creation of the winning subset $X_{winner,j}$, is then discarded from a pool of features used in the next step, and winning subset becomes an ancestor of a deeper subtree.
5. If $j = m$, then stop: the winner subset in step j is the final selected subset of m features. Otherwise, set $j = j + 1$ and continue from step 3.

The forward selection algorithm provides a suboptimal solution, since it does not examine all possible subsets of features. The backward selection algorithm requires more intensive computations than the forward selection. Despite of similarities, both algorithms may provide different results for the same conditions.

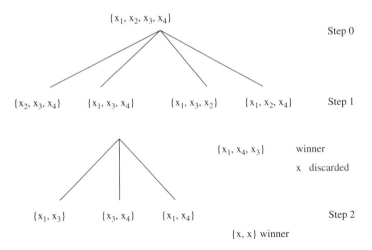

Figure 7.16. A sequential backward search.

If the number m of final features is a unknown a priori, then another optimal search should be employed. Finding the proper number m of features in the final selected feature subset could be realized in a manner similar to the method described earlier for forward selection.

The forward and backward search methods can be combined in several ways, allowing them to cover more feature subsets through increased computations, and thereby to find better suboptimal feature sets. For example, in the so-called **full stepwise search**, operations at each step start as in the backward search. All subsets created by removing one variable from the previous-step pool are evaluated. If the feature selection criterion decrease is below a defined threshold, then a variable is removed. If none of the variables provide a decrease below the threshold, then a variable is added, as in the forward search method.

Based on the concept of **Monte Carlo techniques**, several feature selection methods have been developed using probabilistic search. These methods make probabilistic choices of feature subsets in search of the best subset in a feature space. These random search methods can be used both for open loop and closed loop feature selection algorithms. In general, they do not require that a feature selection criterion must obey the monotonicity condition. Probabilistic algorithms use the inconsistency rate in open loop selection, as a feature goodness criterion and select the feature subset with the minimal number of features that has an inconsistency rate smaller than the predefined threshold. Allowing for features to have some degree of inconsistency opens a way to define the robust feature subset that is not only the best for a given data set but also potentially good for unseen cases (i.e., with generalization ability). The closed loop feature selection scheme has also been proposed for random searches with the ID3 algorithm as classifier.

These probabilistic methods are simple to implement and guarantee finding the best subset of features, if a required number of random trials for subset selection will be performed. These algorithms provide satisfactory results for highly correlated features.

Algorithm: Probabilistic (Monte Carlo) method of feature selection

Given: A data set T_{all} with N_{all} patterns labeled by classes and consisting of n features $X = \{x_1, x_2, \cdots, x_n\}$; a feature subset selection criterion $J_{feature}$ with a defined procedure for its computation based on a limited-size data set $T_{X_{feature}}$; and a maximum number of random subset search trials *max_runs*.

1. Set initially the best-feature subset as equal to an original n-feature set $X_{opt} = X$. Compute the value of the criterion $J_{feature}(X_{feature,0}) = J_{feature}(T_{all})$ for a data set T_{all}.
2. Set $j = 1$ (a search trial number).
3. From all possible 2^n feature subsets, select randomly a distinct subset of features $X_{feature,j}$ (with number of features $1 \leq m_j \leq n$).
4. Create a reduced data set $T_{X_{feature,j}}$ with all N_{all} cases with patterns constituted with m_j features from a subset $X_{feature,j}$.
5. Compute the value of the criterion $J_{feature}(T_{X_{feature,j}})$ for the data set $T_{X_{feature,j}}$.
6. If $J_{feature}(X_{feature,j}) > J_{feature}(X_{feature,j-1})$, then set $X_{best} = X_{feature,j}$ and continue from step 7. Otherwise, continue from step 7.
7. Set $j = j + 1$. If $j \leq max\_runs$, then stop; otherwise, continue from step 3.

The version of the probabilistic algorithm with the open loop inconsistency criterion defined earlier may be formed as follows.

Algorithm: Probabilistic (Monte Carlo) method of open loop feature selection with inconsistency criterion

Given: A data set T_{all} with N_{all} patterns labeled by classes and consisting of n features $X = \{x_1, x_2, \cdots, x_n\}$; a feature subset selection criterion $J_{feature} = J_{inc}$ with a defined procedure for its computation based on a limited-size data set $T_{X_{\text{feature}}}$; a rejection threshold β of feature subset inconsistency; and a maximum number of random subset search trials max_runs.

1. Set an initial value for the best (minimal) feature number as equal to the number n of all original features $m_{\text{best}} = n$. Set initially the best feature subset as equal to an original n-feature set $X_{\text{opt}} = X$.
2. Set $j = 1$ (a search trial number).
3. From all possible 2^n feature subsets, select randomly a distinct subset of features $X_{\text{feature},j}$ (with the number of features $1 \le m_j \le n$).
4. Compute the number of features in the subset

$$X_{feature,j}; \quad m_j = cardinality(X_{feature,j}) \qquad (420)$$

5. Create a reduced data set $T_{X_{\text{feature},j}}$ with all N_{all} cases with patterns consisting of m_j features from a subset $X_{\text{feature},j}$.
6. Compute a value of the inconsistency criterion $J_{\text{inc}}(T_{X_{\text{feature},j}})$ for a data set $T_{X_{\text{feature},j}}$
7. If $m_j < m_{\text{best}}$ and $J_{\text{inc}}((X_{\text{feature},j}) < \beta$, then set $X_{\text{best}} = X_{\text{feature},j}$ and $m_{\text{best}} = m_j$, and continue from step 8. Otherwise, continue from step 8.
8. Set $j = j + 1$. If $j \le max\_runs$, then stop; otherwise, continue from step 3.

Random selection of the feature subset X_{feature} from an original set of all ordered features $x_1, x_2, \cdots, x_n$ can be realized, for example, as follows. First, a random number generator uniformly distributed in $[0,1]$ is executed n times. Each generated number r_i corresponds to one feature x_i. Then, if $r_i > 0.5$, a feature x_i is selected for a subset $X_{\text{feature},j}$; otherwise, it is not selected.

3.1.16. Feature Scaling

Feature scaling (weighting) is used in feature selection. If we consider different discriminatory power of pattern features, feature scaling can be described as assigning continuous weight values from the range $[0, 1]$ to each feature, depending on its impact on prediction. Given an original set of features X forming an n-dimensional pattern $\mathbf{x}$, feature scaling is the transformation of $\mathbf{x}$ into $\mathbf{x}_s$ using the n-dimensional weights vector $\mathbf{w}$ ($w_i \in [0, 1]$):

$$x_{s,i} = w_i x_i, \quad (i = 1, 2, \cdots, n) \qquad (421)$$

In the extreme situation of weight taken only to binary values $w_i \in \{0, 1\}$, feature scaling becomes feature selection. For $w_i = 1$, feature x_i is selected for the final pattern; otherwise, for $w_i = 0$, the feature is removed. Choosing the best feature weights is an optimization problem, which can be as difficult as optimal feature selection. Feature scaling also faces a feature dimensionality problem.

4. Summary and Bibliographical Notes

In this Chapter we have introduced feature selection and feature extraction methods. The most important topics discussed were the unsupervised **Principal Component Analysis** (PCA) and supervised **Fisher's linear discriminant** technique. These can be used for pattern projection, **feature extraction**, and dimensionality reduction. The other unsupervised method covered was the **Independent component analysis** (ICA) used for discovering unknown intrinsic independent variables in the data. ICA estimates unknown **mixing matrix** and **independent components** representing given data. It can be used for **blind source separation** and linear feature extraction and reduction. The **Singular Value Decomposition** method is used often used for extracting features from images and image compression. Feature extraction and compression can be also achieved by using **vector quantization, Fourier transform** and **wavelets**. Vector quantization is the technique of representing (encoding) input data patterns by a smaller finite set of code vectors that approximate the input pattern space. **Learning vector quantization** technique combines supervised learning and vector quantization. **Fourier analysis** can be used to preprocess time-series data and images. **Fourier transform** (both one- and two-dimensional) converts input data into **frequency domain** that provides better understanding and representation of the data. A **fast Fourier transform** is a computationally efficient algorithm for computing a **discrete Fourier transform**. **One-dimensional Fourier transform** is a fundamental technique in processing feature extraction for time-series and speech data, whereas **two-dimensional Fourier transform** is widely used in image processing, including feature extraction. **Wavelets analysis** provides multi-resolution approximation of a time-series signal and images using a fast-decaying oscillating waveform. Wavelets provide localization in time and in a space and a powerful technique for feature extraction from data. **Zernike moments** are used to extract **rotation-invariant** robust features from images using orthogonal moment invariants and orthogonal polynomials.

In **feature selection**, features are divided into **strongly relevant, weakly relevant**, and **irrelevant**. Feature selection methods are divided into **open loop** and **closed loop (wrapper)** methods. All feature selection methods depend on the selection criteria and search method. The **selection criteria** include minimum concept description, mutual information, inconsistency count, rough sets, and interclass separability. In the Chapter we have described **exhaustive search, branch and bound, feature ranking**, and **forward** and **backward** feature selection methods.

Feature selection and extraction methods, including PCA and Fisher's transformations are well presented in [6, 7, 12, 14, 16, 17, 26]. Independent component analysis is nicely described in [11], while the blind source separation problem is described in [4], and applications of ICA are presented in [24, 25]. A description of vector quantization can be found in [9, 15, 18]. Fourier transform and Wavelets are covered in [1, 5, 8, 21]. The Zernike moments were introduced in [28] and later extended in [13, 23]. The open loop feature selection methods include Focus [2] and Relief algorithms [16] and their extensions [3, 6, 12, 17, 22, 26]. The closed loop (wrapper) feature selection methods are described in [3, 6, 10, 12, 22, 23]. The sequential backward selection was introduced in [19], and the forward selection and stepwise methods were presented in [14]. The branch and bound search for feature selection (based on the idea of dynamic programming) is covered in [20, 27].

References

1. Ainsworth, W.A. 1988. *Speech Recognition by Machine*, Peter Peregrinus Ltd., London, UK
2. Almuallim, H., and Dietterich, T.G. 1992. Efficient algorithms for identifying relevant features. *Proceedings of the Ninth Canadian Conference on Artificial Intelligence*, 38–45. Vancouver, Canada
3. Bazan, J.G., Skowron, A., and Swiniarski, R. 2006. Rough sets and vague concept approximation: From sample approximation to adaptive learning, Transactions on Rough Sets V; Journal Subline, Lecture Notes in Computer Science 4100, Springer, Heidelberg, 39–62
4. Bell, A.J., and Sejnowski, T.J. 1995. An information-maximization approach to blind separation and blind deconvolution. *Neural Computation*, 7:1129–1159
5. Burrus, C., Gopinath, R., and Guo, H. 1998. *Introduction to Wavelets and Wavelet Transformations: A Primer*, Prentice Hall
6. Cios, K.J., Pedrycz, W., and Swiniarski, R. 1998. *Data Mining Methods for Knowledge Discovery*, Kluwer
7. Duda, R.O., and Hart, P.E. 2001. *Pattern Recognition and Scene Analysis*, Wiley
8. Fant, C.G. 1973. *Speech Sounds and Features*, MIT Press
9. Gersho, A., and Gray, R. 1992. *Vector Quantization and Signal Compression*, Boston, Kluwer

10. Grzymala-Busse, J.W, Kostek, B., Swiniarski, R., and Szczuka, M. 2004. (Editors-in Chief of a special I volume) Transaction on Rough Sets I. In (Editors-in-Chief Peters, J., and Skowron, A.), Lecture Notes in Computer Sciences on Rough Sets, 3100, Springer, Berlin, New York, pp. 1–404
11. Hyvarinen, A., Karhunen, J., and Oja, E. 2001. *Independent Component Analysis*, John Wiley, New York
12. John, G., Kohavi, R., and Pfleger, K. 1994. Irrelevant features and the subset selection problem. *Proceedings of the Eleventh International Conference on Machine Learning* (ICML-94), 121–129, New Brunswick, NJ
13. Khotanzad, A., Hong, Y.H. 1990. Invariant image recognition by Zernike moments. *IEEE Transactions on Pattern Analysis and Machine Intelligence*, 12(5):489–497
14. Kittler, J. 1986. Feature selection and extraction. In Young, T.Y., and Fu, K.S. (Eds.), *Handbook of Pattern Recognition and Image Processing*, Academic Press, 59–83
15. Kohonen, T. 1997. *Self-Organizing Maps*, Springer
16. Kononenko, I. 1994. Estimating attributes: Analysis and extension of Relief. *Proceedings of European Conference on Machine Learning*, 171–182, Catania, Italy
17. Langley, P. 1994. Selection of relevant features in machine learning. *Proceedings of the AAAI Fall Symposium on Relevance*, 140–144, Orlando, FL
18. Linde, Y., Buzo, A., and Gray, R. 1980. An algorithm for vector quantizer design. *IEEE Transaction on Communications*, 28(1):84–94
19. Marill, T., and Green, D. 1963. On the effectiveness of receptors in recognition systems. *IEEE Transactions on Information Theory*, 9:11–17
20. Narendra, P.M., and Fukunaga, K. 1977. A branch and bound algorithm for feature subset selection. *IEEE Transactions on Computers*, C-26:917–922
21. Rabiner, L.R., and Juang, B.H. 1993. *Fundamentals of Speech Recognition*, Prentice Hall, Englewood Cliffs, N.J.
22. Skowron, A., Swiniarski, R., Synak, P., and Peters, J.F. 2004. Approximation Spaces and Information Granulation. Tsumoto, S., Slowinski, R., and Komorowski, J. (Eds.) *Rough Sets and Current Trends in Computing*, Proceedings of 4th International Conference, RSCTC 2004, Uppsala, Sweden, Springer, pp. 116–126
23. Swiniarski, R. 2004. Application of Zernike Moments, Independent Component Analysis, and Rough and Fuzzy Classifier for Hand-Written Character Recognition. In Klopotek, M.K., Wierzchon, S., and Trojanowski, K. (Eds.), *Intelligent Information Processing and Web Mining*. Proceedings of the International IIS:IIPWM'04 Conference. Zakopane, Poland, May 17–20, Springer, pp. 623–632
24. Swiniarski, R., Lim Hun Ki, Shin Joo Heon and Skowron, A. 2006. Independent Component Analysis, Principal Component Analysis and Rough Sets in Hybrid Mammogram Classification. *Proceedings of the 2006 International Conference on Image Processing, Computer Vision, & Pattern Recognition*, volume II, 640–645, Las Vegas
25. Swiniarski, R., and Skowron, A. 2004. Independent Component Analysis and Rough Sets in Face Recognition. In Grzymala-Busse, J., Kostek, B., Swiniarski, R., and Szczuka, M. (Editors-in Chief of a special I volume) Transaction on Rough Sets I. In (Editors-in-Chief Peters, J., and Skowron, A.), Lecture Notes in Computer Sciences on Rough Sets, 3100, Springer, Berlin, New York, pp. 392–404
26. Swiniarski, R. and Skowron, A. 2003. Rough sets methods in feature selection and recognition. *Pattern Recognition Letters*, 24(6):883–849
27. Yu, B., and Yuan, B. 1993. A more efficient branch and bound algorithm for feature selection. *Pattern Recognition*, 26(6):883–889
28. Zernike, F. 1934. Beugungstheorie des schneidenverfahrens und seimer verbesserten form, der phasenkontrastmethode, *Physica*, 1:689–706

5. Exercises

1. Let us consider the data set containing patterns with three nominal attributes {MONITOR, OS, CPU} labeled by two categorical classes c_1 = Poor, c_2 = Good (Table 7.4). For this data set (decision table):

(a) Compute a set of strongly relevant attributes (for the entire pattern **x**) with respect to a class attribute d, using the idea of core defined by rough sets theory.
(b) Compute sets of weakly relevant attributes (for a whole pattern **x**) with respect to a class attribute d using the idea of reduct defined by rough sets theory.
(c) For selected minimal reduct, design decision rules using the rough sets method.
(d) Find a minimal set of attributes describing all concepts in the data set using exhaustive search.

2. For a data set from Table 7.4, find an optimal set of attributes, using the open loop scheme with individual feature ranking, using the criteria of

   (a) mutual information
   (b) inconsistency count

3. Let us consider a data set containing 20 cases (Table 7.5). Each case is composed of the four feature patterns $\mathbf{x} \in \mathbb{R}^4$ labeled by categorical classes $c_1 = 1$ and $c_2 = 2$ (10 patterns in each class). A pattern's attributes take on real values.
   For the considered data set, find the best feature subset, using the open loop feature selection method, with the following feature selection criteria:

   (a) mutual information
   (b) inconsistency count
   (c) interclass separability

   while applying

   (a) exhaustive search
   (b) branch and bound search
   (c) sequential forward search

4. For a data set from Table 7.5, find the best feature subset, using the closed loop feature selection method, with the relative frequency of classification errors as a feature selection criterion. Use the k-nearest neighbors classifier for feature evaluation and also as a final classifier. For performance evaluation of classifiers, use the holdout error counting method (with partion of the data set used for design and the other portion held out for testing).
5. For a data set from Table 7.5, the provide principal component analysis (PCA) (global for all patterns for all classes):

Table 7.4. Example of the decision table PC.

Object	Condition attributes			Decision attributes
U	C			D
PC	MONITOR	OS	CPU	d
x_1	Color	DOS	486	Good
x_2	Color	Windows	Pentium	Good
x_3	Monochrome	DOS	Pentium	Poor
x_4	Color	Windows	486	Good
x_5	Color	DOS	386	Poor
x_6	Monochrome	Windows	486	Good
x_7	Monochrome	Windows	Pentium	Good
x_8	Color	Windows	386	Good

Table 7.5. A data set with real-valued attributes.

x_1	x_2	x_3	x_4	Class
0.8	0.7	1.1	0.5	1
1.3	0.6	0.9	1.3	1
1.9	1.7	0.3	0.6	1
2.0	0.1	0.3	1.8	1
1.1	1.6	0.1	1.9	1
0.1	0.2	2.1	2.3	1
2.2	2.4	0.3	1.3	1
2.1	1.9	0.5	2.9	1
1.3	2.7	0.3	2.2	1
2.9	3.2	0.8	0.1	1
5.2	2.5	10.6	5.5	2
7.8	9.5	12.2	6.5	2
4.5	7.6	3.9	2.3	2
8.9	6.2	2.9	8.3	2
4.2	5.4	11.3	9.3	2
3.2	8.7	2.5	15.9	2
6.6	6.3	10.3	12.2	2
9.9	2.2	6.8	15.1	2
12.8	4.2	9.8	4.1	2
4.9	9.2	4.8	7.9	2

(a) Compute the covariance matrix
(b) Find an optimal Karhunen-Loéve transform
(c) Transform the original patterns to full-size principal component space. Compute the inverse transform
(d) Transform the original patterns to the reduced principal component space (consider $m = 1, 2, 3$). Compute the inverse transform. Compute the reconstruction error: a) numerically (as a sum of squared errors); b) as a sum of trailing eigenvalues (remove the least significant principal components).

6. For a data set from Table 7.5,

   (a) Provide principal component analysis (PCA) with transformation of data into the full-size principal component space. For the full-size PCA feature vector, compute and evaluate the k-nearest neighbors classifier.
   (b) Select the first m principal components ($m = 4, 3, 2, 1$) as a reduced feature vector. For this feature pattern, design and evaluate the k-nearest neighbors classifier.

7. Generate samples for two time signals $\sin(0.2*x)$ and $\sin(2*x)$, for x from 0 to 10π with the step $\pi/100$. Mix the signals linearly using the mixing matrix

$$x = 0 : \pi/100 : 10*\pi; \qquad H = \begin{bmatrix} 0.1 & 0.9 \\ 0.9 & 0.1 \end{bmatrix}$$

Separate sources using ICA method. Solve the problems

(a) Without whitening.
(b) With whitening and with reduction of the dimension to 1.
(c) With whitening implemented as PCA, and with reduction of the dimension to 1.

8. Design a Matlab program to perform vector quantization on a gray-scale image using a 4×4 pixel block as a quantization unit. Unfold the pixel block as a 16-element block vector **x** by concatenating the subsequent rows of the block. Form the training set T_{tra} containing subsequent block vectors for all blocks of an image (for example, considering the sequence of blocks for an image from the left to the right, etc.), design your optimal codebook using all block vectors from the training data set. Use the generalized Lloyd algorithm to find the optimal codebook. Select the size of codebook (say, $M = 128$). Then quantize the image represented by the training set T_{tra} using your codebook. Find the quantization quality using a PSNR measure. Provide quantization of the image for codebook size equal to 32, 64, and 256. Reconstruct an image from the compressed representation. Compare the results. Provide quantization experiments and compare the resulting quality of quantization.

    As a result of quantization of an image, we obtain the codebook with M codevectors labeled by M indices. The quantization result of the considered image is the vector of encoding indices T_{ind}, which entry contains an index of encoding codevector representing the corresponding block vector in the set T_{tra}. In order to find the reconstructed image from T_{ind} each encoding codevector for a given block is folded into an image block (through row-by-row folding).

9. Synthesize (or get from the Internet) a data set of gray-scale images of handwritten digits $(0, 1, \cdots, 9)$ (at least 10 instances per digit).

    (a) Write in MatLab language (or in another language) a program implementing the computation of complex Zernike moments.
    (b) Extract from each image the Zernike moments with maximal order (6,6) and (12, 12). Design Zernike patterns and construct supervised training and testing sets (with 10 classes).
    (c) Design and test the following classifiers: k-nearest neighbors, error back-propagation neural networks, and learning vector quantization.
    (d) Apply principal component analysis (PCA) to obtained supervised data sets. Project all patterns into principal component space, and reduce the projected patterns to a heuristically selected number. Design and test the classifiers listed in the previous question.
    (e) Form the training and test sets containing raw patterns composed from digit images as concatenated columns. Apply principal component analysis (PCA) to such raw data sets. Project all patterns into the principal component space, and reduce the length of the projected patterns to a heuristically selected number. Design and test the handwritten digit recognitions using classifiers listed in the previous question.

10. Get (from the Internet) a data set of gray-scale images of human faces (with at least 10 classes, and at least 10 instances of face for a class). Extract the face image features and form a pattern. Compare different methods of feature extraction and pattern forming and their impact on classifier accuracy. Write the required operations in Matlab language. Consider the following methods of feature extraction:

    (a) Principal Component Analysis (PCA)
    (b) Independent Component Analysis (ICA)
    (c) Singular Values Decomposition (SVD)
    (d) Zernike Moments
    (e) Two-dimensional Fourier transform (power spectrum features)
    (f) Synthetic features (see previous Chapters) derived from the normalized power spectrum map of two-dimensional Fourier transform (power spectrum features)
    (g) Haar wavelets
    (h) Morlet wavelets

8

Discretization Methods

In this Chapter, we introduce unsupervised and supervised methods for discretization of continuous data attributes. Discretization is one of the most important, and often required, preprocessing methods. Its overall goal is to reduce the complexity of the data for further data mining tasks.

1. Why Discretize Data Attributes?

Often data are described by attributes that assume **continuous** values. If the number of continuous values of the attributes/features is huge, model building for such data can be difficult and/or highly inefficient. Moreover, many data mining algorithms, as described in Part 4 of this book, operate only in discrete search/attribute spaces. Examples of the latter are decision trees and rule algorithms, for which discretization is a necessary preprocessing step, not just a tool for reducing the data and the subsequently generated model complexity.

The goal of **discretization** is to reduce the number of values a continuous attribute assumes by grouping them into a number, n, of intervals (bins).

Two key problems associated with discretization are how to choose the number of intervals (bins), and how to decide on their width. Discretization can be performed with or without taking class information (if available) into account. Analogously to unsupervised vs. supervised learning methods (which require class information), discretization algorithms are divided into two main categories: **unsupervised** and **supervised**. If class information exists (i.e., if we have access to training data) a discretization algorithm should take advantage of it, especially if the subsequently used learning algorithm for model building is supervised. In this case, a discretization algorithm should maximize the interdependence between the attribute values and the class labels. An additional benefit of using class information in the discretization process is that it minimizes the (original) information loss. Figure 8.1 illustrates a trivial case for one attribute, using the same width for all intervals. The top of Figure 8.1 shows a case where a user decided to choose $n = 2$ intervals, without using information about class membership of the data points. The bottom of Figure 8.1 shows the grouping of attribute values into $n = 4$ intervals, while taking into account the class information; the four intervals better discretize the data for a subsequent classification into two classes, represented by white and black ovals. In the first case, by choosing just two intervals, the user made it more difficult for a classifier to distinguish between the classes (since the instances from different classes were grouped into the same intervals). With real data, it would be unusual to have such nicely distributed attribute values; most often there would be a mixture of data points from several classes in each interval, as illustrated in Figure 8.2.

Discretization of continuous attributes is most often performed one attribute at a time, independent of other attributes. This approach is known as **static attribute discretization**. On the other end of the spectrum is **dynamic attribute discretization**, where all attributes are discretized

1. Why Discretize Data Attributes?

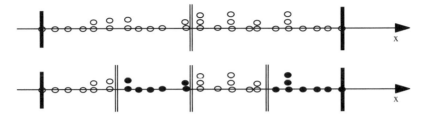

Figure 8.1. Deciding on the number of intervals without using (top) and after using (bottom) class information.

simultaneously while taking into account the interdependencies among them. Still another way to look at discretization algorithms is to consider whether the partitions produced by them apply only to localized regions of the example/instance space. In this case, they are called **local** (like the one performed by the C4.5 algorithm, where not all features are discretized). On the other hand, when all attributes are discretized, they produce $n_1 \bullet n_2 \bullet \ldots n_d$ regions, where n_i is the number of intervals of the ith attribute; such methods are called **global**.

Discretization transforms continuous attribute/feature values into a finite number of intervals and associates with each interval a **discrete** value. Discrete values can be **nominal** (e.g. white, red) or **numerical** (e.g., 1, 2, 3), although for coding purposes nominal values are always converted into numerical ones. Any discretization process consists of two basic steps:

- first, the number of discrete intervals needs to be chosen. This is usually done by the user, although a few discretization algorithms are able to do it on their own. There also exist some heuristic rules that help the user to decide on the number of intervals.
- second, the width (boundary) of each interval (given the range of values of an attribute) must be determined, which is often done by a discretization algorithm itself.

Since all discretization algorithms but one supervised algorithm (outlined later in the Chapter) are *static* and all but one (k-means) are *global* we shall use only the following categorization to describe them:

- **unsupervised** (or class-blind), which discretize attributes without taking into account class information
- **supervised** (or class-aware), which discretize attributes while using interdependence between the known class labels and the attribute values

As we will learn in Part 4 of this book, most of the model-building algorithms are strongly influenced by the number of intervals into which the attributes are divided. One such effect is efficiency of model-building, and another the ability of the model to generalize on new data. The larger the number of intervals, the larger the chance for a model to overfit the data (see Chapter15).

To summarize: on the one hand, a discretization algorithm, used as a preprocessing step, should generate as few discrete intervals as possible. On the other hand, however, too few intervals may hide information about the relationship between the class variable (if known) and the interval variable. The latter situation is especially true when the attribute values are not distributed

Figure 8.2. Distribution of points belonging to three categories over attribute X.

evenly, in which case a large amount of important original information may be lost. In practice, discretization algorithms search for a number of discrete intervals as a tradeoff between these two goals.

2. Unsupervised Discretization Algorithms

Unsupervised discretization algorithms are the simplest to use and implement. They only require the user to specify the number of intervals and/or how many data points should be included in any given interval.

The following heuristic is often used to choose intervals: *the number of intervals for each attribute should not be smaller than the number of classes (if known)*. The other popular heuristic is to choose the number of intervals, n_{Fi}, for each attribute, F_i, ($i = 1, \ldots, n$, where n is the number of attributes), as follows:

$$n_{Fi} = M/(3^*C)$$

where M is the number of training examples and C is the number of known categories.

A description of two unsupervised algorithms follows. We assume that, in general, the user supplies a set of numbers, representing the number of intervals into which each attribute/feature is to be discretized:

$$N = \{n_{F1}, \ldots, n_{Fi}, \ldots, n_{Fn}\}$$

2.1. Equal-Width Discretization

This algorithm first finds the minimum and maximum values for each feature, F_i, and then divides this range into a number, n_{Fi}, of user-specified, equal-width intervals.

2.2. Equal-Frequency Discretization

This algorithm determines the minimum and maximum values of the attribute, sorts all values in ascending order, and divides the range into a user-defined number of intervals, in such a way that every interval contains the same number of sorted values.

3. Supervised Discretization Algorithms

As we will learn in Part 4 of this book (particularly in Chapter 12), the goal of machine learning algorithms is to generate models of the data that well approximate concepts/hypotheses represented (described) by the data. In other words, their goal is to discover relation between the **class variable** (such as the conclusion (THEN part) of a production rule) and the attribute/feature variable (such as the condition (IF part) of a rule). In a similar manner a supervised discretization problem can be formalized in view of the class-attribute interdependence, as described below.

3.1. Information-Theoretic Algorithms

Many supervised discretization algorithms have their origins in **information theory**. A supervised learning task requires having a training dataset consisting of M examples, where each example belongs to only one of the S classes. Let F indicate a continuous attribute/feature. We say that there exists a discretization scheme D on F that discretizes the continuous attribute F into n discrete intervals, bounded by the pairs of numbers:

$$D : \{[d_0, d_1], (d_1, d_2], \ldots, (d_{n-1}, d_n]\} \tag{1}$$

3. Supervised Discretization Algorithms

where d_0 is the minimal value and d_n is the maximal value of attribute F, and the values in Equation (1) are arranged in ascending order.

These values constitute the boundary set for discretization D:

$$\{d_0, d_1, d_2, \ldots, d_{n-1}, d_n\}$$

Each value of attribute F can be assigned into only one of the n intervals defined in Equation (1). The membership of each value, within a certain interval, for attribute F may change with a change of the discretization D. The class variable and the discretization variable of attribute F are treated as random variables defining a two-dimensional frequency matrix, called the **quanta matrix**, as shown in Table 8.1.

In the table, q_{ir} is the total number of continuous values belonging to the ith class that are within interval $(d_{r-1}, d_r]$. M_{i+} is the total number of objects belonging to the ith class, and M_{+r} is the total number of continuous values of attribute F that are within the interval $(d_{r-1}, d_r]$, for $i = 1, 2 \ldots, S$ and $r = 1, 2, \ldots, n$.

Example: Let us assume there are three classes, four intervals, and 33 examples, distributed as illustrated in Figure 8.2. For these data we construct the corresponding quanta matrix, shown in Table 8.2.

The values shown in Table 8.2 have been calculated as follows. Total number of values:

$$M = \sum_{r=1}^{L_j} q_{+r} = \sum_{i=1}^{c} q_{i+}$$

$$M = 8 + 7 + 10 + 8 = 33$$

$$M = 11 + 9 + 13 = 33$$

Number of values in the First interval:

$$M_{+r} = \sum_{i=1}^{c} q_{ir}$$

$$M_{+\text{first}} = 5 + 1 + 2 = 8$$

Table 8.1. Two-dimensional quanta matrix for attribute F and discretization scheme D.

Class	Interval					Class Total
	$[d_0, d_1]$	...	$(d_{r-1}, d_r]$	...	$(d_{n-1}, d_n]$	
C_1	q_{11}	...	q_{1r}	...	q_{1n}	M_{1+}
$\vdots$	$\vdots$	...	$\vdots$	...	$\vdots$	$\vdots$
C_i	q_{i1}	...	q_{ir}	...	q_{in}	M_{i+}
$\vdots$	$\vdots$	...	$\vdots$	...	$\vdots$	$\vdots$
C_S	q_{S1}	...	q_{Sr}	...	q_{Sn}	M_{S+}
Interval Total	M_{+1}	...	M_{+r}	...	M_{+n}	M

Table 8.2. Quanta matrix corresponding to Figure 8.2.

Classes	Intervals				Total
	First	Second	Third	Fourth	
White	5	2	4	0	11
Grey	1	2	2	4	9
Black	2	3	4	4	13
Total	8	7	10	8	33

Number of values in the White class:

$$M_{i+} = \sum_{r=1}^{n} q_{ir}$$

$$M_{\text{white}+} = 5+2+4+0 = 11$$

We now calculate several statistics from the above quanta matrix. The estimated joint probability of the occurrence that attribute F values are within interval $D_r = (d_{r-1}, d_r]$ and belong to class C_i is calculated as follows:

$$p_{ir} = p(C_i, D_r|F) = \frac{q_{ir}}{M} \quad (2)$$

(For the example: $p_{\text{white,first}} = 5/33 = 0.24$)

The estimated class marginal probability that attribute F values belong to class C_i, p_{i+}, and the estimated interval marginal probability that attribute F values are within the interval $D_r = (d_{r-1}, d_r]$ p_{+r}, are:

$$p_{i+} = p(C_i) = \frac{M_{i+}}{M} \quad (3)$$

(For the example: $p_{\text{white}+} = 11/33$)

$$p_{+r} = p(D_r|F) = \frac{M_{+r}}{M} \quad (4)$$

(For the example: $p_{+\text{first}} = 8/33$)

The **Class-Attribute Mutual Information** between the class variable C and the discretization variable D for attribute F given the 2-D frequency matrix is defined as

$$I(C, D|F) = \sum_{i=1}^{S} \sum_{r=1}^{n} p_{ir} \log_2 \frac{p_{ir}}{p_{i+} p_{+r}} \quad (5)$$

(For the example: $I(C, D : vj) = 5/33^* \log((5/33)/(11/33^*8/33)) + \ldots + 4/33^* \log((4/33)/(13/33)^*8/33))$)

Similarly, the **Class-Attribute Information** and **Shannon's entropy** are defined, respectively, as

$$INFO(C, D|F) = \sum_{i=1}^{S} \sum_{r=1}^{n} p_{ir} \log_2 \frac{p_{+r}}{p_{ir}} \quad (6)$$

(For the example: INFO(C, D : vj) = $5/33^* \log((8/33)/(5/33)) + \ldots + 4/33^* \log((8/33)/(4/33))$)

$$H(C, D|F) = \sum_{i=1}^{S} \sum_{r=1}^{n} p_{ir} \log_2 \frac{1}{p_{ir}} \quad (7)$$

(For the example: H(c : vj) = $5/33^* \log(1/(5/33)) + \ldots + 4/33^* \log(1/(4/33))$)

Given Equations (5), (6) and (7), the **Class-Attribute Interdependence Redundancy** criterion (CAIR, or R) and the **Class-Attribute Interdependence Uncertainty** criterion (CAIU, or U) are defined as

$$R(C, D|F) = \frac{I(C, D|F)}{H(C, D|F)} \quad (8)$$

$$U(C, D|F) = \frac{INFO(C, D|F)}{H(C, D|F)} \quad (9)$$

The CAIR criterion measures the interdependence between classes (the larger its value, the better correlated are the class label and the discrete intervals) and the discretized attribute. CAIR is independent of the number of class labels and of the number of unique values of the continuous attribute. The same holds true for the CAIU criterion, but with a reverse relationship. The CAIR criterion is used in a CADD algorithm, described later, that uses a user-specified number of intervals, initializes the discretization intervals using a maximum entropy discretization method, and uses the significance test for selection of a proper confidence interval, all of which add to its complexity. The CAIM algorithm, on the otherhand, described in detail below, avoids these disadvantages.

3.1.1. CAIM: Class-Attribute Interdependency Maximization Algorithm

The **Class-Attribute Interdependency Maximization algorithm** works in a top-down manner: it divides one of the existing intervals into two new intervals, using a criterion function that results in achieving the optimal class-attribute interdependency after the split. It starts with the entire interval $[d_o, d_n]$ and maximizes interdependence between the continuous attribute and its class labels in order to automatically generate a small number of discrete intervals. The discretization criterion used in the CAIM algorithm is described next.

The CAIM criterion (note that the criterion's name is the same as the algorithm's name) measures dependency between the class variable C and the discretization variable D for attribute F, for a given quanta matrix, and is defined as follows:

$$CAIM(C, D|F) = \frac{\sum_{r=1}^{n} \frac{\max_r^2}{M_{+r}}}{n} \quad (10)$$

where n is the number of intervals, r iterates through all intervals, i.e., $r = 1, 2, \ldots, n$, max_r is the maximum value among all q_{ir} values (maximum in the rth column of the quanta matrix), $i = 1, 2, \ldots, S$, and M_{+r} is the total number of continuous values of attribute F that are within the interval $(d_{r-1}, d_r]$.

The CAIM criterion is a heuristic measure defined to quantify the interdependence between classes and the discretized attribute. It has the following properties:

- the larger the value of the CAIM, the higher the interdependence between the class labels and the intervals. The larger the number of values belonging to class C_i within a particular interval (if the number of values belonging to C_i within the interval is the largest, then C_i is called the *leading class*), the higher the interdependence between C_i and the interval. The goal of maximizing the interdependence can be translated into the goal of achieving the largest possible number of values that belong to a leading class, within all intervals. CAIM maximizes the number of values belonging to a leading class by using the max_i operation. CAIM achieves the highest value when all values within a particular interval belong to the same class, for all intervals; then $max_r = M_{ri}$ and $CAIM = M/n$.
- CAIM assumes real values in the interval $[0, M]$, where M is the number of values of attribute F.
- CAIM generates a discretization scheme where each interval potentially has the majority of its values grouped within a single class label
- the squared max_i value is divided by the M_{ri} in order to account for the (negative) impact that values belonging to classes other than the leading class have on the discretization scheme. The more such values, the bigger the value of M_{ri}, which decreases the value of the CAIM criterion.
- scales the max_r^2 value to avoid overflow error during calculations. To do so, we calculate $\frac{max_r^2}{M_{ri}}$ as $\frac{max_r}{M_{ri}} max_r$

– since CAIM favors discretization schemes with smaller numbers of intervals the summed value is divided by the number of intervals n.

The value of the CAIM criterion is calculated with a single pass over the quanta matrix. The optimal discretization scheme can be found by searching over the space of all possible discretization schemes to find the one with the highest value of the CAIM criterion. Since such a search for the globally optimal CAIM value would be highly combinatorial, the algorithm uses a greedy approach to search for an approximation of the optimal value by finding locally maximal values of the criterion. Although this approach does not guarantee finding the global maximum it is computationally inexpensive.

The CAIM algorithm, like all discretization algorithms, consists of two steps: (1) initialization of the candidate interval boundaries and the corresponding initial discretization scheme, and (2) the consecutive additions of a new boundary that results in the locally highest value of the CAIM criterion. The pseudocode of the CAIM algorithm follows.

Given: Data consisting of M examples, S classes, and continuous attributes F_i
For every F_i DO:

Step1.
1.1 find the maximum (d_n) and minimum (d_0) values of F_i
1.2 form a set of all distinct values of F_i in ascending order, and initialize all possible interval boundaries B with minimum, maximum and all the midpoints of all the adjacent pairs in the set
1.3 set the initial discretization scheme as $D : \{[d_0, d_n]\}$, set GlobalCAIM=0

Step2.
2.1 initialize $k = 1$
2.2 tentatively add an inner boundary, which is not already in D, from B, and calculate the corresponding CAIM value
2.3 after all the tentative additions have been tried accept the one with the highest value of CAIM
2.4 if (CAIM > GlobalCAIM or $k < S$), then update D with the boundary accepted in step 2.3 and set GlobalCAIM=CAIM; else terminate
2.5 set $k = k + 1$ and go to step 2.2

Result: Discretization scheme D.

The algorithm starts with a single interval that covers all values of an attribute and then divides it iteratively. From all possible division points that are tried (with replacement) in Step 2.2., it chooses the division boundary that gives the highest value of the CAIM criterion. The algorithm uses the heuristic that every discretized attribute should have at least the number of intervals equal to the number of classes.

The running time of the algorithm is log-linear, namely, $O(M \log(M))$. Tables 8.4 and 8.5 illustrate the performance of the algorithm on several datasets. All but the *smo* dataset can be obtained from the UC Irvine ML repository at http://www.ics.uci.edu/~mlearn/MLRepository.html, while the *smo* can be obtained from the StatLib at http://lib.stat.cmu.edu. A summary description of the eight datasets is shown in Table 8.3.

A comparison of the results achieved by the CAIM algorithm and six other unsupervised and supervised discretization algorithms (described later in this Chapter), on the eight datasets, is shown in Table 8.4. The goodness of discretization is evaluated by three measures: (a) the CAIR criterion value, (b) the number of generated intervals, and (c) the execution time.

As mentioned above, one of the goals of data preprocessing is to make it easier for the subsequently used machine learning algorithm to generate a model of the data. In Tables 8.5 and 8.6 we

Table 8.3. Properties of datasets used for comparing discretization algorithms.

Properties	Datasets							
	Iris	sat	thy	wav	ion	smo	Hea	pid
# of classes	3	6	3	3	2	3	2	2
# of examples	150	6435	7200	3600	351	2855	270	768
# of training / testing examples	10 × cross-validation	10 × cross-validation	10 × cross-validation	10 × cross-validation	10 × cross-validation	10 × cross-validation	10 × cross-validation	10 × cross-validation
# of attributes	4	36	21	21	34	13	13	8
# of continuous attributes	4	36	6	21	32	2	6	8

Table 8.4. Comparison of discretization algorithms using eight datasets (bolded entries indicate the best results).

Criterion	Discretization Method	Dataset															
		iris	std	sat	std	thy	std	wav	std	ion	std	smo	std	hea	std	pid	std
CAIR mean value through all intervals	Equal Width	0.40	0.01	0.24	0	0.071	0	0.068	0	0.098	0	0.011	0	0.087	0	0.058	0
	Equal Frequency	0.41	0.01	0.24	0	0.038	0	0.064	0	0.095	0	0.010	0	0.079	0	0.052	0
	Paterson-Niblett	0.35	0.01	0.21	0	0.144	0.01	**0.141**	0	0.192	0	0.012	0	0.088	0	0.052	0
	Maximum Entropy	0.30	0.01	0.21	0	0.032	0	0.062	0	0.100	0	0.011	0	0.081	0	0.048	0
	CADD	0.51	0.01	**0.26**	0	0.026	0	0.068	0	0.130	0	**0.015**	0	0.098	0.01	0.057	0
	IEM	0.52	0.01	0.22	0	0.141	0.01	0.112	0	**0.193**	0.01	0.000	0	0.118	0.02	0.079	0
	CAIM	**0.54**	0.01	**0.26**	0	**0.170**	0.01	0.130	0	0.168	0	0.010	0	**0.138**	0.01	**0.084**	0.01
# of intervals	Equal Width	16	0	252	0	126	0.48	630	0	640	0	22	0.48	56	0	106	0
	Equal Frequency	16	0	252	0	126	0.48	630	0	640	0	22	0.48	56	0	106	0
	Paterson-Niblett	48	0	432	0	45	0.79	252	0	384	0	17	0.52	48	0.53	62	0.48
	Maximum Entropy	16	0	252	0	125	0.52	630	0	572	6.70	22	0.48	56	0.42	97	0.32
	CADD	16	0.71	246	1.26	84	3.48	628	1.43	536	10.26	22	0.48	55	0.32	96	0.92
	IEM	**12**	0.48	430	4.88	28	1.60	91	1.50	113	17.69	**2**	0	**10**	0.48	17	1.27
	CAIM	**12**	0	**216**	0	**18**	0	**63**	0	**64**	0	6	0	12	0	**16**	0

Table 8.5. Comparison of accuracies achieved by the CLIP4 and C5.0 algorithms (bolded values indicate the best results).

Algorithm	Discretization Method	iris		sat		thy		wav		ion		smo		hea		pid	
		acc	std	acc	std	acc	std	acc	std	acc	std	acc	std	acc	std	acc	std
CLIP4	Equal Width	88.0	6.9	77.5	2.8	91.7	1.9	68.2	2.2	86.9	6.4	68.6	2.0	64.5	10.1	65.5	6.5
	Equal Frequency	91.2	7.6	76.3	3.4	95.7	2.4	65.4	2.9	81.0	3.7	68.9	2.8	72.6	8.2	63.3	6.3
	Paterson-Niblett	87.3	8.6	75.6	3.9	97.4	0.6	60.9	5.2	**93.7**	3.9	68.9	2.7	68.5	13.6	72.7	5.1
	Maximum Entropy	90.0	6.5	76.4	2.7	97.3	0.9	63.5	2.9	82.9	4.8	68.7	2.8	62.6	9.8	63.4	5.1
	CADD	**93.3**	4.4	**77.5**	2.6	70.1	13.9	61.5	3.4	88.8	3.1	68.8	2.5	72.2	11.4	65.5	4.2
	IEM	92.7	4.9	77.2	2.7	**98.8**	0.5	75.2	1.7	92.4	6.9	66.9	2.6	75.2	8.6	72.2	4.2
	CAIM	92.7	8.0	76.4	2.0	97.9	0.4	**76.0**	1.9	92.7	3.9	**69.8**	4.0	**79.3**	5.0	**72.9**	3.7
C5.0	Equal Width	94.7	5.3	86.0	1.6	95.0	1.1	57.7	8.2	85.5	6.4	69.2	5.4	74.7	5.2	70.8	2.8
	Equal Frequency	94.0	5.8	85.1	1.5	97.6	1.2	57.5	7.9	81.0	12.4	70.1	1.7	69.3	5.7	70.3	5.4
	Paterson-Niblett	94.0	4.9	83.0	1.0	97.8	0.4	74.8	5.6	85.0	8.1	70.1	3.2	**79.9**	7.1	71.7	4.4
	Maximum Entropy	93.3	6.3	85.2	1.5	97.7	0.6	55.5	6.2	86.5	8.8	70.2	3.9	73.3	7.6	66.4	5.9
	CADD	93.3	5.4	86.1	0.9	93.5	0.8	56.9	2.1	77.5	11.9	70.2	4.7	73.6	10.6	71.8	2.2
	IEM	**95.3**	4.5	84.6	1.1	99.4	0.2	**76.6**	2.1	92.6	2.9	69.7	1.6	73.4	8.9	**75.8**	4.3
	CAIM	**95.3**	4.5	86.2	1.7	98.9	0.4	72.7	4.2	89.0	5.2	**70.3**	2.9	76.3	8.9	74.6	4.0
	Built-in	92.7	9.4	**86.4**	1.7	**99.8**	0.4	72.6	3.6	87.0	9.5	70.1	1.3	76.8	9.9	73.7	4.9

Table 8.6. Comparison of the number of rules/leaves generated by the CLIP4 and C5.0 algorithms (bolded values indicate the best results).

Algorithm	Discretization Method	iris		sat		thy		wav		ion		smo		pid		hea	
		#	std	#	std	#	std	#	std	#	std	#	std	#	std	#	std
CLIP4	Equal Width	4.2	0.4	47.9	1.2	**7.0**	0.0	**14.0**	0.0	**1.1**	0.3	20.0	0.0	7.3	0.5	7.0	0.5
	Equal Frequency	4.9	0.6	47.4	0.8	**7.0**	0.0	**14.0**	0.0	1.9	0.3	19.9	0.3	7.2	0.4	6.1	0.7
	Paterson-Niblett	5.2	0.4	42.7	0.8	**7.0**	0.0	**14.0**	0.0	2.0	0.0	19.3	0.7	**1.4**	0.5	7.0	1.1
	Maximum Entropy	6.5	0.7	47.1	0.9	**7.0**	0.0	**14.0**	0.0	2.1	0.3	19.8	0.6	7.0	0.0	**6.0**	0.7
	CADD	4.4	0.7	45.9	1.5	**7.0**	0.0	**14.0**	0.0	2.0	0.0	20.0	0.0	7.1	0.3	6.8	0.6
	IEM	4.0	0.5	44.7	0.9	**7.0**	0.0	**14.0**	0.0	2.1	0.7	18.9	0.6	3.6	0.5	8.3	0.5
	CAIM	**3.6**	0.5	45.6	0.7	**7.0**	0.0	**14.0**	0.0	1.9	0.3	**18.5**	0.5	1.9	0.3	7.6	0.5
C5.0	Equal Width	6.0	0.0	348.5	18.1	31.8	2.5	69.8	20.3	32.7	2.9	**1.0**	0.0	249.7	11.4	66.9	5.6
	Equal Frequency	4.2	0.6	367.0	14.1	56.4	4.8	56.3	10.6	36.5	6.5	**1.0**	0.0	303.4	7.8	82.3	0.6
	Paterson-Niblett	11.8	0.4	243.4	7.8	15.9	2.3	**41.3**	8.1	18.2	2.1	**1.0**	0.0	58.6	3.5	58.0	3.5
	Maximum Entropy	6.0	0.0	390.7	21.9	42.0	0.8	63.1	8.5	32.6	2.4	**1.0**	0.0	306.5	11.6	70.8	8.6
	CADD	4.0	0.0	346.6	12.0	35.7	2.9	72.5	15.7	24.6	5.1	**1.0**	0.0	249.7	15.9	73.2	5.8
	IEM	**3.2**	0.6	466.9	22.0	34.1	3.0	270.1	19.0	12.9	3.0	**1.0**	0.0	**11.5**	2.4	**16.2**	2.0
	CAIM	**3.2**	0.6	332.2	16.1	**10.9**	1.4	58.2	5.6	**7.7**	1.3	**1.0**	0.0	20.0	2.4	31.8	2.9
	Built-in	3.8	0.4	287.7	16.6	11.2	1.3	46.2	4.1	11.1	2.0	1.4	1.3	35.0	9.3	33.3	2.5

show the classification results of two machine learning algorithms, achieved after the data were discretized by different discretization algorithms. The algorithms are a decision tree C5.0 and a rule algorithm CLIP4, which are described in Chapter 12. The classification results are compared using two measures: the accuracy (Table 8.5) and the number of the generated rules (Table 8.6).

Note: Since the decision tree algorithm has a built-in front-end discretization algorithm, and thus is able to generate models from continuous attributes, we show its performance while it generated rules from raw data against the results achieved when discretized data were used.

As we see in Table 8.5 the best accuracy, on average, for the two machine learning algorithms is achieved for data discretized with the CAIM algorithm. Table 8.6 compares the number of rules generated by the two machine learning algorithms and indicates that the CAIM algorithm, by generating discretization schemes with small numbers of intervals, reduces the number of rules generated by the CLIP4 algorithm, and the size of the generated decision trees.

3.1.2. χ^2

A supervised learning χ^2 test (**Chi^2 test**) is used in the CAIR/CADD algorithm, but can also be used on its own for discretization purposes.

In the χ^2 test, we use the decision attribute, so it is a supervised discretization method. Let us note that any interval **Boundary Point** (BP) divides the feature values from the range $[a, b]$, into two parts, namely, the left boundary point $L_{BP} = [a, \text{BP}]$ and the right boundary point $R_{BP} = (\text{BP}, b]$; thus $n = 2$. Using statistics, we can measure the degree of independence between the partition defined by the decision attribute and defined by the interval BP. For that purpose, we use the χ^2 test as follows:

$$\chi^2 = \sum_{r=1}^{2} \sum_{i=1}^{C} \frac{(q_{ir} - E_{ir})^2}{E_{ir}}$$

where E_{ir} is the expected frequency of feature F_{ir}:

$$E_{ir} = \frac{q_{+r} \, q_{i+}}{M}$$

If either q_{+r} or q_{i+} is zero, then E_{ir} is set to 0.1. It holds that if the partitions defined by a decision attribute and by an interval boundary point BP are independent, then

$$P(q_{i+}) = P(q_{i+}|L_{BP}) = P(q_{i+}|R_{BP})$$

for any class, which means that $q_{ir} = E_{ir}$ for any $r \in [1, 2]$ and $i \in [1, \ldots, C]$, and $\chi^2 = 0$. Conversely, if an interval boundary point properly separates objects from different classes the value of the χ^2 test for this particular BP should be very high. This observation leads to the following heuristic: *retain interval boundaries with corresponding high values of the χ^2 test and delete those with small corresponding values.*

Example: Let us calculate χ^2 test values for data shown in the contingency tables shown in Table 8.7.

In Table 8.7 the boundary points BP1 through BP4 separate 14 data points for one feature (the points belong to two classes) into two intervals each (Int.1 and Int.2) in four different ways. The χ^2 values corresponding to the partitions shown in Table 8.7 are (from top left to bottom right) 2.26, 7.02, 0.0, and 14.0, respectively.

In order to use the χ^2 test for discretization, the following procedure can be used. First, the user chooses the χ^2 value, based on the desired significance level and degree of freedom using statistical tables. Then, the χ^2 values are calculated for all adjacent intervals. Next, the smallest of the calculated χ^2 values is found and compared with the user-chosen value: if it is smaller,

Table 8.7. Four hypothetical scenarios for discretizing an attribute.

	BP1				BP2		
Class	Int.1	Int.2	Total	Class	Int.1	Int.2	Total
C1	4	4	8	C1	7	1	8
C2	1	5	6	C2	1	5	6
Total	5	9	14	Total	8	6	14
	BP3				BP4		
Class	Int.1	Int.2	Total	Class	Int.1	Int.2	Total
C1	4	4	8	C1	8	0	8
C2	3	3	6	C2	0	6	6
Total	7	7	14	Total	8	6	14

then the two corresponding intervals are merged; otherwise, they are kept. The process is repeated until the smallest value found is larger than the user-specified value. The procedure needs to be repeated for each feature separately. There are also several other ways, not covered here, in which the χ^2 test can be used for discretization of features.

3.1.3. Maximum Entropy Discretization

Let us recall that in order to design an effective discretization scheme, one needs to choose:

– *Initial discretization.* One way of doing this is to start with only one interval bounded by the minimum and the maximum values of the attribute. The optimal interval scheme is found by successively adding the candidate boundary points. On the other hand, a search can begin with all the boundary points as candidates for the optimal interval scheme, and then their number can be reduced by elimination. The easiest way to chooe the candidate boundary points, however, is to take all the midpoints between any two nearby values of a continuous feature; alternatively, the user may use some heuristic to specify their number.
– *Criteria for a discretization scheme.* As mentioned above, CAIM, CAIR, and CAIU criteria are good measures of the interdependence between the class variable and the interval variable, and thus all can be used as discretization criteria.

Let T be the set of all possible discretization schemes, with their corresponding quanta matrices. The goal of the maximum entropy discretization is to find a $t^* \in T$ such that

$$H(t^*) \geq H(t) \quad \forall t \in T$$

where H is **Shannon's entropy** as defined before. This method is intended to ensure maximum entropy with minimum loss of information. To calculate the maximum entropy (ME) for one row in the quanta matrix (one class), a discretization scheme must be found that makes the quanta values as even as possible. However, for a general multiclass problem, discretization based on maximum entropy for the quanta matrix can be highly combinatorial. To avoid this situation, the problem of maximizing the total entropy of the quanta matrix is approximated by maximizing the marginal entropy. Then boundary improvement (by successive local perturbation) is performed to maximize the total entropy of the quanta matrix. The ME algorithm after Ching et al. is presented below.

Given: A training data set consisting of M examples and C classes.
For each feature, DO:

1. Initial selection of the interval boundaries:

   a) Calculate the heuristic number of intervals $= M/(3*C)$
   b) Set the initial boundary so that the sums of the rows for each column in the quanta matrix distribute as evenly as possible to maximize the marginal entropy

2. Local improvement of the interval boundaries

   a) Boundary adjustments are made in increments of the ordered, observed unique feature values to both the lower boundary and the upper boundary for each interval
   b) Accept the new boundary if the total entropy is increased by such an adjustment
   c) Repeat the above until no improvement can be achieved

Result: Final interval boundaries for each feature.

The ME discretization seeks the discretization scheme that preserves the information about a given data set. However, since it hides information about the class-feature interdependence, it is not very helpful for subsequently used ML algorithms, which have to find the class-feature interdependence on their own.

3.1.4. Class-Attribute Interdependence Redundancy Discretization (CAIR)

In order to overcome the ME problem of not retaining the interdependence relationship between the class variable and attribute variable, the CAIR measure can be used as a criterion for the discretization algorithm. Since the problem of maximizing CAIR, if we are to find an optimal number of intervals, is highly combinatorial, a heuristic is used to find a local optimization solution, as outlined in the following pseudocode.

Given: A training data set consisting of M examples and C classes.

1. Interval initialization

   a) Sort in an increasing order the continuous-valued attributes
   b) Calculate the number of intervals from the heuristic $(M/(3*C))$
   c) Perform ME discretization on the sorted values to obtain initial intervals
   d) Form the quanta matrix corresponding to those initial intervals

2. Interval improvement

   a) Based on the existing boundaries, tentatively eliminate each inner boundary in turn, and calculate its corresponding CAIR
   a) After all tentative elimination has been tried on each existing boundary, accept the one with the largest corresponding value of CAIR
   b) Update the boundaries, and repeat this step until there is no increase of CAIR

3. Interval elimination
   At this step the statistically insignificant (redundant) intervals are consolidated. This is done by using the χ^2 test in the following manner:

   a) Perform the following test:

   $$R(C:F_j) \geq \frac{\chi^2}{2 \cdot L \cdot H(C:F_j)}$$

   where χ^2 is the χ^2 value at a certain significance level specified by the user, L is the total number of values in two adjacent intervals, and H is the entropy for these intervals

b) If the test is significant at the specified level, then perform the same test for the next pair of intervals
c) If not, one of the intervals from the pair is redundant, and they are consolidated into one by eliminating the separating boundary

Result: Final interval boundaries for each feature.

The CAIR criterion was used in the CADD algorithm to select a minimum number of intervals without significantly reducing CA interdependence information. The problems with CADD are that it uses the heuristic in selecting the initial number of intervals, that the use of ME to initialize the interval boundaries may result in a very poor discretization scheme, although this outcome is remedied by the boundary perturbation scheme at step 2, and that the confidence level for the χ^2 test needs to be specified by the user. All these factors increase the computational complexity of the algorithm.

3.2. Other Supervised Discretization Algorithms

Below we describe two conceptually simple supervised discretization algorithms that are normally used for other purposes. One is a clustering algorithm and the other a decision tree. Later we outline a dynamic supervised discretization algorithm.

3.2.1. K-means clustering for discretization

The **K-means algorithm**, the widely used clustering algorithm, is based on the minimization of a performance index defined as the sum of the squared distances of all vectors, in a cluster, to its center. The K-means algorithm is described in detail in Chapter 9. Below, we restate it expressly for the purpose of discretization of an attribute, namely, for finding the number and the boundaries of intervals.

Given: A training data set consisting of M examples and C classes, and a user-defined number of intervals n_{Fi} for feature F_i

1. Do class c_j for $(j = 1, \ldots, C)$
2. Choose $K = n_{Fi}$ as the initial number of cluster centers. Initially, the first K values of the feature can be selected as the cluster centers.
3. Distribute the values of the feature among the K cluster centers, based on the minimal distance criterion. As a result, feature values will cluster around the updated K cluster centers.
4. Compute K new cluster centers such that for each cluster the sum of the squared distances from all points in the same cluster to the new cluster center is minimized
5. Check whether the updated K cluster centers are the same as the previous ones, if yes go to step 1; otherwise, go to Step 3

Result: The final boundaries for the single feature that consist of the minimum value of the feature, midpoints between any two nearby cluster prototypes for all classes, and the maximum value of the feature.

The outcome of the algorithm, in an ideal case, is illustrated, for one feature, in Figure 8.3. The behavior of the K-means algorithm is strongly influenced by the number of cluster centers (which the user must choose), the choice of initial cluster centers, the order in which the samples are taken, and geometric properties of the data. In practice, before specifying the number of intervals for each feature, the user might draw the frequency plot for each attribute so that the "guessed" number of intervals (clusters) is as correct as possible. In Figure 8.3 we have correctly "guessed" the number of clusters to be 6, and therefore the result is good. Otherwise, we might have gotten a number of intervals that did not correspond to the true number of clusters present

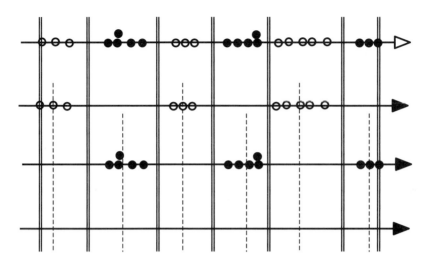

Figure 8.3. Illustration of the K-means discretization algorithm.

in a given attribute (what we are in fact doing is one-dimensional clustering). The problem of choosing the correct number of clusters is inherent to all clustering algorithms. To avoid it, one can cluster the feature values into several different numbers of intervals (clusters) and then calculate some measure of the goodness of clustering (see the discussion of cluster validity measures in Chapter 9). Then, one can choose the number of intervals that corresponds to the optimal value of the used cluster validity measure.

3.2.2. One-level Decision Tree Discretization

In order to better understand the one-level decision tree algorithm the reader is encouraged to read about **decision trees** in Chapter 12. The one-level decision tree algorithm by Holte can be used for feature discretization. It is known as **One-Rule Discretizer**, or **1RD**. It greedily divides the feature range into a number of intervals, using the constraint that each interval must include at least the user-specified minimum number of continuous values (statistics tells us that this number should not be less than 5). The 1RD algorithm starts with initial partition into the intervals, each containing the minimum number of values, and then moves the initial partition boundaries, by adding the feature values, so that each interval contains a strong majority of objects from one decision class. The process is illustrated in Figure 8.4.

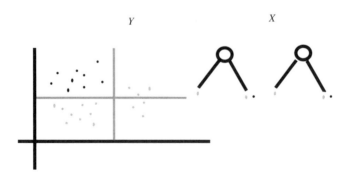

Figure 8.4. Illustration of the 1RD algorithm for 2D data. On the right, we see a discretization using features X and Y, respectively.

3.2.3. Dynamic Attribute Discretization

Few algorithms for dynamic attribute selection exist, and their applicability to large data sets is limited because of the high computational costs. In spite of this situation we outline below one **supervised dynamic discretization** algorithm. Discretization performed by a supervised dynamic discretization algorithm, in an ideal case, may essentially eliminate the need for the subsequent design of a classifier, as illustrated in Figure 8.5, which shows the hypothetical outcome of the algorithm.

From the outcome of such discretization, we may design a classifier (although one that does not recognize all the training instances correctly) in terms of the following rules, using both features:

IF $x_1 = 1$ AND $x_2 = I$ THEN class = MINUS (covers 10 minuses)
IF $x_1 = 2$ AND $x_2 = II$ THEN class = PLUS (covers 10 pluses)
IF $x_1 = 2$ AND $x_2 = III$ THEN class = MINUS (covers 5 minuses)
IF $x_1 = 2$ AND $x_2 = I$ THEN class = MINUS MAJORITY CLASS
                                        (covers 3 minuses & 2 pluses)
IF $x_1 = 1$ AND $x_2 = II$ THEN class = PLUS MAJORITY CLASS
                                        (covers 2 pluses & 1 minus)

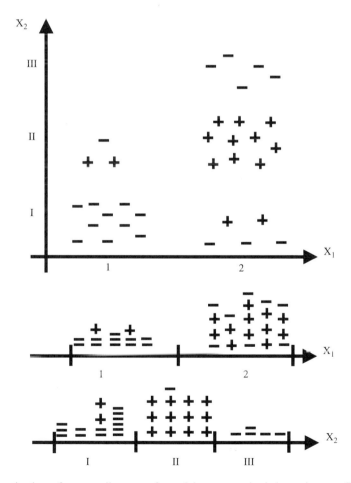

Figure 8.5. Discretization of two attributes performed in a supervised dynamic way. Top: training data instances (their class memberships). Bottom: discretization of the first attribute into two intervals and of the second attribute also into two intervals.

3. Supervised Discretization Algorithms

We note that the discretized feature x_2 alone can be used to design a reasonably accurate classifier if the following rules are specified:

IF $x_2 = $ I   THEN class = MINUS MAJORITY CLASS
(covers 10 minuses & 2 pluses)
IF $x_2 = $ II   THEN class = PLUS MAJORITY CLASS
(covers 10 pluses & 1 minus)
IF $x_2 = $ III   THEN class = MINUS (covers 5 minuses)

By defining the above, simpler classifier, we have in fact used the most discriminatory feature (see Chapter 7), identified by the dynamic discretization algorithm. In specifying both classifiers, we have used the terms PLUS/MINUS MAJORITY CLASS to indicate that the corresponding rules do not cover only one class but a mixture of both. As we will see in Chapter 12, to avoid overfitting of the training data, we accept such rules for large data sets.

Below, after Gama et al., we outline one such supervised dynamic discretization algorithm. It uses training data to build hypotheses for possible discretizations of all continuous attributes into a number of intervals for each attribute. Then it uses test data to validate/choose the best hypothesis. The algorithm uses an A* search to determine the number of intervals for each attribute.

The size of the search space for the A* algorithm is $r_1 \bullet r_2 \bullet \ldots r_d$ regions, where r_i is the number of continuous values for the ith attribute. A state is defined as a collection of vectors $(n_1, n_2 \ldots n_d)$, where n_i is the number of intervals for attribute i. The most general discretization would be into one interval for each attribute, i.e., $(1, 1, \ldots 1)$, and the most specific would be to use all original continuous values for all attributes, i.e. $(r_1, r_2 \ldots r_d)$. The goal of the search is to find the optimal value for each attribute, according to some objective function. The objective function for the A* algorithm is defined as $o(n) = g(n) + d(n)$, where n denotes a state. The function g measures the distance to the goal state, and function d measures the depth of the search tree, or distance from the initial state. The function g is defined using the test data set, that is defined in terms of the error committed on the test data using the current discretization scheme for all attributes. For example, for four attributes, it could be (2,3,5,4). The function d is defined as

$$d(n) = \frac{1}{c} \sum_{i=1}^{l} \log(n_i)$$

where c is a user-specified scaling constant used to control the influence of function d on the objective function. The objective function favors smaller over larger number of intervals for each attribute.

Before one can calculate the value of function g (the error on test data), two operations must be performed for a current candidate hypothesis, say, (2,3,5,4):

– both the training and test data must be discretized into 2, 3, 5, and 4 intervals, respectively
– to achieve this outcome one can use any of the discretization algorithms described in this Chapter (we know the number of intervals for each attribute).

Once we have discretized both data sets, we compute the error on the test data. If it is smaller than the current best value, we accept the new discretization. Using our example, if the current state is (2,3,5,4) and we decide to increase the number of intervals for each attribute by one (our initial state was (1,1,1,1)), and the current node has four children nodes ((3,3,5,4), (2,4,5,4), etc.), the decision of which of these nodes to expand will depend on the value of the objective function: only the node with the smallest value will be expanded.

The stopping criteria for the algorithm are as follows: if the value of the objective function is zero (no error on test data), then stop: optimal discretization has been found. The other criterion is that if after a few iterations (say, 3) there is no improvement of the objective function value, then increase the number of intervals by one. If, again, after a few iterations there is no improvement, then stop: optimal discretization has been found.

From the above outline, we note that this supervised dynamic discretization algorithm, although interesting, is computationally expensive. That is why dynamic algorithms are not used in practice.

3.2.4. Paterson and Niblett Algorithm

Finally, we briefly describe an algorithm that is incorporated into C4.5 and C5.0 algorithms as a preprocessing step, used before building a decision tree, to discretize continuous data. For feature F_i that has a number, r, of continuous values, first the training examples are sorted on the values of feature F_i. Then they are split, on threshold R, into two intervals: those for which $F_i \leq R$ and those for which $F_i > R$. To find the threshold R, all the $r-1$ possible divisions on feature F_i are examined. For each possible value of R, the **information gain** is computed, and the threshold corresponding to the largest value of information gain, or **gain ratio**, is chosen as the threshold to discretize this feature (see Chapter 12 for information about decision trees).

4. Summary and Bibliographical Notes

In this Chapter we introduced two **unsupervised** and several **supervised discretization algorithms** [1, 2, 4, 5, 6, 8, 9, 10, 11]. We also outlined one **dynamic discretization algorithm** [7] and performed comparison between several state-of-the-art algorithms [3, 10]. Discretization algorithms are one of the most important preprocessing tools. Depending on the goal of preprocessing and the type of data, we use either one type of algorithm or another. If the goal is to build a model of the data in a situation where training data are available, one should use a supervised discretization algorithm. If the data are very large then one is constrained to using only the simplest algorithms. The most important topics discussed in this Chapter are the notion of the **quanta matrix** and the **information-theoretic approach** to discretization.

References

1. Cios, K.J., Pedrycz, W., and Swiniarski, R. 1998. *Data Mining Methods for Knowledge Discovery*, Kluwer
2. Ching, J.Y., Wong, A.K.C., and Chan, K.C.C. 1995. Class-dependent discretization for inductive learning from continuous and mixed-mode data. *IEEE Transactions on PAMI*, 17:641–651
3. Cios, K.J., and Kurgan, L. 2004. CLIP4: Hybrid inductive machine learning algorithm that generates inequality rules. *Information Sciences*, 163(1–3): 37–83
4. Doughherty, J., Kohavi, R., and Sahami, M. 1995. Supervised and unsupervised discretization of continuous features. In: Machine Learning: Proceedings of the 12th International Conference, Prieditis, A., and Russell S. (Eds.)
5. Fayyad, U.M., and Irani, K.B. 1992. On the handling of continuous-valued attributes in decision tree generation. *Machine Learning*, 8:87–102
6. Fayyad, U.M., and Irani, K.B. 1993. Multi-interval discretization of continuous-valued attributes for classification learning. In: *Proceedings of the 13th Inernationa lJoint Conference on Artificial Intelligence*, Morgan-Kaufmann, 1022–1027
7. Gama, J., Torgo, L., and Soares, C. 1998. Dynamic discretization of continuous attributes. In: *Proceedings of the Sixth Ibero-American Conference on Artificial Intelligence*, 160–169
8. Holte, R.C. 1993. Very simple classification rules perform well on most commonly used data sets. *Machine Learning*, 11:63–90
9. Kerber, R. 1992. Chimerge: discretization of numeric attributes. In: *Proceedings of the 10th National Conference on Artificial Intelligence*, MIT Press, 123–128

10. Kurgan, L., and Cios, K.J., 2004. CAIM discretization algorithm. *IEEE Transactions on Knowledge and Data Engineering*, 16(2):145–153
11. Paterson, A., and Niblett, T.B. 1987. *ACLS Manual*. Edinburgh Intelligent Terminals, Ltd.

5. Exercises

1. What are the basic categories of discretization algorithms?
2. What is a quanta matrix?
3. Implement and run a one-level decision tree to discretize the iris and pid data sets (see Table 8.3).
4. Implement and run the CAIM algorithm on the iris and pid data sets (see Table 8.3).
5. Implement and run the dynamic discretization algorithm on the iris and pid data sets (see Table 8.3).

Part 4

Data Mining: Methods for Constructing Data Models

9
Unsupervised Learning: Clustering

In this Chapter, we introduce the concept of clustering, present the basic terminology, offer a commonly encountered taxonomy of clustering algorithms, and discuss in detail some representative algorithms. We cover essential issues of scalable clustering that offers an insight into the issues of handling large data sets.

1. From Data to Information Granules or Clusters

Making sense of data has been an ongoing quest within various types of research communities in almost every practical endeavor that deals with collected experimental evidence. The age of information technology, whose eminent manifestation is a vast amount of data, has amplified this quest and made it even more challenging. The collections of large quantities of data *anytime* and *anywhere* have become the predominant reality of our lives.

Within this context, clustering arises as a remarkably rich conceptual and algorithmic framework for data analysis and interpretation. In a nutshell, clustering is about **abstraction**–discovering structure in collections of data. The task of clustering is challenging both conceptually and computationally. As explained in Chapter 4, the term clustering is often used as a synonym for **unsupervised learning** (but we need to remember that another key unsupervised learning technique is association rules). As the name clustering implies, it is anticipated that a suitable, unsupervised algorithm is capable of discovering structure on its own by exploring **similarities** or differences (such as distances) between individual data points in a data set under consideration. This highly intuitive and appealing guideline sounds deceptively simple: cluster two data points if they are "close" to each other and keep doing the same by exploring the distances between newly formed clusters and the remaining data points. The number of different strategies for cluster formation is enormous, and a great many approaches try to determine what "similarity" between elements in the data means. Different clustering algorithms address various facets and properties of clusters. Their computational aspects are of paramount importance, and we need to become cognizant of them at the very beginning, in particular with reference to scalability issues. Let us stress that dividing N data (patterns) into c clusters (groups) gives rise to a huge number of possible **partitions**, which is expressed in the form of the Stirling number:

$$\frac{1}{c!}\sum_{i=1}^{c}(-1)^{c-i}\binom{c}{i}i^N \qquad (1)$$

To illustrate the magnitude of the existing possibilities, let us consider $N = 100$ and $c = 5$, which sounds like a fairly limited problem. Even in this case, we end up with over 10^{67} partitions. Obviously, we need to resort to some optimization techniques the ones known as *clustering methods*.

2. Categories of Clustering Algorithms

Clustering techniques can be divided into three main categories:

1. Partition – based clustering, sometimes referred to as objective function-based clustering
2. Hierarchical clustering
3. Model-based (a mixture of probabilities) clustering.

    The clustering principles for each of these categories are very different which implies very different style of processing and resulting formats of the results. In **partition based clustering**, we rely on a certain objective function whose minimization is supposed to lead us to the "discovery" of the structure existing in the data set. While the algorithmic setup is quite appealing and convincing (the optimization problem could be well formalized), one is never sure what type of structure to expect and hence what should be the most suitable form of the objective function. Typically, in this category of the methods, we predefine the number of clusters and proceed with the optimization of the objective function. There are some variants in which we also allow for successive splits of the clusters, a process that leads us to a dynamically adjusted number of clusters. The essence of **hierarchical clustering** lies in the successive development of clusters; we begin either with successive splits (starting with a single cluster that is an entire data set) or with individual points treated as initial clusters, which we and keep merging (this process leads us to the concept of agglomerative clustering). The essential feature of hierarchical clustering concerns a suitable choice of a distance function and a means to express the distance between data and patterns. These features, in essence, give rise to a spectrum of various clustering methods (single linkage, complete linkage, etc.). In **model-based clustering**, as the name itself stipulates, we assume a certain probabilistic model of the data and then estimate its parameters. In this case, we refer to a so-called mixture density model where we assume that the data are a result of a mixture of c sources of data. Each of these sources is treated as a potential cluster.

3. Similarity Measures

Although we have discussed the concept of **similarity** (or distance) in Chapter 4, it is instructive to cast these ideas in the setting of clustering. In the case of continuous features (variables) one can use many distance functions as similarity measures (see Table 9.1). Each of these distances comes with its own geometry (such as hyperspheres and hyperboxes). As will become clear later on, the choice of distance function implies some specific geometry of the clusters formed.
    In the case of binary variables, we usually do not use distance as a similarity measure. Consider two binary vectors **x** and **y**, that are two strings of binary data:

$$\mathbf{x} = [x_1 x_2 \ldots x_n]^T$$
$$\mathbf{y} = [y_1 y_2 \ldots y_n]^T$$

compare them coordinate-wise and then count the number of occurrences of specific combinations of 0's and 1's:

a) when x_k and y_k are both equal to 1
b) when $x_k = 0$ and $y_k = 1$
c) when $x_k = 1$ and $y_k = 0$
d) when x_k and y_k are both equal to 0

    These four combinations of numbers can be organized into a 2×2 co-occurrence matrix to show how the two strings are "close" to each other. Note that since the strings are composed of

Table 9.1. Selected distance functions between patterns **x** and **y**.

Distance function	Formula and comments
Euclidean distance	$d(\mathbf{x}, \mathbf{y}) = \sqrt{\sum_{i=1}^{n}(x_i - y_i)^2}$
Hamming (city block) distance	$d(\mathbf{x}, \mathbf{y}) = \sum_{i=1}^{n}|x_i - y_i|$
Tchebyschev distance	$d(\mathbf{x}, \mathbf{y}) = \max_{i=1,2,\ldots,n}|x_i - y_i|$
Minkowski distance	$d(\mathbf{x}, \mathbf{y}) = \sqrt[p]{\sum_{i=1}^{n}(x_i - y_i)^p}, p > 0$
Canberra distance	$d(\mathbf{x}, \mathbf{y}) = \sum_{i=1}^{n} \frac{|x_i - y_i|}{x_i + y_i}$, x_i and y_i are positive
Angular separation	$d(\mathbf{x}, \mathbf{y}) = \frac{\sum_{i=1}^{n} x_i y_i}{\left[\sum_{i=1}^{n} x_i^2 \sum_{i=1}^{n} y_i^2\right]^{1/2}}$ Note: similarity measure expresses the angle between the unit vectors in the direction of **x** and **y**

0's and 1's, we encounter four combinations; these are represented in the tabular format shown below. For instance, the first row and the first column corresponds to the number of times 1s' occur in both strings (equal to a):

	1	0
1	a	b
0	c	d

Evidently, the zero nondiagonal entries of this matrix indicate ideal matching (the highest similarity). Based on these four entries, there are several commonly encountered measures of similarity of binary vectors. The simplest is the matching coefficient defined as

$$\frac{a+d}{a+b+c+d} \qquad (2)$$

Another measure of similarity, Russell and Rao's, takes the following form:

$$\frac{a}{a+b+c+d} \qquad (3)$$

The Jacard index is more focused since it involves cases in which both inputs assume values equal to 1:

$$\frac{a}{a+b+c} \qquad (4)$$

The Czekanowski index is practically the same, but by adding the weight factor of 2 it emphasizes the coincidence of situations when both entries in **x** and **y** assume values equal to 1:

$$\frac{2a}{2a+b+c} \qquad (5)$$

For binary data, the Hamming distance could be another viable alternative. It is essential to be aware of the existence of numerous approaches to defining similarity measures used in various application-oriented settings.

When the features assume p discrete values, we can express the level of similarity/matching by counting the number of situations in which the values of the corresponding entries of **x** and **y** coincide. If this occurs r times, the pertinent measure could be in the following form, where n denotes the dimension of the corresponding vectors:

$$d(\mathbf{x}, \mathbf{y}) = \frac{n-r}{n} = 1 - \frac{r}{n} \qquad (6)$$

4. Hierarchical Clustering

Hierarchical clustering algorithms produce a graphical representation of data. The construction of graphs (these methods reveal structure by considering each individual pattern) is done in two modes: the bottom-up and top-down. In the **bottom-up mode**, also known as the **agglomerative** approach, we treat each pattern as a single-element cluster and then successively merge the closest clusters. At each pass of the algorithm, we merge the two clusters that are the closest. The process repeats until we get to a single data set (cluster) or reach a predefined threshold value. The **top-down approach**, also known as the **divisive** approach, works in the opposite direction. We start with the entire set, treat it as a single cluster, and keep splitting it into smaller clusters. Considering the nature of the top-down and bottom-up processes, these methods are quite often computationally inefficient, except possibly in the case of binary patterns.

The results of hierarchical clustering are represented in the form of a **dendrogram**. A dendrogram is defined as a binary tree with a distinguished root that has all the data items at its leaves. An example of a dendrogram is shown in Figure 9.1, along with distance values guiding the process of successive merging of the clusters. Depending upon the distance value, we produce a sequence of nested clusters.

Dendrograms are visually appealing graphical constructs that help us understand how difficult it is to merge two clusters. The nodes (represented in the form of small dots) located at the bottom of the graph correspond to the patterns/data points ($a, b, c, \ldots$). While moving up in the graph, we merge the points that are the closest in terms of some assumed similarity function. For instance, the distance between g and h is the smallest, and thus these two are merged. The distance scale shown at the right-hand side of the graph helps us visualize distance between the clusters. Moving upwards, the clusters get larger.

Thus, at any level of the graph (see the dotted line in Figure 9.1), we can explicitly enumerate the content of the clusters. For instance, following the dotted line, we end up with three clusters, that is $\{a\}$, $\{b, c, d, e\}$, and $\{f, g, h\}$. This process implies a simple stopping criterion: given a certain threshold value of the distance, we stop merging the clusters once the distance between

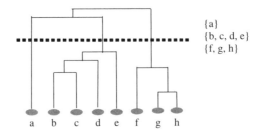

Figure 9.1. A dendrogram as a visualization of structure in the data.

them exceeds the threshold. In other words, merging two quite distinct structures (where their distinctiveness is expressed via the distance value) does not seem to be a good idea.

An important issue arises as to the way in which one can measure distance between two clusters. Note that we have discussed how to express distance between two patterns. Here, since each cluster may contain many patterns, distance computations are not that obvious and certainly not unique. Consider clusters A and B as illustrated in Figure 9.2. Let us describe the distance (between A and B) by $d(A, B)$ and denote the number of patterns in A and B by n_1 and n_2, respectively.

We have several ways of computing the distance between two such clusters, as described below.

Single link method (see Figure 9.2 (a)) the distance $d(A, B)$ is based on the minimal distance between the patterns belonging to A and B. It is computed as

$$d(A, B) = \min_{\mathbf{x} \in A, \mathbf{y} \in B} d(\mathbf{x}, \mathbf{y}) \qquad (7)$$

In essence, this distance is a radically "optimistic" mode, where we involve the closest patterns located in different clusters. The clustering method based on this distance is one of the most commonly used.

Complete link method (see Figure 9.2 (b)) This approach is on the opposite end of the spectrum, since it is based on the farthest distance between two patterns in two clusters:

$$d(A, B) = \max_{\mathbf{x} \in A, \mathbf{y} \in B} d(\mathbf{x}, \mathbf{y}) \qquad (8)$$

Group average link (see Figure 9.2 (c)) in contrast to the two previous approaches, in which the distance is determined on the basis of extreme values of the distance function, in this method we calculate the average between the distances as computed between each pair of patterns, with one pattern from each cluster:

$$d(A, B) = \frac{1}{\text{card}(A)\text{card}(B)} \sum_{\mathbf{x} \in A, \mathbf{y} \in B} d(\mathbf{x}, \mathbf{y}) \qquad (9)$$

Obviously, the computations are more intensive, but they reflect a general tendency between the distances computed for individual pairs of patterns.

One can develop other ways of expressing the distance between clusters A and B. For instance, we can calculate the Hausdorff distance between two sets of patterns:

$$d(A, B) = \max\{\max_{\mathbf{x} \in A} \min_{\mathbf{y} \in B} d(\mathbf{x}, \mathbf{y}), \max_{\mathbf{y} \in B} \min_{\mathbf{x} \in A} d(\mathbf{x}, \mathbf{y})\} \qquad (10)$$

To illustrate the essence of distance computations between a data point and a cluster we will use the following example.

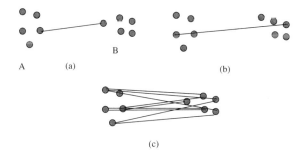

Figure 9.2. Two clusters A and B and several ways of computing the distance between them: (a) single link; (b) complete link; (c) group average link. Data are denoted by small circles.

262 4. Hierarchical Clustering

Example: Let cluster A consist of three points (1,3), (2,3), and (1.5, 0.5). The distance between **x** and A, d(**x**,A) is computed using formulas (7)–(9). In the computations we use the Euclidean, Hamming and Tchebyschev distance, respectively. The results are displayed in a series of graphs shown in Figure 9.3.

An interesting general formula for expressing various agglomerative clustering approaches is known as the Lance-Williams recurrence formula. It expresses the distance between cluster A and B and the cluster formed by merging these two (which gives rise to cluster C):

$$d_{A \cup B,C} = \alpha_A d_{A,C} + \alpha_B d_{B,C} + \beta d_{A,B} + \gamma |d_{A,C} - d_{B,C}| \tag{11}$$

with adjustable values of the parameters $\alpha_A(\alpha_B)$, β, and γ. The choice of values for these parameters implies a certain clustering method, as shown in Table 9.2.

With reference to the algorithmic considerations, hierarchical clustering can be realized in many different ways. One interesting alternative comes in the form of the Jarvis-Patrick (JP) algorithm. For each data point, we form a list of k nearest neighbors (where the neighborhood is expressed in terms of some predefined distance function).

We allocate two points to the same cluster if either of the following conditions is satisfied:

a) the points are within each other's list of nearest neighbors
b) the points have at least k_{min} nearest neighbors in common.

Here k and k_{min} are two integer parameters whose values are specified in advance before building the clusters. The JP algorithm has two evident advantages. First, it allows for nonconvex

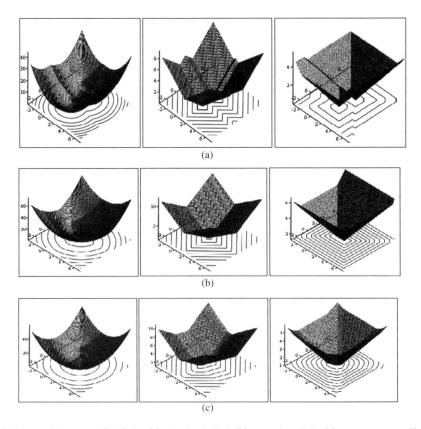

Figure 9.3. Plots of distance d(x,A) for (a) the single link (b) complete link, (c) group average link, and the Euclidean, Hamming, and Tchebyshev distances, from left to right, respectively.

Table 9.2. Values of the parameters in the Lance-Williams recurrence formula and the resulting agglomerative clustering; n_A, n_B and n_C denote the number of patterns in the corresponding clusters.

Clustering method	$\alpha_A(\alpha_B)$	β	γ
Single link	1/2	0	$-1/2$
Complete link	1/2	0	1/2
Centroid	$\dfrac{n_A}{n_A+n_B}$	$-\dfrac{n_A n_B}{(n_A+n_B)^2}$	0
Median	1/2	$-1/4$	0

clusters. This point becomes obvious: the data point a may be clustered with b and b could be grouped with c, and subsequently, the points a and c which do not seem to be related to each other end up in the same cluster. Second, the algorithm is nonparametric. Since it is based upon the ordering of the distances (ranking) it is less sensitive to potential outliers.

5. Objective Function-Based Clustering

The key design challenge in objective function-based clustering is the formulation of an objective function capable of reflecting the nature of the problem so that its minimization reveals meaningful structure (clusters) in the data.

Objective function-based clustering looks for a data structure through minimization of some **performance index** (called also **objective function**). Our anticipation is that a proper choice of the objective function and its minimization would help reveal a "genuine" structure in the data.

There are many possibilities for formulating the objective function and various ways of organizing the optimization activities. In what follows, we discuss several representative algorithms by emphasizing the very nature of the underlying methods. The representation of the structure (clusters) is provided in two ways, namely, as a collection of representatives (prototypes) and as a partition matrix. Let us denote the prototypes by $\mathbf{v}_1, \mathbf{v}_2, \ldots, \mathbf{v}_c$. The partition matrix $U = [u_{ik}]$ consists of c rows and N columns whose entries describe allocation of the corresponding data to the consecutive clusters.

5.1. K-Means Algorithm

The minimum variance criterion is one of the most common options that help organize the data. It comes with a clear and intuitive motivation: the **prototypes** of a large number of data should be such that they minimize a dispersion of data around them. Having N patterns in $\mathbf{R}^n$ and assuming that we are interested in forming c clusters, we compute the sum of dispersions between the patterns and a set of prototypes $\mathbf{v}_1, \mathbf{v}_2, \ldots, \mathbf{v}_c$:

$$Q = \sum_{i=1}^{c} \sum_{k=1}^{N} u_{ik} ||\mathbf{x}_k - \mathbf{v}_i||^2 \qquad (12)$$

where $||\ ||^2$ being Euclidean distance between $\mathbf{x}_k$ and $\mathbf{v}_i$. The important component in the above sum is the partition matrix $U = [u_{ik}], i = 1, 2, \ldots, c,$ $k = 1, 2, \ldots, N$ whose role is to allocate the patterns to the clusters. The entries of U are binary. Pattern k belongs to cluster i when $u_{ik} = 1$. The same pattern is excluded from the cluster when u_{ik} is equal to 0.

5. Objective Function-Based Clustering

Partition matrices satisfy the following conditions:

- each cluster is nontrivial, i.e., it does not include all patterns and is nonempty:

$$0 < \sum_{k=1}^{N} u_{ik} < N, i = 1, 2, \ldots, c \tag{13}$$

- each pattern belongs to a single cluster

$$\sum_{i=1}^{c} u_{ik} = 1, k = 1, 2, \ldots, N \tag{14}$$

The family of partition matrices (binary matrices satisfying these two conditions) will be denoted by **U**. As a result of minimization of Q, we construct the partition matrix and a set of prototypes. Formally, we express this construct in the following way, which is an optimization problem with constraints:

$$\text{Min } Q \text{ with respect to } \mathbf{v}_1, \mathbf{v}_2, \ldots, \mathbf{v}_c \text{ and } U \in \mathbf{U} \tag{15}$$

There are a number of approaches for this optimization. The most common is K-Means, a well established way to cluster data.

The flow of the main optimization activities in **K-Means clustering** can be outlined in the following manner:

Start with some initial configuration of the prototypes $\mathbf{v}_i, i = 1, 2, \ldots, c$ (e.g., choose them randomly)

- iterate
- construct a partition matrix by assigning numeric values to U according to the following rule

$$u_{ik} = \begin{cases} 1, & \text{if } d(\mathbf{x}_k, \mathbf{v}_i) = \min_{j \neq i} d(\mathbf{x}_k, \mathbf{v}_j) \\ 0, & \text{otherwise} \end{cases} \tag{16}$$

- update the prototypes by computing the weighted average, which involves the entries of the partition matrix

$$\mathbf{v}_i = \frac{\sum_{k=1}^{N} u_{ik} \mathbf{x}_k}{\sum_{k=1}^{N} u_{ik}} \tag{17}$$

until the performance index Q stabilizes and does not change, or until the changes are negligible.

Partition matrices form a vehicle to illustrate the structure of the patterns. For instance, the matrix formed for $N = 8$ patterns split into $c = 3$ clusters is shown as follows:

$$U = \begin{bmatrix} 1 & 0 & 0 & 1 & 0 & 1 & 0 & 1 \\ 0 & 1 & 1 & 0 & 0 & 0 & 0 & 0 \\ 0 & 0 & 0 & 0 & 1 & 0 & 1 & 0 \end{bmatrix}$$

Each row describes a single cluster. Thus we have the following arrangement: the first cluster consists of patterns $\{1, 4, 6, 8\}$, the second involves a set of patterns $\{2, 3\}$, and the third one covers the remaining patterns, that is $\{5, 7\}$

Graphical visualization of the partition matrix (data structure) can be shown in the form of a star or radar diagram as shown in Figure 9.4.

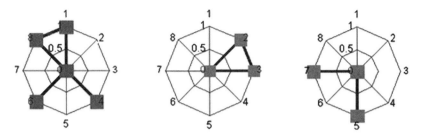

Figure 9.4. Star diagram as a graphical representation of the partition matrix for three clusters.

5.2. Growing a Hierarchy of Clusters

Objective function-based clustering can be organized in some **hierarchical topology** of clusters that helps us reveal a structure in a successive manner while reducing the required computing effort. The crux of this approach is as follows: we start with a fairly small number of clusters, say, 3–5. Given this low number, the optimization may not be very demanding. For each cluster formed here, we determine the value of the associated objective function, say the one given by expression (4) and computed for each cluster separately. Denote it by Q_i. The cluster with the highest value of Q_i is a candidate for further splitting (structural refinement). We cluster the data belonging to this cluster into c groups and next compute the values of the objective function for each of these groups. Again, we find the cluster with the highest objective function and proceed with its refinement (further splits). This process is repeated until we reach the point where the values of the objective functions for each cluster have fallen below a predefined threshold or we have exceeded the maximal number of clusters allowed in this process. The growth of the cluster tree (Figure 9.5) depends on the nature of the data. In some cases, we can envision a fairly balanced growth. In others, the growth could concentrate upon portions of the data where some newly split clusters are subject to further consecutive splits.

The advantage of this stepwise development process is that instead of proceeding with a large number of clusters in advance (which carries a significant computing cost associated with this grouping), we successively handle the clusters that require attention due to their heterogeneity (dispersion).

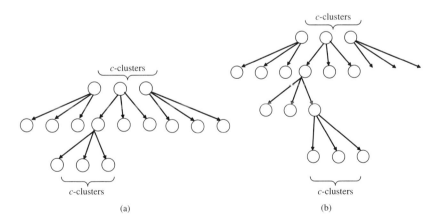

Figure 9.5. Growing a tree of clusters (a) balanced growth where most clusters at the higher level are split; (b) imbalanced growth, in which some clusters are subject to consecutive splits.

5.3. Kernel–based Clustering

The structure present in original data could be quite complicated – a situation that poses a genuine challenge to a variety of clustering techniques. One interesting alternative to address this problem is to elevate the original data $\mathbf{x}, \mathbf{y}, \mathbf{z}\ldots$ to a higher dimensional space M (where typically $M >> N$) in anticipation that the structure arising there could be made quite simple. This possibility could create an advantage when clustering the data in this new space. Consider a certain mapping $\phi(\mathbf{x})$. By applying it to all data to be clustered, e.g., $\mathbf{x}_1, \mathbf{x}_2, \ldots, \mathbf{x}_N$ we end up with $\phi(\mathbf{x}_1), \phi(\mathbf{x}_2), \ldots, \phi(\mathbf{x}_N)$, which are now located in the extended (augmented) space and are subject to clustering. As an illustration of the concept, let us focus on the following objective function involving distances $||.||$ formulated in some augmented space:

$$Q = \sum_{i=1}^{c} \sum_{k=1}^{N} u_{ik}^{m} ||\varphi(\mathbf{x}_k) - \varphi(\mathbf{v}_i)||^2$$

The optimization of Q is realized with respect to the partition matrix U and the prototypes $\phi(\mathbf{v}_1), \phi(\mathbf{v}_2), \ldots, \phi(\mathbf{v}_c)$ located in this new space. This task could appear to be more complicated given the fact that now the patterns are located in a highly dimensional space. Hopefully, there are ways around it that alleviate this potential difficulty. The use of the Mercer theorem allows us to express a scalar product of the transformed data as a **kernel function** $K(\mathbf{x}, \mathbf{y})$ so that

$$K(\mathbf{x}, \mathbf{y}) = \phi(\mathbf{x})^T \phi(\mathbf{y})$$

Obviously, there is a direct correspondence between the kernel function and the mapping ϕ. Here also originates the name of the kernel-based clustering itself.

If we use Gaussian kernels of the form

$$K(\mathbf{x}, \mathbf{y}) = exp(-||\mathbf{x} - \mathbf{y}||^2/\sigma^2)$$

then the distance present in the objective function transforms into the far more manageable form

$$||\phi(\mathbf{x}_k) - \phi(\mathbf{v}_i)||^2 = 2 - K(\mathbf{x}_k, \mathbf{v}_i)$$

Once this transformation has been noted, the derivations of the complete algorithm are straightforward. These will become obvious when we present the details of Fuzzy C-Means (FCM) clustering.

5.4. K-medoids Algorithms

Before we move on to the essence of the **K-medoids clustering** algorithm, we note that the prototypes of the clusters are formed on the basis of all elements, and if any data point is affected by noise, this affects the prototypes. Furthermore, the prototype is not one of the elements of the data, which again could be treated as an undesired phenomenon. To alleviate these difficulties, we resort to representatives of the data that are more robust. These come under the term *medoids*. To explain the concept, we start with the concept of a median.

The **median**, med $\{x_1, x_2, \ldots, x_N\}$, of real numbers is an **ordered statistic** that expresses a "central" element in the set. Assume that the above set of data is ordered, say, $x \leq, x_2, \ldots \leq x_N$. The median is defined as follows: (a) median $= x_{(N+1)/2}$ if N is odd and (b) median $= (x_{N/2} + x_{N/2+1})/2$ if N is even. It is worth stressing that the median is a robust estimator: the result does not depend on noisy data. To illustrates, let us imagine that x_1 assumes a value that is far lower than the rest of the data. x_1 is therefore an *outlier* and, as such, should not impact the result. The median

Figure 9.6. Median of dataset $\{x_1, x_2, \ldots, x_N\}$. Note its robustness property which manifests as no change even in the presence of one or more outlier(s) – see the situation represented on the right-hand side.

reflects this situation – note that its value has not been changed (see Figure 9.6). Evidently, in this case because of the outlier, the mean of this data set is pushed far down in comparison with the previous case.

One could easily demonstrate that the median is a solution to the following optimization problem:

$$\min_{ii} \sum_{k=1}^{N} |x_k - x_{ii}| = \sum_{k=1}^{N} |x_k - \text{med}| \tag{18}$$

where *med* denotes the median. Interestingly, in the above objective function we encounter a Hamming distance, which stands in sharp contrast with the Euclidean distance we found in the *K*-means algorithm. In other words, some objective functions promote the **robustness** of clustering.

In addition to the evident robustness property (highly desirable), we note that the prototype is one of the elements of the data. This is not the case in *K*-means clustering where the prototype is calculated from averaging and hence does not have any interpretability.

In general, for *n*-dimensional data, the objective function governing the clustering into *c* clusters is written in the form

$$Q = \sum_{i=1}^{c} \sum_{k=1}^{N} u_{ik} \|\mathbf{x}_k - \mathbf{v}_i\|^2 = \sum_{i=1}^{c} \sum_{k=1}^{N} \sum_{j=1}^{n} u_{ik} |x_{k_j} - v_{ij}| \tag{19}$$

Finding the minimum of this function leads to the centers of the clusters. Nevertheless, this optimization process can be quite tedious. To avoid this type of the optimization process, other arrangements are considered. They come under the name of specialized clustering techniques including Partitioning Around Medoids (PAM) and Clustering LARge Applications (CLARA).

The underlying idea of the PAM algorithm is to represent the structure in the data by a collection of **medoids** – a family of the most centrally positioned data points. For a given collection of medoids, each data point is grouped around the medoid to which its distance is the shortest. The quality of the produced clustering formed from the collection of medoids is quantified by taking the sum of distances between the medoids and the data belonging to the corresponding cluster represented by the specific medoid. PAM starts with an arbitrary collection of elements treated as medoids. At each step of the optimization, we make an exchange between a certain data point and one of the medoids, assuming that the swap results in improvement in the quality of the clustering. This method has some limitations with reference to the size of the dataset. Experimental results demonstrate that PAM works well for small datasets with a small number of clusters, for example, 100 data points and 5 clusters. To deal with larger datasets, the method has been modified to sample the dataset rather than operating on all data. The resulting method, called CLARA (Clustering LARge Applications), draws the sample, applies PAM to this sample and finds the medoids. CLARA draws multiple samples and produces the best clustering as the output of the clustering.

268 5. Objective Function-Based Clustering

In summary, note that the K-medoids algorithm and its variants offer two evident and highly desirable features:

a) **robustness** and
b) **interpretability** of the prototype (as one of the elements of the dataset).

However it also comes with a high computational overhead, which needs to be phased into the overall knowledge discovery process.

5.5. Fuzzy C-Means Algorithm

In the above described "standard" clustering methods, we assumed that clusters are well-delineated structures, namely, that a data point belongs to only one of the clusters. While this assumption sounds mathematically appealing, it is not fully reflective of reality and comes with some conceptual deficiencies. Consider the two-dimensional data set shown in Figure 9.7.

How many clusters can we distinguish? "Three" seems like a sound answer. If so, we assume $c = 3$ and run a clustering algorithm, the two data points situated between the two quite dense and compact clusters have to be assigned to one of the clusters. K-means will force them to be assigned somewhere. While this assignment process is technically viable, the conceptual aspect is far from being fully accepted. Such points may be difficult to assign using Boolean logic, therefore we would do better to flag them out by showing their partial membership (belongingness) to both clusters. A split of membership of $1/2 - 1/2$ or $0.6 - 0.4$ or the like could be more reflective of the situation presented by these data. To adopt this line of thought, we have to abandon the concept of two-valued logic (0–1 membership). Doing so brings us to the world of fuzzy sets, described in Chapter 4. The allocation of the data point becomes a matter of degree – the higher the membership value, the stronger its bond to the cluster. At the limit, full membership (a degree of membership of 1) indicates that the data point is fully allocated to the cluster. Lower values indicate weaker membership in the cluster. The most "unclear" situation occurs when the membership in each cluster is equal to 1/c that is the element is shared among all clusters to the same extent. Membership degrees are indicative of "borderline" elements: their membership in the clusters is not obvious, and this situation is easily flagged for the user/data analyst.

The concept of **partial membership** in clusters is the cornerstone of fuzzy clustering. In line with objective function-based clustering, we introduce the concept of a fuzzy partition matrix. In contrast to the binary belongingness of elements to individual clusters, we now relax the condition of membership by allowing the values of u_{ik} to be positioned anywhere in [0,1]. The two other

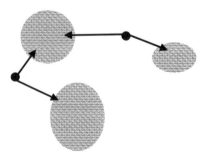

Figure 9.7. Three compact clusters and two isolated data points inbetween them: a two-valued logic challenge and the emergence of fuzzy sets.

fundamental conditions remain the same as before, as given by equations (13) and (14). The **objective function** incorporates the partition matrix and comes in the form of the double sum:

$$Q = \sum_{i=1}^{c} \sum_{k=1}^{N} u_{ik}^m \, ||\mathbf{x}_k - \mathbf{v}_i||^2 \qquad (20)$$

The elements of the partition matrix come with a fuzzification coefficient (m) whose values are greater than 1. The optimization of equation (20) is completed with respect to the prototypes and the partition matrix. Without going into computational details, we offer a general iterative scheme of **Fuzzy C-Means clustering**, in which we successively update the partition matrix and the prototypes.

Initialize: Select the number of clusters (c), stopping value (ε), and fuzzification coefficient (m). The distance function is Euclidean or weighted Euclidean. The initial partition matrix consists of random entries satisfying equations (13)–(14).

Repeat
    update prototypes

$$\mathbf{v}_i = \frac{\sum_{k=1}^{N} u_{ik}^m \mathbf{x}_k}{\sum_{k=1}^{N} u_{ik}^m} \qquad (21)$$

    update partition matrix

$$u_{ik} = \frac{1}{\sum_{l=1}^{c} \left(\frac{||\mathbf{x}_k - \mathbf{v}_i||}{||\mathbf{x}_k - \mathbf{v}_j||} \right)^{2/(m-1)}} \qquad (22)$$

Until a certain stopping criterion has been satisfied

The **stopping criterion** is usually taken to be the distance between two consecutive partition matrices, $U(\text{iter})$ and $U(\text{iter}+1)$. The algorithm is terminated once the following condition is satisfied:

$$\max_{ik} |u_{ik}(\text{iter}+1) - u_{ik}(\text{iter})| \leq \varepsilon \qquad (23)$$

We may use for instance $\varepsilon = 10^{-6}$. Note that the above expression is nothing but a Tchebyshev distance. The role of the parameters of this clustering is the same as that already discussed in the case of K-means. The new parameter here is the fuzzification coefficient, which did not exist in the previous algorithms for the obvious reason that the entries of the partition matrix were only taken to be 0 or 1. Here the fuzzification coefficient plays a visible role by affecting the shape of the membership functions of the clusters. Some snapshots of membership functions are shown in Figure 9.8.

Notably, the values close to 1 yield almost Boolean (binary) membership functions.

5.6. Model-based Algorithms

In this approach, we assume a certain probabilistic model of the data and then estimate its parameters. This structure, which is highly intuitive comes under the name of *mixture density*. We assume that the data are a result of a **mixture** of c sources of data that might be thought of as clusters. Each component of this mixture is described by some conditional probability density function (pdf), $p(\mathbf{x}|\boldsymbol{\theta}_i)$, characterized by a vector of parameters $\boldsymbol{\theta}_i$. The prior probabilities

270   5. Objective Function-Based Clustering

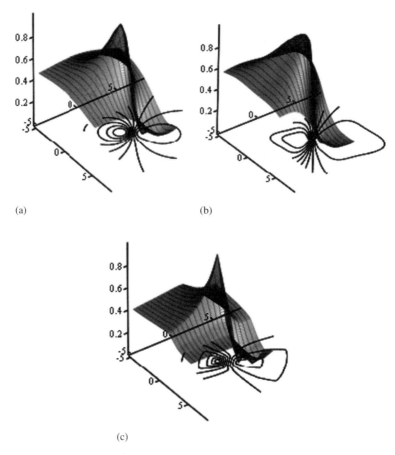

Figure 9.8. Plots of membership functions for several values of the fuzzification coefficient (m) used in the objective function (a) $m = 1.2$; (b) $m = 2.0$; (c) $m = 3.0$.

$p_1, p_2, \ldots, p_c$ of clusters are given. Under these assumptions, the model is additive and comes in the form of mixture densities:

$$p(\mathbf{x}|\boldsymbol{\theta}_1, \boldsymbol{\theta}_2, \ldots, \boldsymbol{\theta}_c) = \sum_{i=1}^{c} p(\mathbf{x}|\boldsymbol{\theta}_i) p_i \qquad (24)$$

Given the nature of the model, we also refer to p_1, p_2, .. and p_c as mixing parameters. To build the model, one has to estimate the parameters of the contributing pdfs. To do so we have to assume that $p(\mathbf{x}, \boldsymbol{\theta})$ is identifiable which means that if $\boldsymbol{\theta} \neq \boldsymbol{\theta}'$ then there exists an $\mathbf{x}$ such that $p(\mathbf{x}|\boldsymbol{\theta}) \neq p(\mathbf{x}|\boldsymbol{\theta})$. The standard approach used to discover the clusters is to carry out **maximum likelihood estimation**. In essence, this estimate maximizes the expression

$$P(\mathbf{X}|\boldsymbol{\theta}) = \prod_{k=1}^{N} p(\mathbf{x}_k|\boldsymbol{\theta}) \qquad (25)$$

One should know that the above optimization problem is not straightforward, especially with high-dimensional data.

5.7. Scalable Clustering Algorithms

5.7.1. Density-Based Clustering (DBSCAN)

As the name indicates, **density-based clustering** methods rely on the formation of clustering on the basis of the density of data points. This approach follows a very intuitive observation: if in some region of the feature space the data are located close to each other, then their density is high, and hence they form a cluster. On the other hand, if there are some points in some region and their density is low, they are most likely potential outliers not associated with the majority of the data.

This appealing observation constitutes the rationale behind the algorithms belonging to the category of density-based clustering. The DBSCAN method is one of the common representatives of this category, with OPTICS, DENCLUE, and CLIQUE following the same line of thought.

To convert these intuitive and compelling ideas into the algorithmic environment, we introduce the notion of the ε-neighborhood. Given some data $\mathbf{x}_k$, the ε-neighborhood, denoted by $N_\varepsilon(\mathbf{x}_k)$, is defined as:

$$N_\varepsilon(\mathbf{x}_k) = \{\mathbf{x} | d(\mathbf{x}, \mathbf{x}_k) \leq \varepsilon\} \tag{26}$$

Note that the form of the distance function implies the geometry of the ε-neighborhood. Obviously, higher values of ε produce larger neighborhoods. For the neighborhood of $\mathbf{x}_k$, we can count the number of data points falling within it. We introduce another parameter, N_Pts, that tells us how many data points fall within a neighborhood. If the neighborhood of $\mathbf{x}_k$ is highly populated by other data, that is

card $(N_\varepsilon(\mathbf{x}_k)) \geq$ N_Pts

we say that $\mathbf{x}_k$ satisfies a core point condition. Otherwise, we label $\mathbf{x}_k$ as a border point.

We say that $\mathbf{x}_i$ is $\mathbf{x}_k$ density-reachable with parameters ε and N_Pts if:

(a) $\mathbf{x}_i$ belongs to $N_\varepsilon(\mathbf{x}_k)$, and
(b) card $(N_\varepsilon(\mathbf{x}_k)) \geq$ N_Pts

Figure 9.8 illustrates this type of **reachability**. Note that the $\mathbf{x}_i$ could also be density-reachable from $\mathbf{x}_k$ by a chain of other data points:

$$\mathbf{x}_{k+1}, \mathbf{x}_{k+2}, \ldots, \mathbf{x}_{i-1} \tag{27}$$

such that $\mathbf{x}_{k+1}$ is density reachable from $\mathbf{x}_k$, $\mathbf{x}_{k+2}$ is density reachable from $\mathbf{x}_{k+1}$, etc. This type of transitivity is again illustrated in Figure 9.9.

The property of density reachability becomes the crux of the underlying clustering algorithm. We form the clusters on the basis of density reachability, and all data belonging to the same cluster are those that are density reachable.

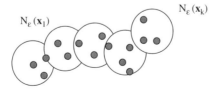

Figure 9.9. The concept of density reachability and its transitive character when a sequence of data is involved.

Given this background rationale, the generic DBSCAN algorithm consists of the following sequence of steps.

Set up the parameters of the neighborhood, e and N_Pts:

(a) arbitrarily select a data point, say, x_k
(b) find (retrieve) all data that are density reachable from x_k
(c) if x_k is a core point, then the cluster has been formed (all points are density reachable from x_k)
(d) otherwise, consider x_k to be a border point and move on to the next data point

The sequence (a) – (d) is repeated until all data points have been processed.

If we take a closer look at the clustering mechanisms, we can see that the concept of density-based reachability focuses on the formation of groups in a local way. In essence, each data point (regarded as a potential core) "looks" at its surroundings, which are formed by its neighborhood of some predefined size (ε). This situation offers a broad variety of potential geometric shapes of clusters that could be formed in this manner. The elements that are not density reachable are treated as outliers. Obviously, as in any other mechanism of unsupervised learning, one should be aware of the number of parameters whose values directly impact the performance of clustering and the form of the results. Three of these play an essential role. The distance function, as already mentioned, defines the geometry of the formed neighborhood. The size of the neighborhood and the number of points, N_Pts, affect the granularity of the search. These latter two are related; a higher value of ε requires higher values of N_Pts. If we consider very small values of ε, the DBSCAN starts building a significant number of clusters. With increasing values of ε, far fewer clusters are formed, more data points are regarded as outliers, and a more general structure is revealed. In essence, these parameters offer a significant level of flexibility, yet the choice of suitable values of the parameters becomes a data-dependent task. The computational complexity of the method is O (NlogN).

5.7.2. Cure

The essence of CURE (Clustering Using Representatives) is to exploit the concept of scattered points in clusters. Let us contrast this method with the two extreme clustering techniques. In centroid-based clustering (such as, e.g., K-means), we use single elements (prototypes) to represent the clusters. In hierarchical clustering, on the other hand, at the beginning of the process all points are representative for the clustering process. In the CURE algorithm, we choose a collection of scattered data in each cluster. The intent of these scattered points is to reflect the shape of the clusters. During the process of clustering, the scattered points shrink toward the center (mean). The speed of shrinking is controlled by the damping coefficient (called the shrinking factor), which assumes values between 0 and 1. If the value of the shrinking factor is 1, then we end up with K-means clustering. At the other extreme, with the shrinking factor equal to 0 (where no shrinking occurs), we end up with hierarchical clustering. Since each cluster is represented by a collection of points, the method is less sensitive to outliers as shrinking helps eliminate their potential impact. For the realization of CURE in the presence of large datasets, the method implements the following process. First, it draws a random sample of data. Assuming that the sample is large enough, one could anticipate that this sample could reflect the overall structure. Next CURE partitions the random sample and clusters the data in each partition.

6. Grid - Based Clustering

The ideas and motivation behind **grid-based clustering** are appealing and quite convincing. Clustering reveals a structure at some level of generality, and we are interested in describing such structures in the language of generic geometric constructs like **hyperboxes** and their combinations.

These result in a highly descriptive shape of the structure. Hence it is appropriate to start by defining a grid in the data space and then processing the resulting hyperboxes. Obviously, each hyperbox is described in terms of some statistics of data falling within its boundaries, yet in further processing we are not concerned with dealing with individual data points. By avoiding this detailed processing, we save computing time. Let us start with a simple illustrative example. Consider a collection of quite irregular clusters as shown in Figure 9.10.

It is very likely that partition-based clustering could have problems with this dataset due to the diversity of geometric shapes of the clusters. Likewise, to carry out clustering, we have to identify the number of clusters in advance, which poses a major challenge.

Grid-based clustering alleviates these problems by successively merging the elements of the grid. The boxes combined together give rise to a fairly faithful description of the clusters, and this otcome becomes possible irrespective of the visible diversity in the geometry of the shapes.

Let us move into a more formal description of grid–based clustering by introducing the main concepts. The key notion is a family of hyperboxes (referred to as *blocks*) that are formed in the data space. Let us denote these by $B_1, B_2, \ldots, B_p$. They satisfy the following requirements: (a) B_i is nonempty in the sense that it includes some data points, (b) the hyperboxes are disjoint, that is $B_i \cap B_j = \emptyset$ if $i \neq j$, and (c) a union of all hyperboxes covers all data that is $\bigcup_{i=1}^{p} B_i = X$ where $X = \{x_1, x_2, \ldots, x_N\}$. We also require that such hyperboxes "cover" some maximal number (say, b_{max}) of data points. Hyperboxes differ between themselves with respect to the number of data they cover. To measure the property of how well a certain hyperbox reflects the data, we compute a density index that is computed as the ratio of the number of data falling within the hyperbox and its volume.

Next, the clustering algorithm clusters the blocks B_i, (and, in essence, the data) into a nested sequence of nonempty and disjoint collections of hyperboxes. The hyperboxes with the highest density become the centers of clusters of hyperboxes. The remaining hyperboxes are afterwards clustered iteratively based on their density index, thereby building new cluster centers or merging with existing clusters. We can merge only those hyperboxes that are adjacent to a certain cluster ("neighbor"). A neighbor search is conducted, starting at the cluster center and inspecting adjacent blocks. If a neighbor block is found, the search proceeds recursively with this hyperbox.

Algorithmically, grid-based clustering comprises the following fundamental phases:

- Formation of the grid structure
- Insertion of data into the grid structure
- Computation of the density index of each hyperbox of the grid structure
- Sorting the hyperboxes with respect to the values of their density index
- Identification of cluster centres (viz. the hyperboxes of the highest density)
- Traversal of neighboring hyperboxes and the merging process

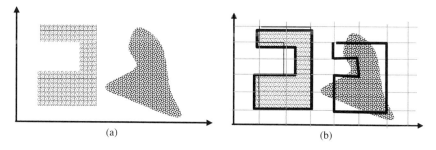

Figure 9.10. (a) A collection of geometrically different clusters; (b) and a grid structure formed in the data space along with clusters being built by merging the adjacent boxes of the grid structure.

One should be aware that since clustering moves only towards merging hyperboxes, the choice of grid (which becomes a prerequisite for the overall algorithm) does deserve careful attention. A grid which is rough may not help capture the details of the structure in the data. A grid that is too detailed produces a significant computational overhead.

In the literature, most studies focus on cases that involve a significant number of data (N) yet they shy away from discussing data of high dimensionality. This tendency is perhaps understandable since grid-based clustering becomes particularly beneficial in these cases (the processing of the hypeboxes abstracts all computing from the large number of the individual data points).

To summarize, let us highlight the outstanding features of grid-based clustering:

- The grid-based clustering algorithm scans the data set only once and in this way can potentially handle large data sets.
- By considering basic building blocks the method could handle a broad range of possible geometric shapes of clusters.
- Since grid-based clustering is predominantly concerned with the notion of density, it is helpful in handling clusters of arbitrary shapes. Similarly, although we rely on the density of points, we can detect potential outliers.

7. Self-Organizing Feature Maps

Objective function-based clustering forms one of the main optimization paradigms of data discovery. To put it in a broader perspective, we switch to an alternative arising in neural networks (see Chapter 13), namely **self-organizing feature maps**. This alternative will help us to contrast the underlying optimization mechanisms and to look at formats of results generated by different clustering methods.

The concept of a Self-Organizing feature Map (SOM) was originally developed by Kohonen. As emphasized in the literature, SOMs are regarded as neural networks composed of a grid of artificial neurons that attempt to show highly dimensional data in a low-dimensional structure, usually in the form of a two- or three-dimensional map. To make such visualization meaningful, one ultimate requirement is that the low-dimensional representation of the originally high-dimensional data has to preserve the **topological properties** of the data set.

In a nutshell, this requirement means that two data points (patterns) that are close to each other in the original feature higher-dimensional space should retain this similarity (or closeness or proximity) in their representation (mapping) in the reduced, lower-dimensional space. Similarly, two distant patterns in the original feature space should retain their distant locations in the lower-dimensional space. By being more descriptive, SOM acts as a **computer eye** that helps us gain insight into the structure of the data and observe relations occurring between patterns that were originally located in a high dimensional space by showing those relations in a low-dimensional, typically two- or three-dimensional space. In this way, we can confine ourselves to a two-dimensional map that apparently preserves all the essential relations between the data as well as the dependencies between the individual variables. In spite of many variations, the generic SOM architecture (as well as the learning algorithm) remains basically the same. Below we summarize the essence of the underlying self-organization algorithm that realizes a certain form of unsupervised learning.

Before proceeding with detailed computations, we introduce necessary notation. We assume, as usual, that the data are vectors composed of "n" real numbers, viz. they are elements of $\mathbf{R}^n$. The SOM is a collection of linear neurons organized in the form of a two-dimensional grid (array), as shown in Figure 9.11.

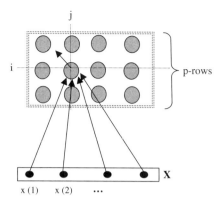

Figure 9.11. A basic topology of the SOM constructed as a grid of identical neurons.

In general, the grid may consist of p rows and r columns; quite commonly, we use the square array of $p \times p$ neurons. Each neuron is equipped with modifiable connections $\mathbf{w}(i, j)$ where these form an n-dimensional vector of connections:

$$\mathbf{w}(i, j) = [w_1(i, j) w_2(i, j) \ldots w_n(i, j)]$$

Note that the pair of indexes $(i0, j0)$ refers to the location of this neuron on the map (array of neurons). It completes computing of the distance function $\|.\|$ between its connections and the input vector $\mathbf{x}$:

$$y(i, j) = \|\mathbf{w}(i, j) - \mathbf{x}\| \qquad (28)$$

The distance function can be any of those discussed earlier. In particular, one could consider Euclidean distance or its weighted version. The same input $\mathbf{x}$ is fed to all neurons. The neuron with the shortest distance between the input vector and its own weight vector becomes activated and is called the winning neuron. Let us denote the coordinates of the neuron by $(i0, j0)$. More precisely, we have

$$(i0, j0) = \arg\min_{(i,j)} \|\mathbf{w}(i, j) - \mathbf{x}\| \qquad (29)$$

The winning neuron best matches (responds to, is similar to) the input vector $\mathbf{x}$. As a winner of the competition, it is rewarded by being allowed to modify its weight so that it becomes positioned even closer to the input. The **learning rule** is read as follows (see Chapter 13, Sec. 3.2.1 for more details):

$$\mathbf{w\_new}(i0, j0) = \mathbf{w}(i0, j0) + \eta(\mathbf{x} - \mathbf{w}(i0, j0)) \qquad (30)$$

where η denotes a learning rate, $\eta > 0$. The higher the learning rate is, the more intensive the updates of the weight are. In addition to the changes of weight of the winning neuron, we often allow its neighbors (the neurons located at the consecutive coordinates of the map) to update their weights as well. This influence is quantified via a **neighbor function** Φ (i, j, i0, j0). In general, this function satisfies two intuitively appealing conditions:

a) it attains a maximum equal to 1 for the winning node, i = i0, j = j0, $\Phi(i0, j0, i0, j0) = 1$ and
b) when the node is apart from the winning node, the value of the function gets lower (the updates are smaller). Evidently, there are also nodes where the neighbor function goes to zero and the nodes are not affected.

Considering the above, we rewrite Equation (30) in the following form:

$$\mathbf{w\_new}(i, j) = \mathbf{w}(i0, j0) + \eta \Phi(i, j, i0, j0)(\mathbf{x} - \mathbf{w}(i, j)) \qquad (31)$$

The typical neighbor function comes in the form

$$\Phi(i, j, i0, j0) = \exp(-\beta((i - i0)^2 + (j - j0)^2)) \qquad (32)$$

with parameter β usually assuming small positive values.

Expression (31) applies to all the nodes (i, j) of the map. As we iterate (update) the weights, the neighborhood function shrinks: at the beginning of the updates, we start with a large region, and when the learning settles down, we start reducing the size of the neighborhood. For instance, one may think of a linear decrease of neighborhood size. To emphasize this relationship, we can use the notation $\Phi(iter, i, j, i0, j0)$ where *iter* denotes consecutive iterations of the learning scheme.

Either the number of iterations is specified in advance or the learning terminates once there are no significant changes in the weights of the neurons.

Some conditions for successful learning of the SOM neural network algorithm are as follows:

- The training data set must be sufficiently large since self-organization relies on statistical properties of data.
- Proper selection of the neighbor function will assure that only the weights of the winning neuron and its neighborhood neurons are locally adjusted
- The radius, and thus the size, of the winning neighborhood must monotonically decrease with learning time (in successive iterations)
- The amount of weight adjustment for neurons in a winning neighborhood depends on how close they are to the input.

Another approach to designing a SOM network is to use heuristics and simulation-based findings:

- Since the weights of the SOM network have to approximate the probability density of input vectors $p(\mathbf{x})$, it is advisable to have visual inspection of this distribution first. This can be done by using Sammon **nonlinear projection** as an initial step before designing the SOM network (we elaborate on this idea later on).
- The size of the 2D array of neurons should be large enough to accommodate all the clusters that can be present in the data. An array that is too small may allow discovery of only the coarse clustering structure (smaller clusters may be put together). For smaller problems, the number of neurons in a 2D array should be approximately equal to the number of input vectors.
- Since good accuracy of statistical modeling requires a large number of samples (say 100,000), in practice, for a limited size of input data sets, we may use the data set several times repetitively, either randomly or by cycling the data in the same sequence through the network.
- To enhance the impact coming some patterns that are known to be important, we can simply present them to the network a large number of times, or increase their learning rate nd/or neighbor function $\Phi(i, j, i0, j0)$.
- Patterns with missing values can be still used for training, with the missing value being replaced by the average of that feature
- when it is desired in some applications such as, monitoring of real-time measurements, to map some nominal data in a specific location on the map (for instance, in the middle) we can copy these patterns as initial values for the neurons in the desired map location, and then keep their learning rate low during successive adjustments.

Example: This example shows the use of the SOM algorithm for finding the hierarchical structure of pattern vectors after Kohonen. It is an interesting example of SOM self-organization and global ordering as shown for five-dimensional patterns. The labeled training set, shown in Table 9.2 contains 32 input vectors with categories labeled by letters A, B, …,Z and digits 1,2,…,6. The SOM neural network has been composed with its 5 inputs fully connected by weights with 70 neurons.

The neurons have been arranged in a rectangular 7×10 array with hexagonal neighborhood geometry:

The SOM network has been trained using only unlabeled patterns selected randomly. After 10,000 iterations, the weights of the network converged to the stable values. Then the trained network was calibrated by presenting to it known labeled patterns (x_i, class$_j$), the classes being shown as feature x_6 in Table 9.2.

```
A B C D E
F
G
H K L M N O P Q R
I   S   W
J   T   X 1 2 3 4 5 6
    U   Y
    V   Z
```

For each labeled pattern, the array response was computed and the neuron with the strongest response was labeled by the pattern class. For example, when the pattern (2,0,0,0,0 ; B) was presented to the network, the left upper neuron in the array had the dominating strongest response was labeled as class B. It was shown that only 32 neurons were activated during calibration, whereas the remaining 38 neurons remained unlabeled, as is shown below.

This structure represents a global order, achieved through local interactions and weights adjustments, and reflects a *minimum spanning tree* relationship among the input patterns as shown below.

This example shows the power of self-organization. The minimum spanning tree is a mathematical technique, originating in graph theory. Assume that the data patterns from a given set will be presented as a planar graph, where each vertex represents one data pattern (vector). In addition, assume that one assigns to each edge, connecting two vertices a weight equal to the distance between the vertices. The minimum spanning tree is the spanning tree of the pattern vertices for which the sum of its connected edges is minimal. In another words, the minimum spanning tree for the data vectors is a corresponding planar tree that connects all data patterns to their closest neighbors such that total length of the tree edges is minimal. Analyzing the data in

7. Self-Organizing Feature Maps

Table 9.2. Kohonen's five-dimensional example.

x_1	x_2	x_3	x_4	x_5	Class
1	0	0	0	0	A
2	0	0	0	0	B
3	0	0	0	0	C
4	0	0	0	0	D
5	0	0	0	0	E
3	1	0	0	0	F
3	2	0	0	0	G
3	3	0	0	0	H
3	4	0	0	0	I
3	5	0	0	0	J
3	3	1	0	0	K
3	3	2	0	0	L
3	3	3	0	0	M
3	3	4	0	0	N
3	3	5	0	0	O
3	3	6	0	0	P
3	3	7	0	0	P
3	3	8	0	0	R
3	3	3	1	0	S
3	3	3	2	0	T
3	3	3	3	0	U
3	3	3	4	0	V
3	3	6	1	0	W
3	3	6	2	0	X
3	3	6	3	0	Y
3	3	6	4	0	Z
3	3	6	2	1	1
3	3	6	2	2	2
3	3	6	2	3	3
3	3	6	2	4	3
3	3	6	2	5	5
3	3	6	2	6	6

Table 9.2, we find that the Euclidean distances between subsequent neighboring patterns differ by a value of one. For instance, for patterns $\mathbf{x}_A$ and $\mathbf{x}_B$

$$||\mathbf{x}_A - \mathbf{x}_B|| = 1 < ||\mathbf{x}_A - \mathbf{x}_i||, \quad i = C, D, \ldots, Z, 1, \ldots, 6$$

SOM and FCM are complementary, and so are their advantages and shortcomings. FCM requires the number of groups (clusters) to be defined in advance. It is guided by a certain performance index (objective function), and the solution comes in the clear form of a certain partition matrix. In contrast, SOM is more user oriented. There is no number of clusters (group) that needs to be specified in advance.

As emphasized very clearly so far, SOM provides an important tool for visualization of high-dimensional data in a two- or three-dimensional space. The preservation of distances is crucial to this visualization process.

The same idea of distance preservation is a cornerstone of the Sammon's **projection method**; however, the realization of this concept is accomplished in quite a different manner. This nonlinear projection attempts to preserve topological relations between patterns in the original and the reduced spaces by preserving the interpattern distances. Sammon's projection algorithm minimizes an error defined as the difference between patterns in the original and reduced feature spaces. More formally, let $\{\mathbf{x}_k\}$ be the set of L n-dimensional vectors $\mathbf{x}_k$ in the original feature space $\mathbf{R}^n$, and let $\{\mathbf{y}_k\}$ be the set of L corresponding m-dimensional vectors $\mathbf{y}$ in the reduced low-dimensional space $\mathbf{R}^m$, with $m << n$. Most often we take $m = 2$. Let us denote by $d(\mathbf{x}_i, \mathbf{x}_j)$ the distance (usually Euclidean) between two vectors in the original feature space. Denote by $d(\mathbf{y}_i, \mathbf{y}_j)$ the distance between two vectors in the (reduced) projected feature space. Sammon's algorithm determines the projection such that it minimizes the following **distortion measure** J_d, for all L patterns, defined as follows

$$J_d = \frac{1}{\sum_{i=1, i\neq j}^{L}\sum_{j=1, j\neq i}^{L} d(\mathbf{x}_i, \mathbf{x}_j)} \sum_{i=1, i\neq j}^{L}\sum_{j=1, j\neq i}^{L} \frac{(d(\mathbf{x}_i, \mathbf{x}_j) - d(\mathbf{y}_i, \mathbf{y}_j))^2}{d(\mathbf{x}_i, \mathbf{x}_j)}$$

This criterion, called Sammon's stress, expresses how well all interpattern distances are preserved in the projection into a lower-dimensional feature space. Minimization procedures, such as gradient descent, can be used to find optimal projections that minimize this distortion criterion. To avoid getting stuck in a local minimum, one may start from different, random, initial configurations, or add noise. Since for every iteration step, $L(L-1)/2$ inter-pattern distances need to be computed, this algorithm becomes impractical for large number of patterns. Other techniques for minimization, such as evolutionary programming, can also be used for finding a solution and possibly even finding the global minimum. Unfortunately, Sammon's algorithm does not provide an explicit function describing the relationship between pattern vectors in both the original and projected spaces. Consequently after finding the optimal projection for a given set of L patterns, the algorithm does not exhibit a generalization capability. Thus, for new data, the minimization procedure must be rerun from scratch to accommodate both old and new data.

8. Clustering and Vector Quantization

Briefly speaking, **vector quantization** concerns various means of data compression, which is essential when dealing with storage and transmission of images, audio files, multimedia information and so forth (see Chapter 7). There are two important criteria, namely, quality of reconstruction (here we are interested in minimal quantization error) and high compression rate (so that we can store and transmit only a small portion of the original data to faithfully reconstruct the original source). An overall scheme for such processing is shown in Figure 9.12.

At the encoding end, we represent data through prototypes. These are usually referred to as a **codebook**. Any input datum is then captured in terms of the elements of the codebook and the index of the representative that matches it best is transmitted or stored. Formally, we can explain the flow of processing in the following manner:

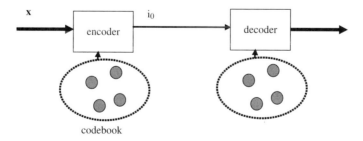

Figure 9.12. The principle of vector quantization. The encoder and decoder both use the same codebook; in transmission, rather than sending a multidimensional vector $\mathbf{x}$, one transmits an index of the best prototypes (i_0) which are used at the decoding end (hence the result of decoding is one of the elements of the codebook).

Encoding: determine the best representative (prototype) of the codebook and store (transmit) its index i_0, $i_0 = \arg\min_i ||\mathbf{x} - \mathbf{v}_i||$ where $\mathbf{v}_i$ denotes the ith prototype.

Decoding: recall the best prototype given the transmitted index (i_0).

Clustering as we have discussed here, is geared towards the discovery of structure in the data. There are some similarities and differences between the processes of vector quantization and clustering.

Clustering is an integral part of vector quantization. As a matter of fact, in the formation of the codebook we directly use various tools of clustering, such as K-Means.

The apparent differences concern key objectives. In vector quantization, we are interested in the quality of reconstruction (recall) while in clustering, cast in the framework of data mining, interpretability becomes very important. For instance, to maintain a low reconstruction error we might need to consider two very close elements of the codebook (their use helps us keep the error low enough). The same two prototypes, considered in the context of data mining may not be retained since their closeness makes them almost the same (conceptually redundant) from the point of view of data interpretation.

9. Cluster Validity

Since clustering is one of the two key unsupervised learning techniques, we should proceed very carefully with the assessment of its results. Are the generated clusters (along with their representation in the form of prototypes, partition matrices, dendrograms, etc.) reflective of the true nature of the data? This is a fundamental issue that permeates all clustering pursuits and profoundly impacts the practical usefulness of the technique. We should be fully cognizant of the fact that, while in essence being unsupervised, clustering is subconsciously endowed with some implicit components of supervision. The selection of these components impacts the character and quality of the results delivered by any clustering algorithm. We focus on the two main components present in almost any algorithm no matter what its nature and algorithmic details (SOM does not require a priori specification of the number of clusters, but its interpretation could be fairly complicated)

The choice of a **similarity measure** plays a primordial role in the search in the data space. We have discussed many types of such measures. We indicated that each distance measure used implies a certain geometry in the data space. As a consequence, the clustering technique endowed with specific distance is searching for structure in data that conforms to this specific geometry (such as hyperspheres in the case of using Euclidean distance). In the latter example, regardless of the "real" structure (shape) of clusters (and we never know it), an algorithm would searches for spherical clusters only. In short, we predispose the clustering right up front to search for clusters of some specific shape. This shortcoming is very much problem dependent. First, one could envision that due to the normality of data distribution, it is very likely that the structure could quite well conform to this geometric model of the clusters. In the case when a more complicated geometry of the clusters is encountered, we can envision that clustering with more clusters may take care of this problem (see Figure 9.13). In essence, one could adopt Euclidean distance as a fairly general and reasonable model of real shape in the data. The price is a larger number of clusters when there is a higher geometric diversity that could be resolved by asking for more clusters to be generated. The advantage is in the numeric treatment, since the Euclidean distance facilitates the optimization aspects of the clustering techniques.

The weighted Euclidean distance is an example of modified distance, which takes into consideration substantial differences in the ranges of the variables. Some other distances such as the Mahalanobis, come with an increased geometric flexibility but are associated with a high price of computing inverse of the corresponding covariance matrices (and thus are often quite prohibitive).

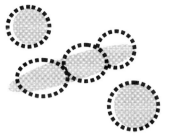

Figure 9.13. Diverse structure/shape of data in which clustering using the Euclidean distance can result in a larger number of spherical clusters (three clusters are required to represent the elongated cluster).

The number of clusters (c) that must be specified by users a priori (except when using SOM) is an extremely important parameter affecting the clustering outcome, since it explicitly determines a level of detail of the overall search for structure. On the one hand, we focus on the ensuing analysis and thus may anticipate that a certain range of the number of clusters (say between 3 and 8) does exist in the data. For example, in doing market analysis, we can easily guess the approximate number of categories of customers or at least contemplate a range of possible numbers of valid clusters.

In general, however, users must guess a priori the number of clusters that a given algorithm will be asked to generate. Normally, the user would guess the range of the number of clusters and then use **cluster validity measures** to assess which particular number of clusters best reveals the true structure in the data.

Two categories of tasks fall under the umbrella of what is referred to as **cluster validity** – a suite of methodologies and algorithms that offer us some mechanisms to validate clustering results.

There are many measures, called **cluster validity indices**, whose values relate to the number of clusters generated, and thus are used to judge the clusters detected in the data and to assess the quality of the structure revealed in this manner. In what follows, we present some of these measures and explain their motivation and computational details. In any case, we must always remember that the success of the clustering validity index depends upon the characteristics of the data and the clustering algorithm being used.

In spite of the diversity of clustering algorithms, we can spell out two fundamental and intuitively appealing requirements to which clusters should adhere.

Compactness. This property expresses how close the elements in a cluster are. For instance, consider a variance of the elements: the lower the value of the variance, the higher the compactness of the cluster. Since we are interested in compact clusters, low values of compactness are desirable. Likewise, we can calculate distances between the elements belonging to a cluster (intra cluster distances).

Separability. For this property, we evaluate how distinct the clusters are. An intuitive way of expressing separability is to compute inter cluster distances. Since we strive for high compactness and high separability, a structure should be characterized by small values of intra cluster distances and large values of inter cluster distances.

The realization of this observation comes in the form of the **Davies-Bouldin index**. To determine its value, we compute the within scatter distance for the ith cluster:

$$s_i = \frac{1}{\text{card}(\Omega_i)} \sum_{x \in \Omega_i} ||\mathbf{x} - \mathbf{v}_i||^2 \qquad (33)$$

with $||.||$ being some distance function. Here Ω_i denotes the ith cluster. We introduce the following distance between the prototypes of the clusters:

$$d_{ij} = ||\mathbf{v}_i - \mathbf{v}_j||^2 \tag{34}$$

which can serve as the inter cluster distance between the two clusters. Now, we define the ratio

$$r_i = \max_{j, j \neq i} \frac{s_i + s_j}{d_{ij}} \tag{35}$$

and sum its values over the clusters arriving at the following sum:

$$r = \frac{1}{c} \sum_{i=1}^{c} r_i \tag{36}$$

The "optimal" ("correct") number of clusters (c) is the number for which the value of r attains its minimum. Note that the minimum of r favors minimal values of the nominator of r_i and maximal values of the denominator in this expression. This is really what makes the clusters compact and well separated.

In the same vein of the minimization of scattering with a cluster or the maximization of inter cluster distances comes the Dunn separation index. In this construct, we first define a **diameter of the cluster**:

$$\Delta(\Omega_i) = \max_{\mathbf{x}, \mathbf{y} \in \Omega_i} ||\mathbf{x} - \mathbf{y}|| \tag{37}$$

Then the inter cluster distance is expressed as

$$\delta(\Omega_i, \Omega_j) = \min_{\mathbf{x} \in \Omega_i, \mathbf{y} \in \Omega_j} ||\mathbf{x} - \mathbf{y}|| \tag{38}$$

The **Dunn separation index** is formed as follows:

$$r = \min_i \min_{j, j \neq c} \frac{\delta(\Omega_i, \Omega_j)}{\max_k \Delta(\Omega_k)} \tag{39}$$

The values of r are maximized with respect to the number of the clusters.

The **Xie-Benie index** relates to fuzzy clustering and realizes the same concept as described above. Here we arrive at the following expression:

$$r = \frac{\sum_{k=1}^{N} \sum_{i=1}^{c} u_{ik}^m ||\mathbf{x}_k - \mathbf{v}_i||^2}{N \{ \min_{i \neq j} ||\mathbf{v}_i - \mathbf{v}_j||^2 \}} \tag{40}$$

Note that the "optimal" number of clusters will result in the lowest values of r yet the index may show some monotonicity when the number of clusters is quite close to the number of the data.

The concepts of scattering and compactness of clusters could be realized in different ways. One quite commonly encountered approach relies on forming a sound compromise between these two concepts. Let us introduce the average scattering

$$\text{av\_scatter} = \frac{1}{c} \sum_{i=1}^{c} \frac{s_i}{s} \tag{41}$$

where s_i denotes a measure of scattering for the ith cluster while s stands for the scattering reported for the entire data set. The expression for the separation between clusters reads as follows

$$\text{separation} = \frac{D_{\min}}{D_{\max}} \sum_{i=1}^{c} \left(\sum_{j=1}^{c} ||\mathbf{v}_i - \mathbf{v}_j|| \right)^{-1} \quad (42)$$

where $D_{\min} = \min_{i,j=1,2,\ldots,c} ||\mathbf{v}_i - \mathbf{v}_j||$ and $D_{\max} = \max_{i,j=1,2,\ldots,c} ||\mathbf{v}_i - \mathbf{v}_j||$. The validity index (SD) is taken as a weighted combination of the average scattering and separation, that is

$$\text{SD} = \alpha \text{ av\_scatter} + \text{separation} \quad (43)$$

where α is used to strike a sound balance between the two components of the index. By monitoring the values of SD treated as a function of c and determining its lowest value, we arrive at the plausible (or "optimal") number of clusters.

With reference to fuzzy clustering, there are several interesting cluster validity indicators. The first of these are based exclusively on the values of the partition matrix, while the third one takes into consideration the data as well as the prototypes. The **partition coefficient** P_1 is built upon the entries of the partition matrix:

$$P_1 = \sum_{i=1}^{c} \sum_{k=1}^{N} u_{ik}^2 \quad (44)$$

The values assumed by this index are in $[1/c, 1]$. At the extreme, if the data belong to a single cluster, viz. u_{ik} equals 1 for some i, then P_1 is equal to 1. On the other hand, if we encounter an equal distribution of membership grades (so it is likely that the data set does not exhibit any evident strongly manifested structure), $u_{ik} = 1/c$ then P_1 is equal to 0. When searching for the number of clusters, we look at the plot of P_1 versus c and choose the number of clusters for which this plot exhibits some "knee" – a place where a substantial change in the values of the index occurs.

The **partition entropy** is defined as:

$$P_2 = -\frac{1}{N} \sum_{i=1}^{c} \sum_{k=1}^{N} u_{ik} \log_a u_{ik} \quad (45)$$

where $a > 0$. The values of P_2 are confined to the interval $[0, \log_a c]$. Again, the most suitable number of clusters is determined by inspecting the character of the relationship P_2 treated as a function of c.

The third validity index not only takes into account the partition matrix but also involves the data as well as the prototypes,

$$P_3 = \sum_{i=1}^{c} \sum_{k=1}^{N} u_{ik}^m (||\mathbf{x}_k - \mathbf{v}_i||^2 - ||\mathbf{v}_i - \mathbf{v}||^2) \quad (46)$$

The first component of P_3 expresses a level of compactness of the clusters (dispersion around the prototypes) while the second one is concerned with the distances between the clusters (more specifically, their prototypes) and the mean of all data ($\mathbf{v}$).

In general, while these validity indexes are useful, in many cases they may produce inconclusive results. Given this effect, one should treat them with a big grain of salt by understanding that they offer only some guidelines and do not decisively point at the unique "correct" number of clusters. Also remember that when the number of clusters approaches the number of data points the usefulness of these indices is limited. Hopefully, in practice the number of clusters will always be a small fraction of the size of the dataset. In many cases, the user (decision-maker) could also provide some general hints in this regard.

10. Random Sampling and Clustering as a Mechanism of Dealing with Large Datasets

Large data sets require very careful treatment. No matter what clustering technique one might envision, it may be impractical due to the size of the data set. The communication overhead could be enormous, and the computing itself could be quite prohibitive. However, a suitable solution to the problem can be identified trough the use of **sampling**. There are two fundamental conceptual and design issues in this respect. First, we should know what a random sample means and how large it should be to become representative of the dataset and the structure we are interested in revealing. Second, once sampling has been completed, one has to know how to perform further clustering. Let us note that some of these mechanisms have been already mentioned in the context of clustering algorithms, such as CURE.

10.1. Random Sampling of Datasets

In a nutshell, we are concerned with a random draw of data from a huge dataset so that the clustering completed for this subset (samples) leads to meaningful results (that could be "similar" to those obtained when we run the clustering algorithm on the entire dataset). Given the clustering, we could rephrase the question about size of the random sample so that the probability of deforming the structure in the sample (for example, by missing some clusters) is low. Let us note that the probability of missing some cluster Ω is low if the sample includes a fraction of the data belonging to this cluster, say f*card(Ω) with $f \in [0, 1]$. Recall that card(Ω) stands for the number of data in Ω. The value of f is dependent on the geometry of the clusters and their separability. In essence, we can envision that if the structure of data is well defined, the clusters will be condensed and well-separated and then the required fraction f could be made substantially lower. Let us determine the size s of the sample such that that the probability that the sample contains fewer than $f$card(Ω) is less than δ. Let Z_j be a binary variable of value 1 if the jth data belongs to the cluster Ω and 0 otherwise. Assume that Z_j are 0-1 random variables treated as independent Bernoulli trials such that $P(Z_j = 1) = \text{card}(\Omega)/N, j = 1, 2, \ldots s$. The number of data in the sample that belong to cluster Ω is then expressed as the sum $Z = \sum_{i=1}^{s} Z_i$. Its expected value is $\mu = E(Z) = s\text{card}(\Omega)/N$. If we use the Chernoff bounds for the independent Poisson trials $Z_1, Z_2, \ldots, Z_s$ we find the following probability bounds:

$$P[Z < (1-\varepsilon)\mu] < \exp(-\mu\varepsilon^2/2) \qquad (47)$$

In other words, the probability that the number of data falls below the expected count m by more than $\varepsilon\mu$ is lower than the right–hand expression (47). We require that the probability that this number falls below $f$card(Ω) should be no more than d. In other words, we require that the following holds:

$$P(Z < f\text{card}(\Omega)) < \delta \qquad (48)$$

Let us rewrite the expression in the following equivalent format:

$$P[Z < (1 - (1 - f\operatorname{card}(\Omega)/\mu))] < \delta \qquad (49)$$

With the use of the Chernoff bound, the above inequality for the probability holds if:

$$\exp\left(-\frac{\mu\left(1 - \frac{f\operatorname{card}(\Omega)}{\mu}\right)^2}{2}\right) \leq \delta \qquad (50)$$

Given that $\mu = s\operatorname{card}(\Omega)/N$, we solve the above equation with respect to s we obtain the following inequality:

$$s \geq fN + \frac{N}{\operatorname{card}(\Omega)}\ln\left(\frac{1}{\delta}\right) + \frac{N}{\operatorname{card}(\Omega)}\sqrt{\left(\ln\left(\frac{1}{\delta}\right)\right)^2 + 2f\operatorname{card}(\Omega)\ln\left(\frac{1}{\delta}\right)} \qquad (51)$$

In other words (51) holds if the sample size is not lower than the number provided above. If we neglect some very small clusters (which may not have played any significant role), it has been shown that the sample size is independent from the original number of data points.

Clustering based on the random samples of the data

The sampling of data is a viable alternative to the dimensionality problem. A random sample is drawn and, on that basis, we discover a structure of the data. To enhance the reliability of the results, another option is to cluster the prototypes developed for each sample. This overall scheme is illustrated in Figure 9.14.

We draw a random sample, complete clustering and return a collection of the prototypes. Denote these by $\mathbf{v}_i[ii]$ with the index ii denoting the ii^{th} random sample. Then all the prototypes are again clustered in this way, reconciling the structures developed at the lower level. Since the number of elements to cluster at this level is far lower than in the sample itself, the clustering at the higher level of the structure does not require any substantial computing.

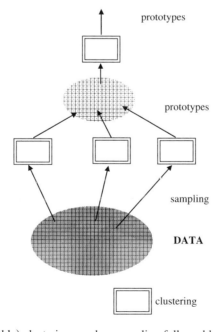

Figure 9.14. A two-level (double) clustering: random sampling followed by clustering of the prototypes.

11. Summary and Biographical Notes

In this Chapter, we have covered clustering, which occupies a predominant position in unsupervised learning and data mining. There are several compelling reasons. For this first, clustering techniques play an important role in revealing structure in data, basically without supervision. This feature of clustering is valuable given the fact that we do not know the structure in data and thus any mechanism that could help develop some insights into it is highly desirable. We presented a spectrum of clustering methods and elaborated on their conceptual properties, computational aspects and scalability. We addressed the issue of validity of clustering, stressing that although several cluster validity indexes are available, their outcomes should be treated with caution. The treatment of huge databases through mechanisms of sampling and distributed clustering was discussed as well. The latter two approaches are essential for dealing with huge datasets.

Clustering has been an area of very intensive research for several decades; the reader may consult classic texts in this area such as [1, 9, 11, 14, 17, 18, 20, 35]. The literature on fuzzy clustering is also abundant starting with the book by Bezdek [3]; interesting and useful references are [4, 7, 8, 12, 13, 16, 21, 22]. Cluster validity issues are presented in [10, 35, 36, 37]. The concept of self-organizing feature maps is well presented by Kohonen [23, 24, 25, 26]. Hierarchical clustering is covered in [31].

References

1. Anderberg, M.R. 1973. *Cluster Analysis for Applications*, Academic Press
2. Babu, G.P., and Murthy, M.N. 1994. Clustering with evolutionary strategies. *Pattern Recognition*, 27, 321–329
3. Bezdek, J.C. 1981. *Pattern Recognition with Fuzzy Objective Function Algorithms*, Plenum Press
4. Bezdek, J.C, Coray, C.R., Guderson, R., and Watson, J. 1981. Detection and characterization of cluster substructure, *SIAM Journal of Applied Mathematics*, 40: 339–372
5. Bezdek, J.C., Keller, J., Krishnampuram, R., and Pal, N.R. 1999. *Fuzzy Models and Algorithms for Pattern Recognition and Image Processing*, Kluwer Academic Publishers
6. Dave, R.N. 1990. Fuzzy shell clustering and application to circle detection in digital images, *International Journal of General Systems*, 16: 343–355
7. Dave, R.N. 1991.Characterization and detection of noise in clustering. *Pattern Recognition Letters*, 12, 657–664
8. Dave, R.N., and Bhaswan, K. 1992. Adaptive c-shells clustering and detection of ellipses, *IEEE Transactions on Neural Networks*, 3: 643–662
9. Devijver, P.A., and Kittler, J. (Eds.). 1987. *Pattern Recognition Theory and Applications*, Springer-Verlag
10. Dubes, R. 1987. How many clusters are the best? – an experiment. *Pattern Recognition*, 20(6): 645–663
11. Duda, R.O., Hart, P.E., and Stork, D.G. 2001. *Pattern Classification*, 2nd edition, John Wiley
12. Dunn, J.C. 1974. A fuzzy relative of the ISODATA process and its use in detecting compact well-separated clusters, *Journal of Cybernetics*, 3(3): 32–57
13. Frigui, H., and Krishnapuram, R. 1996. A comparison of fuzzy shell clustering methods for the detection of ellipses, *IEEE Transactions on Fuzzy Systems*, 4: 193–199
14. Fukunaga, K. 1990. *Introduction to Statistical Pattern Recognition*, 2nd edition, Academic Press
15. Girolami, M. 2002. Mercer kernel-based clustering in feature space. *IEEE Transactions on Neural Networks*, 13(3): 780–784
16. Hoppner, F., Klawonn, F., Kruse, R., and Runkler, T. 1999. *Fuzzy Cluster Analysis*, John Wiley
17. Jain, A.K., Murthy, M.N., and Flynn, P.J. 1999. Data clustering: A review, *ACM Computing Survey*, 31(3): 264–323
18. Jain, A.K., Duin, R.P.W., and Mao, J. 2000. Statistical Pattern recognition: a review, *IEEE Transactions on Pattern Analysis and Machine Intelligence*, 22(1): 4–37

19. Jarvis, R.A., and Patrick, E.A. 1973. Clustering using a similarity measure based on shared near neighbors, *IEEE Transactions on Computers*, C22(11): 1025–1034
20. Kaufmann, L., and Rousseeuw, P.J. 1990. *Finding Groups in Data: An Introduction to Cluster Analysis*, John Wiley
21. Kersten, P.R. 1999. Fuzzy order statistics and their applications to fuzzy clustering, *IEEE Transactions on Fuzzy Systems*, 7(7): 708–712
22. Klawonn, F., and Keller, A. 1998. Fuzzy clustering with evolutionary algorithms, *International Journal of Intelligent Systems*, 13: 975–991
23. Kohonen, T. 1982. Self-organized formation of topologically correct feature maps, *Biological Cybernetics*, 43: 59–69
24. Kohonen, T. 1989. *Self-organization and Associative Memory*, Springer Verlag
25. Kohonen, T. 1995. *Self-organizing Maps*, Springer Verlag
26. Kohonen, T., Kaski, S., Lagus, K., and Honkela, T. 1996. Very large two-level SOM for the browsing of newsgroups, In: *Proceedings of ICANN96, Lecture Notes in Computer Science*, 1112, Springer, 269–274.
27. Krishnapuram, R., and Keller, J. 1993. A possibilistic approach to clustering, *IEEE Transactions on Fuzzy Systems*, 1(1993): 98–110
28. Krishnapuram, R., and Keller, J. 1996. The possibilistic C-Means algorithm: insights and recommendations, *IEEE Transactions on Fuzzy Systems*, 4: 385–393
29. Mali, K., and Mitra, S. 2002. Clustering of symbolic data and its validation, In: Pal, N.R., and Sugeno, M. (Eds.), *Advances in Soft Computing – AFSS 2002*, Springer Verlag, 339–344
30. Michalewicz, Z. 1992. *Genetic Algorithms + Data Structures = Evolution Programs*, Springer Verlag
31. Pedrycz, A., and Reformat, M. 2006. Hierarchical FCM in a stepwise discovery of structure in data, *Soft Computing*, 10: 244–256
32. Roth, V., and Steinhage, V. 1999. Nonlinear discriminant analysis using kernel functions, In: Solla, S., Leen, T.K., and Muller, K.R. (Eds.), *Advances in Neural Information Processing Systems*, MIT Press, 568–574.
33. Sammon, J.W. Jr. 1969. A nonlinear mapping for data structure analysis. *IEEE Transactions on Computers*, 5: 401–409
34. Xie, X.L., and Beni, G. 1991. A validity measure for fuzzy clustering, *IEEE Transactions on Pattern Analysis and Machine Intelligence*, 13: 841–847
35. Webb, A. 2002. *Statistical Pattern Recognition*, 2nd edition, John Wiley
36. Windham, M.P. 1980. Cluster validity for fuzzy clustering algorithms, *Fuzzy Sets & Systems*, 3: 1–9
37. Windham, M.P. 1982. Cluster validity for the fuzzy C-Means clustering algorithms, *IEEE Transactions on Pattern Analysis and Machine Intelligence*, 11: 357–363
38. Vapnik, V.N. 1998. *Statistical Learning Theory*, John Wiley
39. Vesanto, J., and Alhoniemi, A. 2000. Clustering of the self-organizing map, *IEEE Transactions on Neural Networks*, 11: 586–600

12. Exercises

1. In grid-based clustering, form a grid along each variable of d intervals. If we are concerned with n dimensional data, how many hyperboxes are needed in total? If the data set consists of N data points, elaborate on the use of grid-based clustering vis-à-vis the ratio of the number of hyperboxes and the size of the data set. What conclusions could be derived? Based on your findings, offer some design guidelines.
2. Think of possible advantages of using the Tchebyschev distance in clustering versus some other distance functions (hint: think about the interpretability of the clusters by referring to the geometry of this distance).
3. In hierarchical clustering, we use different ways of expressing distance between clusters, which lead to various dendrograms. Elaborate on the impact of this approach on the shape of the resulting clusters.

4. Calculate the similarity between the two binary strings [1 1 0 0 1 1 0 0] and [00 11 11 1 0]. Compare differences between the results.
5. Run a few iterations of the K-Means for the toy dataset (1.0, 0.2) (0.9, 0.5) (2.0, 5.0) (2.1, 4.5) (3.1, 3.2) (0.9, 1.3). Select the number of clusters and justify your choice. Interpret the results.
6. Elaborate on the main differences between clustering and vector quantization.
7. Consider the same data set clustered by two different clustering algorithms and thus yielding two different partition matrices. How could you describe the clusters obtained by one method by using the clusters formed by another one?
8. The weights of a small 2×2 SOM developed for some three-dimensional data are as follows:

(1,1): (0.3 -1.5 2.0) (1,2): (0.0 0.5 0.9)
(2,1): (4.0 2.1 -1.5) (2,2): (0.8 -4.0 -1.0).

Where would you locate the inputs $\mathbf{a} = [0.2 \, 0.6 \, 1.1]$ and $\mathbf{b} = [2.2 \, -1.5 \, -1.2]$?
If you were to assess confidence of this mapping, what could you say about $\mathbf{a}$ and $\mathbf{b}$?
9. In the hierarchical buildup of clusters governed by some objective function, we split the data into a very limited number of clusters and then proceed with a series of successive refinements. Consider that you allow for p phases of usage of the algorithm while using "c" clusters at each level. Compare computing costs of this clustering method with the clustering realized at a single level when using cp clusters.

10

Unsupervised Learning: Association Rules

In this Chapter, we cover the second key technique of unsupervised learning, namely, association rules. The first technique, clustering, was covered in Chapter 9. We discuss both the algorithms and data from which association rules are generated.

1. Introduction

Association rules mining is another key **unsupervised** data mining method, after clustering, that finds interesting **associations** (relationships, dependencies) in large sets of data **items**. The items are stored in the form of **transactions** that can be generated by an external process, or extracted from relational databases or data warehouses. Due to good **scalability** characteristics of the association rules algorithms and the ever-growing size of the accumulated data, association rules are an essential data mining tool for extracting knowledge from data. The discovery of interesting associations provides a source of information often used by businesses for decision making. Some application areas of association rules are market-basket data analysis, cross-marketing, catalog design, loss-leader analysis, clustering, data preprocessing, genomics, etc. Interesting examples are personalization and recommendation systems for browsing web pages (such as Amazon's recommendations of related/associated books) and the analysis of genomic data.

Market-basket analysis, one of the most intuitive applications of association rules, strives to analyze customer buying patterns by finding associations between items that customers put into their baskets. For instance, one can discover that customers buy milk and bread together, and even that some particular brands of milk are more often bought with certain brands of bread, e.g., multigrain bread and soy milk. These and other more interesting (and previously unknown) rules can be used to maximize profits by helping to design successful marketing campaigns, and by customizing store layout. In the case of the milk and bread example, the retailer may not offer discounts for both at the same time, but just for one; the milk can be put at the opposite end of the store with respect to bread, to increase customer traffic so that customers may possibly buy more products. A number of interesting associations can be found in the market basket data, as illustrated in Figure 10.1.

This Chapter provides background and explains which data are suitable for association rule mining, what kinds of association rules can be generated, how to generate association rules, and which rules are the most interesting. The explanations are supported by easy-to-understand examples. Upon finishing this Chapter, the reader will know how to generate and interpret association rules.

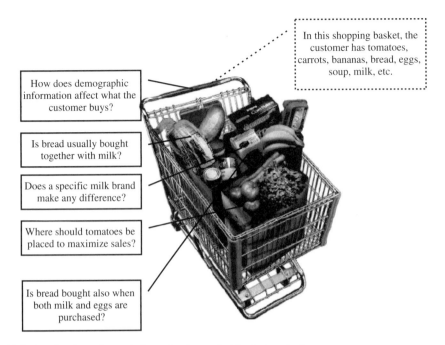

Figure 10.1. Application of association rules in market-basket analysis.

2. Association Rules and Transactional Data

Continuing our example of market-basket analysis, we represent each product in a store as a Boolean variable, which represents whether an item is present or absent. Each customer's basket is represented as a Boolean vector, denoting which items are purchased. The vectors are analyzed to find which products are frequently bought together (by different customers), i.e., **associated** with each other. These cooccurrences are represented in the form of **association rules**:

LHS ⇒ RHS [support, confidence]

where the left-hand side (LHS) implies the right-hand side (RHS), with a given value of support and confidence.

Support and **confidence** are used to measure the quality of a given rule, in terms of its usefulness (strength) and certainty. Support tells how many examples (transactions) from a data set that was used to generate the rule include items from both LHS and RHS. Confidence expresses how many examples (transactions) that include items from LHS also include items from RHS. Measured values are most often expressed as percentages. An association rule is considered interesting if it satisfies minimum values of confidence and support, which are to be specified by the user (domain expert). The following examples are used to illustrate the concepts.

Example: An association rule that describes customers who buy milk and bread.

buys (x, milk) ⇒ buys (x, bread) [25%, 60.0%]

The rule shows that customers who buy milk also buy bread. The direction of the association, from left to right, shows that buying milk "triggers" buying bread. These items are bought together in 25% of store purchases (transactions), and 60% of the baskets that include milk also include bread.

Example: An association rule describing graduate students might read as follows:

major(x, Computer Engineering) AND takes_course(x, Advanced Data Analysis and Decision Making) ⇒ level(x, PhD) [1%, 75%]

The rule has two items in the LHS and one item in the RHS. Association rules can include multiple items in their LHS. The rule in this example states that students who major in Computer Engineering and who take Advanced Data Analysis and Decision Making course are at the Ph.D. level with 1% support and 75% confidence. Again, the support shows that 1% of all students in the database satisfy this association. At the same time, among those who major in Computer Engineering and who take Advanced Data Analysis and Decision Making courses, 75% are the Ph.D. students.

Association rules are derived when data describe events (items in a transaction) that occur at the same time or in close proximity. The two main types of association rules are **single-dimensional** and **multidimensional**. The former refers to one dimension, such as *buy* in the first example, while the latter refers to more than one dimension, as in the second example where we have three dimensions: *major*, *takes_course*, and *level*. Both of these types of association rules could also be categorized as **Boolean** or **quantitative**. The former concerns the presence or absence of an item, while the latter considers quantitative values, which are partitioned into item intervals. An example of a multidimensional quantitative rule follows.

Example: An association rule that describes the finding that young adult customers who earn below 20K buy bread.

age(x, "(18, 25)") ∧ income(x, "<20K") ⇒ buy(x, bread) [0.5%, 50.0%]

The quantitative items such as *age* and *income* are discretized (see Chapter 8).

Association rules can be also categorized as **single-level** and **multilevel**. The former operate on a single level of abstraction, while the latter are based on items that can be expressed at different levels in a **hierarchy** (see Chapter 6). The example below shows a multilevel association rule (which can be contrasted with the single-level rule given in the first example).

Example: A multilevel association rule that describes customers buying skim milk and large white bread.

buys (x, skim_milk) ⇒ buys (x, large_white_bread) [2.5%, 60.0%]

In the above rule, the *milk* and *bread* items are subdivided into different kinds that constitute a hierarchy. For instance, *bread* can be divided into *white* and *wheat*, and each of these two types can be subdivided into *small*, *medium*, and *large*.

In what follows, we describe data and methods that can be used to generate single-dimensional (single-level) Boolean association rules. Next, we discuss how to extend these basic algorithms to address multidimensional, multilevel, and quantitative rules.

2.1. Transactional Data

Input data for an association-rule mining algorithm are provided in the **transactional form**. Each record (example) should consist of a transaction ID and information about all items (in a consistent format) that constitute the transaction, as shown in Figure 10.2.

An example of transactional data that concern the market-basket example for a grocery store is shown in Table 10.1.

Two example association rules that can be found by visual analysis of the above table are

$$\text{Beer} \Rightarrow \text{Eggs}$$

$$\text{Apples} \Rightarrow \text{Celery}$$

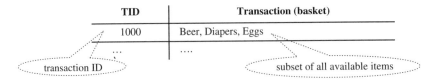

Figure 10.2. Example transaction.

Table 10.1. Example of transactional data..

TID	Transaction (basket)
1000	Apples, Celery, Diapers
2000	Beer, Celery, Eggs
3000	Apples, Beer, Celery, Eggs
4000	Beer, Eggs

The transactional data can be obtained directly from a transactional database (which could be stored by a grocery or a retail store), or alternatively, these data can be obtained by a simple transformation of relational data. A relational table describing patients of a heart clinic (see Chapter 3) can be transformed to the transactional form, as shown in Figure 10.3.

Given a suitable transactional database, in which each transaction is a list of items (say, items being purchased by a customer in a single visit to a store), we aim at finding association rules that relate the presence of one set of items with another set of items. In the following, we formally define basic concepts used to describe an association mining algorithm.

2.2. Basic Concepts

Let $I = \{i_1, i_2, \ldots, i_m\}$ be a **set of items** and D be the **set of transactions** (transactional data set) where each **transaction** $T \subseteq I$ is associated with an **identifier** TID and m is the number of items. Let A and B be two sets of items. A transaction T is said to contain A if and only if $A \subseteq T$. An **association rule** is an implication in the form $A \Rightarrow B$ where $A \subset I$, $B \subset I$, and $A \cap B = \emptyset$

The interestingness of an association rules describes how significant the rule is with respect to D. Two measures are used to quantify the interestingness of a rule:

patient (relational)

patient ID	name	age	sex	chest pain type	defect type	diagnosis
P1	Konrad Black	31	male	1	normal	absent
P2	Konrad Black	26	female	4	fixed	present
P3	Anna White	56	female	2	normal	absent
...	...	...	...	...	...	...

patient (transactional)

TID	Transaction (basket)
1000	name=Konrad Black, age=31, sex=male, chest_pain_type=1, defect type=normal, diagnosis=absent
2000	name=Konrad Black, age=26, sex=female, chest_pain_type=4, defect type=fixed, diagnosis=present
3000	name=Anna White, age=56, sex=female, chest_pain_type=2, defect type=normal, diagnosis=absent
...	...

Figure 10.3. Transformation from relational to transactional data.

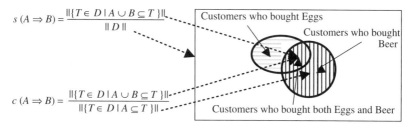

Figure 10.4. Computing support and confidence for the association rule Beer $\Rightarrow$ Eggs.

- **Support**, which indicates the frequency (probability) of the entire rule with respect to D. It is defined as ratio of the number of transactions containing A and B to the total number of transactions (the probability of both A and B cooccurring in D):

$$\text{support}(A \Rightarrow B) = P(A \cup B) = \frac{||\{T \in D | A \cup B \subseteq T\}||}{||D||}$$

- **Confidence**, which indicates the strength of implication in the rule. It is defined as ratio of the number of transactions containing A and B to the number of transactions containing A (conditional probability of B given A):

$$\text{confidence}(A \Rightarrow B) = P(B|A) = \frac{||\{T \in D | A \cup B \subseteq T\}||}{||\{T \in D | A \subseteq T\}||}$$

Figure 10.4 illustrates these concepts using the rule Beer $\Rightarrow$ Eggs as an example.

Rules that satisfy both the minimum support threshold and the minimum confidence threshold are called **strong association rules**, as explained in Figure 10.5. In Figure 10.6, we give another example to better explain the introduced concepts and definitions.

A set of items is referred to as an **itemset** (in data mining, the term itemset is used instead of item set). An itemset that contains k items is referred to as a **k-itemset**. For instance, {Beer, Eggs} is a 2-itemset. The **support count** (also known as frequency, occurrence frequency, or count) of

---

For the given transactions, find all rules such that LHS = $\{A, B\}$, RHS = $\{C\}$, with minimum support = 50% and minimum confidence = 50%.

TID	Transactions
1000	A, B, C
2000	A, C
3000	A, D
4000	B, E, F

Support is the probability that a transaction contains $\{A, B, C\}$, and *confidence* is the conditional probability that a transaction containing $\{A, B\}$ also contains C.
Rule $A \wedge B \Rightarrow C$ [support 25%, confidence 100%] does not satisfy the minimum confidence. Two (shorter) strong association rules are generated as:
 $A \Rightarrow C$ [support 50%, confidence 66.6%]
 $C \Rightarrow A$ [support 50%, confidence 100%]

Figure 10.5. Strong association rules.

294 2. Association Rules and Transactional Data

For the given transactions	
TID	Transactions
1000	Apples, Celery, Diapers
2000	Beer, Celery, Eggs
3000	Apples, Beer, Celery, Eggs
4000	Beer, Eggs

find I, T for TID = 2000, *support* (Beer $\Rightarrow$ Eggs), and *confidence* (Beer $\Rightarrow$ Eggs)
$I = \{$Apples, Beer, Celery, Diapers, Eggs$\}$
$T = \{$Beer, Celery, Eggs$\}$
support (Beer $\Rightarrow$ Eggs) = 75%
confidence (Beer $\Rightarrow$ Eggs) = 100%

Figure 10.6. Example transactional data and related association rule concepts.

an itemset is the number of transactions in D that contain the itemset. A **frequent itemset** is an itemset that satisfies a minimum support level, i.e., the support count of this frequent itemset is greater than or equal to the product of the minimum support and the total number of transactions in D. The number of transactions required for an itemset to satisfy minimum support is called the **minimum support count**. The set of frequent k-itemsets is commonly denoted as L_k.

In simple terms, the generation of association rules boils down to generation of frequent itemsets. Note, that this generation may be difficult if, for a given transactional data set, the number of items, m, is large.

Since we are interested only in strong association rules, i.e., those that satisfy minimum support and minimum confidence (user-defined parameters), we generate these only from frequent itemsets. Given an itemset, say, {Beer, Eggs}, we can generate four association rules:

$\Rightarrow$ Beer, Eggs
Beer $\Rightarrow$ Eggs
Eggs $\Rightarrow$ Beer
Eggs, Beer $\Rightarrow$

The examples below provide more insight into the process of generation of association rules from frequent itemsets. More details are given in Sec. 3.3.

Example: Given the transactions from Figure 10.6 and minimum support of 50%, the frequent itemset {Beer, Eggs, Celery} can be generated. Requiring minimum confidence of 60%, this itemset can be transformed into the following association rule:

Beer $\wedge$ Eggs $\Rightarrow$ Celery [50%, 66%]

The rule can be read as
IF customer buys Beer and Eggs THEN the probability of buying Celery is 66%.

Example: Given the following two itemsets:

{Beer, Eggs} with support 75%
{Beer, Celery, Eggs} with support 50%

we can generate the following rule:
IF customer buys Beer and Eggs THEN the probability of buying Celery is 66%.

3. Mining Single Dimensional, Single-Level Boolean Association Rules

The following four steps are used to generate single-dimensional association rules:

1. Prepare input data in the transactional format.
2. Choose items of interest, i.e., itemsets.
3. Compute support counts to evaluate whether selected itemsets are frequent, i.e., whether they satisfy minimum support.
4. Given the frequent itemsets, generate strong association rules that satisfy the minimum confidence by computing the corresponding conditional probabilities (counts).

Since the frequent itemsets used to generate the association rules satisfy the minimum support, the generated rules also satisfy the minimum support.

The computational performance (scalability) of an association-rule mining algorithm is determined by the second and third steps above. The remaining steps are much less computationally expensive. Therefore, the following discussion concentrates on these two steps.

3.1. The Naïve Algorithm

The simplest way to compute frequent itemsets is to consider all possible itemsets, compute their support, and check whether they are higher than the minimum support threshold. A naive algorithm for generation of frequent itemsets (Steps 2 and 3 above) is shown in Figure 10.7.

Given that 2^m itemsets must be searched and n transactions must be scanned each time, this algorithm requires $O(2^m n)$ tests. This number grows exponentially with the number of items, and thus for larger problems the computations would take an unacceptably long time. Since 2^m is causing the problem, we need to find a way to reduce the number of tests. The itemsets that we can safely assume will not produce frequent itemsets do not need to be tested. This reasoning resulted in the development of the Apriori algorithm that is discussed next.

3.2. The Apriori Algorithm

The Apriori algorithm uses prior knowledge about an important property of frequent itemsets—hence its name. The **Apriori property** of an itemset says that all nonempty subsets of a frequent itemset must also be frequent. In other words, if a given itemset is not frequent (if it does not satisfy the minimum support threshold), then any superset of this itemset will also be not frequent, because it cannot occur more frequently than the original itemset. A proof follows:

Given n transactions, suppose A is a subset of i transactions (itemset).
If $A' \subset A$, then A' is a subset of $i' \leq i$ transactions.
Thus, if $i/n <$ minimum support, then i'/n is also $<$ minimum support.

```
n = |D|
for each subset s of |
  {
    counter = 0;
    for each transaction T in D
      {
        if s is a subset of T;
          counter = counter + 1; }
    if minimum support ≤ counter / n
      add s to frequent itemsets; }
```

Figure 10.7. Naïve algorithm for generation of frequent itemsets.

3. Mining Single Dimensional, Single-Level Boolean Association Rules

The simplest superset of an itemset is the itemset with one more added item. The Apriori property is an **antimonotone property**, i.e., if a set cannot satisfy a property, all of its supersets will also fail the same test.

The Apriori property is used to reduce the number of itemsets that must be searched to find frequent itemsets. The association-rule mining algorithm, the **Apriori algorithm**, performs the iterative search through itemsets, starting with 1-itemsets, through 2-itemsets, 3-itemsets, etc. In general, it finds and processes k-itemsets based on the exploration of $(k-1)$-itemsets. Using the Apriori property, the Apriori algorithm

– first finds all 1-itemsets
– next, finds among them a set of frequent 1-itemsets, L_1
– next extends L_1 to generate 2-itemsets
– next finds among these 2-itemsets a set of frequent 2-itemsets, L_2
– and repeats the process to obtain L_3, L_4, etc.

Based on the Apriori property, in each iteration, k-itemsets that do not satisfy the minimum support are removed and only the remaining k-itemsets are used to generate itemsets for the next, $k+1$, iteration. This process substantially reduces the number of itemsets that must be checked if they are frequent. The algorithm is shown in Figure 10.8.

The only unknown in implementing the Apriori algorithm is how to perform generation of C_k, which is a set of k-itemsets based on L_{k-1}. These k-itemsets are checked against the minimum support to derive L_k. The C_k is generated in two steps:

1. For each frequent itemset FI from L_{k-1}, find each item i that does not belong to FI, but belongs to some other frequent $(k-1)$-itemset in L_{k-1}. Add i to FI to create a k-itemset. Remove duplicate k-itemsets after all additions for all $(k-1)$-itemsets are finished.

    Example: Generation of frequent 2-itemsets from frequent 1-itemsets.
    Frequent 1-itemsets include $\{A\}$, $\{B\}$, $\{C\}$.
    The 2-itemsets generated based on Step 1 are $\{A, B\}$, $\{A, C\}$, $\{B, A\}$, $\{B, C\}$, $\{C, A\}$, and $\{C, B\}$. After elimination of duplicates, the following 2-itemsets are left: $\{A, B\}$, $\{A, C\}$, and $\{B, C\}$.

2. If frequent $(k-1)$-itemsets from L_{k-1} have $(k-2)$-items in common, then create a k-itemset by adding the two different items to $(k-2)$ common items.

    Example: Generation of frequent 4-itemsets from overlapping frequent 3-itemsets.
    For two 3-itemsets $\{A, B, C\}$ and $\{A, B, D\}$ the resulting 4-itemset is $\{A, B, C, D\}$.

```
n = |D|
L₁ = {frequent 1-itemsets}
for (k = 2; L_{k-1} is not empty; k++)
{
    C_k is generated as k-itemset candidates from L_{k-1};
    for each transaction T in D
    {
        C_t = subset(C_k, T);       // k-itemsets that are subsets of T
        for each k-itemset c in C_t
        c.count++; }
    L_k = {c in C_k such that c.count ≥ minimum support; }}
Result: the frequent itemsets are the union of all L_k
```

Figure 10.8. The Apriori algorithm for the generation of frequent itemsets.

3.3. Generating Association Rules from Frequent Itemsets

The last step of the four that are used to generate single-dimensional association rules is to generate association rules from frequent itemsets. The association-rule mining algorithm requires the generation of strong rules, i.e., those that satisfy both minimum confidence and minimum support. The minimum support level is guaranteed by using frequent itemsets, and thus we need only to (1) generate the rules and (2) prune those rules that do not satisfy the minimum confidence.

The confidence can be defined based on the corresponding support values as follows:

$$\text{confidence}(A \Rightarrow B) = P(B|A) = support\_count(A \cup B)/support\_count(A)$$

where $support\_count(A \cup B)$ is the number of transactions in D containing the itemset $A \cup B$, and $support\_count(A)$ is the number of transactions in D containing the itemset A.

Based on this formula, each frequent itemset FI is used to generate association rules in two steps:

1. Generate all nonempty subsets of items, Y, of FI
2. For each Y, output the rules "Y $\Rightarrow$ (FI − Y)" if $support\_count(FI)$ / $support\_count(Y) \geq$ minimum confidence threshold.

To demonstrate the Apriori algorithm in action, we generate association rules from the transactional data given in Figure 10.1, as is shown in Figure 10.9.

3.4. Improving Efficiency of the Apriori Algorithm

The Apriori algorithm was further modified to improve its efficiency (computational complexity). Below we briefly explain the most important improvements.

- **Hashing** is used to reduce the size of the candidate k-itemsets, i.e., itemsets generated from frequent itemsets from iteration $k–1$, C_k, for $k > 1$. For instance, when scanning D to generate L_1 from the candidate 1-itemsets in C_1, we can at the same time generate all 2-itemsets for each transaction, hash (map) them into different buckets of the hash table structure, and increase the corresponding bucket counts. A 2-itemset whose corresponding bucket count is below the support threshold cannot be frequent, and thus we can remove it from the candidate set C_2. In this way, we reduce the number of candidate 2-itemsets that must be examined to obtain L_2.
- **Transaction removal** removes transactions that do not contain frequent itemsets. In general, if a transaction does not contain any frequent k-itemsets, it cannot contain any frequent $(k + 1)$ itemsets, and thus it can be removed from the computation of any frequent t-itemsets, where $t > k$.
- **Data set partitioning** generates frequent itemsets based on the discovery of frequent itemsets in subsets (partition) of D. The method has two steps:

  1. Division of the transactions in D into s nonoverlapping subsets and the mining frequent itemsets in each subset. Given a minimum support threshold (*minimum_support*) for D, then the minimum itemset support for a subset equals *minimum_support*number_transactions_in_this_subset*. Based on this support count, all frequent itemsets (for all k) in each subset, referred to as **local frequent itemsets**, are found. A special data structure, which for each itemset records the TID of the transactions that contains the items in this itemset, is used to find all local frequent k-itemsets, for all $k = 1, 2, 3, \ldots$, in just one scan of D. The frequent local itemsets may or may not be frequent in D, but any itemset that is potentially frequent in D must be frequent in at least one subset. Therefore, local frequent itemsets from all subsets become candidate itemsets for D. The collection of all local frequent itemsets is referred to as **global candidate itemsets** with respect to D.

298   3. Mining Single Dimensional, Single-Level Boolean Association Rules

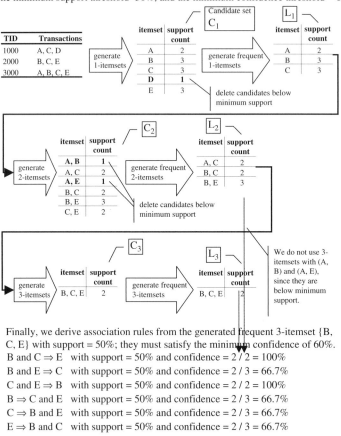

Finally, we derive association rules from the generated frequent 3-itemset {B, C, E} with support = 50%; they must satisfy the minimum confidence of 60%.

B and C $\Rightarrow$ E   with support = 50% and confidence = 2 / 2 = 100%
B and E $\Rightarrow$ C   with support = 50% and confidence = 2 / 3 = 66.7%
C and E $\Rightarrow$ B   with support = 50% and confidence = 2 / 2 = 100%
B $\Rightarrow$ C and E   with support = 50% and confidence = 2 / 3 = 66.7%
C $\Rightarrow$ B and E   with support = 50% and confidence = 2 / 3 = 66.7%
E $\Rightarrow$ B and C   with support = 50% and confidence = 2 / 3 = 66.7%

Figure 10.9. Example generation of association rules using the Apriori algorithm.

2. Computation of frequent itemsets for D based on the global candidate itemsets. One scan of D is performed to find out which of the global candidate itemsets satisfy the support threshold.

The size and number of subsets is usually set so that each of the subsets can fit into the main computer memory. Figure 10.10 illustrates the data set partitioning procedure.

– **Sampling** generates association rules based on a sampled subset of transactions in D. In this case, a randomly selected subset S of D is used to search for the frequent itemsets. The generation of frequent itemsets from S is more efficient (faster), but some of the rules that would have been generated from D may be missing, and some rules generated from S may not be present in D, i.e., the "accuracy" of the rules may be lower. Usually the size of S is selected so that the transactions can fit into the main memory, and thus only one scan of the data is required (no paging). To reduce the possibility that we will miss some of the frequent itemsets from D when generating frequent itemsets from S, we may use a lower support threshold for S

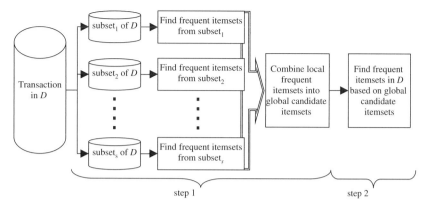

Figure 10.10. Generation of frequent itemsets using data set partitioning.

as compared with the support threshold for D. This approach is especially valuable when the association rules are computed on a very frequent basis.

- **Mining frequent itemsets without generation of candidate itemsets**. One of the main limiting aspects of the Apriori algorithm is that it can still generate very large number of candidate itemsets. For instance, for 10,000 1-itemsets, the Apriori algorithm generates approximately 10,000,000 candidate 2-itemsets and has to compute and store their occurrence frequencies. When a long frequent pattern is generated, say with 100 items, the Apriori algorithm generates as many as 2^{100} candidate itemsets. The other limiting aspect is that the Apriori algorithm may need to repeatedly scan the data set D to check frequencies of a large set of candidate itemsets—a process that is especially transparent when mining long itemsets, i.e., $n+1$ scans is required where n is the length of the longest itemset.

To address these issues, a **divide-and-conquer** method, which decomposes the overall problem into a set of smaller tasks, is used. The method, referred to as **frequent-pattern growth** (FP-growth), compresses the set of frequent (individual) items from D into a **frequent pattern tree** (FP-tree). The tree preserves complete information about D with respect to the frequent pattern mining, is never larger than D, and reduces the irrelevant information, i.e., infrequent items are removed. The FP-tree, instead of D, is scanned to find frequent itemsets. Next, this compacted frequent-item-based data set is divided into a set of conditional data sets, each associated with one frequent item, and each of these conditional data sets is mined separately. A more detailed discussion of the FP-growth algorithms is beyond the scope of this textbook.

Several studies that have considered the performance of frequent itemset generation indicate that the FP-growth algorithm is efficient and scalable for mining both short and long frequent itemsets, and is about an order of magnitude faster than the Apriori algorithm. The FP-growth algorithm is faster than another method known as the **tree-projection** algorithm, which recursively projects D into a tree of projected databases. Intuitively, this improvement is due to avoiding generation of candidate itemsets, using compact data structure and elimination of repeated scans of D. Instead of generation of candidate itemsets, the FP-tree is built and simple counts are computed.

3.5. Finding Interesting Association Rules

As suggested by the example given in Figure 10.9, and depending on the minimum support and confidence values, the user may generate a large number of rules to analyze and assess. The question then is how one can filter out the rules that are potentially the most interesting. First, whenever a rule is interesting (or not) it can be evaluated either objectively or subjectively.

3. Mining Single Dimensional, Single-Level Boolean Association Rules

The ultimate subjective user's evaluations cannot be quantified or anticipated; they are different for different users. That is why objective **interestingness measures**, based on the statistical information present in D, were developed in order to remove uninteresting rules before presenting them to the user.

The **subjective** evaluation of association rules often boils down to checking whether a given rule is unexpected (i.e., surprises the user) and actionable (i.e., the user can do something useful based on the rule). More specifically, the rules are categorized as

- **useful**, when they provide high-quality, actionable information, e.g., diapers $\Rightarrow$ beers
- **trivial**, when they are valid and supported by data, but useless since they confirm well-known facts, e.g., milk $\Rightarrow$ bread
- **inexplicable**, when they concern valid and new facts but cannot be utilized, e.g., grocery_store $\Rightarrow$ milk_is_sold_as_often_as_bread

In most cases, the confidence and support values associated with each rule are used as an **objective** measure to select the most interesting rules. Rules that have confidence and support values higher than other rules are preferred. Although this simple approach works in many cases, we will show that sometimes rules that have high confidence and support may be uninteresting and even misleading. Therefore, an additional quality measure is used. This measure is not used to generate the rules (support and confidence are sufficient for this purpose) but is helpful for selecting interesting rules.

Let us assume that a transactional data set concerning a grocery store contains milk and bread as the frequent items. The data show that on a given day 2,000 transactions were recorded and among these, in 1,200 transactions the customers bought milk, in 1,650 transactions the customers bought bread, and in 900 transactions the customers bough both milk and bread. Given a minimum support threshold of 40% and a minimum confidence threshold of 70%, the "milk $\Rightarrow$ bread [45%, 75%]" rule would be generated. On the other hand, due to low support and confidence values, the "milk $\Rightarrow$ not bread [15%, 25%]" rule would not be generated. At the same time, the latter rule is by far more "accurate," while the first may be misleading. This phenomenon is explained using the **contingency matrix** shown in Table 10.2, which provides corresponding support counts for all (four) combinations of the two Boolean items.

The probability of buying bread is 82.5%, while the confidence of milk $\Rightarrow$ bread is lower and equals 75%. In other words, bread and milk are negatively associated, i.e., buying one results in a decrease in buying the other. Obviously, using this rule would not be a wise decision. The confidence value is only an estimate of the conditional probability of itemset B (bread) given A (milk).

The alternative approach to evaluating the interestingness of association rules is to use measures based on correlation. For an $A \Rightarrow B$ rule, the itemset A is **independent** of the occurrence of the itemset B if $P(A \cup B) = P(A)P(B)$. Otherwise, itemsets A and B are **dependent** and correlated as events. The **correlation measure** (also referred to as lift and interest), which is defined between

Table 10.2. Support count values for the milk $\Rightarrow$ bread example.

	milk	not milk	total
bread	900	750	1650
not bread	300	50	350
total	1200	800	2000

itemsets *A* and *B* (but can be easily extended to more than two itemsets), is defined as

$$\text{correlation}(A, B) = \frac{P(A \cup B)}{P(A)P(B)}$$

If the correlation value is less than 1, then the occurrence of *A* is **negatively correlated** (inhibits) the occurrence of *B*. If the value is greater than 1, then *A* and *B* are **positively correlated**, which means that the occurrence of one implies (promotes) the occurrence of the other. Finally, if the correlation equals 1, then *A* and *B* are **independent**, i.e., there is no correlation between these itemsets.

Based on the contingency matrix shown in Table 10.2, the correlation value for milk $\Rightarrow$ bread equals $0.45/(0.6*0.825) = 0.45/0.495 = 0.91$. This result shows that milk and bread are negatively correlated, and thus the corresponding association rule should not be used. On the other hand, for the milk $\Rightarrow$ not bread rule, the correlation equals $0.15/(0.6*0.175) = 0.15/0.105 = 1.43$. In this case, there is a relatively strong positive correlation between these two itemsets. This example demonstrates that the correlation cannot be successfully captured using support and confidence values.

4. Mining Other Types of Association Rules

The simplest form of an association rules is a Boolean, single-level, single-dimensional rule, which we described in Sec. 10.3 above. In what follows, we describe methods for mining multilevel, multidimensional, and quantitative association rules.

4.1. Multilevel Association Rules

In applications where items form a hierarchy, it may be difficult to find strong association rules at the low level of abstraction due to sparsity of data in the multidimensional space. Strong association rules usually can be found at the higher level of a hierarchy, but they often represent already known, commonsense knowledge. For instance, the milk $\Rightarrow$ bread rule is likely to have strong support, but it is trivial. At the same time, the rule skim_milk $\Rightarrow$ large_white_bread may be useful, but it may have weak support. The corresponding concept hierarchy is shown in Figure 10.11.

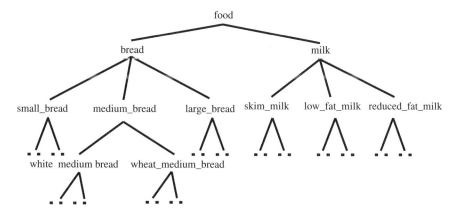

Figure 10.11. Example of a concept hierarchy concerning food.

4. Mining Other Types of Association Rules

An association mining algorithm should be able to generate and traverse between the rules at different levels of abstraction. **Multilevel association rules** are generated by performing a top-down, iterative deepening search. In simple terms, we first find strong rules at the high level(s) in the hierarchy, and then search for lower-level "weaker" rules. For instance, we first generate the milk $\Rightarrow$ bread rule and then concentrate on finding rules that concern breads of different sizes and milks with different fat content.

There are two main families of methods for multilevel association-rule mining:

- Methods based on **uniform support**, where the same minimum support threshold is used to generate rules at all levels of abstraction. In this case, the search for rules is simplified, since we can safely assume that itemsets containing item(s) whose ancestors (in the hierarchy) do not satisfy the minimum support are also not frequent. At the same time, it is very unlikely that items at the lower level of abstraction occur as frequently as those at the higher levels. Consequently, if the minimum support threshold is too high, the algorithm will miss potentially useful associations at lower levels of abstraction. On the other hand, if the threshold is too low, a very large number of potentially uninteresting associations would be generated at the higher levels.
- Methods based on **reduced support**, an approach that addresses the drawbacks of uniform support. In this case, each level of abstraction is furnished with its own minimum support threshold. The lower the abstraction level, the smaller the corresponding threshold. Figure 10.12 illustrates the difference between the uniform and reduced support methods.

If we assume a minimum support threshold value of 10% and use the uniform support method, then *milk* and *reduced_fat_milk* items are frequent and the *skim_milk* and *low_fat_milk* items are removed. In the case of the reduced support method, if the minimum support for the lower abstraction level is lowered by half, *milk*, *reduced_fat_milk*, and *low_fat_milk* items would all be considered frequent. There are three approaches to search for multilevel associations using the reduced support method:

- The **level-by-level independent** method, in which a breadth-first search is performed, i.e., each node in the hierarchy is examined, regardless of whether or not its parent node is found to be frequent.
- The **level-cross-filtering by single item** method, in which an item at a given level in the hierarchy is examined only if its parent at the preceding level is frequent. In this way, a more specific association is generated from a general one. For example, in Figure 10.12, if the minimum support threshold were set to 25%, then *reduced_fat_milk*, *low_fat_milk*, and *skim_milk* would not be considered.
- The **level-cross-filtering by k-itemset** method, in which a k-itemset at a given level is examined only if its parent k-itemset at the preceding level is frequent.

Once the multilevel association rules have been generated, some of them may be redundant due to the ancestor relationship between their items. For example, the following two rules are redundant:

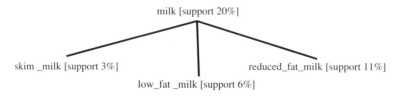

Figure 10.12. Multilevel association-rule mining using uniform and reduced support methods.

milk ⇒ large_bread [10%, 80%]
skim_milk ⇒ large_bread [2%, 75%]

The latter rule doe not provide useful information in the context of the former rule. The former rule is called **ancestor** of the latter rule, since it can be obtained by replacing an item in the latter rule by using their ancestor in the concept hierarchy. A rule is considered **redundant** if its support is similar to its "expected" value, based on the ancestor of the rule. Continuing our example with the two rules shown above, since the former rule has support of 10% and about one quarter of the sold *milk* is *skim_milk*, the expected support for the *skim_milk* is about 2.5%. The actual support for the latter rule is 2%, which means that the rule is redundant. As a consequence, only the former rule, which is more general, is kept.

4.2. Multidimensional and Quantitative Association Rules

The Boolean, single-dimensional association rules consider only one predicate and only the presence/absence of an item. In contrast, **multidimensional association rules** consider multiple predicates, and **quantitative association rules** consider items (attributes) that may assume a range of values instead of just two values.

The motivation for multidimensional association rules stems from data warehouses, which store multidimensional data. In this case, items correspond to features that describe different dimensions, and thus are associated with different predicates. An example in Sec. 2 showed a multidimensional rule that uses three predicates: *major*, *takes_course*, and *level*. Such a rule is referred to as an **interdimension association rule**, since none of the predicates is repeated. In contrast, the **hybrid-dimension association rule** incorporates some of the predicates multiple times (see the following example).

Example: An association rule that describes graduate students.

major(x, Computer Engineering) AND takes_course(x, Advanced Data Analysis and Decision Making) AND takes_course(x, Data Mining) ⇒ level(x, PhD) [0.9%, 90%]

Some of the data attributes are **discrete** (categorical) and **continuous**, in contrast to Boolean attributes, which are used in generic association mining. All attribute types can be categorized as **nominal**, in which case there is no ordering between their values, and **ordinal**, in which case some ordering exists. An example discrete ordinal attribute is *age*, while an example discrete nominal attribute is *major*. The continuous attributes must be preprocessed before being used for association mining. The preprocessing boils down to discretization, which divides the entire range of the attribute values into subintervals and associates a discrete value with each of the intervals. For instance, *age* can be divided into subintervals [0, 9], [9, 18], [18, 30], [30, 65], [65, 120], and these intervals can be associated with the following discrete values: *child*, *teenager*, *young_adult*, *adult*, *senior*. The discretization can be performed manually, based on an associated attribute hierarchy and/or an expert's knowledge. In the case of continuous ordinal attributes, discretization can also be performed automatically (see Chapter 8). A **quantitative association rule** uses discrete and/or discretized continuous items (attributes); see the example shown in Sec. 2.

The multidimensional and quantitative association rules can be generated using the same algorithms as the Boolean, single-dimensional rules. These include the Apriori algorithm and its modifications, such as hashing, partitioning, sampling, etc. The main difference is in how the input transactional data, is prepared. Namely, the continuous attributes must be discretized and the algorithm must search through all relevant attributes (attribute-value pairs; see Sec. 2.1) together, instead of searching just one attribute (predicate).

5. Summary and Bibliographical Notes

In this Chapter, we introduced **association rules** and **association rule mining** algorithms. The most important topics discussed in this Chapter are the following.

- Association rule mining is one of the two key **unsupervised** data mining methods that finds interesting **associations** (correlation) relationships within a large set of **items**.
- The items are stored using **transactions**.
- The association rules can be categorized as **single-dimensional** and **multidimensional**, **Boolean** and **quantitative**, and **single-level** and **multilevel**.
- The interestingness of association rules is measured using **support**, **confidence**, and **correlation**.
- The association rules are generated from **frequent itemsets**, which in turn are generated from transactions.
- The most popular algorithm for the generation of association rules is the **Apriori algorithm**.
- Several modifications to this algorithm have been proposed to improve its efficiency. These include **hashing, transaction removal, data set partitioning, sampling**, and **mining frequent itemsets without generation of candidate itemsets**.
- **Multilevel association rules** are generated using **uniform support**-based and **reduced support**-based methods.

A number of introductory-level textbooks provide material concerning **association rules** and **association rule mining**, such as [10, 12, 19]. Association rule mining was first proposed by Agrawal, Imielinski, and Swami [2], while the **Apriori algorithm** was introduced in [3, 4, 14]. The subsequent improvements to the Apriori algorithm, described in this Chapter, include **hashing** [15], **removal of transactions** [3, 8, 15], **partitioning** [16], **sampling** [20], and **mining itemsets without candidate generation** [9].

Multilevel association rule mining was introduced in [8, 17]. Mining for **quantitative association rules** includes approaches based on rule clustering [13], x-monotone regions [7], and partial completeness [18]. **Multidimensional association rules** were studied in [11]. Finally, the **interestingness** of the association rules was reported in [1, 5, 6].

References

1. Aggrawa, C., and Yu, P. 1999. A new framework for itemset generation. *Proceedings of the ACM Symposium on Principles of Database Systems*, Seattle, USA, 18–24
2. Agrawal, R., Imielinski, T., and Swami, A. 1993. Mining association rules between sets of items in large databases. *Proceedings of the 1993 ACM-SIGMOD International Conference on Management of Data*, Washington, USA, 207–216
3. Agrawal, R., and Srikant, R. 1994. Fast algorithm for mining association rules. *Proceedings of the 1994 International Conference on Very Large Databases*, Santiago, Chile, 487–499
4. Agrawal, R., Mannila, H., Srikant, R., Toivonen, H., and Verkamo, A. 1996. Fast discovery of association rules. In Fayyad, U., Piatesky-Shapiro, G., Smyth, P., and Uthurusamy, R. (Eds.), *Advances in Knowledge Discovery and Data Mining*, AAAI Press/The MIT Press, Menlo Park, CA, 307–328
5. Brin, S., Motwani, R., and Silverstein, C. 1997. Beyond market basket: generalizing association rules to correlations. *Proceedings of the 1997 ACM-SIGMOD International Conference on Management of Data*, Tuscon, USA, 265–276
6. Chen, M., Han, J., and Yu, P. 1996. Data mining: an overview from a database perspective. *IEEE Transactions on Knowledge and Data Engineering*, 8:866–883
7. Fukuda, T., Morimoto, Y., Morishita, S., and Tokuyama, T. 1996. Data mining using two-dimensional optimized association rules: scheme, algorithms and visualization. *Proceedings of the 1996 ACM-SIGMOD International Conference on Management of Data*, Montreal, Canada, 13–23
8. Han, J., and Fu, Y. 1995. Discovery of multiple-level association rules from large databases. *Proceedings of the 1995 International Conference on Very Large Databases*, Zurich, Swizerland, 420–431

9. Han, J., Pei, J., and Yin, Y. 2000. Mining frequent patterns without candidate generation. *Proceedings of the 2000 ACM-SIGMOD International Conference on Management of Data*, Dallas, USA, 1–12
10. Han, K., and Kamber, M. 2001. *Data Mining: Concepts and Techniques*, Morgan Kaufmann, San Fransico, USA
11. Kamber, M., Han, J., and Chiang, J. 1997. Metarule-guided mining of multi-dimensional association rules using data cubes. *Proceedings of the 1997 International Conference on Knowledge Discovery and Data Mining*, Newport Beach, CA, USA, 207–210
12. Kantardzic, M. 2002. *Data Mining: Concepts, Models, Methods, and Algorithms*, Wiley-IEEE Press, Piscataway, NJ, USA
13. Lent, B., Swami, A., and Widom, J. 1997. Clustering association rules. *Proceedings of the 1997 International Conference on Data Engineering*, Birmingham, England, 220–231
14. Mannila, H., Toivonen, H., and Verkamo, A. 1994. Efficient algorithms for discovering association rules. *Proceedings of the AAAI'94 Workshop on Knowledge Discovery in Databases*, Seattle, Washington, USA, 181–192
15. Park, J., Chen, M., and Yu, P. 1995. An effective hash-based algorithm for mining association rules. *Proceedings of the 1995 ACM-SIGMOD International Conference on Management of Data*, San Jose, CA, USA, 175–186
16. Savasere, A., Omiecinski, E., and Navathe, S. 1995. An efficient algorithm for mining association rules in large databases. *Proceeding of the 1995 International Conference on Very Large Databases*, Zurich, Switzerland, 432–443
17. Srikant, R., and Agrawal, R. 1995. Mining generalized association rules. *Proceedings of the 1995 International Conference on Very Large Databases*, Zurich, Switzerland, 407–419
18. Srikant, R., and Agrawal, R. 1996. Mining quantitative association rules in large relational tables. *Proceedings of the 1996 ACM-SIGMOD International Conference on Management of Data*, Montreal, Canada, 1–12
19. Tan, P-N., Steinbach, M., and Kumar, V. 2005. *Introduction to Data Mining*, Pearson Addison Wesley
20. Toivonen, H. 1996. Sampling large databases for association rules. *Proceedings of the 1996 Conference on Very Large Databases*, Bombay, India, 134–145

6. Exercises

1. Generate frequent itemsets for the following transactional data. Assume that the minimum support threshold equals 40%.

TID	Transaction (basket)
1000	C, S, B, M
2000	E, R, D, B
3000	R, K, D, M
4000	C, R, B, D, M
5000	K, B, D, M
6000	R, B, D, M, S
7000	K, B, D, M

2. For the following transactional data, generate frequent itemsets and strong association rules. Assume a minimum support threshold equal to 33.3% and a minimum confidence threshold equal to 60%. Sort the resulting strong rules by their confidence and determine their corresponding support and confidence values.
3. The following transactional data include four transactions. Assuming that the minimum support threshold equals 60% and the minimum confidence threshold equals 80% find all frequent

TID	Transaction (basket)
1000	A, C, D
2000	A, D
3000	B, C, D
4000	B, C
5000	B, C
6000	B, C, D

TID	Transaction (basket)
1000	A, B, D, F
2000	A, B, C, D, E
3000	A, B, C, E
4000	A, B, D

itemsets. Generate all strong association rules of the form "x and $y \Rightarrow y$" where x, y, and z are the corresponding items, i.e., A, B, C, D, E, and F.

4. The number of generated association rules depends on the number of generated frequent itemsets. Assuming that, for a given transactional data set, the number of items, m, is large, estimate how many frequent itemsets can be generated from m items. Consider the case when all candidate itemsets satisfy the minimum support level and when, on average, half of the candidate itemsets satisfy the minimum support level.
5. Use WEKA software (available for free at http://www.cs.waikato.ac.nz/ml/weka/) to generate association rules. Run WEKA with the nominal data set *weather.nominal*, which is included with the software. You can initially use the default values of confidence and support, but you should also try several other combinations. Analyze and interpret the discovered rules.
6. Describe the FP-growth algorithm for mining association algorithms. Use transactions from Exercise 3 to demonstrate how it works. Briefly contrast this algorithm with a tree-projection algorithm.
7. Based on the contingency matrix below, a minimum support threshold equal to 30%, and a minimum confidence threshold equal to 60%, verify whether the "Beer $\Rightarrow$ Diapers" association rule is strong. Describe what kind of correlation (independent, positively correlated, negatively correlated) exists between *beer* and *diapers*.

	Beer	Not beer	Total
Diapers	200	50	250
Not diapers	100	150	250
Total	300	200	500

8. Come up with a simple example (different than the example used in the text) demonstrating that items that constitute a strong association rule can be negatively correlated.
9. Research the topic of *incremental association rules*. Create a bibliography of related literature and write a short paper that discusses the background, motivation, and related algorithms.

11

Supervised Learning: Statistical Methods

In this Chapter, we cover the fundamentals of statistical supervised learning. The discussion addresses two dominant groups of topics in this domain namely, Bayesian methods and regression models. The Bayesian approach is concerned with processing probabilistic information represented in the form of prior probabilities and conditional probabilities. The regression models develop relationships between variables on the basis of available data and make use of assumptions about the statistical nature of such data and the character of possible errors.

1. Bayesian Methods

In this section, we present the fundamentals of Bayesian methods. We first introduce a simple Bayesian technique for a two-class pattern classification and then generalize it for multifeature and multiclass classification. We also discuss classifier design based on discriminant functions for normally distributed probabilities of patterns. Furthermore, we describe major estimation techniques of probability densities used in Bayesian inference. Finally, we discuss a Probabilistic Neural Network (PNN) regarded as a hardware implementation of kernel-based probability density estimation and Bayesian classification.

1.1. Introduction

Statistical processing based on the Bayes decision theory is a fundamental technique for pattern recognition and classification. It is based on the assumption that the classification of patterns (the decision problem) is expressed in probabilistic terms. The statistical characteristics of patterns are expressed as known probability values that describe the random nature of patterns and their features. These probabilistic characteristics are mostly concerned with a priori probabilities and conditional probability densities of patterns and classes. The Bayes decision theory provides a framework for statistical methods for classifying patterns into classes based on probabilities of patterns and their features.

1.2. Basics of Bayesian Methods

Let us assume an experiment involving recognition of two kinds of birds flying over a garden: an eagle and a hawk. The goal of recognition is to discriminate between these two classes of birds. We assume that an observer has collected information about birds for a long time. Furthermore, the two types of birds emerge over time in some random sequence. However, for a concrete

1. Bayesian Methods

instance of a bird, we know that it can be only an eagle or a hawk. We say that a state of nature, which is a bird flying over a garden, has only two distinct states: either it is "an eagle" or "a hawk":

$$\text{States of nature } C = \{\text{"an eagle"}, \text{"a hawk"}\}$$

We can call the state of nature, namely a bird flying over a garden, as C. This is a random variable taking two discrete values c : c_1 for an eagle and c_2 for a hawk:

$$\text{Values of } C = \{c_1, c_2\} = \{\text{"an eagle"}, \text{"a hawk"}\}$$

We can also use numerical coding for the two states of nature, for example,

$$c_1 = 0 \text{ (or 1) represents an eagle (class1)}$$
$$c_2 = 1 \text{ (or 2) represents a hawk (class2)}$$

We may imagine that, for an observer, a bird is perceived as an object, an image, or a pattern. This pattern may be then analyzed by an observer who sees it as a feature pattern, embracing some measurable characteristic of a bird, like size, speed, or lightness level.

Since birds fly over a garden in a random way, we consider a variable C representing a bird as a random variable. We can also consider that this random variable C represents an object as a subject of recognition. Features describing a bird will also have random values.

The goal of classification, based on Bayesian decision theory, is to classify objects based on statistical information about objects in such a way as to minimize the probability of misclassification. The classification capability when dealing with new objects depends on prior statistical information gathered from a population of previously seen randomly appearing objects. The level of accuracy (or misclassification level) with which we can predict the class of a new incoming object depends on the amount of statistical measurements and the nature of knowledge gathered prior to the current experiment. The statistical characteristics of previously seen objects can be expressed in terms of probabilities concerning the objects and their features. When the statistics for previous observations have been calculated based on a limited size sample, then the probabilities can be estimated.

For recognizing an eagle / hawk, we may assume that among the large number N of prior observations it was concluded that a fraction n_{eagle} of them belonged to a class c_1 ("an eagle") and a fraction n_{hawk} belonged to a class c_2 ("a hawk") (with $n_{\text{eagle}} + n_{\text{hawk}} = N$). Then we consider the **a priori probability (prior)** $P(c_1)$ that the next bird will belong to the class c_1 (will be an eagle) and $P(c_2)$ that the next bird will belong to the class c_2 (will be a hawk). The probability $P(c_i)(i = 1, 2)$ corresponds to the fraction n_{c_i} of birds in an ith class, in the limit of an infinite numbers of birds observed:

$$P(c_i) = \lim_{N \to \infty} \frac{n_{c_i}}{N}, i = 1, 2 \qquad (1)$$

For a large number of observed objects, the a priori probability $P(c_i)$ can be estimated by

$$\hat{P}(c_i) = \frac{n_{c_i}}{N} \qquad (2)$$

$P(c_i)$ denotes the (unconditional) probability that an object belongs to class c_i, without the help of any further information about the object.

The a priori probabilities $P(c_1)$ and $P(c_2)$ represent our prior initial knowledge (in statistical terms) about how likely it is that an eagle or a hawk may emerge even before a bird physically appears. For example, if a prior observation shows that an eagle appears four times as often as a hawk, we may have $P(c_1) = P(\text{"an eagle"}) = 0.8$ and $P(c_2) = P(\text{"a hawk"}) = 0.2$.

Suppose that we are forced to make a classification decision for a new bird, which will appear in the next observation, without our having seen it yet. So far the only statistical knowledge about the type of bird is the prior probabilities $P(c_1)$ and $P(c_2)$. Based on this limited knowledge about probabilities, our natural and best decision is to assign the next bird to the class having the higher a priori probability. Thus, we can write a classification decision knowing $P(c_1)$ and $P(c_2)$ as follows:

$$\text{"Assign a bird to a class } c_1 \text{ if } P(c_1) > P(c_2);$$
$$\text{otherwise, assign a bird to a class } c_2\text{"} \qquad (3)$$

If $P(c_1) = P(c_2)$, then both classes have the same chance of occurrence.

For a particular object and a particular classification decision, the probability of classification error is as follows:

$$P(\text{classification\_error}) = \begin{cases} P(c_2) & \text{if we decide } C = c_1 \\ P(c_1) & \text{if we decide } C = c_2 \end{cases} \qquad (4)$$

We see that the probability of a classification error is minimized for the classification decision of the form "select c_1 if $P(c_1) > P(c_2)$, and c_2 otherwise." Selecting a class with a bigger probability value produces a smaller probability of classification error.

1.3. Involving Object Features in Classification

In order to increase classification accuracy, we provide an additional measure of values for some selected features characterizing an object and allowing for better discrimination between one class from another. For classification of an eagle and a hawk, we can assume that a measure of a bird feature, like size, will help to distinguish the two. We anticipate that the average eagle is bigger than the average hawk. According to nature, bird size assumes values in a random way.

We denote a random feature of bird size by x and call it a **feature variable** of an object or a feature. Frequently, the feature variable is assumed to be a continuous random variable taking continuous values from a given range. The variability of a random variable x can be expressed in probabilistic terms. The probability distribution for this variable depends on the state of nature.

We represent a distribution of a random variable x by the **class conditional probability density function (the state conditional probability density function)**

$$p(x|c_i), i = 1, 2 \qquad (5)$$

The probability $p(x|c_i)$ is the probability density function for a value x of a random feature variable given that the object belongs to class c_i (the state of nature is c_i).

For birds, the class conditional probability density functions, for a value x of a feature variable representing bird size, are as follows:

$$p(x|c_1) \text{ for an eagle, } p(x|c_2) \text{ for a hawk} \qquad (6)$$

Examples of probability densities are shown in Figure 11.1. We can provide a similar interpretation for a hawk, that is, $p(x|c_2)$. The conditional probability density functions $p(x|c_1)$ and $p(x|c_2)$ represent distribution of variability of eagle and hawk sizes. It is quite intuitive that knowledge about bird size may help in classifying them into different categories.

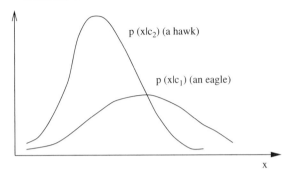

Figure 11.1. An example of probability densities.

The probability density function $p(x|c_i)$ is also called the **likelihood of a class** c_i with respect to the value x of a feature variable. This conditional probability density function suggests that the likelihood that an object belongs to class c_i is bigger if $p(x|c_i)$ is larger.

Let us define the joint probability density function $p(c_i, x)$ to be a probability density that an object is in a class c_i and has a feature variable value x. Let us also define the conditional probability function $P(c_i|x)(i = 1, 2)$, which specifies the probability that the object class is c_i given that the measured value of a feature variable is x. The probability $P(c_i|x)$ is called the **a posteriori (posterior)** probability; its value depends on the a posteriori fact that a feature variable took a concrete value x. We have

$$\sum_{i}^{2} P(c_i|x) = 1 \qquad (7)$$

From probability theory (see Appendix B), we know that the following relations hold for a priori and a posteriori probabilities:

$$p(c_i, x) = P(c_i|x)p(x), \quad i = 1, 2$$
$$p(c_i, x) = p(x|c_i)P(c_i), \quad i = 1, 2 \qquad (8)$$

The term $p(x)$ represents **an unconditional probability density function** for a feature variable x. We know that:

$$p(x) = \sum_{i=1}^{2} p(x|c_i)P(c_i) = p(x|c_1)P(c_1) + p(x|c_2)P(c_2) \qquad (9)$$

Rearranging the above relationships results in **Bayes' rule (Bayes' theorem)**:

$$P(c_i|x) = \frac{p(x|c_i)P(c_i)}{p(x)}, \quad i = 1, 2 \qquad (10)$$

Since $p(x) = \sum_{i=1}^{2} p(x|c_i)P(c_i)$, we can also write

$$P(c_i|x) = \frac{p(x|c_1)P(c_i)}{\sum_{i=1}^{2} p(x|c_i)P(c_i)}, \quad i = 1, 2 \qquad (11)$$

The probability density function $p(x)$ plays the role of a scaling factor, ensuring that a posteriori probabilities sum to $1 (P(c_1|x) + P(c_2|x) = 1)$.

The importance of Bayes' rule is that it allows us to reverse conditions in conditional probabilities. Specifically, the conditional probability $P(c_i|x)$ can be expressed in terms of the a priori probability function $P(c_i)$, together with the class conditional probability density function $p(x|c_i)$. In practice, it is difficult to find $P(c_i|x)$, whereas $p(x|c_i)$ and $P(c_i)$ can be estimated easily by sampling from prior observations.

Now we can formulate a classification decision for a new emerging bird, having a measure of its feature (size) x. Let us assume that we know the a posteriori conditional probabilities $P(c_i|x)(i=1,2)$ from prior observations, as well as their feature variable x. With this knowledge we provide the following classification decision for a new observed bird:

"Assign class c_i to a bird that has the largest value of the a posteriori conditional probability $P(c_i|x)$, for a given x."

This statistical classification rule is best in the sense of minimizing the probability of misclassification (the probability of classification error). This rule is called **Bayes' decision (Bayesian classification rule)**.

We see that for a given bird with measured feature value x, the conditional probability of the classification error is:

$$P(\text{classification\_error}|x) = \begin{cases} P(c_2|x) & \text{if we decide } C = c_1 \\ P(c_1|x) & \text{if we decide } C = c_2 \end{cases} \quad (12)$$

For a given object and a given x, we can easily find that the probability of classification error is minimized if we decide according to Bayes' rule.

Because different objects may emerge in a random sequence, each having a different value of x, we need to show that Bayes' classification rule minimizes the average probability of error (for all possible objects). The average probability of error is defined as

$$P(\text{classification\_error}) = \int_{-\infty}^{\infty} P(\text{classification\_error}, x)\, dx$$

$$= \int_{-\infty}^{\infty} P(\text{classification\_error}|x) p(x)\, dx \quad (13)$$

We see that if the classification decision ensures that for each x the probability $P(\text{classification\_error}|x)$ is as small as possible, then the whole integral will be as small as possible, which guarantees its minimization. This outcome implies that choosing Bayes' classification rule guarantees minimization of the average probability of classification error. For the two-class example, Bayes' rule yields the following value for x conditional probability of classification error:

$$P(\text{classification\_error}|x) = \min\{P(c_1|x), P(c_2|x)\} \quad (14)$$

Since the probability function $P(c_i|x)(i=1,2)$ is difficult to estimate, we express it by the class conditional probability density functions $p(x|c_i), p(x)$, and the probability $P(c_i)$ using Bayes' theorem. Then we form the following Bayes' classification rule:

$$\text{"Decide a class } c_1 \text{ if } \frac{p(x|c_1)P(c_1)}{p(x)} > \frac{p(x|c_2)P(c_2)}{p(x)}$$

$$\text{and a class } c_2 \text{ otherwise."} \quad (15)$$

From the rule, we see that the probability density function $p(x)$ is not important for a classification decision (since it is only a scaling factor ensuring that a posteriori probabilities sum to 1). Hence, we express Bayes' classification rule for the two-class example in the equivalent form:

"Decide a class c_1 if $p(x|c_1)P(c_1) > p(x|c_2)P(c_2)$

and a class c_2 otherwise (if $p(x|c_2)P(c_2) > p(x|c_1)P(c_2)$)."

Since the probabilities $p(x|c_i)$ and $P(c_i)$ can be estimated by sampling the previous observations, Bayes' rule is considered practically relevant. We note that both $p(x|c_i)$ and $P(c_i)$ are important for statistical classification and guarantee minimization of the probability of classification error.

Example: Let us consider a bird classification problem with $P(c_1) = P(\text{"an eagle"}) = 0.8$ and $P(c_2) = P(\text{"a hawk"}) = 0.2$ and known probability density functions $p(x|c_1)$ and $p(x|c_2)$. Assume that, for a new bird, we have measured its size $x = 45$ cm and for this value we computed $p(45|c_1) = 2.2828 \cdot 10^{-2}$ and $p(45|c_2) = 1.1053 \cdot 10^{-2}$. Thus, the classification rule predicts class c_1 ("an eagle") because $p(x|c_1)P(c_1) > p(x|c_2)P(c_2)$ ($2.2828 \cdot 10^{-2} \cdot 0.8 > 1.1053 \cdot 10^{-2} \cdot 0.2$). Let us assume that we have known an unconditional density $p(x)$ value to be equal to $p(45) = 0.3$. The probability of classification error is

$$P(\text{classification\_error}|x) = \min\{P(c_1|x), P(c_2|x)\}$$
$$= \min\{\frac{p(x|c_1)P(c_1)}{p(x)}, \frac{p(x|c_1)P(c_2)}{p(x)}\}$$
$$= \min\{0.0754, 0.07\} = 0.07$$

1.4. Bayesian Classification – General Case

We generalize Bayes' classification rule for objects belonging to more than two classes and for measured characteristics of objects having more than one feature variable. Furthermore, we generalize Bayes' rule by introducing a classification loss function that is more general than the probability of classification error. This approach will be supported by introducing the possibility of deciding classification actions other than the physical states of nature of an object (such as **rejection** or **unknown**).

As we know from Chapter 7, features may be of different types. Here we consider real-valued features of an object as n-dimensional column vector $\mathbf{x} \in \mathbb{R}^n$:

$$\mathbf{x} = \begin{bmatrix} x_1 \\ x_2 \\ \vdots \\ x_n \end{bmatrix} \quad (16)$$

where $x_i \in \mathbb{R}$ is an ith element of a feature vector. A feature vector $\mathbf{x}$ represents an object and is regarded as a point in n-dimensional space. The concrete realization of a vector $\mathbf{x}$ represents an instance of a measurement for the given object. For example, for an eagle / hawk we can measure two features, size and weight (which form the two-dimensional feature vector $\mathbf{x} \in \mathbb{R}^2$)

$$\mathbf{x} = \begin{bmatrix} x_1 \\ x_2 \end{bmatrix} = \begin{bmatrix} \text{size} \\ \text{weight} \end{bmatrix} \quad (17)$$

1.4.1. Bayes' Classification Rule for Multiclass Multifeature Objects

Objects can be classified into more than two classes. Generally, we assume that an object may belong to l distinct classes (l distinct states of nature):

$$C = \{c_1, c_2, \cdots, c_l\} \tag{18}$$

For example, we may classify birds into four classes: eagle, hawk, stork, or owl.

Let us express Bayes' theorem and Bayes' classification rule for multiclass and multifeature classification problems. We denote by $P(c_i)$ the **a priori probability** that the next object will belong to class c_i. The probability $P(c_i)(i = 1, 2, \cdots, l)$ corresponds to the fraction of birds in the ith class, in the limit of an infinite numbers of birds observed.

The **class conditional probability density function** (the **state conditional probability density function**) is denoted for all l classes by $p(\mathbf{x}|c_i)(i = 1, 2, \cdots, l)$. The joint probability density function is denoted by $p(c_i, \mathbf{x})$ and is the probability density that an object is in class c_i and has feature vector value $\mathbf{x}$. The conditional probability function $P(c_i|\mathbf{x})(i = 1, 2, \cdots, l)$ specifies the probability that an object class is c_i given that the measured value of a feature vector is $\mathbf{x}$. The probability $P(c_i|\mathbf{x})$ is the **a posteriori (posterior)** probability, the value of which depends on the a posteriori fact that a feature vector took the value $\mathbf{x}$. We have

$$\sum_{i=1}^{l} P(c_i|\mathbf{x}) = 1 \tag{19}$$

From probability theory, we know that the following relations hold for a priori and a posteriori probabilities:

$$p(c_i, \mathbf{x}) = P(c_i|\mathbf{x})p(\mathbf{x}), \quad i = 1, 2, \cdots, l$$
$$p(c_i, \mathbf{x}) = p(\mathbf{x}|c_i)P(c_i), \quad i = 1, 2, \cdots, l \tag{20}$$

The term $p(\mathbf{x})$ represents the **unconditional probability density function** for a feature vector $\mathbf{x}$. We know that

$$p(\mathbf{x}) = \sum_{i=1}^{l} p(\mathbf{x}|c_i)P(c_i) = p(\mathbf{x}|c_1)P(c_1) + \cdots + p(\mathbf{x}|c_l)P(c_l), \quad i = 1, 2, \cdots, l \tag{21}$$

Rearranging the above equations results in **Bayes' theorem**:

$$P(c_i|\mathbf{x}) = \frac{p(\mathbf{x}|c_i)P(c_i)}{p(\mathbf{x})}, \quad i = 1, 2, \cdots, l \tag{22}$$

Since $p(\mathbf{x}) = \sum_{i=1}^{l} p(\mathbf{x}|c_i)P(c_i)$, we can also write

$$P(c_i|\mathbf{x}) = \frac{p(\mathbf{x}|c_i)P(c_i)}{\sum_{i=1}^{l} p(\mathbf{x}|c_i)P(c_i)}, \quad i = 1, 2, \cdots, l \tag{23}$$

For multiclass objects with a multidimensional feature vector, Bayes' classification rule can be stated as follows:

"*Given an object with a corresponding feature vector value* $\mathbf{x}$,

assign an object to a class c_j

with the highest a posteriori conditional probability $P(c_j|\mathbf{x})$."

1. Bayesian Methods

In other words:

> "For a given object with a given value **x**
> of a feature vector, assign an object to class c_j when
> $$P(c_j|\mathbf{x}) > P(c_i|\mathbf{x}), \quad i = 1, 2, \cdots, l; \; i \neq j\text{"} \tag{24}$$

The conditional probability $P(c_i|\mathbf{x})$ is difficult to ascertain; however, using Bayes' theorem, we express it in terms of $p(\mathbf{x}|c_i)$, $P(c_i)$ and $p(\mathbf{x})$:

> "A given object, with a given value **x** of a feature vector,
> can be classified as belonging to class c_j when
> $$\frac{p(\mathbf{x}|c_j)P(c_j)}{p(\mathbf{x})} > \frac{p(\mathbf{x}|c_i)P(c_i)}{p(\mathbf{x})}, \quad i = 1, 2, \cdots, l; \; i \neq j\text{"} \tag{25}$$

After canceling the scaling probability $p(\mathbf{x})$ from both sides, we obtain the following form of the Bayes classification rule:

> "Assign an object with a given value **x**
> of a feature vector to class c_j when
> $$p(\mathbf{x}|c_j)P(c_j) > p(\mathbf{x}|c_i)P(c_i), \quad i = 1, 2, \cdots, l; \; i \neq j\text{"} \tag{26}$$

In the sense of the minimization of the probability of classification, Bayes' error rule is the theoretically optimal classification rule. In other words, there is no other classification rule that yields lower values of the classification error.

1.5. Classification that Minimizes Risk

In some applications, like medical diagnosis, the criterion of minimization of the classification error may not be the most desirable. To incorporate the fact that misclassifications of some classes are more costly than others, we define a classification that is based on a minimization criterion that involve a loss regarding a given classification decision for a given true state of nature.

Let us assume that, based on the measured feature vector **x**, a classification decision yields class c_j, whereas the object's true class is c_i. We define the **loss** function:

$$L(\text{decision\_class}_j \mid \text{true\_class}_i) \tag{27}$$

as cost (penalty, weight) due to the fact of assigning an object to class c_j when in fact the true class is c_i. We denote a loss function (a function of a classification decision) $L(\text{decision\_class}_j|\text{true\_class}_i)$ by L_{ij}. For l-class classification problems, the loss elements form a square $l \times l$ **loss matrix L**:

$$\mathbf{L} = \begin{bmatrix} L_{11} & L_{12} & \cdots & L_{1l} \\ L_{21} & L_{22} & \cdots & L_{2l} \\ \vdots & \vdots & \ddots & \vdots \\ L_{l1} & L_{l2} & \cdots & L_{ll} \end{bmatrix} \tag{28}$$

Selection of the loss matrix depends on the specificity of the application and expertise available in a given domain.

Let us consider classification of an object with measured feature vector value $\mathbf{x}$. We do not know true class of this object – it can be any number of the set of possible classes $\{c_1, c_2, \cdots, c_l\}$. Assume that $P(c_i|\mathbf{x})$ is the conditional probability that the object belongs to a true class c_i based on $\mathbf{x}$. We define the **expected** (average) conditional loss associated with making a decision that the object belongs to a class c_j, when in fact the object may belong to another class c_i ($i = 1, 2, \cdots, l;\ i \neq j$), as

$$R(c_j|\mathbf{x}) = \sum_{i=1}^{l} L(decision\_class_j \mid true\_class_i) P(c_i|\mathbf{x}) \tag{29}$$

or in a short form as

$$R_j = \sum_{i=1}^{l} L_{ij} P(c_i|\mathbf{x}) \tag{30}$$

The expected loss $R_j = R(c_j|\mathbf{x})$ is called **conditional risk**. This risk, associated with a decision c_j, is conditioned on the realization $\mathbf{x}$ of a feature vector. If, for a given new object, we measure a feature vector value as $\mathbf{x}$, then we can minimize the conditional risk by taking a classification decision $c_j (j \in \{1, 2, \cdots, l\})$, which minimizes the conditional risk criterion R_j.

A classification rule (classification decision) assigns a class for each realization $\mathbf{x}$ of a feature vector. Now we define the overall risk as the expected loss associated with a given classification decision, considered for all possible realizations $\mathbf{x}$ of a n-dimensional feature vector from $\mathbf{R_x}$ as

$$R = \int_{\mathbf{R_x}} R(c_j|\mathbf{x})\, d\mathbf{x} = \int_{\mathbf{R_x}} \sum_{i=1}^{l} L_{ij} P(c_i|\mathbf{x})\, d\mathbf{x} \tag{31}$$

where the integral is calculated for an entire feature vector space $\mathbf{R_x}$.

The overall risk R can be considered as a classification criterion for minimizing risk related to a classification decision. From the definition of risk, we see that if a classification decision c_j is made so that it guarantees that the conditional risk $R(c_j|\mathbf{x})$ is as small as possible for each realization of a feature vector $\mathbf{x}$, then overall risk is also minimized. This result leads to the generalization of Bayes' rule for minimization of probability of the classification error.

Let us now state a classification problem as deciding which class a new object belongs to, for a given realization of a feature vector $\mathbf{x}$, provided that minimization of a risk R is our minimization criterion.

For this general classification problem, we have the following Bayes' classification rule: "For a given object with a given value $\mathbf{x}$ of a feature vector, evaluate all conditional risks for all possible classes $c_j (j = 1, 2, \cdots, l)$,

$$R(c_j|\mathbf{x}) = \sum_{i=1}^{l} L_{ij} P(c_i|\mathbf{x}), \quad j = 1, 2, \cdots, l \tag{32}$$

and choose a decision c_j for which the conditional risk $R(c_j|\mathbf{x})$ is minimal:

$$R(c_j|\mathbf{x}) < R(c_k|\mathbf{x}), \quad k = 1, 2, \cdots, l;\quad k \neq j \tag{33}$$

The minimal overall risk R obtained as a result of the application of Bayes' classification decision is called the **Bayes risk**.

1. Bayesian Methods

The definition of the conditional risk $R(c_i|\mathbf{x})$ of Bayes' classification rule can also be written in the form:

"For a given object with a given value $\mathbf{x}$ of a feature vector,

choose a decision (a class) c_j for which

$$\sum_{i=1}^{l} L_{ij} P(c_i|\mathbf{x}) < \sum_{i=1}^{l} L_{ik} P(c_i|\mathbf{x}), \quad k = 1, 2, \cdots, l, \ k \neq j" \tag{34}$$

From Bayes' theorem, we know that

$$P(c_i|\mathbf{x}) = \frac{p(\mathbf{x}|c_i) P(c_i)}{p(\mathbf{x})} \tag{35}$$

Thus, we can rewrite Bayes' rule as

"Choose a decision (a class) c_j for which

$$\sum_{i=1}^{l} L_{ij} \frac{p(\mathbf{x}|c_i) P(c_i)}{p(\mathbf{x})} < \sum_{i=1}^{l} L_{ik} \frac{p(\mathbf{x}|c_i) P(c_i)}{p(\mathbf{x})}, \quad k = 1, 2, \cdots, l, \ k \neq j" \tag{36}$$

Since $p(\mathbf{x})$ as the scaling factor appears on both sides of the inequality, we can write a practical form of Bayes' classification rule:

"Choose a decision (a class) c_j for which

$$\sum_{i=1}^{l} L_{ij} p(\mathbf{x}|c_i) P(c_i) < \sum_{i=1}^{l} L_{ik} p(\mathbf{x}|c_i) P(c_i), \quad k = 1, 2, \cdots, l, \ k \neq j" \tag{37}$$

1.5.1. Bayesian Classification Minimizing the Probability of Error

Frequently a classification problem is associated with deciding to which class $c_i (i = 1, 2, \cdots, l)$ a new object belongs. This decision is a prediction of a true state of nature. Here we exclude from the classification rule decisions the category of not being a true state of nature (like a decision "unknown"). We can assume that wrong classification decisions are equally costly for all classes. Thus, if an object is classified as belonging to a class c_j when the true class is c_i, then when $j = i$, the classification is correct; otherwise, if $j \neq i$, we have a classification error. We do not weigh a loss concerning misclassification of each state of nature; rather we are interested only in judging whether a classification is correct or incorrect. Formally, we define the symmetrical **zero-one** conditional loss function as

$$L_{ij} = L(\text{decision\_class}_j \mid \text{true\_class}_i) = \begin{cases} 0 & \text{for } i = j \\ 1 & \text{for } i \neq j \end{cases} \tag{38}$$

The loss function assigns a loss of 0 for correct classification and a loss of 1 for any kind of classification error (regardless of class type). All errors are equally costly.

It is easy to notice that for the zero-one loss function the conditional risk $R(c_j|\mathbf{x})$ criterion is the same as the average probability of classification error:

$$R(c_j|\mathbf{x}) = \sum_{i=1}^{l} L_{ij} P(c_i|\mathbf{x}) = \sum_{i=1, \ i \neq j}^{l} P(c_i|\mathbf{x}) = 1 - P(c_j|\mathbf{x}) \tag{39}$$

where $P(c_j|\mathbf{x})$ is the conditional probability that the classification decision c_j is correct for a given $\mathbf{x}$.

From Equation (30), we know that Bayes' classification rule provides a classification decision that minimizes the conditional risk $R(c_j|\mathbf{x})$. We can observe that in order to minimize the conditional risk, we have to find a classification decision c_j that minimizes $1 - P(c_j|\mathbf{x})$, which means maximizing the posteriori conditional probability $P(c_j|\mathbf{x})$. This leads to minimization by the Bayes' classification rule of the average probability of the classification error:

"*Choose a decision (a class) c_j for which*

$$P(c_j|\mathbf{x}) > P(c_k|\mathbf{x}), \quad k = 1, 2, \cdots, l, \ i \neq j"$$

For this statement of classification, it is easy to find a classification rule that guarantees minimization of the average probability of classification error (i.e., the **error rate**). An average probability of classification error is thus used as a criterion of minimization for selecting the best classification decision.

If a loss matrix **L** is chosen in a way that assigns a loss of 1 for all incorrectly decided classes and a loss of 0 for correct classes ($L_{ii} = 0$), yielding a loss matrix

$$\mathbf{L} = \begin{bmatrix} 0 & 1 & 1 & 1 \\ 1 & 0 & 1 & 1 \\ \vdots & \vdots & \ddots & \vdots \\ 1 & 1 & 1 & 0 \end{bmatrix} \tag{40}$$

then minimization by the generalized Bayes' classification rule of the risk criterion is equivalent to minimization by Bayes' rule of the average probability of classification error.

1.5.2. Generalization of the Maximum Likelihood Classification

For multiclass multifeature objects, we define the generalized likelihood ratio for classes c_i and c_j:

$$\frac{p(\mathbf{x}|c_j)}{p(\mathbf{x}|c_i)}, \quad j, i = 1, 2, \cdots, l; \quad j \neq i \tag{41}$$

and the generalized threshold value as

$$\theta_{ji} = \frac{(L_{ij} - L_{ii})P(c_i)}{(L_{ji} - L_{jj})P(c_j)} \tag{42}$$

Thus the **maximum likelihood** classification rule, which minimizes the risk, can be interpreted as choosing a class for which the likelihood ratio exceeds all other threshold values for other classes:

"*Decide class c_j if*

$$\frac{p(\mathbf{x}|c_j)}{p(\mathbf{x}|c_i)} > \theta_{ji}, \quad i = 1, 2, \cdots, l; \ i \neq j" \tag{43}$$

For symmetric loss functions $L_{ij} = L_{ji}$ with $L_{ii} = 0$ (no loss for correct decisions), we can obtain the following classification rule:

"*Decide a class c_j if*

$$\frac{p(\mathbf{x}|c_j)}{p(\mathbf{x}|c_i)} > \frac{P(c_i)}{P(c_j)} \quad i = 1, 2, \cdots, l; i \neq j" \tag{44}$$

1. Bayesian Methods

1.6. Decision Regions and Probability of Errors

We will analyze classification error regardless of the type of classifier (Bayes' or other). Generally, the task of a classifier is to assign each instance **x** of a feature vector of a considered object to one of the possible classes $c_i (i = 1, 2, \cdots, l)$. In other words, that a classifier divides the feature space into l disjoint **decision regions** (decision subspaces) $R_1, R_2, \cdots, R_l$. The region R_i is a subspace such that each realization **x** of a feature vector of an object falling into this region will be assigned to a class c_i (see Figure 11.2).

In general, regions $R_i (i = 1, 2, \cdots, l)$ need not be contiguous and may be divided into several disjoint subregions; however, all vectors in these subregions will belong to the same class c_i.

The regions intersect, and boundaries between adjacent regions are called **decision boundaries** (**decision surfaces**) because classification decisions change across boundaries.

The task of a classifier design is to find classification rules that will guarantee division of a feature space into optimal decision regions $R_1, R_2, \cdots, R_l$ (with optimal decision boundaries) that will minimize a selected classification performance criterion. These optimal decision boundaries do not guarantee an ideal, errorless classification, but do guarantee that the average error will be minimal according to the chosen minimization criterion.

We will address this difficult classification problem of finding optimal decision regions and decision boundaries by analyzing a two-class classification with n-dimensional feature vector **x**. Assume that a feature space was divided (possibly not optimally) into two distinct regions, namely R_1, corresponding to class c_1, and R_2, corresponding to class c_2. We have two scenarios in which classification error occurs. The first occurs when a measured feature vector value **x** falls into region R_1, yielding a decision c_1, when the true class is c_2. The second occurs when a feature vector realization **x** falls into region R_2, yielding a decision c_2, when the true class is c_1. Since both errors are mutually exclusive, we calculate the total probability of each kind of classification error as

$$P(\text{classification\_error}) = P(\mathbf{x} \in R_1, c_2) + P(\mathbf{x} \in R_2, c_1)$$
$$= P(\mathbf{x} \in R_1 | c_2) P(c_2) + P(\mathbf{x} \in R_2 | c_1) P(c_1) \qquad (45)$$

Using Bayes' theorem, we find

$$P(\text{classification\_error}) = \int_{R_1} p(\mathbf{x}|c_2) P(c_2) d\mathbf{x} + \int_{R_2} p(\mathbf{x}|c_1) P(c_1) d\mathbf{x} \qquad (46)$$

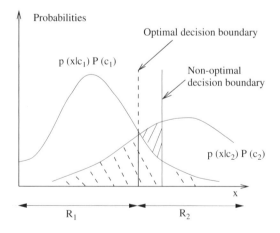

Figure 11.2. Decision boundaries.

Now we form optimal decision regions R_1 and R_2 (thus defining a classification rule) that maximize the above probability $P(\text{classification\_error})$. Let us consider an object with a given realization of a feature vector $\mathbf{x}$. The probability of classification error is minimized if for $p(\mathbf{x}|c_1)P(c_1) > p(\mathbf{x}|c_2)P(c_2)$ we choose region R_1 such that $\mathbf{x}$ will be in region R_1, because this guarantees a smaller contribution to the probability of classification error (minimization criterion). This outcome is equivalent to minimization by Bayes' classification rule of the average probability of classification error by choosing a class with the largest a posteriori probability.

An optimal classification formula based on minimization of the probability of classification error can be generalized for a multiclass multiple feature classification. For multiclass problems, it is more convenient to define the probability of the correct classification decision. For objects being classified into l distinct classes $c_1, c_2, \cdots, c_l$, with n-dimensional feature vector $\mathbf{x} \in \mathbb{R}^n$, we have the following average probability of correct classification of new objects based on value $\mathbf{x}$ of a feature vector:

$$P(\text{classification\_correct}) = \sum_{i=1}^{l} P(\mathbf{x} \in R_i, c_i)$$

$$= \sum_{i=1}^{l} P(\mathbf{x} \in R_i | c_i) P(c_i)$$

$$= \sum_{i=1}^{l} \int_{R_i} p(\mathbf{x}|c_i) P(c_i) d\mathbf{x} \qquad (47)$$

where R_i is a decision region associated with a class c_i.

Classification problems can be stated as choosing a decision region R_i (thus defining a classification rule) that maximize the probability $P(\text{classification\_correct})$ of correct classification being an optimization criterion. Thus, this criterion is maximized by choosing a region R_i such that each feature vector value $\mathbf{x}$ is assigned to a class for which an integral $\int_{R_i} p(\mathbf{x}|c_i) P(c_i) d\mathbf{x}$ is maximal (for all $\mathbf{x}$). This is equivalent to Bayes' classification rule, which guarantees the best possible partition of a feature space into decision regions from the point of view of maximization of the average probability of correct classification.

1.7. Discriminant Functions

Classifiers can be designed in many ways using different paradigms and different optimal design criteria. In the previous sections, we have discussed statistical classifiers based on Bayes' decisions, which use the comparison of relative values of the probabilities concerning objects and their features. One of the most classical forms of a classifier design is to express the class selection problem, based on values $\mathbf{x}$ of a feature vector, in a canonical form using a set of explicitly defined **discriminant functions**:

$$d_i(\mathbf{x}), \quad i = 1, 2, \cdots, l \qquad (48)$$

Each discriminant is associated with a specific class $c_i (i = 1, 2, \cdots, l)$. This discriminant-type classifier assigns an object with a given value $\mathbf{x}$ of a feature vector to a class c_j if

$$d_j(\mathbf{x}) > d_i(\mathbf{x}), \quad \text{for all} \quad i = 1, 2, \cdots, l; \ i \neq j \qquad (49)$$

In other words, the classifier assigns to an object a class for which the corresponding discriminant value is the largest.

1. Bayesian Methods

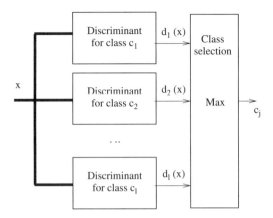

Figure 11.3. The discriminant classifier.

The discriminant classifier can be designed as a system (see Figure 11.3) containing a collection of discriminants $d_i(\mathbf{x})(i = 1, 2, \cdots, l)$ associated with each class $c_i(i = 1, 2, \cdots, l)$ along with a module that selects the discriminant with the largest value as the recognized class:

$$d_i(\mathbf{x}), \quad i = 1, 2, \cdots, l$$
$$\max(d_i(\mathbf{x})), \quad i = 1, 2, \cdots, 1 \tag{50}$$

For a discriminant function-based classifier, the classification rule can be expressed as follows: Given an object with a given value $\mathbf{x}$ of a feature vector:

(1) Compute numerical values of all discriminant functions for $\mathbf{x}$:

$$d_i(\mathbf{x}), \quad i = 1, 2, \cdots, l \tag{51}$$

(2) Choose a class c_j as a prediction of true class for which a value of the associated discriminant function $d_j(\mathbf{x})$ is the largest:

$$\text{Select a class } c_j \text{ for which } d_j(\mathbf{x}) = \max(d_i(\mathbf{x})), \quad i = 1, 2, \cdots, 1$$

The discriminant function can be defined in several ways based on the classification optimization criterion.

Let us define discriminant functions for those cases in which Bayesian classification minimizes the probability of classification error. For Bayesian classifiers, the natural choice for the discriminant function is the a posteriori conditional probability $P(c_i|\mathbf{x})$:

$$d_i(\mathbf{x}) = P(c_i|\mathbf{x}), \quad i = 1, 2, \cdots, l \tag{52}$$

Using Bayes' theorem, we find more practical versions of the discriminant function:

$$d_i(\mathbf{x}) = \frac{p(\mathbf{x}|c_i)P(c_i)}{p(\mathbf{x})}, \quad i = 1, 2, \cdots, l \tag{53}$$

$$d_i(\mathbf{x}) = p(\mathbf{x}|c_i)P(c_i), \quad i = 1, 2, \cdots, l \tag{54}$$

Note that only the relative values of the discriminant functions are important for class determination. Thus, any monotonically increasing function $f(d_i(\mathbf{x}))$ of $d_i(\mathbf{x})$ will provide an identical classification decision. Hence, we find the equivalent form of the Bayesian discriminant by taking a **natural logarithmic** function $\ln d_i(\mathbf{x})$ for the above discriminant:

$$d_i(\mathbf{x}) = \ln p(\mathbf{x}|c_i) + \ln P(c_i), \quad i = 1, 2, \cdots, l \quad (55)$$

All of the above discriminant functions provide equivalent classification.

By choosing a discriminant function $d_i(\mathbf{x})$ for each class c_i, the classification rule is defined as well. This means that a feature space is divided into l distinct decision regions $R_i (i = 1, 2, \cdots, l)$. Each of these is associated with the decision stating that if a feature vector value $\mathbf{x}$ falls into region R_j, then an object is classified as belonging to a class c_j.

Therefore, if, for a given value $\mathbf{x}$ of a feature vector, $d_j(\mathbf{x}) > d_i(\mathbf{x})$ for all $i = 1, 2, \cdots, l; i \neq j$, then $\mathbf{x}$ is in the corresponding region R_j, and a classification decision assigns a new object to a class c_j. Discriminant functions define the decision boundaries that separate the decision regions. Generally, the decision boundaries are defined by neighboring decision regions when the corresponding discriminant function values are equal. The decision boundaries define a surface in the feature space where classification decisions change. For the contiguous neighboring regions R_j and R_i, the decision boundary that separates them can be found by equaling the associated discriminant functions:

$$d_j(\mathbf{x}) = d_i(\mathbf{x}) \quad (56)$$

The decision boundaries are unaffected by the increasingly monotonic transformation of discriminant functions.

Generally, for risk that minimizes Bayes' classifier, discriminant functions can be defined as

$$d_i(\mathbf{x}) = -R(c_i|\mathbf{x}), \quad i = 1, 2, \cdots, l \quad (57)$$

A minus sign in the discriminant definition has been added because the maximal value of a discriminant will be achieved for the minimal conditional risk $R(c_i|\mathbf{x})$.

1.7.1. Bayesian Discriminant Functions for Two Classes

For two-class (c_1 and c_2) classification, we define two discriminant functions, namely, $d_1(\mathbf{x})$ and $d_2(\mathbf{x})$. These discriminant functions define two decision regions R_1 and R_2 in a feature space that is divided by the decision boundary where discriminant functions are equal:

$$d_1(\mathbf{x}) = d_2(\mathbf{x}) \quad (58)$$

and where a classification decision changes from class c_1 to c_2. An object with the feature vector $\mathbf{x}$ is classified to class c_1 (belonging to the decision region R_1) if $d_1(\mathbf{x}) > d_2(\mathbf{x})$. Otherwise, it is assigned to class c_2 (belonging to the decision region R_2). We see that for two-class classification a feature space is divided into two distinct regions; thus, instead of choosing a classifier with two discriminant functions, we design a **dichotomizer** with a single discriminant function:

$$d(\mathbf{x}) = d_1(\mathbf{x}) - d_2(\mathbf{x}) \quad (59)$$

For a given $\mathbf{x}$, the dichotomizer computes the value of a single discriminant function $d(\mathbf{x})$ and assigns a class, based on the sign of this value.

For discriminant functions selected on the basis of Bayes' rule, we find the following dichotomizers for two-class recognition:

$$d(\mathbf{x}) = P(c_1|\mathbf{x}) - P(c_2|\mathbf{x}) \quad (60)$$

$$d(\mathbf{x}) = p(\mathbf{x}|c_1)P(c_1) - p(\mathbf{x}|c_2)P(c_2) \quad (61)$$

or

$$d(\mathbf{x}) = \ln \frac{p(\mathbf{x}|c_1)}{p(\mathbf{x}|c_2)} + \ln \frac{P(c_1)}{P(c_2)} \qquad (62)$$

1.7.2. Quadratic and Linear Discriminants Derived from the Bayes Rule

Under the assumption of multivariate normality of feature vector distribution densities within the classes, a quadratic discriminant can be derived from the Bayes rule. With a further assumption about the equality of the covariance matrices computed for the features of the patterns for each of the classes, a simple linear discriminant can be derived from the Bayes rule. These discriminants can be designed based on estimation of class probabilities and within-class covariance matrices. This leads to a practical statistical classifier design for given data.

In the previous section, we have seen the following form of the Bayesian discriminant:

$$d_i(\mathbf{x}) = \ln p(\mathbf{x}|c_i) + \ln P(c_i|\mathbf{x}), \quad i = 1, 2, \cdots, l \qquad (63)$$

A classifier based on these discriminants assigns an object with a given feature vector $\mathbf{x}$ to a class c_j for which the value of the discriminant function is the largest.

Quadratic Discriminant. Let us assume a multivariate normal Gaussian distribution of the feature vector $\mathbf{x}$ within each class. Thus, each component of the feature vector has a Gaussian (normal) distribution within a class. The vector form of the normal or Gaussian distribution of the probability density function $p(\mathbf{x}|C_i)$, for the feature vector $\mathbf{x}$ within a class C_i, is given by the expression

$$p(\mathbf{x}|c_i) = \frac{1}{(2\pi)^{\frac{n}{2}}|\mathbf{\Sigma}_i|^{\frac{1}{2}}} \exp\left[-\frac{1}{2}(\mathbf{x}-\boldsymbol{\mu}_i)^T \mathbf{\Sigma}_i^{-1} (\mathbf{x}-\boldsymbol{\mu}_i)\right] \qquad (64)$$

where $\boldsymbol{\mu}_i$ is the mean vector of the ith class feature vector, $\mathbf{\Sigma}_i$ is the ith class feature vector covariance matrix, $|\mathbf{\Sigma}_i|$ is the determinant of the covariance matrix, and n is the dimension of the feature vector $\mathbf{x}$. Substituting Equation (64) into Equation (63) gives the following form of the discriminant function:

$$d_i(\mathbf{x}) = \ln \frac{1}{(2\pi)^{\frac{n}{2}}|\mathbf{\Sigma}_i|^{\frac{1}{2}}} \exp\left[-\frac{1}{2}(\mathbf{x}-\boldsymbol{\mu}_i)^T \mathbf{\Sigma}_i^{-1} (\mathbf{x}-\boldsymbol{\mu}_i)\right] + \ln P(c_i),$$
$$i = 1, 2, \cdots, l \qquad (65)$$

Expanding the logarithm gives:

$$d_i(\mathbf{x}) = -\frac{1}{2}\ln|\mathbf{\Sigma}_i| - \frac{1}{2}(\mathbf{x}-\boldsymbol{\mu}_i)^T \mathbf{\Sigma}_i^{-1}(\mathbf{x}-\boldsymbol{\mu}_i) - \frac{n}{2}\ln 2\pi + \ln P(c_i),$$
$$i = 1, 2, \cdots, l \qquad (66)$$

Finally, eliminating the constant element $\frac{n}{2}\ln 2\pi$ leads to the following expression:

$$d_i(\mathbf{x}) = -\frac{1}{2}\ln|\mathbf{\Sigma}_i| - \frac{1}{2}(\mathbf{x}-\boldsymbol{\mu}_i)^T \mathbf{\Sigma}_i^{-1}(\mathbf{x}-\boldsymbol{\mu}_i) + \ln P(c_i),$$
$$i = 1, 2, \cdots, l \qquad (67)$$

We see that the above discriminant is a quadratic function of the feature vector $\mathbf{x}$ for a given $P(c_i)$ and $\mathbf{\Sigma}_i$. It is called **a quadratic discriminant**. Decision boundaries between classes i and

j along which we have equalities $d_i(\mathbf{x}) = d_j(\mathbf{x})$ are hyperquadratic functions in n-dimensional feature space (hyperspheres, hyperellipsoids, hyperparaboloids, etc.).

A classifier based on the quadratic Bayesian discriminants can be constructed as follows:

Given: A pattern $\mathbf{x}$. Values of state conditional probability densities $p(\mathbf{x}|c_i)$ and the a priori probabilities $P(c_i)$ for all classes $i = 1, 2, \cdots, l$.

(1) Compute values of the mean vectors $\boldsymbol{\mu}_i$ and the covariance matrices $\boldsymbol{\Sigma}_i$ for all classes $l = 1, 2, \cdots, l$ based on the training set.

(2) Compute values of the discriminant function for all classes

$$d_i(\mathbf{x}) = -\frac{1}{2}\ln|\boldsymbol{\Sigma}_i| - \frac{1}{2}(\mathbf{x} - \boldsymbol{\mu}_i)^T \boldsymbol{\Sigma}_i^{-1}(\mathbf{x} - \boldsymbol{\mu}_i) + \ln P(c_i),$$

$$i = 1, 2, \cdots, l \quad (68)$$

(3) Choose a class c_j as a prediction of true class for which a value of the associated discriminant function $d_j(\mathbf{x})$ is largest:

$$\text{Select a class } c_j \text{ for which } d_j(\mathbf{x}) = \max(d_i(\mathbf{x})), \quad i = 1, 2, \cdots, l$$

Result: The predicted class.

Linear Discriminant: Equal Within-Class Covariance Matrices. An important linear discriminant form can be derived assuming equal covariances for all classes $\boldsymbol{\Sigma}_i = \boldsymbol{\Sigma} (i = 1, 2, \cdots, l)$. In this case in the above quadratic discriminant function, the term $\ln|\boldsymbol{\Sigma}_i| = \ln|\boldsymbol{\Sigma}|$ is class independent and thus can be dropped, yielding the discriminant form:

$$d_i(\mathbf{x}) = -\frac{1}{2}(\mathbf{x} - \boldsymbol{\mu}_i)^T \boldsymbol{\Sigma}^{-1}(\mathbf{x} - \boldsymbol{\mu}_i) + \ln P(c_i), \quad i = 1, 2, \cdots, l \quad (69)$$

Let us expand the quadratic form in the above discriminant:

$$\frac{1}{2}(\mathbf{x} - \boldsymbol{\mu}_i)^T \boldsymbol{\Sigma}^{-1}(\mathbf{x} - \boldsymbol{\mu}_i)$$

$$= \frac{1}{2}\mathbf{x}^T \boldsymbol{\Sigma}^{-1}\mathbf{x} - \frac{1}{2}\mathbf{x}^T \boldsymbol{\Sigma}^{-1}\boldsymbol{\mu}_i - \frac{1}{2}\boldsymbol{\mu}_i^T \boldsymbol{\Sigma}^{-1}\mathbf{x} + \frac{1}{2}\boldsymbol{\mu}_i^T \boldsymbol{\Sigma}^{-1}\boldsymbol{\mu}_i$$

Because $\boldsymbol{\Sigma}$ is symmetric, its inverse is also symmetric, and hence $\frac{1}{2}\mathbf{x}^T \boldsymbol{\Sigma}^{-1}\boldsymbol{\mu}_i = \frac{1}{2}\boldsymbol{\mu}_i^T \boldsymbol{\Sigma}^{-1}\mathbf{x}$. In addition, we see that the term $\frac{1}{2}\mathbf{x}^T \boldsymbol{\Sigma}^{-1}\mathbf{x}$ is class independent; therefore, it may be eliminated, yielding a linear form of discriminant functions:

$$d_i(\mathbf{x}) = \boldsymbol{\mu}_i^T \boldsymbol{\Sigma}^{-1}\mathbf{x} - \frac{1}{2}\boldsymbol{\mu}_i^T \boldsymbol{\Sigma}^{-1}\boldsymbol{\mu}_i + \ln P(c_i), \quad i = 1, 2, \cdots, l \quad (70)$$

The above discriminant, as a linear function of a feature vector $\mathbf{x}$, is called a **linear discriminant function**. Decision boundaries between classes i and j, for which $d_i(\mathbf{x}) = d_j(\mathbf{x})$, are pieces of hyperplanes in n-dimensional feature space.

For two-class classification with two-dimensional feature vectors the decision boundary between classes is a line (see Figure 11.4).

The classification process using linear discriminants can be stated as follows:

(1) Compute, for a given $\mathbf{x}$, numerical values of discriminant functions for all classes:

$$d_i(\mathbf{x}) = \boldsymbol{\mu}_i^T \boldsymbol{\Sigma}^{-1}\mathbf{x} - \frac{1}{2}\boldsymbol{\mu}_i^T \boldsymbol{\Sigma}^{-1}\boldsymbol{\mu}_i + \ln P(c_i), \quad i = 1, 2, \cdots, l \quad (71)$$

1. Bayesian Methods

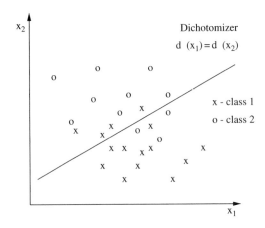

Figure 11.4. Two-class two-feature pattern classification.

(2) Choose a class c_j for which a value of the discriminant function $d_j(\mathbf{x})$ is largest:

Select a class c_j for which $d_j(\mathbf{x}) = \max(d_i(\mathbf{x}))$, $i = 1, 2, \cdots, 1$

Example: Let us assume that the following two-feature patterns $\mathbf{x} \in \mathbb{R}^2$ from two classes $c_1 = 0$ and $c_2 = 1$ have been drawn according to the Gaussian (normal) density distribution:

x_1	x_2	class
1	2	0
2	2	0
2	3	0
3	1	0
3	2	0

x_1	x_2	class
6	8	1
7	8	1
8	7	1
8	8	1
7	9	1

From this set of patterns, we estimate, for each class separately, the mean vectors and covariance matrices. The mean values are equal to:

$$\hat{\boldsymbol{\mu}}_1 = \frac{1}{5}\sum_{i=1}^{5} \mathbf{x}^{\text{class}1,i} = \begin{bmatrix} 2.2 \\ 2.0 \end{bmatrix}$$

$$\hat{\boldsymbol{\mu}}_2 = \frac{1}{5}\sum_{i=1}^{5} \mathbf{x}^{\text{class}2,i} = \begin{bmatrix} 7.2 \\ 8.0 \end{bmatrix}$$

where $\mathbf{x}^{\text{class}j,i}$ denotes the ith pattern from class j. The estimate for the covariance matrix for patterns from class 1 can be computed as

$$\hat{\boldsymbol{\Sigma}}_1 = \frac{1}{5-1}\sum_{i=1}^{N}(\mathbf{x}^{\text{class}1,i} - \hat{\boldsymbol{\mu}}_1)(\mathbf{x}^{\text{class}1,i} - \hat{\boldsymbol{\mu}}_1)^T$$

$$= \frac{1}{4}[\begin{bmatrix} 1-2.2 \\ 2-2 \end{bmatrix}[1-2.2, 2-2] + \begin{bmatrix} 2-2.2 \\ 2-2 \end{bmatrix}[2-2.2, 2-2]$$

$$+ \begin{bmatrix} 2-2.2 \\ 3-2 \end{bmatrix}[2-2.2, 3-2] + \begin{bmatrix} 3-2.2 \\ 1-2 \end{bmatrix}[3-2.2, 1-2]$$

$$+ \begin{bmatrix} 3-2.2 \\ 2-2 \end{bmatrix}[3-2.2, 2-2]]$$

$$= \frac{1}{4}\begin{bmatrix} 2.8 & -1.0 \\ -1.0 & 2.00 \end{bmatrix}$$

$$= \begin{bmatrix} 0.7 & -0.25 \\ -0.25 & 0.5 \end{bmatrix}$$

Similarly, we have

$$\hat{\mathbf{\Sigma}}_2 = \begin{bmatrix} 0.7 & -0.25 \\ -0.25 & 0.5 \end{bmatrix}$$

We see that the estimates of the symmetric covariance matrices for both classes are equal, $\hat{\mathbf{\Sigma}}_1 = \hat{\mathbf{\Sigma}}_2 = \hat{\mathbf{\Sigma}}$, and thus the simpler linear version of discriminants can be used:

$$d_i(\mathbf{x}) = \hat{\boldsymbol{\mu}}_i^T \hat{\mathbf{\Sigma}}^{-1} \mathbf{x} - \frac{1}{2}\hat{\boldsymbol{\mu}}_i^T \hat{\mathbf{\Sigma}}^{-1} \hat{\boldsymbol{\mu}}_i + \ln P(c_i), \quad i = 1, 2$$

We compute

$$\hat{\mathbf{\Sigma}}^{-1} = \begin{bmatrix} 1.7391304 & 0.8695652 \\ 0.8695652 & 2.4347826 \end{bmatrix}$$

$$\hat{\boldsymbol{\mu}}_1^T \hat{\mathbf{\Sigma}}^{-1} = [5.5652174, \quad 6.7826087]$$

$$\hat{\boldsymbol{\mu}}_2^T \hat{\mathbf{\Sigma}}^{-1} = [19.478261, \quad 25.739130]$$

$$\frac{1}{2}\hat{\boldsymbol{\mu}}_1^T \hat{\mathbf{\Sigma}}^{-1} \hat{\boldsymbol{\mu}}_1 = \frac{1}{2} \cdot 25.808696$$

$$\frac{1}{2}\hat{\boldsymbol{\mu}}_2^T \hat{\mathbf{\Sigma}}^{-1} \hat{\boldsymbol{\mu}}_2 = \frac{1}{2} \cdot 346.15652$$

and $\ln P(c_1) = \ln P(c_2) = \ln 0.5 = -0.6931$. This leads to the following linear discriminant functions for both classes:

$$d_1(\mathbf{x}) = 5.5652174 x_1 + 6.7826087 x_2 - 22.9043 - 0.6931$$

$$d_2(\mathbf{x}) = 19.478261 x_1 + 25.739130 x_2 - 173.0833 - 0.6931$$

We can also find a dichotomizer for this two-class classification representing the Bayes decision boundary between the two classes (see Figure 11.5):

$$d(\mathbf{x}) = d_1(\mathbf{x}) - d_2(\mathbf{x}) = 0$$

$$d(\mathbf{x}) = -13.9130 x_1 - 18.9565 x_2 + 150.1790 = 0$$

or

$$x_2 = -0.7339 x_1 + 7.9219$$

which is (for equal covariance matrices) a line separating clusters of patterns belonging to the two classes. Patterns "above" ($d(\mathbf{x}) < 0$) the decision boundary $d(\mathbf{x})$ are classified as belonging to the class c_2 and pattern "below" ($d(\mathbf{x}) > 0$) as belonging to the class c_1 (see Figure 11.5). We see that the data patterns do not overlap and that ideal classification of these training patterns is possible by the linear discriminants and linear decision boundary.

Let us test how the designed linear discriminant classifies a new pattern $\mathbf{x}^{11} = [4, 1]^T$.

1. Bayesian Methods

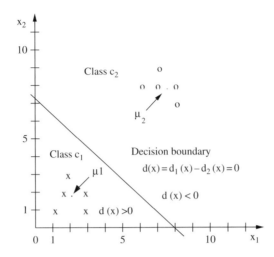

Figure 11.5. Two-class two-feature pattern dichotomizer.

We compute values of two discriminants for two classes, namely, $d_1([4, 1]^T) = 5.4461$ and $d_2([4, 1]^T) = -51.3416$. Then we assign $\mathbf{x}^{11}$ to class c_1, since $d_1([4, 1]^T) > d_2([4, 1]^T)$. For the other new pattern $\mathbf{x}^{12} = [6, 7]^T$, we have $d_1([6, 7]^T) = 57.2722$ and $d_2([6, 7]^T) = 123.2671$. Then we assign the new pattern $\mathbf{x}^{12} = [6, 7]^T$ to the class c_2, since $d_1([6, 7]^T) < d_2([6, 7]^T)$.

We can also use a dichotomizer for classifying new patterns. For the pattern $\mathbf{x}^{11} = [4, 1]^T$, we have $d([4, 1]^T) = 75.5705 > 0$, indicating the class c_1 (the pattern "below" the decision boundary). For $\mathbf{x}^{12} = [6, 7]^T$, the dichotomizer value is $d([6, 7]^T) = -65.9945 < 0$, indicating the class c_2 (the pattern located "above" the decision boundary).

The linear discriminant function can also be presented in the form of a linear neural network, that is,

$$d_i(\mathbf{x}) = \mathbf{w}_i^T \mathbf{x} + w_{i0}, \quad i = 1, 2, \cdots, l \tag{72}$$

where the weight vector is defined as

$$\mathbf{w}_i = \mathbf{\Sigma}^{-1} \boldsymbol{\mu}_i \tag{73}$$

and the threshold is determined by

$$w_{i0} = -\boldsymbol{\mu}_i^T \mathbf{\Sigma}^{-1} \boldsymbol{\mu}_i + \ln P(c_i) \tag{74}$$

For contiguous regions R_i and R_j for classes i and j, the decision boundary between these classes is governed by the linear relationship

$$d_i(\mathbf{x}) - d_j(\mathbf{x}) = \mathbf{b}\mathbf{x} + w_{ij0} \tag{75}$$

where

$$\mathbf{b} = (\mathbf{w}_i - \mathbf{w}_j)^T = (\boldsymbol{\mu}_i \mathbf{\Sigma}^{-1} - \boldsymbol{\mu}_j \mathbf{\Sigma}^{-1})^T \tag{76}$$

$$w_{ij0} = w_{i0} - w_{j0}$$
$$= -\boldsymbol{\mu}_i^T \mathbf{\Sigma}^{-1} \boldsymbol{\mu}_i + \ln P(c_i) + \boldsymbol{\mu}_j^T \mathbf{\Sigma}^{-1} \boldsymbol{\mu}_j - \ln P(c_j) \tag{77}$$

The hyperplane, separating classes i and j, is generally not orthogonal to the line passing through the means $\boldsymbol{\mu}_i$ and $\boldsymbol{\mu}_j$. For nonequal a priori probabilities $P(c_i)$, the separating hyperplane is translated away from the more likely mean.

Minimum Mahalanobis Distance Classifier. If we assume equal covariance matrices for all classes $\Sigma_i = \Sigma (i = 1, 2, \cdots, l)$, and equal a priori probabilities for all classes $P(c_i) = P$, then we can ignore in the discriminant equation the term $\ln P(c_i)$. This gives a simple form of the discriminant (after neglecting $\frac{1}{2}$):

$$d_i(\mathbf{x}) = -(\mathbf{x} - \boldsymbol{\mu}_i)^T \Sigma^{-1}(\mathbf{x} - \boldsymbol{\mu}_i), \quad i = 1, 2, \cdots, l \tag{78}$$

We observe that a classifier based on the above discriminants assigns an object with a given value $\mathbf{x}$ of a feature vector to a class j for which the squared Mahalanobis distance $(\mathbf{x} - \boldsymbol{\mu}_j)^T \Sigma^{-1}(\mathbf{x} - \boldsymbol{\mu}_j)$ of $\mathbf{x}$ to the mean vector $\boldsymbol{\mu}_j$ is smallest (because this approach gives the largest value of discriminant). In other words, a classifier selects the class c_j for which a value $\mathbf{x}$ is nearest, in the sense of Mahalanobis distance, to the corresponding mean vector $\boldsymbol{\mu}_j$. This classifier is called **a minimum Mahalanobis distance classifier**.

We find the following linear version of the minimum Mahalanobis distance classifier (by expanding the squared Mahalanobis distance):

$$d_i(\mathbf{x}) = \boldsymbol{\mu}_i^T \Sigma^{-1} \mathbf{x} - \frac{1}{2} \boldsymbol{\mu}_i^T \Sigma^{-1} \boldsymbol{\mu}_i, \quad i = 1, 2, \cdots, l \tag{79}$$

The above discriminant is a linear function of a feature vector $\mathbf{x}$.

A minimum Mahalanobis distance classifier algorithm can be constructed as follows.

Given: The mean vectors for all classes $\boldsymbol{\mu}_i (i = 1, 2, \cdots, l)$ and a given value $\mathbf{x}$ of a feature vector.

(1) Compute numerical values of the Mahalanobis distances $(\mathbf{x} - \boldsymbol{\mu}_i)^T \Sigma^{-1}(\mathbf{x} - \boldsymbol{\mu}_i) (i = 1, 2, \cdots, l)$ between $\mathbf{x}$ and means $\boldsymbol{\mu}_i$ for all classes.
(2) Choose a class c_j as a prediction of true class, for which the value of the associated Mahalanobis distance attains the minimum:

$$(\mathbf{x} - \boldsymbol{\mu}_j)^T \Sigma^{-1}(\mathbf{x} - \boldsymbol{\mu}_j) = \min_i (\mathbf{x} - \boldsymbol{\mu}_i)^T \Sigma^{-1}(\mathbf{x} - \boldsymbol{\mu}_i), \quad i = 1, 2, \cdots, l \tag{80}$$

Result: The predicted class.

For a minimum Mahalanobis distance classifier, the mean vectors act as l prototypes for l classes, and a new feature vector value $\mathbf{x}$ is matched to the prototypes. The predicted class is assigned to a class with the nearest mean to $\mathbf{x}$ in the sense of the Mahalanobis distance.

Linear Discriminant for Statistically Independent Features. The discriminant function can be simplified if we assume that the covariance matrices for all classes are equal and that features are statistically independent. In this case, each feature has the same variance Σ^2, and the covariance matrix becomes diagonal:

$$\Sigma = \Sigma^2 \mathbf{I} \tag{81}$$

where $\mathbf{I}$ is an $n \times n$ identity matrix (having ones only on the main diagonal, and zeros elsewhere else). We see that $|\Sigma_i| = \sigma^{2n}$ and $\Sigma_i^{-1} = \sigma^{-2} \mathbf{I}$, which leads to the simple form of the discriminant:

$$d_i(\mathbf{x}) = -\frac{\|\mathbf{x} - \boldsymbol{\mu}_i\|^2}{2\Sigma^2} + \ln P(c_i), \quad i = 1, 2, \cdots, l \tag{82}$$

where $\|\cdot\|$ denotes the Euclidean norm

$$\|\mathbf{x} - \boldsymbol{\mu}_i\| = \sqrt{(\mathbf{x} - \boldsymbol{\mu}_i)^T (\mathbf{x} - \boldsymbol{\mu}_i)} \tag{83}$$

1. Bayesian Methods

Notice that in the above discriminant the squared distance between a feature vector and a mean vector is normalized by the variance σ^2 and biased by $P(c_i)$. Thus, we find that, for a feature vector value $\mathbf{x}$ equally distant from the two mean vectors of two different classes, the class with the bigger a priori probability $P(c_j)$ will be chosen.

In fact, we do not need to compute the Euclidean distances for the above discriminants. Note that

$$(\mathbf{x}-\boldsymbol{\mu}_i)^T(\mathbf{x}-\boldsymbol{\mu}_i) = \mathbf{x}^T\mathbf{x} - 2\boldsymbol{\mu}_i^T\mathbf{x} + \boldsymbol{\mu}_i^T\boldsymbol{\mu}_i \tag{84}$$

which yields the quadratic discriminant formula

$$d_i(\mathbf{x}) = -\frac{1}{2\sigma^2}\left[\mathbf{x}^T\mathbf{x} - 2\boldsymbol{\mu}_i^T\mathbf{x} + \boldsymbol{\mu}_i^T\boldsymbol{\mu}_i\right] + \ln P(c_i), \quad i=1,2,\cdots,l \tag{85}$$

However, since the term $\mathbf{x}^T\mathbf{x}$ is the same for all classes, it can be dropped, which gives the linear discriminant formula

$$d_i(\mathbf{x}) = \frac{1}{\sigma^2}\boldsymbol{\mu}_i^T\mathbf{x} - \frac{1}{2\sigma^2}\boldsymbol{\mu}_i^T\boldsymbol{\mu}_i + \ln P(c_i), \quad i=1,2,\cdots,l \tag{86}$$

This linear discriminant can be written in the "neural network" style as a linear threshold machine:

$$d_i(\mathbf{x}) = \mathbf{w}_i^T\mathbf{x} + w_{i0} \tag{87}$$

where the weight vector is defined as

$$\mathbf{w}_i = \frac{1}{\sigma^2}\boldsymbol{\mu}_i \tag{88}$$

and the threshold weight (a bias in the ith direction) is

$$w_{i0} = -\frac{1}{2\sigma^2}\boldsymbol{\mu}_i^T\boldsymbol{\mu}_i + \ln P(c_i) \tag{89}$$

The decision surfaces for the above linear discriminants are pieces of hyperplanes defined by equations $d_i(\mathbf{x}) - d_j(\mathbf{x})$. For contiguous regions R_i and R_j for classes i and j, with the highest a posteriori probabilities, the decision boundaries between these classes are governed by the linear equation

$$d_i(\mathbf{x}) - d_j(\mathbf{x}) = \mathbf{b}\mathbf{x} + w_{ij0} \tag{90}$$

where

$$\mathbf{b} = (\mathbf{w}_i - \mathbf{w}_j)^T = \frac{1}{\sigma^2}(\boldsymbol{\mu}_i - \boldsymbol{\mu}_j)^T \tag{91}$$

and

$$w_{ij0} = -\frac{1}{2\sigma^2}\left(\boldsymbol{\mu}_i^T\boldsymbol{\mu}_i - \boldsymbol{\mu}_j^T\boldsymbol{\mu}_j\right) + \ln P(c_i) - \ln P(c_j) \tag{92}$$

The above decision hyperplane is orthogonal to the vector $\boldsymbol{\mu}_i - \boldsymbol{\mu}_j$ (orthogonal to the line linking means $\boldsymbol{\mu}_i$ and $\boldsymbol{\mu}_j$ of classes i and j). For equal a priori probabilities $P(c_i)$ and $P(c_j)$, the intersection point of the line joining the means and the decision hyperplane is situated in the middle of the segment joining the two means $\boldsymbol{\mu}_i$ and $\boldsymbol{\mu}_j$. For nonequal a priori probabilities $P(c_i)$, the intersection point is away from the larger probability (toward the less likelihood mean). We see that for a small variance $\boldsymbol{\Sigma}^2$, compared with a distance between means $\|\boldsymbol{\mu}_i - \boldsymbol{\mu}_j\|$, the influence of a priori probabilities on the position of the decision boundary is small.

Minimum Euclidean Distance Classifier. If we can assume that the covariance matrices for all classes are equal and that features are statistically independent, then we also have $\boldsymbol{\Sigma} = \boldsymbol{\Sigma}^2 \mathbf{I}$. If, additionally, we have equal a priori probabilities for all classes $P(c_i) = P$, then we can ignore the term $\ln P(c_i)$. This produces a straightforward form of discriminants:

$$d_i(\mathbf{x}) = -\frac{\|\mathbf{x} - \boldsymbol{\mu}_i\|^2}{2\sigma^2}, \quad i = 1, 2, \cdots, l \tag{93}$$

or (neglecting the term $\boldsymbol{\Sigma}^2$, which is unimportant for discrimination):

$$d_i(\mathbf{x}) = -2\sigma^2, \quad i = 1, 2, \cdots, l \tag{94}$$

We see that a classifier based on the above discriminants assigns an object with a given value $\mathbf{x}$ of a feature vector to a class j for which the Euclidean distance $\|\mathbf{x} - \boldsymbol{\mu}_j\|$ of $\mathbf{x}$ to the mean vector $\boldsymbol{\mu}_j$ is the smallest (because this gives the largest value of the discriminant). In other words, a classifier selects the class c_j of which a value $\mathbf{x}$ is nearest to the corresponding mean vector $\boldsymbol{\mu}_j$. This classifier is called **a minimum distance classifier or a minimum Euclidean distance classifier**.

We also find the following linear version of the minimum distance classifier (expanding the Euclidean norm and neglecting the term $\boldsymbol{\Sigma}^2$):

$$d_i(\mathbf{x}) = \boldsymbol{\mu}_i^T \mathbf{x} - \boldsymbol{\mu}_i^T \boldsymbol{\mu}_i, \quad i = 1, 2, \cdots, l \tag{95}$$

The above minimum Euclidean distance discriminant is a linear function of a feature vector $\mathbf{x}$ (a linear discriminant, a linear classifier).

Below we give a pseudocode for a minimum Euclidean distance classifier.

Given: Mean vectors for all classes $\boldsymbol{\mu}_i (i = 1, 2, \cdots, l)$ and a given value $\mathbf{x}$ of a feature vector.

(1) Compute numerical values of Euclidean distances between $\mathbf{x}$ and means $\boldsymbol{\mu}_i$ for all classes:

$$\|\mathbf{x} - \boldsymbol{\mu}_i\| = \sqrt{\sum_{k=1}^{n} (x_k - \mu_{k,i})^2}, \quad i = 1, 2, \cdots, l \tag{96}$$

(2) Choose a class c_j as a prediction of true class for which a value of the associated Euclidean distance is smallest:

$$\|\mathbf{x} - \boldsymbol{\mu}_j\| = \min \|\mathbf{x} - \boldsymbol{\mu}_i\|, \quad i = 1, 2, \cdots, l \tag{97}$$

Result: The predicted class.

Limitations of Bayesian Normal Discriminant. The use of the assumptions of multivariate normality within classes, and equal covariance matrices between classes, leads to the linear discriminant form of the Bayes rule. If the assumptions are true, the linear discriminant is equivalent to the optimal classifier from the point of view of minimal probability of the classification error. Obviously, in real-world situations, these assumptions are satisfied only approximately. However, even if the assumptions are not valid, the linear discriminant has shown good performance. Due to its simple structure, the linear discriminant tends not to overfit the training data set, which may lead to stronger generalization ability for unseen cases (see Chapter 15).

1.8. Estimation of Probability Densities

We can design an optimal Bayesian classifier, in the sense of minimizing the probability of classification error, by knowing the priori probabilities $P(c_i)$ and class conditional probability densities $p(\mathbf{x}|c_i)$ for all classes $c_i (i = 1, 2, \cdots, l)$. In practice, we almost never have sufficient knowledge about the probabilistic characteristics of the real patterns under consideration. This knowledge is usually very vague, uncertain, and incomplete. Our partial knowledge about the probabilistic nature of recognized objects is derived from a limited, finite set of observed a priori patterns/objects (a limited sample).

Due to the limited number of a priori observed objects, we try to estimate a priori probabilities and to state conditional probability densities. This estimation should be optimal according to the well-defined estimation criterion. The computed estimates can then be further used instead of true probabilities in Bayesian classifier design. However, we should remember that estimates only approximately represent true probabilities, and therefore these estimates may be a source of classification uncertainty.

Estimates of a priori probabilities can be expressed as the following ratios:

$$\hat{P}(c_i) = \frac{n_{c_i}}{N}, \quad i = 1, 2, \cdots, l \tag{98}$$

where n_{c_i} is the number of objects in the ith class of a limited sample, and N is the total number of objects in a sample.

In Bayesian processing, the estimation of the class conditional probability densities $p(\mathbf{x}|c_i)$ is a major concern. Three main approaches are used in probability density estimation.

– **Parametric methods** (with the assumption of a specific functional form of a probability density function).
– **Nonparametric methods** (without the assumption of a specific functional form of a probability density function).
– **Semiparametric method** (a combination of parametric and nonparametric methods).

Parametric methods assume that a specific functional form (for example, a specific algebraic formula) of the probability density function with a certain number of parameters is given as a model of the probability density. Thus, the task of estimation in parametric methods is to estimate the optimal values of parameters.

Nonparametric methods do not assume a specific functional form of the probability density function, and this function is derived entirely from given data.

The semiparametric method is a combination of parametric and nonparametric methods, and assumes a very general functional form of probability densities and a possibly variable number of adaptively adjusted parameters.

The estimation of the probability density can be done in both supervised and unsupervised modes. Probability density estimation techniques also include neural network methods.

1.8.1. Parametric Methods of Probability Density Estimation

Let us assume that we have a priori observations of objects and corresponding patterns as a limited-size sample containing N patterns belonging to l classes $c_i (i = 1, 2, \cdots, l)$:

$$X = \{\mathbf{x}^1, \mathbf{x}^2, \cdots, \mathbf{x}^N\} \tag{99}$$

Having patterns labeled by the corresponding classes, we can split the set of all patterns X according to a class into l disjoint sets:

$$X_{c_1}, X_{c_2}, \cdots, X_{c_l}, \quad \bigcup_{i=1,2,\cdots,c_l} X_i = X \tag{100}$$

Each set X_{c_i} contains N_i patterns belonging to class c_i. We assume that patterns from each set X_{c_i} have been drawn independently according to the class conditional probability density function $p(\mathbf{x}|c_i)$ ($i = 1, 2, \cdots, l$). Assume that the parametric form of the class conditional probability density is given as a function

$$p(\mathbf{x}|c_i, \boldsymbol{\theta}_i) \qquad (101)$$

where $\boldsymbol{\theta}_i = (\theta_{i1}, \theta_{i2}, \cdots, \theta_{im})^T$ is the m-dimensional vector of parameters for the ith class. For example, if the probability density has a normal (Gaussian) form:

$$p(\mathbf{x}|c_i) = p(\mathbf{x}|c_i, \boldsymbol{\theta}_i) = \frac{1}{(2\pi)^{\frac{n}{2}}|\boldsymbol{\Sigma}_i|^{\frac{1}{2}}} \exp\left[-\frac{1}{2}(\mathbf{x}-\boldsymbol{\mu}_i)^T \boldsymbol{\Sigma}_i^{-1}(\mathbf{x}-\boldsymbol{\mu}_i)\right] \qquad (102)$$

then the parameter vector can be constituted of the elements of the ith class mean vector $\boldsymbol{\mu}_i$ and the covariance matrix elements $\boldsymbol{\Sigma}_i$:

$$\boldsymbol{\theta}_i = (\boldsymbol{\mu}_i^T, \boldsymbol{\Sigma}_{i1}, \boldsymbol{\Sigma}_{i2}, \cdots, \boldsymbol{\Sigma}_{in})^T \qquad (103)$$

where $\boldsymbol{\Sigma}_{ij}$ denotes the jth row of a covariance matrix $\boldsymbol{\Sigma}_i$. For the one-dimensional feature vector for the normal distribution, the parameter vector contains a mean and a variance:

$$\boldsymbol{\theta}_i = \begin{bmatrix} \mu_i \\ \Sigma_i^2 \end{bmatrix} \qquad (104)$$

Since we assumed that patterns for each class were generated independently, the parameter estimation of probability density can be provided in a similar way, independently for each class. We may simplify the notation assuming that we estimate the probability density of a certain class. This allows us to omit the class index, denoting $p(\mathbf{x}|c_i, \boldsymbol{\theta}_i)$ by $p(\mathbf{x}|\boldsymbol{\theta})$.

1.8.2. The Maximum Likelihood Estimation of Parameters

The maximum likelihood optimal parameter estimation problem can be stated as follows. Assume that we are given a limited-size set of N patterns $\mathbf{x}^i$ for a given class generated independently:

$$X = \{\mathbf{x}^1, \mathbf{x}^2, \cdots, \mathbf{x}^N\} \qquad (105)$$

We also assume that we know a parametric form $p(\mathbf{x}|\boldsymbol{\theta})$ of a conditional probability density function. The task of estimation is to find the optimal (the best according to the used criterion) value of the parameter vector $\boldsymbol{\theta}$ of a given dimension m.

Since patterns $\mathbf{x}^i$ are drawn independently according to distribution $p(\mathbf{x}|\boldsymbol{\theta})$, the joint probability density of all patterns from the data set X can be written as a product of the probabilities for each pattern separately:

$$L(\boldsymbol{\theta}) = p(X|\boldsymbol{\theta}) = \prod_{i=1}^{N} p(\mathbf{x}^i|\boldsymbol{\theta}) \qquad (106)$$

The probability density $L(\boldsymbol{\theta})$ is a function of a parameter vector $\boldsymbol{\theta}$ for a given set of patterns X. It is called the **likelihood** of θ for a given set of patterns X.

The function $L(\boldsymbol{\theta})$ can be chosen as a criterion for finding the optimal estimate of $\boldsymbol{\theta}$. Given this optimal criterion, such an estimation is called the **maximum likelihood estimation** of parameters $\boldsymbol{\theta}$. In this technique, the optimal value $\hat{\boldsymbol{\theta}}$ of a parameter vector is chosen to maximize the criterion $L(\boldsymbol{\theta})$. This can be viewed as choosing the $\boldsymbol{\theta}$ that most likely matches a given observed pattern set.

1. Bayesian Methods

For computational reasons, in maximizing the likelihood estimation criterion, it is more convenient to minimize the negative natural logarithm of the likelihood $L(\boldsymbol{\theta})$:

$$J(\boldsymbol{\theta}) = -\ln L(\boldsymbol{\theta}) = -\sum_{i=1}^{N} \ln p(\mathbf{x}^i|\boldsymbol{\theta}) \qquad (107)$$

and the optimal estimate is chosen by minimizing $J(\boldsymbol{\theta})$ (corresponding to the maximizing likelihood $L(\boldsymbol{\theta})$). However, since the negative $\ln$ is a monotonic function, both criteria are equivalent.

For the differentiable function $p(\mathbf{x}^i|\boldsymbol{\theta})$, the minimum value of $\boldsymbol{\theta}$ can be found from the set of m equations (obtained by making equal to $\mathbf{0}$ the gradient of $J(\boldsymbol{\theta})$ for $\boldsymbol{\theta}$):

$$\frac{\partial}{\partial \boldsymbol{\theta}} J(\boldsymbol{\theta}) = \begin{bmatrix} \frac{\partial J(\boldsymbol{\theta})}{\partial \theta_1} \\ \frac{\partial J(\boldsymbol{\theta})}{\partial \theta_2} \\ \vdots \\ \frac{\partial J(\boldsymbol{\theta})}{\partial \theta_m} \end{bmatrix} = \frac{\partial}{\partial \boldsymbol{\theta}} \left[-\sum_{i=1}^{N} \ln p(\mathbf{x}^i|\boldsymbol{\theta}) \right]$$

$$= \begin{bmatrix} \frac{\partial [-\sum_{i=1}^{N} \ln p(\mathbf{x}^i|\boldsymbol{\theta})]}{\partial \theta_1} \\ \frac{\partial [-\sum_{i=1}^{N} \ln p(\mathbf{x}^i|\boldsymbol{\theta})]}{\partial \theta_2} \\ \vdots \\ \frac{\partial [-\sum_{i=1}^{N} \ln p(\mathbf{x}^i|\boldsymbol{\theta})]}{\partial \theta_m} \end{bmatrix} = \mathbf{0} \qquad (108)$$

For the normal form of a probability density function $N(\boldsymbol{\mu}, \boldsymbol{\Sigma})$ with unknown parameters $\boldsymbol{\mu}$ and $\boldsymbol{\Sigma}$ constituting vector $\boldsymbol{\theta}$, we can find a close algebraic formula for maximum likelihood estimates of the parameter vector:

$$\hat{\boldsymbol{\mu}} = \frac{1}{N} \sum_{i=1}^{N} \mathbf{x}^i \qquad (109)$$

$$\hat{\boldsymbol{\Sigma}} = \frac{1}{N} \sum_{i=1}^{N} (\mathbf{x}^i - \hat{\boldsymbol{\mu}})(\mathbf{x}^i - \hat{\boldsymbol{\mu}})^T \qquad (110)$$

These equations show that the maximum likelihood estimate $\hat{\boldsymbol{\mu}}$ for the mean $\boldsymbol{\mu}$ is the sample mean (the average for a given set of patterns). Similarly, the maximum likelihood estimate $\hat{\boldsymbol{\Sigma}}$ of the covariance matrix $\boldsymbol{\Sigma}$ is the sample arithmetic average of the N matrices (outer products) $(\mathbf{x}^i - \hat{\boldsymbol{\mu}})(\mathbf{x}^i - \hat{\boldsymbol{\mu}})^T$.

We show how to find the estimates for the normal form of the probability density

$$p(x) = \frac{1}{(2\pi\sigma^2)^{\frac{1}{2}}} \exp\left[-\frac{1}{2\sigma^2}(x-\mu)^2\right] \qquad (111)$$

for a one-dimensional pattern vector x with two unknown parameters $\theta_1 = \mu$ and $\theta_2 = \Sigma^2$ forming a two-dimensional feature vector $\boldsymbol{\theta}$:

$$\boldsymbol{\theta} = \begin{bmatrix} \theta_1 \\ \theta_2 \end{bmatrix} \begin{bmatrix} \mu \\ \sigma^2 \end{bmatrix} \qquad (112)$$

The maximum likelihood estimation criterion in this case can be written as

$$J(\boldsymbol{\theta}) = \frac{1}{N} \sum_{i=1}^{N} \left[\frac{1}{2} \ln 2\pi\theta_2 + \frac{1}{2\theta_2}(x^i - \theta_1)^2 \right] \qquad (113)$$

The gradient of this criterion with respect to $\boldsymbol{\theta}$ equal to zero leads to the solution of the maximum likelihood estimates for the parameters:

$$\hat{\theta}_1 = \hat{\boldsymbol{\mu}} = \frac{1}{N}\sum_{i=1}^{N} x^i \tag{114}$$

$$\hat{\theta}_2 = \hat{\sigma}^2 = \frac{1}{N}\sum_{i=1}^{N}(x^i - \hat{\boldsymbol{\mu}})^2 \tag{115}$$

These results can be similarly extended for the multivariate case.

Theory shows that the maximum likelihood estimate of a covariance matrix $\boldsymbol{\Sigma}$ is biased, i.e., the expected value of $\hat{\boldsymbol{\Sigma}}$ is not equal to $\boldsymbol{\Sigma}$. For example, for the one-dimensional pattern, the expected value of $\boldsymbol{\Sigma}$,

$$E[\hat{\sigma}] = \frac{N-1}{N}\sigma^2 \tag{116}$$

(where $\boldsymbol{\Sigma}^2$ is the true variance) is not equal to σ. For $N \to \infty$, the bias disappears. The unbiased estimate can be obtained in the form

$$\hat{\boldsymbol{\Sigma}} = \frac{1}{N-1}\sum_{i=1}^{N}(\mathbf{x}^i - \hat{\boldsymbol{\mu}})(\mathbf{x}^i - \hat{\boldsymbol{\mu}})^T \tag{117}$$

1.8.3. Nonparametric Methods of Probability Density Estimation

In many problems, the functional form of probability density function (for example, the formula for normal distribution) is not known in advance. More general methods of probability density estimation seek, based on existing data, the functional form of the probability density function as well as numerical values of its parameters. Some techniques, like histograms, do not provide the close algebraic formula for density, but do provide precomputed numerical values for densities. Neural networks are also used for density estimation. These estimation methods are called **nonparametric**.

We will introduce the following nonparametric techniques: histogram, kernel-based method, k-nearest neighbors, and nearest neighbors.

General Idea of Nonparametric Probability Density Estimation. Suppose that a limited set of N samples (patterns) $X = \{\mathbf{x}^1, \mathbf{x}^2, \cdots, \mathbf{x}^N\}$ is independently generated for a given class according to an unknown probability density function $p(\mathbf{x})$. Our intent is to determine an estimate $\hat{p}(\mathbf{x})$ of a true probability density $p(\mathbf{x})$ based on the available limited-size sample.

The probability that a new pattern $\mathbf{x}$, drawn according to the unknown probability density $p(\mathbf{x})$, will fall inside a region R is given by

$$P = \int_{\tilde{\mathbf{x}} \in R} p(\tilde{\mathbf{x}})\, d\tilde{\mathbf{x}} \tag{118}$$

For a small region and for continuous $p(\mathbf{x})$, with almost the same values within a region R, we can (for a pattern lying inside R) write the following approximation:

$$P = \int_{\tilde{\mathbf{x}} \in R} p(\tilde{\mathbf{x}})\, d\tilde{\mathbf{x}} \approx p(\mathbf{x})V \tag{119}$$

where V is the volume of a region R ($V = \int_{\tilde{\mathbf{x}} \in R} d\tilde{\mathbf{x}}$). Based on this approximation, we see that we can estimate $p(\mathbf{x})$ for a given $\mathbf{x}$ knowing the probability P that $\mathbf{x}$ will fall in R.

To find P, we first note that the probability that for N sample patterns set k of them will fall in a region R is given by the binomial law

$$P_{k \text{ from } N} = \frac{N!}{k!(N-k)!} P^k (1-P)^{N-k} \quad (120)$$

The mean of the fraction of patterns k/N falling into a region R is $E[k/N] = P$, with variance around the mean $E[(k/N - P)^2] = P(1-P)/N$. Because the binomial distribution $P_{k \text{ from } N}$ for k sharply peaks around a mean for $N \to \infty$, we assume that the ratio k/N is a sound estimate of the probability P:

$$P \approx \frac{k}{N} \quad (121)$$

Combining Equations (119) and (120), we can find the following approximation for a probability density function for a given pattern $\mathbf{x}$:

$$p(\mathbf{x}) \approx \hat{p} = \frac{k}{NV} \quad (122)$$

To endure that the above region-based probability density estimation will converge to true $p(\mathbf{x})$, we need a decision about the proper selection of region R. The choice of R should be an optimal compromise between contradicting requirements. First, in order to be sure that the approximation $P \approx k/N$ holds, region R must be large, causing P to be large and thus ensuring that the binomial distribution will sharply peak around a mean. Second, the approximation $P = p(\mathbf{x})V$ is more accurate if R (and thus V) is smaller, guaranteeing that $p(\mathbf{x})$ remains almost constant in the integration region. The best estimate requires finding an optimal region R.

There are two major techniques for probability density estimation based on region selection and counting the number of patterns falling in the region:

– Kernel-based methods
– K-nearest neighbors methods

In the kernel-based method, region R (and thus the volume V) is fixed and a number k of patterns falling in the region is counted from the data.

In the k-nearest neighbors method, the number k of patterns is predetermined. Based on this value, a region is determined from the data with a corresponding volume. It has been shown that both techniques converge to the true probability density in the limit of infinite N ($N \to \infty$), provided that the region volume shrinks with increasing N, and k grows with N.

Both techniques require training data to be present during the estimation of the probability density value $p(\mathbf{x})$ for a new pattern $\mathbf{x}$.

Kernel-based Method and Parzen Window. One of the most straightforward techniques of probability density estimation $p(\mathbf{x})$ for a new pattern $\mathbf{x}$, based on a given training set T_{tra} containing N patterns $\mathbf{x}^1, \mathbf{x}^2, \cdots, \mathbf{x}^N$, is called the **kernel-based method**. This method is based on fixing around a pattern vector $\mathbf{x}$ a region R (and thus a region volume V) and counting a number k of given training patterns falling in this region by using a special kernel function associated with the region. Such a kernel function is also called a **Parzen window**.

One choice of a fixed region for an n-dimensional pattern $\mathbf{x}$ is an n-dimensional hypercube with side length equal to h, centered around point $\mathbf{x}$. The volume of this hypercube is

$$V = h^n \quad (123)$$

We find an analytic formula for calculating the number k of patterns from a training set falling within the hypercube, with the help of the following **kernel (window) function**:

$$\psi(\mathbf{y}) = \begin{cases} 1 & \text{for } |y_i| \leq 1/2, \ i = 1, 2, \cdots, n \\ 0 & \text{otherwise} \end{cases} \quad (124)$$

This kernel function (a **hypercube-type Parzen window**) corresponds to a unit hypercube (with a side length $h = 1$) centered at the origin. It allows us to decide whether a given pattern falls inside the unit hypercube. This decision may be easily extended for the hypercube with side h. We see that for the pattern $\mathbf{x}^i$, the value of kernel function $\psi((\mathbf{x} - \mathbf{x}^i)/h)$ is equal to 1 if a pattern $\mathbf{x}^i$ falls within the hypercube with a side h centered on the point $\mathbf{x}$, and it is equal to 0 otherwise.

For an N-sample-pattern set, the total number of patterns falling within the hypercube centered around $\mathbf{x}$ is given by

$$k = \sum_{i=1}^{N} \psi\left(\frac{\mathbf{x} - \mathbf{x}^i}{h}\right) \quad (125)$$

Substituting the above into Equation (121) gives the estimate of the probability density function for a given state $\mathbf{x}$ based on a given training set:

$$\hat{p}(\mathbf{x}) = \frac{1}{N} \sum_{i=1}^{N} \frac{1}{h^n} \psi\left(\frac{\mathbf{x} - \mathbf{x}^i}{h}\right) = \frac{1}{N h^n} \sum_{i}^{N} \psi\left(\frac{\mathbf{x} - \mathbf{x}^i}{h}\right) \quad (126)$$

The hypercube-type kernel function method resembles the histogram technique for a probability density estimation. However, instead of fixed bins defined in advance, the kernel method locates hypercube cells at new pattern points. The hypercube kernel methods, like a histogram, suffer from a discontinuity of the estimate on boundaries of hypercubes. The kernel-based methods, in general, require that a training set of N patterns be available for the density estimate of new patterns as a reference pattern set.

In order to obtain a smooth estimate of a probability density, we use other forms of a kernel function $\psi(\mathbf{x})$. A kernel function must satisfy two conditions:

$$\psi(\mathbf{y}) \geq 0 \quad (127)$$

$$\int_{\text{all } \mathbf{y}} \psi(\mathbf{y}) \, d\mathbf{y} = 1$$

One of the natural choices can be the radial symmetric multivariate Gaussian (normal) kernel:

$$\psi(\mathbf{y}) = \frac{1}{(2\pi)^{n/2}} \left(\exp -\frac{\|\mathbf{y}\|^2}{2}\right) \quad (128)$$

where $\|\mathbf{y}\| = \sqrt{\mathbf{y}^T \mathbf{y}}$, which for $\mathbf{y} = \frac{\mathbf{x} - \mathbf{x}^i}{h}$ gives the following estimate:

$$\hat{p}(\mathbf{x}) = \frac{1}{(2\pi)^{n/2} h^n N} \sum_{i=1}^{N} \exp\left(\frac{-\|\mathbf{x} - \mathbf{x}^i\|^2}{2h^2}\right)$$

$$= \frac{1}{(2\pi)^{n/2} h^n N} \sum_{i=1}^{N} \exp\left(\frac{-(\mathbf{x} - \mathbf{x}^i)^T (\mathbf{x} - \mathbf{x}^i)}{2h^2}\right) \quad (129)$$

The term $\|\mathbf{x} - \mathbf{x}^i\| = ((\mathbf{x} - \mathbf{x}^i)^T (\mathbf{x} - \mathbf{x}^i))^{1/2}$ is the Euclidean distance between patterns $\mathbf{x}$ and $\mathbf{x}^i$.

If the class-dependent $p(\mathbf{x}|c_k)$ probability density is estimated, then in the simple approach, the designer has to separate from the training set all patterns belonging to the same class c_k and use only these patterns to compute an estimate. Let us denote the number of patterns from the kth class by N_k, and the ith reference pattern from the training set from the class c_k by $\mathbf{x}^{k,i}$. Then the class conditional probability density is given by the expression

$$\hat{p}(\mathbf{x}|c_k) = \frac{1}{N_i} \sum_{i=1}^{N_k} \frac{1}{h_k^n} \psi\left(\frac{\mathbf{x} - \mathbf{x}^{k,i}}{h_k}\right) \tag{130}$$

where h_k is the class-dependent smoothing parameter. The probability density is computed separately for each class. For the Gaussian kernel, we have

$$\hat{p}(\mathbf{x}|c_k) = \frac{1}{(2\pi)^{n/2} h_k^n N_k} \sum_{i=1}^{N_k} \exp\left(\frac{-\|\mathbf{x} - \mathbf{x}^{k,i}\|^2}{2h_k^2}\right) \tag{131}$$

The design problem with a kernel-based estimation is associated with the selection of a kernel function and a "window size." For example, for a Parzen window that is a rectangular cube, this problem relates to choosing the size of a side h of a hypercube. A proper selection of h, called a **smoothing parameter**, is crucial for obtaining a good estimate of true density. For large h, an estimate may be too smooth. A value of h that is too small may result in too much detail and noise from the data. The selection of h depends on the character of the data and could be completed experimentally. Let us note that the fixed size of a region limits kernel-based estimation methods.

A simple design approach assumes an equal smoothing parameter for all classes. We can also try to tune distinct smoothing parameters for each class. The generalization ability of the kernel-based density estimation depends on the training set and on smoothing parameters. We can also select a reduced training set that is a subset of an original training set. Kernel-based estimation is memory and computation intensive. It requires availability of the training set during the estimation of density for new patterns. To reduce this burden, we consider some reduction of the training data. For example, the patterns from the same class of the training set can be clustered. Then, instead of taking all patterns, we can use only centers of clusters as reference training patterns.

***K*-nearest Neighbors.** A method of probability density estimation with variable size regions is the *k*-nearest neighbors method. In the *k*-nearest neighbors method, a number of patterns k within a region is fixed, whereas a region size (and thus a volume V) varies depending on the data. In this technique, the estimation of probability density for a given new pattern $\mathbf{x}$, based on a given training set T_{tra} of N patterns, is provided in the following way. First, a small n-dimensional sphere is located in the pattern space centered at the point $\mathbf{x}$. Then a radius of this sphere is extended until the sphere contains exactly the fixed number k of patterns from a given training set. Then an estimate of the probability density $\hat{p}(\mathbf{x})$ for $\mathbf{x}$ is computed as

$$\hat{p}(\mathbf{x}) = \frac{k}{NV} \tag{132}$$

The *k*-nearest neighbors methods lead to an estimate that in fact is not a true probability density, since its integral does not converge over pattern space.

Modifications of the *k*-nearest neighbors estimation may include selecting as a reference pattern only some part of the training set or clustering training data set and using only cluster centers as reference patterns.

K-nearest Neighbors Classification Rule. The k-nearest neighbors probability density estimation method can be simply modified as the **k-nearest neighbors classification rule**. Let us assume that we are given a training set T_{tra} of N patterns $\mathbf{x}^1, \mathbf{x}^2, \cdots, \mathbf{x}^N$ labeled by l classes and containing N_i patterns for the ith class $c_i (i = 1, 2, \cdots, l;\ \sum_i^l N_i = l)$. A new pattern $\mathbf{x}$ is assigned to the class c_j most frequently appearing within the k-nearest neighbors for $\mathbf{x}$. In other words for a given $\mathbf{x}$, the first k-nearest neighbors from a training set should be found (regardless of a class label) based on a defined pattern distance measure. Then, among the selected k nearest neighbors, numbers n_i of patterns belonging to each class c_i are computed. The predicted class c_j assigned to $\mathbf{x}$ corresponds to a class for which n_j is the largest (a winner class in voting among k-nearest neighbors).

We can combine the k-nearest neighbors method with the Gaussian classification rule, assigning to $\mathbf{x}$ a class for which the conditional a posteriori probability $P(c_j|\mathbf{x})$ is largest. Using the k-nearest neighbors method, we have the following estimate (approximation) of a class conditional probability density in the region containing the k-nearest neighbors of $\mathbf{x}$:

$$\hat{p}(\mathbf{x}|c_i) = \frac{n_i}{N_i V} \tag{133}$$

For the unconditional probability density,

$$\hat{p}(\mathbf{x}) = \frac{K}{NV} \tag{134}$$

We also can find an approximation for the a priori probability:

$$\hat{P}(c_i) = \frac{n_i}{N} \tag{135}$$

From Bayes' theorem, we have

$$P(c_i|\mathbf{x}) = \frac{p(\mathbf{x}|c_i)P(c_i)}{p(\mathbf{x})} \approx \frac{n_i}{k} \tag{136}$$

The classification rule, which minimizes the probability of misclassification error, selects the predicted class c_j for a pattern $\mathbf{x}$ for which the ratio $\frac{n_j}{k}$ is the largest.

Nearest Neighbor Classification Rule. The simple version of the k-nearest neighbors classification is for a number of neighbors k equal to one. This **nearest neighbor** classification assigns for a new state $\mathbf{x}$ a class of nearest neighbor patterns from a given training set $\mathbf{x}^1, \mathbf{x}^2, \cdots, \mathbf{x}^N$.

The algorithm for the nearest neighbors classification is stored as follows.

Given: A training set T_{tra} of N patterns $\mathbf{x}^1, \mathbf{x}^2, \cdots, \mathbf{x}^N$ labeled by l classes. A new pattern $\mathbf{x}$.

(1) Compute for a given $\mathbf{x}$ the nearest neighbor $\mathbf{x}_j$ from a whole training set based on the defined pattern distance measure *distance* $(\mathbf{x}, \mathbf{x}^i)$.
(2) Assign to $\mathbf{x}$ a class c_j of nearest neighbors to $\mathbf{x}$.

Result: The predicted class.

1.8.4. Semiparametric Methods of Probability Density Estimation

In the previous sections, we discussed parametric and nonparametric methods of probability density estimation. Parametric methods assume a known form of probability density function that is global for the entire data set, and the task of estimation is to find unknown parameters of a known function based on existing data. The disadvantage of parametric methods is choosing one density function form without the possibility of local fitting to specific regions of data. Nonparametric methods allow more general forms of density function but suffer from large dimensionality of the required parameters, depending on the data size. They also require the presence of the entire data set for estimation of the density of new patterns. For large data sets these techniques are time consuming and memory consuming.

The combination of parametric and nonparametric methods gives the foundation for **semiparametric** methods. Two frequently used semiparametric methods are

– Functional approximation
– Mixture models (mixtures of probability densities)

These methods are concerned with the formation of a density estimation model regared as a linear or nonlinear parametric combination of known basis functions or probability density functions (for example, normal densities) localized in certain regions of data.

For semiparametric methods, the estimation task is to find forms of basis functions or component densities and their numbers and parameters, along with the mixing parameter values used in the model. For the semiparametric model of density, the computation of an estimate for a new pattern requires only an algebraic (or other) form of the density model instead of all the data.

A major advantage of semiparametric models is their ability to precisely fit component functions locally to specific regions of a feature space, based on discoveries about probability distributions and their modalities from the existing data.

Functional Approximation. This method assumes approximation of density by the linear combination of m basis functions $\phi_i(\mathbf{x})$:

$$\hat{p}(\mathbf{x}) = g(\mathbf{x}, \mathbf{a}) = \sum_{i=1}^{m} a_i \phi_i(\mathbf{x}) \tag{137}$$

where $a_i (i = 1, 2, \cdots, m)$ are elements of a parameter vector $\mathbf{a} \in \mathbb{R}^m$. An example is a symmetric radial basis function for which an approximation of density can be proposed by the following smooth continuous scalar function defined in $\mathbb{R}$:

$$g(\mathbf{x}, \mathbf{a}) = \sum_{i=1}^{m} a_i \phi_i(\|\mathbf{x} - \mathbf{x}_i\|)$$

$$= a_1 \phi_1(\|\mathbf{x} - \mathbf{x}_1\|) + a_2 \phi_2(\|\mathbf{x} - \mathbf{x}_2\|) + \cdots + a_m \phi_m(\|\mathbf{x} - \mathbf{x}_m\|) \tag{138}$$

where $\|\mathbf{x} - \mathbf{x}_i\|$ is a distance between two vectors in $\mathbb{R}^n$. We see that an approximating function $g(\mathbf{x}, \mathbf{a})$ is constructed as a linear combination of radial basis functions $\phi_i(\|\mathbf{x} - \mathbf{x}_i\|)$ ($i = 1, 2, \cdots, m$) centered around m center vectors $\mathbf{x}_i$. Let us consider a smooth continuous scalar radial basis function (a kernel) $\phi_c(\|\mathbf{x} - \mathbf{x}_c\|)$ defined in $[0, 1]$ and centered around a center $\mathbf{x}_c \in \mathbb{R}^n$. The basis function is radially symmetric. Thus, a function value is identical for vectors $\mathbf{x}$ that lie at the fixed radial distance from the center of the basis function $\mathbf{x}_c$. There are different approaches for selecting basis functions. However, the most commonly used basis function is the Gaussian radial function (Gaussian kernel):

$$\phi_c(\|\mathbf{x} - \mathbf{x}_c\|) = \phi_c(\|\mathbf{x} - \mathbf{x}_c\|, \boldsymbol{\Sigma}_c)$$

$$= \exp\left[-\frac{(\mathbf{x} - \mathbf{x}_c)^T (\mathbf{x} - \mathbf{x}_c)}{2\boldsymbol{\Sigma}_c^2}\right] \tag{139}$$

with two parameters such as $\mathbf{x}_c$ being a center of the Gaussian radial function, and $\boldsymbol{\Sigma}_c^2$ being the normalization parameter for the Gaussian kernel. Furthermore, $\|\mathbf{x} - \mathbf{x}_c\|^2 = (\mathbf{x} - \mathbf{x}_c)^T (\mathbf{x} - \mathbf{x}_c)$ is a squared Euclidean distance between vectors $\mathbf{x}$ and $\mathbf{x}_c$. The normalization parameter $\mathbf{\Sigma}$ has the same form as the standard deviation in the Gaussian normal probability distribution, although it is estimated in a different way. The Gaussian kernel values are in the range from 0 to 1. This means that the closer the vector $\mathbf{x}$ is to the center $\mathbf{x}_c$ of the Gaussian kernel, the larger is the function value. The goal of functional approximation of density is to find, for selected basis functions with given corresponding parameters, an optimal value $\hat{\mathbf{a}}$ of a parameter vector according to the assumed optimization criterion. For example, it can be read as:

$$J(\mathbf{a}) = \int_{\text{all } x} [p(\mathbf{x}) - \hat{p}(\mathbf{x})]^2 \, d\mathbf{x} \tag{140}$$

Substitution of an approximating function $g(\mathbf{x}, \mathbf{a})$ for $\hat{p}(\mathbf{x}, \mathbf{p})$ gives

$$J(\mathbf{a}) = \int_{\text{all } x} [p(\mathbf{x}) + \sum_{i=1}^{m} a_i \phi_i(\mathbf{x})]^2 \, d\mathbf{x} \tag{141}$$

Finding optimal parameters requires minimization of the criterion $J(\mathbf{a})$, with necessary conditions for the minimum given by setting to zero the values of the criterion gradient for $\mathbf{a}$:

$$\frac{\partial J(\mathbf{a})}{\partial \mathbf{a}} = \mathbf{0} \tag{142}$$

or

$$\frac{\partial J(\mathbf{a})}{\partial a_j} = 0, \quad j = 1, 2, \cdots, m \tag{143}$$

We find

$$\frac{\partial J(\mathbf{a})}{\partial a_j} = 2 \int_{\text{all } x} \left[p(\mathbf{x}) - \sum_{i=1}^{m} a_i \phi_i(\mathbf{x}) \right] \phi_j(\mathbf{x}) \, d\mathbf{x} = 0 \tag{144}$$

which gives

$$\int_{\text{all } x} \phi_j(\mathbf{x}) p(\mathbf{x}) \, d\mathbf{x} = \int_{\text{all } x} \phi_j(\mathbf{x}) \left[\sum_{i=1}^{m} a_i \phi_i(\mathbf{x}) \right] d\mathbf{x} \tag{145}$$

By definition, $\int_{\text{all } x} \phi_j(\mathbf{x}) p(\mathbf{x}) \, d\mathbf{x}$ is an expected value $E[\phi_j(\mathbf{x})]$ of a function $\phi_j(\mathbf{x})$. Rearrangement of the above equations gives the set of m linear equations for optimal parameters $a_1, a_2, \cdots, a_m$:

$$\sum_{i=1}^{m} a_i \int_{\text{all } x} \phi_j(\mathbf{x}) \phi_i(\mathbf{x}) \, d\mathbf{x} = E[\phi_j(\mathbf{x})], \quad j = 1, 2, \cdots, m \tag{146}$$

The solution of the above equations requires knowledge of a density $p(\mathbf{x})$. However, one can use the following approximation:

$$E[\phi_j(\mathbf{x})] = \int_{\text{all } x} \phi_j(\mathbf{x}) p(\mathbf{x}) \, d\mathbf{x} \approx \frac{1}{N} \sum_{k=1}^{N} \phi_j(\mathbf{x}^k) \tag{147}$$

where N is a number of patterns. Thus, we have

$$\sum_{i=1}^{m} p_i \int_{\text{all } x} \phi_j(\mathbf{x}) \phi_i(\mathbf{x}) \, d\mathbf{x} = \frac{1}{N} \sum_{k=1}^{N} \phi_j(\mathbf{x}^k), \quad j = 1, 2, \cdots, m \tag{148}$$

This set of m linear equations for a_i can be solved for a given m basis function $\phi_j(\mathbf{x})$. If the orthonormal basis functions are used, which satisfy the conditions

$$\int_{\text{all } x} \phi_j(\mathbf{x})\phi_i(\mathbf{x}) \, d\mathbf{x} = \begin{cases} 1 & \text{for } i = j \\ 0 & \text{for } i \neq j \end{cases} \tag{149}$$

then we can obtain the following optimal estimates for parameters:

$$\hat{a}_j = \frac{1}{N}\sum_{k=1}^{N} \phi_j(\mathbf{x}^k), \quad j = 1, 2, \cdots, m \tag{150}$$

The above batch formula for optimal parameter calculation requires knowledge of all N patterns. The sequential (iterative) formula is

$$\hat{a}_j^{N+1} = \frac{1}{N+1}\left[N\hat{a}_j^N + \phi_j(\mathbf{x}^{N+1})\right] \tag{151}$$

where $\hat{a}_j^N$ and $\hat{a}_j^{N+1}$ are optimal coefficients obtained for N and $N+1$ pattern samples, respectively. Knowing the optimal values of parameters and basis functions, we have a functional approximation of probability density:

$$\hat{p}(\mathbf{x}) = \sum_{i=1}^{m} \hat{a}_i \phi_i(\mathbf{x}) \tag{152}$$

The design procedure for the functional approximation of density first requires the selection of the number and forms of basis functions $\phi(\mathbf{x})$, along with their parameters.

The algorithm for the functional approximation of a probability density is as follows:

Given: A training set T_{tra} of N patterns $\mathbf{x}^1, \mathbf{x}^2, \cdots, \mathbf{x}^N$. The m orthonormal radial basis functions $\phi_i(\mathbf{x})(i = 1, 2, \cdots, m)$, along with their parameters.

(1) Compute the estimates of unknown parameters

$$\hat{a}_j = \frac{1}{N}\sum_{k=1}^{N} \phi_j(\mathbf{x}^k), \quad j = 1, 2, \cdots, m \tag{153}$$

(2) Form the model of the probability density as a functional approximation

$$\hat{p}(\mathbf{x}) = \sum_{i=1}^{m} \hat{a}_i \phi_i(\mathbf{x}) \tag{154}$$

Result: The probability density of patterns.

Mixture Models (Mixtures of Probability Densities). These models are based on linear parametric combination of known probability density functions (for example, normal densities) localized in certain regions of data. One of the natural choices for a mixture model is the following linear mixture distribution:

$$p(\mathbf{x}) = p(\mathbf{x}|\theta) = p(\mathbf{x}|\boldsymbol{\theta}, \mathbf{P}) = \sum_{i=1}^{m} p_i(\mathbf{x}|\boldsymbol{\theta}_i)P_i \tag{155}$$

where $p_i(\mathbf{x}|\boldsymbol{\theta}_i)$ is the ith **component density** with parameter vector $\boldsymbol{\theta}_i$, m is a number of component densities, and P_i is the ith **mixing parameter**. The terms $\boldsymbol{\theta}$ and $\mathbf{P}$ denote the component densities parameter vector and the mixing parameter vector, respectively. Finally, Θ denotes a parameter vector composed of θ and $\mathbf{P}$. To simplify notation, we describe the mixture model as

$$p(\mathbf{x}) = \sum_{i=1}^{m} p_i(\mathbf{x}) P_i \tag{156}$$

The above linear mixture distribution resembles the definition of the unconditional probability density. Indeed, we shall consider P_i as the a priori probability that the pattern $\mathbf{x}$ has been generated by the i_{th} component density $p_i(\mathbf{x})$ of the mixture. Furthermore, we select P_i to satisfy

$$\sum_{i=1}^{m} P_i = 1$$

$$0 \leq P_i \leq 1 \tag{157}$$

and assume normalization of component densities:

$$\int_{\text{all } \mathbf{x}} p_i(\mathbf{x}) d\mathbf{x} = 1 \tag{158}$$

which then can be compared with class conditional densities. Generation of a new pattern by the above mixture model is as follows. First, one i^{th} component density is randomly selected with probability P_i. Then a pattern $\mathbf{x}$ is generated according to the selected probability density $p_i(\mathbf{x})$.

Most frequently mixtures of normal Gaussian densities are used for probability density modeling. Mixture models play an important role in the design of radial basis neural networks (see Chapter 13) and mixture of experts.

1.8.5. Distance Between Probability Densities and the Kullback-Leibler Distance

The goal of density estimation procedures is to find a density model as close as possible to the true density. We can define distance $d(p(\mathbf{x}), \hat{p}(\mathbf{x}))$ between two densities, with true density $p(\mathbf{x})$ and its approximate estimate $\hat{p}(\mathbf{x})$. The negative natural logarithm of the likelihood function $L = p(X) = \prod_{i=1}^{N} p(\mathbf{x}^i)$ for N patterns $\mathbf{x}^1, \mathbf{x}^2, \cdots, \mathbf{x}^N$ is

$$-\ln L = -\sum_{i=1}^{N} \ln p(\mathbf{x}^i) \tag{159}$$

For the model $\hat{p}(\mathbf{x})$, the average of the negative log-likelihood per pattern, in the limit as a number of patterns go to infinity, can be expressed as an expected value

$$E[-\ln L] = -\lim_{N \to \infty} \frac{1}{N} \sum_{i=1}^{N} \ln \hat{p}(\mathbf{x}^i) = -\int_{\text{all } \mathbf{x}} p(\mathbf{x}) \ln \hat{p}(\mathbf{x}) \, d\mathbf{x} \tag{160}$$

The above expectation may be considered as a measure telling us how closely the model and the true density agree. For identical densities $p(\mathbf{x}) = \hat{p}(\mathbf{x})$, the expected value is

$$-\int_{\text{all } \mathbf{x}} p(\mathbf{x}) \ln p(\mathbf{x}) \, d\mathbf{x} \tag{161}$$

which is the **entropy** of $p(\mathbf{x})$. Subtraction of this entropy from the expectation $E[-\ln L]$ gives a measure of the distance between $p(\mathbf{x})$ and $\hat{p}(\mathbf{x})$:

$$d(p(\mathbf{x}), \hat{p}(\mathbf{x})) = -\int_{\text{all } \mathbf{x}} p(\mathbf{x}) \ln \frac{\hat{p}(\mathbf{x})}{p(\mathbf{x})} \, d\mathbf{x} \tag{162}$$

and is called the **Kullback-Leibler distance**.

1.9. Probabilistic Neural Network

The kernel-based probability density estimation and the optimal Bayesian classification rule can be bases for designing a **probabilistic neural network** (PNN). The optimal Bayes' classification rule is stated as follows: Given values of state conditional probability densities $p(\mathbf{x}|c_k)$ and a priori probabilities $P(c_k)$ for all classes $k = 1, 2, \cdots, l$:

"*Assign an object with pattern vector* $\mathbf{x}$ *to a class* c_j *when*

$$p(\mathbf{x}|c_j)P(c_j) > p(\mathbf{x}|c_k)P(c_k), \quad k = 1, 2, \cdots, l; \quad k \neq j$$"

This rule requires knowledge of probabilities $p(\mathbf{x}|c_k)$ and $P(c_k)$ for all classes $c_k (k = 1, 2, \cdots, l)$. The kernel-based method provides the following estimate of a probability density function $p(\mathbf{x}|c_k)$ for a given state $\mathbf{x}$ from the class c_k based on given N_k training patterns $\mathbf{x}^{k,i}$ from class c_k:

$$\hat{p}(\mathbf{x}|c_k) = \frac{1}{N_k h_k^n} \sum_{i=1}^{N_k} \psi\left(\frac{\mathbf{x} - \mathbf{x}^{k,i}}{h_k}\right) \tag{163}$$

where $\psi(\mathbf{y})$ is the selected kernel function (a Parzen window) and h_k is a class-dependent smoothing parameter.

The probabilistic neural network is, in fact, a hardware implementation of the kernel-based method of density estimation and Bayesian optimal classification (providing minimization of the average probability of the classification error).

The PNN is a static network with feedforward architecture (see also Chapter 13). It is composed of four layers: an **input layer**; two hidden layers, a **pattern layer** and a **summation layer** (density estimates); and an **output layer (decision layer)** (see Figure 11.6).

An input layer (weightless) consists of n neurons (units), each receiving one element $x_i (i = 1, 2, \cdots, n)$ of the n-dimensional input pattern vector $\mathbf{x}$.

The PNN network is designed based on the specific training set of patterns T_{tra} used for computing the probability densities for each class. We assume that the training set T_{tra} contains N patterns, with N_k patterns for each class $c_k (k = 1, 2, \cdots, l) \left(\sum_k^l N_k = N\right)$.

A pattern layer consists of N neurons (units, nodes), each representing one reference pattern from the training set T_{tra}. Each neuron in the pattern layer receives n inputs (outputs of each input layer unit — the entire input pattern $\mathbf{x}$). Since the number N of units in the pattern layer is equal to the number of reference patterns in the training set, the topology of the PNN depends on the number of patterns in the training set. The N neurons in the pattern layer are divided into l groups (subsets) $PN_k (k = 1, 2, \cdots, l)$, each containing N_k neurons with reference patterns $\mathbf{x}^{k,i}$ belonging to the same class c_k.

The transfer function of the pattern layer neuron implements a kernel function $\Psi\left(\frac{\mathbf{x} - \mathbf{x}^{k,i}}{h_k}\right)$ (a Parzen window). This neuron uses two parameters: the ith reference pattern $\mathbf{x}^{k,i}$ from the training set from the given class c_k, and the kernel smoothing parameter h_k for a given class c_k. In some applications, the smoothing parameter h is set to be the same for all classes. Generally, the pattern neuron does not use an output activation function. The output of the pattern neuron is directly attached (with a weightless connection) to the summation-layer neuron representing the same class c_k as a class of a reference pattern of the considered-pattern neuron.

The weightless second hidden layer is the summation layer. The number of neurons in the summation layer is equal to the number of classes l. Each ith neuron in the summation layer corresponds to the calculation of an estimate of a probability density $p(\mathbf{x}|c_k)$ for the kth class $c_k (k = 1, 2, \cdots, l)$. The transfer function of the summation neuron is a sum $\sum$. The kth summation-layer neuron, associated with the kth class c_k, receives N_k outputs from all pattern-layer neurons (belonging to the group PN_k) having as a parameter an exemplary pattern $\mathbf{x}^{k,i}$ from class c_k.

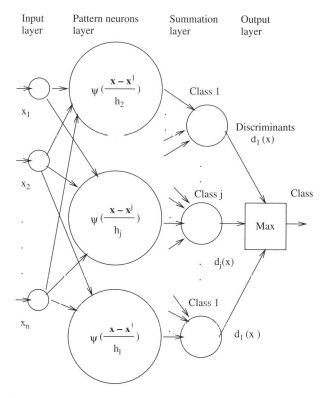

Figure 11.6. Probabilistic neural network.

The output activation function of the summation layer neuron is generally equal to $\frac{1}{N_k h_k^n}$ but may be modified for different kernel functions. The smoothing parameter h_k is generally set for each class separately. The final output of the kth neuron of the summation layer is equal to the estimate $\hat{p}(\mathbf{x}|c_k)$ of a class conditional probability density for a given input pattern $\mathbf{x}$ and a given class c_k.

The output layer is the classification decision layer that implements Bayes' classification rule by selecting the largest value $\hat{p}(\mathbf{x}|c_k)\hat{P}(c_k)$ and thus decides a class c_j for the pattern $\mathbf{x}$. The output unit receives l outputs from the summation-layer neurons representing the probability density estimates $\hat{p}(\mathbf{x}|c_k)$ for each class c_k. The weights for the output unit are the estimates $\hat{P}(c_k)$ of a priori probabilities for each class c_k. Thus, the output of the kth summation-layer neuron is multiplied by probability $\hat{P}(c_k)$. For Bayesian classification involving risk factors for each class, the weights of the output units may be additionally multiplied by the risk term. The output classification unit may be implemented as a winner-take-all neural network (see Chapter 9).

The design of the probabilistic neural network (PNN) requires a training set and values of a priori class probabilities $P(c_k)(k = 1, 2, \cdots, l)$ for all l classes. For a given training set T_{tra} (considered as a reference pattern set) containing N patterns with n features belonging to l classes, the architecture of the PNN is formed as follows.

The network consists of four layers: input, pattern, summation, and output. The number of network inputs is equal to the number n of features in the pattern vector. The pattern neuron layer will contain N pattern neurons, each implementing the selected kernel function. The number of pattern neurons is equal to the number of reference patterns in the training set T_{tra}. Each pattern neuron will take as a parameter the ith reference pattern vector $\mathbf{x}^{k,i}$ from the training set T_{tra} (from a given class c_k) and a smoothing parameter h_k set for a class c_k to which the reference pattern belongs. The N pattern neurons are divided into l groups (subsets) $PN_k (k = 1, 2, \cdots, l)$, each containing N_k neurons with reference patterns $\mathbf{x}^{k,i}$ belonging to the same class c_k.

The summation layer will contain l weightless summation neurons, each for one class $c_k (k = 1, 2, \cdots, l)$. The output activation function of the summation-layer neurons requires two parameters: the number N_k of patterns from class c_k in the training set, and a class-dependent smoothing parameter h_k. The output of the summation-layer neuron (after an output activation function) represents the probability density estimate for a given input pattern from a given class.

The output layer contains one decision unit. It may be, for example, implemented as a winner-take-all competitive network or in some other selected way. The weights of the output units are set to values equal to the estimates $\hat{P}(c_k)$ of a priori probabilities for each class. The single direct k^{th} input (after exceeding the weight) is the Bayes' discriminant function $d_k(\mathbf{x})$ for a given input pattern $\mathbf{x}$ and a class c_k.

The PNN network is fully connected between the input and the pattern layers. The connections between the pattern layer and the summation layer are sparse. The pattern-layer neuron associated with the reference pattern from class c_k is connected only to the summation-layer neuron associated with this class.

The training of the PNN network is in fact part of the network design, since it does not involve trial and error learning. The whole training set T_{tra} is presented to the PNN only once. Each training set pattern is loaded to the pattern layer neuron as a reference parameter.

The designer has to select a kernel function and smoothing parameters $h_k (k = 1, 2, \cdots, l)$, usually by trial and error. A simple design approach assumes equal smoothing parameters for all classes $h = h_k (k = 1, 2, \cdots, l)$, whereas one may obtain better results by choosing different smoothing parameters for each class.

One suggestion for a class-dependent smoothing parameter selection is based on the discovery of the average minimum distances between patterns in the reference training set T_{tra} belonging to the same class. Let us denote by mdp_k^i the minimum distance between a pattern $\mathbf{x}^{k,i}$ from the training set and the nearest pattern within the same class c_k. Then the average minimum distance within class k is

$$mpd_{\text{average},k} = \frac{1}{N_k} \sum_{i=1}^{N_i} mpd_k^i \qquad (164)$$

where N_k denotes the number of patterns from class c_k in the training set. One choice for the between-pattern distance measure could be the Euclidean distance. With an average minimum distance computed for patterns of the kth class, the smoothing parameter for class c_k can be chosen as

$$h_k = a \cdot mpd_{\text{average},k} \qquad (165)$$

where a is a constant that has to be found experimentally. Since the PNN is memory and computation intensive, some design methods select only some subset of the training set for reference patterns only, which reduces the computation burden.

Pattern Processing. Processing of patterns by the already-designed PNN network is performed in the feedforward manner. The input pattern is prxesented to the network and processed forward by each layer. The resulting output is the predicted class.

Probabilistic Neural Network with the Radial Gaussian Kernel. The most natural choice for a kernel function is a radial symmetric multivariate Gaussian (normal) kernel:

$$\psi \left(\frac{\mathbf{x} - \mathbf{x}^{k,i}}{h_k} \right) = \frac{1}{(2\pi)^{n/2} h_k^n N_k} \exp \left(-\frac{\|\mathbf{x} - \mathbf{x}^{k,i}\|^2}{2h_k^2} \right) \qquad (166)$$

where $\mathbf{x}^{k,i}$ denotes the ith training set pattern from a k class c_k. For the above Gaussian kernel, the kth neuron in the pattern layer implements, for the reference pattern $\mathbf{x}^{k,i}$ from the class c_k, the function

$$v_{k,i} = -\frac{\|\mathbf{x} - \mathbf{x}^{k,i}\|^2}{2h_k^2} = \frac{1}{h_k^2}\left[\mathbf{x}^T\mathbf{x}^{k,i} - \frac{1}{2}\left(\mathbf{x}^T\mathbf{x} + (\mathbf{x}^{k,i})^T\mathbf{x}^{k,i}\right)\right] \quad (167)$$

For the Gaussian kernel, the pattern neuron will have the exponential output activation function $\exp(v_{k,i})$. The term $\frac{1}{(2\pi)^{n/2}h_k^n}$ is moved to the summing layer as altering the output activation function. The kth output neuron of the summation layer, forming a probability density for the kth class, will have the following output activation function:

$$\frac{1}{(2\pi)^{n/2}h_k^n N_k} \quad (168)$$

PNN with the Radial Gaussian Normal Kernel and Normalized Patterns. A probabilistic neural network with a radial Gaussian normal kernel may be simplified if all training set reference patterns and input patterns are normalized to have unit length $\sum_{i=1}^{n} x_i^2 = 1$, $\mathbf{x}^2\mathbf{x} = 1$. The normalization of a pattern is given by the formula $\mathbf{x}/\|\mathbf{x}\|$ (where $\|\mathbf{x}\| = \sqrt{\mathbf{x}^T\mathbf{x}}$). For the normalized patterns, the transfer function for the pattern-layer neuron becomes:

$$\frac{1}{h_k^2}\left[\mathbf{x}^T\mathbf{x}^{k,i} - \frac{1}{2}\left(\mathbf{x}^T\mathbf{x} + (\mathbf{x}^{k,i})^T\mathbf{x}^{k,i}\right)\right] = \frac{1}{h_k^2}\left(\mathbf{x}^T\mathbf{x}^{k,i} - 1\right) = \frac{1}{h_k^2}\left(z_{k,i} - 1\right) \quad (169)$$

where $z_{k,i} = \mathbf{x}^T\mathbf{x}^{k,i}$.

Normalization of patterns allows for a simpler architecture of the pattern-layer neurons, containing here also input weights and an exponential output activation function. The transfer function of a pattern neuron can be divided into a neuron's transfer function and an output activation function. The n weights of the ith pattern neuron from the kth class (forming a weight vector $\mathbf{w}^{k,i}$) are set by the values of elements of an ith reference pattern from the kth class: $w_j^{k,i} = x_j^{k,i} (j = 1, 2, \cdots, n)$ ($\mathbf{w}^{k,i} = \mathbf{x}^{k,i}$). The pattern neuron realizes the sum of input pattern elements multiplied by the corresponding weights (the outer product $(\mathbf{w}^{k,i})^T$):

$$z_{k,i} = \sum_{j=1}^{n} w_j^{k,i} x_j$$

$$z_{k,i} = (\mathbf{w}^{k,i})^T \mathbf{x} \quad (170)$$

The pattern-neuron output activation function has the form $\exp(z_{k,i} - 1)/h_k^2$. Other layers of the network are similar to the PNN with Gaussian kernel, discusses above for nonnormalized patterns.

Training of the PNN networks with the Gaussian kernel and normalized patterns is part of network design and does not involve trial and error learning. The whole training set T_{tra} is presented to the PNN only one time. Each training set pattern is assigned as the weight vector of the corresponding pattern layer neuron.

The processing of a new input pattern in the already-designed PNN network first requires pattern normalization (as a front-end operation). Then the normalized pattern is presented to the network and subsequently processed by each layer. The resulting output will be the predicted class.

1.10. Constraints in Classifier Design

Classifiers based on the optimal Bayesian classification rule are designed based on known class conditional probabilities $p(\mathbf{x}|c_i)$ ($i = 1, 2, \cdots, l$) and a priori class probabilities $P(c_i)$ ($i = 1, 2, \ldots, l$). The probabilities characterize the probabilistic nature of real phenomena generating classified patterns. Let us assume that, based on an analysis of existing a priori gathered data (or based on known statistical characteristics of the generated patterns), we can find precisely the class conditional probability densities and a priori class probabilities. Then we can design the ideal optimal classfier based on Bayes' rule. This type of optimal design assumes as an optimization criterion the average probability of classification error for all population of generated patterns. For the ideal optimal Bayes classifier, we do not need to analyze the performance of the designed classifier for unseen patterns with the identical underlying probability characteristic. We know that an ideal optimal classifier will guarantee minimization of the average probability of the classification error. However, ideal situations rarely exist in real life. The class conditional probability densities and a priori class probabilities are normally not known a priori and must be estimated in some optimal way based on a specific subset of a finite (often severely limited) data set gathered a priori. We assume that this specific subset of data (a training set) well represents patterns generated by a physical phenomenon and that we can assume that these patterns are drawn according to the characteristic of underlying-phenomenon probability density. Furthermore, the average probability of classification error is difficult to calculate for real data.

In real-life situations, we instead seek feasible suboptimal solutions of Bayesian classifier design. The major task and difficulty in this suboptimal design is the estimation of class conditional probabilities $p(\mathbf{x}|c_i)(i = 1, 2, \cdots, l)$ based on a limited sample. The samples are frequently collected randomly, not using a well-planned experimental procedure; thus, the assumptions that they are a good representation of nature and are satisfactory to grasp pattern probability density, are often not satisfied. Specifically, in the design of Bayesian classifiers, particularly in the design of Gaussian normal classifiers, a frequently made assumption about the normal form of probability densities governing the generation of patterns is not necessarily true.

Another difficulty in probability density estimation arises from the fact that the estimation techniques discussed earlier require selecting (heuristically) design parameters from a limited pattern sample. This, in practice, leads to a suboptimal solution.

2. Regression

To store, interpret, and process massive data, we first intuitively look for simple data models. In the design of optimal models, the most prominent approach is the minimization **method of least squares of errors (LS)**, which uses as a criterion a sum of squares of modeling errors. We begin with the concept of regression analysis and the least squares method for discovering linear models of data. Then we present simple linear regression analysis, including linear model design for **two-dimensional** data. Later we present the mathematically elegant and easy-to-understand matrix version of simple linear regression. One of the mathematical attractions of linear models is their ability to find closed analytical formulas for model parameters, optimal in the sense of minimum least squares of errors.

2.1. Data Models

Mathematical models are useful approximate representations of phenomena that generate data and may be used for prediction, classification, compression, or control design. A mathematical model of biological or chemical processes, dynamic systems, etc. may be obtained from the principles and laws of physics that govern the process behavior. However, in different data processing scenarios, there is a demand to find a mathematical model by processing existing data without employing the law of physics governing data generating phenomena. These types of models are called "black-box" models. In the design of a black-box type of model, one tries to fit the data by using well-known mathematical functions, for example, the function defining a line.

Often, when a model of data is designed based on a sample from a given population, this kind of data analysis and model discovery is called **regression analysis**.

Model design consists of finding a model structure, computing optimal values of model parameters, and assessing the model quality (see Chapter 15).

A model structure relates to the type (and for some models the order) of mathematical formulas that describe the system behavior (for example, equations of lines and planes, difference equations, differential equations, Boolean functions, etc.) Depending on the model structure, we can divide regression models into the following categories:

- Simple linear regression
- Multiple linear regression
- Neural network-based linear regression
- Polynomial regression.
- Logistic regression
- Log-linear regression
- Local piecewise linear regression.
- Nonlinear regression (with a nonlinear model)
- Neural network-based nonlinear regression

We can also distinguish static and dynamic models. A **static model** produces outcomes based only on the current input. Such a model has no internal memory. In contrast, a **dynamic model** produces outcomes based on the current input and the past history of the model behavior. Such a model has internal memory represented by the **internal state**.

Models are designed based on given data. Although data may represent the entire population, in reality, the designer often has access only to a sample of the population. Frequently a data space is surprisingly "empty", since a sample might be very small. For example, suppose that during an experiment we have measured N values of the observable entity y for N different values of known conditions x. We may consider measuring patient temperature y in time scale x, or observing a reaction (output) y of a process as a result of providing an input signal. One instance of the experiment is represented by a pair (x^i, y^i) of numerical values of an observation y^i for a given condition x^i. In most cases, we say that y is the dependent variable, whereas x is an independent variable.

Regression analysis is applied to the given data set that is the random sample from a certain population (see Figure 11.7). Most of the time, both independent and dependent variables are continuous valued (real valued).

As a result of experiment, or of gathering the statistical data, we collect N pairs of the experimental data set named T_{orig}:

$$T_{\text{orig}} = \{(x^1, y^1), (x^2, y^2), \cdots, (x^N, y^N)\} = \{(x^i, y^i), \ i = 1, 2, \ldots, N\} \qquad (171)$$

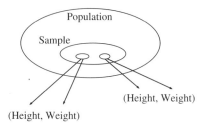

Figure 11.7. Data gathering.

2. Regression

where the superscript i denotes a case number. Each ith pair of data (x^i, y^i) represents one i^{th} data point (one i^{th} case). This pair is composed of the ith value x^i of the independent variable x and the corresponding (resulting) value y^i of the dependent variable y.

Regression analysis is a statistical method used to discover the relationship between variables and to design a data model that can be used to predict variable values based on other variables (see Figure 11.8 and 11.9). Generally, the discovered model can be nonlinear. Linear regression belongs to the statistical methods used to determine the linear relationship and linear data model for two or more variables. A simple linear regression is the most common technique which discovers linear association (relations) and a linear model between two generally random variables. A simple linear regression attempts to find the linear relationship between two variables, x and y, and to discover a linear model, i.e., a line equation $y = b + ax$, which is the best fit to given data in order to predict values of data. This modeling line is called the **regression line** of y on x and the equation of that line is called a **regression equation** (**regression model**). Linear regression attempts to discover the best linear model representing the given dataset, based on a defined performance criterion. This process corresponds to finding a line which is the best fit to a given dataset containing a sample population.

Although linear regression is generally considered between two random variables, in typical approaches linear regression analysis provides a prediction of a **dependent variable** y based on an **independent variable** x. The independent variable x is also called the **input variable**, **predictor variable**, or **explanatory variable**, whereas the dependent variable y is called the **outcome**, **observation**, **response**, or **criterion variable**.

Frequently, values of the independent variable x are fixed, and the corresponding values of the dependent variable y are subject to random variation.

The term independent variable means that the value of the variable can be chosen arbitrarily, and consequently, the dependent variable is an effect of the input variable. There is a causal dependency of the dependent variable on the independent variable (an input-output model concept).

One of the results of regression analysis is a data model. The discovered regression model can be used for prediction of real values of the dependent variable. Note that not all values of y from T_{orig} will occur exactly on the regression line.

In regression analysis, only the dependent variable y has to be considered as a random sample. The independent variable x does not need to be a random sample and can be chosen arbitrarily.

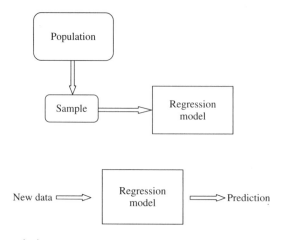

Figure 11.8. Regression analysis.

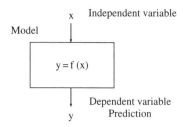

Figure 11.9. Prediction.

Data Visualization: Scatter Plot. An easy way to present and examine given two-dimensional data is through a scatter plot. A scatter diagram contains a plot of the points from a given data set onto a rectangular Cartesian coordinate system. The horizontal axis represents the independent variable x, and the vertical axis represents the dependent variable y (see Figure 11.10, where x represents Height and y Weight).

One point of a scatter plot depicts one pair of data (x^i, y^i). A scatter plot helps visually to find any relationship that may exist among the data. It is also useful in analyzing the data and the errors in linear regression analysis. For multiple regression with two independent variables, it is possible to visualize data in a three-dimensional scatter plot. Additional insight into regression analysis provides overlapping plots of scatter data points with a regression line (see Figure 11.11).

2.2. Simple Linear Regression Analysis

As an example of linear regression, we consider discovering the relationship and simultaneous changes of height and weight with age for a certain population. Let us assume that data have been gathered from the population of students in San Diego in the age range from 9 to 14. Here, age can be represented by the independent variable x, and weight by the dependent variable y.

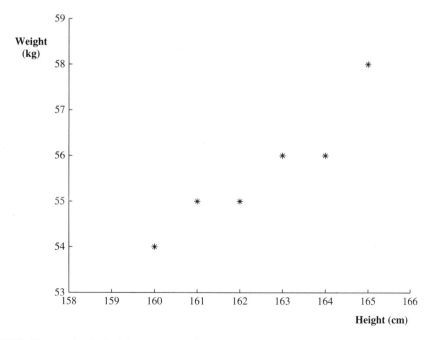

Figure 11.10. Scatter plot for height versus weight data.

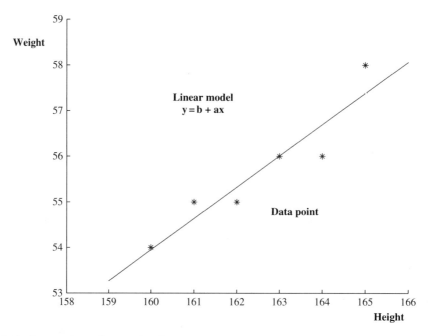

Figure 11.11. Scatter plot and regression line.

Regression analysis discovers the relation between height and weight and shows "how change in height can influence the change in weight." The discovered regression model can be used to predict a weight based on the height of a student. The linear regression postulates that the linear function (dependency)

$$y = b + ax \qquad (172)$$

is an adequate model representing the relationship between x and y. The structure of the model is the straight line $y = b + ax$, where a is the slope and b is the intercept of the line. Coefficients a and b are model parameters.

We try to find a linear model (line) of best fit to the data containing pairs of values for height and weight of a sample of students. The regression line is that of y on x, because we are predicting values of y based on the corresponding values of x.

Table 11.1 shows a randomly selected sample of six students from the considered population.

Figure 11.12 shows a scatter plot and the best-fitting line as the linear regression model $y = b + ax$ with parameter values $a = 0.6857$ and $b = -55.7619$. The prediction (regression) error is the difference between the real value of y and the predicted value computed using the regression model $e = y - (b + ax)$. The vector of errors for all cases from the data is $\mathbf{e} = [0.0476 \quad 0.3619 \quad -0.3238 \quad -0.0095 \quad -0.6952 \quad 0.6190]^T$. The sum of squared errors is equal to $J = 1.10484$. As will be explained later, this sum is the minimal value of the performance criterion.

Linear regression is a fundamental data mining tool used in a variety of domains to discovery linear relations between variables and to design a linear variable based on known values of one

Table 11.1. Relation between height and weight: sample data.

Height	x	160	161	162	163	164	165
Weight	y	54	55	55	56	56	58

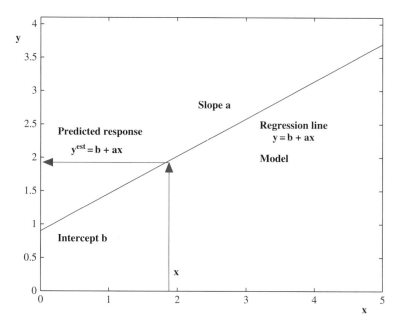

Figure 11.12. Regression model.

or more independent variables. Linear models are approximate representations of real relations between variables, since data are rarely ideally linear. We must remember that global linear regression models may not be adequate for "nonlinear" data. Despite these limitations, linear regression, due to its simplicity, is widely used in almost all domains of science and engineering.

Assumptions. A linear model is designed based on given data as a sample of a certain population. It can be considered to be a finite set of pairs of values (x^i, y^i) of x and y drawn according to a given probability distribution from a population. This limited knowledge about the true relationship between x and y is used in the model-building process. Linear regression analysis is carried out on the basis of the following assumptions:

1) The observations y^i ($i = 1, \cdots, N$) are random samples and are mutually independent.
2) The regression error terms (the difference between the predicted value and the true value) are also mutually independent, with the same distribution (normal distribution with zero mean) and constant variances.
3) The distribution of the error term is independent of the joint distribution of explanatory variables. It is also assumed that unknown parameters of regression models are constants.

Simple Linear Regression Analysis. Simple linear regression analysis assumes a linear form of the data model $y = b + ax$ for a given data set. It comprises evaluation of basic statistical characteristics of data, capturing the linear relationship between two variables: the independent variable x and the dependent variable y (through correlation analysis). Furthermore, it provides an estimation of the optimal parameters of a linear model (mostly by using the least squares method; see Figure 11.13) and assesses model quality and generalization ability to predict the outcome for new data (also see Chapter 15).

2. Regression

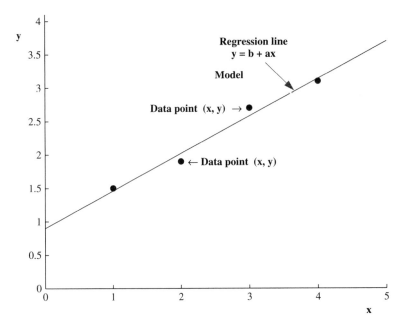

Figure 11.13. Scatter plot and optimal regression model.

Model Structure. For a given set T_{orig} of N data points $\{(x^i, y^i), \quad i = 1, 2, \cdots, N\}$, we form the hypothesis that the possibility exists to discover a data model as the functional relationship between y and x as expressed by a function $y = f(x)$.

Generally, a function $f(x)$ could be nonlinear in x (see Figure 11.14). For example,

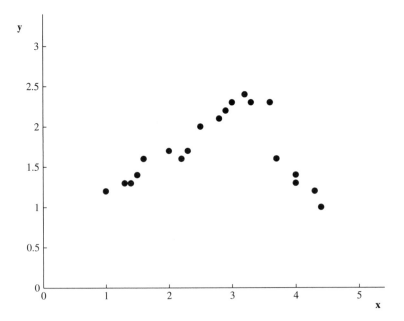

Figure 11.14. Nonlinear data.

$$y = f(x) = a_0 + a_1 x + a_2 x^2 + a_3 x^3 \tag{173}$$

However, several structural forms of this function can be considered including a line equation

$$y = f(x) = b + ax \tag{174}$$

as a simple data model.

A Model in Multiple Linear Regression. In a multiple regression, we have data with n input independent variables, denoted as a vector of independent variables $\mathbf{x} = [x_i, x_2, \cdots, x_n]^T (\mathbf{x} \in \mathbb{R}^n)$ and one (generally many) dependent output variable y. Here, one data pair is composed as $(\mathbf{x}^i, y^i)$. The relations between variables and the multiple regression model can be discovered from data containing multiple independent variables.

One can try to best fit the data by using the linear function

$$y = a_0 + a_1 x_1 + a_2 x_2 + \cdots + a_n x_n \tag{175}$$

where $\mathbf{a} = [a_0, a_1, \cdots, a_n]^T$ is a parameter vector $\mathbf{x} \in \mathbb{R}^n$). For $n = 3$, the linear regression model represents a plane. For $n > 3$, we are concerned with a certain hyperplane.

Generally, the multiple regression model can be a function (generally nonlinear) of $\mathbf{x}$: $f(\mathbf{x})$:

$$y = f(\mathbf{x}) = a_0 + a_1 \phi_1(\mathbf{x}) + a_2 \phi_2(\mathbf{x}) + \cdots + a_m \phi_m(\mathbf{x}) \tag{176}$$

where $a_i (i = 0, 1, 2, \cdots, m)$ are parameters and $\phi_i(\mathbf{x})(i = 1, 2, \cdots, m)$ are generally nonlinear functions of $\mathbf{x}$.

Regression Errors. Because a model is an approximation of the reality, computation of a predicted outcome will come with certain error. Let us consider the ith data point (x^i, y^i) from the data set T_{orig}, with values x^i and y^i of the independent and dependent variable, respectively; the predicted variable value computed by the model $y^{i,\text{est}} = \hat{b} + \hat{a} x^i$ lies on the regression line. This value is different than the true data value y^i from the data point (x^i, y^i). This difference e = real-value − predicted-value, i.e,

$$e_i = y^i - f(x^i) = y^i - y^{i,\text{est}} = y^i - (\hat{b} + \hat{a} x^i) \tag{177}$$

is the **regression error** also called the **residual** or **modeling error**.

The residuals have an easy interpretation on a scatter plot with an overlapping plot of the regression line (see Figure 11.15).

The error is the vertical distance from the regression line to the point on the scatter plot.

Performance Criterion – Sum of Squared Errors. The most natural and commonly encountered performance criterion is **the sum of squared errors** (the residuals sum of squares):

$$J(\text{parameters}) = \sum_{i=1}^{N} (e^i)^2 = \sum_{i=1}^{N} (y^i - f(x^i))^2 \tag{178}$$

where (x^i, y^i) is the ith data point from a data set, x^i is the ith instance of the input independent variable, y^i is the ith instance of the output dependent variable, and $f(x)$ is a model. A scalar value of the performance criterion computed for the entire data set is used as a measure for comparing

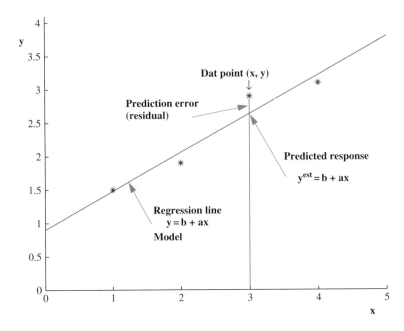

Figure 11.15. Regression errors.

model quality in procedures of searching to discover the best regression model for the data. For the model which is best fit to data the residual sum of squares is minimum.

For multiple regression, with a vector **x** of n-input independent variables the sum of squared errors performance criterion is defined as

$$J(\text{parameters}) = \sum_{i=1}^{N}(e^i)^2 = \sum_{i=1}^{N}(y^i - f(\mathbf{x}^i))^2 \qquad (179)$$

where $(\mathbf{x}^i, y^i)$ is the ith data point from a data set, $\mathbf{x}^i$ is the ith instance of the input independent variables vector, and $f(\mathbf{x})$ is a model.

A performance criterion lets us compare models for different values of parameters. The minimization technique in regression that uses as a criterion the sum of squared error – **method of least squares or errors** (LSE) or, in short, the **method of least squares**.

We assume that regression errors are random values with normal distribution and mean equal to zero.

Basic Statistical Characteristics of Data. Regression analysis requires computation of the basic statistical characteristics of data and variables. We need to find the characteristic of initial data: mean, variance, and standard deviation of the values of both variables x and y. We also need to compute sums of squared variations of the dependent variable y due to both regression analysis and model imperfection.

The estimates of the mean and the variance are major statistical descriptors of data clusters. The mean of N samples can be estimated as

$$\bar{x} = \frac{1}{N}\sum_{i=1}^{N} x^i, \quad \bar{y} = \frac{1}{N}\sum_{i=1}^{N} y^i \qquad (180)$$

The mean is the center of mass of data related to x. The center of the data is defined as

$$\left(\frac{\sum_{i=1}^{N} x^i}{N}, \frac{\sum_{i=1}^{N} y^i}{N}\right) \tag{181}$$

The identified optimal regression line passes through the center of the data.
  The variance is defined as

$$\sigma_x^2 = \frac{1}{N}\sigma_{i=1}^{N}(x^i - \bar{x})^2, \quad \sigma_y^2 = \frac{1}{N}\sum_{i=1}^{N}(y^i - \bar{y})^2 \tag{182}$$

The variance is the second moment around the mean.
  The standard deviation is the square root of the variance:

$$\sigma_x = \sqrt{\Sigma_x^2}, \quad \sigma_y = \sqrt{\sigma_y^2} \tag{183}$$

In statistics, the covariance between two real-valued random variables x and y, with expected values $E(x) = \bar{x}$ and $E(y) = \bar{y}$, is defined as

$$\operatorname{cov}(x, y) = E((x - \bar{x})(y - \bar{y})) \tag{184}$$

where E denotes the expected value.
  In calculation of covariance for the N data values (x^i, y^i), the following numerical estimate is used:

$$\operatorname{cov}(x, y) = \frac{\sum_{i=1}^{N}(x^i - \bar{x})(y^i - \bar{y})}{N} \tag{185}$$

Covariance indicates the extent to which two random variables covary. For example, if the analysis of two technology stocks shows that they are affected by the same industry trends, their prices will tend to rise or fall together. Then they covary. Covariance and correlation measure such tendencies.
  Let us recall useful equations for the sum of squared values containing N data points (x^i, y^i) about their respective mean values $\bar{x}$ and $\bar{y}$:

$$S_x = \sum_{i=1}^{N}(x^i - \bar{x})^2 = \sum_{i=1}^{N}(x^i)^2 - N\bar{x}^2, \quad S_y = \sum_{i=1}^{N}(y^i - \bar{y})^2 = \sum_{i=1}^{N}(y^i)^2 - N\bar{y}^2 \tag{186}$$

$$S_{xy} = \sum_{i=1}^{N}(x^i - \bar{x})(y^i - \bar{y}) = \sum_{i=1}^{N} x^i y^i - N\bar{x}\bar{y} \tag{187}$$

Sum of Squared Variations in y Caused by the Regression Model. Due to imperfection of the regression model, the dependent variable y will fluctuate around the regression model. The following sums of squared variations in y are useful in regression analysis.
  We write that the total sum of the squared variations in y,

$$S_{\text{total}} = \sum_{i=1}^{N}(y^i - \bar{y})^2 \tag{188}$$

is equal to the sum of the squared deviations (due to regression and explained by the regression model):

$$S_{ye} = \sum_{i=1}^{N}(y^{i,\text{est}} - \bar{y})^2 \tag{189}$$

plus the sum of squared unexplained variations (due to errors):

$$S_{re} = \sum_{i=1}^{N}(y^i - y^{i,\text{est}})^2 \tag{190}$$

The total sum of squared variations in y is expressed by:

$$S_{\text{total}} = S_{ye} + S_{re} \tag{191}$$

$$S_{\text{total}} = \sum_{i=1}^{N}(y^i - \bar{y})^2 = \sum_{i=1}^{N}(y^{i,\text{est}} - \bar{y})^2 + \sum_{i=1}^{N}(y^i - y^{i,\text{est}})^2$$

These formulas are used to define important regression measures (for example, the correlation coefficient).

Computing Optimal Values of the Regression Model Parameters. For a known structure of the linear regression model $y = b + ax$, optimal model parameters values have to be computed based on the given data set and the defined performance criterion. Finding values of parameters is the static minimization problem. A designer tries to define a model goodness criterion for judging what values of parameters guarantee the best fit of a model to data. These parameters are optimal in the sense of the given performance criterion. The most famous is the criteria of the sum of squared regression errors.

There are few methods for estimation of optimal model parameter values. The most commonly used methods are the following:

– The analytical offline method, which is based on least squares of errors and uses closed mathematical formula for optimal model parameter values
– The analytical recursive offline method, which is based on the least squares of errors criterion
– Searching iteratively optimal model parameters which is based on gradient descent methods
– Neural network-based regression, which uses mechanisms of learning

In the next sections, we derive closed formulas for the optimal values of parameters for the linear model $y = b + ax$.

Simple Linear Regression Analysis, Linear Least Squares, and Design of a Model. Let us assume that the data set $T_{\text{orig}} = \{(x^i, y^i), (i = 1, 2, \cdots, N)\}$ is available. One choice for the regression function can be a linear function representing a line on the (x, y) plane:

$$y = b + ax \tag{192}$$

where a is a slope and b is an intercept.

The general linear model structure is

$$y = f(x) = a_0 + a_1 \phi_1(x) \tag{193}$$

with a_0 representing intercept b (bias), a_1 representing slope a, and $\phi_1(x) = x$. This line equation is the linear mathematical model for the data, where a and b are model parameters. If we have decided that the line equation will be the mathematical model of data, then we have to find optimal values of the model parameters a and b by using a selected model performance criterion. The most obvious choice is the sum of squared errors.

Sometimes we anticipate the "linear nature" of a process that generates experimental data, but due to the nature of a population, measurement errors, or other disturbances, the measured data not necessary fall on the line $y = b + ax$. Therefore, we try to find the line with parameters a and b that will best fit the experimental data. Let us assume that we will now compute, using the model $y = b + ax$, the predicted value (estimated response) y^{est} for the known value y^i of the independent variable x from the pair of data $(x^i, y^i)(i = 1, 2, \cdots, N)$. The line equations will not be satisfied ideally for all pairs (i.e., $y^i - (b + ax^i) = 0$). We will instead obtain a set of equations that includes error terms ($e^i = y^i - f(x^i) = y^i - (b + ax^i)$), representing the difference between the true data y^i, obtained from the measurement, and the data approximated by the model, $f(x^i) = b + ax^i$ for the known x^i:

$$
\begin{aligned}
y^1 &= (b + ax^1) + e^1 \\
y^2 &= (b + ax^2) + e^2 \\
&\vdots \\
y^i &= (b + ax^i) + e^i \\
&\vdots \\
y^N &= (b + ax^N) + e^N
\end{aligned}
\tag{194}
$$

where $e^i = y^i - (b + ax^i)$ is the error for the ith pair of the data.

To answer the question of which regression line is the "best" fit for the experimental data, we must define the performance criterion that will allow us to objectively judge, in a quantitative way, the quality of fit. The criterion is usually defined as the sum of the squares of all errors:

$$
J(a, b) = \sum_{i=1}^{N} [y^i - f(x^i)]^2 = \sum_{i=1}^{N} [y^i - (b + ax^i)]^2
\tag{195}
$$

The criterion $J(a, b)$ for a linear model is the function of the parameters a and b. For the given a and b and a given set of experimental data, the criterion $J(a, b)$ takes a numerical scalar value. We can compare the values of $J(a, b)$ for different pairs of parameters (a, b) for given experimental data $\{(x^i, y^i), (i = 1, 2, \cdots, N)\}$. In other words, we can compare models.

Figure 11.16 shows the performance curve, representing values of the performance criterion versus values of one parameter a for the model $y = ax$, which is the line passing through the origin, with the second parameter b (the intercept) equal to zero ($b = 0$). We can see that the performance curve has a minimum.

Now, to solve the problem of the best fit of the experimental data $\{(x^i, y^i), (i = 1, 2, \cdots, N)\}$ by the line $y = b + ax$, we should find parameters values $(\hat{a}, \hat{b})$ minimizing the criterion $J(a, b)$:

$$
\hat{a}, \hat{b} = \arg \min_{\hat{a}, \hat{b}} J(a, b) \sum_{i=1}^{N} [y^i - (b + ax^i)]^2
\tag{196}
$$

Parameters $(\hat{a}, \hat{b})$ are called the **optimal parameters** for the criterion $J(a, b)$ and the data $\{(x^i, y^i), (i = 1, 2, \cdots, N)\}$. For the optimal parameters, the criterion will take the minimum value $J(\hat{a}, \hat{b})$.

We know from calculus that to make $J(a, b)$ a minimum the following necessary conditions for the minimum of a function must be satisfied:

$$
\frac{\partial J(a, b)}{\partial a} = 0, \quad \frac{\partial J(a, b)}{\partial b} = 0
\tag{197}
$$

We can find the partial derivatives of $J(a, b)$ as

$$
\frac{\partial J(a, b)}{\partial a} = \sum_{i=1}^{N} -2[y^i - (b + ax^i)](x^i) = 0
$$

$$
\frac{\partial J(a, b)}{\partial b} = \sum_{i=1}^{N} (-2)[y^i - (b + ax^i)] = 0
\tag{198}
$$

2. Regression

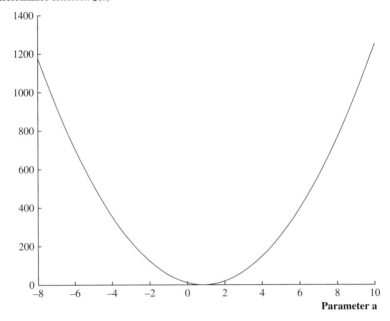

Figure 11.16. Performance criterion curve.

As a result, we have obtained two linear equations of two unknowns a and b, called the **normal equations**. We can rewrite these equations in the form

$$\left(\sum_{i=1}^{N}(x^i)^2\right)a + \left(\sum_{i=1}^{N}x^i\right)b = \sum_{i=1}^{N}x^i y^i$$

$$\left(\sum_{i=1}^{N}x^i\right)a + Nb = \sum_{i=1}^{N}y^i \qquad (199)$$

The solution of the above equations may be easily found as

$$\hat{a} = \frac{1}{d}\left(N\sum_{i=1}^{N}x^i y^i - \sum_{i=1}^{N}x^i \sum_{i=1}^{N}y^i\right)$$

$$\hat{b} = \frac{1}{d}\left(\sum_{i=1}^{N}(x^i)^2 \sum_{i=1}^{N}y^i - \sum_{i=1}^{N}x^i \sum_{i=1}^{N}x^i y^i\right)$$

$$d = N\sum_{i=1}^{N}(x^i)^2 - \left(\sum_{i=1}^{N}x^i\right)^2 \qquad (200)$$

We can also write the solution as

$$\hat{a} = \frac{N\sum_{i=1}^{N}x^i y^i - \sum_{i=1}^{N}x^i \sum_{i=1}^{N}y^i}{N\sum_{i=1}^{N}(x^i)^2 - \left(\sum_{i=1}^{N}x^i\right)^2} \qquad (201)$$

$$\hat{b} = \frac{\sum_{i=1}^{N}(x^i)^2 \sum_{i=1}^{N}y^i - \sum_{i=1}^{N}x^i \sum_{i=1}^{N}x^i y^i}{N\sum_{i=1}^{N}(x^i)^2 - \left(\sum_{i=1}^{N}x^i\right)^2} \qquad (202)$$

For the application of the least square method a procedure for simple linear regression is given below

Given: The number N of experimental observations, and the set of the N experimental data points $\{(x_i, y_i), i = 1, 2, \ldots, N\}$.

(1) Compute the statistical characteristics of the data (Equations (186)–(187)).
(2) Compute the estimates of the model optimal parameters using Equations (201) and (202).
(3) Assess the regression model quality indicating how well the model fits the data. Compute

   (a) Standard error of estimate (Equation (177))
   (b) Correlation coefficient r
   (c) Coefficient of determination r^2

Result: Optimal model parameter values $\hat{a}$ and $\hat{b}$.

Once the linear regression model has been found, we need a measure of model quality. This measure should tell us how well the model fits real data. For example, for a simple linear model, a correlation coefficient and a coefficient of determination can be considered to indicate a model quality. As will be shown in the next sections, these statistical measures of model quality are based on linear correlation analysis.

Example: Let us consider a data set composed of four pairs ($N = 4$) of the experimental data, with one independent variable x and one dependent variable y (see Table 11.2).
The optimal parameters of the regression line $y = b + ax$ are found as

$$d = 4 \sum_{i=1}^{4} (x^i)^2 - \left(\sum_{i=1}^{4} x^i y^i \right)^2 = 4 \cdot 30 - 10^2 = 20$$

$$\hat{a} = \frac{1}{d} \left(4 \sum_{i=1}^{4} x^i y^i - \sum_{i=1}^{4} x^i \sum_{i=1}^{4} y_i \right) = \frac{1}{20} (4 \cdot 25.8 - 10 \cdot 9.2) = 0.56$$

and

$$\hat{b} = \frac{1}{d} \left(\sum_{i=1}^{4} (x^i)^2 \sum_{i=1}^{4} y^i - \sum_{i=1}^{4} x_i \sum_{i=1}^{4} x^i y^i \right) = \frac{1}{20} (30 \cdot 9.2 - 10 \cdot 25.8) = 0.9$$

The resulting regression line $y = 0.9 + 0.56x$ represents the best model of the data with respect to the minimum of the least squares of regression errors. We see that, for a given set of pairs of data (x^i, y^i) ($i = 1, 2, 3, 4$), the optimal regression line provides a data fit with the following regression errors (regression residuals):

$$e_1 = y^1 - (\hat{b} + \hat{a}x^1) = 1.5 - (0.9 + 0.56 \cdot 1) = 0.04$$

Table 11.2. Sample of four data points.

x	y
1.0	1.5
2.0	1.9
3.0	2.7
4.0	3.1

$$e_2 = y^2 - (\hat{b} + \hat{a}x^2) = 1.9 - (0.9 + 0.56 \cdot 2) = -0.12$$

$$e_3 = y^3 - (\hat{b} + \hat{a}x^3) = 2.7 - (0.9 + 0.56 \cdot 3) = 0.12$$

$$e_4 = y^4 - (\hat{b} + \hat{a}x^4) = 3.1 - (0.9 + 0.56 \cdot 4) = -0.04$$

The corresponding value of the performance criterion is

$$J(\hat{a}, \hat{b}) = J(0.56, 0.9) = (0.04)^2 + (-0.12)^2 + (0.12)^2 + (-0.04)^2 = 0.0304$$

Optimal Parameter Values in the Minimum Least Squares Sense. For the given set T containing N data points (x^i, y^i), the regression parameter – the slope b and the intercept a of the line equation (regression equation) $y = b + ax$ that best fits the data in the sense of the minimum of least square errors – can be found by Equations (201) and (202).

As we may recall, the required conditions for a valid linear regression are as follow:

- The error term $e = y - (b + ax)$ is normally distributed.
- The error variance is the same for all values of x.
- Error are independent of each other.

Quality of the Linear Regression Model and Linear Correlation Analysis.. Having computed the parameter of the linear model, we now would like to evaluate a measure that will tell us about the quality of a model. In other words, we want to know how well a regression line represents the true relationship between points in the data. The assessment of model quality raises the following questions:

- Is a linear regression model appropriate for the data set at hand?
- Does a linear relationship actually exist between the two variables x and y, and what is its strength?

The concept of linear correlation analysis (with the resulting correlation coefficient) can be used as a measure of how well the trends predicted by the values follow the trends in the training data. Similarly, the coefficient of determination can be used to measure how well the regression line fits the data points.

Correlation of Variables. Correlation describes the strength, or degree, of linear relationship. That is, correlation lets us specify to what extent two variables vary together. Correlation analysis is used to assess the simultaneous variability of a collection of variables. The relationships among variables in a correlation analysis are generally not directional.

Autocorrelation. Autocorrelation is correlation over time in which the level of dependency observed at some time period affects the level of response in the next time period. Autocorrelation plays an important role in processing time series (temporal sequential data).

Correlation Coefficient. The correlation coefficient is a measure of the quality of least squares fitting the training data. The linear **correlation coefficient** is a statistical measure of the strength and direction of a linear relationship between two variables. The linear correlation coefficient is also known as the product-moment coefficient of correlation or Pearson's correlation. In other words, the correlation coefficient r is a measure of how well the predicted values computed by linear regression model "fit" the true data. The correlation coefficient, r, is a measure of the reliability of the linear relationship between values of the independent variable x and the dependent variable y. The correlation coefficient takes scalar real values from the range $-1 \leq r \leq 1$. A value of $r = 1$ indicates an exact linear relationship between x and y. Values of r close to 1 indicate excellent linear reliability. If the correlation coefficient is relatively far away from 1, the predictions based on the linear relationship are less reliable.

The correlation coefficient is defined as the covariance of the independent variable x and the dependent variable y divided by the product of the standard deviations of x and y. Given a data set containing N data points (x^i, y^i), the correlation coefficient r is calculated as

$$r = \frac{S_{xy}}{\sqrt{S_x}\sqrt{S_y}} = \frac{\frac{\sum_{i=1}^{N}(x^i-\bar{x})(y^i-\bar{y})}{N}}{\sqrt{\frac{\sum_{i=1}^{N}(x^i-\bar{x})^2}{N}}\sqrt{\frac{\sum_{i=1}^{N}(y^i-\bar{y})^2}{N}}} \quad (203)$$

The correlation coefficient can also be computed using the formula

$$r = \frac{N\sum_{i=1}^{N}(x^i y^i) - \sum_{i=1}^{N} x^i \sum_{i=1}^{N} y^i}{\sqrt{N\sum_{i=1}^{N}(x^i)^2 - (\sum_{i=1}^{N} x^i)^2}\sqrt{N\sum_{i=1}^{N}(y^i)^2 - (\sum_{i=1}^{N} y^i)^2}} \quad (204)$$

It can be easily observed that if there is no relationship between the values predicted by the model and the actual true values, the correlation coefficient is 0 or is very low (the predicted values are no better than random numbers). As the strength of the relationship between the predicted values and the actual values increases, so does the correlation coefficient.

The signs $+$ and $-$ are used to distinguish positive linear correlations and negative linear correlations, respectively. Positive values of the correlation coefficient indicate a relation between x and y such that, as values of x increase, then the values for y increase as well. If x and y have a strong positive linear correlation, then r is close to $+1$. For the perfect positive fit, when all data points lie exactly on a straight line, the value of r is equal to $+1$. Negative values of r indicate a relationship between x and y such that when values for x increase, the values for y decrease. For a strong negative linear correlation x and y, the value of r is close to -1. If there is no linear correlation (or a weak linear correlation) between x and y, then the value for r is zero or near zero (indicating that there is a random, nonlinear relationship between the two variables).

Coefficient of Determination. In linear regression analysis, we also use another statistical measure, the **coefficient of determination**, denoted by r^2. The **coefficient of determination** r^2 measures how well the regression line represents (fits) the data.

The coefficient tells us how much of the variation in y is explained by the linear relationship with x. Formally, the coefficient of determination computes the proportion of the variance (fluctuation) of one variable (y) that is predicted based on the other variable (x). The coefficient of determination can be defined as

- the percent of variation in the dependent variable y that can be explained by the regression equation,
- the explained variation in y divided by the total variation, or
- the square of r (correlation coefficient).

Every sample has some variation in it. The total variation is made up of two parts, namely the part that can be explained by the regression equation and the part that cannot. The ratio of the explained variation to the total variation is a measure of how well the regression line represents the given dataset. If the regression line passes perfectly through every point on the scatter plot, this model is able to explain all the variation. The further the regression line is from the points, the less the model is able to explain (see Figure 11.17).

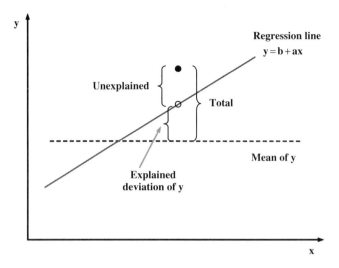

Figure 11.17. Explained and unexplained variation in y.

The coefficient of determination takes real values from the range $0 \leq r^2 \leq 1$ and denotes the strength of the linear association between x and y. The coefficient of determination represents the percent of the data that is closest to the optimal regression line (the optimal linear model) that best fits the data.

Example: If the coefficient of correlation has the value $r = 0.9327$, then the value of the coefficient of determination is $r^2 = 0.8700$. It can be understood that 87% portion of the total variation in y can be explained by the linear relationship between x and y, as it is described by the optimal regression model of the data. The remaining portion 13% of the total variation in y remains unexplained.

Let us recall the sums of squared deviations of regression variables. We can write that the total sum of the squared variations in y, $\sum_{i=1}^{N}(y^i - \bar{y})^2$, is equal to the sum of the squared deviations of the regression line (explained by regression model line), $\sum_{i=1}^{N}(y^{i,\text{est}} - \bar{y})^2$, plus the sum of squared unexplained variations (errors) $\sum_{i=1}^{N}(y^i - y^{i,\text{est}})^2$. The total sum of squared variations in y is expressed by:

$$\sum_{i=1}^{N}(y^i - \bar{y})^2 = \sum_{i=1}^{N}(y^{i,\text{est}} - \bar{y})^2 + \sum_{i=1}^{N}(y^i - y^{i,\text{est}})^2 \tag{205}$$

The coefficient of determination r^2 is the proportion of a sample variance of a dependent variable that is "explained" by the independent variables when a linear regression is provided.

The coefficient of determination can be computed as follows:

$$r^2 = \frac{\text{explained variation}}{\text{total variation}} = \frac{\text{explained sum of squares}}{\text{total sum of squares}}$$

$$= \frac{\sum_{i=1}^{N}(y^{i,\text{est}} - \bar{y})^2}{\sum_{i=1}^{N}(y^i - \bar{y})^2} \tag{206}$$

where y^i is the value for the dependent variable y, $\bar{y}$ is the average of the dependent value, and $y_{i,\text{est}}$ are predicted values for the dependent variable (the predicted values are calculated using the regression equation):

$$r^2 = \frac{S_{xy}^2}{S_x S_y} \tag{207}$$

Given a data set containing N data points (x^i, y^i), the coefficient of determination r^2 is calculated as

$$r^2 = \frac{\left[N \sum_{i=1}^{N} (x^i y^i) - \sum_{i=1}^{N} x^i \sum_{i=1}^{N} y^i\right]^2}{\left[N \sum_{i=1}^{N} (x^i)^2 - \left(\sum_{i=1}^{N} x^i\right)^2\right]\left[N \sum_{i=1}^{N} (y^i)^2 - \left(\sum_{i=1}^{N} y^i\right)^2\right]} \tag{208}$$

If $r^2 = 0$, then there is no linear relation between x and y. If $r^2 = 1$, then there is a perfect match of points and the regression line.

Matrix Version of Simple Linear Regression Based on Least Squares Method. The least squares regression of the experimental data may be written using a matrix form. Let us consider linear regression of experimental data $\{(x^i, y^i), i = 1, 2, \cdots, N\}$, using the more general linear form of the model:

$$y = f(x) = a_0 + a_1 \phi_1(x) \tag{209}$$

with a_0 representing the intercept b (bias), a_1 representing the slope a, and ϕ being a function of x. We denote the set of experimental data as two N-element column vectors:

$$\mathbf{x} = \begin{bmatrix} x^1 \\ x^2 \\ \vdots \\ x^N \end{bmatrix}, \quad \mathbf{y} = \begin{bmatrix} y^1 \\ y^2 \\ \vdots \\ y^N \end{bmatrix} \tag{210}$$

where $\mathbf{x} \in \mathbb{R}^N$ and $\mathbf{y} \in \mathbb{R}^N$ are composed of continuous valued elements (taking on real values). Vector $\mathbf{x}$ contains values of the independent variable x and vector $\mathbf{y}$ values of the dependent variable y.

We define a two-element model parameter vector as

$$\mathbf{a} = \begin{bmatrix} a_0 \\ a_1 \end{bmatrix} = \begin{bmatrix} b \\ a \end{bmatrix} \tag{211}$$

For a simple line equation $b + ax$, we have $\phi_1(x) = x$. The data vector representing one pair of experimental data (x^i, y^i) can be defined as a 1×2 row vector:

$$\phi_i = [1 \phi_1(x^i)] = [1 \ x^i] \tag{212}$$

The value 1 in the data vector allows us to consider bias in the line model $y = b + ax$.

The $N \times 2$ data matrix for all N experimental data can be defined as

$$\Phi = \begin{bmatrix} \phi_1 \\ \phi_2 \\ \vdots \\ \phi_N \end{bmatrix} = \begin{bmatrix} 1 & \phi_1(x^1) \\ 1 & \phi_1(x^2) \\ \vdots & \vdots \\ 1 & \phi_1(x^N) \end{bmatrix} = \begin{bmatrix} 1 & x^1 \\ 1 & x^2 \\ \vdots & \vdots \\ 1 & x^N \end{bmatrix} \tag{213}$$

2. Regression

The matrix form of the model description (the estimation of $\hat{y}$) for all N experimental data points is written as follows:

$$\hat{\mathbf{y}} = \mathbf{\Phi}\,a = \begin{bmatrix} 1 & \phi_1(x^1) \\ 1 & \phi_1(x^2) \\ \vdots & \vdots \\ 1 & \phi_i(x^N) \end{bmatrix} \begin{bmatrix} a_0 \\ a_1 \end{bmatrix} = \begin{bmatrix} 1 & x^1 \\ 1 & x^2 \\ \vdots & \vdots \\ 1 & x^N \end{bmatrix} \begin{bmatrix} b \\ a \end{bmatrix} \quad (214)$$

The error vector $\mathbf{e} \in \mathbb{R}^N$, whose elements represent the difference $y^i - (b + ax^i)$ between the observed value y^i and the corresponding value computed by the model $b + ax^i$, is defined as

$$\mathbf{e} = \begin{bmatrix} e^1 \\ e^2 \\ \vdots \\ e^N \end{bmatrix} = \begin{bmatrix} y^1 - (b + ax^1) \\ y^2 - (b + ax^2) \\ \vdots \\ y^N - (b + ax^N) \end{bmatrix} \quad (215)$$

The regression error in the matrix form is written as

$$\mathbf{e} = \mathbf{y} - \hat{\mathbf{y}} = \mathbf{y} - \mathbf{\Phi}\,a \quad (216)$$

The performance criterion of the least squares may now be expressed, in the compact vector form, as a function of the parameter vector $\mathbf{a}$:

$$J(\mathbf{a}) = \sum_{i=1}^{N}(e^i)^2 = \mathbf{e}^T\mathbf{e} = (\mathbf{y} - \mathbf{\Phi}a)^T(\mathbf{y} - \mathbf{\Phi}a) = ||\mathbf{e}||_2 \quad (217)$$

where $||\mathbf{e}||_2$ is the Euclidan norm

$$\mathbf{e} = \mathbf{y} - \hat{\mathbf{y}} = \mathbf{y} - \mathbf{\Phi}a \quad (218)$$

and $\hat{\mathbf{y}} = \mathbf{\Phi}a$. Finally, we obtain

$$J(\mathbf{a}) = (\mathbf{y} - \mathbf{\Phi}a)^T(\mathbf{y} - \mathbf{\Phi}a) = \mathbf{y}^T\mathbf{y} - \mathbf{y}^T\mathbf{\Phi}a - \mathbf{a}^T\mathbf{\Phi}^T\mathbf{y} + \mathbf{a}^T\mathbf{\Phi}^T\mathbf{\Phi}a \quad (219)$$

We can compare the values of $J(\mathbf{a})$ for different pairs of parameters for the given experimental data $\{(x^i, y^i), (i = 1, 2, \cdots, N)\}$. Now, to solve the problem of the best fit of the experimental data by the linear model

$$y^i = b + ax^i, \quad i = 1, 2, \cdots, N \quad (220)$$

we search the value of the parameter vector $\hat{\mathbf{a}}$ for which the performance (minimization) criterion $J(\mathbf{a}) = \mathbf{e}^T\mathbf{e}$ is minimal. The parameters ($\hat{\mathbf{a}}$) are called **optimal parameters** for the criterion $J(\mathbf{a})$ and the data $\{(x^i, y^i), (i = 1, 2, \cdots, N)\}$. For the optimal parameters, the criterion takes the minimum value $\hat{J}(\hat{\mathbf{a}})$:

$$\hat{\mathbf{a}} = \arg\min_{\mathbf{a}} J(\mathbf{a}) = \arg\min_{\mathbf{a}}(\mathbf{y} - \mathbf{\Phi}a)^T(\mathbf{y} - \mathbf{\Phi}a) \quad (221)$$

We know from calculus that in order to make $J(\mathbf{a})$ a minimum, the following necessary condition for the minimum of a function must be satisfied:

$$\frac{\partial J(\mathbf{a})}{\partial \mathbf{a}} = \frac{\partial}{\partial \mathbf{a}}[\mathbf{y}^T\mathbf{y} - \mathbf{y}^T\mathbf{\Phi}a - \mathbf{a}^T\mathbf{\Phi}^T\mathbf{y} + \mathbf{a}^T\mathbf{\Phi}^T\mathbf{\Phi}a] = 0 \quad (222)$$

After some derivations, we obtain

$$\frac{\partial J(\mathbf{a})}{\partial \mathbf{a}} = -\mathbf{y}^T \mathbf{\Phi} - \mathbf{\Phi}^T \mathbf{y} + 2\mathbf{\Phi}^T \mathbf{\Phi} a \qquad (223)$$

Knowing that $\mathbf{y}^T \mathbf{\Phi} = \mathbf{y}\mathbf{\Phi}^T$, we eventually obtain the **normal equation**:

$$-\mathbf{\Phi}^T \mathbf{y} + \mathbf{\Phi}^T \mathbf{\Phi} a = 0 \qquad (224)$$

This equation solves for the unique minimum solution for an optimal parameter vector $\mathbf{a} = \hat{\mathbf{a}}$,

$$\hat{\mathbf{a}} = (\mathbf{\Phi}^T \mathbf{\Phi})^{-1} \mathbf{\Phi}^T \mathbf{y} \qquad (225)$$

if the matrix of the second derivatives of a performance criterion (called Hessian)

$$\frac{\partial J^2(\mathbf{a})}{\partial \mathbf{a}^2} = 2(\mathbf{\Phi}^T \mathbf{\Phi}) \qquad (226)$$

is positive definite.

The value of the criterion for the optimal parameter vector is

$$J\hat{\mathbf{a}} = \frac{1}{2}(\mathbf{y} - \mathbf{\Phi}\hat{\mathbf{a}})^T (\mathbf{y} - \mathbf{\Phi}\hat{\mathbf{a}}) \qquad (227)$$

The matrix $(\mathbf{\Phi}^T \mathbf{\Phi})^{-1}$ is called the **pseudoinverse** of the $\mathbf{\Phi}$ if the matrix $\mathbf{\Phi}^T \mathbf{\Phi}$ is nonsingular. The invertible matrix $\mathbf{\Phi}^T \mathbf{\Phi}$ is called **the excitation matrix**. If the Hessian of the performance criterion is positive definite, the solution for the optimal parameters gives a minimum value of the performance criterion $J(\mathbf{a})$.

The error vector $\mathbf{e} \in \mathbb{R}^N$, whose elements represent the difference $y^i - (\hat{b} + \hat{a}x^i)$ between the observed value y^i and the corresponding values computed by the model $\hat{b} + \hat{a}x^i$, is

$$\mathbf{e} = \begin{bmatrix} e^1 \\ e^2 \\ \vdots \\ e^N \end{bmatrix} = \begin{bmatrix} y^1 - (\hat{b} + \hat{a}x^1) \\ y^2 - (\hat{b} + \hat{a}x^2) \\ \vdots \\ y^N - (\hat{b} + \hat{a}x^N) \end{bmatrix} \qquad (228)$$

The regression error for the model with the optimal parameter vector can be written as

$$\mathbf{e} = \mathbf{y} - \hat{\mathbf{y}} = \mathbf{y} - \mathbf{\Phi}\hat{\mathbf{a}} \qquad (229)$$

with elements representing regression errors.

Example: Let us consider again the dataset shown in Table 11.3.

Table 11.3. Sample data set.

x	y
1	1.5
2	1.9
3	2.7
4	3.1

2. Regression

We will fit the linear regression model $y = b + ax$ to the data. First, we compose vectors $\mathbf{x}$ and $\mathbf{y}$:

$$\mathbf{x}^T = [1\ 2\ 3\ 4],\ \mathbf{y}^T = [1.5\ 1.9\ 2.7\ 3.1]$$

We have $\phi_1(x) = x$, and the data vector representing one pair of experimental data (x^i, y^i) as the 1×2 row vector is

$$\phi_i = [1\ \phi_1(x^i)] = [1\ x^i]$$

Eventually, we compose the data matrix

$$\mathbf{\Phi} = \begin{bmatrix} \phi_1 \\ \phi_2 \\ \phi_3 \\ \phi_4 \end{bmatrix} = \begin{bmatrix} 1 & \phi_1(x^1) \\ 1 & \phi_1(x^2) \\ 1 & \phi_1(x^3) \\ 1 & \phi_1(x^4) \end{bmatrix} = \begin{bmatrix} 1 & x^1 \\ 1 & x^2 \\ 1 & x^3 \\ 1 & x^4 \end{bmatrix} = \begin{bmatrix} 1 & 1 \\ 1 & 2 \\ 1 & 3 \\ 1 & 4 \end{bmatrix}$$

The optimal model parameters can be found as

$$\hat{\mathbf{a}} = (\mathbf{\Phi}^T \mathbf{\Phi})^{-1} \mathbf{\Phi}^T \mathbf{y}$$

$$= \left(\begin{bmatrix} 1 & 1 & 1 & 1 \\ 1 & 2 & 3 & 4 \end{bmatrix} \begin{bmatrix} 1 & 1 \\ 1 & 2 \\ 1 & 3 \\ 1 & 4 \end{bmatrix} \right)^{-1} \begin{bmatrix} 1 & 1 & 1 & 1 \\ 1 & 2 & 3 & 4 \end{bmatrix} \begin{bmatrix} 1.5 \\ 1.9 \\ 2.7 \\ 3.1 \end{bmatrix}$$

$$= \begin{bmatrix} 0.9 \\ 0.56 \end{bmatrix}$$

where $\hat{a}_1 = 0.9$ is a bias (b) and $\hat{a}_2 = 0.56$ is the slope (a). We have obtained the optimal linear regression model $y = 0.56x + 0.9$. The residuals (regression errors) are

$$\mathbf{e} = \begin{bmatrix} 0.0400 \\ -0.1200 \\ 0.1200 \\ -0.0400 \end{bmatrix}$$

2.3. Multiple Regression

The multiple regression analysis is the statistical technique of exploring the relation (association) between the set of n independent variables that are used to explain the variability of one (generally many) dependent variable y.

The simplest multiple regression analysis refers to data with two independent variables x_1 and x_2, one dependent variable y, and the selected linear regression model:

$$y = f(x_1, x_2) = a_0 + a_1 x_1 + a_2 x_2 \tag{230}$$

where a_0, a_1, and a_2 are model parameters. This model represents a plane in three-dimensional space (x_1, x_2, y). As the graphical interpretation of multiple linear regression, we can imagine fitting the regression plane to data points obtained as a **three-dimensional** scatter plot.

Generally, for linear multiple regression with n independent variables $x_1, x_2, \cdots, x_n$, denoted as the n-dimensional vector $\mathbf{x} \in \mathbb{R}^n$, and the $n+1$ parameter vector $a_0, a_1, \cdots, a_n$, denoted as the $(n+1)$-dimensional parameter vector $\mathbf{a} \in \mathbb{R}^{n+1}$,

$$\mathbf{x} = \begin{bmatrix} x_1 \\ x_2 \\ \vdots \\ x_n \end{bmatrix},\ \mathbf{a} = \begin{bmatrix} a_0 \\ a_1 \\ \vdots \\ a_n \end{bmatrix} \tag{231}$$

the linear model can be defined as

$$y = f(x_1, x_2, \cdots, x_n) = a_0 + a_1 x_1 + a_2 x_2 + \cdots, + a_n x_n \qquad (232)$$

Using vector notation, we may also write

$$y = f(\mathbf{x}) = \mathbf{a}^T \mathbf{xa}, \qquad \text{where} \quad \mathbf{xa} = [1, x_1, x_2, \cdots, x_n]^T \qquad (233)$$

This regression model is represented by a **hyperplane** in $(n+1)$-dimensional space.

Because the observations $y^i (i = 1, \cdots, N)$ are random samples, they are mutually independent. Hence, the error terms are also mutually independent, with the same distribution, zero mean, and constant variances. The distribution of the error term is independent of the joint distribution of explanatory variables. We also assume that unknown parameters of regression models are constants. Proper computations require that the inequality $N \geq n$ holds.

Geometrical Interpretation: Regression Errors. The multiple regression analysis can be geometrically interpreted as follows. We can imagine that N data cases $(\mathbf{x}^i, y^i)$ are represented as N points in $(n+1)$-dimensional space. The goal of multiple regression is to find a hyperplane in the $(n+1)$-dimensional space that will best fit the data. We can solve this minimization problem by using the method of least squares, with the performance criterion being the sum of squared errors

$$J(\mathbf{a}) = \sum_{i=1}^{N} \left(y^i - \left(a_0 + a_1 x_1^i + \cdots, a_n x_n^i \right) \right)^2 \qquad (234)$$

or by using vector notation:

$$J(\mathbf{a}) = \sum_{i=1}^{N} \left(y^i - \mathbf{a}^T \mathbf{xa} \right)^2, \qquad \text{where} \quad \mathbf{xa} = [1, x_1, x_2, \cdots, x_n]^T \qquad (235)$$

A multiple regression error for the ith case $(\mathbf{x}^i, y^i)$ is the difference between the predicted value $f(\mathbf{x}^i)$ (which is on the hyperplane) and the true value y^i from the ith case. In other words, an error is the "vertical" distance between the ith point and the point on the plane that represents the predicted value. The optimal hyperplane guarantees minimization of the sum of squares of errors. The error variance and standard error of the estimate in multiple regression are

$$S_e^2 = \frac{\sum_{i=1}^{N} (y^i - y^{i,\text{est}})^2}{N - n - 1}, \quad S_e = \sqrt{\frac{\sum_{i=1}^{N} (y^i - y^{i,\text{est}})^2}{N - n - 1}} \qquad (236)$$

where y^i is the value of the dependent variable for the ith data case, $\hat{y}^{i,\text{est}}$ is the corresponding value estimated from the regression model equation, N is the number of cases in the data set, and n is the number of independent variables. The standard error of the estimate informs us about the variability of the estimation error and can be used to compare models.

Degree of Freedom. The denominator $N - n - 1$ in the previous equation tells us that in multiple regression with n independent variables, the standard error has $N - n - 1$ degrees of freedom. The degree of freedom has been reduced from N by $n + 1$ because $n + 1$ numerical parameters $a_0, a_1, a_2, \cdots, a_n$ of the regression model have been estimated from the data.

2.4. General Least Squares and Multiple Regression

Gauss proposed the least squares principle in the eighteenth century in the following way: "the sum of squares of the differences between the actually observed and computed values multiplied by numbers that measure the degree of precision is a minimum." The least squares method has been widely used ever since.

Let us assume that we have carried out experiments in a physical system and have recorded the scalar outputs y as a result of some signals (variables) $x_1, \cdots, x_n$ associated with the process input, output, environment, or internal states that have influenced the system behavior. Let us form an n-element vector constituted with all the considered input variables (independent variables) $\mathbf{x} = [x_1, x_2, \cdots, x_n]^T$, $\mathbf{x} \in \mathbb{R}^n$. Let us assume that we have recorded the set $\{(\mathbf{x}^i, y^i), i = 1, 2, \cdots, N\}$ of N experimental observations. We are searching for the regression model of the data given in the general form of the weighted sum of m functions:

$$y^i = a_0 + a_1 \phi_1(\mathbf{x}^i) + a_2 \phi_2(\mathbf{x}^i) + \cdots + a_m \phi_m(\mathbf{x}^i) \tag{237}$$

where $\phi_k(\mathbf{x}^i), k = 1, 2, \cdots, m$, is the known function of input-variables vector $\mathbf{x}$. In the regression model, $a_0, a_1, a_2, \cdots, a_m$ represent model parameters denoted by the $(m+1)$-element parameter vector $\mathbf{a} = [a_0, a_1, a_2, \cdots, a_m]^T$, $\mathbf{a} \in \mathbb{R}^{m+1}$. The m, representing the number of functions $\phi_k(\mathbf{x}^i)$, is called the **order of the model**. Proper computations require that the inequality $N \geq m$ holds. We want to find the values of the regression model parameters $\mathbf{b}$, which provide that the values $\hat{y}^i$, computed by the model for the given value of the input-variables vector $\mathbf{x}^i$ for the ith case, agree as closely as possible with the measured values y^i of the variable y. To answer the question of which model parameters "best" fit the experimental data, we define the performance criterion (performance index) that will allow us to objectively judge, in a quantitative way, the quality of a fit. As we have already discussed, the most natural criterion is the sum of squares of all errors for all data cases $e^i = y^i - \hat{y}^i$:

$$J(\mathbf{a}) = \sum_{i=1}^{N}(e^i)^2 = \sum_{i=1}^{N}(y^i - \hat{y}^i)^2$$

$$= \sum_{i=1}^{N}[y^i - (a_0 + a_1 \phi_1(\mathbf{x}^i) + a_2 \phi_2(\mathbf{x}^i) + \cdots + a_m \phi_m(\mathbf{x}^i))]^2 \tag{238}$$

The performance criterion for given regression variables, $J(\mathbf{a})$ is the function of the model parameters denoted as vector $\mathbf{a}$. For the given values of the parameters and the given set of experimental data $\{(\mathbf{x}^i, y^i), (i = 1, 2, \cdots, N)\}$, the criterion $J(\mathbf{a})$ takes a scalar numerical value. Additionally, for the given parameters, the model will be responsible for the following regression errors (residues) for the given pair of experimental data $(\mathbf{x}^i, y^i)(i = 1, 2, \cdots, N)$:

$$\begin{aligned} e^i &= y^i - \hat{y}^i \\ &= y^i - [a_0 + a_1 \phi_1(\mathbf{x}^i) + a_2 \phi_2(\mathbf{x}^i) + \cdots + a_m \phi_m(\mathbf{x}^i)], i = 1, 2, \cdots, N \end{aligned} \tag{239}$$

Let us define the $(m+1)$-dimensional row **data vector**

$$\phi_i = [1 \; \phi_1(\mathbf{x}^i) \; \phi_2(\mathbf{x}^i), \cdots, \phi_m(\mathbf{x}^i)] \tag{240}$$

where the first element equal to 1 allows us to consider bias in the model.

The set of the process-N observed outputs (values of the dependent variable) is denoted by the N-element column vector $\mathbf{y}$, and the errors are represented by the N-element column vector $\mathbf{e}$:

$$\mathbf{y} = [y^1, y^2, \cdots, y^N]^T \quad \mathbf{e} = [e^1, e^2, \cdots, e^N]^T \tag{241}$$

Finally, the set of all functions $\phi_k(\mathbf{x}^i)$, $(k = 1, 2, \cdots, m; i = 1, 2, \cdots, N)$ for all N cases of recorded experiment is denoted by the $N \times (m+1)$ **data matrix (design matrix)**:

$$\boldsymbol{\Phi} = \begin{bmatrix} 1 & \phi_1(\mathbf{x}^1) & \phi_2(\mathbf{x}^1) & \cdots & \phi_m(\mathbf{x}^1) \\ 1 & \phi_1(\mathbf{x}^2) & \phi_2(\mathbf{x}^2) & \cdots & \phi_m(\mathbf{x}^2) \\ \vdots & \vdots & \vdots & \ddots & \vdots \\ 1 & \phi_1(\mathbf{x}^N) & \phi_2(\mathbf{x}^N) & \cdots & \phi_m(\mathbf{x}^N) \end{bmatrix} = \begin{bmatrix} \phi_1^T \\ \phi_2^T \\ \vdots \\ \phi_N^T \end{bmatrix} \quad (242)$$

The matrix form of the model description for all N experimental data is written as follows:

$$\hat{\mathbf{y}} = \begin{bmatrix} 1 & \phi_1(\mathbf{x}^1) & \phi_2(\mathbf{x}^1) & \cdots & \phi_m(\mathbf{x}^1) \\ 1 & \phi_1(\mathbf{x}^2) & \phi_2(\mathbf{x}^2) & \cdots & \phi_m(\mathbf{x}^2) \\ \vdots & \vdots & \vdots & \ddots & \vdots \\ 1 & \phi_1(\mathbf{x}^N) & \phi_2(\mathbf{x}^N) & \cdots & \phi_m(\mathbf{x}^N) \end{bmatrix} \begin{bmatrix} a_0 \\ a_1 \\ \vdots \\ a_m \end{bmatrix} = \boldsymbol{\Phi}\mathbf{a} \quad (243)$$

The performance criterion of the least squares may now be expressed in the compact vector form:

$$J(\mathbf{a}) = \sum_{i=1}^{N} (e^i)^2 = \mathbf{e}^T \mathbf{e} = (\mathbf{y} - \boldsymbol{\Phi}a)^T(\mathbf{y} - \boldsymbol{\Phi}a) = \|\mathbf{e}\|_2 \quad (244)$$

where

$$\mathbf{e} = \mathbf{y} - \hat{\mathbf{y}} = \mathbf{y} - \boldsymbol{\Phi}a \quad (245)$$

are modeling errors (residuals) and $\hat{\mathbf{y}} = \boldsymbol{\Phi}a$ is computed by the regression model as a vector of predicted values of the dependent variable.

Finally, we have

$$J(\mathbf{a}) = (\mathbf{y} - \boldsymbol{\Phi}a)^T(\mathbf{y} - \boldsymbol{\Phi}a) = \mathbf{y}^T\mathbf{y} - \mathbf{y}^T\boldsymbol{\Phi}\mathbf{a} - \mathbf{a}^T\boldsymbol{\Phi}^T\mathbf{y} + \mathbf{a}^T\boldsymbol{\Phi}^T\boldsymbol{\Phi}\mathbf{a} \quad (246)$$

We can compare the values of $J(\mathbf{a})$ for different sets of parameters for given experimental data $\{(\mathbf{x}^i, y_i), (i = 1, 2, \cdots, N)\}$. Now, to solve the problem of the best fit by the linear regression model

$$y_i = a_0 + a_1\phi_1(\mathbf{x}^i) + a_2\phi_2(\mathbf{x}^i) + \cdots + a_m\phi_m(\mathbf{x}^i) = \phi_i^T \mathbf{a}, \quad i = 1, 2, \cdots, N \quad (247)$$

we should search for the value of the parameter vector $\hat{\mathbf{a}}$ for which the performance criterion $J(\mathbf{a}) = \mathbf{e}^T\mathbf{e}$ is minimal. Parameters ($\hat{\mathbf{a}}$) are called the **optimal parameters** for the criterion $J(\mathbf{a})$ and the data $\{(\mathbf{x}^i, y^i), (i = 1, 2, \cdots, N)\}$. For the optimal parameters, the criterion will take the minimum value $\hat{J}(\hat{\mathbf{a}})$:

$$\hat{\mathbf{a}} = \arg\min_{\mathbf{a}} J(\mathbf{a}) = \arg\min_{\mathbf{a}} (\mathbf{y} - \boldsymbol{\Phi}a)^T(\mathbf{y} - \boldsymbol{\Phi}a) \quad (248)$$

To produce a minimum of $J(\mathbf{a})$, we must satisfy the following necessary condition for the minimum of a function:

$$\frac{\partial J(\mathbf{a})}{\partial \mathbf{a}} = \frac{\partial}{\partial \mathbf{a}} [\mathbf{y}^T\mathbf{y} - \mathbf{y}^T\boldsymbol{\Phi}\mathbf{a} - \mathbf{a}^T\boldsymbol{\Phi}^T\mathbf{y} + \mathbf{a}^T\boldsymbol{\Phi}^T\boldsymbol{\Phi}\mathbf{a}] = 0 \quad (249)$$

After some derivations, we obtain

$$\frac{\partial J(\mathbf{a})}{\partial \mathbf{a}} = -\mathbf{y}^T\boldsymbol{\Phi} - \boldsymbol{\Phi}^T\mathbf{y} + 2\boldsymbol{\Phi}^T\boldsymbol{\Phi}\mathbf{a} \quad (250)$$

Knowing that $\mathbf{y}^T\boldsymbol{\Phi} = \mathbf{y}\boldsymbol{\Phi}^T$, we eventually obtain the **normal equation**

$$\boldsymbol{\Phi}^T\mathbf{y} + \boldsymbol{\Phi}^T\boldsymbol{\Phi}\mathbf{a} = 0 \tag{251}$$

whose solution gives us the value of the optimal parameter vector $\hat{\mathbf{a}}$:

$$\hat{\mathbf{a}} = (\boldsymbol{\Phi}^T\boldsymbol{\Phi})^{-1}\boldsymbol{\Phi}^T\mathbf{y} \tag{252}$$

The value of the performance criterion (a sum of squared errors) for the optimal parameter vector is

$$J(\hat{\mathbf{a}}) = (\mathbf{y} - \boldsymbol{\Phi}\hat{\mathbf{a}})^T(\mathbf{y} - \boldsymbol{\Phi}\hat{\mathbf{a}}) \tag{253}$$

The matrix $(\boldsymbol{\Phi}^T\boldsymbol{\Phi})^{-1}$ is called the **pseudo-inverse** of matrix $\boldsymbol{\Phi}$ if the matrix $\boldsymbol{\Phi}^T\boldsymbol{\Phi}$ is nonsingular. The invertible matrix $\boldsymbol{\Phi}^T\boldsymbol{\Phi}$ is called the **excitation matrix**. If the Hessian (the matrix of the second derivatives of the performance criterion)

$$\frac{\partial J^2(\mathbf{a})}{\partial \mathbf{a}^2} = 2(\boldsymbol{\Phi}^T\boldsymbol{\Phi}) \tag{254}$$

is positive definite, the normal equation solves for the unique minimum solution for an optimal parameter vector $\mathbf{a} = \hat{\mathbf{a}}$

$$\hat{\mathbf{a}} = (\boldsymbol{\Phi}^T\boldsymbol{\Phi})^{-1}\boldsymbol{\Phi}^T\mathbf{y} \tag{255}$$

Practical, Numerically Stable Computation of the Optimal Model Parameters. The solution for the optimal least-squares parameters is almost never computed from the equation $\hat{\mathbf{a}} = (\boldsymbol{\Phi}^T\boldsymbol{\Phi})^{-1}\boldsymbol{\Phi}^T\mathbf{y}$ due to its poor numerical performance in cases when the matrix $\boldsymbol{\Phi}^T\boldsymbol{\Phi}$ (the covariance matrix) is ill conditioned. The most successful remedies rely on various matrix decomposition methods, such as the prominent LU/LDL^T, QR, Cholesky, or UDU^T decompositions (see Appendix A).

For the known decompositions we have the following solutions

Method	Decomposition	Solution of least-squares
QR	$\boldsymbol{\Phi} = \mathbf{QR}$	$\hat{\mathbf{a}} = \mathbf{R}^{-1}\mathbf{Q}^T\mathbf{y}$
LU	$\boldsymbol{\Phi}^T\boldsymbol{\Phi} = \mathbf{LU}$	$\hat{\mathbf{a}} = \mathbf{U}^{-1}\mathbf{L}^{-1}(\boldsymbol{\Phi}^T\mathbf{y})$
LDL^T	$\boldsymbol{\Phi}^T\boldsymbol{\Phi} = \mathbf{LDL}^T$	$\hat{\mathbf{a}} = (\mathbf{L}^T)^{-1}\mathbf{D}^{-1}\mathbf{L}^{-1}(\boldsymbol{\Phi}^T\mathbf{y})$
Cholesky	$\boldsymbol{\Phi}^T\boldsymbol{\Phi} = \mathbf{GG}^T$	$\hat{\mathbf{a}} = (\mathbf{G}^T)^{-1}\mathbf{G}^{-1}(\boldsymbol{\Phi}^T\mathbf{y})$

2.5. Assessing the Quality of the Multiple Regression Model

The quality of the multiple regression model informs us how well the multiple regression hyperplane (regression model) fits the data. The following measures can be used to assess the quality of the multiple regression model:

– Standard error of estimate
– Coefficient of multiple determination
– Multiple correlation coefficient
– C_p statistic

The Coefficient of Multiple Determination, R^2. The fit of the multiple regression model to the data can be assessed by the coefficient of multiple determination.

We recall that in the simple linear regression analysis, the coefficient of correlation r between two (generally random) variables x (explanatory) and y (dependent) is a numerical measure of the relation (association) between these two variables. The square of the coefficient of correlation defines the coefficient of determination r^2, which indicates the portion of variance in the dependent variable y that is accounted for by the variation in the independent variable x.

A multiple regression counterpart of the coefficient of determination r^2 is the coefficient of multiple determination, R^2.

Let us recall definitions of squared variations. We remember that the total sum of the squared variations in y,

$$S_{\text{total}} = \sum_{i=1}^{N}(y^i - \bar{y})^2 \qquad (256)$$

is equal to the sum of

(1) the sum of squared deviations of the regression hyperplane (due to regression; explained by the regression model), and
(2) the sum of squared unexplained variations (due to errors).

Thus, the total sum of squared variations in y is expressed by

$$S_{\text{total}} = S_{ye} + S_{re}$$

$$S_{\text{total}} = \sum_{i=1}^{N}(y^i - \bar{y})^2 = \sum_{i=1}^{N}(y^{i,\text{est}} - \bar{y})^2 + \sum_{i=1}^{N}(y^i - y^{i,\text{est}})^2 \qquad (257)$$

The coefficient of multiple determination R^2 is the percent of the variance in the dependent variable that can be explained by all of the independent variables taken together.

The coefficient of determination can be computed using the formula:

$$R^2 = \frac{\text{explained sum of squares}}{\text{total sum of squares}}$$

or

$$R^2 = \frac{S_{ye}}{S_{\text{total}}} = \frac{\sum_{i=1}^{N}(y^{i,\text{est}} - \bar{y})^2}{\sum_{i=1}^{N}(y^i - \bar{y})^2} \qquad (258)$$

where y^i is the values for the dependent variable y, $\bar{y}$ is the average of the dependent value, and $y^{i,\text{est}}$ are the predicted values for the dependent variable (the predicted values are calculated using the regression equation).

Coefficient R^2 always increases as the number of model design parameters increases. Adjusted R^2 uses the number of design parameters plus a constant that are used in the model and the number of data points N in order to correct the statistic of this coefficient in situations when unnecessary parameters are used in the model structure.

We define the adjusted coefficient of multiple determination R^2_{adj} as

$$R^2_{\text{adj}} = 1 - \frac{N-1}{(N-n-1)}(1-R^2) = 1 - \frac{S_{re}/(N-n-1)}{S_{\text{total}}/(N-1)} \qquad (259)$$

Table 11.4. Three-dimensional data.

x_1	x_2	y
1	4	6
2	5	7
3	8	5
4	2	8

C_p Statistic. To compare multiple regression models C_p, a statistic measure is used:

$$C_p = \frac{\left(1 - R_n^2\right)(N - np)}{\left(1 - R_{np}^2\right)} - (N - 2(n+1)) \qquad (260)$$

where n is the number of independent variables, np is the total number of model parameters (including the intercept), R_n^2 is the coefficient of multiple determination for a regression model with n independent variables, and R_{np}^2 is the coefficient of multiple determination for a regression model with all np estimated parameters.

When comparing alternative regression models, the designer aims to choose models whose values of C_n is close to or below $(n+1)$.

Multiple Correlation. Another measure of model quality is the multiple correlation coefficient R, which computes the amount of correlation between more than two variables. One variable is the dependent variable, and the others are independent variables. A value of R can be found as the positive square root of R^2 (coefficient of multiple determination).

The multiple correlation coefficient, denoted by R, is a measure of the strength of the linear relationship between the dependent variable y and the set of independent variables $x_1, x_2, \ldots, x_n$. The multiple correlation coefficient is a measure of how well the regression equation (hyperplane) fits the data. A value of R close to 1 indicates that the fit is very good. A value near zero indicates that the model is not a good approximation of the data and cannot be efficiently used for prediction.

Example: Let us consider a multiple linear regression analysis for the data set containing $N = 4$ cases, composed with one dependent variable y and two independent variables x_1 and x_2 (see Table 11.4).

Figure 11.18 shows the scatter plot of data points in three-dimensional space (x_1, x_2, y).

We fit to the data the linear multiple regression model, representing a plane $y = f(\mathbf{x}) = a_0 + a_1 x_1 + a_2 x_2$ in the three dimensional space (x_1, x_2, y). First, we compose the four element vector $\mathbf{y}$ of four values of the dependent-variable y (as the third column from the data set):

$$\mathbf{y} = \begin{bmatrix} 6 \\ 7 \\ 5 \\ 8 \end{bmatrix}$$

and the two-element independent variable vector $\mathbf{x} = [x_1, x_2]^T (\mathbf{x} \in \mathbb{R}^2)$. Then we define the three-element model parameter vector

$$\mathbf{a} = \begin{bmatrix} a_0 \\ a_1 \\ a_2 \end{bmatrix}$$

and the data vector

$$\phi_i = [1 \, \phi_1(\mathbf{x}^i) \, \phi_2(\mathbf{x}^i)] = [1 \, x_1 \, x_2]$$

where $\phi_1(\mathbf{x}) = x_1$, $\phi_2(\mathbf{x}) = x_2$. An auxiliary entry of the vector of the data vector set to 1 in the data vector allows us to consider a bias a_0 in the plane model $y = f(\mathbf{x}) = a_0 + a_1 x_1 + a_2 x_2$. Then we form the data matrix

$$\mathbf{\Phi} = \begin{bmatrix} 1 & 1 & 4 \\ 1 & 2 & 5 \\ 1 & 3 & 8 \\ 1 & 4 & 2 \end{bmatrix}$$

The optimal model parameters can be found as

$$\hat{\mathbf{a}} = (\mathbf{\Phi}^T \mathbf{\Phi})^{-1} \mathbf{\Phi}^T \mathbf{y}$$

$$= \left(\begin{bmatrix} 1 & 1 & 1 & 1 \\ 1 & 2 & 3 & 4 \\ 4 & 5 & 8 & 2 \end{bmatrix} \begin{bmatrix} 1 & 1 & 4 \\ 1 & 2 & 5 \\ 1 & 3 & 8 \\ 1 & 4 & 2 \end{bmatrix} \right)^{-1} \begin{bmatrix} 1 & 1 & 1 & 1 \\ 1 & 2 & 3 & 4 \\ 4 & 5 & 8 & 2 \end{bmatrix} \begin{bmatrix} 1.5 \\ 1.9 \\ 2.7 \\ 3.1 \end{bmatrix} = \begin{bmatrix} 0.9 \\ 0.56 \\ 3.1 \end{bmatrix}$$

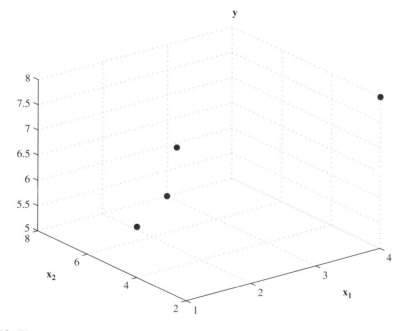

Figure 11.18. The scatter plot for three-dimensional data.

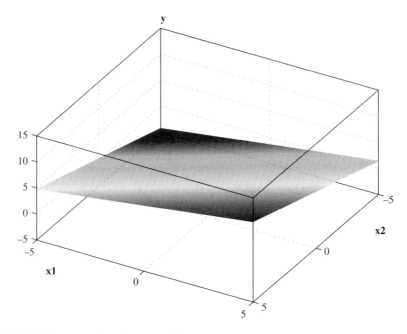

Figure 11.19. Multiple regression, plane model.

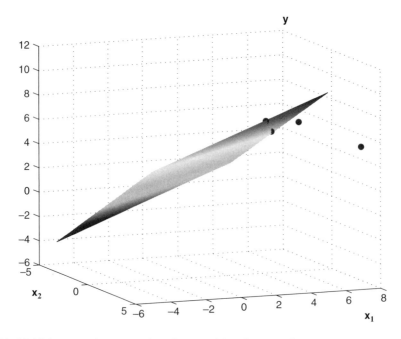

Figure 11.20. Multiple regression, regression plane model and scatter plot.

Now we have obtained the optimal model $y = 3.1 + 0.9x_1 + 0.56x_2$, with the optimal values of parameters $\hat{a}_1 = 0.9$, $\hat{a}_2 = 0.56$, and $\hat{a}_3 = 3.1$. The optimal regression model is the equation of the plane (Figure 11.19) in (x_1, x_2, y) space.

Figure 11.20 shows the plane as a linear regression model for three-dimensional data overlapped by a scatter plot of data. The residuals (errors) are

$$\mathbf{e} = \begin{bmatrix} 0.0400 \\ -0.1200 \\ 0.1200 \\ -0.0400 \end{bmatrix}$$

The criterion value for the optimal parameters is 0.016

3. Summary and Bibliographical Notes

In this Chapter, we have discussed the key concepts of supervised learning realized in the setting of statistical methods. The fundamental nature of these methods stems from the fact that they fully exploit probabilistic knowledge about data. This becomes particularly visible in case of Bayesian methods, which exploits the concept of conditional probabilities and prior probabilities — all of which encapsulate statistical characteristics of the data. We showed that Bayesian classifiers are optimal in the sense that for given probabilistic characteristics of underlying data, the resulting classifier produces prediction with the lowest possible probability of classification error. While the relevance of this fundamental result has to be fully acknowledged, we should remember that the empirical quality of such classifiers depends on the quality of estimation of the underlying probability density functions. This, in turn, emphasizes the role of effective estimation procedures, which have been discussed in detail.

Undoubtedly, regression models constitute one of the fundamental modeling methods. In the presence of several assumptions about the statistical characteristics of data (including normality of data), one can endow the models with several mechanisms of assessment of the quality of the models, including the generation of confidence intervals. While linear regression models can serve as a sound basis, there is a wealth of generalizations that are capable of handling nonlinear relationships and the statistical nature of data.

Bayesian reasoning and classification schemes were discussed from a variety of standpoints and within a diversity of numerous applications; the fundamentals are covered in [2, 3, 5, 7, 13]. Estimation issues are studied in [17, 21]. Various issues of categorical data analysis are presented in [20]. Regression analysis has been a vast and much diversified area of fundamental and applied research. Starting from the seminal study by Gauss [6], there is a long list of excellent references, say, [1, 9, 22]. Statistical aspects of pattern recognition and classification in particular are lucidly exposed in [8, 10, 11, 12, 14, 15, 23, 25]. The statistical aspects of neurocomputing along with some comparative studies are discussed in [4, 16, 18, 19] while probabilistic neural networks are presented in [24]. In [7], some linkages are discussed between statistical approaches and the processing of information granules, in particular with fuzzy sets.

References

1. Aitchison, J., and Dunsmore, I.R. 1975. *Statistical Prediction Analysis*, Cambridge University Press
2. Bernardo, J.M., and Smith, A.F.M. 1994. *Bayesian Theory*, Wiley
3. Besag, J., Green, P., Higdon, D., and Mengersen, K. 1995. Bayesian computation and stochastic systems. *StatSci*, 10:3–66
4. Bishop, C.M. 1995. *Neural Networks for Pattern Recognition*, Oxford Press
5. Bolstad, William M. 2004. *Introduction to Bayesian Statistics*, John Wiley

6. Gauss C.F. 1809. Theoria Motus Corporum Coelestium in Sectionibus Conicis Solem Ambientum.
7. Cios, K.J., Pedrycz, W., and Swiniarski, R. 1998. *Data Mining Methods for Knowledge Discovery*, Kluwer
8. Devijver, P.A., and Kittler, J. 1982. *Pattern Recognition: A Statistical Approach*, Prentice Hall
9. Draper, N.R., and Smith, H. 1996. Applied Regression Analysis Wiley Series in Probability and Statistics
10. Duda, R.O., Hart, P.E., and Stork D.G. 2001. *Pattern Classification*, Wiley
11. Fu, K.S. 1982. *Syntactic Pattern Recognition and Applications*, Prentice Hall
12. Fukunaga, K. 1990. *Introduction to Statistical Pattern Recognition*, Academic Press
13. Gelman, A., Carlin, J., Stern, H., and Rubin, D. 1995. *Bayesian Data Analysis*, Chapman and Hall
14. Hastie, T., and Tibshirani, R. 1994. Discriminant analysis by Gaussian mixtures. Technical report, AT&T Bell Laboratories
15. Hastie, T., and Tibshirani, R. 1996. Discriminant analysis by Gaussian mixtures. *JRSSB*, 58:158–176
16. Holmstrom, L., Koistinen, P., Laaksonen, J., and Oja, E. 1996. Comparison of Neural and Statistical Classifiers – Theory and Practice. Research Report A13, Rolf Evalinna Institute, University of Helsinki, Finland
17. Kullback, S. 1959. *Information Theory and Statistics*, Dover Publications
18. Mackay, D.J.C. 2003. *Information theory, inference, and learning algorithms*, Cambridge University Press
19. Michie, D., Spiegelthalter, D.J., and Taylor, C.C. (Eds.). 1994. *Machine Learning, Neural and Statistical Classification*, Ellis Horwood
20. Myers, R.H. 1986. Classical and Modern Regression with Applications, Boston, MA: Duxbury Press.
21. Parzen, E. 1962. On estimation of a probability density function and mode. *Annals of Mathematical Statistics*, 33:1065–1076
22. Rawlings, J.O. 1988. *Applied Regression Analysis: A Research Tool*, Pacific Grove, CA: Wadsworth and Brooks/Cole Advanced Books and Software
23. Ripley, B.D. 1996. *Pattern Recognition and Neural Networks*, Cambridge University Press
24. Specht, D.F. 1990. Probabilistic neural networks. *Neural Networks*, 3(1):109–118
25. Webb, A. 1999. *Statistical Pattern Recognition*, Arnold

4. Exercises

1. Consider a two-category classification problem with a scalar feature x and a priori probabilities $P(c_1) = 0.3$ and $P(c_2) = 0.7$. Assume that probability density functions for each classes $p(x|c_1)$ and $p(x|c_2)$ are known. A new object has been measured, and we obtained $x = 35$. For this value of x, we have computed $p(35|c_1) = 2.2785 \cdot 10^{-2}$ and $p(35|c_2) = 1.0953 \cdot 10^{-2}$. Classify a new object using Bayes' rule. Compute the probability of classification error for a classification decision taken.
2. Consider the two-category scalar feature classification problem with patterns drawn for each class independently according to the normal probability density distributions $p(x|c_i) = N(\boldsymbol{\mu}_i, \boldsymbol{\Sigma}_i)(i = 1, 2)$, with the parameter values $\boldsymbol{\mu}_1 = 1$, $\boldsymbol{\sigma}_1 = 1$ for class 1 and $\boldsymbol{\mu}_2 = 2$, $\boldsymbol{\sigma}_1 = 2$ for class 2. Compute the Bayes decision boundary.
3. Generate, using a random number generator, 30 scalar feature patterns for the 1 having the normal Gaussian probability density $p(x|c_1) = \frac{1}{(2\pi\Sigma^2)^{\frac{1}{2}}} \exp\left[-\frac{1}{2\sigma^2}(x-\boldsymbol{\mu})^2\right]$ with parameters $\boldsymbol{\mu}_1 = 10$ and $\boldsymbol{\sigma}_1 = 5$. Similarly, generate 20 scalar feature patterns for class 2 with $\boldsymbol{\mu}_2 = 15$ and $\boldsymbol{\sigma}_2 = 4$. Compute the mean and standard deviation for each class. Assuming these normal forms of probability densities of patterns within a class, find:
   (a) The quadratic discriminant functions $d_1(x)$ and $d_2(x)$ for each class
   (b) The Bayes decision boundary $d(x) = d_1(x) - d_2(x)$ between two classes
   (c) After assigning a class to patterns $x = 18$, $x = 8$, $x = 22$ using Bayes' classification rule, the probability of classification error for these pattern and classification decisions
   (d) The average probability of classification error for this classification problem

4. Consider a two-category, two-feature classification problem with patterns drawn for each class independently according to the normal probability density distributions $p(\mathbf{x}|c_i) = N(\boldsymbol{\mu}_i, \boldsymbol{\Sigma}_i)(i = 1, 2)$ with the following values of parameters: $\boldsymbol{\mu}_1 = [1, 1]^T$, $\boldsymbol{\Sigma}_1 = \mathbf{I}$ for the class 1 and $\boldsymbol{\mu}_2 = [2, 2]^T$, $\boldsymbol{\Sigma}_2 = \sigma^2 \mathbf{I}(\sigma = 2)$ for class 2. Compute the Bayes decision boundary.

5. Let us assume that the following two-feature patterns $\mathbf{x} \in \mathbb{R}^2$ from two classes $c_1 = 0$ and $c_2 = 1$ have been drawn according to the Gaussian normal density distribution shown below:

Class 1	x_1	x_2	class
	1	2	0
	2	2	0
	2	3	0
	3	1	0
	3	2	0
	3	9	0
	3	8	0
	7	6	0
	7	10	0

Class 2	x_1	x_2	class
	6	8	1
	7	8	1
	8	7	1
	8	8	1
	7	9	1
	6	1	1
	6	5	1
	9	9	1
	1	1	1

From this set of patterns, compute for each class separately the estimate mean vectors and covariance matrices. Design Bayesian discriminants and the decision boundary (dichotomizer) between two classes. Plot on a (x_1, x_2) plane the pattern points and dichotomizer. Classify the new patterns $[0.5, 0.5]^T$ and $[7, 3]^T$ using discriminants and a dichotomizer.

Modify the pattern sets for each class from the above tables, adding independently to each feature of each pattern the random noise that has a the normal Gaussian distribution with parameters $\mu = 0.0$ an $\sigma = 1$. Design the Bayesian discriminants and the dichotomizer based on these patterns and test them for the new patterns $[0.6, 0.7]^T$ and $[0.4, 0.8]^T$. Also, test these classifiers for new patterns obtained by modification of the pattern sets for each class from the above tables, by adding independently to each feature of each pattern the random noise having the normal Gaussian distribution with parameters $\mu = 0.0$ an $\sigma = 2$.

6. Let us assume that the following three-feature patterns $\mathbf{x} \in \mathbb{R}^3$ from two classes $c_1 = 0$ and $c_2 = 1$ have been drawn according to the Gaussian normal density distribution:

Class 1	x_1	x_2	x_3	class
	1	2	2	0
	2	2	2	0
	2	3	2	0
	3	1	2	0
	3	2	2	0

Class 2	x_1	x_2	x_3	class
	6	8	6	1
	7	8	8	1
	8	7	7	1
	8	8	8	1
	7	9	8	1

From this set of patterns, compute, for each class separately, the estimated mean vectors and covariance matrices. Design the Bayesian discriminants and the decision boundary (dichotomizer) between the two classes. Plot in (x_1, x_2, x_3) space the pattern points and dichotomizer. Classify new patterns $[0.5, 0.5, 1]^T$ and $[7, 9, 10]^T$ using discriminants and dichotomizer.

Modify the pattern sets for each class from the above tables, adding independently to each pattern the random noise that has a normal multivariate normal Gaussian distribution with parameters $\boldsymbol{\mu} = [0.0, 0.0, 0.0]^T$ and $\boldsymbol{\Sigma} = \text{diag}[\sigma_{ii}]$, with $\sigma_{11} = 1$, $\sigma_{22} = 0.5$, and $\sigma_{33} = 1.5$.

Design the Bayesian discriminants and dichotomizer based on these patterns and test them for the new patterns $[0.6, 0.7, 9.0]^T$ and $[0.4, 0.8, 0.9]^T$. Also test these classifiers for new patterns obtained by adding independently to each pattern the random noise having the normal multivariate normal Gaussian distribution with parameters $\boldsymbol{\mu} = [0.0, 0.0, 0.0]^T$ and $\boldsymbol{\Sigma} = \text{diag}[\sigma_{ii}]$, with $\sigma_{11} = 1.5$, $\sigma_{22} = 2.0$, and $\sigma_{33} = 1.0$.

7. Let us consider a two-feature (two-dimensional) probability density function that is a sum of two normal distributions (mixture of Gaussian distributions)

$$p(\mathbf{x}) = P_1 p_1(\mathbf{x}) + P_2 p_2(\mathbf{x})$$

where

$$p_i(\mathbf{x}) = \frac{1}{(2\pi)^{\frac{n}{2}} |\boldsymbol{\Sigma}_i|^{\frac{1}{2}}} \exp\left[-\frac{1}{2}(\mathbf{x} - \boldsymbol{\mu}_i)^T \boldsymbol{\Sigma}_i^{-1} (\mathbf{x} - \boldsymbol{\mu}_i)\right], i = 1, 2$$

with $P_1 = 0.4$ and $P_2 = 0.6$. The parameters for the component normal densities (with statistically independent features) are $\boldsymbol{\mu}_1 = [10, 10]^T$, $\boldsymbol{\Sigma}_1 = \sigma_1^2 \mathbf{I}$ with $\sigma_1 = 4$, and $\boldsymbol{\mu}_2 = [20, 20]^T$, $\boldsymbol{\Sigma}_2 = \sigma_2^2 \mathbf{I}$ with $\sigma_2 = 2$. Compute and plot the mixture density values for the ranges $x_1 \in [-40, 60]$ and $x_2 \in [-40, 60]$. Compute and plot the loci (contour) of pattern vectors with constant probability density values $p(\mathbf{x}) = 0.2$. Compute thee Euclidean and the squared Mahalanobis distance from the mean $\boldsymbol{\mu}_1$ and $\boldsymbol{\mu}_2$ to one selected vector on the calculated loci.

8. For the patterns from Exercise 7, estimate a probability density function for all patterns regardless of class type. Estimate the probability density functions for each class separately and classify new patterns based on the k-nearest neighbors classification rule. Consider as neighbors members $k = 2, 3, 4, 5, 6, 7, 8$. Classify also new patterns based on the nearest neighbor classification rule ($k = 1$).

9. Generate randomly 300 scalar feature patterns for class 1 having the mixture density distribution from Exercise 7 with the parameters $\mu_1 = 10$, $\sigma_1 = 2$ and $\mu_2 = 20$, $\sigma_2 = 1$. Similarly, generate randomly 300 scalar feature patterns for class 2 with the parameters $\mu_1 = 12$, $\sigma_1 = 2$ and $\mu_2 = 25$, $\sigma_2 = 2$. Based on the generated data, design the following classifiers:

(a) k-nearest neighbors (provide experiments for different values of k)
(b) nearest neighbor

Test the designed classifiers for the new patterns $x = 18$, $x = 7$, $x = 28$. Generate randomly an additional 100 patterns for class 1 and an additional 100 patterns for class 2 for the above parameters and densities, constituting a validation data set. Test the quality of the designed classifiers for this validation set. Generate randomly an additional 100 patterns for class 1 and an additional 100 patterns for class 2 for the above parameters and densities, constituting a testing data set. Using trial and error, find the best number k of neighbors guaranteeing minimalization of the number of misclassifications of patterns from the testing set.

10. In the table shown below variables x and y represent height and weight of a student, respectively, and z represents time achieved by the student in a 100 [m] race competition.

Height	x	160	161	162	163	164	165	166	167	168	189	170
Weight	y	54	55	55	56	57	56	56	57	56	58	59
100m Time	z	14.3	14.2	15.1	15.1	13.0	14.2	11.7	13.3	12.6	13.2	12.9

(a) Perform linear regression analysis for the independent variable x and dependent variable z.

i Compute all the necessary data set characteristics.
ii Compute the optimal linear regression model $z = b + ax$. Find optimal parameters of the model using the least squares method. Use closed formulas for optimal values of the parameters.
iii Calculate regression errors and plot their values. Is the error distribution normal?
iv Compute the minimal value of the performance criterion.
v Evaluate regression model quality by computing standard error of the estimate, correlation coefficient, and coefficient of determination. Explain meaning of these measures.
vi Create a scatter plot of data points overlapped with the plot of the regression line. Calculate regression errors.
vii Calculate the predicted value of the dependent variable for the input independent variable $x = 165.5$.

(b) Perform multiple linear regression analysis for the two independent variables x and y, and one dependent variable z. Repeat operations i–vii with the following changes: in ii change model $x = b + ax$ into a plane model $z = a_0 + a_1 x + a_2 y$, in vi the regression line into the regression plane, and substitute in vii $x = 165.65$ with $x = 165.5$ and $y = 12.9$.

12

Supervised Learning: Decision Trees, Rule Algorithms, and Their Hybrids

In this Chapter, we describe representative algorithms of the two key supervised inductive machine learning techniques, namely, decision trees and rule algorithms. We present both basic and more advanced versions of decision trees, such as the ID3, C4.5, ID5R, and 1RD algorithms. Note that all decision tree algorithms are based on the fundamental concept learning algorithm originally proposed by Hunt. On the other hand, most rule algorithms have their origins in the set-covering problem first tackled by Michalski. The rule algorithms are represented by the DataSqueezer algorithm and the hybrid algorithms by the CLIP4 algorithm that combines the best characteristics of decision trees and rule algorithms.

1. What is Inductive Machine Learning?

Learning means different things to different people and depends on the area of investigation. For instance, the understanding of learning in the cognitive sciences is different than that in psychology or education or computational intelligence. How shall we then understand learning in the context of data mining? First, we narrow it down to the notion of **machine learning** (ML), which means that machines/computers perform learning, not people. There are several understandings of ML even in this much narrower domain. *We define learning as the ability of an agent (like an algorithm) to improve its own performance based on past experience*. In other words, learning is the ability of a "trained" algorithm to make predictions. Past experience is usually provided to the machine learning algorithm in the form of historical data, stored in a database describing some domain. Note that most of the data mining techniques described in Part IV of this book do "learn" in some way.

The big advantage of inductive machine learning algorithms, as described in this Chapter, over other methods is that the models generated by them, expressed in the form of decision trees or IF… THEN… rules, are explicit and can be validated, modified, learned from, or used for training novices in a given domain. These advantages are in contrast to the "black box" nature of neural networks (and other methods) that have difficulty making sense of, say, weights and connections in neural networks, parameters in a classifier, and so on. Therefore, inductive ML is quite often the preferred methodology in fields like medical diagnosis, or control, where a decision maker wants to understand and validate the generated rules against his/her own domain knowledge.

To deal with large quantities of data, inductive machine learning algorithms (as with any other DM algorithms) should be scalable. Machine learning researchers devote significant effort to reducing complexity of inductive ML algorithms, from as high as $O(n^4)$, where *n* is the number of

1. What is Inductive Machine Learning?

examples, to recent state-of-the-art algorithms that are nearly linear, more accurately $O(n \log n)$. We will discuss a range of such algorithms in this Chapter and describe some of them in detail.

Another desirable characteristic of inductive machine learning algorithms is their ability (or inability) to deal with incomplete data. Many real datasets have records that include missing values due to a variety of reasons, such as manual data entry errors, incorrect measurements, equipment errors, etc. It is common, for example in medicine, to encounter datasets that have about half their values missing. Thus, a desirable property of any ML algorithm is its robustness to missing and noisy data.

In this Chapter, we are concerned only with supervised inductive machine learning algorithms. The key concept in inductive ML is that of a **hypothesis**, which is generated by a given algorithm. A hypothesis approximates some **concept.** A concept can mean anything, say, a concept of a planet, of a class/category, of a result of a soccer game, etc. We generally assume that only a **teacher/oracle** knows the true meaning of a concept and describes that concept, by means of examples, to a **learner** whose task is to generate a hypothesis (one or more) that best approximates the concept. For instance, the concept of a *category* can be described to a learner in terms of input-output pairs such as (high fever, pneumonia), or (weak, pneumonia). Although we often make the assumption that the terms *concept* and *hypothesis* are equivalent, this is not quite correct. The reason is that the learner receives from a teacher only a finite set of examples that describe the concept, and thus the hypotheses about the concept can only approximate the concept (for which the teacher may have possibly an infinite number of examples). In practice, the learner receives only a limited number of examples about the concept, since generation of a large number of training data pairs can be expensive and time consuming. Nevertheless, while being aware of this distinction, we will use these two terms interchangeably. Most often, historical data about a concept are stored in a database, along with each object's classification, called a **decision attribute**, the latter performing the role of the teacher. Thus, pneumonia in the above examples is a decision attribute.

Since hypotheses in inductive ML are often described in terms of production IF... THEN... rules, we will also use the term **rule** (and later add still another "equivalent" term: **cover**) to mean *hypothesis*. The formal definition of a cover is given later in the Chapter.

To summarize, *inductive ML algorithms generate hypotheses that approximate concepts; hypotheses are often expressed in the form of production rule; and we say that the rules "cover" the examples.*

Building a model of some domain represented by historical data can be done in two basic ways. First, we can run statistical analyses of the available information and data to estimate the assumed model's parameters. Second, if only a very limited amount of data and information about the domain is available, the model can be built by using model-free techniques like inductive machine learning, neural networks, etc. Let us note that most inductive ML (and neural networks) algorithms are unstable, that is, small changes in input data may result in big changes in their outputs.

We often think of ML as a process of building a model of some domain, a representation of which is stored in a database. Any **inductive ML process** is broken into two phases:

– The **learning phase**, where the algorithm analyzes the data and recognizes similarities among data objects. The result of this analysis is the generation of a tree, or equivalent to it set of production rules.
– The **testing phase**, where the rules are evaluated on new data, and some performance measures are computed.

When many hypotheses (rules) are generated from data, there is usually a need, for comprehensibility and other reasons, to select only a few of them. This selection can be accomplished by using heuristics like Occam's razor (in ML this means choosing the simplest/shortest rules describing a concept) or a minimum-description-length principle; see discussion of these and similar techniques

Chapter 12 Supervised Learning: Decision Trees, Rule Algorithms, and Their Hybrids

in Chapter 15. On the other hand, the user may use his or her background knowledge of the domain to select the best hypotheses; this type of learning is known as **constructive learning**. The user can also reason from first principles, such as laws of physics and mathematical theorems, or can make a final decision based on his or her knowledge of the problem to make sure that the problem constraints, some perhaps known only to the user, are satisfied.

The complexity and nature of a domain from which examples are drawn determines the difficulty of an inductive ML process. When only a few features/attributes describe the training examples, the learning task is much easier than when the examples are highly dimensional. When many irrelevant features describe the examples, the ML algorithms must be able to distinguish between the relevant and irrelevant features. Another issue is the amount of noise present in the data, both in the features describing the examples (pattern vectors) and/or in the description of the classes; fortunately, several inductive ML algorithms cope well with noise. Finally, we very often assume that the data describing a certain phenomenon do not change over time, which is not true in many domains; few inductive ML algorithms can deal with the latter problem.

Inductive machine learning can also be seen as a search problem, where the task is to search for hypotheses that best describe the concept. We can imagine starting with some initial hypothesis and then searching for one that covers as many examples as possible. We say that *an example is covered by a rule when it satisfies all conditions of the IF part of the rule*. The search spaces can be huge and we may have to keep track of all possible solutions.

Still another view of inductive ML, for classification purposes, is to see it as a problem of designing a classifier, i.e., finding boundaries that encompass only examples belonging to a given class. Those boundaries can either partition the entire sample space or leave parts of it unassigned to either of the classes (the more frequent outcome).

1.1. Categorization of ML algorithms

There are many ways to categorize ML algorithms. At the highest level, they are divided into **supervised** and **unsupervised ML algorithms**; we focus only on the former in this Chapter. If we consider the inference methods used by inductive ML algorithms, we can divide them into **inductive** versus **deductive**; we address here only inductive ML algorithms. At another level, we have algorithms that are either **incremental** or **nonincremental**. In nonincremental learning, all the training examples are presented simultaneously (as a batch file) to the algorithm. In contrast, in incremental learning the examples are presented one at a time, and the algorithm improves its performance based on these new examples without the need of retraining (see Chapter 3). A large number of the existing ML systems are nonincremental. At still another level, some ML algorithms can deal with structured data while other deal with unstructured data.

In supervised learning, known **as inductive ML** or **learning from examples**, the user, serving the role of the teacher, must provide examples describing each concept/class. When there is no a priori knowledge of classes, supervised learning algorithms can still be used if the data have a clustering structure. In this situation, we first run a clustering algorithm to reveal natural groupings in the data, and the user must label these before an inductive ML algorithm is used (see Chapter 9).

Any supervised learning algorithm must be provided with a **training data set**. Let us assume that such a set S consists of M training data pairs, belonging to C classes:

$$S = \{(x_i \in c_j) \mid i = 1, \ldots, M; j = 1, \ldots, C\}$$

Training data pairs are often called examples, where x_i is an n-dimensional pattern vector whose components are called features/attributes, and c_j is a known class.

The mapping function f, $c = f(x)$, is not known. The training data represent information about some domain, with the frequently used assumption that the features represent only properties of the examples but not the relationships between the examples. A supervised machine learning

1. What is Inductive Machine Learning?

algorithm then searches the space of all possible hypotheses, H, for the hypothesis (one or more) that best estimates this unknown mapping function f. Within the framework of classification, we can think about the hypothesis as being (part of) a classifier.

ML algorithms find hypotheses by discovering common features (attributes) of examples that represent a class. The resulting hypotheses (concept approximations) can be written as production (IF… THEN…) rules. When the rules are used on new examples unseen in training, they should be able to predict the class membership of these examples.

Depending on the level of "supervision", there exists a range of supervised learning paradigms. On one end, there is **rote learning,** in which the system is "told" the correct rules. In **learning by analogy,** the system is taught the correct response to a similar, but not identical, task; the system must adapt the previous response by generating a new rule applicable to the new situation. A similar type of learning is **case-based learning** in which the learning system stores all the cases it has studied so far, together with their outcomes. When a new case is encountered, the system tries to adapt to this new case the stored previous behavior. In **explanation-based learning,** the system analyzes a set of example solutions in order to determine why each was successful (or not). After these explanations are generated, they are used for solving new problems. **Inductive machine learning** is another type of learning that constitutes the main topic of this Chapter. On the other end of the spectrum is **unsupervised learning**, such as clustering and association rule mining, which are described in Chapters 9 and 10. There is, however, a difference between clustering, as used in taxonomy, and clustering as used within a framework of ML: namely, classical clustering is best suited for handling numerical data. In ML we use the term **conceptual clustering** to differentiate the process from classical clustering, since the former is able to deal with nominal data.

Conceptual clustering consists of two tasks: finding clusters in a given data set; and characterization, which generates a concept description for each cluster found by clustering. Each of the clustering algorithms described in Chapter 9 can be used for clustering. Characterization, on the other hand, belongs to supervised machine learning methods. Conceptual clustering can be thought of as a hybrid combining unsupervised and supervised approaches to learning.

It is worth noting that any supervised ML algorithm can be transformed into an unsupervised one. For example, if data patterns are described by n features, we can run the supervised algorithm n times, each time with a different feature playing the role of the decision attribute that we are trying to predict. The result will be n models/classifiers of data, with one of them being possibly correct.

1.1.1. Inductive versus Deductive Learning

Deduction and induction are two basic techniques for inferring new information from data. **Deduction** infers information that is a logical consequence of the information stored in the data. It is provably correct if the data describing some domain are correct. **Induction**, on the other hand, infers generalized information, or knowledge, from the data by searching for regularities among the data. It is correct for the given data but merely plausible outside of the data. A vast majority of ML algorithms use induction. Learning by induction is essentially a search process for a correct rule, or a set of rules, guided by the provided training examples, and is the main topic of this Chapter.

1.1.2. Structured versus Unstructured Algorithms

In learning hypotheses (approximations of concepts) there is a need to choose a description language. This choice may limit the domains in which a given algorithm can be used. There are two basic types of domains: **structured** and **unstructured**. An example belonging to an unstructured domain is usually described by an <attribute, value> pair. Description languages applicable to unstructured domains include decision trees, production rules, and decision tables (as used in rough sets). Decision trees and production rules have similar representation power; they are described in detail later in this Chapter. For domains having an internal structure, often a first-order calculus language can be used. However, because of the generally huge hypothesis space and high memory requirements, a relatively small number of ML algorithms use first-order logic for model building. Other languages used for structured concept descriptions are semantic networks and frames. They are, however, much more difficult to handle than decision trees or rules.

1.2. Generation of Hypotheses

We shall illustrate the problems involved in inductive machine learning by means of an example that will be carried on throughout this Chapter to illustrate various algorithms.

First, let us define an *information system* (IS):

$$IS = <S, Q, V, f>$$

where

S is a finite set of examples, $S = \{e_1, e_2, \ldots, e_M\}$, where M is the number of examples
Q is a finite set of features, $Q = \{F_1, F_2, \ldots, F_n\}$, where n is the number of features
$V = \cup V_{F_j}$ is a set of feature values where V_{F_j} is the domain of feature $F_j \in Q$
$v_i \in V_{F_j}$ is a value of feature F_j
$f = S \times Q \to V$ is an information function satisfying
$f(e_i, F_i) \in V_{F_j}$ for every $e_i \in S$ and $F_j \in Q$

The set S is often called the *learning/training data set*, that is a subset of the entire universe (known only to the teacher/oracle), which is defined as a Cartesian product of all feature domains $V_{F_j} (j = 1, 2 \ldots n)$.

Example: The scenario for our example is as follows. A travel agency collected data about its customers in terms of the following attributes (features): type of phone call received (F1), language fluency (F2), ticket type sought (F3), and age (F4). The agency also knew who bought, or did not buy, a ticket, which constitutes a decision attribute (F5). Table 12.1 shows values for each attribute/feature (and its numerical encoding). Ticket type was discretized according to the trip distance: Short (1–250 miles), Local (251–1000) and Long (> 1000). Age was discretized into Very young (18–25 years), Young (26–40), Middle aged (41–60), Old (61–80), and Very old (> 80). Discretization was done to make it "easier" for a ML algorithm to learn (it will become apparent later why), and the encoding was done so that a computer could handle this (now numerical) data. Note, that we have five examples of the "Buy" decision (we shall call these Positive examples) and four examples of the "Not buy" decision (called Negative examples). Although the naming of the examples is arbitrary, the following rule is often used: those examples for which we are to generate a hypothesis are called Positive. Thus, in our example case we want to generate a model of customers who "Buy" a ticket. Similarly, when dealing with multiclass problems, we shall call examples of class of interest Positive (for which we generate a model) and all the rest Negative. As a result, for multiclass problems we must generate several sets of hypotheses, one for each class.

Let us remember that available learning/training examples almost never cover the universe of all possible examples (we receive only what the teacher/data owner gives us). Thus, the generated rules need to be "general" enough (defined later) so that they are able to perform sufficiently well on the unseen new examples.

1. What is Inductive Machine Learning?

Table 12.1. Travel agency data.

S	F1	F2	F3	F4	F5
Example	Type of call	Lang. fluency	Ticket type	Age	Decision attribute
E1	Local (1)	Fluent (1)	Long (3)	Very young (1)	Buy (1)
E2	Local (1)	Fluent (1)	Local (1)	Old (4)	Buy (1)
E3	Long dist. (2)	Not fluent (3)	Short (2)	Very old (5)	Buy (1)
E4	Intern. (3)	Accent (2)	Long (3)	Very old (5)	Buy (1)
E5	Local (1)	Fluent (1)	Short (2)	Middle (3)	Buy (1)
E6	Local (1)	Not fluent (3)	Short (2)	Very young (1)	Not buy (2)
E7	Intern. (3)	Fluent (1)	Short (2)	Middle (3)	Not buy (2)
E8	Intern. (3)	Foreign (4)	Long (3)	Young (2)	Not buy (2)
E9	Local (1)	Not fluent (3)	Long (3)	Middle (3)	Not buy (2)

The information system, based on Table 12.1 data, is thus as follows:

$S = \{e_1, \ldots e_9\}, M = 9$
$Q = \{F_1, \ldots F_5\}, n = 5$
$V_{F1} = \{\text{Local, Long distance, International}\}$
$V_{F2} = \{\text{Fluent, Accent, Not fluent, Foreign}\}$
$V_{F3} = \{\text{Local, Short, Long}\}$
$V_{F4} = \{\text{Very young, Young, Middle, Old, Very old}\}$
$V_{F5} = \{\text{Buy, Not buy}\}$

By visual analysis of the data shown in Table 12.1 we can easily generate the following rule:

IF (F1 = 1) AND (F2 = 1) THEN F5 = 1
or put in words:
IF Type of call = Local AND Language fluency = Fluent THEN Decision = Buy

Let us notice that this mentally generated rule covers three out of five positive examples; in fact, it is a good rule, since it covers the majority of the positive examples. Such rules are called **strong rules**.

The goal of inductive machine learning algorithms is to generate rules (hypotheses) like the one we have just shown in an automatic way. After learning, the generated rules must be tested on unseen examples to assess their predictive power. If the rules fail to correctly classify majority of the test examples (to calculate this outcome we assume that we know their "true" classes) the learning phase is repeated by using several of the procedures described in Chapter 15.

The common characteristic of machine learning algorithms is their ability to, often almost perfectly, to cover/classify training examples, which may lead to overfitting of the data (see Chapter 15). A trivial example of overfitting would be to generate five rules to describe the five positive examples. The rules would be the examples themselves, so the first overfitting rule would be:

$$\text{IF } F_1 = 1 \text{ AND } F_2 = 1 \text{ AND } F_3 = 3 \text{ AND } F_4 = 1 \text{ THEN } F_5 = 1$$

and so on. Obviously, if the rules were that specific they would probably perform very poorly on new examples, i.e., they would commit several errors while assigning the examples to the classes.

As stated above, the goodness of the generated rules needs to be evaluated by testing these rules on new data. It is important to establish a balance between the rules' **generalization** and **specialization** in order to generate a set of rules that have good predictive power. *A more general rule is one that covers more positive training examples.* A specialized rule, on the other hand, may cover, in an extreme case (like the one shown above), only one example. Suppose we have already generated a set of rules. The rules can be generalized, or specialized, by using the following common techniques:

- *Replacing constants with variables* (a more general rule is obtained by replacing constants, in rules that have the same outcome, by a variable, and thus merging them into one rule). For example, the three rules:

    IF $F_1 =$ krystyna   THEN student
    IF $F_1 =$ konrad   THEN student
    IF $F_1 =$ karol   THEN student

    can be replaced by a more general rule:

    IF $F_1 =$ Soic   THEN student

    where the first three names are constants (instances) of the variable (Last name) Soic (we follow here first-order logic notation).
- *Using disjuncts for rule generalization and conjuncts for rule specialization.*
  For example

    IF $F_1 = 1$ AND $F_2 = 5$ THEN class1

    is a more specialized rule (uses the conjunct AND) than the two rules

    IF $F_1 = 1$ THEN class1
    (OR)
    IF $F_2 = 5$ THEN class1

    The last two rules are in disjunction, i.e., are connected by a disjunct OR (OR is never written explicitly; thus any collection of production rules is always ORed).
- *Moving up in a hierarchy for generalization.* If there is a known hierarchy in a given domain, the generalization can be performed by replacing the conditions involving the knowledge at the lower level by the common conditions involving the knowledge at the higher level.
  For example

    IF $F_1 =$ canary   THEN class $=$ sing
    IF $F_1 =$ nightingale   THEN class $=$ sing

    can be replaced (if hierarchy of "songbirds" exists) by a rule
    IF $F_1 =$ songbird   THEN class $=$ sing
- *Chunking*. This technique is based on the assumption that, given a goal, every problem encountered and solved on the way to this goal can be treated as a subgoal. Then we can generate rules for each of the subgoals, and put them together.

In the following sections, we concentrate on three families of inductive machine learning algorithms, namely, decision trees, rule algorithms, and hybrid, and show in detail how they approach and solve the task of generating hypotheses (building models) from training data.

The representation of the knowledge format (see Chapter 5) used in representing hypotheses generated from a set of examples is a key element that distinguishes between the three families of inductive machine learning algorithms. In some applications, it is more important to understand the used knowledge structure, and to find its relation to the data from which that structure is generated, than simply to show that the system performs well on test data. In other words, the generated knowledge-representation structure should be simple and easy to comprehend. In terms of knowledge representation, inductive machine learning algorithms are of two basic types: those that use rules and those that use decision trees. There are also some learning algorithms that use decision graphs. Note that a decision tree is a special case of a decision graph.

In what follows, we first describe decision tree algorithms, then rule algorithms, and finally a hybrid algorithm that combines the best characteristics of the two.

2. Decision Trees

Let us recall that a **decision tree** consists of nodes and **branches** connecting the nodes. The nodes located at the bottom of the tree are called **leaves** and indicate classes (see Figure 12.1). The top node in the tree, called the **root**, contains all training examples that are to be divided into classes. All nodes except the leaves are called **decision nodes**, since they specify decision to be performed at this node based on a single feature. Each decision node has a number of **children nodes**, equal to the number of values that a given feature assumes. All decision tree algorithms are based on Hunt's fundamental algorithm of **concept learning**. This algorithm embodies a method used by humans when learning simple concepts, namely, finding key distinguishing features between two categories, represented by positive and negative (training) examples. Hunt's algorithm is based on a **divide and conquer** strategy. The task is to divide the set S, consisting of n examples belonging to c classes, into disjoint subsets that create a **partition** of the data into subsets containing examples from one class only. The following pseudocode summarizes the algorithm.

Given: A set of training examples, S.

1. Select the most discriminatory (significant) feature
2. Split the entire set, S, located at the root of the tree, into several subsets using the selected feature. The number of children nodes originating from the root is equal to the number of values the selected feature takes on.
3. Recursively find the most significant feature for each subset generated in step 2 and then split it top-down into subsets. If each subset contains examples belonging to one class only (a leaf node), then stop; otherwise go to step 3.

Result: The decision tree from which classification rules can be extracted.

Based on Hunt's algorithm, Quinlan developed an algorithm called ID3 (Interactive Dichotomizer3), in which he used **Shannon's entropy** as a criterion for selecting the most significant/discriminatory feature:

$$Entropy(S) = \sum_{i=1}^{c} -p_i \cdot \log_2(p_i)$$

where p_i is the proportion of the examples belonging to the ith class.

The uncertainty in each node is reduced by choosing the feature that most reduces its entropy (via the split). To achieve this result, **Information Gain** that measures expected reduction in entropy caused by knowing the value of a feature F_j, is used:

$$Information\ Gain(S, F_j) = Entropy(S) - \sum_{v_i \in V_{F_j}} \frac{|S_{v_i}|}{|S|} \cdot Entropy(S_{v_i})$$

where V_{F_j} is a set of all possible values of feature F_j and S_{v_i} is a subset of S, for which feature F_j has value v_i.

Information gain is used to select the best feature (reducing the entropy by the largest amount) at each step of growing a decision tree. To compensate for the bias of the information gain for cases with many outcomes, a measure called the **Gain Ratio** is used:

$$Gain\ Ratio(S, F_j) = \frac{Information\ Gain(S, F_j)}{Split\ Information(S, F_j)}$$

where

$$Split\ Information(S, F_j) = \sum_{i=1}^{C} \frac{|S_i|}{|S|} \cdot \log_2 \left(\frac{|S_i|}{|S|} \right)$$

Split Information *is the entropy of S with respect to values of feature* F_j. In a situation when two or more features have the same value of information gain the feature that has the smaller number of values is selected. Use of the gain ratio results in the generation of smaller trees. The pseudocode of the discrete ID3 algorithm follows.

Given: A set of training examples, S.

1. Create the root node containing the entire set S
2. If all examples are positive, or negative, then stop: decision tree has one node
3. Otherwise (the general case)
   Select feature F_j that has the largest *Information Gain* value
   For each value v_i from the domain of feature F_j:

   (a) add a new branch corresponding to this best feature value v_i, and a new node, which stores all the examples that have value v_i for feature F_j
   (b) if the node stores examples belonging to one class only, then it becomes a leaf node; otherwise below this node add a new subtree, and go to step 3

Result: A decision tree.

This algorithm uses **inductive bias** during learning, namely, it prefers small over large decision trees. A decision tree in fact consists of disjunctions of conjunctions of the feature values, and thus can be represented by an equivalent set of IF… THEN… rules, as is illustrated later.

The information gain and the gain ratio are defined using the concept of entropy to measure chaos/order in the data. Entropy achieves a maximum of 1, in two-category problems, when the data contain an equal number of positive and negative examples; it has a value of 0 when the data consist only of examples of one type.

Other measures can also be used for finding the most discriminatory feature for building decision trees. Let us first write the entropy function in a slightly different form so that it can be easily compared with other measures:

$$Entropy(n) = \sum_{i=1}^{c} -p(i|n) \cdot \log_2 p(i|n)$$

where $p(i|n)$ denotes the fraction of examples from class i at node n.

One measure that can be used instead of entropy is the **misclassification error,** defined as

$$MissclassError(n) = 1 - \max_i(p(i|n))$$

Another popular measure, called **Gini index**, is defined as

$$Gini(n) = 1 - \sum_{i=1}^{c} p^2(i|n)$$

Note that these two measures are similar to the entropy measure because all three attain maximum value (0.5 for these two and 1 for entropy) for data with an equal number of positive and negative examples (the worst case). All three measures attain the minimum value of 0 for data subsets containing only one type of examples (the best/desired case). The values differ

2. Decision Trees

for any other mixture of positive and negative examples, but all three measures indicate the same trend, namely, they show decreasing values when data become more homogenous after performing a split on a feature/attribute that most reduces the "chaos" in the data.

Example: We illustrate the algorithm using the travel agency data shown in Table 12.1. With the entropy measure, the initial entropy for the entire data set (five positive and four negative examples) is calculated as follows:

$$Entropy(S) = \sum_{i=1}^{c} -p_i \cdot \log_2(p_i)$$

$$Entropy(S) = -\frac{5}{9}\log_2\frac{5}{9} - \frac{4}{9}\log_2\frac{4}{9} = -\log_2\frac{1}{2} = 0.9911$$

Next, we calculate the following for attribute F1 and each of its values:

$$Entropy(S, F1_{Local}) = -\frac{3}{5}\log_2\frac{3}{5} - \frac{2}{5}\log_2\frac{2}{5} = 0.971$$

$$Entropy(S, F1_{LongDistance}) = -\frac{1}{1}\log_2\frac{1}{1} - 0 = 0$$

$$Entropy(S, F1_{International}) = -\frac{1}{3}\log_2\frac{1}{3} - \frac{2}{3}\log_2\frac{2}{3} = 0.9183$$

Having the above values we calculate the information gain for feature F1 as

$$Gain(S, F1) = 0.9911 - \{\frac{5}{9} \cdot 0.9710 + \frac{1}{9} \cdot 0 + \frac{3}{9} \cdot 0.9183\} = 0.1545$$

We perform similar calculations for features F2, F3, and F4, and choose the "best" feature (in terms of the largest value of the information gain). This best feature (F4) is used at the root of the tree to split the examples into subsets, which are stored at the child nodes. Next, at each child node we perform the same calculations to find the best feature at this level of the decision tree, and we keep splitting the examples until the leaf nodes contain examples belonging to one class only. Figure 12.1 illustrates the generated decision tree for the data in Table 12.1.

From this tree, we can write the following four rules that cover examples belonging to the category Buy:

IF (F4 = 1) AND (F2 = Fluent(1))   THEN F5 = Buy
IF (F4 = 3) AND (F3 = Local(1))   THEN F5 = Buy
IF (F4 = 4) AND (F2 = Fluent(1))   THEN F5 = Buy
IF (F4 = 5)   THEN F5 = Buy (covers 2 examples)

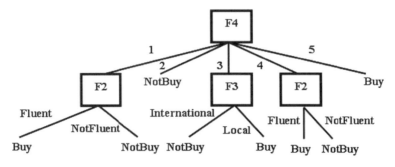

Figure 12.1. A decision tree built using the travel agency example.

Similarly, we can write rules that cover/describe the category NotBuy. Note that in practice, for two-category problems, we only need to write one set of rules (usually for the category Positive (Buy)), since if the new example does not match any of the rules for class Positive, it is recognized as belonging to category Negative (NotBuy).

IF (F4 = 1) AND (F2 = Not Fluent(3))   THEN F5 = Not buy (2)
IF (F4 = 2)   THEN F5 − Not buy (2)
IF (F4 = 3) AND (F3 = International(3)) THEN F5 = Not buy (2)
IF (F4 = 4) AND (F2 = Not Fluent(3))   THEN F5 = Not buy (2)

The generated tree is 100% correct, i.e., it correctly recognizes all positive examples in the training data. Also notice that none of the generated rules is the same as the rule that we have found by mentally analyzing the data.

An important constraint of decision trees is that they can be generated only from data described by discrete attributes (although some algorithms appear to deal with continuous data by using an internal front-end discretization algorithm) and only from two-category data. How then can we use them when the data contain examples belonging to several categories? In that case, we generate several different decision trees, each distinguishing one category versus the other categories put together. For example, in the case of three categories, we would generate a tree for the first category (called positive) versus the other two categories treated as one (negative), then for the second category (now treated as positive) versus the remaining two, etc. When a new example comes, it is checked versus the rules generated for each category separately, and a decision is made based on the best match.

In the case when attributes describing the data are continuous, we first need to discretize them using one of the many existing discretization methods described in Chapter 8. Let us note that some decision tree algorithms, like C4.5, have a built-in front-end discretization algorithm, so to the user it appears that the algorithm operates on continuous data (while in fact it does not). The better strategy, however, is to use different discretization algorithms (those that take into account class information and those that do not), since depending on which discretization algorithm is used, some of the subsequently generated decision trees will be better than others.

Decision trees become incomprehensible when their size grows. For example, application of the ID3 algorithm to the chess end game generated a very complex and large tree that could not be understood even by chess experts. To remedy this problem pruning techniques (see below) are used to make decision trees more comprehensible. Decision tree algorithms also do not easily support incremental learning. Although ID3 still works when examples are supplied one at a time, it grows a new tree from scratch each time a new example is supplied. To remedy this situation, an algorithm called ID5R was proposed by Utgoff that does not grow a new tree for each new example but restructures the existing tree to make it consistent with all previously used examples, which are retained but not reprocessed.

To avoid the problem of overfitting (see Chapter 15) decision trees are pruned down in such a way that there is no significant loss of classification accuracy. Pruning techniques can be divided into **prepruning** and **postpruning**, depending on when the pruning occurs during the growth process of the tree. *In prepruning the growth of the tree stops when it is determined that no attribute will significantly increase the information gain in the process of classifying the data.* Postpruning, on the other hand, involves already-constructed trees. Often, the complexity of the tree is compared with the observed loss in classification accuracy in order to make a decision about how much of the tree (how many branches) should be eliminated.

The extreme case of prepruning was introduced by Holte, who proposed an algorithm called 1RD for building decision trees that are only one level deep. Interestingly, he has shown that

392   2. Decision Trees

the classification performance of such trees is comparable with other more complex algorithms (similarly to using the Naïve Bayes algorithm; see Chapter 11). It is interesting to note that the 1RD algorithm can be also used for feature discretization (see Chapter 8).

Some more advanced decision tree algorithms allow for growing trees from data containing missing values, use windowing techniques to deal with large data sets, use techniques like boosting to improve performance, etc. **Windowing** simply means that the training data is divided into subsets, called windows, from which several decision trees are grown. Then the best rules, extracted from all the trees, are chosen according to the lowest predicted error rate.

Decision trees and rule algorithms always create decision boundaries that are parallel to the coordinate axes defined by the features. In other words, they create hypercube decision regions in high-dimensional spaces. An obvious remedy for this shortcoming is to have an algorithm that is able to place a decision boundary at any angle. This idea is used in the Continuous ID3 (CID3) neural network algorithm, which uses the entropy minimization technique to place hyperplanes for solving a given classification problem. During the process of minimizing the entropy, CID3 also generates its own topology. It starts with just one neuron and adds new neurons and/or new hidden layers until a given problem is solved (when the entropy value is reduced to a small value close to zero). Minimization of entropy is used for "best" placement of separating hyperplanes, in terms of their orientation and position. Unlike the ID3 algorithm, which creates many classification "boxes", CID3 uses a much smaller number of hyperplanes to achieve the same goal. The top part of Figure 12.2 illustrates the result of a decision tree algorithm used on a two-class (indicated by pluses and minuses), two-feature problem. Although CID3 can place hyperplanes at any angle it does not guarantee finding an optimal solution, like the one shown in Figure 12.2. However, the number of the generated hyperplanes is smaller than the number of parallel hyperplanes that would be generated by the ID3 algorithm.

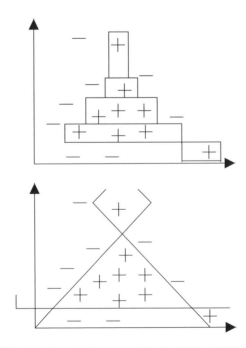

Figure 12.2. Comparison of decision boundaries created by the ID3 and CID3 algorithms.

The advantages and disadvantages of decision trees can be summarized as follows:

- they reveal relationships between the rules (which can be written out from the tree), thereby making it easy to see the structure of the data
- they produce rules that best describe all the classes in the training data set,
- they are computationally simple
- they may generate very complex (long) rules, which are very hard to prune
- they generate a large number of corresponding rules that can become excessively large unless pruning techniques are used to make the rules more comprehensible, and
- they require large amounts of memory to store the entire tree for deriving the rules.

3. Rule Algorithms

In this section we describe **rule algorithms**, sometimes referred to as **rule learners**. Rule induction (generation) is distinct from the generation of decision trees. While it is trivial to write a set of rules given a decision tree (as we have shown above), it is more complex to generate rules directly from data. The rules, however, have many advantages over decision trees. Namely, they are easy to comprehend; their output can be easily written in first-order logic format, or directly used as a knowledge base in expert systems; the background knowledge can be easily added into a set of rules; and they are modular and independent, i.e., a single rule can be understood without reference to other rules. Independence means that, in contrast to rules written out from decision trees, they do not share any common parts/attributes (partial paths in the decision tree). Although in some applications rules perform better than decision trees, their disadvantage is that they do not show relationships between the rules (since the rules are independent).

Table 12.2 compares inductive ML algorithms in terms of their complexity. The first three listed are decision tree algorithms, the next is a rule algorithm, and the last two are hybrid algorithms. Here we concentrate on rule algorithms and describe in detail a DataSqeezer algorithm. Later we shall describe a hybrid algorithm, CLIP4. The algorithms described in this Chapter are indicated by bold letters in Table 12.2.

3.1. DataSqueezer Algorithm

Let us denote a training dataset by D, consisting of s examples and k attributes. The subsets of positive examples, D_P, and negative examples, D_N, satisfy these properties:

$$D_P \cup D_N = D, \quad D_P \cap D_N = \emptyset, \quad D_N \neq \emptyset, \text{ and } \quad D_P \neq \emptyset$$

Figure 12.3 shows the pseudocode of the algorithm. Vectors and matrices are denoted by capital letters, while elements of vectors/matrices are denoted with the same name but use small letters.

Table 12.2. Complexity of some inductive ML algorithms.

Algorithm	Complexity
ID3	O(n)
C4.5 rules	$O(n^3)$
C5.0	O(n log n)
DataSqeezer	**O(n log n)**
CN2	$O(n^2)$
CLIP4	$O(n^2)$

3. Rule Algorithms

Given: POS, NEG, k (number of attributes), n (number of examples).

Step1.
1.1 G_{POS} = DataReduction(POS, k);
1.2 G_{NEG} = DataReduction(NEG, k);
Step2.
2.1   Initialize RULES = []; i = 1;
     // where $rules_i$ denotes i^{th} rule stored in RULES
2.2   create LIST = list of all columns in G_{POS}
2.3   within every G_{POS} column that is on LIST, for every nonmissing value a from selected column j   compute sum, s_{aj}, of values of $gpos_i[k+1]$ for every row i, in which a appears and multiply s_{aj}, by the number of values the attribute j has
2.4   select maximal s_{aj}, remove j from LIST, add "j = a" selector to $rules_i$
2.5.1 **if** $rules_i$ does not describe any rows in G_{NEG}
2.5.2   **then** remove all rows described by $rules_i$ from G_{POS}, i = i + 1;
2.5.3   **if** G_{POS} is not empty go to 2.2, **else** terminate
2.5.4 **else** go to 2.3

Result: RULES describing POS.

DataReduction (D, k)   // data reduction procedure for D = POS or D = NEG
DR.1   Initialize G = []; i = 1; tmp = d_1; g_1 = d_1; $g_1[k+1]$ = 1;
DR.2.1 **for** j = 1 to N_D // for positive/negative data; N_D is N_{POS} or N_{NEG}
DR.2.2 **for** kk = 1 to k   // for all attributes
DR.2.3   **if** ($d_j[kk] \neq tmp[kk]$ **or** $d_j[kk]$ = '*')
DR.2.4   **then** tmp[kk] = '*';   // '*' denotes missing" do not care" value
DR.2.5 **if** (number of non missing values in tmp ≥ 2)
DR.2.6   **then** g_i = tmp; $g_i[k+1]$++;
DR.2.7 **else** i++; g_i = d_j; $g_i[k+1]$ = 1; tmp = d_j;
DR.2.8 **return** G;

Figure 12.3. Pseudocode of the DataSqueezer algorithm

The matrix of positive examples is denoted as POS and their number as N_{POS}; similarly NEG denotes the matrix of negative examples and their number is N_{NEG}. The POS and NEG matrices are formed by using all positive and negative examples, where examples are represented by rows, and features/attributes by columns. Positive examples, POS, and negative examples, NEG, are denoted in the DataReduction procedure by $d_i[j]$ values, where $j = 1, \ldots, k$ is the column number, and i is the example number (row number in matrix D, which denotes either POS or NEG). The algorithm also uses matrices that store intermediate results (G_{POS} for the POS, and G_{NEG} for the NEG), which have k columns. Each cell of the G_{POS} matrix is denoted by $gpos_i[j]$, where i is a row number and j is a column number; and similarly for the G_{NEG} matrix each cell is denoted by $gneg_i[j]$. The G_{POS} stores a reduced subset of the data from the POS matrix, and the G_{NEG} stores a reduced subset of the data from the NEG matrix. The need for and meaning of this reduction is explained later. The G_{NEG} and G_{POS} matrices have an additional $(k+1)^{th}$ column that stores the number of examples from the NEG and POS matrices, described by a particular row in the G_{NEG} and G_{POS}, respectively. For example, $gpos_2[k+1]$ stores the number of examples from the POS that are described by the second row in the G_{POS} matrix. Below we describe the two steps of the pseudocode in more detail.

Step 1: The algorithm performs data reduction to generalize information stored in the original data. Data reduction is performed via the use of a prototypical concept learning procedure based on the FindS algorithm of Mitchell. It is performed for both positive and negative data and results

Chapter 12 Supervised Learning: Decision Trees, Rule Algorithms, and Their Hybrids   395

in generation of the G_{POS} and G_{NEG} matrices. This reduction is related to the least generalization. The main difference is that FindS performs the least generalization multiple times for the entire positive set through a beam search strategy, while the DataSqueezer performs it only once in a linear fashion by generalizing consecutive examples. It also generalizes the negative data.

Step 2: DataSqeezer generates rules by a greedy hill-climbing search on the reduced data. A rule is generated by using the search procedure starting with an empty rule, and adding selectors (selectors are the values that attributes assume; see the formal definition of a selector in section 4.1.1) until the termination criterion is satisfied. The rule that is being generated consists of selectors generated using the G_{POS}, and is checked against the G_{NEG}. If the rule covers any data in the G_{NEG}, a new selector is added to make it more specific, and thus better able to distinguish between positive and negative data. The maximum depth of the search is equal to the number of features. Next, the examples covered by the generated rule are removed, and the process is repeated. The algorithm is very simple to implement since it requires use of a table as the only data structure.

Example: In Figure 12.4 we illustrate how the process works by using our travel agency data (see Table 12.1). POS $= 5$, NEG $= 4$, $k = 4$, and $n = 9$.

In Step 1, the G_{POS} and G_{NEG} tables are computed. First the DataReduction(POS, k) call in line 1.1 is executed. In line DR.1 the variables, G=[] and $d_1 = $ tmp $= $ [Local Fluent Long Very-young], are initialized. After computing DR.2.1÷DR.2.4 for $j = 2$, tmp $= $ [Local Fluent **], and after computing DR.2.5 and DR.2.6, $g_1 = $ [Local Fluent **2]. After executing for the loop for $j = 3$, tmp does not change and $g_1 = $ [Local Fluent **3]. For $j = 4$, tmp $= $ [****] and thus is based on DR.2.7, $g_2 = $ [Long-dist Not-fluent Short Very-old 1] and tmp $= $ [Long-dist Not-fluent Short Very-old]. For $j = 5$, tmp $= $ [*** Very-old] and thus based on DR.2.7, $g_3 = $ [Intern Accent Long Very-old 1] and tmp $= $ [Intern Accent Long Very-old], and the process is terminated. As the result $G = G_{POS} = [g_1\ g_2\ g_3]^T$. Analogous computations are performed when DataReduction(NEG, k) is called in line 1.2.

The last column of G_{POS} and G_{NEG} gives the number of rows from POS and NEG, respectively, which a particular row in G_{POS} and G_{NEG} describes. The "*" stands for "do not care" value. For instance, the first row in G_{POS} covers the first three rows in POS.

In Step 2, rules are generated using the G_{POS} and G_{NEG} tables. A rule is generated by incrementally adding selectors using the G_{POS} table. A selector with the highest summed value from the last columns is chosen and added incrementally until a rule will not describe any rows in the G_{NEG} table. Next, the rows described by the generated rule are removed, and the process repeats. Detailed computations follow. First we initialize RULES $= $ [] and LIST $= $ [1, 2, 3, 4]. Next, in line 2.3, s_{aj} values are computed as $s_{Local1} = 9$ (since *Local* has a summed value of 3 in the last column in G_{POS} and the total number of values for j $= 1$, which is *Type of call*, is 3), $s_{Long-dist.1} = 3$, $s_{Intern.1} = 3$, $s_{fluent2} = 12$, $s_{Not-fluent2} = 4$, $s_{Accent2} = 4$, $s_{Short3} = 3$, $s_{Long3} = 3$, $s_{Very-old4} = 8$. In line 2.4, $s_{fluent2}$ is selected, LIST $= $ [1, 3, 4], and $rules_1 = $ [Lang fluency $= $ fluent].

Next, in line 2.5.1, $rules_1$ is verified to describe second row in G_{NEG}, and thus we iterate back to 2.3. Since the second column has been removed from LIST, in line 2.4 we will select S_{Local1}, LIST $= $ [3, 4], and $rules_1 = $ [Lang fluency $= $ fluent, Type of Call $= $ local]. Since $rules_1$ does not cover any row in G_{NEG}, in line 2.5.2, $G_{POS} = $ [Long-dist Not-fluent Short Very-old; Intern. Accent Long Very-old], and since it is not empty (line 2.5.3), the algorithm iterates back to line 2.2. Again, LIST $= $ [1, 2, 3, 4] and $s_{Long-dist.1} = 3$, $s_{Intern.1} = 3$, $s_{Not-fluent2} = 4$, $s_{Accent2} = 4$, $s_{Short3} = 3$, $s_{Long3} = 3$, and $s_{Very-old4} = 8$. Thus $rules_2 = $ [Age $= $ Very-old] and LIST $= $ [1, 2, 3] are computed in line 2.4. Since $rules_2$ does not describe any rows in G_{NEG}, we remove the corresponding rows in line 2.5.2 from G_{POS}, which becomes empty, and the algorithm terminates.

As result of the above operations, the following two rules are generated (see Figure 12.4), which cover all five POS training examples:

396   3. Rule Algorithms

    IF LangFluency = Fluent AND TypeofCall = Local   THEN Buy (covers 3 examples)
    IF Age = Very old   THEN Buy (covers 2 examples)

or as the algorithm would write it

    IF F2=1 AND F1=1   THEN F5=1   (covers 3 examples)
    IF F4=5   THEN F5=1   (covers 2 examples)

Analogous computations can be performed assuming that POS stores *Not buy* decisions and NEG stores *Buy* decisions. The reader is encouraged to perform the computations manually, and should generate the following rules:

    IF F1=1 AND F2=3   THEN F5=2   (covers 2 examples)
    IF F1=3 AND F4=3   THEN F5=2   (covers 1 example)
    IF F4=2   THEN F5=2   (covers 1 example)

The DataSqueezer algorithm handles data with missing values very well. The algorithm uses all available information while ignoring missing values, i.e., these are handled as "do not care" values, as was illustrated above. In fact, the algorithm is robust to a very large number of missing values and was shown to successfully generate rules even from data that have up to 60% of missing values.

DataSqueezer, like all algorithms covered in this Chapter, handles only discrete-valued numerical and nominal attributes (the latter are encoded into numerical values). The generated rules are independent of the encoding scheme, since, during the rule induction process, no distances are calculated between feature values.

DataSqueezer uses two thresholds that can be set by the user (by default, they are set to zero).

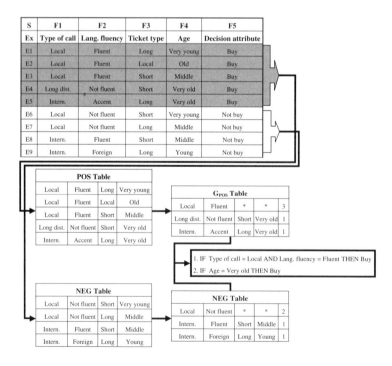

Figure 12.4. Illustration of the rule generation process by the DataSqueezer algorithm.

The **Pruning Threshold** is used to prune very specific rules. The rule generation process is terminated if the first selector added to $rule_i$ has s_{aj} value equal to or smaller than the threshold's value. The algorithm induces rules by selecting maximal s_{aj} values (selectors that cover most of the POS examples) and removes examples that have already been covered. This approach has the benefit of leaving small subsets of positive examples that store different examples from the majority already covered; these (possibly outliers) can be filtered out using this threshold.

The **Generalization Threshold** is used to allow for rules that cover a small amount of negative data. It relaxes the requirement from line 2.5.1 (Figure 12.3) and allows for acceptance of rules that cover negative examples; the number is equal to or smaller than this threshold. It is a useful mechanism in the case of data with overlapping classes, or in the case of inconsistent examples present in the training dataset (examples that should have been but were not eliminated during preprocessing).

Both thresholds are set as percentages of the POS data size and thus are scalable. As a rule of thumb, they are usually set to small values.

DataSqueezer generates a set of rules for each class. Due to the rule generation process and use of the thresholds, conflicts in the generated rules may arise. Two outcomes are possible, namely, a new example may be assigned to a particular class or it may be left unclassified. To resolve possible conflicts the following is done:

- All rules that cover a given example are found. If no rules cover it, then it is left unclassified. This situation may occur for examples with missing values for all attributes used by the rules.
- For every class, the goodness of rules describing the class and covering the example is summed. The example is assigned to the class with the highest value. In case of a tie, the example is left unclassified. The goodness value for each rule is equal to the percentage of the POS examples that it covers.

This classification procedure is identical to the procedure used in the CLIP4 algorithm and is described in detail in the next section.

Let us note that according to the operation of the DataSqeezer algorithm all unclassified examples are treated as incorrect classifications (since they are not covered by the model's rules). As such, they contribute toward lowering the algorithm's classification accuracy. This outcome is in contrast to the C5.0 algorithm that uses the **default hypothesis**, which states that if an example is not covered by any rule it is assigned to the class with the highest frequency (the default class) in the training data. Consequently, each example is always classified; however, this mechanism may lead to significant but artificial improvement in terms of the accuracy of the model. In the extreme case, for highly **skewed datasets** (where one of the classes has significantly larger number of training examples), it may lead to generation of the default hypothesis as the only rule. Skewness is a common problem in data mining; for instance with fraud detection systems, frequently one must deal with data that is skewed 100:1 against examples of fraud, which are obviously the most important examples to detect. To deal with skewness, one can resample the data by oversampling the minority class and undersampling the majority class.

In what follows, we compare DataSqueezer with decision trees and hybrid algorithms, covered in this Chapter, in terms of accuracy and rule complexity. The 22 datasets used for benchmarking consist of from 151 to 200,000 training examples and from 15 to 500,000 testing examples. The number of attributes ranges from 5 to 61, and the number of classes is between 2 and 10. All datasets are available from either the UCI repository at http://www.ics.uci.edu/~mlearn/MLRepository.html, or from the StatLog repository at http://lib.stat.cmu.edu.

All datasets are summarized in Table 12.3 (for the *ipums* dataset, the years of 1970, 1980, and 1990 were used as classes). Table 12.4 shows both the accuracy and the number of rules generated and their complexity (measured by the number of selectors used). The comparison is with C5.0 (proprietary implementation of decision trees) and CLIP4. Some tests were performed using 10-fold cross-validation.

Table 12.3. Description of the datasets used for the benchmarking tests.

#	Abbr.	Set name	Size	#class	#attrib.	Test data	#	Abbr.	Set name	Size	#class	#attrib.	Test data
1	**adult**	Adult	48842	2	14	16281	12	**led**	LED display	6000	10	7	4000
2	**bcw**	Wisconsin breast cancer	699	2	9	10CV	13	**pid**	PIMA Indian diabetes	768	2	8	10CV
3	**bld**	BUPA liver disorder	345	2	6	10CV	14	**sat**	StatLog satellite image	6435	6	37	2000
4	**bos**	Boston housing	506	3	13	10CV	15	**seg**	Image segmentation	2310	7	19	10CV
5	**cid**	Census-income	299285	2	40	99762	16	**smo**	Attitude smoking restr.	2855	3	13	1000
6	**cmc**	Contraceptive method	1473	3	9	10CV	17	**spect**	SPECT heart imaging	267	2	22	187
7	**dna**	StatLog DNA	3190	3	61	10CV	18	**tae**	TA evaluation	151	3	5	10CV
8	**forc**	Forest cover	581012	7	54	565892	19	**thy**	Thyroid disease	7200	3	21	3428
9	**hea**	StatLog heart disease	270	2	13	10CV	20	**veh**	StatLog vehicle silhouette	846	4	18	10CV
10	**ipum**	IPUMS census	233584	3	61	70076	21	**vot**	Congressional voting rec	435	2	16	10CV
11	**kdd**	Intrusion (kdd cup 99)	805050	40	42	311029	22	**wav**	Waveform	3600	3	21	3000

Table 12.4. The accuracy comparison.

Data set	C5.0	CLIP4	DataSqueezer		
			accuracy	sensitivity	specificity
bcw	94 (±2.6)	95 (±2.5)	94 (±2.8)	92 (±3.5)	98 (±3.3)
bld	68 (±7.2)	63 (±5.4)	68 (±7.1)	86 (±18.5)	44 (±21.5)
bos	75 (±6.1)	71 (±2.7)	70 (±6.4)	70 (±6.1)	88 (±4.3)
cmc	53 (±3.4)	47 (±5.1)	44 (±4.3)	40 (±4.2)	73 (±2.0)
dna	94	91	92	92	97
hea	78 (±7.6)	72 (±10.2)	79 (±6.0)	89 (±8.3)	66 (±13.5)
led	74	71	68	68	97
pid	75 (±5.0)	71 (±4.5)	76 (±5.6)	83 (±8.5)	61 (±10.3)
sat	86	80	80	78	96
seg	93 (±1.2)	86 (±1.9)	84 (±2.5)	83 (±2.1)	98 (±0.4)
smo	68	68	68	33	67
tae	52 (±12.5)	60 (±11.8)	55 (±7.3)	53 (±8.4)	79 (±3.8)
thy	99	99	96	95	99
veh	75 (±4.4)	56 (±4.5)	61 (±4.2)	61 (±3.2)	88 (±1.6)
vot	96 (±3.9)	94 (±2.2)	95 (±2.8)	93 (±3.3)	96 (±5.2)
wav	76	75	77	77	89
MEAN (stdev)	**78.5**(±14.4)	**74.9**(±15.0)	**75.4**(±14.9)	74.6 (±19.1)	83.5 (±16.7)
adult	85	83	82	94	41
cid	95	89	91	94	45
forc	65	54	55	56	90
ipums	100	–	84	82	97
kdd	92	–	96	12	91
spect	76	86	79	47	81
MEAN all (stdev)	**80.4**(±14.1)	**75.6**(±14.8)	**77.0**(±14.6)	71.7 (±23.0)	80.9 (±19.0)

Table 12.4 shows the sensitivity and specificity for DataSqueezer. Note that for multi-class problems, the sensitivity and specificity are computed for each class separately, and only the average values are shown. Results for the experiments with 10-fold cross-validation also show standard deviations. Table 12.5 shows a comparison of the rules complexity.

The number of selectors per rule gives a better indication of the complexity of the rules than just their sheer number. It is very small for all algorithms except CLIP4, where the larger number is caused by the inequality format of rules generated. An important factor when discussing difference in accuracy is that, as mentioned above, C5.0 uses the default hypothesis, while both CLIP4 and DataSqueezer do not, which obviously lowers their accuracy.

The use of the default hypothesis may artificially increase the accuracy of the generated rules but may also lead to generation of the default hypothesis as the only "artificial" rule, and may have no practical value, since no data model (rules) for any of the classes is generated. For example, for the *smo* dataset, the CLIP4 and DataSqueezer algorithms generated "real" rules while the C5.0 generated only default hypotheses. We also note that use of a default hypothesis may also lower the average number and complexity of the generated rules.

4. Hybrid Algorithms

From the two well-known **hybrid inductive machine learning algorithms** shown in Table 12.5 we describe below the CLIP4 algorithm. CLIP4 was chosen because the CN2 algorithm uses entropy to guide the process of rule generation, in a fashion very similar to the process used in building decision

4. Hybrid Algorithms

Table 12.5. The number of rules and rule complexity.

	C5.0			CLIP4			DataSqueezer		
Data set	mean # rules	mean # select	# select / rule	mean # rules	mean # select	# select / rule	mean # rules	mean # select	# select / rule
bcw	16	16	1.0	4	122	30.5	4	13	3.3
bld	14	42	3.0	10	272	27.2	3	14	4.7
bos	18	68	3.8	10	133	13.3	20	107	5.4
cmc	48	184	3.8	8	61	7.6	20	70	3.5
dna	40	107	2.7	8	90	11.3	39	97	2.5
hea	10	21	2.1	12	192	16.0	5	17	3.4
led	20	79	4.0	41	189	4.6	51	194	3.8
pid	10	22	2.2	4	64	16.0	2	8	4.0
sat	96	498	5.2	61	3199	52.4	57	257	4.5
seg	42	181	4.3	39	1170	30.0	57	219	3.8
smo	0	0	0	18	242	13.4	6	12	2.0
tae	12	33	2.8	9	273	30.3	21	57	2.7
thy	7	15	2.1	4	119	29.8	7	28	4.0
veh	37	142	3.8	21	381	18.1	24	80	3.3
vot	4	6	1.5	10	52	5.2	1	2	2.0
wav	30	119	4.0	9	85	9.4	22	65	3.0
MEAN	25.3	95.8	2.9	16.8	415.3	18.9	21.2	77.5	3.4
stdev	(±23.9)	(123.5)	(±1.4)	(±16.3)	(±789.1)	(±12.7)	(±19.8)	(±80.3)	(±0.9)
adult	54	181	3.3	72	7561	105.0	61	395	6.5
cid	146	412	2.8	19	1895	99.7	15	95	6.3
forc	432	1731	4.0	63	2438	38.7	59	2105	35.7
Ipums	75	197	2.6	–	–	–	108	1492	13.8
kdd	108	354	3.3	–	–	–	26	409	15.7
spect	4	6	1.5	1	9	9.0	1	9	9.0
MEAN all	55.6	200.6	2.9	21.2	927.4	28.4	27.7	261.1	6.5
stdev	(±92.3)	(±368.6)	(±1.2)	(±21.8)	(±1800.6)	(±28.2)	(±27.6)	(±520.2)	(±7.4)

trees (see Sec. 2). The CLIP4 algorithm, on the other hand, uses a very different mechanism for rule generation, namely, an integer linear programming.

4.1. CLIP4 Algorithm

The CLIP4 (Cover Learning (using) Integer Programming) algorithm is a hybrid that combines ideas (like its predecessors the CLILP2 and CLIP3 algorithms) of two families of inductive ML algorithms, namely, rule algorithms and decision tree algorithms. More precisely, CLIP4 uses a rule-generation schema similar to Michalski's AQ algorithms, as well as the tree-growing technique to divide training data into subsets at each level of a (virtual) decision tree in a fashion similar to Quinlan's algorithms. The main difference between CLIP4 and the two families of algorithms is CLIP4's extensive use of the set covering (SC) problem and the novel solution constituting its core operation, which is performed several times to generate the rules. Specifically, the SC algorithm is used to select the most discriminating features, to grow new branches of the tree, to select data subsets from which CLIP4 generates the least overlapping rules, and to generate final rules from the (virtual) tree leaves, which store subsets of the data.

An important characteristic feature that distinguishes CLIP4 from the vast majority of ML algorithms is that it generates production rules that involve inequalities. This results in generating a small number of compact rules, especially in domains where attributes have large number of values and where the majority of them are correlated with the target class. In this case, algorithms using equalities would generate hundreds of complex rules. In the next section, we describe a novel SC algorithm developed expressly for use in CLIP4.

4.1.1. CLIP4's Set-Covering Algorithm

Several key operations performed by CLIP4 are modeled and solved by the **set covering** algorithm, which is a simplified version of **integer programming** (IP). In general, IP is used for optimization of a function, subject to a large number of constraints. Several simplifications are made to the IP model to transform it into the SC problem: the function that is the subject of optimization has all its coefficients set to one; their variables are binary, $x_i = \{0, 1\}$; the constraint function coefficients are also binary; and all constraint functions are greater than or equal to one. The SC problem is NP-hard, and thus only approximate solution can be found.

Before we describe the SC algorithm we shall introduce the problem (see Figure 12.5) and present a solution to it. First, we transform the IP problem into the binary matrix (BIN) representation that is obtained by using the variables and constraint coefficients. BIN's columns correspond to variables (features/attributes) of the optimized function; its rows correspond to function constraints (examples).

CLIP4 finds the solution of the SC problem in terms of selecting a minimal number of columns that have the smallest total number of 1's. This outcome is obtained by minimizing the number of 1's that overlap among the columns and within the same row. The solution consists of a binary vector composed of the selected columns. All rows for which there is a value of 1 in the matrix, in a particular column, are assumed to be "covered" by this column.

To obtain the solution we use the SC algorithm (developed for the CLIP4 algorithm), which is summarized by the following pseudocode.

Given: BINary matrix. *Initialize*: Remove all empty (inactive) rows from the BINary matrix; if the matrix has no 1's, then return error.

1. Select active rows that have the minimum number of 1's in rows – *min-rows*
2. Select columns that have the maximum number of 1's within the *min-rows* – *max-columns*
3. Within *max-columns* find columns that have the maximum number of 1's in all active rows – *max-max-columns*. If there is more than one *max-max-column*, go to Step 4., otherwise go to Step 5.
4. Within *max-max-columns* find the first column that has the lowest number of 1's in the inactive rows
5. Add the selected column to the solution
6. Mark the inactive rows. If all the rows are inactive then terminate; otherwise go to Step 1.

Result: Solution to the SC problem.

$$\begin{aligned} &\text{Minimize}: \\ &x_1 + x_2 + x_3 + x_4 + x_5 = Z \\ &\text{Subject to}: \\ &x_1 + x_3 + x_4 \geq 1 \\ &x_2 + x_3 + x_5 \geq 1 \\ &x_3 + x_4 + x_5 \geq 1 \\ &x_1 + x_4 \geq 1 \\ &\text{Solution}: \\ &Z = 2, \quad \text{when } x_1 = 1, x_2 = 0, x_3 = 1, x_4 = 0, x_5 = 0 \end{aligned}$$

$$\begin{aligned} &\text{Minimize}: \\ &x_1 + x_2 + x_3 + x_4 + x_5 = Z \\ &\text{Subject to}: \\ &\begin{bmatrix} 1,0,1,1,0 \\ 0,1,1,0,1 \\ 0,0,1,1,1 \\ 1,0,0,1,0 \end{bmatrix} \cdot \begin{bmatrix} x_1 \\ x_2 \\ x_3 \\ x_4 \\ x_5 \end{bmatrix} \geq 1 \end{aligned}$$

Figure 12.5. Simplified set-covering problem and its solution (on the left); and in BIN matrix form (on the right)

402   4. Hybrid Algorithms

In the above pseudocode, an active row is a row not covered by the partial solution, and an inactive row is a row already covered by the partial solution.

Example: To illustrate how the SC algorithm works we use here a slightly different example than the one shown in Figure 12.5. The difference is that some matrix rows (constraints) are repeated, as shown in Figure 12.6. The solution, also shown in Figure 12.6 consists of the second and fourth columns, which have no overlapping 1's in the same rows.

Before we describe the CLIP4 algorithm in detail, let us first introduce key notations (these are the same as in DataSqueezer except that they use different letters to make them consistent with the original articles cited at the end of the Chapter). The set of all training examples is denoted by S. A subset of positive examples is denoted by S_P and the subset of negative examples by S_N. S_P and S_N are represented by matrices whose rows represent examples and whose columns

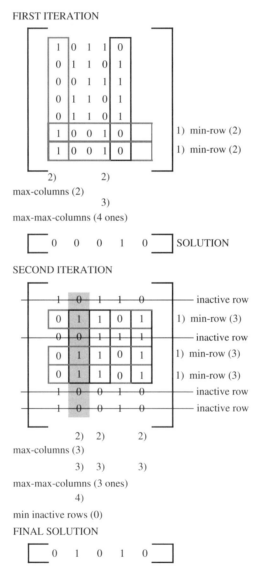

Figure 12.6. Solution of the SC problem using the SC algorithm.

correspond to attributes. The matrix of positive examples is denoted as POS and their number by N_{POS}. Similarly for the negative examples, we have matrix NEG and number N_{NEG}. The following properties are satisfied for the subsets:

$$S_P \cup S_N = S, \quad S_P \cap S_N = \emptyset, \quad S_N \neq \emptyset, \text{ and } \quad S_P \neq \emptyset$$

The examples are described by a set of K attribute-value pairs:

$$e = \wedge_{j=1}^{K}[a_j \# v_j]$$

where a_j denotes the jth attribute with value $v_j \in d_j$, and # is a relation ($\neq, =, <, \approx, \leq$, etc.), where K is the number of attributes. An example e consists of a set of selectors $s_j = [a_j \neq v_j]$.
The CLIP4 algorithm generates rules in the form:

$$\text{IF}(s_1 \wedge \ldots \wedge s_m) \quad \text{THEN class} = \text{class}_i$$

where all selectors are only in the form $s_i = [a_j \neq v_j]$, that is use only inequalities.

The positive examples from the matrix POS are described by a set of values $\text{pos}_i[j]$, where $j = 1, \ldots, K$ is the column number and i is the example number (the row number in the POS matrix). The negative examples are described similarly by a set of $\text{neg}_i[j]$ values. CLIP4 uses binary matrices (BIN) that are composed of K columns, filled with either 1 or 0 as values. Each element of the BIN matrix is denoted by $\text{bin}_i[j]$, where i is a row number and j is a column number. These matrices are the result of operations performed by CLIP4, which are modeled and solved using the SC algorithm described above. The pseudocode of the CLIP4 algorithm is provided in Figure 12.7. The algorithm consists of three phases that are explained below.

Phase I: The POS data is partitioned into subsets of similar data in a decision-tree-like manner. Each node represents a data subset. Each level of the tree is built using one negative example to find selectors that distinguish between all positive examples and the negative example. The selectors are used to create new branches of the tree. During tree growth, we use pruning to eliminate noise from the data (described later) and to avoid excessive growth, which reduces execution time.

Explanation of lines 2–23: The tree is grown in a top-down manner. At the ith tree level, N_i subsets, represented by matrices $\text{POS}_{i,j}$ ($j = 1, \ldots, N_i$), are generated using N_{i-1} subsets from the previous tree level and the single negative example neg_i. Each subset, represented by the matrix $\text{POS}_{i,j}$ (examples that constitute this matrix's elements are denoted as $\text{pos}^{i,j}$), is transformed into a BIN matrix of the same size as the $\text{POS}_{i,j}$ matrix, using the neg_i example, and then modeled and solved using the SC method. The solution is used to generate subsets for the next tree level, represented by the $\text{POS}_{i+1,j}$ matrices.

CLIP4 grows a virtual "tree", since any iteration consists of only one level: the most recent bottom tree level. Moreover, this tree is pruned so that even the one level that is kept has at most a few nodes. This results in high memory efficiency. The data splits are made by generating not just one "best" division, based on some "best" feature (say in terms of the highest information gain, as in decision trees), but a set of divisions based on any feature that distinguishes between the positive and the one negative example. This mechanism of partitioning data assures generation of the general rules.

Phase II: A set of terminal subsets (tree leaves) is selected using two criteria. First, large subsets are preferred over small ones (in accordance with Occam's razor, which hopefully will result in the rules that are "strong" and general), while the accepted subsets (between them) must cover the entire POS data. Second, we want to use the completeness criterion. To that end, we first perform a back-projection of one of the selected positive data subsets using the entire NEG data,

404   4. Hybrid Algorithms

```
         N₀=1; Create POS₀,₁ consisting of entire Sₚ, Create NEG consisting of entire Sₙ    // Initialize
1   for negᵢ, i=1 to N_NEG do {                                                              // PHASE I STARTS
2       for j=1 to Nᵢ₋₁ do {                                                                 //for each POSᵢ₋₁,ⱼ matrix
3           for k=1 to K do {                                                                //create new BINⱼ matrix
4               for l=1 to number of POSᵢ₋₁,ⱼ rows do {
5                   if posₗⁱ⁻¹,ʲ[k] = negᵢ[k] then binₗʲ[k] = 0;
6                   if posₗⁱ⁻¹,ʲ[k] ≠ negᵢ[k] then binₗʲ[k] = 1;
7                   if posₗⁱ⁻¹,ʲ[k] = '*' then binₗʲ[k] = 0;                                 // missing value encountered
8                   if negᵢ[k] = '*' then binₗʲ[k] = 0; }}                                   // missing value encountered
9           SOLⱼ = SolveSCProblem(BINⱼ);}
10      PruneMatrices(POSᵢ₋₁,ⱼ, SOLⱼ, j=1,...,Nᵢ₋₁);
11      ApplyGeneticOperators(POSᵢ₋₁,ⱼ, SOLⱼ, j=1,...,Nᵢ₋₁);
12      Nᵢ=1;                                                                                //counter for POSᵢ₊₁ matrices
13      for j=1 to Nᵢ₋₁ do {                                                                 //for each POSᵢ,ⱼ matrix
14          if POSᵢ₋₁,ⱼ was not pruned or redundant then {
15              for k=1 to K do {                                                            //through entire solution vector
16                  if SOLⱼ[k]=1 then {                                                      //then create new POS_{i,Ni} matrix
17                      for l=1 to number of POSᵢ₋₁,ⱼ rows do {
18                          if posₗⁱ⁻¹,ʲ[k]≠negᵢ[k] then add posₗⁱ⁻¹,ʲ to POS_{i,Ni} matrix; }}
19                  Nᵢ=Nᵢ+1; }}}
20      for j=1 to Nᵢ do {                                //for each POSᵢ matrix check if it large enough to be not considered as a noise
21          if number of rows of POSᵢ,ⱼ < NoiseThreshold then { remove POSᵢ,ⱼ from the tree; Nᵢ=Nᵢ-1;}}
22      EliminateRedundantMatrices(POSᵢ,ⱼ j=1,...,Nᵢ); }
23  Create BIN matrix that consist of N_{N_NEG} columns and N_POS rows, and fill with zeros    // PHASE II STARTS

24  for i=1 to N_{N_NEG} do {                                                                // for all tree leaves
25      for j=1 to N_POS do {
26          for k=1 to number of rows of POS_{N_NEG,i} do { if posⱼ⁰,⁰ = posₖ^{N_NEG,i} then binⱼ[i]=1;}}}
27  SOL = SolveSCProblem (BIN);                                                              // select best leaf node subsets
28  for i=1 to N_{N_NEG} do {                                                                // through entire solution vector
29      create BINᵢ=NEG;
30      if SOL[i]=1 then {                                                                   // back-project POS_{Nneg,i} matrix
31          for j=1 to K do {
32              for k=1 to N_NEG do {
33                  if negₖ[j]='*' then binₖⁱ[j]=0; else {
34                      for l=1 to number of rows of POS_{N_NEG,i} do { if posₗ^{N_NEG,i}[j]=negₖ[j] then binₖⁱ[j]=0; }}}}
35      for j=1 to K do {                                                                    // convert BINᵢ values to binary
36          for k=1 to N_NEG do { if binₖⁱ[j]≠0 then binₖⁱ[j]=0;}}
37      SOLᵢ = SolveSCProblem (BINᵢ);
38      for j=1 to K do {                                                                    // start generation of i-th rule
39          if SOLᵢ[j]=1 then {                                                              // add selectors to the rule
40              for k=1 to N_NEG do { if binₖⁱ[j]=1 then add "aⱼ≠negₖ[j]" selector to the Ruleᵢ; }}}}
                                                                                             // PHASE III STARTS
41  best#covered=0; previous_best#covered=0;                                                 // holds # ex. covered by the rule
42  while best#covered≥0 do {                                                                //until best rules are accepted
43      for i=1 to N_{N_NEG} do {                                                            // through all generated rules
44          coversᵢ=N_POS;                                                                   // # examples covered by Ruleᵢ
45          for j=1 to N_POS do {                                                            // for all examples in POS
46              for k=1 to K do {
47                  for l=1 to number of selectors in Ruleᵢ do {
48                      if aₗ=k and vₖ=posⱼ⁰,¹[k] then { coversᵢ=coversᵢ-1; j=j+1; }}}}      // example not covered by Ruleᵢ
49          if best#covered<coversᵢ then best#covered=coversᵢ; best_rule=Ruleᵢ; }
50      N_POS = N_POS – best#covered;
51      if N_POS<StopThreshold or N_POS=0 then STOP;
52      POS₀,₁ = POS₀,₁ – examples covered by best_rule;
53      if (best#covered<previous_best#covered/2 and best#covered<N_POS/2) then best#covered=-1;    // multiple rules
54      previous_best#covered_examples=best#covered_examples; }
```

Figure 12.7. Pseudocode of the CLIP4 algorithm

and then convert the resulting matrix into a BIN matrix and solve it using the SC method. The solution is used to generate a rule, and the process is repeated for every selected positive data subset. Explanation of lines 24–41: $N_{N_{NEG}}$ is the number of tree leaves. The binary BIN matrix is used to select the minimal set of leaves that covers the entire POS data. Back-projection results in a matrix BIN_i that has 1's for the selectors that can distinguish between positive and negative examples. The back-projection generates one matrix for every terminal subset ($POS_{N_{NEG},i}$). It is computed from the NEG matrix, by setting a value from NEG to zero if the same value appears in the corresponding column in the $POS_{N_{NEG},i}$ matrix; otherwise it is left unchanged. The ith rule is generated by solving the SC problem for BIN_i and adding a selector for every 1 that is in any column indicated by the solution.

The rule induction generates a rule directly from two sets of data (NEG data and the selected subset of positive data). It does not require the use of logic or storing and traversing the entire tree, as in decision trees, but only a comparison of values between the two matrices. Thus, the only two data structures required to generate the rules are lists (vectors) and matrices.

Phase III: A set of best rules is selected from the generated rules. Rules that cover the most POS examples are chosen as possibly the most general. If there is a tie between the "best rules", the rule that uses the minimal number of selectors is chosen. Explanation of lines 42–55: To find the number of examples covered by a rule, we first find examples not covered by the rule, and subtract these from the total number of examples. We do this because the rules consist of selectors involving inequalities. The variable called $covers_i$ records the number of positive examples covered by the ith rule. After a rule is accepted, the positive examples covered by it are removed from the POS matrix (line 53).

In this phase, more than one rule can be generated. The rules are accepted in order from the strongest to the weakest. We use a heuristic for accepting multiple rules, which states that the next rule is accepted when it covers at least half the number of examples covered by the previously accepted rule and at least half the number of positive examples not covered by any rule so far. CLIP4 generates multiple rules in a single sweep through the data. The thresholds used in the pseudocode in Figure 12.7 are described later.

The basic characteristics of the algorithm are:

- completeness, meaning that the generated set of rules describes all positive examples
- consistency, meaning that rules do not describe any of the negative examples
- use of a minimal number of selectors in the generated rules.

CLIP4 is memory efficient (because it stores only the bottom level of the tree), generates strong rules (because of its mechanism of data partitioning and use of heuristics for rules acceptance), and is fast (because of pruning and accepting multiple rules in one iteration). The algorithm is also robust to noisy and missing-value data, as shown later in the Chapter.

Example: Below we illustrate how CLIP4 works using again our travel agency example. The data from Table 12.1 are stored in two matrices POS and NEG.

$$POS = \begin{bmatrix} 1,1,3,1 \\ 1,1,1,4 \\ 2,3,2,5 \\ 3,2,3,5 \\ 1,1,2,3 \end{bmatrix} \quad NEG = \begin{bmatrix} 1,3,2,1 \\ 3,1,2,3 \\ 3,4,3,2 \\ 1,3,3,4 \end{bmatrix}$$

406   4. Hybrid Algorithms

Phase I: Using the first negative example [1,3,2,1] and matrix POS, we obtain the following binary matrix:

$$BIN = \begin{bmatrix} 0,1,1,0 \\ 0,1,1,1 \\ 1,0,0,1 \\ 1,1,1,1 \\ 0,1,0,1 \end{bmatrix} \quad SOL = \begin{bmatrix} 1,1,0,0 \\ 0,1,0,1 \\ 0,0,1,1 \end{bmatrix}$$

From the BIN matrix we obtain, using the SC algorithm, three equally good solutions, stored in matrix SOL. We always choose the first one (going from left to right while analyzing the examples), so the accepted solution is:

$$SOL = [1,1,0,0]$$

Next, we split the matrix POS on the just obtained solution vector, looking at the appropriate values of features 1 and 2 (indicated by the two 1s in vector SOL). The corresponding values of feature 1 in matrix POS are 1, 2, and 3, but the neg_1 example has a value 1 in the first feature and must therefore be different from it. Thus we obtain:

$$F1 \neq 1 \; POS1 = \begin{bmatrix} 2,3,2,5 \\ 3,2,3,5 \end{bmatrix}$$

Similarly for feature 2: It too has values 1,2 and 3 in matrix POS, but since the neg_1 has a value of 3 it must be different from it:

$$F2 \neq 3 \; POS2 = \begin{bmatrix} 1,1,3,1 \\ 1,1,1,4 \\ 3,2,3,5 \\ 1,1,2,3 \end{bmatrix}$$

We have just used the SC solution to find the most discriminative pairs *(feature, its value)*, using information about the values in matrix NEG to split the matrix POS into subsets.

Phase II: After repeating the process illustrated above, at the end of Phase I we end up with just two matrices (the leaf nodes of the virtual decision tree) the matrix numbers 8 and 9 are used for tracking purposes and are not important:

$$POS8 = \begin{bmatrix} 2,3,2,5 \\ 3,2,3,5 \end{bmatrix} \quad POS9 = \begin{bmatrix} 1,1,3,1 \\ 1,1,1,4 \\ 1,1,2,3 \end{bmatrix} \quad POS = \begin{bmatrix} 1,1,3,1 \\ 1,1,1,4 \\ 2,3,2,5 \\ 3,2,3,5 \\ 1,1,2,3 \end{bmatrix}$$

Next, we generate the TM matrix. Its number of columns is equal to the number of matrices at the end of Phase I. It is obtained by comparing (sliding down) the examples of POS8 and POS9 with (over) the entire matrix POS (shown on the right for the convenience of the reader). If the feature values of the examples in POS8 or in POS9 are the same as in the matrix POS, then we put a 1 in the same row of TM as in the covered example in POS. A 1 in the first column of TM indicates that this particular example is covered by POS 8; a 1 in the second column indicates that it is covered by POS9. Thus, examples 3 and 4 from POS are covered by POS8, while examples 1, 2, and 5 are covered by POS9. Otherwise, we put a 0 in the column. The purpose of constructing a TM matrix is to decide how many of the matrices (subsets) generated at the end of Phase I are

Chapter 12 Supervised Learning: Decision Trees, Rule Algorithms, and Their Hybrids

needed to cover, between them, all examples from matrix POS (in our case, both are needed, but this is a special case; in general, from several matrices only a few will be needed).

$$TM = \begin{bmatrix} 0,1 \\ 0,1 \\ 1,0 \\ 1,0 \\ 0,1 \end{bmatrix}$$

$$SOL = [1,1]$$

Next, we again use the SC algorithm to find the solution (i.e., how many subsets are needed), and the SOL vector tells us that both are needed.

We then create a back-projection NEG matrix, obtained by comparing all matrices at the end of Phase I with the entire matrix NEG (shown here only for POS9). This step ensures that none of the NEG examples is covered.

$$POS9 = \begin{bmatrix} 1,1,3,1 \\ 1,1,1,4 \\ 1,1,2,3 \end{bmatrix} \quad NEG = \begin{bmatrix} 1,3,2,1 \\ 3,1,2,3 \\ 3,4,3,2 \\ 1,3,3,4 \end{bmatrix}$$

$$backproj \quad NEG = \begin{bmatrix} 0,3,0,0 \\ 3,0,0,0 \\ 3,4,0,2 \\ 0,3,0,4 \end{bmatrix} \begin{bmatrix} 0,1,0,0 \\ 1,0,0,0 \\ 1,1,0,1 \\ 0,1,0,1 \end{bmatrix}$$

$$SOL = [1,1,0,0]$$

The matrix *backproj* NEG is obtained by sliding POS9 over the entire matrix NEG and keeping only the values from matrix NEG that are different from the POS9 examples (note that this approach is the opposite of what we have done before). Then this matrix is converted into a BIN matrix, and again the SC algorithm is invoked to find a solution.

From this solution and from the *backproj* NEG matrix (in general matrices), we are now ready to generate the first rule:

IF (F1 $\neq$ 3) AND (F2 $\neq$ 3) AND (F2 $\neq$ 4) THEN F5 = Buy (covers examples e1, e2, and e5)

By the same process, using POS8, we generate one more rule:

IF (F4 $\neq$ 1) AND (F4 $\neq$ 3) AND (F4 $\neq$ 2) AND (F4 $\neq$ 4) THEN F5 = Buy (covers examples e3 and e4)

Phase III: Using the CLIP4 heuristic we choose only the first rule and remove from matrix POS all examples covered the first rule. Next, we repeat the entire process on the reduced matrix POS:

$$POS' = \begin{bmatrix} 2,3,2,5 \\ 3,2,3,5 \end{bmatrix}$$

After going again through all phases of the algorithm, we generate just one rule:

IF (F4 $\neq$ 1) AND (F4 $\neq$ 3) AND (F4 $\neq$ 2) AND (F4 $\neq$ 4) THEN F5 = Buy

408   **4. Hybrid Algorithms**

Note that we have already generated this rule, but such an outcome does not have to be the case. For the final outcome, in two iterations, the algorithm generates a set of rules that covers all positive examples and none of the negative:

IF (F1≠3) AND (F2≠3) AND (F2≠4)   THEN F5=Buy
IF (F4=5)   THEN F5=Buy

Note that by using "reverse" logic (knowing the feature values for attribute F4) it is possible to convert the second rule into the equalities rule, as written above. Verbally, the two rules say:

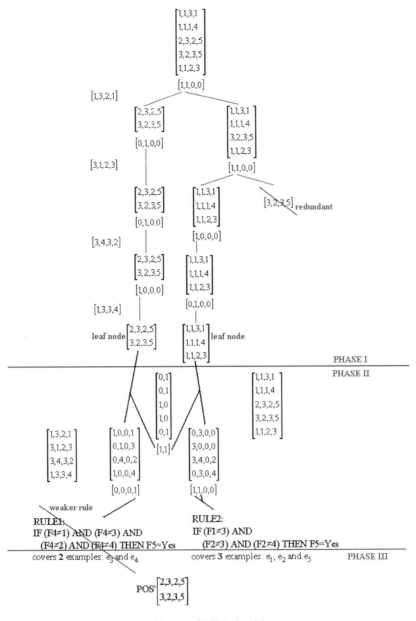

Figure 12.8. Summary of the process used by the CLIP4 algorithm.

Chapter 12 Supervised Learning: Decision Trees, Rule Algorithms, and Their Hybrids

IF Call ≠ International AND Language Fluency ≠ Bad
  AND Language Fluency ≠ Foreign   THEN Buy
IF Customer is 80 years or older   THEN Buy

We illustrate the process used by CLIP4 in Figure 12.8. A high-level view of the same example is shown in Figure 12.9. Both Figure illustrate how CLIP4 performs tree generation, data partitioning, and rule generation.

4.1.2. Handling of Missing Values

Data with attribute missing values are often encountered. If the number of examples with missing values is small, as a percentage of the data, we can simply discard them in a preprocessing step. However, in many real situations – for example, in medical data – the number of examples (records) with missing values can exceed 50% of the data. In such real situations, we might use all examples and all attributes with values and omit from our computations only these attributes with missing values. The CLIP4 algorithm does not discard any examples with missing values nor does it fill them with any computed values. It simply takes advantage of the available values in the examples and ignores the missing values.

Attributes with missing values influence the first two phases of the algorithm. They are removed from processing by filling in the corresponding cell in the BIN matrix cell with a 0 when the new branch of the tree in Phase I is generated, and when the back-projection in Phase II is performed. The example shown in Table 12.6 is used to show how rules are generated from examples with missing values. The mechanism used in CLIP4 to deal with missing attribute

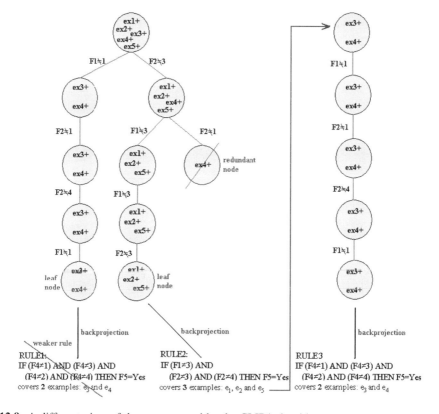

Figure 12.9. A different view of the process used by the CLIP4 algorithm.

Table 12.6. Data with missing values.

Ex. #	F1	F2	F3	F4	Class
1	1	2	3	*	1
2	1	3	1	2	1
3	*	3	2	5	1
4	3	3	2	2	1
5	1	1	1	3	1
6	3	1	2	5	2
7	1	2	2	4	2
8	2	1	*	3	2

values assures that they are not used in generated rules and that all existing attribute values, even in examples with missing attributes, are used during the rule generation process. The advantage of this mechanism is that the user simply supplies data to the algorithm and obtains rules without needing to use some statistical methods to calculate missing values. This approach works well under the assumption that the remaining (complete) portion of the data is sufficient to infer the correct model.

Example: Data shown in Table 12.6 (slightly modified travel agency data) illustrate how CLIP4 deals with missing values. The missing values are denoted by "*" and Class 1 indicates the POS class. CLIP4 generates two rules:
  IF F1 $\neq$ 3 AND F1 $\neq$ 2 AND F3 $\neq$ 2   THEN class 1
  IF F2 $\neq$ 2 AND F2 $\neq$ 1   THEN class 1

The first rule covers examples 1, 2, and 5, while the second covers examples 2, 3, and 4. Between them they cover all positive examples, including those with missing values, and none of the negative examples. The rules overlap since both cover the second example.

4.1.3. Handling of Continuous Attributes and Classification

As mentioned above, a machine learning algorithm can deal with continuous values in two ways:

– by processing continuous attributes itself, or
– by performing a front-end discretization, which is usually performed prior to the rule induction process (as in decision tree algorithms).

CLIP4 uses the second approach by using the supervised discretization algorithms such as CAIM and others described in Chapter 8. CLIP4 classifies as example by using sets of rules generated for all classes. Two classification outcomes are possible: an example is assigned to a particular class, or it is left unclassified. To classify an example, two principles apply:

– All rules that cover the example are found. If no rules cover the example then it remains unclassified. Such a situation may occur if the example has missing values for all attributes used by the rules.
– For every class, goodness of the rules describing a particular class and covering the example is summed. The example is assigned to a class that has the highest summed value. If there is a tie, then the example is left unclassified. For each generated rule, a goodness value, equal to the percentage of the positive training examples that it covers, is calculated. For instance, if an example is covered by two rules from class 1 with corresponding goodness values of 15% and 20%, but the example is also covered by two rules from class 2 with goodness values 50% and 10%, then it would be classified to Class 2 (the sum of goodness values for Class 1 is 35% vs. 60% for Class 2).

4.1.4. Thresholds

CLIP4 uses three thresholds (by default these are all 0) to perform tree pruning and to remove "noisy" examples:

The **Noise Threshold** determines which nodes (possibly containing noisy positive examples) are pruned from the tree grown in Phase I. The threshold prunes every node that contains fewer examples than its value.

The **Pruning Threshold** is used to prune nodes from the generated tree. It uses a goodness value, identical to the fitness function described later, to perform selection of the nodes. The threshold selects the first few nodes with the highest fitness value and removes the remaining nodes from the tree.

The **Stop Threshold** stops the algorithm when fewer than the threshold number of positive examples remains uncovered. CLIP4 generates rules by partitioning the data into subsets containing similar examples, and removes examples covered by already generated rules. This approach has the advantage of eliminating small subsets of positive examples (which contain examples different than the majority already covered) from the subsequent rule generation process. If the user were to know the amount of noise in the data then the threshold could be set to this value (e.g., 5%). The noise and stop thresholds are specified as a percentage of the size of positive data and thus are easily scalable.

4.1.5. Use of Genetic Operators to Improve Accuracy on Small Training Data

Genetic algorithms (GAs) are search algorithms that mimic the natural selection process. They start with an initial population of N individuals/ chromosomes in the search space, determine the chances of survival of these individuals, and evolve the population to retain individuals with the highest corresponding value of a fitness function, while eliminating weaker individuals. CLIP4 uses this idea to improve the accuracy of generated rules. Its genetic module works by exploiting a single loop through a number of evolving populations. The loop consists of establishing the initial population of individuals and then selecting the new population from the old population, altering and evaluating the new population, and replacing the old one with the new. These operations are performed until a termination criterion is satisfied. CLIP4 uses the GA in Phase I to enhance the partitioning of the data and to obtain more "general" leaf node subsets. The components of the genetic module are as follows:

– *Population* and *individual*

An individual/chromosome is defined as a node in the tree and consists of the $POS_{i,j}$ matrix (the jth matrix at the ith tree level) and $SOL_{i,j}$ (the solution to the SC problem obtained from the $POS_{i,j}$ matrix). A population is defined as a set of nodes at the same level of the tree.

– *Encoding* and *decoding scheme*

There is no need for encoding using the individuals defined above since GA operators are used on the binary $SOL_{i,j}$ vector.

– *Selection* of the new population

The initial population is the first tree level that consists of at least two nodes. CLIP4 uses the following fitness function to select the most suitable individuals for the next generation:

$$fitness_{i,j} = \frac{number\ of\ examples\ that\ constitute\ POS_{i,j}}{number\ of\ subsets\ that\ will\ be\ generated\ from\ POS_{i,j}\ at\ (i+1)^{th}\ tree\ level}$$

The fitness value is calculated as the number of rows of the $POS_{i,j}$ matrix divided by the number of 1's from the $SOL_{i,j}$ vector. The fitness function has high values for the tree nodes that consist of a large number of examples with a low branching factor. These two properties

412   4. Hybrid Algorithms

influence the generalization ability of the rules and the speed of the algorithm. The mechanism for selecting individuals for the next population is as follows:

- All individuals are ranked using the fitness function
- Half of the individuals with the highest fitness are automatically selected for the next population (they will branch to create nodes for the next tree level)
- The second half of the next population is generated by matching the best with the worst individuals (the best with the worst, the second best with the second worst, etc.) and applying GA operators to obtain new individuals (new nodes in the tree). This mechanism promotes the generation of new tree branches that contain large number of examples.

Example: An illustration of the selection mechanism used in the CLIP4 algorithm is shown in Figure 12.10.

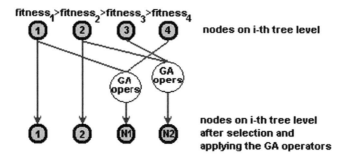

Figure 12.10. Selection mechanism performed by the GA module.

The GA module uses crossover and mutation operators. Both are applied only to the $SOL_{i,j}$ vectors. The resulting $ChildSOL_{i,j}$ vector, together with the $POS_{i,j}$ matrix of the parent with the higher fitness value, constitutes the new individual. The selection of the $SOL_{i,j}$ matrix assures that the resulting individual is consistent with CLIP4's way of partitioning data. The crossover operator is defined as

$$ChildSOL_i = \max(Parent1SOL_i, Parent2SOL_i)$$

where $Parent1SOL_i$ and $Parent2SOL_i$ are the ith values of $SOL_{i,j}$ vectors of the two parent nodes.

CLIP4 uses mutation with 10% probability to flip a value in the ChildSOL vector to 1, regardless of the existing value. For particular data, the probability of mutation can be established experimentally. Each 1 in the ChildSOL generates a new branch, except for 1's taken from the $SOL_{i,j}$ of the parent with higher fitness value, which are discarded because they would generate branches redundant with the branches generated by the parent.

Example: An example of the crossover operation is shown in Figure 12.11.

The termination criterion checks whether the bottom level of the tree has been reached. The entire evolution process of the GA module is shown in Figure 12.12.

The GA module of the CLIP4 algorithm is used when only a small data set a small portion of the search space is available to that covers generate rules that potentially cover a not yet described portion of the state space.

Chapter 12 Supervised Learning: Decision Trees, Rule Algorithms, and Their Hybrids 413

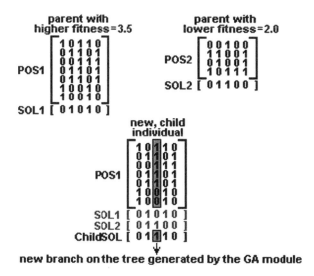

Figure 12.11. Example of the crossover performed by the GA module.

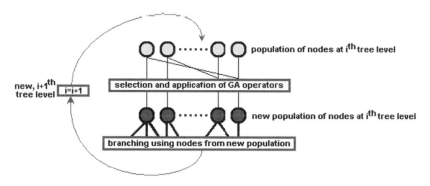

Figure 12.12. The evolution process performed by the GA module.

4.1.6. Pruning

CLIP4 uses prepruning to prune the tree during its process of tree generation. The prepruning stops the learning process even if some positive examples have still not been covered and while some negative examples are still covered. More precisely, CLIP4 prunes the tree grown in Phase I as follows:

– First, it selects a number (defined by the pruning threshold) of best nodes on the ith tree level. The selection is performed based on the goodness criterion identical to the fitness function described later. Only the selected nodes are used to branch into new nodes, and are passed to the $(i+1)^{th}$ tree level.
– Second, all redundant nodes that resulted from the branching process are removed. Two nodes are redundant if one mode contains positive examples that are identical to each other, or forms a subset of positive examples of the other node. The node with the smaller number of examples is pruned first.
– Third, after the redundant nodes are removed, each new node is evaluated using the noise threshold. If the node contains fewer examples than the number specified by the noise threshold, then the node is pruned.

414   4. Hybrid Algorithms

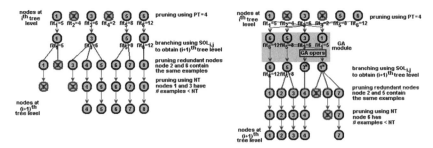

Figure 12.13. Example of pruning while generating the $(i+1)^{th}$ tree level from the ith level. Left: without the GA; right: with GA.

The prepruning method used in the CLIP4 algorithm avoids some disadvantages of classical prepruning. Namely, it never allows negative examples to be covered by the rules, and it ensures the completeness condition of the rule generation process. This type of prepruning increases the accuracy of the generated rules and lowers the complexity of the algorithm. An example of the prepruning process is shown in Figure 12.13.

4.1.7. Feature and Selector Ranking

The CLIP4 algorithm not only generates a data model in terms of rules but also ranks attributes and selectors. This aspect of the algorithm can be used for feature selection (see Chapter 7). Let us repeat here that feature selection is a process of finding a subset of original features that is optimal according to some defined criterion; in the case of classification, the criterion is to obtain the highest predictive accuracy. CLIP4 ranks attributes and selectors by assigning to them a goodness value that quantifies their relevance to a particular learning task. These rankings provide additional insight into the information hidden in the data.

The goodness of each attribute and selector is computed by means of a set of generated rules for a given class. All attributes with goodness value greater than zero are strongly relevant to the classification task. Strong relevancy means that the attribute cannot be removed from the attribute set without decreasing the accuracy of classification. The other attributes can be removed from the data. The attribute and selector goodness values are computed in the following manner:

- Each generated rule has a goodness value equal to the percentage value of the positive training examples it covers. Recall that each rule consists of one or more (attribute, value of the selector) pairs.
- Each selector has a goodness value equal to the goodness of the rule from which it comes. The goodness of the same selectors from different rules is summed, and then scaled to the (0,100) range; with 100 being the highest. Scaling of the goodness values is necessary because otherwise the summed goodness for a particular selector can grow to over 100, while only the ratio to the goodness's of other selectors is important.
- For each attribute, the sum of the scaled goodness of all attribute selectors is computed and then divided by the number of attribute values to obtain the goodness of the attribute.

Example: The following example illustrates attribute and selector ranking. Suppose we have two-category data, described by five attributes a1 through a5, that assume the values $a1 = \{1, 2, 3\}$, $a2 = \{1, 2, 3\}$, $a3 = \{1, 2\}$, $a4 = \{1, 2, 3\}$, and $a5 = \{1, 2, 3, 4\}$, while $a6 = \{1, 2\}$ is the decision attribute. Suppose CLIP4 generated the following rules from the data:

IF a5 ≠ 2 and a5 ≠ 3 and a5 ≠ 4   THEN class = 1 (covers 46% (29/62) positive examples)
IF a1 ≠ 1 and a1 ≠ 2 and a2 ≠ 2 and a2 ≠ 1   THEN class = 1 (covers 27% (17/62) positive examples)
IF a1 ≠ 1 and a1 ≠ 3 and a2 ≠ 3 and a2 ≠ 1   THEN class = 1 (covers 24% (15/62) positive examples)
IF a1 ≠ 2 a1 ≠ 3 and a2 ≠ 2 and a2 ≠ 3   THEN class = 1 (covers 14% (9/62) positive examples)

Using the information about attribute values we can obtain equality rules:

IF a5 = 1   THEN class = 1 (covers 46% (29/62) positive examples)
IF a1 = 3 and a2 = 3   THEN class = 1 (covers 27% (17/62) positive examples)
IF a1 = 2 and a2 = 2   THEN class = 1 (covers 24% (15/62) positive examples)
IF a1 = 1 and a2 = 1   THEN class = 1 (covers 14% (9/62) positive examples)

We calculate the goodness values for the selectors, as shown below in descending order (we can then calculate the goodness of attributes):

(a5, 1);   goodness 46
(a1, 3) and (a2, 3);   goodness 27
(a1, 2) and (a2, 2);   goodness 24
(a1, 1) and (a2, 1);   goodness 14

In order to show their relative goodness, they are scaled to the 0–100 range:

(a5, 1);   goodness 100
(a1, 3) and (a2, 3);   goodness 58.7
(a1, 2) and (a2, 2);   goodness 52.2
(a1, 1) and (a2, 1);   goodness 30.4

Thus, for attribute a1, we have the following selectors and goodness values:
(a1,3) with goodness 58.7, (a1,2) with goodness 52.2, and (a1,1) with goodness 30.4. From this we calculate the goodness of the first attribute a1 as follows:

$$(58.7 + 52.2 + 30.4)/3 = 47.1$$

We calculate the goodness of a2 similarly. For attribute a5, however, we have the following selectors and their goodness values: (a5,1) with goodness 100, and (a5,2) until (a5,4) with goodness of 0. Thus its goodness is calculated as

$$(100 + 0 + 0 + 0)/4 = 25.0$$

The remaining attributes, namely a3, a4, and a6 all have a goodness value of 0 because they were not used in the generated rules.

CLIP4 picked attributes a2 and a1, with the highest goodness value, and attribute a5, with goodness above zero only for the selector (a5,1). This allows removing attributes a3, a4, and a6 as irrelevant. The selector ranking gives additional insight into the a5 attribute, namely, that only its value of 1 is strongly relevant to the target concept.

The feature and selector ranking performed by CLIP4 algorithm can be used as follows:

– To select only relevant attributes (features). The user can discard all attributes that have a goodness of 0 and still have a correct (as equally accurate) model of the data.
– To provide additional insight into data properties. The selector ranking can help in analyzing the data in terms of the relevance of the selectors to the classification task.

5. Summary and Bibliographical Notes

In this Chapter, we described three types of **inductive machine learning** algorithms: **decision trees**, **rule algorithms**, and their **hybrids**. We have shown in detail how these algorithms work so that readers can easily reimplement them or come up with extensions/improvements. We have also presented a performance comparison based on several data sets and have discussed their advantages and disadvantages. The reader is referred to [4, 9, 12, 14, 16] for a general discussion of inductive machine learning. For a discussion of decision trees, the reader may consult [1, 4, 9, 10, 11, 17, 18], while for information on rule algorithms, the reader is referred to [4, 7, 8, 13, 15] and for information on hybrid algorithms to [2–6].

References

1. Cios, K.J., and Liu, N. 1992. Machine learning in generation of a neural network architecture: a Continuous ID3 approach. *IEEE Transactions on Neural Networks*, 3(2):280–291
2. Cios, K.J., and Liu, N. 1995. An algorithm which learns multiple covers via integer linear programming. Part I – The CLILP2 Algorithm. *Kybernetes*, 24: 29–50
3. Cios, K.J., Wedding, D.K. and Liu, N. 1997. CLIP3: cover learning using integer programming. *Kybernetes*, 26(4–5):513–536
4. Cios, K.J., Pedrycz, W., and Swiniarski, R. 1998. *Data Mining Methods for Knowledge Discovery.* Kluwer
5. Cios, K.J., and Kurgan, L. 2004. CLIP4: Hybrid inductive machine learning algorithm that generates inequality rules. *Information Sciences*, 163(1–3):37–83
6. Clark, P., and Niblett, T. 1989. The CN2 algorithm. *Machine Learning*, 3:261–283
7. Cohen, W. 1995. Fast effective rule induction. *Proceedings of the 12th International Conference on Machine Learning*, 115–123, Lake Tahoe, CA
8. Cohen, W., and Singer, Y. 1999. A simple, fast and effective rule learner. *Proceedings of the 16th National Conference on Artificial Intelligence*, 335–342
9. Han, J., and Kamber, M. 2001. *Data Mining: Concepts and Techniques*, Morgan Kaufmann
10. Holte, R.C. 1993. Very simple classification rules perform well on most commonly used data sets. *Machine Learning*, 11:63–90
11. Hunt, E.B., Marin, J., and Stone, P.J. 1966. *Experiments in Induction*, Academic Press
12. Kodratoff, Y. 1988. *Introduction to Machine Learning*, Morgan-Kaufmann
13. Kurgan, L., Cios, K.J., and Dick, S. 2006. Highly scalable and robust rule learner: performance evaluation and comparison, *IEEE Transactions on Systems Man and Cybernetics*, Part B, 36(1):32–53
14. Langley, P. 1996. *Elements of Machine Learning*, Morgan-Kaufmann
15. Michalski, R.S. 1969. On the quasi-minimal solution of the general covering problem. In: *Proceedings of the 5th International Symposium on Information Processing (FCIP 69)*, Bled, Yugoslavia, A3:25–128
16. Mitchell, T.M. 1997. *Machine Learning*, McGraw-Hill
17. Quinlan, J.R. 1993. *C4.5 Programs for Machine Learning*, Morgan-Kaufmann
18. Utgoff, P.E. 1989. Incremental induction of decision trees. *Machine Learning*, 4:161–186

6. Exercises

1. What is inductive machine learning?
2. What is the basic method of generating rules from data?
3. Do we always have to choose the "best" attribute on which to split the data while building decision trees?
4. What is pruning?
5. Calculate by hand all information gains, grow decision tree, and extract rules for class 1 for the training data shown in Table 12.1 using the ID3 algorithm.

6. Find by hand the rule(s) covering class 1, from the data shown in Table 12.1, using the CLIP4 algorithm. Illustrate graphically the process of generating the rules.
7. Implement and run on the three MONK data sets the following algorithms:

   a) ID3
   b) DataSquezer
   c) CLIP4

   The MONK data can be found at http://www.ics.uci.edu/~mlcarn/MLSummary.html

13
Supervised Learning: Neural Networks

In this Chapter, we introduce the general ideas behind artificial neural network (NN) algorithms. First, we discuss biological neurons and then move on to describe their artificial neuron models, the first component of any NN. Then we describe two other key components, namely, learning rules used for updating connections (weights/synapses) between the neurons, and network topologies. The three elements play a pivotal role in designing any NN. Next, we cover in detail the supervised Radial Basis Function (RBF) network, motivated by its usefulness in data mining and knowledge discovery tasks. Note that an unsupervised Kohonen's Self-Organizing Feature Maps (SOM) has already been described in Chapter 9 in our discussion of clustering algorithms; SOM is often seen as a NN (hardware) implementation of a clustering algorithm. Another category of NN covered in this book (see Chapter 11) concerns probabilistic neural networks.

1. Introduction

Interest in **artificial neural networks** arose from the realization that the human brain is, and for a long time will be, the best recognition and modeling device. So it is no surprise that researchers have tried over the last fifty years to mimic the brain's operation. Early modeling attempts were very crude, not least because of the then much poorer understanding of its operations. However, more recent NN algorithms have been able to model specific regions of the brain very accurately and to solve a variety of tasks outside the modeling realm. These advances are due in a large part to rapid progress in neuroscience, from a better understanding of how specific regions of the brain organize, reorganize, learn, and function to a better understanding at a lower level of how individual **neurons** function and operate. The discoveries of neuroscience are closely followed by computational scientists who accurately model operations of biological neurons and develop **learning rules** for updating connections between neurons. Other reasons for advances in NN are the ever greater speed and memory of computers and progress in parallel and distributed computing. Still, even DARPA, after spending millions of dollars on research in automated image recognition, has realized that human operators are still better, in terms of their speed and accuracy in recognizing images, than the most powerful computer systems. Because of that realization, research efforts now focus on better understanding of brain operations (in neuroscience) and on developing artificial software and hardware systems to mimic those operations – the approach known as **biomimetics**. Very appealing characteristics for computational problem-solving include the brain's learning and generalization abilities such as self-organization, robustness, and repeatability of basic computations. In this Chapter, we cover neuron models that closely mimic biological neurons and the learning rules that are applicable to networks of such neurons.

Artificial neural networks, even those that use a very simple neuron model, have the attractive property of being **universal approximators** – that is, they can approximate any mapping function between known inputs and corresponding known outputs (collectively known as training data), to any desired degree of accuracy (meaning the error of approximation can be made arbitrarily small). This property is very useful in many applications. However, if we train a NN to fit the training data too closely so that the model is overfitted by requiring a very small error the network may have poor predictive ability on new data.

In many domains, NNs have been successfully used for a variety of knowledge discovery tasks. One such task is clustering, a key unsupervised method in data mining, exploratory data analysis, and data compression. An example of a NN performing clustering is Kohonen's SOM network, which is described in Chapter 9. NNs have also demonstrated great success in solving problems of image recognition and compression. Since more and more information is now collected in terms of images, NNs that can process images, for example, for diagnostic or decision-making purposes, are becoming increasingly important.

To design any type of NN, the following key elements, described later in detail, must be determined:

- the **neuron model**(s) used for computations
- a **learning rule**, used to update the weights/synapses associated with connections between the neurons in a network
- the **topology** of a network, which determines how the neurons are arranged and interconnected

The first two elements must be specified a priori. However, topology can be determined in a variety of ways: it can be static (defined a priori), or it can be determined in a dynamic way, namely, by adding neurons and/or hidden layers as needed to solve a problem. Such NNs are known as **ontogenic**. The static topology of many types of networks is often determined by trial and error. However, the topology of some NNs, like RBFs, is determined to a large extent by their type.

The ability to learn, via the use of a learning rule in a network to update its connections, is an indispensable characteristic of any neural network algorithm. Probabilistic neural networks lack this ability and are not described here but rather in Chapter 11. In this book, we cover three types of NN, namely, SOM, probabilistic, and RBF. RBFs are the main topic of this Chapter, along with neuron models and learning rules.

2. Biological Neurons and their Models

All **neuron models** (called neurons or nodes, for short) resemble biological neurons to some extent. The degree of this resemblance is an important distinguishing factor between different neuron models. The choice of a model depends to some extent on the purpose for which it will be used. The most accurate neuron models mimic all the key characteristics of biological neurons, most importantly their **temporal spiking** nature (meaning that a neuron generates a train of electrical spikes after it is activated); membrane potential; sodium, potassium, and calcium channels; threshold accommodation; refractory periods; etc. The cost of high accuracy, however, is high model complexity and thus networks comprised of such neurons usually do not exceed a few thousand neurons. An example of a biologically accurate model is the Hodgkin-Huxley neuron model. Other models, like the **integrate-and-fire model**, are less complex but also more loosely biologically correct. At the other end of the spectrum, in terms of complexity, is the McCulloch-Pitts model, the first ever model of a biological neuron that is very simple and preserves only a couple of features of a biological neuron, namely, that after receiving the inputs it sums them up and, if the sum is above a specified threshold, "fires" (generates) a single output. Because of its simplicity, networks built of such neurons could be quite large.

The first learning rules for networks of artificial neurons were developed for simple neuron models. More recently, however, learning rules are being developed for networks that use **spiking neuron** models, and these have been shown to outperform networks consisting of simple neuron models in tasks like autonomous image analysis. With parallel NN implementations, based on spiking neurons and the learning rules that operate on them, they will be used more frequently for solving difficult problems because of their self-organizing properties.

2.1. Biological Neurons

Figure 13.1 shows a sketch of a biological **neuron**. The **nucleus** (body of the neuron) receives the stimuli (inputs) via its **dendrites** at the contact points called **synapses** (shown in black), sums up these stimuli and, if the sum is above its membrane threshold potential, it generates a **postsynaptic potential** (a **train of spikes**) and sends it along its **axon** to other neurons, which are connected to its terminals again via synapses (in NN terminology called **weights**). The frequency of the generated spikes depends on the strength of the stimuli. The synapses between the neurons are plastic (modifiable) and determine the influence of the incoming signals on the neuron's "firing" (generating a spike).

A neuron that feeds into other neurons (mainly through its axon, but also thorough other connections) is called a **presynaptic neuron**, and a neuron that receives inputs from presynaptic neurons is called a **postsynaptic neuron**.

In what follows, we first introduce a spiking neuron model, since it closely resembles a biological neuron, and then the McCulloch-Pitts simple neuron model.

2.2. Spiking Neuron Models

Two types of neuron models most closely resemble biological neurons, namely, **conductance** models (like the Hodgkin-Huxley model) and **integrate-and-fire** models (like the MacGregor model). In what follows we describe the integrate-and-fire model and then the **very simple spiking neuron** model, which is less accurate but retains key functions of biological neurons.

Conductance models describe the dynamics of the neuron's membrane potential and its spike generation in great detail by mimicking the biological neurons' response and by modeling several ion channels, like potassium, calcium, and sodium; postinhibitory rebounds; spiking activity; and the multi-compartmental structure of a neuron (its nucleus, dendrites, and axon are modeled separately). Consideration of the compartmental geometry of the neuron increases the model's accuracy but also considerably raises its computational complexity. In many applications, such

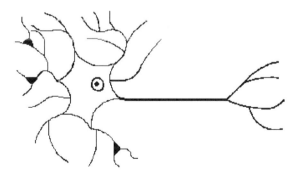

Figure 13.1. Sketch of a biological neuron.

422   **2. Biological Neurons and their Models**

as the modeling of learning processes, spike-generation modeling can be simplified; although it decreases degree of biological similarity, it allows for the study of much larger NNs from a more "global" perspective.

Integrate-and-fire neuron models provide some simplification of spike generation while accounting for the membrane potential and other essential neuron properties. MacGregor's model closely models the biological neuron behavior in terms of its membrane potential, potassium channel response, refractory properties, and adaptation to stimuli. Instead of modeling each individual channel (except for potassium), however, it imitates the resulting neuron's excitatory and inhibitory properties. For accurate modeling, it is always crucial to preserve the representation of the spiking nature of the biological neurons. Biological neurons, both excitatory (those that help other neurons to fire) and inhibitory (those that prevent other neurons from firing), generate a series of action potentials (also called a train

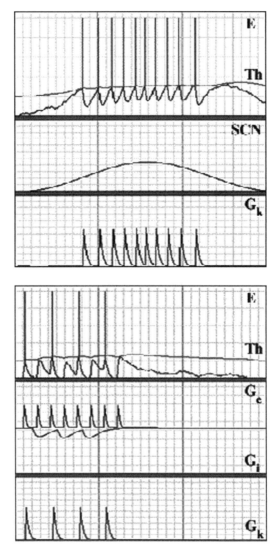

Figure 13.2. Neuron responses to an external artificial stimulation, SCN, (top), and to natural stimulation with excitatory, G_e, and inhibitory, G_i, spikes (bottom), where E is membrane potential, Th is threshold, and G_K is potassium channel conductance.

of spikes, or just spikes) in response to stimulation. The spikes' duration and frequency depend on the character and strength of the stimulus. The McGregor integrate-and-fire model is described by Equations 1–4, while Figure 13.2 illustrates its operations.

Spike generation:

$$S = \begin{cases} 1 & E \geq T_h \\ 0 & E < T_h \end{cases} \tag{1}$$

Refractory properties:

$$\frac{dG_K}{dt} = \frac{-G_K + B \cdot S}{T_{GK}} \tag{2}$$

Threshold accommodation:

$$\frac{dT_h}{dt} = \frac{-(T_h - T_{h0}) + c \cdot E}{T_{th}} \tag{3}$$

Transmembrane potential:

$$\frac{dE}{dt} = \frac{-E + G_K \cdot (E_K - E) + G_e \cdot (E_e - E)}{T_{mem}} + \frac{G_i \cdot (E_i - E) + SCN + N}{T_{mem}} \tag{4}$$

where the following holds:

Transmembrane potentials:

$E = V - V_r \qquad E_K = V_K - V_r$
$E_i = V_i - V_r \qquad E_e = V_e - V_r$
V is the membrane potential
V_r is the membrane resting potential
V_K is the potassium resting potential
V_i is the inhibitory resting potential
V_e is the excitatory resting potential

Transmembrane conductances:

$G_K = g_K / G$
$G_i = g_{si} / G$
$G_e = g_{se} / G$
g_K is the potassium resting conductance
g_{si} is the inhibitory resting conductance
g_{se} is the excitatory resting conductance

G is the membrane resting conductance
$SCN = SC/G$ is the current through a membrane
SC is the current injected into cell
T_{GK} is the time constant for G_K decay
T_h is the threshold value
T_{h0} is the resting value of threshold
T_{th} is the time constant for decay of threshold
T_{mem} is the membrane time constant
$T_{mem} = C/G$
C is the for the second training phase capacitance
c determines rise of threshold $c \in [0, 1]$
B is the postfiring potassium increment
N is the noise
S is the flag showing neuron firing $\{0, 1\}$

The neuron's membrane potential changes according to the input signals (incoming spikes). The spikes enter the neuron through synaptic connections, thereby, increasing the synaptic conductance. This results in postsynaptic potential changes. There are two types of synaptic connections in this model: **excitatory** and **inhibitory**. The type of synaptic connection depends upon the presynaptic neuron. The weighted sum of all excitatory and inhibitory synaptic conductances yields the excitatory and inhibitory stimulus values, respectively. If the excitatory stimulus is too weak or the inhibitory stimulus too strong, the membrane potential cannot reach the threshold, and the neuron does not fire. If the stimulus is strong enough for the membrane potential to reach the threshold, the neuron fires (generates a spike train traveling along the axon). For a short time, immediately after the spike generation, the neuron is incapable of responding to any additional stimulation. This time interval is referred to as the **absolute refractory period**. Following the absolute refractory period is an interval known as the **relative refractory period**. During the relative refractory period, the neuron can only respond to very strong stimulation. Depending on the neuron properties, it may accommodate to the stimulating signals, requiring a stronger stimulus for the generation of consecutive spikes.

The **Very Simple Spiking Neuron** (VSSN) model, presented below, does not require solving any differential equations, as opposed to integrate-and-fire model. However, it retains these key functions of biological neurons: a) that the neurons can form multiple time-delayed connections with varying strengths to other neurons, b) that communication between the neurons occurs by slowly decaying intensity pulses, which are transmitted to connecting neurons, c) that the firing rate is proportional to the input intensity, and d) that firing occurs only when the input intensity is above a firing threshold.

The VSSN model is described by a set of simple equations, which can be broken down into input (Equation 5), internal operation (Equations 6–11), and output (Equations 12–13).

$$I_c(t) = \sum_m \min_m \left(\sum_n W_n \cdot O_n(t - t_n), I_{clamp_m} \right) \tag{5}$$

$$F_a(t) = \begin{Bmatrix} t - \hat{F} \leqslant t_r & 1 \\ else & 0 \end{Bmatrix} \tag{6}$$

$$\hat{F}_L(t) = \begin{Bmatrix} F_a = 0 \text{ AND } I_c(t) > T_v(t) & t \\ else & \hat{F} \end{Bmatrix} \tag{7}$$

$$T_v(t) = \begin{Bmatrix} F_a = 1 & T2 \\ else & TRF(t) \end{Bmatrix} \tag{8}$$

$$\text{TRF}(t) = \begin{cases} \frac{T1-T2}{S_{trf}} + t_r \geq t - F_L & T1 \\ \text{else} & (t - F_L - t_r)S_{trf} + T2 \end{cases} \quad (9)$$

$$O_{\text{update}}(t) = \begin{cases} \hat{F}_L(t) \neq \hat{F}\ 1, \hat{F} = \hat{F}_L(t) \\ \text{else} \qquad 0 \end{cases} \quad (10)$$

$$Buffer_{\text{update}}(x) = \begin{cases} x \neq 0\ i = x \ldots x + w \rightarrow Buffer_i = tri\left(i - \hat{F}\right) \\ \text{else} \qquad\qquad Buffer \end{cases} \quad (11)$$

$$tri(t) : \begin{cases} \left(\frac{t}{w} \geq 0\right) \cap \left(\frac{t}{w} \leq 1\right) a\left(1 - \frac{t}{w}\right) \\ \text{else} \qquad 0 \end{cases} \quad (12)$$

$$O(t) = Buffer_t \quad (13)$$

where:

I_c	is the input intensity at time t
W_n	is the weight of connection n
$O_n(t)$	is the output for n at time t
t_n	is the delay for connection n
I_{clamp_n}	is the maximum allowed intensity for connection group m
$F_a(t)$	determines if the neuron is in the *absolute refractory period*
t_r	is the Absolute Recovery time
$\hat{F}$	is the time of last firing
$\hat{F}_L(t)$	returns new value of $\hat{F}$ if the firing threshold is exceeded
$T_v(t)$	is the value of variable threshold
$TRF(t)$	is the threshold recovery function
T2	is the high threshold
T1	is the initial threshold
S_{trf}	is the slope of the TRF
$O_{\text{update}}(t)$	determines whether to update the output *Buffer*
$Buffer_{\text{update}}(x)$	overlays new output into the *Buffer*
Buffer	is the buffer containing all output values up to a network specified maximum delay
$Buffer_i$	is an index into buffer for time i
$tri(x)$	is the triangular output function used to describe synaptic potentials
w	is the width of the triangular output
α	is the amplitude of triangular pulse
$O(t)$	is the output at time t

Below we summarize the VSSN's operation. Note that in the model all information is represented as an abstract intensity. The input intensity is an instantaneous measurement of activity (Equation 5), the connections can be sub-grouped, and the subgroups can be clamped to a maximum allowed intensity. All synaptic potentials are positive while inhibition is realized by setting the weighting parameter to a negative value. All connections can also have a delay. Internally, the first thing to check for is whether the model is in a period of absolute recovery (Equation 6). In the VSSN model, inputs are summed but the result is ignored if Equation 6 is true or that the neuron is in the *absolute refractory period*.

The next thing to check for is whether the firing condition has been met. Equation 7 evaluates a threshold for firing conditions, where the input intensity is compared against a variable firing threshold; if the input is greater than the firing threshold and the model is not in an absolute refractory period then the current time is returned, which indicates that a firing event has occurred.

The variable threshold is returned by Equation 8 that returns the high threshold, or T2, if in an absolute refractory period, or else returns the value of the Threshold Recovery Function (TRF). The TRF is a simple linear decay from T2 to T1 (Equation 9). Equation 9 results in a firing frequency that is proportional to input. The VSSN uses instantaneous intensity measurement and a linear equation to achieve firing frequency proportional to the input.

The next step is to use the firing result to generate an output. Equation 10 returns the current time if an output needs to be generated, or 0 if not. The VSSN maintains an output buffer that is as long as the maximum allowed delay in the network. When an output is generated the resulting intensities are added into the buffer (Equation 11). The VSSN uses triangular pulses to represent output intensities (Equation 12).

Using the triangular equation as an approximation of the α function vastly simplifies output calculations in the VSSN model. The final step in a simulation is to distribute the individual output results to all members of the network in a computationally efficient manner. The VSSN uses a buffer to retrieve output intensities. This approach is very efficient as the retrieval of an output value is simply an indexed lookup into the output buffer (Equation 13).

The VSSN model scales linearly as connections are added and the model is independent of the amount of activity on those connections, which gives it a significant performance advantage over the integrate-and-fire model.

Graphically, the operation of the VSSN model is illustrated in Figure 13.3 that shows how an example input generates four action potentials (not shown) resulting in four excitatory post synaptic potentials (shown). The output spacing (frequency) is proportional to the input because of the TRF (bold line Figure 13.3a). It also shows how output intensities are modeled as triangular pulses, and how, when more than one output overlaps, the intensities are added linearly. The VSSN model assumes that there is an infinite supply of a "neurotransmitter" available and that any event resulting in firing will cause an equal and additional generation of synaptic potential.

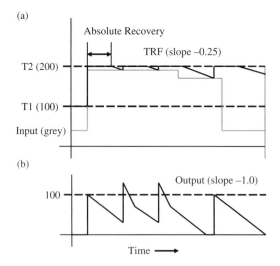

Figure 13.3. The VSNN's operational details showing how an input affects the model. (a) Internally the model uses two thresholds (T1 and T2) to determine when to fire based on the net input. The TRF determines the time varying fire threshold used to approximate the input to firing frequency observed in biological neurons. (b) Externally all outputs are represented as right triangles, and when two or more outputs overlap they are added linearly.

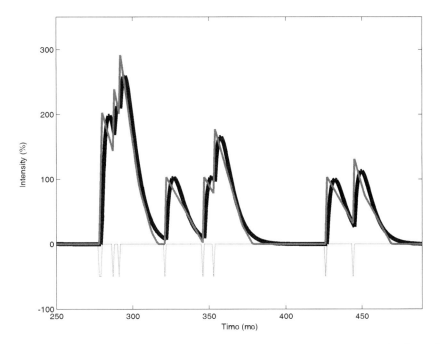

Figure 13.4. A comparison between synaptic potentials produced by the VSSN model (gray line) and the integrate-and-fire model (thick black line) based on potentials generated times following a Poisson distribution (dashed negative going line).

Figure 13.4 compares the synaptic potentials generated by both the VSSN model (with right triangular potentials) and the integrate-and-fire model (with alpha functions). A Poisson distribution determines the release of synaptic potentials. The firing times are shown in Figure 13.4 as the dashed negative spike train. Alpha function parameters are such that the maximum output occurs at 5 ms and then decays back to zero in about 20 ms. The parameters of a triangular function are matched to the alpha function by minimizing the squared differences between the outputs of the functions using a differential evolution algorithm. The Figure shows that overall there is good agreement between the two waveforms.

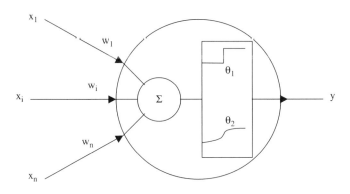

Figure 13.5. Artificial neuron model.

2.3. Simple Neuron Models

Simple models of a neuron ignore almost all biological properties except for their "firing" characteristic (generating an output). In 1943, McCulloch and Pitts were the first to propose such a model, shown in Figure 13.5. It performs two simple operations: a dot product summation of w_i and x_i, followed by a threshold output activation function operating on this sum (essentially normalizing the output so that it is within either the 0-1 range or the -1 to $+1$ range). The two operations are depicted in Figure 13.5 as a small circle (with the Σ sign inside) and as a step function (within a rectangle). The neuron receives inputs, x_i, from other neurons and compares the sum with the threshold, θ, to produce an output, y, which is sent to other neurons after scaling by an output activation function, f. This process is described by Equation (14):

$$y = f\left(\sum_{i=1}^{n} w_i\, x_i - \theta_1\right) \tag{14}$$

where n is the dimension of the input vector x, and w_i are the connection weights representing the synapses.

Through the use of a two-layer ensemble of such neurons, together with a simple learning rule (see below), it is possible to solve many linearly separable pattern recognition problems. However, a seemingly simple but far-reaching replacement of the discontinuous threshold, like the step function, with a continuous threshold, like the sigmoid, has made it possible to compute error gradients and come up with a learning rule (called backpropagation learning) for multilayer networks. A sigmoid function (shown in Figure 13.5 below the step function) is defined by Equation (15):

$$f(x) = \left(1 + \exp\left(-\frac{x}{\theta_2}\right)\right)^{-1} \tag{15}$$

where x is the net input to the sigmoid and θ_2 represents the steepness of the sigmoid curve.

Although this simple model of a neuron, when used in NNs, has proven successful in thousands of applications, it does not model well its biological counterpart and is thus not used for brain modeling purposes.

3. Learning Rules

All **learning rules** are based to a lesser or greater degree on Donald Hebb's (1949) famous observation from his experiments: if a presynaptic neuron i repeatedly fires a postsynaptic neuron j, then the synaptic strength between the two is increased. Let us note that it was computational scientists who expressed this observation as a learning rule for artificial neural networks. Below we describe learning rules for networks of spiking neurons first and then learning rules for networks using simple neuron models.

3.1. Learning Rules for Spiking Neuron Models

According to Hebb, the adjustment of the strength of synaptic connections between neurons takes place every time the postsynaptic neuron fires. If firing occurs, the synaptic weight values are updated according to Equation (16). The learning rate α controls the amount of adjustment; it can assume any value, with 0 meaning that there is no learning. To keep the synaptic connection strength bounded, a sigmoid function is used to produce a smoothly shaped learning curve:

$$w_{ij}(t+1) = sig(w_{ij}(t) + \alpha_{+-} \cdot PSP_{ij}(t)) \tag{16}$$

where t is time; w_{ij} is the synaptic weight between neurons i and j; α_{+-} is the learning rate, which can have different value for positive and negative adjustment; PSP_{ij} is the postsynaptic potential value for the connection between neurons i and j; and $sig(x)$ is sigmoid function.

Typically, learning rules for spiking neuron model networks use artificial functions to evaluate the amount of synaptic strength adjustment for each synapse. The **Synaptic Time-Delayed Plasticity** (STDP) rule is described by Equation (17) and is illustrated in Figure 13.6 (on the right):

$$STDP(\Delta t) = \begin{cases} \alpha_+ \exp(-\Delta t/\tau_+) & if \quad \Delta t > 0 \\ -\alpha_- \exp(\Delta t/\tau_-) & if \quad \Delta t \leqslant 0 \end{cases} \quad (17)$$

where Δt is the time between pre-and postsynaptic neuron firings; α_{+-} is the learning rate; and τ_{+-} is the time constant.

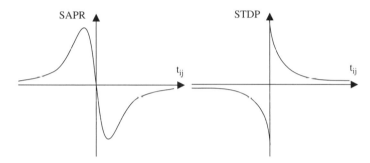

Figure 13.6. Comparison of learning functions: SAPR and STDP.

In contrast to rules like the STDP, the **Synaptic Activity Plasticity Rule** (SAPR) uses actual synaptic dynamics to evaluate the amount of adjustment, as shown in Figure 13.6 (on the left). The advantage of the SAPR over other learning rules for spiking neural models is that it uses the actual value present in each synapse instead of using an artificial function. Modification of the synaptic weights between pre- and postsynaptic neurons takes place every time the postsynaptic neuron fires. When firing occurs, all the neuron's incoming synapses are evaluated, and their synaptic strength is adjusted depending on the particular synapse type and its recent activity. The amount of adjustment is proportional to the contribution of a particular synapse to the neuron's firing. If a particular excitatory presynaptic neuron spike arrives before the postsynaptic neuron fires, the related synapse has a positive contribution, and thus its synaptic strength increases by an amount proportional to the current postsynaptic potential (PSP) value. When an excitatory presynaptic neuron's spike arrives after the postsynaptic neuron fires, it has no contribution to the recent firing and its strength is decreased by an amount proportional to the current PSP value.

There is no explicit equation or function shape for synaptic strength adjustment in the SAPR. The adjustment only approximates a possible function plot by using a general PSP shape. Figure 13.6 shows just one example of a learning function for excitatory and inhibitory synapses; the actual shape depends on the particular synapse parameters, current synaptic strength, and learning rate used. In contrast to the STDP function, the SAPR function is continuous, has a finite range of values, and mimics the shape based on biological observations.

Arrival of the action potential at the synaptic connection changes the synaptic conduction that elicits synaptic current alteration and thus results in the **postsynaptic potential** (PSP) modification. The PSP can be either **excitatory** (EPSP) or **inhibitory** (IPSP) and is proportional to the synaptic conductance change. These changes directly affect the neuron's membrane potential. The synaptic conductance between neurons i and j is obtained from Equation (18).

3. Learning Rules

$$g_{ij}(t) = \frac{t - \Delta t_{ij}}{T_{ij}} \cdot \exp\left(1 - \frac{t - \Delta t_{ij}}{T_{ij}}\right) \quad (18)$$

where g_{ij} is the synaptic conductance; t is the time; Δt_{ij} is the propagation time; and T_{ij} is the synapse time constant.

Figure 13.7 shows the normalized shapes of the EPSP and IPSP.

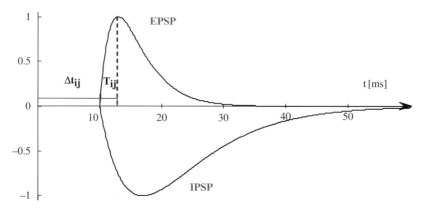

Figure 13.7. EPSP and IPSP normalized shapes.

3.2. Learning Rules for Simple Neuron Models

Below we describe three simple learning rules.

3.2.1. Hebbian and anti-Hebbian learning

Hebb's postulate, mentioned above, has been translated into a simple formula known as **Hebb's rule**:

$$w_{ij}(t+1) = w_{ij}(t) + \eta \, y_j(t) x_i(t)$$

where $\eta \in [0, 1]$ is a learning rate and t an iteration (or time) step. By the use of this rule, starting from some initial randomly set weight vector, the resulting final weight vector will point in the direction of the maximum variance in the data. A variation of the above rule is when the two neurons do not fire together frequently enough; then the synaptic strength between the two is decreased (the penalty for not firing). This outcome is known as anti-Hebbian learning rule:

$$w_{ij}(t+1) = w_{ij}(t) - \eta \, y_j(t) x_i(t)$$

The problem associated with Hebbian learning is the unbounded growth/decrease of the weights, which, however, can be easily alleviated by normalization, or imposing a saturation value for the weights.

3.2.2. Winner-Take-All or Competitive Learning

Competitive learning is another biologically inspired simple learning rule. The neurons compete among themselves for activation, and only the winning neuron, w_i, is able to fire and at the same time strengthen its weight according to:

$$w_i(t+1) = w_i(t) + \eta(t)(x - w_i(t))$$

Often the neighboring neurons are also allowed to modify their weights; they use the same learning rule but with their own weights, w_j. The winning neuron is selected as the one for which the distance between w_i and x is minimal. The result of this competition is that the winning neuron's weight vector moves even closer to the input vector x. If the training data are normalized and consist of several clusters, then the winning neuron weight vectors will point to cluster centers.

3.2.3. Simple Error-Correction Learning

Error-correction is probably the first learning rule that was used in very simple neural networks called Perceptrons:

$$w_i(t+1) = w_i(t) + \eta(d-y) x_i$$

where both y and $d \in (0, 1)$. This rule says that the error measured as the difference between the actual output y and the desired (known) output d guides the process of updating the weights. This rule is applicable to a single-layer feedforward NN.

4. Neural Network Topologies

In most cases, neural **network topology** needs to be specified, to a smaller or larger extent, by the network designer. The exceptions are **ontogenic neural networks** that generate their own topology on the fly (as needed to solve a problem), which use criteria like entropy to guide their self-design. NN topology can be looked upon as a directed graph, with neurons as nodes and the weighted interneuron connections as arcs. In this view, NN topologies can be divided into two broad categories:

– **feedforward**, with no loops and connections within the same layer
– **recurrent**, with possible feedback loops

The Kohonen SOM network, Radial Basis Function (RBF) network and probabilistic networks are examples of feedforward networks, while a Hopfield network (not covered) is an example of a recurrent network. These two basic topologies are illustrated in Figure 13.8. Let us notice that although the network shown at the top is composed of three **layers** – input, hidden, and output – the input layer only distributes data to the hidden layer without performing any calculations. To make this distinction, inputs are indicated as black dots and "true" neurons as empty circles. Later in the Chapter we will discuss RBF topology in greater detail.

5. Radial Basis Function Neural Networks

Radial Basis Function (RBF) networks have been shown to be useful in data mining tasks because of their many desirable characteristics. First, the time required to train RBF networks is much shorter than the time required for most other types of NNs. Second, their topology is relatively simple to determine. Third, unlike most of the supervised learning NN algorithms that find only a local optimum, the RBFs find a global optimum. Like other NNs they are universal approximators, which means they can approximate any continuous function to any degree of accuracy, given a sufficient number of hidden-layer neurons.

RBFs originated mainly in the field of **regularization theory**, which was developed as an approach to solving **ill-posed problems**. Ill-posed problems are those that do not fulfill the criteria of well-posed, i.e., *a problem is well-posed when a solution to it exists, is unique, and depends on initial data.* An example of an ill-posed problem is exactly that presented by model-building, given the training data; namely, we want to find a mapping function between

432   5. Radial Basis Function Neural Networks

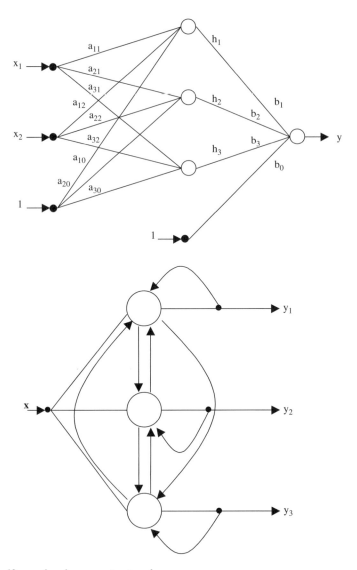

Figure 13.8. Feedforward and recurrent networks.

known inputs and the corresponding known outputs. The problem is ill-posed because it has an infinite number of solutions. One approach to solving an ill-posed problem is to **regularize** it by introducing some constraints in order to restrict the search space. In an approximation task of finding a mapping function between input-output pairs, *regularization means smoothing of the approximation function*.

Let us now recall the idea of approximating a function. Given a finite set of training pairs (x, y), we are interested in finding a mapping function, f, from an n-dimensional input space to an m-dimensional output space. The function f does not have to pass through all the data points but needs to "best" approximate the "true" function f': $y = f'(x)$. In general, the approximation of the (true) function, given by training data points, can be achieved by the function f:

$$f: y = f(x, w)$$

where w is a parameter called a weight within the framework of NN. The problem with approximating a function (given the training data points) is that all three functions shown in Figure 13.9 approximate it. Clearly, however, the approximation by a function that is closest to the data points, as indicated by the dotted line, is better than the other function indicated by the heavy line. Here the notion of regularization/smoothness comes into play, namely, the function that is closest to the training data points is a smooth function.

How do we choose the smoothest function from many possible approximations, and how close to the data should it be? For this purpose, we use a smoothness function, $G(f(x))$, with the property of having larger values for less smooth functions but achieving minimum for the smoothest function. There are several ways to design such a function, but most involve calculating derivatives. For example, we can calculate the first derivative of a function (which amplifies oscillations and is thus indicative of smoothness) and then square it and sum it up:

$$G(f(x)) = \int_R f'(x)^2 \, dx$$

The behavior of this function is shown in Figure 13.10. The top of the Figure shows the first derivatives of the two functions that approximate the true function; the heavy line corresponds to the nonsmooth function shown in Figure 13.9. The bottom of the Figure shows the squares of the first derivatives. The area under the bottom curves gives the values of $G(f(x))$ for the two functions. It is easy to see that the smooth function correctly identifies the smoothest of the two approximating functions.

The general problem of approximating a function for given training data can then be formulated as one that minimizes the following function:

$$H(f(x)) = \sum_i^N (y_i - f(x_i))^2 + \lambda G(f(x)$$

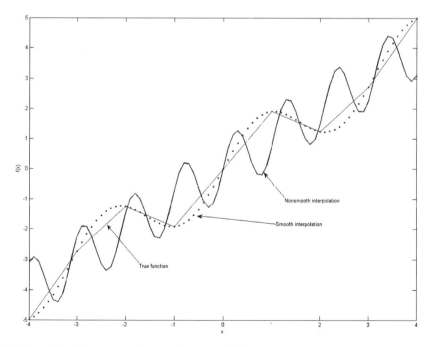

Figure 13.9. Smooth and non-smooth approximations of data points.

434 5. Radial Basis Function Neural Networks

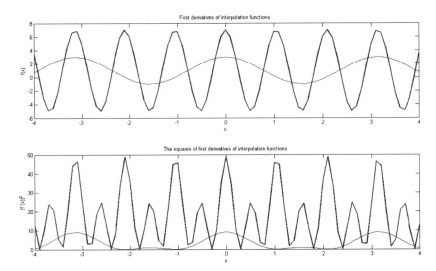

Figure 13.10. First derivatives of the two approximating functions shown in Figure 13.9 (top) and their squares (bottom) used for calculating the smoothness functional $G(f(x))$.

where (x_i, y_i) is one of N training data pairs, $f(x_i)$ indicates the actual value, and y_i indicates the desired value. The first part of this functional minimizes the error (empirical risk) between the data and the approximating function $f(x)$, while the second part forces it to be a smooth function. In other words, the first part enforces closeness to the data, and the second its smoothness. λ is a regularization parameter: the smaller it is, the smaller is the smoothness of the approximating function $f(x)$. If $\lambda = 0$ then the function goes through all training data points (resulting in overfitting the data, not interpolating/approximating it).

With the minimizing function in hand, the task is to find the approximating function $f(x)$ that minimizes it. We do not show here the derivations for the solution but just state that it takes the form:

$$f(x) = \sum_i^N w_i \Phi(x; c_i) + w_0$$

where $\Phi(x; c_i)$ is called a **basis function**. If the basis function is Gaussian and uses the Euclidean metric to calculate the similarity/distance between x and c_i, then the above equation defines a **Radial Basis Function** (RBF) that can be interpreted/implemented as a neural network. Such a Gaussian basis function can be rewritten as shown below. We note that it is a linear combination of non-linear functions Φ:

$$y = f(x) = \sum_i^N w_i \Phi(\|x - x_i\|) = \sum_i^N w_i \Phi(v_i)$$

where $v_i = \|x_i - c_i\|$, N is the number of neurons in the hidden layer centered at points c_i, and the $\|.\|$ is a **metric** used to calculate the distance between vectors x_i and c_i. We will see later than in the simplest RBF realization, the c_i's are equal to the x_i's. Numerous distance measures and basis functions can be used for the calculation of v and $\Phi(v)$, respectively, as described later.

Realization of radial basis functions within the framework of NN was first proposed in the late 1990s. Figure 13.11 shows a simple RBF network with three neurons in the hidden layer, and one output neuron. Outputs of the hidden layer are denoted as o_1, o_2, and o_3. The term **radial** means that the basis functions are radially symmetric, i.e., function values are identical for vectors x

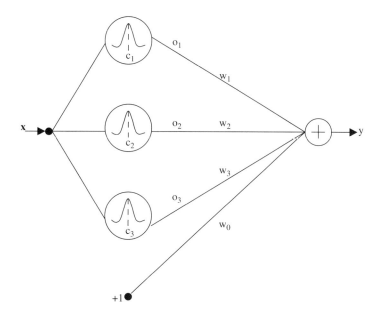

Figure 13.11. Simple, one-output RBF network.

that lie at a constant distance from the basis function center, c_i. Later we describe methods for determining the centers.

An RBF network topology always consists of three layers of neurons: an input layer, a hidden layer, and an output layer. The input layer only distributes the input vectors to the hidden layer. Note that there are no weights associated with the connections between the input larger and the hidden layer. The hidden layer differs from other NNs since each neuron in an RBF network represents a data "cluster" that is centered at a particular point, with a given radius. Each neuron in the hidden layer calculates the distance v_i between the input vector and its radial basis function center c_i, and transforms it, via a basis function T, to calculate the output of a neuron, o_i. This output is multiplied by a weight value w_i and fed into the output layer. The output layer consists of neuron(s) that linearly combine the outputs of the hidden layer to calculate the output value(s).

When designing an RBF network, before its training, several key parameters must be determined:

(a) topology, i.e., determination of the number of hidden layer neurons (and their centers), and determination of the number of output neurons
(b) selection of the similarity/distance measures
(c) determination of the hidden-layer neuron radii
(d) selection of basis functions
(e) neuron models used in the hidden and output layers

After we have described the selection of the RBF parameters, we will discuss the confidence measures used to calculate confidence in the output of an RBF network on new unseen data. We will also show that the task of function approximation performed by an RBF can also be achieved by fuzzy basis functions. This will allow for interpretation of the RBF network in terms of the equivalent fuzzy production IF... THEN... rules. The latter are very useful in a knowledge discovery process, since the rules are easily understood, while the weights and connections of a NN are only a "black box" to the user. In short, the equivalence of RBFs and fuzzy production rules allows for interpretation of what the network has learned in a language that is understandable to the end user.

5.1. Topology

Although RBF topology consists of only three layers, we need to determine how many neurons (nodes) should be used in a hidden layer.

The number of "neurons" in the input-layer is defined by the dimensionality of the input vector, X. The input layer nodes are not neurons at all: they are just feeding the input vector into all hidden-layer neurons.

The number of neurons in the output layer is determined by the number of categories present in the training data. Thus, for example, for three categories we would have three output neurons, etc. For two-category problems, it is sufficient to have one output neuron (which would be "on" for one category, and "off" for the other category).

5.2. Determining the Number of Hidden Layer Neurons and Their Centers

When designing an RBF network topology we need to determine the number of neurons in a hidden layer and their centers. In general, the number of hidden layer neurons is determined by a **trial and error method**. The simplest method, but applicable only to smaller data sets, is the **Neuron at Data Point** (NADP) method. It uses all input training data points as neurons in a hidden layer; they then become their own centers (each input vector determines the center of the corresponding hidden layer neuron). Such a network will always learn the training data set correctly because there is one neuron (cluster) center for each input training data point. This approach is very simple but it may result in overfitting instead of approximating/interpolating the input-output function.

For large data sets, one method to determine the number of hidden layer nodes (and their centers) is to **cluster** the input data and then to use the cluster prototypes as neuron centers. The pervasive problem associated with clustering, however, is how to decide/find the "correct" number of clusters in the data (and thus the number of hidden layer neurons). All the algorithms described in Chapter 9 can be used for clustering the input data. Another problem associated with clustering is the choice of a similarity/distance measure for grouping the data into clusters; this measure is another key parameter to be decided for an RBF basis function (see discussion below).

Another method to decide on the number of hidden-layer nodes in large data sets is the **Orthogonal Least Squares** (OLS) algorithm, which is a variation of the NADP method. In this approach, the hidden-layer neurons are all preset to a constant radius value while the center points are chosen from the training data. The problem of overfitting the RBF network is not a concern in this approach, since unlike in the set up of the NADP, only a subset of the training points is used. The difficulty with the OLS approach is in selecting the best subset of training points in order to maximize the accuracy of the network. The OLS algorithm addresses this problem by transforming the training data vectors into a set of orthogonal basis vectors. This process allows for calculating the individual contribution of each vector. The user will then select the single training point that will make the greatest contribution to the network's accuracy and add it to the hidden-layer of the network. Once that point is added to the network, the remaining points must be recalculated to determine how they will affect the accuracy of the new network. Again, the point that has the greatest effect on the network's accuracy is selected. This process continues until the calculated values of adding more points to the network's hidden-layer become negligible. The OLS method avoids the problem of overfitting the network, but it is limited to radii of one preset size. The other disadvantage is that the user is required to determine the optimal subset, which for large training data sets may require a prohibitive amount of time.

5.3. Distance Measures

Selecting a distance (similarity) measure is important, as it determines the shape of the formed clusters. One of the most popular is the Euclidean metric:

$$v_i = \sqrt{(X - C_i)^2}$$

where C_i stands for the center vector of the ith hidden layer neuron. The Euclidean metric implies that all the hidden-layer neurons will form hyperspheres. Figure 13.12 illustrates the difference between classification performed by **hyperplane classifiers** (whose decision functions/surfaces are hyperplanes) and **kernel classifiers** (which use higher-order decision functions/surfaces). An example of the first type is feedforward neural networks and of the second is RBF networks.

To accommodate for shapes other than hyperspheres a weighted Euclidean distance can be used to allow for hyperellipsoidally shaped clusters:

$$v_i = \sqrt{\sum_{j=1}^{K} d_j^2 (x_j - c_{ij})^2}$$

where the subscript i indexes an RBF center vector. The subscript j refers to the specific element of the x, c, and d vectors that is being calculated. Each d_j describes how much weight should be attached to the difference in the x_j and the c_{ij} elements. The difficulty is that the user must determine what influence each element of x should have. Determining values of d adds time to the design of the RBF network, since it is usually done by the trial and error method. This approach still implies that all the hyperellipses are of the same size and point in the same direction. The Mahalanobis distance is used to accommodate for any shape of the clusters, but calculating the covariance matrix is computationally very expensive:

$$v_i = \sqrt{(x - c_i)^T M_i^{-1} (x - c_i)}$$

where matrix M_i is the covariance matrix of neuron i.

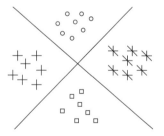

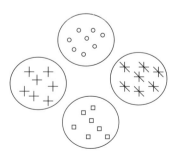

Figure 13.12. Classification outcomes performed by a hyperplane (top) and kernel classifiers (bottom).

5.4. Neuron Radii

The next parameter to be determined for an RBF network is the radii, or σ values, of the centers (clusters), unless the basis function $\Phi(v)$ is not a function of σ.

The simplest approach is to set all radii values to a constant value. The disadvantage of this approach is that some of the clusters may be much larger than others. Another difficulty is that finding the correct value requires using a trial and error method. If the value is too large, there may be a significant cluster overlap, making training difficult. If the value is too small, then most of the input vectors may not fall into any of the neuron centers, and the output of all the basis functions, and the final output of the network, will be small. If computation time is not a concern, the **P-nearest neighbors method** can be used to determine each radius individually. Figure 13.13 illustrates the outcome of the method. Finding the σ value for each neuron requires finding the distances to P nearest center points, where P is an integer with a minimum value of 1 and a maximum value equal to the number of neurons in the hidden layer. The heuristic is to set the value of $P = 2$ so that the distance is computed from the neuron center to all the other neuron centers. Then the P smallest distances are used to compute the node's radius, using the formula:

$$\sigma_i = \sqrt{\frac{1}{P}\sum_{p=1}^{P} v_p}$$

A variation of the above is to calculate distances from the neuron center to all other neurons. The smallest nonzero distance is then selected and multiplied by a scaling value and the center's radius is set to:

$$\sigma_i = \alpha \min_{(v>0)}(v)$$

where the term $\min_{(v>0)}(v)$ is a function that returns the smallest distance value computed that is greater than 0. The purpose of this nonzero criterion is that in some situations (as when NADP is used), it is possible to have two neurons with the same center point. Without the $v > 0$ requirement, the radius of the neuron would be zero. The α term is a scaling value and is in the range of $0.5 \leq \alpha \leq 1.0$. When α is at its smallest, then the clusters will cover the maximum area with no overlap. When it is the largest, the clusters will cover the maximum area with no cluster completely contained inside another.

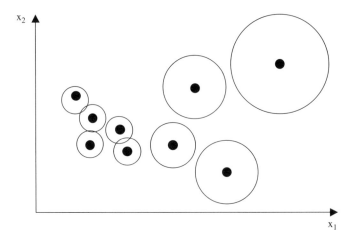

Figure 13.13. Illustration of the *P*-nearest neighbors method.

Another approach, the **class interference algorithm**, is used to prevent RBF neurons from responding too strongly to input vectors from other classes. Figure 13.14 illustrates the concept. In this method, each neuron must be assigned to one of the class types, but a class type may have more than one neuron assigned to it. A possible heuristic is to set every neuron's σ value equal to the distance to the farthest neighboring neuron's center point. Next, a threshold value t is set. This threshold is the maximum allowable output for a basis function $\Phi(v)$ assuming that the input vector X belongs to a different class. If the $\Phi(v)$ exceeds the threshold, then the radius is scaled back by a factor of γ, where $0 < \gamma < 1$. The new radius is calculated using the equation:

$$\sigma_i^+ = \gamma \sigma_i^-$$

The training data are then cycled through the network continuously until no further adjustment of the radii is needed.

5.5. Basis Functions

We have already mentioned a **Gaussian basis function** when introducing the concept of RBFs. A sufficient condition for designing any basis function is that it must be conditionally or strictly positive definite. Finding such functions is time consuming, and therefore a simpler approach is often used that takes into account a relation between strictly positive definite functions and monotonic functions.

A function f(x) is **monotonic** *on $(0, \infty)$ if it is positive for even derivatives and negative for odd derivatives*, or $f(x) \geq 0$, $f'(x) \leq 0$, $f''(x) \geq 0$, etc., for $x \in (0, \infty)$. A function $f(x)$ is monotonic on $(0, \infty)$ if and only if $f(x^2)$ is positive definite. In other words, if $f(x)$ is monotonic, then $f(x^2)$ could be used as a basis function. The set of available basis functions can be extended by determining that if the first derivative of a function $f(x)$ is monotonic, then $f(x)$ can also be used as a radial basis function.

In the equations that follow, the σ term represents the radius of the neuron. When neurons do not use radii, such as when Mahalanobis distance is used, then the σ terms are set to 1.0 to simplify the calculations. By far the most popular basis function used in RBFs is the Gaussian:

$$\Phi(v) = \exp^{-(v^2/\sigma^2)}$$

The thin plate spline function is another popular function that, unlike the Gaussian, is independent of the radius of the neuron:

$$\Phi(v) = v^2 \ln(v)$$

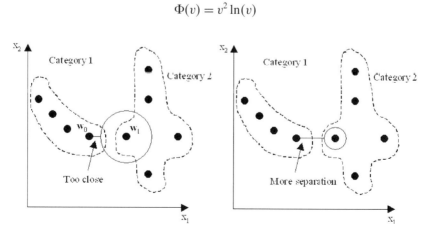

Figure 13.14. Illustration of the class interference method.

5. Radial Basis Function Neural Networks

It may happen in calculations that an input vector is identical with the neuron's center point. Since in this case the output value of the function cannot be determined, we set $\Phi(v) = 0.0$ when $v = 0.0$. An interesting property of the thin plate spline function is that between the values of $v = 0.0$ and $v = 1.0$ the output has a negative value, while for values greater than 1.0 the output is positive and approaches infinity. Still another basis function is the quadratic, defined as

$$\Phi(v) = (v^2 + \sigma^2)^\beta$$

Although this function is valid in the range $0 < \beta < 1$, the β value is most often set to $\beta = 1/2$.

The inverse quadratic is defined below for $\beta > 0$. It has the property of converging toward zero as the distance v increases:

$$\Phi(v) = (v^2 + \sigma^2)^{-\beta}$$

The cubic spline, like the thin plate spline is only a function of distance v, and not of the radius:

$$\Phi(v) = v^3$$

Since there is no requirement that the basis function of an RBF network be nonlinear, the linear spline basis function is used in some applications:

$$\Phi(v) = v$$

5.6. Calculating Output Values of a Hidden Layer

After each hidden-layer neuron calculates its $\Phi(v)$ term, the output neuron calculates the dot product of two vectors of length N, $\phi = [\phi(v_1), \phi(v_2), \ldots, \phi(v_N)]$ and $w = [w_1, w_2, \ldots, w_N]$

$$y = \sum_{i=1}^{N} w_i \Phi(v_i)$$

After the output layer calculates its output, the next operation is to transform the result into a form expected by the user. In the case when we want the RBF network to determine whether an input vector belongs to one of C classes, then the network is trained so that if an instance of a specific class enters the network, the result for that output neuron is 1 while the rest of the output neurons return a 0. In practice, to increase the speed of training, we train the outputs to be 0.9 and 0.1; for example, for the three-output network they will be 0.9, 0.1 and 0.1.

5.7. Training of the Output Weights

The only training that is required in the RBF network is to find the weight values between the hidden layer and the output layer. This can be done in various ways. A popular one is to minimize the mean absolute error:

$$\Delta_i = |d_i - y_i|$$

where d_i is the desired output and y_i is the actual output. After the error is determined, the network then adjusts the output weight values, using the following learning rule:

$$w(t+1) = w(t) + \eta \Phi_j(v_i) \Delta_i$$

The η value is a learning constant that has a value ranging from 0 to 1, but is generally set to a value closer to 0.0 than to 1.0. The $\phi_j(v_i)$ is the basis function output from the jth neuron in the hidden layer. Training entails sending all T training vectors into the network and continuously readjusting the weights. The network then computes the mean absolute error:

$$error = \frac{1}{T}\sum_{i=1}^{T}|\Delta_i|$$

If the error is below some acceptable tolerance value then the training ends; otherwise the data are resent through the network and the weights are adjusted again until either the network reaches tolerance or until the error stops decreasing significantly with each training cycle.

5.8. Neuron Models

Two different neuron models are used in RBF networks: one in the hidden layer and another in the output layer. The model/operation of the neuron in the hidden layer is determined by the type of the basis function used (most often Gaussian) and the similarity/distance measure used to calculate the distance between the input vector and the hidden-layer center.

The neuron model used in the output layer can be linear (i.e., it sums up the outputs of the hidden layer multiplied by the weights, to produce the output) or sigmoidal (i.e., produces output in the range 0–1). The second type of an output layer neuron is most often used for classification problems, where, during training, the outputs are trained to be either 0 or 1 (in practice, 0.1 and 0.9, respectively).

After having described the design parameters for the RBF networks, we state the pseudocode of the RBF algorithm.

Given: Training data pairs, stopping error criteria, sigmoidal output neuron, Gaussian basis function, Euclidean metric, and learning rule.

1. Determine the network topology (e.g., by clustering the data)
2. Choose the neuron radii (e.g., by setting all to a constant value)
3. Randomly initialize the weights between the hidden and output layer
4. Present the input vector, x, to all hidden layer neurons
5. Each hidden-layer neuron computes the distance, v_i, between x and its center point, c_i
6. Each hidden-layer neuron calculates its basis function output, $\Phi(v)$, using the radii values
7. Each $\Phi(v)$ value is multiplied by a weight value, w, and is fed to the output neuron(s)
8. The output neuron(s) sums all the values and outputs the sum as its answer
9. For each output that is not correct, adjust the weights by using a learning rule
10. If all output values are below some specified error value, then stop; otherwise go to Step 8
11. Repeat Steps 3 through 9 for all remaining training data pairs

Result: A trained RBF network.

Example: This example (see Figure 13.15) illustrates the use of the RBF algorithm, after it has been trained, on two test data points. During training (not shown) dozens of training data pairs, for each of the classes, are used to come up with the center points. The generated network has three neurons in the hidden layer, and one output neuron.
The hidden layer neurons are centered at points A(3, 3), B(5, 4), and C(6.5, 4), and their respective radii are 2, 1, and 1. The weight values between the hidden and output layer were calculate to be 5.0167, −2.4998, and 10.0809, respectively. We now test the network with two new input vectors x_1 (4.5, 4) and x_2 (8, 0.5). The RBF outputs are calculated as follows.
In Step 5: Each neuron in the hidden layer calculates the distance from the input vector to its cluster center by using, the Euclidean distance. For the first test vector x_1

$$d(A, x_1) = ((4.5, 4), (3, 3)) = 1.8028$$

5. Radial Basis Function Neural Networks

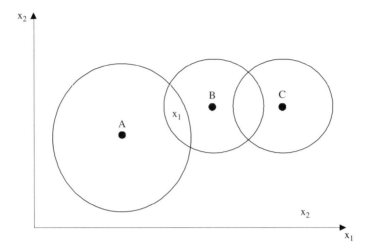

Figure 13.15. Radii and centers of three hidden-layer neurons, and two test points x_1 and x_2.

$$d(B, x_1) = ((4.5, 4), (5, 4)) = 0.5000$$
$$d(C, x_1) = ((4.5, 4), (6.5, 4)) = 2.0000$$

and for the x_2

$$d(A, x_2) = 5.5902 \quad d(B, x_2) = 4.6098 \quad d(C, x_2) = 3.8079$$

In Step 6: Using the distances and their respective radii, the output of the hidden-layer neuron, using the Gaussian basis function, is calculated as follows. For the first vector

$$\exp(-(1.8028/2)2) = 0.4449$$
$$\exp(-(0.5000/1)2) = 0.7788$$
$$\exp(-(2.0000/1)2) = 0.0183$$

and for the second

$$0.0004, \quad 0.0000, \quad 0.0000$$

In Step 7: The outputs are multiplied by the weights, and the result is as follows:

$$\text{for } x_1: \quad 2.2319, -1.9468, 0.1844$$
$$\text{and for } x_2: 0.0020, 0.0000, \quad 0.0000$$

In Step 8: The output neuron sums these values to produce its output value:

$$\text{for } x_1: \quad 0.4696$$
$$\text{and for } x_2: 0.0020$$

Are these outputs correct? We will postpone the answer to this question until we have described the confidence measures. Let us just note that NNs do not give a correct output for a test data vector that lies outside of the range of the data on which the network was trained; the network will always produce some output, but we will not know whether it is correct or not. The purpose of using confidences measures is to flag test data points for which the generated network output is not reliable.

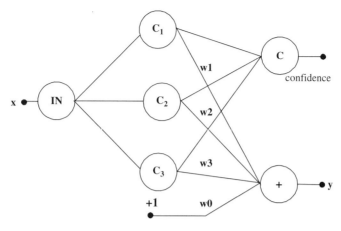

Figure 13.16. RBF network with reliability measures outputs.

5.9. Confidence Measures

RBF networks, like other supervised NNs, give good results for unseen input data that are similar/close to any of the training data points but give poor results for unfamiliar (very different) input data. Being able to flag such data is very important. If some outputs (corresponding to test input data) were flagged as unreliable, then we would discard such unreliable output (treat it as "do not know" output). Such flagging would decrease the number of false positives/negatives at the expense of lowering the overall accuracy (since a smaller number of data points would be classified).

The first method to calculate confidence in the output of the RBF network, known as a Parzen window (see Chapter 11), was based on estimating the density of an unknown probability distribution function. It was calculated along with the RBF output. This method, however, is computationally expensive and is not covered here.

Instead, we introduce a simple method to evaluate confidence in the RBF network output by calculating **certainty factors** and using them as a **confidence measure**. CFs are calculated from the input vector's proximity to the hidden layer's neuron centers, as shown in Figure 13.16.

Certainty factors are generated in RBF networks by using the $\phi(v)$ values. When the output $\phi(v)$ of a hidden layer neuron is near 1.0, then this indicates that a point lies near the center of a cluster, which means that the new data point is very familiar/close to that particular neuron. A value that is near 0.0 will lie far outside of a cluster, and thus the data are very unfamiliar to that particular neuron. If all the neurons have a small $\phi(v)$ value, then the network's certainty in the answer is small, and our confidence in the result is also small. Since N values, one $\phi(v)$ for each neuron in the hidden layer, are generated, the question arises how an overall confidence measure should be determined. The first thought would be to choose the maximum $\phi(v)$ value and consider it as the certainty factor:

$$CF = \max \Phi(v_i)$$

From a certainty factor point of view, the disadvantage of this approach is that it would only take into account the maximum $\phi(v)$ value in the determination of the certainty factor, which might result in an instance in which two or more of the hidden-layer neurons are reasonably close to an input vector but neither is very close to it. Therefore, the calculated certainty factor may be lower than expected. In order to factor in the certainty of other neurons, the following equation is used to calculate the certainty:

$$CF_i = CF_{i-1} + (1 - CF_{i-1}) \max_i (\Phi(v))$$

where i is the number of neurons that are used to determine the certainty factor. In general, the value of i should be set to N (the number of neurons in the hidden layer), but it can be set to a lower value if computational time is critical. The term CF_{i-1} refers to the certainty factor of the network using only i-1 neurons to calculate the confidence. The value of CF_0 (the certainty when no neurons in the hidden layer are used) is defined to be 0.0. Lastly, the term $\max_i(\phi(v))$ is defined as the i-th largest value of $\phi(v)$. So, $\max_1$ means the largest value of $\phi(v)$ and $\max_2$ the second largest, etc. The pseudocode for calculating certainty factors is given below.

Given: An RBF network with an additional output neuron that calculates the certainty value.

1. For a given input vector, calculate the outputs, $\Phi(v)$, of the hidden-layer nodes
2. Sort them in descending order
3. Set CF = 0.0
4. For all outputs $\Phi(v)$ calculate:

$$CF(t+1) = CF(t) + (1.0 - CF(t))\max_i \Phi(v)$$

$$CF(t) = CF(t+1)$$

Result: The final value of $CF(t)$ is the generated certainty factor.

Knowing the value of the certainty factor, the user can use some threshold value (say 0.7) to either accept or reject the network's result.

Example: Assume that four $\Phi(v)$ values have been calculated, for a given input vector, as 0.6, 0.1, 0.5, and 0.2, as shown in Table 13.1. The overall network certainty is computed to be 0.86. Note, that incorporating each additional hidden layer neuron increases the certainty factor by a smaller and smaller amount until the network reaches its final certainty factor value. This example of diminishing returns indicates that it is not necessary to actually use $i = N$ to arrive at the optimum value for network certainty. A value less than N will likely give an answer close to the true certainty factor; however, a value of $i = 1$ would greatly underestimate the network's certainty factor.

A disadvantage of certainty factors is that they treat every neuron's output as equal regardless of the number of training points that actually are in that cluster. However, they are very simple to compute, and they yield meaningful results even for high-dimensional vectors and neurons with large radii.

Example: Let us go back to the example illustrated in Figure 13.15. For the first test vector x_1, the value of $CF = .88$ and for the second test vector, $CF = 0.0$. Thus, using a threshold of 0.7, we would accept the first result as reliable and reject the one for test vector x_2, which agrees with our intuition.

5.10. RBF Networks in Knowledge Discovery

How can we use RBF networks in the process of knowledge discovery? In the first place, all NNs, including RBFs, generate **black box models** of data, since the discovered function/relationship between the inputs and outputs is encoded in weights and connections of the trained network.

Table 13.1. Calculation of certainty factors.

i	CF_{i-1}	$1-CF_{i-1}$	$Max_i(f(v))$	CF_i
1	0.00	1.00	0.60	0.60
2	0.60	0.40	0.50	0.80
3	0.80	0.20	0.20	0.84
4	0.84	0.16	0.10	0.86

As such, these networks are not comprehensible by humans. Second, although all but the SOM neural network are supervised, the training data in many domains are not available. In this case, the only technique that can be used to analyze such data is clustering, which has an inherent problem: choosing a correct number of clusters (see Chapter 9). Even if one uses the unsupervised SOM, which finds clusters without the need to specify their number a priori, the difficult problem remains of interpreting the meaning of the found clusters (assigning labels to them).

As we have seen, the RBFs with Gaussian basis functions also group data into "clusters" around each of the neurons in its hidden layer. So the question is how one can use this information to address the second problem. Below we describe a couple of approaches, specific to RBFs, for doing so. One is to show equivalence of RBFs and fuzzy production systems so that the RBF trained network can be written in terms of easy-to-understand rules. The other uses fuzzy production rules and fuzzy context for deciding which part of the data (stored in a database) to focus on.

5.10.1. Rule-Based Indirect Interpretation of the RBF Network

An RBF network is equivalent, under some weak conditions, to a system of **fuzzy production rules**. Note that a fuzzy system (a set of fuzzy rules) can also approximate any continuous function to any degree of accuracy, just like a NN. This equivalence is important for knowledge discovery tasks, since one can first find a model/trend in the data using an RBF network, which can be done relatively quickly and easily, and then find a "corresponding" set of fuzzy rules that are easily comprehensible, as opposed to the weights and connections of an RBF network. Figure 13.17 illustrates this idea.

We thus see that, on the one hand, one can approximate a function using an RBF network with Gaussian basis functions ϕ_i, and on the other hand, make the approximation via fuzzy sets A_i. Since fuzzy sets can assume the same shape as Gaussians we can equate them, i.e., $A_i = \phi_i$. To keep the discussion simple, let us note that the receptive fields form the condition portion of the rules. The conclusion portion, on the other hand, is a numeric value. Given this, a rule is written in the form:

IF x is A_i THEN y is y_i

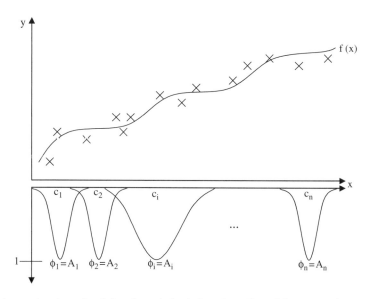

Figure 13.17. Approximation of training data via basis functions Φ_i and fuzzy sets A_i.

where y_i is a numeric representation associated with A_i. The computation of the output of the system of rules depends on the activation of individual A_i. For instance, we can compute y_i as a weighted average of experimental data falling under the support of A_i, namely:

$$y_i = \frac{\sum_{k=1}^{M} A_i(x(k))y(k)}{\sum_{k=1}^{M} A_i(x(k))}$$

where $(x(k), y(k))$ are training data pairs.

A generalization of the previous rule, where the conclusion is also treated as a fuzzy set, has the form:

$$\text{IF } x \text{ is} A_j \text{ THEN } y \text{ is } B_j$$

where A_j and B_j are fuzzy sets. Each rule represents a partial mapping from x into y.

Figure 13.18 illustrates the concept of approximating the function with fuzzy production rules. It is easy to notice that the more fuzzy sets we use, for each variable, the higher the approximation accuracy is. The fuzzy sets for the output variable Y can also be defined as Gaussians, although for the purpose of showing the equivalence of an RBF network with a fuzzy system they are defined as their modal values (single values at the peaks of the Gaussians).

If we analyze the previous rule, we notice that from the topology of a standard RBF network, like the one shown in Figure 13.11, we can also write a similar rule, namely:

$$\text{IF } x \text{ falls within the receptive field } \Phi_i \text{ THEN } w_i$$

However, the value of the weight w_i has no meaning, whereas fuzzy set B_j can have real meaning. A simple solution is to replace the weight w_i with a fuzzy set B_j. Let us comment on the significance of the equivalence of these two topologies for the knowledge discovery process. Once a trend in the data is discovered with the RBF network, we can switch to a set of fuzzy production rules, where the rules have a clear meaning to the user.

In other words, the output variable may take on a specific value for a specific value of the input variable (or a set of values if it is a fuzzy set), that is, the RBF network can be made understandable through the use of the equivalent fuzzy rules system, which is shown in Figure 13.19. Let us note that the seemingly "second" hidden layer is not a layer but is used to normalize the outputs of the

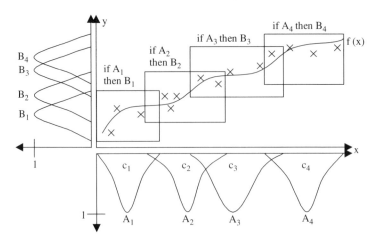

Figure 13.18. Approximation of a function with fuzzy production rules.

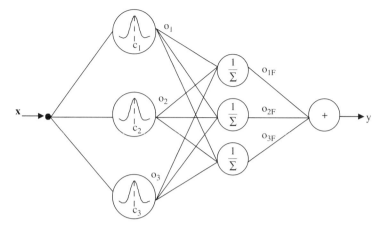

Figure 13.19. Fuzzy production rules system.

first (and only) hidden layer so that the sum $o_1 + o_2 + o_3 = 1$, which is not the case for regular RBFs.

5.10.2. Fuzzy Context RBF Network

Our goal is to make sense of data by designing an RBF network that will result in a neural network topology that is easy to interpret. We know that the hidden layer in the RBF network is formed by a series of radial basis functions. Their outputs are then combined linearly by the neuron(s) at the output layer. By keeping this topology but realizing it in a slightly different way we can accomplish our goal. Recall that when designing an RBF network, we need to define the radial basis function centers. For large data sets one solution is to cluster the data and use the found cluster prototypes as the centers of the receptive fields. We can do something similar, however, through the use of a **fuzzy context**. We define a series of such contexts, also called *data mining windows*, for the output variable(s). *Fuzzy contexts are represented by fuzzy sets and are used to select from a database only the records of interest to the user.* More specifically, only those records whose corresponding field matches the context to a nonzero degree are processed. As shown in Figure 13.20, the fuzzy context, say, *negative small output*, corresponds only to specific input–output pairs to be selected from the entire database.

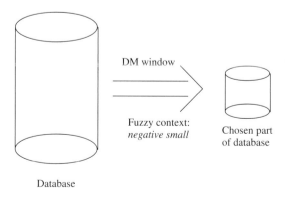

Figure 13.20. Fuzzy context and its data windowing effect on the original training set.

5. Radial Basis Function Neural Networks

As a result, only the selected (input) data becomes "clustered". This leads to **context-oriented clustering** of data, done for each context separately. The collection of such identified cluster centers is then used to form neuron centers for the RBF hidden layer. More specifically, for each context i we cluster pertinent data into c_i clusters. This means that for p contexts, we end up with $c_1, c_2, \ldots c_p$ clusters. The collection of these clusters will define the hidden layer for the RBF network. The outputs of the neurons of this layer are then linearly combined by the output-layer neuron(s).

The output neuron, with a linear transfer function, combines the outputs of the hidden-layer neurons via its weights. The difference, however, when using the fuzzy-context RBF network is that all connections are fuzzy sets and not numbers/weights as in the standard RBF network. Moreover, these fuzzy sets are the fuzzy contexts already specified for the purpose of clustering. This eliminates the need to train the weights between the hidden and output layers. One advantage of using the fuzzy context is shortening of training time, which can be substantial for large databases. More importantly, the network can be directly interpreted as collection of IF…THEN…rules, as we have learned above. Since the connections between the neurons are now defined as fuzzy sets, the result is also a fuzzy set. The computations involve rules of fuzzy arithmetic.

To illustrate the fuzzy context RBF let us denote the output (see Figure 13.21) by y:

$$y = (o_{11} + o_{12} + \ldots + o_{1c1})A_1 + (o_{21} + o_{22} + \ldots + o_{2c2})A_2 + \ldots$$
$$+ (o_{p1} + o_{p2} + \ldots + o_{pcp})A_p$$

where $o_i \in [0, 1]$ are activation levels of the receptive fields associated with the ith context, and As are the fuzzy sets of the context. The calculations are greatly simplified if the contexts are taken as triangular fuzzy numbers (although one could also Gaussians fuzzy numbers or use some other form of a membership function). Then the output of the network (result) is also a triangular number, with the lower and upper bounds, as well as the modal value, being computed separately.

Example: Let us consider the network with two fuzzy contexts specified as triangular fuzzy numbers. Figure 13.21 illustrates the topology of the network and the two contexts. Let us assume that for a given input x, the outputs of the first three hidden-layer neurons corresponding to the first fuzzy A_1 context (negative small) are: 0.0, 0.05, 0.60, and of the next two hidden-layer nodes

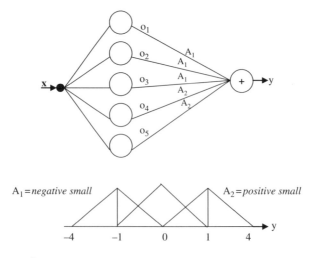

Figure 13.21. Fuzzy sets of context.

corresponding to the second fuzzy A_2 context (positive small) are 0.35 and 0.00. Then we obtain the following:

$$\text{lower bound}: (0+0.05+0.60)(-4)+(0.35+0)0 = -2.60$$
$$\text{modal value}: (0+0.05+0.60)(-1)+(0.35+0)1 = -0.30$$
$$\text{upper bound}: (0+0.05+0.60)0+(0.35+0)4 = 1.40$$

The result is also a triangular fuzzy number, with the modal value of -0.30 and lower and upper bounds of -2.60 and 1.40, respectively. Note that the granularity (see Chapter 5) of the result is usually lower than the granularity of the original fuzzy sets of the context. In particular, if the input x is such that the activation levels of the receptive fields are equally distributed across the contexts, say:

$$\text{first fuzzy context:} \quad 0.0, 0.5, 0.0$$
$$\text{second fuzzy context is:} 0.5, 0.0$$

we obtain a significant spreading effect in the sense that the result is a fuzzy number, with the parameters being averages of the individual contexts. That is:

$$\text{lower bound: } 0.5(-4.0)+(0.5)0 = -2.00$$
$$\text{modal value: } 0.5(-1.0)+(0.5)1 = 0.0$$
$$\text{upper bound: } 0.50+(0.5)4 = 2.00$$

The fuzzy-context RBF network can be easily interpreted as a collection of rules. The condition part of each rule is represented by the hidden-layer cluster prototype, while the conclusion of the rule is formed by the associated fuzzy set of the context. Notice that a receptive field is a fuzzy relation. Each conclusion is a fuzzy set. The rules are thus expressed at the level of lower information granularity, namely, fuzzy relations and fuzzy sets. Each rule has the form:

$$\text{IF input} = \text{hidden layer neuron } r \text{ THEN output} = \text{context } A$$

where A is the corresponding context.

With the outputs of the hidden-layer neurons, being continuous and firing to some degree, the rule with the highest level of firing will be chosen. The result, say, A, is then taken at some confidence level equal to the level of firing of this rule.

6. Summary and Bibliographical Notes

In this Chapter, we introduced the basic building blocks of **neural network algorithms**. Specifically, we discussed the **biological neuron** and its two models: the **spiking neuron model** and a **simple neuron model** [6, 10, 11, 17]. Then we described the **learning rules**, which are used to update the weights between the interconnected neurons, again for networks utilizing the spiking and simple neuron models [16, 17], and we briefly reviewed popular **neural network topologies**.

Then we introduced in detail a neural network that has been shown to be very useful in data mining applications, namely, the **Radial Basis Function,** which is well covered in several publications [1, 2, 3, 4, 5, 8, 9, 14, 15, 18]. We also introduced a simple measure of confidence in an RBF output, calculated based on the outputs of a hidden layer [19]. Next, we talked about the use of RBFs in the knowledge discovery process. The most important characteristic of an RBF in that respect is its equivalence to a **system of fuzzy rules** [7, 8]. Using this feature, RBFs can be used for focusing only on those parts of data that are of greatest interest to the user [12, 13].

References

1. Broomhead, D.S., and Lowe, D. 1988. Multivariable functional interpolation and adaptive networks. *Complex Systems*, 2:321–355
2. Chen, S., Cowan, C.F.N., and Grant, P.M. 1991. Orthogonal least squares learning algorithm for radial basis function network. *IEEE Transactions on Neural Networks*, 2:302–309
3. Cios, K.J., and Pedrycz, W. 1997. Neuro-fuzzy systems. In: *Handbook on Neural Computation*, Oxford University Press, D1.1–D1.8
4. Cios, K.J., Pedrycz, W., and Swiniarski R. 1998. *Data Mining Methods for Knowledge Discovery*, Kluwer
5. Fiesler, E., and Cios, K.J. 1997. Supervised ontogenic neural networks. In: *Handbook on Neural Computation*, Oxford University Press, C1.7
6. Hebb, D.O. 1949. *The Organization of Behavior*, Wiley
7. Jang, R., and Sun, C-T. 1993. Functional equivalence between radial basis function networks and fuzzy inference systems. *IEEE Transactions on Neural Networks*, 4:156–159
8. Kecman, V., and Pfeiffer, B.M. 1994. Exploiting the structural equivalence of learning fuzzy systems and radial basis function neural networks. In: *Proceedings of EUFIT'94*, Aachen. 1:58–66
9. Kecman, V. 2001. *Learning and Soft Computing*, MIT Press
10. Lovelace, J., and Cios, K.J. 2007. A very simple spiking neuron model that allows for efficient modeling of complex systems. *Neural Computation*, (19):1–26
11. McCulloch, W.S., and Pitts, W.H. 1943. A logical calculus of the ideas immanent in nervous activity. *Bulletin of Mathematics and Biophysics*, 5:115–133
12. Pedrycz, W. 1998. Conditional fuzzy clustering in the design of radial basis function neural networks. *IEEE Transactions on Neural Networks*, 9(4):601–612
13. Pedrycz, W., and Vasilakos, A.V. 1999. Linguistic models and linguistic modeling. *IEEE Transactions on Systems, Man, and Cybernetics*, Part C, 29(6):745–757
14. Poggio, T., and Girosi, F. 1990. Networks for approximation and learning. *Proceedings of the IEEE*, 78:1481–1497
15. Poggio, F. 1994. Regularization theory, radial basis functions and networks. In: *From Statistics to Neural Networks: Theory and Pattern Recognition Applications*, NATO ASI Series, 136:83–104
16. Song, S., Miller, K.D., and Abbot, L.F. 2000. Competitive Hebbian learning through spike timing-dependent synaptic plasticity, *Nature Neuroscience*, (3):9
17. Swiercz, W., Cios, K.J., Staley, K., Kurgan, L., Accurso, F., and Sagel, S. 2006. New synaptic plasticity rule for networks of spiking neurons, *IEEE Transactions on Neural Networks*, 17(1):94–105
18. Wang, X-L., and Mendel, J.M. 1992. Fuzzy basis functions, universal approximation and orthogonal least-squares learning. *IEEE Transactions on Neural Networks*, 3:807–814
19. Wedding, D.K. II, and Cios, K.J. 1998. Certainty factors versus Parzen windows as reliability measures in RBF networks. *Neurocomputing*, 19(1–3): 151–165

7. Exercises

1. Given the following training data:

x_1	x_2	x_3	y
1	0	0	1
1	0	1	1
1	1	0	1
0	1	1	0
0	0	1	0
0	1	0	0

and the initial weight vector $\mathbf{w} = [1111]$, use the simple error-correction rule to find, by hand, the final weight vector. First, draw the training data and the required neural network topology. Use the bias term $x_{n+1} = 1$.

2. Derive a learning rule, similar to the simple error-correction rule, but assume a neuron model that uses a sigmoid transfer function with $\theta_2 = 1$. Assume that the error is calculated as $E = (1/2)(d-y)^2$, not as $(d-y)$, to simplify calculations. The rule will have the form:

$$w(t+1) = w(t) - \eta \frac{\delta E}{\delta w}$$

and your task is to calculate this partial derivative (see Figure 13.5).

3. Given the following XOR data:

x_1	x_2	y
0	1	1
1	0	1
0	0	0
1	1	0

a) Use the RBF network (see Figure 13.11) with the Gaussian basis function with $\sigma = 1$, and the NADP method, to calculate the outputs of all the hidden layer neurons for all inputs. Show your results in a tabular form.

b) Define just two neurons in the hidden layer of the RBF network, with the centers equal to the last two training data points, and calculate the same as in a). In addition, draw the data points in the original space of x_1 and x_2 and in the image space of ϕ_1 and ϕ_2.

4. Assume that the data points corresponding to Figure 13.11 are as follows:

x_1	x_2	Category
1	8	1
2	7	1
3	6	1
5	5	1
8	5	2
10	11	2
10	8	2
10	2	2
12	5	2

Use an RBF network, with the NADP method, for selecting the basis function centers with the Gaussian basis function, with equal spreads, and calculate the hidden-layer weights by hand. Show all calculations.

14

Text Mining

In this Chapter, we explain concepts and methods of text mining. We describe information retrieval that provides a platform for performing text mining. The specific topics include linguistic text preprocessing, text similarity measures, approximate search, and the notion of relevance.

1. Introduction

In the previous Chapters, we focused on data mining methods for analysis and extraction of useful knowledge from structured data such as flat file, relational, transactional, and data warehouse data. In contrast, in this Chapter we are concerned with the analysis of **text databases** that consist of mainly semistructured or unstructured data. Text databases usually consist of collections of articles, research papers, e-mails, blogs, discussion forums, and web pages. With the rapid growth of Internet use by both businesses and individuals, these resources are growing in size extremely fast due to the advancements in conversion of paper-based documents into an electronic format and the proliferation of the WWW services.

Semi-structured data are neither completely structured (e.g., a relational table) nor completely unstructured (e.g., free text). A semistructured document contains some structured fields such as title, list of authors, keywords, publication date, category, etc., and some unstructured fields, such as abstract and contents. One of the unique features of text mining is the very large number of features that describe each document. Typically, for a given text database we can extract thousands of features like keywords/concepts/terms. In contrast, when analyzing flat files or a relational table, we find that the number of features typically ranges from a few to a few hundred. The features and the documents are very sparse, i.e., each document will contain only a small subset of all the keywords, and most keywords will occur only in a limited number of documents. Therefore, most of the entries in a binary vector of features, which is often used to represent a document, may be zeros (indicating absence of the features).

Traditional data mining techniques are often not adequate to handle the large size of already large and rapidly growing text databases. Typically, only a small portion of the documents from a text database is relevant to a given user, and thus selection of these documents (and extraction of knowledge from them) is the focus of **text mining**. This area provides users with methodologies to compare documents, rank their importance and relevance, and find patterns and trends across different (sometimes related) documents.

We introduce here the **information retrieval** field, which provides basic functionalities for performing text mining. Next, we concentrate on techniques that are used to transform a semistructured (or unstructured) document into a form suitable for text mining. These include simple **linguistic preprocessing**, such as removal of stop words, stemming and finding synonyms. Finally, the **vector-space model**, which uses a **term frequency matrix** and **tf-idf weighting,** and

453

text similarity measures are described, and a discussion about **latent semantic analysis** and **relevance feedback** improvements to information retrieval systems is presented. Upon finishing this Chapter, the reader will recognize and understand the problems and solutions related to text mining.

2. Information Retrieval Systems

There are three main types of retrieval within the knowledge discovery process framework:

- **data retrieval**, which concerns retrieval of structured data from DBMS and data warehouses (see Chapter 6)
- **information retrieval**, which concerns the organization and retrieval of information from large collections of semistructured or unstructured text-based databases and the WWW
- **knowledge retrieval**, which concerns the generation of knowledge from (usually) structured data, as thoroughly discussed in Chapter 9 through 13

The ISO 2382/1 standard, published in 1984, defines **information retrieval** (IR) as "[the] actions, methods and procedures for recovering stored data to provide information on a given subject." According to this standard, the actions include text indexing, inquiry analysis, and relevance analysis; the data include text, tables, diagrams, speech, video, etc.; and the information includes the relevant knowledge needed to support problem solving, knowledge acquisition, etc. Here we concentrate on information retrieval in text databases.

A typical IR task is to identify documents that are relevant to the user's input, expressed in terms of keywords or example documents. This approach is in contrast to database systems that focus on efficient processing of queries and transactions over structured data and include additional functionalities such as transaction processing, concurrent control, recovery mechanisms, etc. (see Chapter 6). The IR focuses on a different set of problems that includes handling of semistructured or unstructured documents, **approximate search** based on keywords, and the notion of **relevance**.

2.1. Basic Definitions

The core IR vocabulary includes the following four terms:

- **database**, which is defined as a collection of text documents
- **document**, which consists of a sequence of terms in a natural language that expresses ideas about some topic
- **term**, which is defined as a semantic unit, phrase, or word (or more precisely a root of the word)
- **query**, which is a request for documents that concern a particular topic that is of interest to the user of an IR system

IR systems aim at finding relevant documents in response to the user's query. They match the language of the query with the language of the document. However, the matching of a simple word (term) between a query and a document does not provide a proper solution, since the same word may have many different semantic meanings. This is the **polysemy problem**, in which the same term means different things depending on a particular context. For instance, *make* can be associated with "make a mistake," "make of a car," "make up excuses," and "make-believe". Similarly, for the query *Abraham Lincoln*, the IR system may return documents that contain sentences like "Abraham owns a Lincoln. It is a very comfortable car."—which probably is not what we were looking for. Finally, for a general query such as *what is the funniest movie ever made?*, how would the IR know what the user's preferences are? The latter example demonstrates the difficulties inherent in natural languages. The main mechanisms that address the above issues are the approximate search and the relevance measure.

Most data mining algorithms and DBMSs provide just one "right" answer, e.g., the shortest path from node A to node B in a graph, a diagnosis based on a patient's clinical information, or an answer to the user's SQL query, such as

$$SELECT\ Price\ FROM\ Sales\ WHERE\ Item = "book"$$

In contrast, IR systems provide a number of possibly good answers and let the user choose the right one(s). For instance, the user's query *Lincoln* may return information about Abraham Lincoln, a Lincoln dealership, the Lincoln Memorial, the University of Nebraska-Lincoln, and the Lincoln University in New Zealand. In other words, IR systems do not give just one right answer but perform an **approximate search** that returns multiple, potentially correct answers.

IR systems are concerned with providing information stored in a database. The key element required to implement such systems is to provide measurement of the **relevance** between the documents stored in the database and the user's query, i.e., to measure the relation between the requested information and the retrieved information. IR systems use heuristics to find such relevant information. They find documents that are close to the *right* answer and use heuristics to measure *how close* they come to the *right* answer. The inherent difficulty is that very often we do not know the *right* answer. A partial solution involves using measures of **precision** and **recall** that measure the *accuracy* of IR systems (see Section 2.4).

2.2. Architecture of Information Retrieval Systems

An overall architecture of IR systems is shown in Figure 14.1. In general, these systems first preprocess the database and implement methods to compute relevance between the documents in the preprocessed database and the user's query. Systems that are capable of dealing with semi-structured data annotate individual terms or portions of the document using tags; otherwise, the tagging is skipped and the terms are left without annotations. After a user writes the query,

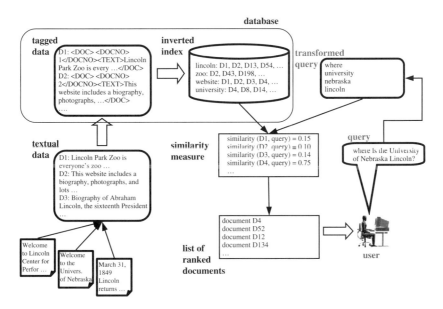

Figure 14.1. Basic architecture of information retrieval systems.

the IR system transforms it to extract important terms that are consistent with terms extracted from the preprocessed documents, and computes similarity (relevance) between the query and the documents based on these terms. Finally, a ranked list of documents, according to their decreasing similarity with the query, is returned.

Each document (and the user's query) is represented using a **bag-of-words model**, which ignores order of the words in the document, the syntactic structure of the document, and individual sentences. The document is transformed into a bag of independent words. The words are stored in a special **search database**, which is organized as an **inverted index**, i.e., it converts the original documents that consist of a set of words into a list of words associated with the corresponding documents where they occur.

2.3. Linguistic Preprocessing

The creation of the inverted index requires **linguistic preprocessing**, which aims at extracting important terms from the document represented as the bag of words. Here, we constrain our discussion to preprocessing of English text only. Term extraction usually includes two main operations:

1. **Removal of stop words**. Stop words are defined as terms that are irrelevant with respect to the main subject of the database, although they may occur frequently in the documents. They include determiners, conjunctions, prepositions, and the like.

    – A **determiner** is a nonlexical element preceding a noun in a noun phrase, and includes articles (a, an, the); demonstratives, when used with noun phrases (this, that, these, those); possessive determiners (her, his, its, my, our, their, your); and quantifiers (all, few, many, several, some, every).
    – A **conjunction** is a part of speech that is used to combine two words, phrases, or clauses together, and includes coordinating conjunctions (for, and, nor, but, or, yet, so), correlative conjunctions (both ... and, either ... or, not (only) ... but (... also)), and subordinating conjunctions (after, although, if, unless, because).
    – A **preposition** links nouns, pronouns, and phrases to other words in a sentence (on, beneath, over, of, during, beside, etc.).
    – Finally, stop words include some custom-defined words that are related to the subject of the database, e.g., for a database that lists all research papers related to heart diseases, *heart* and *disease* should be removed.

2. **Stemming**. The words that appear in documents often have many morphological variants. Therefore, each word that is not a stop word is reduced to its corresponding **stem word** (term), i.e., the words are stemmed to obtain their *root form* by removing common prefixes and suffixes. In this way, we can identify groups of corresponding words where the words in the group are syntactical variants of each other and can collect only one word per group. For instance, the words *disease, diseases,* and *diseased* share a common stem term *disease*, and can be treated as different occurrences of this word.

In short, the search database lists only stemmed terms that are filtered through the stop word list. Optionally, more advanced IR systems perform **part-of-speech tagging**, which associates each filtered term with corresponding categories such as noun, verb, adjective, etc.

The user's **query** consists of words combined by Boolean operators, and, in some cases, specific additional features such as contextual or positional operators. For instance, a query *Abraham Lincoln* is represented as *Abraham* AND *Lincoln* and can include positional operator *Abraham* AND *Lincoln* AND *located in the document's topic*. The query is transformed using the bag-of-words model and the same linguistic preprocessing as the documents themselves. Additionally, queries are often expanded by finding synonyms. **Synonyms** are words that have similar meaning but are different from a morphological point of view. Similarly to stemming, this operation aims at finding a group of these related words. The main difference is that synonyms do not share the stem term, but rather are found based on a thesaurus. For example, if a user would type *heart disease* query, then the query would be expanded to contain all synonyms of *disease* like ailment, complication, condition, disorder, fever, ill, illness, infirmity, malady, sickness, etc.

IR systems do not use linguistic analysis of the semantics of the stored documents and the queries because they use the bag-of-words model and are therefore domain independent. Next, we describe how the similarity (relevance) between the documents and queries is computed.

2.4. Measures of Text Retrieval

Let us suppose that an IR system has returned a set of documents to the user's query. We will define measures that allow us to evaluate how accurate (correct) the system's answer is. Two categories of documents can be found in the entire database: (1) **relevant documents**, which are the documents relevant to the user's query, and (2) **retrieved documents**, which are the documents returned to the user by the system. The relation between the two is illustrated by using the Venn diagram shown in Figure 14.2.

These two types of documents give rise to four possible outcomes, shown in Table 14.1. The two combinations shown in italics, i.e., *retrieved and irrelevant* and *not retrieved and relevant*, represent mistakes, while the bolded combinations, i.e., *X* and **not retrieved and irrelevant** documents, are correct results.

This table is referred to as the **confusion matrix** (for details, see Chapter 15) and is used to define two measures commonly used in text retrieval: precision and recall. For any specific user's query the set of *not retrieved and irrelevant* documents is relatively large, and thus can distort the measures used to evaluate the returned documents. Therefore, both measures are defined using the remaining three outcomes.

Precision evaluates the ability of the IR system to retrieve top-ranked documents that are mostly relevant, and is defined as the percentage of the retrieved documents that are truly relevant to the user's query.

$$\text{Precision} = \frac{X}{retrieved\ documents}$$

Recall, on the other hand, evaluates ability of the IR system to find *all* the relevant items in the database and is defined as the percentage of the documents that are relevant to the user's query

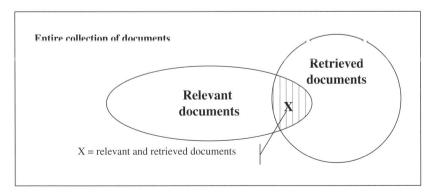

Figure 14.2. Relation between relevant and retrieved documents.

2. Information Retrieval Systems

Table 14.1. Relation between relevant and retrieved documents.

		IR system output	
		Retrieved	Not retrieved
Annotation of documents in the database	Relevant	**X = retrieved and relevant**	**not retrieved and relevant**
	Irrelevant	*retrieved and irrelevant*	**not retrieved and irrelevant**

and that were retrieved.

$$\text{Recall} = \frac{X}{relevant\ documents}$$

Note that, outside the area of information retrieval, recall is better known under the name of *sensitivity* (see Chapter 15).

The quality of the documents returned by the IR system in response to the query is measured by a combination of precision and recall. As is usually the case, a tradeoff between precision and recall needs to be achieved (see Figure 14.3).

Two extreme cases of the two measures are

– very high (about 1) precision and low recall. In this case, the system returns a few documents and almost all of them are relevant, but at the same time a significant number of other relevant documents is missing.
– very high (about 1) recall and relatively low precision. In this case, the system returns a large number of documents that include almost all relevant documents but also include a significant number of unwanted documents.

The normal cases would lie on a solid curve, shown in Figure 14.3. Generally, it is not possible to achieve the ideal case, where all retrieved documents are relevant and no relevant documents are missing in the list returned to the user. An example computation of precision and recall, which shows the tradeoff between these two measures, is provided in Figure 14.4. Assuming that the total number of relevant documents in the entire database is 7, the precision and recall values are computed for different numbers of the returned documents.

Now we explain how the precision and recall are computed based on the user's query and the returned set of documents. The precision is relatively easy to estimate, i.e., the user can analyze each returned document and judge whether it is relevant to his/her query. On the other hand, the

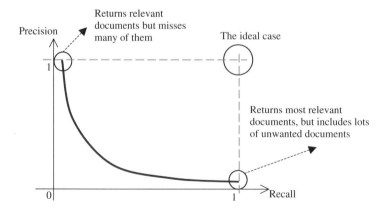

Figure 14.3. Tradeoff between precision and recall.

Total # of relevant documents = 7			
Rank	doc #	Relevant?	Precision (P) / Recall (R)
1	134	yes	P = 1/1 = 1, R = 1/7 = 0.14
2	1987	yes	P = 2/2 = 1, R = 2/7 = 0.29
3	21		
4	8712	yes	P = 3/4 = 0.75, R = 3/7 = 0.43
5	112		
6	567	yes	P = 4/6 = 0.67, R = 4/7 = 0.57
7	810		
8	12	yes	P = 5/8 = 0.63, R = 5/7 = 0.71
9	346		
10	478		
11	7834	yes	P = 6/11 = 0.55, R = 6/7 = 0.86
12	3412		still missing one relevant document, and thus will never reach 100% recall

Figure 14.4. Tradeoff between precision (P) and recall (R).

recall value depends on knowing how many documents in the entire database are relevant to the query. This number is often not available, and thus one of the following techniques can be used to estimate it:

- sampling across the database and performing relevance judgment on the documents
- using different retrieval algorithms on the same database for the same query; the relevant documents are the aggregate of all returned documents

Next, we describe how to measure similarity between two text documents. The similarity measure is necessary to be able to select documents relevant to the user's query, i.e., the user query is also converted into a text document.

2.5. Vector-Space Model

Similarity between text documents is defined based on the bag-of-words representation and uses a **vector-space model** in which each document in the database and the user's query are represented by a multidimensional vector. The dimensions correspond to individual terms in the database. In this model:

- **Vocabulary** is the set of all distinct terms that remain after linguistic preprocessing of the documents in the database, and contains t **index terms**. These "orthogonal" terms form a vector space.
- Each term, i, in either a document or query, j, is given a real-valued **weight** w_{ij}.
- Documents and queries are expressed as t-dimensional vectors $d_j = (w_{1j}, w_{2j}, \ldots, w_{tj})$. We assume that there are n documents in the database, i.e., $j = 1, 2, \ldots, n$

An example of the three-dimensional vector-space model for two documents D_1 and D_2, a user's query Q, and three terms T_1, T_2, and T_3 is shown in Figure 14.5.

In the vector-space model, a database of all documents is represented by a **term-document matrix** (also referred to as **term-frequency matrix**). Each cell in the matrix corresponds to a given weight of a term in a given document (see Figure 14.6). A value of zero means that the term is not present in the document.

Now we explain how the weights w_{ij} are computed. The process involves two complementary elements: the frequency of a term i in document j, and the inverse document frequency of term i.

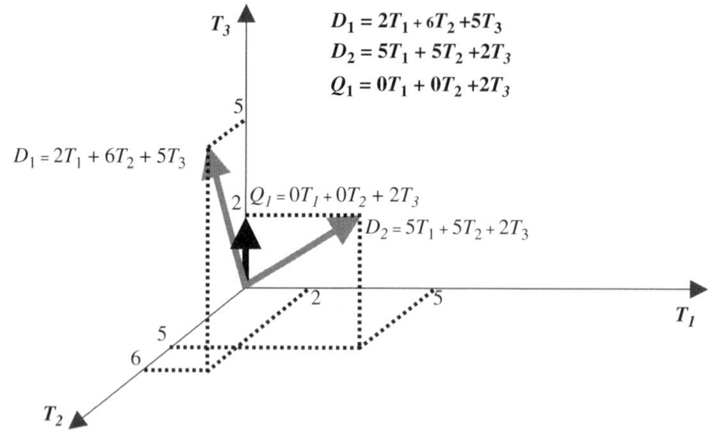

Figure 14.5. Example vector-space model with two documents D_1 and D_2 and query Q_1.

More frequent terms in a document are more important, since they are indicative of the topic of a document. Therefore, the **frequency of a term** i in document j is defined as

$$tf_{ij} = \frac{f_{ij}}{\max_i(f_{ij})}$$

where f_{ij} is the number of times term i occurs in document j. The frequency is normalized by the frequency of the most common term in the document.

The **inverse document frequency** is used to indicate the discriminative power of a term i. In general, terms that appear in many different documents are less indicative for a specific topic. Therefore, the inverse document frequency is defined as

$$idf_i = \log_2(\frac{n}{df_i})$$

where df_i is the document frequency of term i and equals the number of documents that contain term i. $\log_2$ is used to dampen the effect relative to tf_{ij}.

Finally, weights w_{ij} are computed using the **tf-idf measure** (term frequency-inversed document frequency), which is defined as

$$w_{ij} = tf_{ij} \times idf_i$$

The highest weight is assigned to terms that occur frequently in the document j, but rarely in the rest of the database of documents. Although some other ways of determining term weights are

$$\begin{bmatrix} & T_1 & T_2 & \cdots & T_t \\ D_1 & w_{11} & w_{21} & \cdots & w_{t1} \\ D_2 & w_{12} & w_{22} & \cdots & w_{t2} \\ \vdots & \vdots & \vdots & \ddots & \vdots \\ D_n & w_{1n} & w_{2n} & \cdots & w_{tn} \end{bmatrix}$$

Figure 14.6. Example term-document matrix for a database of n documents and t terms.

> Document: "data cube contains x data dimension, y data dimension, and z data dimension"
> (underlines show the "ignored" letters due to linguistic preprocessing)
>
> – The term frequencies are
>
>   data (4), dimension (3), cube (1), contain (1)
>
> – If we assume that the entire collection contains 10,000 documents and the document frequencies of these four terms are
>
>   data (1300), dimension (250), cube (50), contain (3100)
>
> then the following tf-idf weights will be computed:
>
> | | | | |
> |---|---|---|---|
> | data: | tf = 4/4 | idf = log 2(10000/1300) = 2.94 | tf-idf = 2.94 |
> | dimension: | tf = 3/4 | idf = log 2(10000/250) = 5.32 | tf-idf = 3.99 |
> | cube: | tf = 1/4 | idf = log 2(10000/50) = 7.64 | tf-idf = 1.91 |
> | contain: | tf = 1/4 | idf = log 2(10000/3100) = 1.69 | tf-idf = 0.42 |

Figure 14.7. Example computation of tf-idf weights.

also used, the tf-idf weighting is the most popular and has been found to work very well through extensive experimentation.

Figure 14.7 shows an example computation of tf-idf weights for a document that includes the sentence "data cube contains x data dimension, y data dimension, and z data dimension."

The vector-space model and the tf-idf weighting are used to define measure for finding documents that are close to the user's query. There are multiple ways in which this closeness can be measured, e.g., distance between vectors, angle between vectors, distance or angle using projection of the vectors, etc. (see Figure 14.5). In the next section, we describe one of the most popular similarity measures.

2.6. Text Similarity Measures

Text similarity measure is a function that is used to compute the degree of similarity between two vectors. It is used to quantify similarity between the user's query and each of the documents from the database. The measure allows ranking of the documents with respect to their similarity (relevance) to the query. After the documents are ranked, a fixed number of top-scoring documents is returned to the user. Alternatively, a threshold may be used to decide how many documents will be returned. The threshold can be used to control the tradeoff between precision and recall, e.g., a high threshold value will usually result in high precision and low recall.

A popular text-similarity measure is the **cosine measure**, which is computed as the angle between the two vectors. Given the vectors representing document d_j and query q, and t terms extracted from the database, the cosine measure is defined as

$$similarity(\vec{d_j}, \vec{q}) = \frac{\vec{d_j} \cdot \vec{q}}{|\vec{d_j}| \cdot |\vec{q}|} = \frac{\sum_{i=1}^{t}(w_{ij} \cdot w_{iq})}{\sqrt{\sum_{i=1}^{t} w_{ij}^2 \cdot \sum_{i=1}^{t} w_{iq}^2}}$$

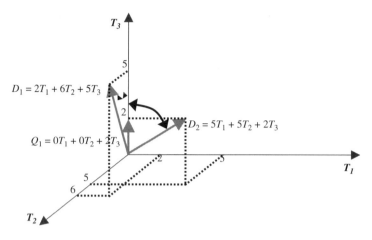

Figure 14.8. Similarity between documents D_1 and D_2 and query Q_1.

Example: For two documents $D_1 = 2T_1 + 6T_2 + 5T_3$ and $D_2 = 5T_1 + 5T_2 + 2T_3$ and query $Q = 0T_1 + 0T_2 + 2T_3$ given in Figure 14.5, the following cosine values are computed:

$$similarity(\vec{D_1}, \vec{Q}) = \frac{(2 \cdot 0 + 6 \cdot 0 + 5 \cdot 2)}{\sqrt{(4 + 36 + 25) \cdot (0 + 0 + 4)}} = \frac{10}{\sqrt{65 \cdot 4}} = 0.62$$

$$similarity(\vec{D_2}, \vec{Q}) = \frac{(5 \cdot 0 + 5 \cdot 0 + 2 \cdot 2)}{\sqrt{(25 + 25 + 4) \cdot (0 + 0 + 4)}} = \frac{4}{\sqrt{54 \cdot 4}} = 0.27$$

The example shows that according to the cosine measure document, D_1 is more similar to the query than document D_2. In fact, the angle between D_1 and Q is much smaller than the angle between D_2 and Q, see Figure 14.8.

This result agrees with common sense. The query has two terms T_3 and no terms T_1 and T_2. Document D_1 has three more terms T_3 than document D_2 and fewer irrelevant terms T_1 and T_2.

3. Improving Information Retrieval Systems

The basic design of the IR systems described in the previous section can be enhanced to increase precision and recall and to improve the term-document matrix. The former issue is often addressed by using the relevance feedback mechanism, while the latter is addressed by using latent semantic indexing.

3.1. Latent Semantic Indexing

Latent semantic indexing aims to improve the effectiveness of the IR system by retrieving documents that are more relevant to a user's query through manipulation of the term-document matrix. The original term-document matrix is often too large for the available computing resources. Additionally, it is also presumed to be noisy, e.g., some anecdotal occurrences of terms should be eliminated, and too sparse with respect to the "true" term-document matrix. Therefore, the original matrix is approximated by a smaller, "de-noisified" matrix.

The new matrix is computed using the **singular value decomposition** (SVD) technique, a well-known matrix theory method. Given a $t \times n$ term-document matrix representing t terms and n documents, the SVD removes some of the rows and columns of the original matrix to reduce it to the size of $k \times k$, where k is usually around a few hundred, while in real IR systems t and n

values are around a few thousand. SVD aims to remove the least significant parts of the original matrix by following these steps:

- Create a term-document matrix.
- Compute the singular value decompositions of the term-document matrix by splitting it into three smaller matrices U, S, and V, where U and V are orthogonal, i.e., $U^T U = I$, and S is a diagonal matrix of singular values. The latter matrix is of size $k \times k$ and constitutes the reduced version of the original term-document matrix.
- For each document, replace its term-document vector by a new one that excludes terms eliminated during SVD
- Store all new vectors and create a multidimensional index to facilitate the search procedures.

SVD is described in details in Chapter 7. Commonly used implementations of SVD can be found in MATLAB and LAPACK.

3.2. Relevance Feedback

Relevance feedback aims to improve the relevance of the returned documents by modifying the user's query. It adds additional terms to the initial query, which may match more relevant documents. The overall process of the relevance feedback system is shown in Figure 14.9. It consists of the following steps:

1. Perform an IR search on the initial query.
2. Obtain feedback from the user as to what documents are relevant and find new terms (with respect to the initial query) from known relevant documents.
3. Add new terms to the initial query to form a new query.
4. Repeat the search using the new query.
5. Return a set of relevant documents based on the new query.
6. User evaluates the returned documents.

The relevance feedback (Step 2 of the above procedure) can be performed either manually or automatically. In the former case, the user manually identifies relevant documents, and new terms are selected either manually or automatically. In the latter case, the relevant documents are identified automatically by using the top-ranked documents, and new terms are selected automatically using the following steps:

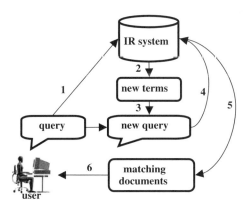

Figure 14.9. The relevance feedback-based IR system.

- Find the N top-ranked documents.
- Identify all terms from the N top-ranked documents.
- Select the feedback terms, say, based on selecting several top terms with respect to maximal values of tf-idf weights and excluding the terms from the original user's query.

The relevance feedback usually improves average precision by increasing the number of "good" terms in the query, but at the same time it requires more computational work and in some case can decrease effectiveness, i.e., one newly added bad term can undo the good caused by lots of good terms.

4. Summary and Bibliographical Notes

In this Chapter we introduced concepts and methods related to **text mining** and **information retrieval**. The most important topics discussed in this Chapter are the following:

- **Information retrieval** (IR) concerns the organization and retrieval of information from large collections of **semistructured** or **unstructured** text-based databases and the WWW.
- **Documents** are processed by the IR systems and consist of a sequence of terms in a natural language that expresses ideas about some topic. A **term** is defined as a semantic unit, phrase, or a root of the word.
- A **query** is the request for documents that concern a particular topic of interest to the IR system user.
- In IR, each document (and the user's query) is represented by using a **bag-of-words model**. The words (terms) are stored in a special **search database**, which is organized as an **inverted index**.
- Creation of the inverted index requires **linguistic preprocessing**, which includes removal of **stop words** and **stemming**, and aims at extracting important terms from document. **Synonyms** are also found to expand the user's queries.
- **Precision** and **recall** (sensitivity) measures are used to evaluate the accuracy of IR systems.
- The similarity between text documents is defined based on the bag-of-words representation and uses a **vector-space model** that is used to derive the **term-document matrix**. The most popular text similarity is the **cosine measure**.
- IR systems increase their accuracy by using a **relevance feedback mechanism** and improve effectiveness by using **latent semantic indexing**.

Introductory-level texts and survey articles concerning **text mining** can be found in [1, 2, 4, 6, 9, 11], while collections of edited articles are in [5, 7, 8]. The **latent semantic indexing** method is described in [3], and related resources can be found on the web site of the Latent Semantic Analysis Group at the University of Colorado at Boulder (http://lsa.colorado.edu/). **Relevance feedback** was proposed in [10, 12].

References

1. Baeza-Yates, R., and Ribeiro-Neto, B. 1999. *Modern Information Retrieval*, Addison Wesley New York, USA
2. Berry, M. 2003. *Survey of Text Mining: Clustering, Classification, and Retrieval*, Springer-Verlag, New York, USA
3. Deerwester, S., Dumais, S., Furnas, G., Landauer, T., and Harshman, R. 1990. Indexing by latent semantic analysis. *Journal of the American Society for Information Science*, 41:391–407
4. Feldman, R., and Sanger, J. 2006. *The Text Mining Handbook: Advanced Approaches in Analyzing Unstructured Data*, Cambridge University Press

5. Franke, J., Nakhaeizadeh, G., and Renz, I. (Eds.). 2003. *Text Mining: Theoretical Aspects and Applications*, Physica-Verlag, Heidelberg
6. Han, J., and Kamber, M. 2000. *Data Mining: Concepts and Techniques*, Morgan Kaufmann
7. Hotho, A., Nürnberger, A., and Paaß, G. 2005. A brief survey of text mining. *GLDV—Journal for Computational Linguistics and Language Technology*, 20(1):19–62
8. Raghavan, P. 1997. Information retrieval algorithms: a survey. *Proceedings of the 1997 ACM-SIAM Symposium on Discrete Algorithms*, New Orleans, USA, 11–18
9. Rijsbergen, C.J. 1990. *Information Retrieval*, Butterworth London, England
10. Rocchio, J. 1971. Relevance feedback in information retrieval. In Salton, G. (Ed.), *The Smart Retrieval System: Experiments in Automatic Document Processing*, Prentice Hall, Upper Saddle River, NJ, USA, 313–323
11. Salton, G., and McGill, M. 1983. *Introduction to Modern Information Retrieval*, McGraw-Hill New York, USA
12. Salton, G., and Buckley, C. 1990. Improving retrieval performance by relevance feedback. *Journal of the American Society for Information Science*, 41(4): 288–297

5. Exercises

1. Define stemming and discuss how it is performed. Are stemming and lemmatization two synonyms or two different terms? If they are different, then explain the differences and discuss which one is more suitable in the context of an information retrieval task.
2. Discuss how XML and related standards could help to perform text mining. Give a list of relevant XML-based standards and briefly describe what they are used for.
3. Investigate how synonyms can be identified. Given an answer to this general question, research how the synonyms can be identified for the MEDLINE® repository (On-line Medical Literature Analysis and Retrieval System), which is the world's biggest repository of medical literature citations and abstracts (hint: visit http://www.nlm.nih.gov/databases/).
4. Research and discuss different text similarity measures and contrast them with the cosine measure, which was introduced in this Chapter.

    (a) Use scientific references to point out the advantages and disadvantages of the other measures when compared with the cosine measure.
    (b) Compute the corresponding similarity values for the two documents and the query given in Section 2.5.

5. Consider a text database that consists of email messages. The emails can be viewed as semistructured documents that include annotated fields such as topic, sender, etc., and unstructured text. Investigate the following:

    (a) How would you store this database to facilitate multidimensional searches (by topic, sender, title, etc.)?
    (b) Assuming that each email in the database could be categorized as *spam*, *normal*, and *priority*, describe how you would design a data mining system that can automatically categorize future emails into these categories. Include information about how this system will be trained and how the emails will be preprocessed.
    (c) Research the literature on existing systems that perform classification as described in question 6b. Give the citations, if any, and briefly describe the design of such systems.

Part 5

Data Models Assessment

15

Assessment of Data Models

In this Chapter, we describe methods for the selection and assessment of models generated from data. Before we describe these methods, an important note needs to be made: It is the user/owner of the data who ultimately approves or disapproves the generated model of the data. The user often does so without using any formal methods but instead judges the usefulness of the generated knowledge (information) using his/her domain knowledge.

1. Introduction

Below we describe the methods to be used by a data miner *before* presenting the selected model, and its assessment, to the user. The user then makes a final decision about whether to accept the model in its current form, and use it for some purpose, or to ask the data miner to generate a new one. The latter outcome is frequently the fate of initial models: they are rejected because it takes several iterations (in the knowledge discovery process loop) and interactions with the user before an acceptable model is found. After all, it is the owner of the data who decided to undertake a data mining project and thus will accept only results that were not previously known to him (although known results validate DM models generated in a "blind" way). The user's expectation of a DM project is to find some new information/knowledge, hidden in the data, so that it can be used for some advantage.

Another note: A data miner assesses the quality of the generated model often by using the same data that were used to generate the model itself (albeit divided into training and testing data sets), while the user depends not only on the DM data and results but also on his or her deep (expert) domain knowledge. In spite of the KDP requirement that the data miner learns as much as possible about the DM domain and the data, his or her knowledge obviously constitutes only a subset of the knowledge of the experts (data owners). Learning knowledge about a domain is not a trivial task. To illustrate this, let us note that the best bioinformaticians would be those with doctoral degrees (and experience) in both molecular biology and computer science, which is rarely the case. Data miners in this example are computer scientists who know only bits and pieces about molecular biology.

Most often, a data miner generates several models of the data and needs to decide which one to accept as the "best" in terms of how well it explains the data and/or its predictive power, before presenting it to the data owner. What are the methods for selecting the "best" model from among several generated? As stated above, the data owner utilizes his **domain knowledge** for **model assessment**. Data miners can do the same but since their knowledge is not as deep as that of the users, they have attempted to formalize the process by coming up with some simple measures, often called **interestingness measures**, which the users supposedly use; later in this Chapter we will discuss some of them. A similar approach would be for data miners to reason from first principles, such as laws of physics, mathematical theorems, etc.; however, this approach is beyond the scope of this book. Instead, in this Chapter we focus on the heuristic, data-reuse (data resampling), and analytic methods for model selection and validation.

2. Models, their Selection, and their Assessment

Let us recall the definition of a **model** before talking about methods for its assessment and selection. We will use the terms *model, classifier*, and *estimator* with the understanding that they mean basically the same thing. A **classifier** is a model of data used for a classification purpose: given a new input, it assigns that input to one of the classes it was designed/trained to recognize. The term **estimator** originated in statistics and will be defined later. The reason we will use all these terms interchangeably is that model assessment methods originated independently in disciplines such as statistics, pattern recognition, and information theory, while using different terminology.

A *model can be defined as a description of causal relationships between input and output variables*. This description can be formulated in several ways and can take the form of a classifier, neural network, decision tree, production rules, mathematical equations, etc. In fact, many data modeling methods were developed in the form of mathematical equations in the area of statistics.

The number of independent pieces of information required for estimating the model is called the model's **degree of freedom**. For instance, if we calculate the mean, it has one degree of freedom. From this definition, especially when the number of degrees of freedom is equal to the number of parameters (the typical case), originates one of the simplest heuristics for model selection: one should choose a **parsimonious model**, one that uses the fewest number of parameters, among several acceptably well-performing models.

When we talk about a model, there is always a **model error** associated with it. Model error is calculated as the difference between the observed/true value and the model output value, and is expressed either as an absolute or squared error between the observed and model output values. When we generate a model of the data, we say that we *fit the model* to the data. However, in addition to fitting the model to the data, we are also interested in using the model for *prediction*. Thus, once we have generated several models and have selected the "best" one, we need to *validate it*, not only for its *goodness of fit* (fit error), but also for its **goodness of prediction** (prediction error). In the neural network and machine learning literature, goodness of prediction is often referred to as the **generalization error**. The latter term ties goodness of prediction into the concepts of overfitting, or underfitting, the data. **Overfitting** *means an unnecessary increase in model complexity*. For example, increasing the number of parameters and the model degrees of freedom, beyond what is necessary increases the model's variance (although we never know what is necessary in the first place: we can only estimate it after assessing its goodness of prediction). **Underfitting** is the opposite notion to overfitting, i.e., too simple a model will not fit the data well. As a corollary/rule of thumb of these definitions, when the available data set is small, we should fit to it a simple model, not a complex one. See more discussion about overfitting and underfitting in the section on the bias-variance dilemma below. In the typical data mining situation, several models are generated. Then the data miner needs to choose one "best" model, an undertaking referred to as **model selection**.

From the above discussion, we note that to evaluate the model, we need to calculate its error. However, the error can be calculated only if training data, meaning known inputs corresponding to known outputs, are available. We also recall from Part IV that data mining methods for building models of data can be divided into three general categories: unsupervised, supervised, and semisupervised. Thus, on the one hand, we can categorize model assessment techniques depending on the nature of the data available (training versus unsupervised (meaning no known outputs are associated with data items)) and as our ability or inability to calculate the errors. On the other hand, we can divide model assessment techniques into groups based on the nature of the methods:

- **Data-reuse**, or resampling, methods (simple split, cross-validation, bootstrap) are very popular in evaluating supervised learning models.
- **Heuristic** (parsimonious model, Occam's razor) methods are not formal, are very simple, but are probably the most frequently used (often subconsciously used) methods.
- **Analytical** (Minimum Description Length, Aikake's Information Criterion, and Bayesian Information Criterion) methods are formal and elegant, but not very practical.
- **Interestingness measures** try to mimic process of model evaluation used by the owner/user of the data; they are relatively new but becoming more popular.

Another possible categorization would be into methods for model selection, methods for assessment of the goodness of fit, and methods for assessment of the goodness of prediction. We will present the model assessment methods in what follows according to the above four-bullet categorization but will use the names of the techniques themselves, since these are better known.

An important note: since the evaluation of models generated by unsupervised (or semisupervised) methods is more difficult than the evaluation of models generated from training data, we have described the quality assessment of the former in the Chapters where they are discussed. As a result, the evaluation of clustering models, done via calculation of various cluster validity measures, is presented in Chapter 9 on clustering because it was easier to explain and understand them there. The same is the case for association rules, covered in Chapter 10, and text mining methods, described in Chapter 14. However, some measures described in the latter two Chapters are also mentioned here for completeness. In what follows, we describe techniques for model validation and model selection.

2.1. The Bias-Variance Dilemma

To evaluate goodness of fit and goodness of prediction of a model, we must be able to calculate error. There are two components of error: bias and variance. **Bias** can be defined as the error that cannot be reduced by increasing the sample size. It is present even if an infinite sample is available; it is a systematic error. Sources of bias are, for example, a measurement error (an experimental error that cannot be removed), a sample error (the sample may not be correctly drawn from the population, and thus may not represent the data correctly), or an error associated with a particular form of an estimator, etc. Bias is calculated as the difference between the estimated expected value and the true value of some parameter:

$$B = E(\bar{p}) - p$$

Its squared value, B^2, is one of the two components (the other is variance, S) of the mean square error, MSE, which calculates the mean square difference between a true value of a parameter, p, and its estimated value, $\bar{p}$:

$$MSE(\bar{p}) = E(\bar{p} - p)^2 = S(\bar{p}) + B^2(\bar{p})$$

where the variance is

$$S^2 = \left(\sum_i (\bar{p}_i - p_i)^2 \right) / (N - 1)$$

for the unbiased estimation of the population variance, or

$$S^2 = \left(\sum_i (\bar{p}_i - p_i)^2\right)/N$$

for the biased one.

Variance can be defined as an additional error (additional to bias) that is incurred given a finite sample (because of sensitivity to random fluctuations). **Biased estimators** have a nonzero bias (meaning that the estimated expected value is different from the true value) while **unbiased estimators** have zero bias of the estimate. An **Estimator** is a method used to calculate a parameter. It is a variable defined as function of the sample values. Examples are a histogram density estimator, which estimates density based on counts per interval/bin; and a Bayesian estimator, which estimates the a posteriori probability from the a priori probability via Bayes' rule. The simplest nonparametric estimator (meaning one that does not depend on complete knowledge of the underlying distribution) is the sample mean that estimates the population mean. The latter, in fact, constitutes the simplest **model** of the data. We know from Part IV of this book that estimators/models of the data are characterized by different degrees of complexity (in the case of the just given examples, the complexity ranges, in ascending order, from a sample mean, through a histogram, to a Bayesian estimator).

From the above definition of the MSE, we see that there is a tradeoff between bias and variance, known as the *bias-variance dilemma*. If a given MSE has large bias, then it has a small accompanying variance, and vice versa. We are interested in finding an estimator/model that is neither too complex (and thus may overfit the data) nor too simple (and thus may underfit the data). Such a model can be found by minimizing the MSE value, with acceptable bias and variance. In Figure 15.1 we show a graph where the *y*-axis represents the values of the bias/variance/MSE, while the *x*-axis represents the complexity/data size of the estimator/model.

Now we shift attention to the goodness-of-fit and goodness-of-prediction, or in other words, to training and test errors. We know from Part IV that it is easy, say, to overtrain a neural network (making it unnecessarily complex by having too many neurons in a hidden layer) by reducing the training error to a very small value. Overtraining usually means that the data have been overfitted.

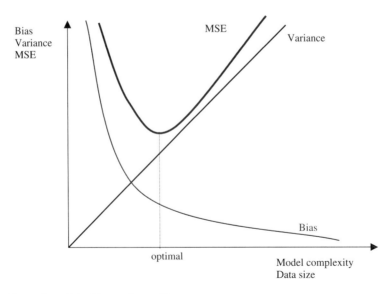

Figure 15.1. Illustration of the bias-variance dilemma.

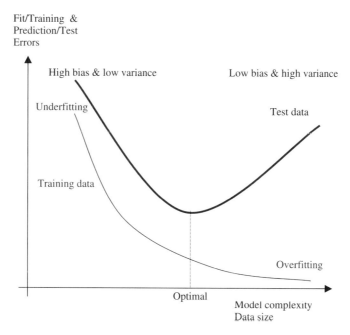

Figure 15.2. Choosing an optimally complex model with acceptable fit and prediction errors.

When such an (over-) trained neural network is used on the test data, its prediction/generalization error is usually large. Again, we are interested in finding a model that is neither too complex nor too simple. Figure 15.2 illustrates the general idea of choosing a model that does not overfit the data, which means that it should perform acceptably well on both training and test data, as measured by the error between the true/desired value and the actual model output value.

3. Simple Split and Cross-Validation

When a large data set composed of known inputs corresponding to known outputs exists, we evaluate the model simply by *splitting the available data into two parts: the* **training** *part, used for fitting the model, and the* **test** *part, used for evaluation of its goodness of prediction.*

The question is how to split the data into these two parts. The best strategy is to do so randomly, using a rule-of-thumb formula that states that about 2/3 (or 1/2) of the data should be used for training. A good approach is to start with about 1/2 of the data for training, but if goodness of prediction on the test data is not acceptable, then to go back and use 2/3 of the data for training. This method is characterized by high bias (because of the small data size) but low variance.

If the results are still unsatisfactory, then we need to use a more expensive computational method, called **cross-validation**, to estimate the goodness of prediction of the model. Informally, we can state that cross-validation is to be used in situations where the data set is relatively small but difficult (meaning that splitting it into just two parts does not result in good prediction).

Let n be the number of data points in the training data set. Let k be an integer index that is much smaller than n. In a **k-fold cross-validation**, we divide the entire data set into k equal-size subsets, and use k-1 parts for training and the remaining part for testing and calculation of the prediction error (goodness of prediction). We repeat the procedure k times, and report the average from the k runs. This method is frequently used in reporting the results in the literature as a **10-fold cross-validation**, when the data set is divided into ten subsets and the final prediction error is calculated as 1/10 times the sum of the ten errors.

In an extreme situation, when the data set is very small, we use **n-fold cross-validation**, which is also known as the **leave-one-out** method. It is computationally expensive and obviously not practical for use on large data sets; it is characterized by low bias accompanied by high variance.

4. Bootstrap

Bootstrap is computationally more expensive than cross-validation. Bootstrap gives a nonparametric estimate of the error of a model in terms of its bias and variance. It works as follows. We draw x samples (in practice we choose x to be between 30 and 100) of equal size from a population consisting of n samples, with the purpose of calculating confidence intervals for the estimates. A strong assumption is made that the available data set of n samples constitutes the population itself. Then, from this "population", x samples of size n (notice that this is the same size as the original data set), called **bootstrap samples**, are drawn, with replacement. Next, we fine-tune the existing model (originally fitted to our data of size n) to each of the x bootstrap samples. Having done this, we can assess the goodness of fit (error) for each of the bootstrap samples, and we can average over the x samples (let us denote by x_i the ith realization of the bootstrap sample, where $i = 1, \ldots, x$) to calculate the bootstrap estimate of the bias and variance as

$$B(t) = \frac{1}{x} \sum_{i}^{x} x_i - t$$

where t is the estimate calculated from the original data:

$$S^2(t) = \sum_{i}^{x} (x_i - t_{ave})^2 / (x-1) \quad \text{where } t_{ave} = \frac{1}{x} \sum_{i}^{x} x_i$$

To use bootstrap for calculating goodness of prediction, we treat the set of bootstrap samples as the training data set, and the original training data set as the test set. Thus, we fit the model to all bootstrap samples, and calculate the prediction error on the original data. The problem with this approach is that prediction error is too optimistic (good), since the bootstrap samples are drawn with replacement. In other words, we calculate the prediction/generalization error on highly overlapping data (with many data items in the test set being the same as the training data). Stated differently, the bootstrap method is a sort of "cheating", similar to a situation where one would report prediction error calculated on the training data. In this respect, cross-validation is better than bootstrap, since we calculate prediction error on a portion of data that was set aside for this purpose. Because of this basic drawback, there exist many, improved, variations of the basic bootstrap method, the most obvious one being its combination with cross-validation.

5. Occam's Razor Heuristic

The most popular heuristic for assessing goodness of a model is **Occam's razor**. It states that *the simplest explanation (model) of the observed phenomena, in a given domain, is most likely to be the correct one*. The idea is very convincing, but there are problems associated with its use in practice. For one, we do not know how to determine "the simplest explanation". Most would intuitively agree with its author, William of Ockham (more often spelled "Occam"), who wrote, "Entia non sunt multiplicanda praeter necessitatem" ('entities are not to be multiplied beyond necessity'). In other words, given several models (specified, for example, in terms of production IF... THEN... rules), a more "compact" model (composed of a smaller number of rules, especially if on average the rules in this model are shorter than in all other models) should be chosen. Having said this, we note that, in many cases, the Occam's razor (in analogy to naïve Bayes) works quite well and can be used as a guiding principle (assuming that there is not enough time to do a more extensive evaluation, or that appropriate data required for calculating other measures do not exist) for selecting one of the generated models (either by supervised or unsupervised model building methods).

Also, worth mentioning is that the Occam's razor heuristic is built into most machine learning algorithms (including those described in this book) – that is, a simpler/shorter/more compact description of the data is preferred over more complex ones. Thus, another problem with this heuristic is that we want to use it for a model assessment, when we have already used it for generating the model itself. Still another issue with Occam's razor is that some researchers hold that it may be entirely incorrect in some situations, the argument being as follows: if we model some process/data known to be very complex, why should a simple model of these data be preferred at all?

Because of the above problems associated with Occam's razor heuristic, researchers came up with analytical ways for assessing model quality. One important such method, the **Minimum Description Length principle**, is only briefly sketched below, since its detailed description would exceed the scope of this book and also since it is not the easiest to use in practice.

6. Minimum Description Length Principle

Rissanen, following the work of Kolmogorov (Kolmogorov complexity), introduced the **Minimum Description Length (MDL) principle**. It was designed to be general and independent of any underlying probability distribution. Rissaneen (1989) wrote, "We never want to make a false assumption that the observed data actually were generated by a distribution of some kind, say, Gaussian, and then go on and analyze the consequences and make further deductions. Our deductions can be entertaining but quite irrelevant... ." This statement is in stark contrast to statistical methods, because the MDL provides a clear interpretation regardless of whether some underlying "true/natural" model of data exists or not.

The basic idea of the MDL principle is connected with a specific understanding/definition of learning (model building). Namely, learning can be understood as finding regularities in the data, where regularity is understood as the ability to compress the data. Thus, we may say that learning can be understood as the ability to compress the data, i.e., to come up with a compact description of the data (the model). This is, in fact, what inductive machine learning, also known as inductive inference, algorithms do. For instance, if we had a data set consisting of 900 items/examples, described by 10 features/attributes, and were able to describe them using only three rules (say, one covering 650 examples, the second 300, and the third 150 examples; notice that the same examples are covered by more than one rule), then what we have built is a very compact model of the data (which could have been chosen by a simple Occam's razor). In the parlance of machine learning, we desire to select the most general model that does not overfit the data. In the parlance of MDL, having a set of models, M, about the data, D, we want to select the model that most compresses the data. Both methods specify the same goal, but are stated using different language. Notice, that the MDL principle has a strong information theory flavor.

In its original form, the MDL principle was stated as follows: If a system can be defined in terms of input and corresponding output data, then in the worst case (longest), it can be described by supplying the entire data set (thus constituting longest/least compressed model of the data). However, if regularities can be discovered in the data, say, expressed in the form of production rules, then a much shorter description is possible, and it can be measured by the MDL principle. *The MDL principle says that the complexity of a theory (model/hypothesis) is measured by the number of bits needed to encode the theory itself, plus the number of bits needed to encode the data using the theory.*

More formally, from a set of models, we choose as the "best" the model that minimizes the following sum: $L(M) + L(D|M)$, where $L(M)$ is the length (in bits) of the description of the model, and $L(D|M)$ is length of the description of data encoded using model M. The basic idea behind this definition can be explained using notions of underfitting and overfitting. We note that it is quite easy to find a complex model, meaning one having large $L(M)$ value, that overfits the data (i.e., with a small $L(D|M)$ value). It is equally easy to find a simple model, with a small $L(M)$ value, that underfits the data but with a large $L(D|M)$ value. Notice the similarity of this approach to the bias-variance dilemma discussed above. What we are thus looking for is a model that constitutes the best compromise between the two cases. Moreover, suppose that we have generated two models that explain/fit the data equally well; then the MDL principle tells us to choose the one that is simpler (for instance, having the smaller number of parameters; recall that such a model is called **parsimonious**), i.e., allows for the most compact description of the data. In this sense, the MDL principle can be seen as a formalization of the Occam's razor heuristic.

7. Akaike's Information Criterion and Bayesian Information Criterion

The **Akaike Information Criterion** (AIC) and the **Bayesian Information Criterion** (BIC) are statistical measures that are briefly sketched below. The reasons for introducing them here are that they make our presentation of assessment methods more comprehensive and shed additional light on other measures introduced in the Chapter. The two measures are used for choosing between models that use different number of parameters, d, and are closely related to each other, as well as to the MDL principle. The idea behind them is as follows. We want to estimate the prediction error, E, and use it for model selection. What we can easily calculate is the training error, TrE. However, since the test vectors do not need to coincide with the training vectors, the TrE is often too optimistic. To remedy this situation we estimate the error of optimism, Eop, and calculate the in-sample (given the sample) error as follows:

$$E = TrE + Eop$$

This general idea underlies both the AIC and the BIC definitions, although the latter has its origins in the Bayesian approach to model selection. The two measures, however, can be shown to be equivalent, although they are biased towards the selection of different models. Specifically, if we have generated a family of models (including the true model), the BIC will choose the correct one as the sample size, N, reaches infinity, while the AIC will choose an overly complex model. For finite samples, however, the BIC will tend to choose models that are too simple; use of the AIC in the latter situation can be a good strategy.

7.1. AIC

The AIC is defined as

$$AIC = -2 log L + 2(d/N)$$

where logL is the maximized log-likelihood, defined as

$$\log L = \sum_{i=1}^{N} \log \overline{P}_\theta(y_i)$$

$P_\theta(Y)$ is a family of densities containing the "true" density, $\overline{\theta}$ is the maximum-likelihood where estimate of θ, and d is the number of parameters in the model.

If we have generated a family of models, which we can tune by α (i.e., TrE(α) and $d(\alpha)$), then we rewrite AIC as

$$AIC(\alpha) = TrE(\alpha) + 2(d(\alpha)/N)S^2$$

where variance S^2 is defined as

$$S^2 = \sum_i (y_i - \bar{y}_i)^2$$

The AIC(α) is an estimate of the test error curve; thus we choose as the optimal model the one that minimizes this function.

7.2. BIC

The BIC can be stated similarly to AIC:

$$BIC = -2 log L + d\ log N$$

We see that the two definitions are equivalent in the sense that the "2" in the second term has been replaced by logN in th BIC definition.

For a Gaussian distribution with variance S^2, we can write BIC as

$$BIC = (N/S^2)(TrE + d/N\ log N)$$

Again, we choose the optimal model as the one corresponding to the minimum value of the BIC. The BIC favors simpler models since it heavily penalizes more complex models. The BIC, in fact, selects the same models that would be selected by using the MDL principle.

8. Sensitivity, Specificity, and ROC Analyses

Let us assume that some underlying "truth" (also known as the gold standard, or hypothesis) exists, in contrast to the MDL principle, which does not require this assumption. This assumption means that training data are available, i.e., known inputs corresponding to known outputs. It further implies that we know the total number of positive examples, P, and the total number of negative examples, N. Then we are able to form a **confusion matrix**, also known as a **misclassification matrix** (the most revealing, but less frequently used, name of the table) or as a **contingency table**. Table 15.1 shows the general form of a confusion matrix that is used for calculating goodness of fit and goodness of prediction errors.

TP is defined as the case in which the test result and gold standard (truth) are both positive; FP is the case in which the test result is positive but the gold standard is negative; TN is the case where both are negative; and FN is the case where the test result is negative but the gold standard is positive. Using Table 15.1 we can calculate the following two key error measures:

Table 15.1. Possible outcomes of a test, given the knowledge of "truth".

	Test	Result	
Truth (Gold standard; Hypothesis)	Positive	Negative	
Positive	True Positive (TP) (no error)	False Negative (FN) (Rejection error, Type I error)	P (total of true positives)
Negative	False Positive (FP) (Acceptance error, Type II error)	True Negative (TN) (no error)	N (total of true negatives)
	Total of Test recognized as Positive	Total of Test recognized as Negative	Total population

$$\text{Sensitivity} = \text{TP}/P = \text{TP}/(\text{TP}+\text{FN}) = \text{Recall}$$

Sensitivity measures how often we find what we are looking for (say, a certain disease). In other words, sensitivity is 1 if all instances of the True class are classified to the True class. In Table 15.1, what we are looking for is $P = (\text{TP}+\text{FN})$, i.e., all the cases in which the gold-standard (disease) is actually positive (but we found only TP of them) – thus the ratio. Sensitivity is also known in the literature under a variety of near-synonyms: TP rate, hit rate, etc. In the information extraction/text mining literature, it is known under the term *recall*.

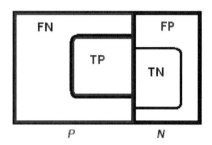

$$\text{Specificity} = \text{TN}/N = \text{TN}/(\text{FP}+\text{TN})$$

Specificity measures how often what we find is what we are not looking for (say, a certain disease); in other words, it measures the ability of a test to be negative when the disease is not present. Specificity is 1 if only instances of the True class are classified to the True class. In Table 15.1, what we find is $N = (\text{FP}+\text{TN})$, i.e., all the cases in which the gold standard is actually negative (but we have found only TN of them).

Stated in a different way, sensitivity measures the ability of a test to be positive when the "disease" is actually present and to be negative when the "disease" is not present. Accuracy (defined below) gives the overall evaluation.

Unfortunately, people often report only **accuracy** instead of calculating both sensitivity and specificity. The reason is that, in a situation where, say, sensitivity is high (say, over 0.9) while specificity is low (say, about 0.3), the accuracy may look acceptable (over 0.6, or 60%), as can be figured out from its definition:

$$\text{Accuracy1} = (\text{TP}+\text{TN})/(P+N)$$

Sometimes a different definition for accuracy is used, one that ignores the number of TN:

$$\text{Accuracy2} = \text{TP}/(P+N)$$

Thus, if only accuracy is reported, the results may look better than they actually are. Because of this one should always calculate and report both the sensitivity and specificity of test results. Consequently, our confidence in results for which only accuracy is reported should be low.

In text mining literature, a still different measure, called **precision**, is used. It is defined as

$$\text{Precision} = \text{TP}/(\text{TP}+\text{FP})$$

where $\text{TP}+\text{FN} = P$ are called **relevant** documents and $\text{TP}+\text{FP}$ are called **retrieved** documents.

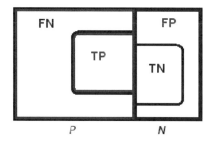

Let us notice the similarity of precision to the Accuracy2 definition: the difference is that now the "universe" is narrowed down to the total of what is found (TP + TN). Below we give an example illustrating the calculation of sensitivity (recall) and specificity.

Example: Table 15.3 illustrates the calculation of sensitivity, specificity, and accuracy, given the results presented as confusion matrix shown in Table 15.2.

Let us now suppose that we have only two classes, say, Normal (N) and Abnormal (A), but the classifier (test result) cannot recognize some examples as belonging to either of the two classes (A or B): the test result is reported as "unknown", as indicated by letter U. How should we construct a confusion matrix for such a case and calculate the three measures? We explain it via the means of the following example.

Example: Suppose we have a data set that consists of 24 examples, consisting of two classes: 15 from class A and 9 from class N. A set of rules (a model) is applied to this test set and gives the following results. For examples from class A, 13 were classified as A, 0 were classified as N, and 2 were classified as U (neither A nor N). For examples from class N, 3 were classified as A, 5 were classified as N, and 1 was classified as U. Now, we form two confusion matrices (Tables 15.4 and 15.5) from which we can calculate the three measures.

Table 15.6 summarizes the results for the above example.

Judging by the accuracy values alone, the results seem to be acceptable, but we notice that the model actually has problems in distinguishing between the two classes, namely, the rules for class

Table 15.2. Misclassification matrix of results on a data set with 27 examples.

True; Gold standard diagnosis	Test		Results		
	Classified as A	Classified as B	Classified as C	Classified as D	
A	4	1	1	1	7
B	0	7	0	0	7
C	0	0	2	1	3
D	0	0	0	10	10
	4	8	3	12	27

Table 15.3. Calculation of sensitivity, specificity, and accuracy.

	Class A	Class B	Class C	Class D
Sensitivity	.57 (4/7)	1.0 (7/7)	.67 (2/3)	1.0 (10/10)
Specificity	1.0 (20/20)	.95 (19/20)	.96 (23/24)	.88 (15/17)
Accuracy	.89 (4+20)/27	.96 (7+19)/27	.93 (2+23)/27	.93 (10+15)/27

Table 15.4. Results for class A.

Class A	Test result positive (A)	Test result negative (N or U)
negative rules (for N)	3	$5+1=6$
positive rules (for A)	13	$0+2=2$

Sensitivity $= TP/(TP+FN) = 13/(13+2) = .867$
Specificity $= TN/(FP+TN) = 6/(3+6) = .667$
Accuracy $= (TP+TN)/(TP+TN+FP+FN) = (13+6)/(13+2+3+6) = .792$

Table 15.5. Results for class N.

Class N	Test result positive (N)	Test result negative (A or U)
negative rules (for N)	5	$3+1=4$
positive rules (for A)	0	$13+2=15$

Sensitivity $= TP/(TP+FN) = 5/(5+4) = .556$
Specificity $= TN/(FP+TN) = 15/(15+0) = 1.0$
Accuracy $= (TP+TN)/(TP+TN+FP+FN) = (5+15)/(5+4+15+0) = .883$

Table 15.6. Summary results for classes A and N.

Results for class	Sensitivity	Specificity	Accuracy
A	.867	.667	.792
N	.556	1.0	.833
Mean	.712	.834	.813

A are too general (although they have high sensitivity, they classify many class N examples as class A), while the rules for class N are too specific (high specificity, but they fail to classify many class N examples as class N); see again Tables 15.4 and 15.5.

Sometimes two other measures, calculated from the confusion matrix, are used. One is called the false discovery rate:

$$\text{False Discovery Rate(FDR)} = FP/(TP+FP)$$

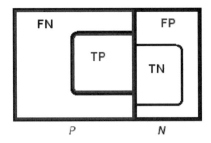

The other one is the **F-measure**, calculated as the harmonic mean of recall/sensitivity and precision:

$$\text{F-measure} = (2 \times \text{Precision} \times \text{Recall})/(\text{Precision} + \text{Recall})$$

Now let us define the *FP rate* as being equal to FP/N similarly (but using "negative" logic) to the *TP rate* (which is equal to TP/P and is better known as sensitivity). Then we can rewrite the specificity as

$$\text{Specificity} = 1 - \text{FP rate}$$

from which we obtain

$$\text{FP rate} = 1 - \text{Specificity}$$

The above definition of specificity leads us to the Receiver Operating Characteristics curves and analyses.

8.1. Receiver Operating Characteristics Analysis

Receiver Operating Characteristics (ROC) analysis is performed by drawing curves in two-dimensional space, with axes defined by the *TP rate* and *FP rate*, or equivalently, by using terms of sensitivity and specificity. That is, the *y*-axis represents $Sensitivity = TP\ rate$, while the *x*-axis represents $1 - Specificity = FP\ rate$, as shown in Figure 15.3.

ROC plots allow for visual comparison of several models (classifiers). For each model, we calculate its sensitivity and specificity, and draw it as a point on the ROC graph. A few comments about the graph shown in Figure 15.3 are necessary. Table 15.1 represents the evaluation of a model/classifier, which when drawn on the ROC graph represents a single point, corresponding to the value (1-specificity, sensitivity), denoted as point *P* in Figure 15.3. The ideal model/classifier would be represented by the location (0,1) on the graph, corresponding to 100% specificity and 100% sensitivity. Points (0,0) and (1,1) represent 100% specificity and 0% sensitivity for the first point, and 0% specificity and 100% sensitivity for the second, respectively. Neither of the latter two points would represent an acceptable model to a data miner. All points lying on the curve connecting the two points ((0,0) and (1,1)) represent random guessing of the classes (equal values of 1-specificity and sensitivity, or in other words, equal values of the TP and FP rates. Stated differently these models/classifiers would recognize equal amounts of TP and FP, with the point *W* at (0.5, 0.5) representing 50% specificity and 50% sensitivity. None of the points on this diagonal curve would be acceptable to a data miner.

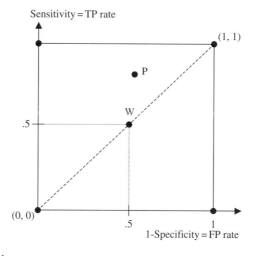

Figure 15.3. An ROC graph.

482   8. Sensitivity, Specificity, and ROC Analyses

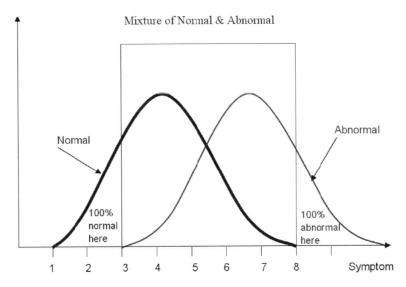

Figure 15.4. Distribution of Normal (healthy) Abnormal (sick) patients.

These observations suggests the strategy for always "operating" in the region above the diagonal line ($y = x$), since in this region the TP rate is higher than the FP rate.

So far we have discussed only single points on the ROC graph. But how can we obtain a curve on the ROC plot corresponding to a classifier, say A? To explain, let us assume that for a (large) class called *Abnormal* (say, a class representing people with some disease), we have drawn a distribution of the examples over some symptom (feature/attribute). And let us assume that we have done the same for a (large) class *Normal* (say, representing people without a disease) over the same symptom. Quite often the two distributions would overlap, as shown in Figure 15.4.

How can we thus distinguish (diagnose) between Normal and Abnormal patients? We do so by choosing a threshold for the symptom. We should keep in mind, however, that regardless of the value we choose, it will always result in having both FP and FN present, as illustrated in Figure 15.5 using an arbitrary threshold of 4.5.

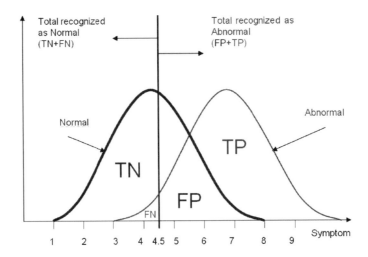

Figure 15.5. Division into Normal and Abnormal patients using a threshold of 4.5.

The use of one threshold corresponds to having a point classifier (and the corresponding confusion matrix), and to drawing just one point in the ROC space (in the above example, point P). By varying the value of the threshold, however, we can obtain several points, which when connected constitute an ROC curve, as shown in Figure 15.6. What we have done is to change the values of one parameter/attribute used in the model/classifier, say, classifier A, to obtain the curve. Some observations about the curve follow. The big question is: Is it possible to choose an "optimal" threshold? If we agree that no matter what we do, it is not possible to distinguish perfectly between Positive and Negative examples (there will always be some FPs), then the answer depends on the user's goal: if the user wants to be very strict, i.e., wants to minimize the number of FPs (knowing that doing so also entails a lower number of TP recognized), then we choose a point on the curve, for classifier A, around the middle but closer to the point (0,0) – for instance, point S in Figure 15.6. If, on the other hand, the goal is to recognize a larger number of TP (knowing that doing so also entails a larger number of FP), then a point on a curve around the middle but closer to the point (1,1) should be chosen – for instance, point D in Figure 15.6. Point D, in fact, corresponds to the situation shown in Figure 15.5 for the threshold of 4.5.

We can perform a similar analysis for classifier B and draw its corresponding ROC curve, also shown in Figure 15.6. Classifier B, in fact, can be similar to classifier A but employ different, or a different number of, symptoms (attributes). Having specified the two curves, the task now is to decide which of the two classifiers constitutes a better model/classifier of the data. We could perform visual analysis: the curve more to the upper left would indicate a better classifier. However, the curves often overlap, as shown in Figure 15.6, and a decision may not be so easy to make. For that purpose the popular method called **Area Under Curve** (AUC) is used. For example, if we look at Figure 15.6, the area under the diagonal curve is 0.5. Thus, we are interested

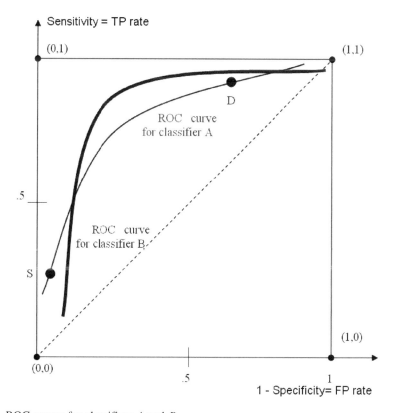

Figure 15.6. ROC curves for classifiers A and B.

in choosing a model/classifier that has maximum area under its corresponding ROC curve: the larger the area, the better performing the model/classifier is. There exists a measure similar to the AUC for assessing the goodness of a model/classifier, known as the **Gini coefficient**, which is defined as twice the area between the diagonal and the ROC curve; the two measures, however, are equivalent, since $\text{Gini} + 1 = 2 \text{ AUC}$.

9. Interestingness Criteria

As we stressed before, it is the owner of data who makes the final decision either to accept or reject the generated data model, depending on its novelty, interestingness, and perceived usefulness.

Similarly to expert systems research carried out in the 1970s, when computer scientists mimicked human decision-making processes by interviewing experts and codifying the rules they use (e.g., how to diagnose a patient) in terms of production rules, data miners undertook a similar effort to formalize the measures (supposedly) used by domain experts/data owners to evaluate models of data.

Such criteria can be roughly divided into those assessing the interestingness of the association rules generated from unsupervised data, and those assessing the interestingness of rules generated by inductive machine learning (i.e., decision trees/production rules) from supervised data. We briefly describe below a few simple measures, starting with the latter.

Let us assume that we have generated a set of production rules or a decision tree. Since we can have hundreds of such rules/branches in a tree, we use a general strategy called **rule refinement** to reduce their number and to focus only on the more important rules (according to some criterion). The strategy is as follows. First, we identify potentially interesting rules, namely, those that satisfy user-specified criteria such as strengths of the rules, their complexity, etc., or that are similar to rules that already satisfy such criteria. Second, we identify a subset of theses rules, called *technically interesting* rules (the name arises from using at this stage more formal methods such as chi-square, AIC, BIC, etc.). In the third step, we remove all but the technically interesting rules. Note that the just-described process of rule refinement is "equivalent" to the selection of the "best" rules carried out by the end user. The processes are similar, since the user ultimately concentrates only on a small number of "best" rules. The difference is that the owner of the data would use heuristic rules, often subconsciously, as a result of his or her deep understanding of a given domain, while in the process described above formal methods are used to achieve the same result. The goal of the process is to present the user with only a few dozen rules at most, possibly selected from hundreds of rules. The user will eventually perform his or her own final selection process and choose maybe a few rules, say, from one dozen presented. In one application (monitoring cystic fibrosis progression), we have generated dozens of strong rules covering at least a specified large number of examples, but the owner of the data selected just one rule, which described a previously unknown relationship in the cystic fibrosis literature. To the user, this was a discovery of a true "knowledge nugget".

Another criterion, probabilistic in nature, is called **interestingness**. Note that several different criteria use this same name, but we describe just one. The criterion is used to assess the interestingness of generated classification rules, one rule at a time. Classification rules, for this purpose, are divided into two types: *discriminant rules* (when evidence, e, implies a hypothesis, h; such rules specify conditions sufficient to distinguish between classes and are the rules most frequently used in practice), and *characteristic rules* (when hypothesis implies evidence; these rules specify the conditions necessary for membership in a class). To assess the interestingness of a characteristic rule, we first calculate sufficiency, S, and necessity, N:

$$S(e \rightarrow h) = \frac{p(e|h)}{p(e|\neg h)}$$

$$N(e \to h) = \frac{p(\neg e|h)}{p(\neg e|\neg h)}$$

where $\to$ stands for implication and the $\neg$ symbol stands for negation. These two measures are used for determining the interestingness of different forms of characteristic rules, namely, $h \to e$, $h \to \neg e$, $\neg h \to e$, and $\neg h \to \neg e$, by using the following relations:

$$IC(h \to e) = \begin{cases} (1 - N(e \to h)) \times p(h), & 0 \le N(e \to h) < 1 \\ 0 & otherwise \end{cases}$$

$$IC(h \to \neg e) = \begin{cases} (1 - S(e \to h)) \times p(h), & 0 \le S(e \to h) < 1 \\ 0 & otherwise \end{cases}$$

$$IC(\neg h \to e) = \begin{cases} (1 - 1/N(e \to h)) \times p(\neg h), & 1 < N(e \to h) < \infty \\ 0 & otherwise \end{cases}$$

and

$$IC(\neg h \to \neg e) = \begin{cases} (1 - 1/S(e \to h)) \times p(\neg h), & 1 < S(e \to h) < \infty \\ 0 & otherwise \end{cases}$$

The calculated values of interestingness are in the range from 0 (min) to 1 (max), and the owner of the data needs to use some threshold (say, .5) to make a decision to retain or remove the rules.

Distance, another criterion, measures the distance between two generated rules at a time in order to find the strongest rules (these with highest coverage). It is defined by:

$$D(R_i, R_j) = \begin{cases} \dfrac{DA(R_i, R_j) + 2DV(R_i, R_j) - 2EV(R_i, R_j)}{N(R_i) + N(R_j)}, & NO(R_i, R_j) = 0 \\ 0 & otherwise \end{cases}$$

where R_i and R_j are rules; $DA(R_i, R_j)$ is the sum of the number of attributes present in R_i but not in R_j, and the sum of attributes present in R_j but not in R_i; $DV(R_i, R_j)$ is the number of attributes in R_i and R_j that have slightly overlapping values (in less than 2/3 of the range); $EV(R_i, R_j)$ is the number of attributes in both rules that have overlapping values (more than 2/3) in the range; $N(R_i)$ and $N(R_j)$ are the number of attributes in each rule; and $NO(R_i, R_j)$ is the number of attributes in R_i and R_j with nonoverlapping values. The criterion calculates values in the range from -1 to 1, indicating strong and slight overlap, respectively. The value of 2 means no overlap. The most interesting rules are those with the highest average distance to the other rules.

10. Summary and Bibliographical Notes

In this Chapter, we explained several **model selection** and **model assessment** methods. We divided these methods into four broad categories: **data reuse, heuristic, formal,** and **interestingness measures**. One should use methods from different categories on the same data. In fact, interestingness measures could and should be used after using methods such as data reuse. The **AIC** and **BIC methods** were discussed for completeness of the presentation and to show their relationship with other methods. As the reader (data miner) has certainly noticed, some of these methods are quite complex for use in large DM projects (although we avoided introducing many other even more computationally expensive methods). So the question arises: Which ones are most frequently used in practice? Certainly the **Occam's razor,** the **resampling** method of **cross-validation,** and **sensitivity** and **specificity** analyses, including **ROC curves,** are the most popular and most often seen in reports on model assessment [2, 3, 6, 7].

Most of the work in the area of model assessment and selection has been done in the areas of statistics and pattern recognition. However, many of the methods, although accurate, are not very practical in data mining undertakings. An excellent treatment of statistical methods can be found in [6, 10, 13, 14, 15]. Some data mining books and reports also briefly discuss the problem of model assessment [1, 5, 8], while interestingness measures are described in [4, 8, 9, 11]. The problem of dealing with missing attribute values is discussed in [12].

References

1. Cios, K.J., Pedrycz, W., and Swiniarski, R. 1998. *Data Mining Methods for Knowledge Discovery*, Kluwer
2. Duda, R.O., Hart, P.E., and Stork, D.G. 2001. *Pattern Classification*, Wiley
3. Frank, I.E., and Todeshini, R. 1994. *The Data Analysis Handbook*, Elsevier
4. Gago, P., and Bentos, C. 1998. A metric for selection of the most promising rules. In Zytkow, J. and Quafafou, M. (Eds.), *Proceedings of the 2nd European Conference on the Principles of Data Mining and Knowledge Discovery*, Nantes, France, September, 19–27
5. Han, J. and Kamber, M. 2001. *Data Mining: Concepts and Techniques*, Morgan Kaufmann
6. Hastie, T., Tibshirani, R., and Friedman, J. 2001. *The Elements of Statistical Learning: Data Mining, Inference and Prediction*. Springer
7. Grunwald, P.D., Myung, I.J., and Pitt, M.A. (Eds.). 2005. *Advances in Minimum Description Length Theory and Applications*, MIT Press
8. Hilderman, R.J., and Hamilton, H.J. 1999. Knowledge discovery and interestingness measures: a survey. *Technical Report CS 99–04*, University of Regina, Saskatchewan, Canada
9. Kamber, M., and Shinghal, R. 1996. Evaluating the interestingness of characteristic rules. In: *Proceedings of the 2nd International Conference on Knowledge Discovery and Data Mining*, Portland, August, 263–266
10. Kecman, V. 2001. *Learning and Soft Computing*, MIT Press
11. Klemettinen, M., Mannila, H., Ronkainen, P., Toivonen, H., and Verkamo, A.I. 1994. Finding interesting rules from large sets of discovered association rules. In: Adam, N.R., Bhargava, B.K., and Yesha, Y. (Eds), *Proceedings of the 3rd International Conference on Information and Knowledge Management*, Gaitersburg, Maryland, 401–407
12. Kurgan, L., Cios, K.J., Sontag, M., and Accurso, F. 2005. Mining the cystic fibrosis data. In: *New Generation of Data Mining Applications*, Kantardzic, M., and Zurada, J. (Eds.), IEEE Press – Wiley, 415–444
13. Moore, G.W., and Hutchins, G.M. 1983. Consistency versus completeness in medical decision-making: exemplar of 155 patients autopsied after coronary artery bypass graft surgery. *Medical Informatics*, London, July-September, 8(3): 197–207
14. Rissanen, J. 1989. *Stochastic Complexity in Statistical Inquiry*, World Scientific
15. Webb, A. 1999. *Statistical Pattern Recognition*, Arnold

11. Exercises

1. Discuss why the quality of generated models must be assessed.
2. Explain the differences between analytical, heuristic, and data reuse/ resampling methods, and interestingness methods.
3. Which methods are applicable for assessing model prediction?
4. Which methods are used for assessing goodness of fit?
5. What methods are used for model selection?
6. After generating a model for your data mining data set, calculate sensitivities and specificities and draw the corresponding ROC curves.

Part 6

Data Security and Privacy Issues

16

Data Security, Privacy and Data Mining

In this Chapter, we introduce the central concepts of security and privacy in data mining. We also relate these concepts to the idea of distributed data mining, where we are concerned with a number of separate databases. The conceptual framework of collaborative clustering and consensus driven clustering is also introduced, along with its main algorithmic aspects and facets and application-oriented realizations and developments. To focus our discussion, we use the framework of fuzzy clustering and Fuzzy C-Means (FCM), in particular since this algorithm is well reported in the literature, comes with a transparent interpretation, and brings with a wealth of case studies and applications. Likewise, it offers a transparent setting in which the ideas of collaborative and distributed clustering can be clearly explained.

1. Privacy in Data Mining

The issue of privacy and security of data emerged at a relatively early stage in the development of data mining. This development is not at all surprising given that all activities of data mining revolve around data and many sensitive issues of accessibility or possible reconstruction of data records exist. To alleviate possible shortcomings along these lines, three main directions have been actively pursued:

- **Data sanitation.** Here the key point is to modify the data so that those data points deemed sensitive cannot be directly data mined. It is anticipated that such modification of data will not significantly impact the main findings given the total volume of data.
- **Data distortion** also known as **data perturbation** or **data randomization**, offers privacy via some modification of individual data. While the distortion affects the values of the individual records, its impact on the discovery and quantification of the main relationships is likely to be still quite negligible.
- **Cryptographic methods**. Here, different techniques from cryptography are considered so that the original data are not revealed during the data mining process. Cryptographic techniques are commonly used in secure multiparty computation in order to allow multiple parties to collaborate in join computing while learning nothing except for the final result of their combined activity. However, we must state that, while attractive on the surface, cryptographic methods come with a high communication and computational overhead, and those costs may be quite prohibitive especially when dealing with large datasets.

2. Privacy Versus Levels of Information Granularity

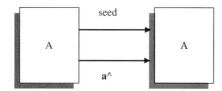

Figure 16.1. Computing the dot product of **a** and **b** by sending a limited number of messages between the sites: message-1: generator seed; k-messages: successive coordinates of $\hat{\mathbf{a}}$.

Interestingly enough, the level of communication required for high-dimensional problems can be radically reduced and still lead to quite acceptable results. An interesting example in this area is **distributed dot product** algorithm. It is well known that computation of the dot product is essential to the determination of the Euclidean distance – a task commonly encountered in numerous classification problems. Let us consider two n-dimensional real vectors, $\mathbf{a} = [a_1 \ a_2 \ldots a_n]^T$ and $\mathbf{b} = [b_1 \ b_2 \ldots b_n]^T$ of high dimensionality, say n, $\dim(\mathbf{a}) = \dim(\mathbf{b}) = n$ that are located at two sites A and B. Note that the Euclidean distance $d(\mathbf{a}, \mathbf{b})$ is directly based on the dot product, namely, $d(\mathbf{a}, \mathbf{b}) = \mathbf{a}^T\mathbf{a} + \mathbf{b}^T\mathbf{b} + \mathbf{a}^T\mathbf{b}$. Our objective is to compute the dot product using a small number of messages being sent between the two sites, see Figure 16.1.

The crux of the method is sending a short k-dimensional ($k << n$) vector instead of the original n-dimensional vectors **a** and **b**. The algorithm of computing $\mathbf{a}^T\mathbf{b}$ works as follows:

- A sends B a seed of the random number generator
- both A and B generate a $k \times$ by n matrix R populated by the entries coming from the random number generator (the generator produces numbers that are generated independently from some fixed distribution with zero mean and finite variance). At the sites computed are the vectors $\hat{a} = R\mathbf{a}$ and $\hat{b} = R\mathbf{b}$
- A sends $\hat{\mathbf{a}}$ to B (k-messages)
- B computes the distance $d(\hat{a}, \hat{b}) = \frac{\hat{a}^T \hat{b}}{k}$.

Depending on the values of k, one could demonstrate experimentally that the accuracy of computing of dot product even for values of k that constitute a low percentage of n is high or quite acceptable. For instance, given the random numbers generated by a uniform distribution $U[-1, 1]$, for k = 10% of the original value of "n" (where $n = 10,000$), the mean error in the calculations of the dot product using the above algorithm is 0.043, and it drops down to 0.026 when the value of k is equal to 30% of the original dimensionality of the vector.

2. Privacy Versus Levels of Information Granularity

While direct access to numeric data is not allowed due to of the privacy constraints, all possible interaction could be realized through some interaction occurring at the higher level of **abstraction** delivered by information granules. In objective function based fuzzy clustering, two important facets of information granulation are conveyed by (a) **partition matrices** and (b) **prototypes**. Partition matrices are, in essence, a collection of fuzzy sets that reflect the nature of the data. They do not reveal detailed numeric information. In this sense, there is no breach of privacy, and partition matrices could be communicated without revealing details about individual data points. Likewise prototypes are reflective of the structure of data and form a summarization of data. Detailed numeric data are hidden behind the prototypes and cannot be reconstructed back to the original form of the individual data points. In both cases, no numeric data are directly made available.

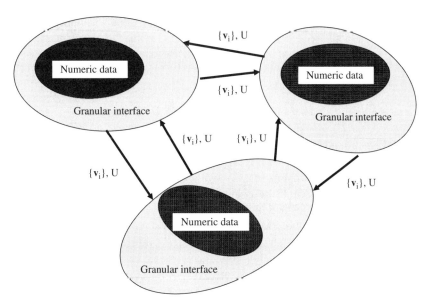

Figure 16.2. A granular interface offering secure communication and formed by the results of the fuzzy clustering (partition matrices U and prototypes v_i): a global view.

The level of **information granularity** is linked with the level of detail and in this sense, when the level of granularity changes occur, a possible leakage of privacy could occur. For instance, in the limit when the number of clusters becomes equal to the number of data points, each prototype is just the data point and no privacy is retained. Obviously, this scenario is quite unrealistic since the structure (the number of clusters) is kept quite condensed when contrasted with all data. The schematic view of privacy offered through information granulation resulting within the process of clustering is illustrated in Figure 16.2. We note here that the granular constructs (either prototypes or partition matrices) build some granular interfaces.

3. Distributed Data Mining

Quite commonly, we encounter situations where databases are distributed rather than centralized. For instance, different outlets of the same company may operate independently and collect data about customers populating their independent databases. These data are not available to others. In banking, each branch may run its own database and such databases could be geographically remote from each other. Individual health institutions could have separate datasets with very limited communication between them. In sensor networks (which have become quite popular given the nature of various initiatives such as intelligent houses, the information highway, etc.), we encounter local databases that operate independently from each other and are inherently distributed. These are also subject to numerous technical constraints (e.g., a fairly limited communication bandwidth, limited power supply, etc.) that significantly reduce the possible interaction between the datasets. Under these circumstances, "standard" data mining activities now face new challenges that need to be addressed. It becomes apparent that processing all data in a centralized manner cannot be realized. On the other hand, the data mining of each of the individual databases could benefit from availability of findings coming from other databases. The technical constraints and privacy issues dictate a certain level of interaction. There are two general modes of interaction, namely **collaborative clustering** and

consensus clustering, both of which aim at the realization of the data mining in the distributed environment. The main difference between the two lies in the level of interaction. Collaborative clustering is positioned on the more active side, where structures are revealed in a more collective manner through some ongoing interaction. Consensus-driven clustering is focused on the reconciliation of the findings when there is no active involvement at the stage of forming clusters.

4. Collaborative Clustering

When dealing with distributed databases, we are often interested in a collaborative style of discovery of relationships that could be common to all the databases. In many scenarios, such collaborative pursuits could be deemed highly beneficial. We could envision a situation where the databases are located in quite remote locations and, given some privacy requirements as well as possible technical constraints, we are not allowed to collect (transfer) all data into a single location and run any centralized algorithm of data mining, say, clustering. On the other hand, at the level of each database, each administrator/analyst involved in data collection, database maintenance, and other activities could easily appreciate the need for some joint activities of data mining. Schematically, we can envision the overall situation as schematically shown in Figure 16.3.

While the collaboration can assume a variety of detailed schemes, two of those are most essential since they are conceptually appealing and practically relevant. We refer to them as the **horizontal** and **vertical** modes of collaboration or in brief horizontal and vertical clustering. More descriptively, given data sets $\mathbf{X}[1], \mathbf{X}[2], .. \mathbf{X}[p]$, where P denotes their number and $\mathbf{X}[ii]$ stands for the ii-th data set (we adhere to the consistent notation of using square brackets to identify a certain data set) in **horizontal clustering** the same objects that are described in *different* feature spaces. For instance, the records for the same collection of patients might be built into each separate medical institution. The schematic illustration of this mode of clustering shown in Figure 16.4 underlines the fact that any possible collaboration can occur at the structural level through the information granules (clusters) built over the data; the clusters are shown in the form of an auxiliary interface layer surrounding the data. The net of directed links shows how the

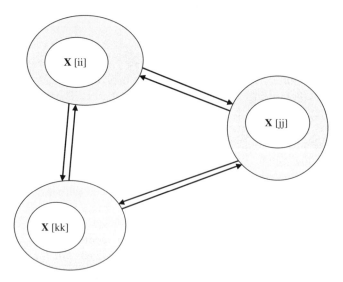

Figure 16.3. A scheme of collaborative clustering involving several datasets interacting at the level of granular interfaces.

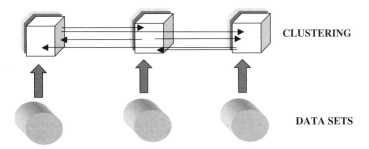

Figure 16.4. A general scheme of horizontal clustering; all communication is realized through some granular interface.

collaboration between different data sets takes place. The width of the links emphasizes the fact that the intensity of collaboration could be different depending upon the data set involved and the intention of the collaborator say, the willingness of some organization to accept findings from external sources).

The mode of **vertical clustering**, Figure 16.5, is complementary (or dual) to the one already presented. Here the data sets are described in the same feature space but deal with *different patterns*.

A number of **hybrid** models of collaboration are available where we encounter data sets with possible links of vertical and horizontal collaboration.

The collaborative clustering exhibits two important features:

- The databases are **distributed** and there is no sharing of their content in terms of the individual records. This restriction is mandated by some privacy and security concerns. Communication between the databases can be realized at higher levels of abstraction which prevent us from any sharing of detailed numeric data.
- Given the existing communication mechanisms, clustering for individual datasets takes into account the structures of other datasets and *actively* engages them in the determination of the clusters; hence the term collaborative clustering

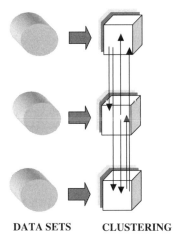

Figure 16.5. A general scheme of vertical clustering; note the "stack" of data sets communicating through some layer of granular communication.

In what follows, we discuss an overall scheme of collaborative clustering by elaborating on some key technical facets that are helpful in better understanding the mechanisms. This discussion will also shed light on the specifics of the overall scheme and its algorithmic underpinnings. To focus our attention, we discuss the algorithmic aspects in the setting of the FCM algorithm (which is helpful given the transparency of the method and the detailed discussion offered in Chapter 9). To be more specific, we focus on the vertical mode of collaboration. Furthermore, let us recall the construct of the proximity matrix that is inherently associated with a partition matrix and will be used in facilitating collaborative clustering when dealing with different numbers of clusters. Given a partition matrix $U = [u_{ik}], i = 1, 2, .., c; k = 1, 2, \ldots, N$, the induced proximity matrix reads as $\text{Prox}(U) = [p_{kl}]k, l = 1, 2, \ldots, N$ where $p_{kl} = \sum_{i=1}^{c} \min(u_{ik}, u_{il})$. An important note: in contrast to partition matrices that depend upon the number of clusters (c), the proximity matrix does not involve the number of clusters and pertains to the data points themselves.

5. The Development of the Horizontal Model of Collaboration

Collaboration during the clustering process deserves careful treatment. We do not know in advance whether the structures emerging (being discovered) at the level of the individual datasets are somewhat compatible and therefore supportive of collaborative activities. It could well be that, in some cases, the inherent structures of datasets are very different, thus preventing any effective collaboration. The fundamental decision is whether to allow some datasets to collaborate or not. This important decision needs to be made up front. One feasible possibility would be to exercise some mechanisms to evaluate consistency of the clusters (structure) at site ii and some other dataset jj. Consider that the fuzzy clustering has been completed separately for each dataset. The resulting structures represented by the prototypes are denoted by $\tilde{v}_1[ii], \tilde{v}_2[ii], \ldots, \tilde{v}_c[ii]$ for the ii-the dataset and $\tilde{v}_1[jj], \tilde{v}_2[jj], \ldots, \tilde{v}_c[jj]$. Consider the ii-th data set. An equivalent representation of its structure comes in the form of the partition matrix. For the ii^{th} dataset, the partition matrix is denoted by $\tilde{U}[ii]$ whose elements are computed on the basis of the prototypes use in the dataset $\mathbf{X}[ii]$.

$$\tilde{u}_{ik}[ii] = \frac{1}{\sum_{j=1}^{c} \left(\frac{\|\mathbf{x}_k - \tilde{v}_i[ii]\|}{\|\mathbf{x}_k - \tilde{v}_j[ii]\|} \right)^{2/(m-1)}} \qquad (1)$$

where $\mathbf{x}_k \in \mathbf{X}[ii]$.

The prototypes of the jj^{th} dataset being available for collaborative purposes when presented to $\mathbf{X}[ii]$ give rise to the partition matrix $\tilde{U}[ii|jj]$ formed for the elements of $\mathbf{X}[ii]$ in the standard manner:

$$\tilde{u}_{ik}[ii|jj] = \frac{1}{\sum_{j=1}^{c} \left(\frac{\|\mathbf{x}_k - \tilde{v}_i[jj]\|}{\|\mathbf{x}_k - \tilde{v}_j[jj]\|} \right)^{2/(m-1)}} \qquad (2)$$

Again, the calculations concern the data points of $\mathbf{X}[ii]$. Refer to Figure 16.6 for the pertinent computing details.

Given the partition matrices $\tilde{U}[ii]$ and $\tilde{U}[ii|jj]$ (**induced partition matrices**) we can check whether they are "compatible" i.e., whether collaboration between the two datasets could be meaningful. We can test whether the histograms of the membership grades of $\tilde{U}[ii]$ and $\tilde{U}[ii|jj]$ are statistically different (that is there is a statistically significant difference). This could be done using e.g., a standard nonparametric test such as χ^2. If the hypothesis of significant statistical difference

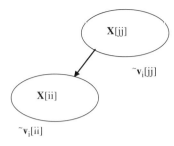

Figure 16.6. Statistical verification of the possibility of collaboration between datasets ii and jj.

between the partition matrices (that is the corresponding structures) is not rejected, then we consider that the ii-th data set can collaborate with the jj^{th} one. Noticeably, the relationship is not reciprocal, so the issue of collaboration between the jj^{th} dataset with the ii^{th} needs to be investigated separately.

5.1. The Augmented Objective Function

The "standard" objective function, minimized at the level of the ii^{th} dataset comes in the well-known form of the double sum, $\sum_{i=1}^{c[ii]} \sum_{k=1}^{M[ii]} u_{ik}^m[ii] \|\mathbf{x}_k - \mathbf{v}[ii]\|^2$. Given that we admit collaboration with the jj^{th} dataset, in the search for structure we take advantage of the knowledge of prototypes representing the jj^{th} dataset and attempt to position the prototypes $\mathbf{v}_1[ii], \mathbf{v}_2[ii], \ldots, \mathbf{v}_c[ii]$ closer to the corresponding prototypes $\mathbf{v}_1[jj], \mathbf{v}_2[jj], \ldots, \mathbf{v}_c[jj]$. This request is reflected in the form of the **augmented objective function** to come in the following format:

$$Q[ii] = \sum_{k=1}^{N[ii]} \sum_{i=1}^{c} u_{ik}^2[ii] d_{ik}^2[ii] + \sum_{\substack{jj=1 \\ jj \neq ii}}^{P} \beta[ii,jj] \sum_{i=1}^{c} \sum_{k=1}^{M[ii]} u_{ik}^2[ii] \|\mathbf{v}_i[ii] - \mathbf{v}_i[jj]\|^2 \qquad (3)$$

The first component is the same as the one guiding the clustering for dataset $\mathbf{X}[ii]$ while the second part reflects the guidance coming from all other datasets that we have identified as potential collaborators (which wa done using the χ^2 test described before previous section). The level of collaboration (which is asymmetric) is guided by the value collaboration coefficient. Its value is chosen based on potential benefits of collaboration as will be discussed in more detail in the next section. More specifically, $\beta[ii,jj]$ is a collaboration coefficient supporting an impact coming from the jj^{th} data set and affecting the structure to be determined in the ii^{th} data set. The number of patterns in the ii^{th} data set is denoted by $N[ii]$. We use a different letter to distinguish between the horizontal and vertical collaboration. The interpretation of (3) is quite obvious: the first term is the objective function directed towards the search for structure in the ii^{th} data set while the second articulates the differences between the prototypes (weighted by the partition matrix of the ii^{th} data set), which have to be made smaller by refining of the partition matrix (or effectively by moving the prototypes in the feature space).

The optimization of $Q[ii]$ involves the determination of the partition matrix $U[ii]$ and the prototypes $\mathbf{v}_i[ii]$. As before, we solve the problem for each data set separately and allow the results interact so that collaboration takes place between the sets. The minimization of the objective function with respect to the partition matrix requires the use of the technique of Lagrange multipliers due to the existence of the standard constraints imposed on the partition matrix. We form an augmented objective function V incorporating the Lagrange multiplier λ and deal with each individual pattern ($t = 1, 2, \ldots, N[ii]$)

$$V = \sum_{i=1}^{c} u_{it}^2[ii] d_{it}^2[ii] + \sum_{\substack{jj=1 \\ jj \neq ii}}^{P} \beta[ii,jj] \sum_{i=1}^{c} u_{it}^2[ii] \|\mathbf{v}_i[ii] - \mathbf{v}_i[jj]\|^2 - \lambda \left(\sum_{i=1}^{c} u_{it} - 1 \right) \qquad (4)$$

5. The Development of the Horizontal Model of Collaboration

Taking the derivative of V with respect to $u_{st}[ii]$ and making it zero, we have

$$\frac{\partial V}{\partial u_{st}} = 2u_{st}[ii]d_{st}^2[ii] + 2\sum_{\substack{jj=1\\jj\neq ii}}^{P}\beta[ii,jj]u_{st}[ii]\|\mathbf{v}_i[ii]-\mathbf{v}_i[jj]\|^2 - \lambda = 0 \tag{5}$$

For notational convenience, let us introduce the shorthand expression

$$D_{ii,jj} = \|\mathbf{v}_i[ii]-\mathbf{v}_i[jj]\|^2 \tag{6}$$

From equation (5) we derive

$$u_{st}[ii] = \frac{\lambda}{2\left(d_{st}^2[ii] + \sum_{\substack{jj=1\\jj\neq ii}}^{P}\beta[ii,jj]D_{ii,jj}\right)} \tag{7}$$

By virtue of the standard normalization condition $\sum_{j=1}^{c} u_{jt}[ii] = 1$, we have

$$\frac{\lambda}{2} = \frac{1}{\sum_{j=1}^{c}\dfrac{1}{d_{jt}^2[ii] + \sum_{\substack{jj=1\\jj\neq ii}}^{P}\beta[ii,jj]D_{ii,jj}}} \tag{8}$$

With the abbreviated notation

$$\varphi[ii] = \sum_{\substack{jj\neq ii}}^{P}\beta[ii,jj]D_{ii,jj} \tag{9}$$

the partition matrix comes in the form

$$u_{st}[ii] = \frac{1}{\sum_{j=1}^{c}\dfrac{d_{st}^2[ii]+\varphi[ii]}{d_{jt}^2[ii]+\varphi[ii]}} \tag{10}$$

For the prototypes, we complete the calculation of the gradient of Q with respect to the coordinates of the prototype $\mathbf{v}[ii]$ and the solve the following system of equations:

$$\frac{\partial Q[ii]}{\partial \mathbf{v}_{st}[ii]} = 0, s = 1,2,..,c; t = 1,2,..n \tag{11}$$

Next we obtain

$$\frac{\partial Q[ii]}{\partial v_{st}[ii]} = 2\sum_{k=1}^{N}u_{sk}^2[ii](x_{kt}-v_{st}[ii]) + 2\sum_{jj\neq ii}^{P}\beta[ii,jj]\sum_{k=1}^{N}u_{sk}^2[ii](v_{st}[ii]-v_{st}[jj]) = 0 \tag{12}$$

In the sequel,

$$v_{st}[ii]\left(\sum_{jj\neq ii}^{P}\beta[ii,jj]\sum_{k=1}^{M[ii]}u_{sk}^2[ii] - \sum_{k=1}^{M[ii]}u_{sk}^2[ii]\right) = \sum_{jj\neq ii}^{P}\beta[ii,jj]\sum_{k=1}^{M[ii]}u_{sk}^2[ii]v_{st}[jj] - \sum_{k=1}^{M[ii]}u_{sk}^2[ii]x_{kt} \tag{13}$$

Finally, we get

$$v_{st}[ii] = \frac{\sum_{\substack{jj \neq ii}}^{P} \beta[ii,jj] \sum_{k=1}^{M[ii]} u_{sk}^2[ii] v_{st}[jj] - 2 \sum_{k=1}^{M[ii]} u_{sk}^2[ii] x_{kt}}{\sum_{\substack{jj \neq ii}}^{P} \beta[ii,jj] \sum_{k=1}^{M[ii]} u_{sk}^2[ii] - \sum_{k=1}^{M[ii]} u_{sk}^2[ii])} \quad (14)$$

An interesting application of vertical clustering occurs when we deal with huge data sets. Instead of clustering them in a single pass, we split them into individual data sets, cluster these separately and then actively reconcile the results through the collaborative exchange of prototypes.

5.2. The Assessment of the Strength of Collaboration

The choice of a suitable **level of collaboration** realized between the datasets through clustering as denoted in (3) by $\beta[ii,jj]$, deserves attention. High values of the collaboration coefficient may lead to some instability of collaboration. Values of this coefficient that are too low may produce a very limited effect of collaboration that could eventually become almost nonexistent. Generally speaking, the values of the collaboration coefficient could be asymmetric that is, $\beta[ii,jj] \neq \beta[jj,ii]$. This outcome is not surprising: we might have a case in which at the level of dataset ii, we are eager to collaborate and quite seriously accept the findings resulting from discovery in dataset jj while the opposite might not be true. Since the values of the collaboration coefficients could be different for any pair of datasets, the optimization of their values could be quite demanding and computationally intensive. To alleviate these shortcomings, let us express the coefficient $\beta[ii,jj]$ as the product $\beta[ii,jj] = \omega f(ii,jj)$ i.e., we view it as a function of the specific datasets under collaboration as calibrated by some constant $\omega(>0)$ whose value does not depend upon the indexes of the datasets. Function $f(ii,jj)$ can be chosen in several ways. In general, we can envision the following intuitive requirement: if the structure revealed at the site of the jj^{th} data set is quite different from the one present at the ii^{th} data set, the level of collaboration could be set up quite low. If there is a high level of agreement between the structure revealed at the jj^{th} set with what has been found so far at the ii^{th} dataset, then $f(ii,jj)$ should assume high values. Given these guidelines, we propose the following form of $f(ii,jj)$

$$f(ii,jj) = 1 - \frac{Q[ii|jj]}{Q[ii] + Q[ii|jj]} \quad (15)$$

Here $Q[ii]$ denotes a value of the objective function obtained for clustering without any collaboration (viz. the partition matrix and the prototypes are formed on the basis of optimization realized for X[ii] only). $Q[ii|jj]$ denotes the value of the objective function computed for the prototypes obtained for **X**[jj] (without any collaboration) and used for data in **X**[ii]; see Figure 16.7.

In essence, the values of $Q[ii]$ and $Q[ii|jj]$ reflect the compatibility of the structures in the corresponding data sets and in this manner tell us about a possible level of successful collaboration. The prototypes obtained for the data set jj which are used to determine the value of the objective function for the ii^{th} data set, could lead to quite comparable values of the objective function if the structure in X[jj] resembles the structure in X[ii]. In this case we envision $Q[ii] < Q[ii|jj]$ yet $Q[ii] \approx Q[ii|jj]$. On the other hand, if the structure in X[jj] is very different that is $Q[ii|jj] >> Q[ii]$, the collaborative impact from what has been established for X[jj] could not be very advantageous. If $Q[ii|jj]$ is close to $Q[ii]$, $f(ii,jj)$ approaches 1/2. When $Q[ii|jj] >> Q[ii]$, the values of $f(ii,jj)$ are close to zero.

Following the process described above, we are left now with a single coefficient (ω) controlling all collaborative activities for all datasets. This outcome is far more practical yet the value of ω needs to be properly selected. Here, several alternatives are possible:

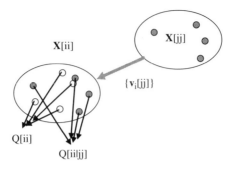

Figure 16.7. Computing the values of $Q[ii|jj]$ realized on the basis of the prototypes computed for $\mathbf{X}[jj]$.

(a) one could monitor the values of the overall objective function (3) during the course of optimization (minimization) The plot of the minimized objective function; oscillations and a lack of convergence. This might suggest that the values of ω are too high (a tight collaboration) and need to be reduced to assure a smooth interaction between the datasets.

(b) we could also look at the differences between the results obtained without collaboration and with collaboration. For instance, the difference between the proximity matrices formed on the basis of the partition matrices constructed for the same data set $X[ii]$ without collaboration and with collaboration, could serve as an indicator of the differences between the results. Such differences could be constrained by allowing for only some limited changes caused by the collaboration.

6. Dealing with Different Levels of Granularity in the Collaboration Process

So far, we have made a strong assumption that the same number of clusters is formed at each individual dataset. This conjecture could well be valid in many cases (since we consider collaboration at the same level of information granularity) but It could also be quite inappropriate in some other cases. To cope with this problem, we need to move our optimization activity to a higher conceptual plane by comparing clustering results at the level of the proximity matrices. When operating at this level of abstraction, we do not need to make any assumptions about a uniform level of granularity occurring across all constructs.

6.1. Consensus-based Fuzzy Clustering

In contrast to collaborative clustering, in which there is an ongoing active involvement of all data sets and the clustering algorithms running on individual datasets are impacted by results from other sites, **consensus-based clustering** focuses mainly on the reconciliation of the individually developed structures. In this sense, building consensus involves the formation of structure based on on the basis of the individual results of clustering developed separately (without any interaction) at the time the clustering algorithm is run. In this section, we are concerned with a collection of clustering methods being run on the same data set. Hence $U[ii]$, $U[jj]$ stand here for the partition matrices produced by the corresponding clustering method. The essential step is to determine some correspondence between the prototypes (partition matrices) formed by each clustering method. Since there has been no prototype interaction during cluster building, no linkages exist between prototypes once the clustering has been completed. The determination of this correspondence is an NP-complete problem and this limits the feasibility of finding an optimal solution. One

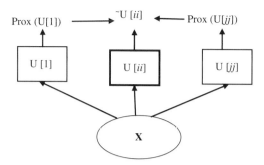

Figure 16.8. A development of consensus-based clustering; the consensus building is focused on the partition matrix generated by the ii-th clustering method, U[*ii*].

way of alleviating this problem is to develop consensus at the level of partition matrices. The use of proximity matrices helps us eliminate the need to identify correspondence between the clusters and handle the cases where different numbers of clusters are used when running a specific clustering method.

Consensus formation is accomplished in the following manner. Given the partition matrices U[*ii*], U[*jj*], etc. which are being developed individually, let us consider consensus building with a focus on U[*ii*]. Given the information about structure in the form of U[*ii*] and other partition matrices U[*jj*], $jj \neq ii$, the implied consensus-driven partition matrix ~U[*ii*] results from a sound agreement with the original partition matrix U[*ii*]. In other words, we would ~U[*ii*] to be as close as possible to U[*ii*]. The minimization of the distance $\|U[ii] - \sim U[ii]\|^2$ could be a viable optimization alternative. Other sources of structural information exist (see Figure 16.8) but here we cannot establish a direct relationship between U[*ii*] (and ~U[*ii*]) and U[*jj*] for the reasons outlined earlier. These difficulties could be alleviated by considering the corresponding induced proximity matrices, say Prox(U[*jj*]). The way in which the **proximity matrix** has been formed relieves us from establishing a correspondence between the rows of the partition matrices (fuzzy clusters) and the number of clusters. In this sense, we may compare Prox (~U[*ii*]) and Prox (U[*jj*]) and search for consensus by minimizing the distance $\|\text{Prox}(\sim U[ii]) - \text{Prox}(U[jj])\|^2$. Considering all sources of structural information, consensus building can the be translated into the minimization of the following problem:

$$\|U[ii] - \sim U[ii]\|^2 + \gamma \sum_{jj \neq ii}^{P} \|\text{Prox}(U[jj]) - \text{Prox}(\sim U[ii])\|^2 \qquad (16)$$

The two components of this expression reflect the two essential sources of information about the structure. The positive weight (scaling) factor (γ) aims at striking a sound compromise within the partition matrix U[ii] associated with the ii^{th} dataset. The result of the consensus reached for the ii^{th} method is the fuzzy partition matrix ~U[*ii*], which minimizes the above performance index (16).

7. Summary and Biographical Notes

We have raised several issues of **data privacy** and **security** in the setting of data mining and have identified several ways of addressing them, including **data sanitation**, **data distortion**, and **cryptographic methods**. In particular, we focused on the role of **information granularity** as a vehicle to carry out collaborative activities (say, clustering) while not releasing detailed numeric data.

Privacy issues in data mining have been raised and carefully discussed in [1, 2, 3, 4, 9, 10, 11, 16, 19, 21, 22]. Various classification and clustering algorithms operating under the constraints of privacy have been presented in [7, 8, 17, 20]. **Distributed clustering** was studied in [12, 17, 18]. **Knowledge-based clustering** along with communication mechanisms realized at the level of information granules and the impact on security was presented in [14, 15].

References

1. Agarwal, R., and Srikant, R. 2000. Privacy-preserving data mining. In *Proceedings. of the ACM SIGMOD Conference on Management of Data*, ACM Press, 439–450
2. Claerhout, B., and DeMoor, G.J.E. 2005. Privacy protection for clinical and genomic data: The use of privacy-enhancing techniques in medicine. *International Journal of Medical Informatics*, 74(2–4): 257–265
3. Clifton, C. 2000. Using sample size to limit exposure to data mining. *Journal of Computer Security* 8(4): 281–307
4. Clifton, C., and Marks, D. 1996. Security and privacy implications of data mining. In *Workshop on Data Mining and Knowledge Discovery*, 15–19, Montreal, Canada
5. Clifton, C., and Thuraisingham, B. 2001. Emerging standards for data mining. *Computer Standards & Interfaces*, 23(3): 187–193
6. Da Silva, J.C., Giannella, C., Bhargava, R., Kargupta, H., and Klusch, M. 2005. Distributed data mining and agents, *Engineering Applications of Artificial Intelligence*, 18(7): 791–807
7. Du, W., and Zhan, Z. 2002. Building decision tree classifier on private data. In Clifton, C., Estivill-Castro, V. (Eds.), *IEEE ICDM Workshop on Privacy, Security and Data Mining, Conferences in Research and Practice in Information Technology*, 14, 1–8, Maebashi City, Japan, ACS
8. Evfimievski, A., Srikant, R., Agrawal, R., and Gehrke, J. 2004. Privacy preserving mining of association rules, *Information Systems*, 29(4): 343–364
9. Johnsten, T., and Raghavan, V.V. 2002. A methodology for hiding knowledge in databases. In Clifton, C., Estivill-Castro, C. (Eds.), IEEE ICDM Workshop on Privacy, Security and Data Mining, Conferences in Research and Practice in Information Technology, 14, 9–17, Maebashi City, Japan, ACS
10. Kargupta, H., Kun, L., Datta, S., Ryan, J., and Sivakumar, K. 2003. Homeland security and privacy sensitive data mining from multi-party distributed resources, *Proceedings 12th IEEE International Conference on Fuzzy Systems, FUZZ '03*, 2: 1257–1260
11. Lindell, Y., and Pinkas, B. 2000. Privacy preserving data mining. In *Lecture Notes in Computer Science*, 1880: 36–54
12. Merugu, S., and Ghosh, J. 2005. A privacy-sensitive approach to distributed clustering, *Pattern Recognition Letters*, 26(4): 399–410
13. Park, B., and Kargupta, H. 2003. Distributed data mining: algorithms, systems, and applications. In Ye, N. (Ed.), *The Handbook of Data Mining*, Lawrence Erlbaum Associates, 341–358
14. Pedrycz, W. 2005. *Knowledge-Based Clustering: From Data to Information Granules*, John Wiley, Hoboken, NJ
15. Pedrycz, W. 2002. Collaborative fuzzy clustering, *Pattern Recognition Letters*, 23(14): 1675–1686
16. Pinkas, B. 2002. Cryptographic techniques for privacy-preserving data mining. *ACM SIGKDD Explorations Newsletter*, 4(2): 12–19
17. Strehl, A., and Ghosh, J. 2002. Cluster ensembles—a knowledge reuse framework for combining multiple partitions. *Journal of Machine Learning Research*, 3: 583–617
18. Tsoumakas, G., Angelis, L., and Vlahavas, I. 2004. Clustering classifiers for knowledge discovery from physically distributed databases, *Data & Knowledge Engineering*, 49(3): 223–242
19. Verykios, V.S., Bertino, E., Fovino, I.N., Provenza, L.P., Saygin, Y., and Theodoridis, Y. 2004. State-of-the-art in privacy preserving data mining. *SIGMOD Record*, 33(1): 50–57
20. Wang, S.L., and Jafari, A. 2005. Using unknowns for hiding sensitive predictive association rules. Proceedings *2005 IEEE International Conference on Information Reuse and Integration*, 223–228
21. Wang, E.T., Lee, G., and Lin, Y.T. 2005. A novel method for protecting sensitive knowledge in association rules mining, *Proceedings of the 29th Annual International Computer Software and Applications Conference (COMPSAC 2005)*, 2: 511–516

22. Wang, K., Yu, P.S., and Chakraborty, S. 2004. Bottom-up generalization: a data mining solution to privacy protection, *Proceedings of the 4th IEEE International Conference on Data Mining, ICDM 2004*, 249–256

8. Exercises

1. Generate a synthetic dataset and run an FCM method on it. Choose a certain number of clusters and a value of the fuzzification coefficient. Treat the resulting partition matrix as a reference (the reference partition matrix). Run the FCM for some other values of the fuzzification coefficient and the number of clusters, obtain the corresponding partition matrix and compare it with the reference partition matrix using the χ^2 test. Assume a certain confidence level. Discuss the results. Elaborate on the impact of the differences between the results (partition matrices) depending upon the number of clusters and the values of the fuzzification coefficient.
2. Rules could be generated in a distributed fashion on the basis of locally available datasets; see the Figure below.

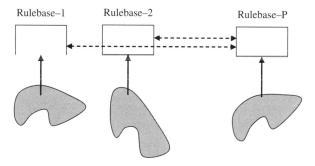

How could you communicate findings as to the local results (rules), and what would be the conditions under which communication and possible collaboration could be possible?
3. Information granules $A_1, A_2, \ldots, A_P$ coming from different sources are sought to be semantically equivalent (viz. conveying similar descriptions of the same concept). How could you characterize these findings in a more synthetic way? Discuss cases when A_i are represented as (a) sets, and (b) fuzzy sets.
4. To avoid disclosure of sensitive data, we could communicate those data as some information granules, say intervals or hypercubes instead of real numbers. This approach could have implications on ensuing processing such as, e.g., classification. As a specific case, consider a nearest neighbor classification rule with the three prototypes representing categories ω_1, ω_2, and ω_3, as illustrated below. Consider the coordinates of the prototypes to be (1,1) (2.5, 3.1), and (1.7, 4.5). What happens if the granularity of the data point (1.5, 3.0) decreases? Refer again to the Figure. The hyperbox built around the data is uniformly expanded along two coordinates. Discuss the relationship between the level of granularity and the class assignment.

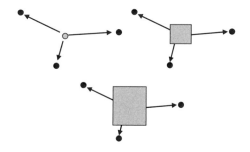

Part 7

Overview of Key Mathematical Concepts

Appendix A
Linear Algebra

In this appendix we provide a review of basic operations on vectors, matrices and linear transformations.

1. Vectors

We begin by defining a mathematical abstraction known as a **vector space**. In linear algebra the fundamental concepts relate to the n-**tuples** and their algebraic properties.

Definition: An ordered n-**tuple** is considered as a sequence of n **terms** $(a_1, a_2, \cdots, a_n)$, where n is a positive integer.

We see that an ordered n-**tuple** has **terms** whereas a set has members.

Example: A sequence (5) is called an ordered 1-**tuple**. A 2-**tuple**, for example (3, 6) (where 6 appears after 3) is called an ordered pair, and 3-**tuple** is called an ordered triple. A sequence (9, 3, 4, 4, 1) is called an ordered 5-**tuple**.

Let us denote the set of all ordered 1-**tuples** of real numbers by $\mathbb{R}$. We will write for example $(3.5) \in \mathbb{R}$.

Definition: The Euclidean n-**space** $\mathbb{R}^n$ (or $\mathbb{R}$) is the set of all ordered n-**tuples** of real numbers. So, $\alpha \in \mathbb{R}^n$ means that α is an ordered n-**tuple** of real numbers. A n-**tuple** is also called **vector** or n-**vector**.

Example: The $(2.5, 6.9, 12.4) \in \mathbb{R}^3$ is 3-**tuple** in 3-dimensional Euclidean space.

Let us consider an Euclidean **space** whose elements are **real** numbers. Loosely speaking, a **vector space** is a set whose elements can be added to one another and whose elements can be multiplied by **real** numbers.

Definition: A linear n-dimensional Euclidean **vector space** over the **real** numbers denoted $\mathbb{R}^n$, is a set of elements (n-**tuples, vectors**) for which the operations of vector **addition** (+) and scalar **multiplication** ($\cdot$) of vectors has been defined such that:

1. $\mathbf{x} + \mathbf{y} \in \mathbb{R}^n$
2. $\alpha \cdot \mathbf{x} \in \mathbb{R}^n$
3. $\mathbf{x} + \mathbf{y} = \mathbf{y} + \mathbf{x} \in \mathbb{R}^n$ (*commutativity*)
4. $\alpha \cdot (\mathbf{x} + \mathbf{y}) = \alpha \cdot \mathbf{x} + \alpha \cdot \mathbf{y}$ (*distributivity*)
5. $(\alpha + \beta) \cdot \mathbf{x} = \alpha \cdot \mathbf{x} + \beta \cdot \mathbf{x}$ (*distributivity*)
6. $(\mathbf{x} + \mathbf{y}) + \mathbf{z} = \mathbf{x} + (\mathbf{y} + \mathbf{z})$ (*associativity*)
7. $(\alpha \beta) \cdot \mathbf{x} = \alpha \cdot (\beta \cdot \mathbf{x})$ (*associativity*)

8. There exists a unique zero vector **0** such that for every $\mathbf{x} \in \mathbb{R}^n$, $\mathbf{x} + \mathbf{0} = \mathbf{0} + \mathbf{x} = \mathbf{x}$
9. For scalars 0 and 1, $0 \cdot \mathbf{x} = \mathbf{0}$ (a vector), and $1 \cdot \mathbf{x} = \mathbf{x}$ (a vector)
10. For each $\mathbf{x} \in \mathbb{R}^n$, there exists a unique element $-\mathbf{x} \in \mathbb{R}^n$ such that $\mathbf{x} + (-\mathbf{x}) = (-\mathbf{x}) + \mathbf{x} = \mathbf{0}$ (a vector)

1.1. Vectors and Vector Operations

As we have mentioned, **a vector** is an *n*-tuple. However, the following notation of a vector is widely used.

A **vector** (more precisely a **column vector**) is an array of elements arranged in a column and denoted by a lower case letter. The **vector's dimension** is given by the number of its elements. For example an *n*-dimensional vector over the **real** numbers, denoted $\mathbf{x} \in \mathbb{R}^n$, is:

$$\mathbf{x} = \begin{bmatrix} x_1 \\ x_2 \\ \vdots \\ x_n \end{bmatrix}, \quad\quad\quad (A.1)$$

where each x_i is a real number.

A vector space can be in the form of *n*-**space**, with its vectors being in the form of *n*-**tuples**.

Example: The vector

$$\mathbf{x} = \begin{bmatrix} 1 \\ 3 \\ 4 \end{bmatrix}$$

represents a vector in three-dimensional Euclidean space $\mathbb{R}^3$, starting at the origin with coordinates $(0, 0, 0)$ and ending at point with coordinates $(1, 3, 4)$.

1.1.1. Geometrical Notation of a Vector

Each vector may be represented in terms of basic components (vectors) parallel to the axes of a Cartesian coordinate system. The basic unit length vectors **i**, **j** starting at the origin **0** and parallel to the axes are called **basis vectors**. For the 2-dimensional (x, y) coordinate system the basis vector $\mathbf{i} = \begin{bmatrix} 1 \\ 0 \end{bmatrix}$ is a vector from the origin $(0, 0)$ to the point $(1, 0)$ on the *x*-axis. The basis vector $\mathbf{j} = \begin{bmatrix} 0 \\ 1 \end{bmatrix}$ is a vector from the origin $(0, 0)$ to the point $(0, 1)$ on the *y*-axis.

The vector **v** from the point P_1 to P_2 in 2D space is described as:

$$\mathbf{v} = a\,\mathbf{i} + b\,\mathbf{j} \quad\quad\quad (A.2)$$

The term $a\,\mathbf{i}$ denotes a vector parallel to *x*-axis and being the **component of a vector v** in **i** direction (parallel to the *x*-axis). The scalar a is the length of this component vector. The component of vector in **i** direction if a is positive is directed to the right, and to the left if a is negative.

The term $b\,\mathbf{j}$ denotes a vector parallel to *y*-axis and being the **component of a vector v** in **j** direction (parallel to the *y*-axis). The scalar b is the length of this component vector. The component of vector in **j** direction if b is positive is directed similarly as *y*-axis, and to opposite to the direction of *y*-axis if a is negative.

The length of vector $\mathbf{v} = a\,\mathbf{i} + b\,\mathbf{j}$ is equal to $||\mathbf{v}|| = \sqrt{a^2 + b^2}$.

Similarly for 3-dimensional space, we have three basic vectors **i**, **j** and **k**, where **k** is unit vector parallel to z-axis, from the origin 0, 0, 0) to the point (0, 0. 0) on z-axis. The basic vectors **i** and **j** are basic unit vectors along the axes x and y.

The **position vector r** from the origin **0** = (0, 0. 0) to a point in space $P(x, y, z)$ is

$$\mathbf{r} = x\,\mathbf{i} + y\,\mathbf{j} + c\,\mathbf{k} \qquad (A.3)$$

The vector $\mathbf{r}_{P_1,P_2}$ from a point $P_1(x_1, y_1, z_1)$ to a point $P_2(x_2, y_2, z_2)$ is described as

$$\mathbf{r}_{P_1,P_2} = (x_2 = x_1)\mathbf{i} + (y_2 - y_1)\mathbf{j} + (z_2 - z_1)\mathbf{k} \qquad (A.4)$$

Two vectors in $\mathbb{R}^n$ **a** and **b** are equal if $a_1 = b_1, a_2 = b_2, \cdots, a_n = b_n$.

1.1.2. Vector Addition
Addition of vectors is defined:

$$\mathbf{x} + \mathbf{y} = \begin{bmatrix} x_1 \\ x_2 \\ \vdots \\ x_n \end{bmatrix} + \begin{bmatrix} y_1 \\ y_2 \\ \vdots \\ y_n \end{bmatrix} = \begin{bmatrix} x_1 + y_1 \\ x_2 + y_2 \\ \vdots \\ x_n + y_n \end{bmatrix} \qquad (A.5)$$

Example:

$$\mathbf{x} + \mathbf{y} = \begin{bmatrix} 2 \\ 6 \\ -5 \end{bmatrix} + \begin{bmatrix} 0 \\ 3 \\ 4 \end{bmatrix} = \begin{bmatrix} 2 \\ 9 \\ -1 \end{bmatrix}$$

The subtraction operation of two vectors is defined similarly. Because a **vector space** is associative under addition, more than two vector may be added (or subtracted) together.

1.1.3. Vector Multiplication by Scalar
A scalar multiplication of a vector is defined:

$$\alpha \cdot \mathbf{x} = \alpha \cdot \begin{bmatrix} x_1 \\ x_2 \\ \vdots \\ x_n \end{bmatrix} = \begin{bmatrix} \alpha \cdot x_1 \\ \alpha \cdot x_2 \\ \vdots \\ \alpha \cdot x_n \end{bmatrix}, \qquad (A.6)$$

where α is a **real** scalar.

Example:

$$\alpha \cdot \mathbf{x} = 2 \cdot \begin{bmatrix} 4 \\ 5 \\ 0 \end{bmatrix} = \begin{bmatrix} 8 \\ 10 \\ 0 \end{bmatrix}$$

The division operation of a scalar and a vector is defined similarly. More than two vectors may be multiplied (or divided) together.

The operations of vector addition and vector subtraction are defined for vectors **of the same dimension only** and can **not** be performed upon vectors of differing dimensions.

1. Vectors

1.1.4. Zero Vector
The **zero** vector **sometimes denoted 0** is a vector having all elements equal to zero, e.g., the 2-dimensional **0** vector:

$$\mathbf{x} = \begin{bmatrix} 0 \\ 0 \end{bmatrix} \tag{A.7}$$

1.1.5. Unit Vector
Any vector with unit length (along the coordinate axes) is called a **unit vector**. The basic vectors **i, j,** and **k** are unit vectors. For example $||\mathbf{i}|| = ||1 \cdot \mathbf{i} + 0 \cdot \mathbf{j} + 0 \cdot \mathbf{k}|| = \sqrt{1^2 + 0^2 + 0^2}$.

1.1.6. Direction
For nonzero vector **v** the **direction** is a unit vector obtain by dividing a vector **v** by its length

$$direction\ of\ \mathbf{v} = \frac{\mathbf{v}}{||\mathbf{v}||} \tag{A.8}$$

1.1.7. Vector Transpose
The transpose of a column vector **x** yields a row vector $\mathbf{x}^T$

$$\mathbf{x}^T = \begin{bmatrix} x_1 \\ x_2 \\ \vdots \\ x_n \end{bmatrix}^T = \begin{bmatrix} x_1\ x_2\ \cdots\ x_n \end{bmatrix} \tag{A.9}$$

thus T denotes the **transposition operation**.

Example:

$$\mathbf{x}^T = \begin{bmatrix} 4 \\ 8 \\ 1 \end{bmatrix}^T = \begin{bmatrix} 4\ 8\ 1 \end{bmatrix}$$

In this book the word **vector** is taken to mean a **column vector**. **Row vectors** are always denoted by the transpose of a column vector.

1.1.8. Vector Average
An average of r vectors $\mathbf{x}_i (i = 1, 2, \cdots, r)$ **of the same dimension** n may be defined as:

$$\mathbf{z} = \frac{1}{r}(\mathbf{x}_1 + \mathbf{x}_2 + \cdots + \mathbf{x}_r)$$

$$= \frac{1}{r} \cdot \left(\begin{bmatrix} x_{1_1} \\ x_{1_2} \\ \vdots \\ x_{1_n} \end{bmatrix} + \begin{bmatrix} x_{2_1} \\ x_{2_2} \\ \vdots \\ x_{2_n} \end{bmatrix} + \cdots + \begin{bmatrix} x_{r_1} \\ x_{r_2} \\ \vdots \\ x_{r_n} \end{bmatrix} \right) \tag{A.10}$$

1.1.9. Inner Product

The **inner** or **dot** product of two vectors **x** and **y** of the same dimension is a **scalar** defined by:

$$\mathbf{x}^T \cdot \mathbf{y} = (\mathbf{x}, \mathbf{y}) = x_1 y_1 + x_2 y_2 + \cdots + x_n y_n = \sum_{i=1}^{n} x_i y_i \qquad (A.11)$$

Note that the inner product of vector **x** and **y** requires that a transposed vector **x** be multiplied by the **y** vector. Sometimes the inner product is denoted simply by juxtaposition of the vectors x and y, for example, as $<\mathbf{x}, \mathbf{y}>$ or $(\mathbf{x}, \mathbf{y})$.

Example: The inner product of two vectors $\mathbf{x} = \begin{bmatrix} 4 \\ 1 \\ 7 \end{bmatrix}$ and $\mathbf{y} = \begin{bmatrix} 0 \\ 2 \\ -3 \end{bmatrix}$

$$\mathbf{x}^T \mathbf{y} = \begin{bmatrix} 4 & 1 & 7 \end{bmatrix}^T \begin{bmatrix} 0 \\ 2 \\ -3 \end{bmatrix} = 4 \cdot 0 + 1 \cdot 2 + 7 \cdot (-3) = 19$$

Inner Product as a Mapping

We can see that an inner product is in fact specific kind of mapping. For each natural number **n**, the inner product is a mapping of $\mathbb{R}^n \times \mathbb{R}^n$ into $\mathbb{R}$.

1.1.10. Orthogonal Vectors

Two vectors **x** and **y** are said to be **orthogonal** if their inner product is equal to zero

$$\mathbf{x}^T \mathbf{y} = 0 \qquad (A.12)$$

here 0 is a scalar.

Example: Two vectors $\mathbf{x} = \begin{bmatrix} 4 \\ 0 \end{bmatrix}$ and $\mathbf{y} = \begin{bmatrix} 0 \\ 2 \end{bmatrix}$ and are orthogonal, since their inner product is equal to zero

$$\mathbf{x}^T \cdot \mathbf{y} = \begin{bmatrix} 4 \\ 0 \end{bmatrix}^T = \begin{bmatrix} 0 & 2 \end{bmatrix} = 4 \cdot 0 + 0 \cdot 2 = 0$$

1.1.11. Vector Norm

The magnitude of a vector may be measure in different ways. One method, called the vector **norm**, is a function from $\mathbb{R}^n$ into $\mathbb{R}$ for **x** an element of $\mathbb{R}^n$. It is denoted $||\mathbf{x}||$ and satisfies the following conditions:

1. $||\mathbf{x}|| \geq 0$, and the equality holds if and only if $\mathbf{x} = \mathbf{0}$
2. $||\alpha \mathbf{x}|| = |\alpha| \cdot ||\mathbf{x}||$, where $|\alpha|$ is the absolute value of scalar α

and is defined as:

$$||\mathbf{x}|| = \sqrt{\mathbf{x}^T \mathbf{x}} = \sqrt{x_1^2 + x_2^2 + \cdots + x_n^2} \qquad (A.13)$$

Example: For the vector $\mathbf{x} = \begin{bmatrix} 4 \\ 3 \end{bmatrix}$ the norm is

$$||\mathbf{x}|| = \sqrt{\mathbf{x}^T \mathbf{x}} = \sqrt{4^2 + 3^2} = 5$$

We observe that the above norm of vector represents the vector length.

1.1.12. Properties of Inner Product of Vectors

We observe the following important properties of the inner product:

1. $\mathbf{x}^T \mathbf{y} = \mathbf{y}^T \mathbf{x}$ (symmetry)
2. $\alpha(\mathbf{x}^T \mathbf{y}) = (\alpha \mathbf{x})^T \mathbf{y} = \mathbf{x}^T (\alpha \mathbf{y})$
3. $(\alpha \mathbf{x} + \mathbf{y})^T \mathbf{z} = \alpha(\mathbf{x}^T \mathbf{z}) + \mathbf{y}^T \mathbf{z}$ (linearity in variable)
4. $\mathbf{x}^T \mathbf{x} = ||\mathbf{x}||^2 > 0 (\mathbf{x} \neq \mathbf{0}); \mathbf{0} \cdot \mathbf{0} = 0$ (positive definiteness)

A vector space with an inner product is called an **inner product space**.

1.1.13. Cauchy-Schwartz Inequality

The upper bound on the inner product is defined by the Cauchy-Schwartz inequality:

$$|\mathbf{x}^T \mathbf{y}| \leq ||\mathbf{x}||\, ||\mathbf{y}||, \tag{A.14}$$

where $|a|$ denotes an absolute value of a.

1.1.14. Vector Metric

In many applications of linear algebra we need some measure (metric) of a distance between vectors.

Given n-dimensional vectors $\mathbf{x}$ and $\mathbf{y}$ we can then define the function $d(\mathbf{x}, \mathbf{y})$ from $\mathbb{R}^n \times \mathbb{R}^n$ to $\mathbb{R}$ which satisfies:

1. $d(\mathbf{x}, \mathbf{y}) \geq 0$, where equality holds if and only if $\mathbf{x} = \mathbf{y}$
2. $d(\mathbf{x}, \mathbf{y}) = d(\mathbf{y}, \mathbf{x})$
3. $d(\mathbf{x}, \mathbf{y}) \leq d(\mathbf{x}, \mathbf{z}) + d(\mathbf{z}, \mathbf{y})$ (triangle inequality)

as:

$$d(\mathbf{x}, \mathbf{y}) = \sqrt{(x_1 - y_1)^2 + (x_2 - y_2)^2 + \cdots + (x_n - y_n)^2} = ||\mathbf{x} - \mathbf{y}|| \tag{A.15}$$

The metric $d(\mathbf{x}, \mathbf{y})$ may be interpresented as a distance between two vectors.

Example: For the vectors $\mathbf{x} = \begin{bmatrix} 4 \\ 3 \end{bmatrix}$ and $\mathbf{y} = \begin{bmatrix} 2 \\ 1 \end{bmatrix}$ the distance (metric) is

$$d(\mathbf{x}, \mathbf{y}) = ||\mathbf{x} - \mathbf{y}|| = \sqrt{(4-2)^2 + (3-1)^2} = \sqrt{8}$$

We see that

$$d(\mathbf{x}, \mathbf{x}) = ||\mathbf{x}|| = \sqrt{\mathbf{x}^T \mathbf{x}} \tag{A.16}$$

1.1.15. Outer Product

The operation $\mathbf{xy}^T$ on n-dimensional vectors $\mathbf{x}$ and $\mathbf{y}$ is called an **outer** or **cross product** and is defined as:

$$\mathbf{xy}^T = \begin{bmatrix} x_1 \\ x_2 \\ \vdots \\ x_n \end{bmatrix} \begin{bmatrix} y_1 & y_2 & \cdots & y_n \end{bmatrix} = \begin{bmatrix} x_1 y_1 & x_1 y_2 & \cdots & x_1 y_n \\ x_2 y_1 & x_2 y_2 & \cdots & x_2 y_n \\ \vdots & \vdots & \ddots & \vdots \\ x_n y_1 & x_n y_2 & \cdots & x_n y_n \end{bmatrix} \tag{A.17}$$

The outer product is a square matrix of dimension $n \times n$.

We also can write

$$\mathbf{xy}^T = \begin{bmatrix} x_1 \\ x_2 \\ \vdots \\ x_n \end{bmatrix} \begin{bmatrix} y_1 & y_2 & \cdots & y_n \end{bmatrix} = \begin{bmatrix} \mathbf{x}y_1 & \mathbf{x}y_2 & \cdots & \mathbf{x}y_n \end{bmatrix} \quad (A.18)$$

Note that the outer product requires that a vector $\mathbf{x}$ to be multiplied by a transposed vector $\mathbf{y}$. Sometimes the outer product is denoted simply by juxtaposition of the vectors x and y.

Example: For the vectors $\mathbf{x} = \begin{bmatrix} 2 \\ 3 \end{bmatrix}$ and $\mathbf{y} = \begin{bmatrix} 4 \\ 5 \end{bmatrix}$ the outer product is the square matrix

$$\mathbf{xy}^T = \begin{bmatrix} \mathbf{x}y_1 & \mathbf{x}y_2 \end{bmatrix} = \begin{bmatrix} 2 \\ 3 \end{bmatrix} \begin{bmatrix} 4 & 5 \end{bmatrix} = \begin{bmatrix} 2 \cdot 4 & 2 \cdot 5 \\ 3 \cdot 4 & 3 \cdot 5 \end{bmatrix} = \begin{bmatrix} 8 & 10 \\ 2 & 15 \end{bmatrix}$$

Note: to perform **inner** or **outer** multiplication of two vectors requires that the vectors be of the same dimension. Thus, one cannot take the **inner** or **outer** product of vectors of differing dimension.

1.2. Geometrical Interpretations of Vectors

1.2.1. Length

Consider the inner product of a vector $\mathbf{x}$ with itself,

$$\mathbf{x}^T\mathbf{x} = x_1^2 + x_2^2 + \cdots + x_n^2 = \sum_{i=1}^{n} x_i^2 \quad (A.19)$$

If $\mathbf{x}$ is 2-dimensional we see that the value above is the length of the vector squared.

Definition: Let $\mathbf{x}$ be any vector, by the **length** of $\mathbf{x}$ we mean the real number

$$||\mathbf{x}|| = \sqrt{\mathbf{x}^T\mathbf{x}} \quad (A.20)$$

For any vector $\mathbf{x}$ we have

1. $||\mathbf{x}|| > 0$ if $\mathbf{x} \neq \mathbf{0}$. Moreover $||\mathbf{0}|| = 0$.
2. $||\alpha\mathbf{x}|| = |\alpha| \, ||\mathbf{x}||$.
3. $\frac{\mathbf{x}}{||\mathbf{x}||}$ is vector whose length is 1.
4. $||\mathbf{x}+\mathbf{y}|| = ||\mathbf{x}||^2 + ||\mathbf{y}||^2 + 2(\mathbf{x}^T\mathbf{y})$
5. $||\mathbf{x}-\mathbf{y}|| = ||\mathbf{x}||^2 + ||\mathbf{y}||^2 - 2(\mathbf{x}^T\mathbf{y})$
6. $|\mathbf{x}^T\mathbf{y}| \leq ||\mathbf{x}||\,||\mathbf{y}||$ \qquad (Cauchy-Schwartz inequality)

Taking the root of this number yields the length of the vector $\mathbf{x}$ and is the same as the vector **norm**:

$$length \; of \; \mathbf{x} = ||\mathbf{x}|| = \sqrt{\mathbf{x}^T\mathbf{x}} = \sqrt{\sum_{i=1}^{n} x_i^2} \quad (A.21)$$

Example: The length of vector $\mathbf{x} = \begin{bmatrix} 2 \\ 3 \\ 4 \end{bmatrix}$ is

$$length \; of \; \mathbf{x} = ||\mathbf{x}|| = \sqrt{\mathbf{x}^T\mathbf{x}} = \sqrt{\sum_{i=1}^{n} x_i^2} = \sqrt{2^2 + 3^2 + 4^2} = \sqrt{29}$$

1.2.2. Angles

Vectors in Euclidean space have a geometrical interpretation. We can find the angle θ between two vectors **x** and **y** as

$$\cos\theta = \frac{\mathbf{x}^T\mathbf{y}}{||\mathbf{x}||\,||\mathbf{y}||} \tag{A.22}$$

Definition: For nonzero vectors **x** and **y**, the real number

$$\theta = arc\cos\frac{\mathbf{x}^T\mathbf{y}}{||\mathbf{x}||\,||\mathbf{y}||} \tag{A.23}$$

is said to be the **angle** between **x** and **y**, denoted as **angle (x, y)**.

Example: For 2-dimensional Euclidean space, the angle between two vectors

$$\mathbf{x} = \begin{bmatrix} 0 \\ 1 \end{bmatrix}, \quad \mathbf{y} = \begin{bmatrix} 1 \\ 1 \end{bmatrix}$$

is $\theta = 45^o$ since

$$\cos\theta = \frac{1}{1\sqrt{2}} = 0.707,$$

where

$$\mathbf{x}^T\mathbf{y} = 1, \quad ||\mathbf{x}|| = 1, \quad ||\mathbf{y}|| = \sqrt{2}$$

From the definition of $\cos\theta$ we can find after easy derivation that

$$\mathbf{x}^T\mathbf{y} = ||\mathbf{x}||\,||\mathbf{y}||\cos(\theta) \tag{A.24}$$

which means that the **inner product** $\mathbf{x}^T\mathbf{y}$ is proportional to the cosine of the vector angle.

1.2.3. Triangle (Cosine) Formula

Let **x** and **y** be any n-vectors. We have a triangle formula for the two vectors:

$$||\mathbf{x}-\mathbf{y}||^2 = ||\mathbf{x}||^2 + ||\mathbf{y}||^2 - 2||\mathbf{x}||\,||\mathbf{y}||\cos\theta \tag{A.25}$$

which may be extended (since $||\mathbf{x}-\mathbf{y}||^2 = (\mathbf{y}-\mathbf{x})^T(\mathbf{y}-\mathbf{x})$)

$$\mathbf{x}^T\mathbf{x} - 2\mathbf{x}^T\mathbf{y} + \mathbf{y}^T\mathbf{y} = \mathbf{x}^T\mathbf{x} + \mathbf{y}^T\mathbf{y} - 2||\mathbf{x}||\,||\mathbf{y}||\cos\theta \tag{A.26}$$

1.2.4. Triangle Inequality

For any two vectors the following **triangle inequality** holds

$$||\mathbf{x}+\mathbf{y}||^2 \leq ||\mathbf{x}|| + ||\mathbf{y}|| \tag{A.27}$$

1.2.5. Orthogonality

Two vectors are said to be **orthogonal** (sometimes denoted $\mathbf{x} \perp \mathbf{y}$) when their inner product $\mathbf{x}^T\mathbf{y}$ vanishes (i.e., is equal to zero). Geometrically, when the angle between two orthogonal vectors is 90° or 270° then

$$\cos \theta = 0 \qquad (A.28)$$

and the **inner product** will vanish.

Example: The vectors

$$\mathbf{x} = \begin{bmatrix} 1 \\ 0 \end{bmatrix}, \quad \mathbf{y} = \begin{bmatrix} 0 \\ 1 \end{bmatrix}$$

are orthogonal since their inner product $\mathbf{x}^T\mathbf{y} = \begin{bmatrix} 1 & 0 \end{bmatrix} \begin{bmatrix} 0 \\ 1 \end{bmatrix} = 1 \cdot 0 + 0 \cdot 1 = 0$ equals zero.

For orthogonal vectors $\mathbf{x}$ and $\mathbf{y}$ we have that $\cos \theta = 0$. The vector that joins vectors $\mathbf{x}$ and $\mathbf{y}$ can be found by the distance between these two vectors.

Example: For $\mathbf{x}$ and $\mathbf{y}$ (as given above) we note that the vector metric of $\mathbf{x}$ and $\mathbf{y}$ is

$$d(\mathbf{x}, \mathbf{y}) = \sqrt{(x_1 - y_1)^2 + (x_2 - y_2)^2} = \sqrt{(1-0)^2 + (0-(-1))^2} = \sqrt{1+1} = \sqrt{2}$$

and for the vector created by subtracting $\mathbf{x}$ and $\mathbf{y}$ we have the vector norm of

$$||\mathbf{x} - \mathbf{y}|| = \sqrt{(x_1 - y_1)^2 + (x_2 - y_2)^2} = \sqrt{(1-0)^2 + (0-(-1))^2} = \sqrt{1+1} = \sqrt{2}$$

thus the vector metric equals the vector norm.

Thus we see that the vector $\mathbf{x} - \mathbf{y}$ describes the hypotenuse of the triangle formed by joining $\mathbf{x}$ and $\mathbf{y}$.

Recall the Pythagorian theorem. We see that, for orthogonal vectors:

$$||\mathbf{x} - \mathbf{y}||^2 = ||\mathbf{x}||^2 + ||\mathbf{y}||^2 \qquad (A.29)$$

1.2.6. Projection of One Vector onto Another

Let us consider two vectors $\mathbf{x}$ and $\mathbf{y}$ with angle θ between them. It is possible to talk about the projection of vector $\mathbf{x}$ onto vector $\mathbf{y}$. The distance

$$z = ||\mathbf{x}|| \cos \theta \qquad (A.30)$$

is called the projection of $\mathbf{x}$ on $\mathbf{y}$ and tells how much $\mathbf{x}$ is pointing in the direction of $\mathbf{y}$. We also observe that

$$z = ||\mathbf{x}|| \cos \theta = ||\mathbf{x}|| \frac{\mathbf{x}^T\mathbf{y}}{||\mathbf{x}||\,||\mathbf{y}||} = \frac{\mathbf{x}^T\mathbf{y}}{||\mathbf{y}||} \qquad (A.31)$$

Note that the projection of one vector onto a vector orthogonal to it is zero.

1. Vectors

1.3. Linear Combination of Vectors

Definition: A **linear combination** (or weighted sum) of m vectors $\mathbf{x}_i \in \mathbb{R}^n (i = 1, 2, \cdots, m)$ is a vector $\mathbf{v} \in \mathbb{R}^n$ such that

$$\mathbf{v} = \alpha_1 \mathbf{x}_1 + \alpha_2 \mathbf{x}_2 + \cdots + \alpha_m \mathbf{x}_m$$

$$= \alpha_1 \begin{bmatrix} x_{11} \\ x_{12} \\ \vdots \\ x_{1n} \end{bmatrix} + \alpha_2 \begin{bmatrix} x_{21} \\ x_{22} \\ \vdots \\ x_{2n} \end{bmatrix} + \cdots + \alpha_m \begin{bmatrix} x_{m1} \\ x_{m2} \\ \vdots \\ x_{mn} \end{bmatrix}$$

$$= \sum_{i=1}^{n} \alpha_i \mathbf{x}_i = \begin{bmatrix} v_1 \\ v_2 \\ \vdots \\ v_n \end{bmatrix} \in \mathbb{R}^n, \tag{A.32}$$

where each $\alpha_i (i = 1, 2, \cdots, m)$ are real scalars.

The linear combination $\mathbf{v}$ is said to be **non-trivial** if the scalars $\alpha_i (i = 1, 2, \cdots, m)$ are not all zero.

Example: For the vectors

$$\mathbf{x}_1 = \begin{bmatrix} 1 \\ 2 \end{bmatrix}, \quad \mathbf{x}_2 = \begin{bmatrix} 3 \\ 2 \end{bmatrix}$$

the linear combination with $\alpha_{1,2} = 2$ produces the vector

$$\mathbf{v} = \alpha_1 \mathbf{x}_1 + \alpha_2 \mathbf{x}_2 = 2 \begin{bmatrix} 1 \\ 2 \end{bmatrix} + 2 \begin{bmatrix} 3 \\ 2 \end{bmatrix} = \begin{bmatrix} 2 \\ 4 \end{bmatrix} + \begin{bmatrix} 6 \\ 4 \end{bmatrix} = \begin{bmatrix} 8 \\ 8 \end{bmatrix}$$

1.3.1. Linearly Independent Vectors

The concept of linearly independent vectors plays a very important role in linear algebra.

Definition: A collection (or set) $M = \{\mathbf{x}_1, \mathbf{x}_2, \cdots, \mathbf{x}_m\}$ of m vectors in $\mathbb{R}^n$, that is, $\{\mathbf{x}_i : \mathbf{x}_i \in \mathbb{R}^n, i = (1, 2, 3, \cdots, m)\}$ is said to be **linearly independent** when their linear combination:

$$\mathbf{v} = \sum_{i=1}^{m} \alpha_i \mathbf{x}_i = \alpha_1 \mathbf{x}_1 + \alpha_2 \mathbf{x}_2 + \cdots + \alpha_m \mathbf{x}_m \tag{A.33}$$

is **0 for only trivial combinations of scalars**, that is, $\alpha_i = 0$ for $i = (1, 2, 3, \cdots, m)$. When the set of m scalars α_i are not all 0 then the linear combination $\mathbf{v}$ can not equal the zero vector $\mathbf{0}$.

Linear independence for a collection M of m vectors means that none of the vectors in the set may be expressed as a linear combination of the others. That is, for all $\mathbf{x}_i \in M$,

$$\mathbf{x}_i \neq \sum_{j=1, i \neq j}^{m} \alpha_j \mathbf{x}_i \tag{A.34}$$

because the set of α's can not be found.

Vector collections that are not linear independent are said to be **linear dependent** which means their linear combination may become $\mathbf{0}$ even if not all the scalars α_i are equal to zero. Also, at least one of the x_i may be expressed as a linear combination of the others.

Example: We can see that the vectors

$$\mathbf{x} = \begin{bmatrix} 1 \\ 0 \end{bmatrix}, \; \mathbf{y} = \begin{bmatrix} 0 \\ 1 \end{bmatrix}$$

are linearly independent since we cannot find nonzero α_1 and α_2 for which

$$\alpha_1 \begin{bmatrix} 1 \\ 0 \end{bmatrix} + \alpha_2 \begin{bmatrix} 0 \\ 1 \end{bmatrix} = \mathbf{0}$$

On the other hand the vectors

$$\mathbf{z} = \begin{bmatrix} 1 \\ 2 \end{bmatrix}, \; \mathbf{u} = \begin{bmatrix} 2 \\ 4 \end{bmatrix}$$

are linearly dependent since

$$\alpha_1 \begin{bmatrix} 1 \\ 2 \end{bmatrix} + \alpha_2 \begin{bmatrix} 2 \\ 4 \end{bmatrix} = \mathbf{0}$$

when $\alpha_1 = -2$ and $\alpha_2 = 1$. And the vector $\mathbf{u}$ may be may be expressed as a linear combination of vector $\mathbf{z}$

$$\mathbf{u} = \begin{bmatrix} 2 \\ 4 \end{bmatrix} = 2 \begin{bmatrix} 1 \\ 2 \end{bmatrix} = 2\mathbf{z}$$

Consider $\mathbb{R}^3$, 3-dimensional Euclidean space. In this space linear dependence is easily visualized. Two vectors $\in \mathbb{R}^3$ are linearly dependent if they both lie on the same line which passes through the origin (0, 0, 0). Three vectors $\in \mathbb{R}^3$ are linearly dependent if they lie on a plane that passes through the origin, and each of the vectors can be expressed as a linear combination of the others. Four vectors $\in \mathbb{R}^3$ are always linearly dependent. Generally speaking, a collection of m vectors in $\mathbb{R}^n$ are always linearly dependent if $m > n$. Random selection of three vectors in $\mathbb{R}^3$ (or higher dimensional space) will usually produce a linear independent collection.

1.4. Subspaces

Definition: A **subspace** of $\mathbb{R}^n$ is a subset V with the property that if $\mathbf{x}$ and $\mathbf{y}$ are any two elements of V and α is any real number, then $\mathbf{x} + \mathbf{y} \in V$ and $\alpha \cdot \mathbf{x} \in V$.

The simplest subspace is the set $V = \{\mathbf{0}\}$, consisting of vector $\mathbf{0}$ alone. Another simple subspace is the set of all multiples of some non-zero vector v – geometrically we see this as all the vectors that lie on a line that passes through the origin and through vector v.

1.4.1. A Spanning Subspace in $\mathbb{R}^n$

Consider a set S of m ($m \leq n$) vectors $\{\mathbf{x}_i \in \mathbb{R}^n \mid i = (1, 2, 3, \cdots, m)\}$ and **all** possible linear combinations of them, e.g., $\mathbf{s} = \alpha_1 \mathbf{x}_1 + \alpha_2 \mathbf{x}_2 + \cdots + \alpha_m \mathbf{x}_m$ where the coefficients $\alpha_i (i = 1, 2, \cdots, m)$ take on all values from $(-\infty, +\infty)$.

Definition: The subspace V of $\mathbb{R}^n$ spanned by a set of vectors $S = \{\mathbf{x}_i \in \mathbb{R}^n \mid i = (1, 2, 3, \cdots, m)\}$ consists of all linear combinations of these vectors and is denoted by

$$span\, \{\mathbf{x}_1, \mathbf{x}_2, \cdots, \mathbf{x}_m\} \qquad (A.35)$$

or

$$[S] \qquad (A.36)$$

This is a linear subspace contained in $\mathbb{R}^n$. Every vector $\mathbf{v}$ in $span\ \{\mathbf{x}_1, \mathbf{x}_2, \cdots, \mathbf{x}_m\}$ can be expressed as some combination of the vectors $\{\mathbf{x}_1, \mathbf{x}_2, \cdots, \mathbf{x}_m\}$, i.e. $\forall \mathbf{v} \in span\ \{\mathbf{x}_1, \mathbf{x}_2, \cdots, \mathbf{x}_m\}$

$$\mathbf{v} = \alpha_1 \mathbf{x}_1 + \alpha_2 \mathbf{x}_2 + \cdots + \alpha_m \mathbf{x}_m \tag{A.37}$$

for some set of coefficients $\alpha_i (i = 1, 2, \cdots, m)$.

Example: In the 3-dimensional vector space $\mathbb{R}^3$ any straight line or any plane which passes through the origin are linear subspaces (spanned on one and two vectors, respectively). It can be shown that three vectors in $\mathbb{R}^3$

$$\mathbf{x}_1 = \begin{bmatrix} 1 \\ 0 \\ 0 \end{bmatrix},\ \mathbf{x}_2 = \begin{bmatrix} 0 \\ 1 \\ 0 \end{bmatrix},\ \mathbf{x}_3 = \begin{bmatrix} -2 \\ 0 \\ 0 \end{bmatrix}$$

always span a plane, while two vectors **may** span a plane but always span a line. For example, as given above, $\mathbf{x}_1$ and $\mathbf{x}_2$ span a plane while $\mathbf{x}_1$ and $\mathbf{x}_3$ span a line.

1.4.2. Basis

Definition: Assume that the set S of m ($m \leq n$) vectors in $\mathbb{R}^n$ spans the linear space $span\ \{\mathbf{x}_1, \mathbf{x}_2, \mathbf{x}_3, \cdots, \mathbf{x}_m\}$ in $\mathbb{R}^n$. The set $S = \{\mathbf{x}_1, \mathbf{x}_2, \mathbf{x}_3, \cdots, \mathbf{x}_m\}$ of vectors is said to be a **basis** for the subspace if the vectors are linearly independent. The **dimension** of the subspace is m the number of elements in the spanning set. The dimension of a subspace is also called the **degree of freedom** of the subspace. The set of m linearly independent vectors in $\mathbb{R}^n$

$$\mathbf{e}_1 = \begin{bmatrix} 1 \\ 0 \\ \vdots \\ 0 \end{bmatrix},\ \mathbf{e}_2 = \begin{bmatrix} 0 \\ 1 \\ \vdots \\ 0 \end{bmatrix},\ \cdots,\ \mathbf{e}_m = \begin{bmatrix} 0 \\ 0 \\ \vdots \\ 1 \end{bmatrix} \tag{A.38}$$

is called the **standard** (or natural, or canonical) basis for the space $\mathbb{R}^n$. A vector space has infinitely many different bases (i.e., a vector space does not have a unique basis). All bases of a vector space contain the same number of vectors. Any **linearly independent** set in $\mathbb{R}^n$ can be extended to a basis by adding more vectors to it if necessary. Any spanning set in $\mathbb{R}^n$ can be reduced to a basis by discarding vectors if necessary. A basis is a maximal and minimal spanning set, i.e., it cannot be made larger without losing linear independence and cannot be made smaller a still span the space.

1.4.3. Hyperplanes and Linear Manifolds

In higher dimensional spaces ($\mathbb{R}^n$, $n \geq 3$), very important linear subspaces are those defined as a linear combination of $n-1$ linearly independent vectors. They are called **hyperplanes** and divide $\mathbb{R}^n$ into two halfspaces. All linear subspaces, including $\mathbb{R}^n$, are called **linear manifolds**, and the set of vectors which define the manifold is said to **span** it.

1.4.4. Orthogonal Spaces

Definition: Two subspaces V and W of the same space $\mathbb{R}^n$ are **orthogonal** if every vector $\mathbf{v} \in V$ is orthogonal to every vector $\mathbf{w} \in W$

$$\mathbf{v}^T \mathbf{w} = 0,\ for\ all\ \mathbf{v}\ and\ \mathbf{w} \tag{A.39}$$

1.4.5. Orthogonal Projections

Definition: For two vectors **x**, **y** elements of the same vector space the vector

$$\mathbf{x}^{py} = \frac{\mathbf{y}^T\mathbf{y}}{||\mathbf{y}||^2} \mathbf{x} = \frac{\mathbf{y}^T\mathbf{x}}{||\mathbf{y}||^2}\mathbf{y} \quad (A.40)$$

is the orthogonal projection of **x** onto **y**, in the sense that the difference vector $\mathbf{x} - \mathbf{x}^{py}$ is orthogonal to **y**. Recall that the angle between the vectors **x** and **y** is

$$\theta = \cos^{-1} \frac{\mathbf{x}^T\mathbf{y}}{||\mathbf{x}||\,||\mathbf{y}||} \quad (A.41)$$

1.4.6. Vector Decomposition

Assume that V is a subspace of $\mathbb{R}^n$.

Theorem: An arbitrary vector $\mathbf{x} \in \mathbb{R}^n$ can be uniquely decomposed into the sum of two vectors

$$\mathbf{x} = \hat{\mathbf{x}} + \tilde{\mathbf{x}}, \quad (A.42)$$

where $\hat{\mathbf{x}} \in V$, and $\tilde{\mathbf{x}}$ is orthogonal to vector space V. In the above theorem vector $\hat{\mathbf{x}}$ is called the **orthogonal projection** of vector **x** on V.

1.4.7. Orthogonal Complement

Definition: The space V^o is called the **orthogonal complement** of V and is the set of all vectors in $\mathbb{R}^n$ which are orthogonal to V.

Thus, in the above theorem $\tilde{\mathbf{x}}$ is called the orthogonal projection on V^o.

Example: Consider the space $\mathbb{R}^3$ with the **standard** basis $\mathbf{e}_i$ ($i = 1, 2, 3$). If the subspace V is defined to be all vectors along the vertical **axis**

$$V = \{\mathbf{x} \in \mathbb{R}^3 \mid \mathbf{x} = \alpha\mathbf{e}_3, \text{ where } \alpha \text{ is a scalar}\},$$

then its orthogonal complement consists of all vectors in the horizontal **plane**

$$V^o = \{\mathbf{x} \in \mathbb{R}^3 \mid \mathbf{x} = \beta\mathbf{e}_1 + \gamma\mathbf{e}_2, \text{ where } \beta \text{ and } \gamma \text{ are scalars}\}$$

Thus, any vector $\mathbf{x} \in \mathbb{R}^3$ may be decomposed on V and V^o, for example

$$\mathbf{x} = \begin{bmatrix} x_1 \\ x_2 \\ x_3 \end{bmatrix} = \mathbf{y} + \mathbf{z} = \begin{bmatrix} 0 \\ 0 \\ x_3 \end{bmatrix} + \begin{bmatrix} x_1 \\ x_2 \\ 0 \end{bmatrix}$$

1.4.8. Projection Theorem

Given vector $\mathbf{x} \in \mathbb{R}^n$ and subspace V of $\mathbb{R}^n$, then

$$\mathbf{x} = \mathbf{x}' + \mathbf{x}'' \quad (A.43)$$

(where $\mathbf{x}' \in V$, $\mathbf{x}'' \in V^o$) and the orthogonal projection $\mathbf{x}'$ has a property that $||\mathbf{x}'||^2$ is minimum.

1. Vectors

1.4.9. Gramm-Smith Orthogonalization

In an **orthogonal basis** of a vector space, every vector in the basis is perpendicular to all the other vectors in the basis. That is, the coordinate axes of the base are mutually orthogonal.

Definition: The set of vectors $\{x_1, x_2, \cdots, x_k\}$ are **orthonormal** if

$$(x_i)^T x_j = \begin{cases} 0 & \text{whenever} \quad i \neq j, \quad \text{giving orthogonality} \\ 1 & \text{whenever} \quad i = j, \quad \text{giving normalization} \end{cases} \quad (A.44)$$

As mentioned before the **standard basis**

$$e_1 = \begin{bmatrix} 1 \\ 0 \\ \vdots \\ 0 \end{bmatrix}, \quad e_2 = \begin{bmatrix} 0 \\ 1 \\ \vdots \\ 0 \end{bmatrix}, \cdots, e_m = \begin{bmatrix} 0 \\ 0 \\ \vdots \\ 1 \end{bmatrix} \quad (A.45)$$

is an important orthogonal basis with perpendicular axes, which can be rotated without changing the right angles at which they meet. For 2-dimensional space the standard basis

$$e_1 = \begin{bmatrix} 1 \\ 0 \end{bmatrix} \text{ and } e_2 = \begin{bmatrix} 0 \\ 1 \end{bmatrix} \quad (A.46)$$

is perpendicular as well as horizontal and vertical. Not often do we initially have an orthogonal basis. However, the Gramm-Smith orthogonalization procedure allows us to construct for any linear space $\mathbb{R}^k$ and starting with a **skewed** basis an orthogonal basis (set of vectors) which is mutually orthogonal and spans the space $\mathbb{R}^k$.

Algorithm: Assume that we have a set of nonzero vectors $\{x^1, x^2, \cdots, x^l\}$ in $\mathbb{R}^k$ ($l \geq k$) which span the space $\mathbb{R}^k$.

1. In construction of a vector basis, one direction may be freely selected, for example $b_1 = x_1$.
2. For x_2 (where x_2 is not in the direction of x^1) we can find the second basis b_2 orthogonal to b_1 by:

$$b_2 = x_2 - \frac{x_2^T b_1}{\|b_1\|^2} b_1 \quad (A.47)$$

The inner product $e_1^T e_2$ is equal to 0, since they are orthogonal. Thus the construction extractions any components of x_1 that are in the x_2 direction. If, on the other hand, x_2 is in the direction of x_1, we just ignore x_2 since x_2 is represented by x_1.

3. Generally, for the next basis we can use the recursive formula

$$b_i = x_i - \sum_{j=1}^{i-1} \frac{x_i^T b_j}{\|b_j\|^2} b_j, \quad (i = 2, \cdots, k) \quad (A.48)$$

where $i = 2, \cdots, k$ and only nonzero b_j must be considered.

When the Gramm-Smith orthogonalization procedure starts with linearly independent vectors $\{x_1, x_2, \cdots, x_k\}$ it ends with orthogonal vectors $\{b_1, b_2, \cdots, b_k\}$ which span $\mathbb{R}^k$.

Example: For three linearly independent vectors in $\mathbb{R}^3$

$$x_1 = \begin{bmatrix} 1 \\ 0 \\ 1 \end{bmatrix}, \quad x_2 = \begin{bmatrix} 1 \\ 0 \\ 0 \end{bmatrix}, \quad x_3 = \begin{bmatrix} 2 \\ 2 \\ 0 \end{bmatrix}$$

We can select

$$\mathbf{b}_1 = \mathbf{x}_1 = \begin{bmatrix} 1 \\ 0 \\ 1 \end{bmatrix}$$

To find b_2 we subtract the elements in direction $\mathbf{b}_1$ from $\mathbf{x}_2$

$$\mathbf{b}^2 = \mathbf{x}_2 - \frac{\mathbf{x}_2^T \mathbf{b}_1}{||\mathbf{b}_1||_2} \mathbf{b}_1 = \begin{bmatrix} 1 \\ 0 \\ 0 \end{bmatrix} - \frac{1}{2} \begin{bmatrix} 1 \\ 0 \\ 1 \end{bmatrix} = \begin{bmatrix} 0.5 \\ 0 \\ -0.5 \end{bmatrix}$$

And lastly the basis $\mathbf{b}_3$ is computed as

$$\mathbf{b}_3 = \mathbf{x}_3 - \frac{\mathbf{x}_3^T \mathbf{b}_1}{||\mathbf{b}_1||_2} \mathbf{x}_1 - \frac{\mathbf{x}_3^T \mathbf{b}_2}{||\mathbf{b}_2||_2} \mathbf{x}_2 = \begin{bmatrix} 2 \\ 2 \\ 0 \end{bmatrix} - \frac{2}{4} \begin{bmatrix} 1 \\ 0 \\ 1 \end{bmatrix} - \frac{1}{0.5} \begin{bmatrix} 0.5 \\ 0 \\ -0.5 \end{bmatrix} = \begin{bmatrix} 0 \\ 1 \\ 0 \end{bmatrix}$$

We see that the resulting basis is orthogonal.

2. Matrices

2.1. Basic Operations

A **matrix** is an $n \times m$ rectangular array of elements with n rows and m columns. Matrices are denoted by capital letters with elements denoted by lower case letters with subscripts.

$$\mathbf{A} = [a_{ij}] = \begin{bmatrix} a_{11} & a_{12} & \cdots & a_{1m} \\ a_{21} & a_{22} & \cdots & a_{2m} \\ \vdots & \vdots & \ddots & \vdots \\ a_{n1} & a_{n2} & \cdots & a_{nm} \end{bmatrix} \quad (A.49)$$

When a matrix $\mathbf{A}$ is defined in $n \times m$ space we denote this by $\mathbf{A} \in C^{n \times m}$, where $n \times m$ is the matrix dimension.

Example:

$$\mathbf{A} = \begin{bmatrix} 2 & 1 & 5 & -0.5 \\ 0 & 6 & 6 & 23 \\ 8 & 7 & 4 & 2 \end{bmatrix}; \quad \mathbf{B} = \begin{bmatrix} 1 \\ 5 \end{bmatrix}; \quad \mathbf{C} = \begin{bmatrix} 1 & 0 & 4 \end{bmatrix}$$

We have the following dimensionality of the above matrices: $\mathbf{A} \in \mathbb{R}^{3 \times 4}$, $\mathbf{B} \in \mathbb{R}^{2 \times 1}$, and $\mathbf{C} \in \mathbb{R}^{1 \times 2}$.

2.1.1. Matrix Addition

Matrix addition is defined for equal dimension matrices, e.g., $n \times m$, as

$$\mathbf{A} + \mathbf{B} = [a_{ij} + b_{ij}] = \begin{bmatrix} a_{11} + b_{11} & a_{12} + b_{12} & \cdots & a_{1m} + b_{1m} \\ a_{21} + b_{21} & a_{22} + b_{22} & \cdots & a_{2m} + b_{2m} \\ \vdots & \vdots & \ddots & \vdots \\ a_{n1} + b_{n1} & a_{n2} + b_{n2} & \cdots & a_{nm} + b_{nm} \end{bmatrix} \quad (A.50)$$

Example:

$$\mathbf{C} = \mathbf{A} + \mathbf{B} = \begin{bmatrix} 1 & 5 \\ 0 & -3 \end{bmatrix} + \begin{bmatrix} 3 & 1 \\ 2 & 7 \end{bmatrix} = \begin{bmatrix} 4 & 6 \\ 2 & 4 \end{bmatrix}$$

2. Matrices

2.1.2. Matrix Multiplication by a Scalar

The matrix **A** may be multiplied by a scalar k as

$$k\mathbf{A} = [ka_{ij}] = \begin{bmatrix} ka_{11} & ka_{12} & \cdots & ka_{1m} \\ ka_{21} & ka_{22} & \cdots & ka_{2m} \\ \vdots & \vdots & \ddots & \vdots \\ ka_{n1} & ka_{n2} & \cdots & ka_{nm} \end{bmatrix} \quad (A.51)$$

Example:

$$\mathbf{B} = 4 \begin{bmatrix} 1 & 5 \\ 0 & -3 \end{bmatrix} = \begin{bmatrix} 4 & 20 \\ 0 & -12 \end{bmatrix}$$

2.1.3. Matrix Multiplication

For so called **conformable** matrices; that is, matrix **A** of dimension $n \times m$ and matrix **B** of dimension $m \times r$ (note that the number of columns in **A** is equal to the number of rows in **B**) the $n \times r$ dimensional matrix **C** is formed by:

$$\mathbf{C} = \mathbf{AB} = [c_{ij}] \quad (A.52)$$

where one element c_{ij} of the resulting matrix c is defined as:

$$c_{ij} = \begin{bmatrix} \vdots & \vdots & \cdots & \vdots \\ a_{i1} & a_{i2} & \ddots & a_{im} \\ \vdots & \vdots & \cdots & \vdots \end{bmatrix}_{n \times m} \begin{bmatrix} \cdots & a_{1j} & \cdots \\ \cdots & a_{2j} & \cdots \\ \cdots & \cdots & \cdots \\ \cdots & \cdots & \cdots \\ \cdots & \cdots & \cdots \\ \cdots & a_{mj} & \cdots \end{bmatrix}_{m \times r}$$

$$= \sum_{k=1}^{m} a_{ik} b_{kj}, \quad i = 1, 2, \cdots, n; \quad j = 1, 2, \cdots, r \quad (A.53)$$

and is obtained by computing the inner product of row i of matrix **A** by column j of matrix **B**.

For example for the 2×3 matrix A and the 3×2 matrix **B** we get **C**, a 2×2 matrix

$$\mathbf{C} = \mathbf{AB} = \begin{bmatrix} a_{11} & a_{12} & a_{13} \\ a_{21} & a_{22} & a_{23} \end{bmatrix} \begin{bmatrix} b_{11} & b_{12} & b_{13} \\ b_{21} & b_{22} & b_{23} \\ b_{31} & b_{32} & b_{33} \end{bmatrix}$$

$$= \begin{bmatrix} a_{11}b_{11} + a_{12}b_{21} + a_{13}b_{31} & a_{11}b_{12} + a_{12}b_{22} + a_{13}b_{32} \\ a_{21}b_{11} + a_{22}b_{21} + a_{23}b_{31} & a_{21}b_{12} + a_{22}b_{22} + a_{23}b_{32} \end{bmatrix}$$

Example:

$$\mathbf{C} = \mathbf{AB} = \begin{bmatrix} 1 & 5 \\ 0 & -3 \end{bmatrix} \begin{bmatrix} 3 & 1 \\ 2 & 7 \end{bmatrix} = \begin{bmatrix} 13 & 36 \\ -6 & -21 \end{bmatrix}$$

2.1.4. Multiplication of Matrix by Vector

For vector $\mathbf{x}$ of dimension m and matrix $\mathbf{A}$ of dimension $n \times m$ the vector dimension m, the vector-matrix product $\mathbf{y} = A\mathbf{x}$ is defined such that the resulting n-dimensional vector $\mathbf{y}$ is defined as

$$y_i = \sum_{j=1}^{m} a_{ij} x_j. \qquad (A.54)$$

For example the 2×3 matrix $\mathbf{A}$ and 3-dimensional vector $\mathbf{x}$ are multiplied as:

$$\mathbf{y} = \mathbf{A}\mathbf{x} = \begin{bmatrix} a_{11} & a_{12} & a_{13} \\ a_{21} & a_{22} & a_{23} \end{bmatrix} \begin{bmatrix} x_1 \\ x_2 \\ x_3 \end{bmatrix} = \begin{bmatrix} a_{11}x_1 + a_{12}x_2 + a_{13}x_3 \\ a_{21}x_1 + a_{22}x_2 + a_{23}x_3 \end{bmatrix}$$

When the number of rows and columns are equal, e.g., an $n \times n$ matrix, we have a square matrix.

Example:

$$\mathbf{x} = \mathbf{A}\mathbf{y} = \begin{bmatrix} 1 & 5 \\ 0 & -3 \end{bmatrix} \begin{bmatrix} 3 \\ 2 \end{bmatrix} = \begin{bmatrix} 13 \\ -6 \end{bmatrix}$$

2.1.5. Equal Matrices

Two matrices $\mathbf{A}$ and $\mathbf{B}$ are equal if and only if all of their corresponding elements are equal $a_{ij} = b_{ij}$ for all $i = 1, 2, \cdots, n$ and $j = 1, 2, \cdots, m$.

2.1.6. Diagonal Matrix

Entries of the matrix starting the top left corner and proceeding to the bottom right corner are said to be on the **main diagonal** of that matrix.

A square matrix having only nonzero elements on the main diagonal and only zero elements otherwise is called a **diagonal** matrix.

Example:

$$\mathbf{A} = \begin{bmatrix} 2 & 0 & 0 & 0 \\ 0 & 1 & 0 & 0 \\ 0 & 0 & 5 & 0 \\ 0 & 0 & 0 & -4 \end{bmatrix}$$

2.1.7. Triangular Matrix

A square matrix of the form

$$\mathbf{A} = \begin{bmatrix} a_{11} & a_{12} & a_{13} & \cdots & a_{1n} \\ 0 & u_{22} & u_{23} & \cdots & a_{2n} \\ 0 & 0 & a_{33} & \cdots & a_{3n} \\ \vdots & \vdots & \vdots & \ddots & \vdots \\ 0 & 0 & 0 & \cdots & a_{nn} \end{bmatrix} \qquad (A.55)$$

is called **upper triangular** (or block triangular).

In a triangular matrix each entry above the main diagonal is 0 (for **upper triangular matrix**), or each entry below the main diagonal is 0 (for **lower triangular matrix**).

2.1.8. Zero Matrix

The matrix in which all element are zero is called zero matrix, denoted $[\mathbf{0}]$.

2.1.9. Identity Matrix

A square matrix with 1's at each position of the main diagonal and 0's in all other positions is called the **identity matrix**, denoted **I**

$$\mathbf{I} = \begin{bmatrix} 1 & 0 & \cdots & 0 \\ 0 & 1 & \cdots & 0 \\ \vdots & \vdots & \ddots & \vdots \\ 0 & 0 & \cdots & 1 \end{bmatrix} = [\delta_{ij}] \quad (A.56)$$

where δ_{ij} is the Kronecker delta defined as

$$\delta_{ij} = \begin{cases} 1 & \text{if } i = j \\ 0 & \text{if } i \neq j \end{cases} \quad (A.57)$$

The result of multiplying a matrix **A** by matrix **I** is equal to matrix **A**, thus multiplication by the identity matrix does not change the original matrix:

$$\mathbf{AI} = \mathbf{IA} = \mathbf{A} \quad (A.58)$$

2.1.10. Matrix Arithmetic

The following rules hold:

1. $\mathbf{A} + \mathbf{B} = \mathbf{B} + \mathbf{A}$ (commutative law of addition)
2. $\mathbf{A} + (\mathbf{B} + \mathbf{C}) = (\mathbf{A} + \mathbf{B}) + \mathbf{C}$ (associative law of addition)
3. $\mathbf{A}(\mathbf{BC}) = (\mathbf{AB})\mathbf{C}$ (associative law of multiplication)
4. $\mathbf{A}(\mathbf{B} + \mathbf{C}) = \mathbf{AB} + \mathbf{AC}$ (distributive law)
5. $\mathbf{A}(\mathbf{B} - \mathbf{C}) = \mathbf{AB} - \mathbf{AC}$
6. $(\mathbf{A} + \mathbf{B})\mathbf{C} = \mathbf{AC} + \mathbf{BC}$
7. $\alpha(\mathbf{B} + \mathbf{C}) = \alpha\mathbf{B} + \alpha\mathbf{C}$
8. $(\alpha + \beta)\mathbf{A} = \alpha\mathbf{A} + \beta\mathbf{A}$
9. $(\alpha\beta)\mathbf{A} = \alpha(\beta\mathbf{A})$
10. $\alpha(\mathbf{AB}) = (\alpha\mathbf{A})\mathbf{B} = \mathbf{A}(\alpha\mathbf{B})$

For square matrix **A** and the integers r, s the following equalities hold

$$\mathbf{A}^r \mathbf{A}^s = \mathbf{A}^{r+s}; \quad (\mathbf{A}^r)^s = \mathbf{A}^{rs} \quad (A.59)$$

2.1.11. Matrix as a n-vectors of Rows or Columns

We can observe that the rows of an $n \times m$ matrix **A** may be regarded as a m-**dimensional vectors** $\mathbf{A}_i^T = [a_{i1}, a_{i2}, \cdots, a_{ij}, \cdots, a_{im}]$ $(i = 1, 2, \cdots, n)$

$$\begin{bmatrix} \mathbf{A}_1^T \\ \mathbf{A}_2^T \\ \vdots \\ \mathbf{A}_n^T \end{bmatrix} \quad (A.60)$$

Similarly, the columns of an $n \times m$ matrix **A** may be regarded as an m-dimensional vectors

$$\mathbf{A}_i = \begin{bmatrix} a_{i1} \\ a_{i2} \\ \vdots \\ a_{ij} \\ \vdots \\ a_{im} \end{bmatrix}, \quad (i = 1, 2, \cdots, n) \quad (A.61)$$

and matrix **A** may be written as

$$\mathbf{A} = \begin{bmatrix} \mathbf{A}_1^T & \mathbf{A}_2^T & \cdots & \mathbf{A}_n^T \end{bmatrix} \quad (A.62)$$

2.2. Determinant of Matrix

One of the most important problems in the matrix theory is associating a single real number with a square matrix. This real number, called the **determinant** of the matrix, is the result of mapping of a matrix into R. The determinant plays a very important role in linear algebra, for example, if a matrix is nonsingular, i.e., it has a nonzero determinant, then we can invert it, but if it is **singular**, i.e., its determinant is equal to zero, then it does not not have an invert.

The determinant of 1×1 matrix $\mathbf{A} = [a]$ is equal to $\det \mathbf{A} = |\mathbf{A}| = a$.

The determinant of the square matrix **A** of dimension 2×2 is:

$$\det \mathbf{A} = |\mathbf{A}| = \det \begin{bmatrix} a_{11} & a_{12} \\ a_{21} & a_{22} \end{bmatrix} = a_{11}(a_{22}) - a_{12}(a_{21}) \quad (A.63)$$

Example:

$$\det \mathbf{A} = |\mathbf{A}| = \det \begin{bmatrix} 1 & 2 \\ 5 & 4 \end{bmatrix} = 1 \cdot 4 - 2 \cdot 5 = -6$$

The determinant of the 3×3 matrix **A** is:

$$\det \mathbf{A} = |\mathbf{A}|$$

$$= a_{11} \det \begin{bmatrix} a_{22} & a_{23} \\ a_{32} & a_{33} \end{bmatrix} - a_{12} \det \begin{bmatrix} a_{21} & a_{23} \\ a_{31} & a_{33} \end{bmatrix} + a_{13} \det \begin{bmatrix} a_{21} & a_{22} \\ a_{31} & a_{32} \end{bmatrix}$$

$$= a_{11}(a_{22}a_{33} - a_{23}a_{32}) - a_{12}(a_{23}a_{31} - a_{21}a_{33}) + a_{13}(a_{21}a_{32} - a_{22}a_{31})$$

$$= a_{11}a_{22}a_{33} - a_{11}a_{23}a_{32} - a_{12}a_{23}a_{31} + a_{12}a_{21}a_{33} + a_{13}a_{21}a_{32} - a_{13}a_{22}a_{31} \quad (A.64)$$

Example:

$$\det \mathbf{A} = |\mathbf{A}| = \det \begin{bmatrix} 2 & 3 & -2 \\ 1 & 4 & 0 \\ 6 & 7 & -8 \end{bmatrix}$$

$$= 2 \det \begin{bmatrix} 4 & 0 \\ 7 & -8 \end{bmatrix} - (-3) \det \begin{bmatrix} 1 & 0 \\ 6 & -8 \end{bmatrix} + (-2) \det \begin{bmatrix} 1 & 4 \\ 6 & 7 \end{bmatrix}$$

$$= 2(-32) - 3(-8) - 2(-17) = -6$$

We generalize computation of a determinant. For a square $n \times n$ matrix:

$$\mathbf{A} = \begin{bmatrix} a_{11} & a_{12} & \cdots & a_{1n} \\ a_{21} & a_{22} & \cdots & a_{2n} \\ \vdots & \vdots & \ddots & \vdots \\ a_{n1} & a_{n2} & \cdots & a_{nn} \end{bmatrix} \quad (A.65)$$

Its determinant is computed as:

$$\det \mathbf{A} = |\mathbf{A}| = a_{11}M_{11} - a_{12}M_{12} + a_{13}M_{13} + \cdots + (-1)^{n-1} a_{1n}M_{1n}, \quad (A.66)$$

where M_{1j}, called a **minor** of entry a_{1j}, is the determinant of the $(n-1) \times (n-1)$ matrix (submatrix) obtained from **A** by deleting its first row and j^{th} column. This matrix is called submatrix of matrix **A**. We see that the determinant has been found from the entries of the first row of matrix **A** and determinants of its submatrices. We call this technique row expansion. A similar procedure can be repeated starting from any other row or any column of matrix (column expansion).

Generally a minor M_{ij} of matrix **A** of entry a_{ij}, is defined to be determinant of the submatrix remaining after removing of the ith row and j^{th} column of **A**.

The sign problem for elements of determinant computation is solved by understanding of an idea of **cofactors** c_{ij} of entry a_{ij} of **A** defined as

$$c_{ij} = (-1)^{i+j} M_{ij} \tag{A.67}$$

Example:

$$\mathbf{A} = \begin{bmatrix} 2 & 3 & -2 \\ 1 & 4 & 0 \\ 6 & 7 & -8 \end{bmatrix}, \quad M_{11} = \det \begin{bmatrix} 4 & 0 \\ 7 & -8 \end{bmatrix} = 4 \cdot (-8) - 0 \cdot 7 = -32$$

and cofactor c_{11} is defined as

$$c_{11} = (-1)^{1+1} M_{11} = (-1)^2 (-32) = -32$$

A sign of cofactors $c_{ij} = \overline{+} M_{ij}$ depends on the position of the entry in relation to the matrix

$$\begin{bmatrix} + & - & + & - & \cdots \\ - & + & - & + & \cdots \\ + & - & + & - & \cdots \\ & & \vdots & & \end{bmatrix}; \ c_{11} = +M_{11}, \ c_{12} = -M_{12}, \ c_{13} = -M_{13}, \cdots \tag{A.68}$$

Let us recall definition of permutation of integers. A **permutation** of a set of integers is some arrangement of these integers but without any repetitions or omissions.

Example: The permutations of set of three integers $\{1, 2, 3\}$ are

$$(1, 2, 3), (1, 3, 2), (2, 1, 4), (2, 4, 1), (4, 1, 2), (4, 2, 1)$$

We say that an inversion in permutation occurs when a larger integer appears before a smaller one.

Definition: Let us consider permutations of numbers $1, 2, 3, \cdots, n$, denoted by $i_1, i_2, \cdots, i_n$. In permutation some numbers following i_1 may be less then i_1. The number of these cases in respect to the position i_1 is called the **number of inversions** in the arrangements pertaining to i_1. Similarly, there is a number of inversion pertaining to each of the other i^{th} position, being the number of indices that come after i^{th} in the permutation and are less than i^{th}. The **index** of the permutation is the sum of all of the number of inversions pertaining to the separate indices.

Example: The permutation

$$(7, 1, 2, 8) \ has \ two \ inversions$$

and permutation

$$(1, 2, 8, 7) \ has \ one \ inversion$$

The permutation

$$(3, 2, 1) \text{ has three inversions}$$

We observe that the determinant is a sum of all signed productions of the form $\pm a_{1i} a_{2j} a_{3k}$, where i, j, and k are permutations of 1, 2, 3 in some order. Half of productions have plus signs and other half minus signs, according to so-called index of permutation. The sign is positive when the index of permutation is even, and negative when the index is odd. Each term in the determinant has n factors where n is the dimension of the square matrix. Each factor of a term has as its second subscript a permutation of the numbers $1, 2, \cdots, n$. Terms that are odd permutations are added and terms that are even permutations are subtracted. Thus, when $n = 3$ we have as permutations, and their implied terms:

Permutation	Index	Signed product
1 2 3	0	$+a_{11}a_{22}a_{33}$
1 3 2	1	$-a_{11}a_{23}a_{32}$
2 3 1	2	$+a_{12}a_{23}a_{31}$
2 1 3	1	$-a_{12}a_{21}a_{33}$
3 1 2	2	$+a_{13}a_{21}a_{32}$
3 2 1	3	$-a_{13}a_{22}a_{31}$

and for $n = 4$:

$1\ 2\ 3\ 4 \rightarrow +a_{11}a_{22}a_{33}a_{44}$
$1\ 2\ 4\ 3 \rightarrow -a_{11}a_{22}a_{34}a_{43}$
$1\ 3\ 2\ 4 \rightarrow +a_{11}a_{23}a_{32}a_{44}$
$1\ 3\ 4\ 2 \rightarrow -a_{11}a_{23}a_{34}a_{42}$
$1\ 4\ 2\ 3 \rightarrow +a_{11}a_{24}a_{32}a_{43}$
$1\ 4\ 3\ 2 \rightarrow -a_{11}a_{24}a_{33}a_{42}$

$2\ 1\ 3\ 4 \rightarrow +a_{12}a_{21}a_{33}a_{44}$
$2\ 1\ 4\ 3 \rightarrow -a_{12}a_{21}a_{34}a_{43}$
$2\ 3\ 4\ 1 \rightarrow +a_{12}a_{23}a_{34}a_{41}$
$2\ 3\ 1\ 4 \rightarrow -a_{12}a_{23}a_{31}a_{44}$
$2\ 4\ 1\ 3 \rightarrow +a_{12}a_{24}a_{31}a_{43}$
$2\ 4\ 3\ 1 \rightarrow -a_{12}a_{24}a_{33}a_{41}$

$3\ 1\ 2\ 4 \rightarrow +a_{13}a_{21}a_{32}a_{44}$
$3\ 1\ 4\ 2 \rightarrow -a_{13}a_{21}a_{34}a_{42}$
$3\ 2\ 4\ 1 \rightarrow +a_{13}a_{22}a_{34}a_{41}$
$3\ 2\ 1\ 4 \rightarrow -a_{13}a_{22}a_{31}a_{44}$
$3\ 4\ 1\ 2 \rightarrow +a_{13}a_{24}a_{31}a_{42}$
$3\ 4\ 2\ 1 \rightarrow -a_{13}a_{24}a_{32}a_{41}$

$4\ 1\ 2\ 3 \rightarrow +a_{14}a_{21}a_{32}a_{43}$
$4\ 1\ 3\ 2 \rightarrow -a_{14}a_{21}a_{33}a_{42}$
$4\ 2\ 3\ 1 \rightarrow +a_{14}a_{22}a_{33}a_{41}$
$4\ 2\ 1\ 3 \rightarrow -a_{14}a_{22}a_{31}a_{43}$
$4\ 3\ 1\ 2 \rightarrow +a_{14}a_{23}a_{31}a_{42}$
$4\ 3\ 2\ 1 \rightarrow -a_{14}a_{23}a_{32}a_{41}$

For the square matrix A of dimension $n \times n$ a **determinant** is defined as the following scalar:

$$\det A = |A| = \sum_{i=1}^{n} \sum_{j=1, j \neq i}^{n} \cdots \sum_{l=1, l \neq i, l \neq j, \cdots, l \neq k}^{n} \text{sign } a_{1i} a_{2j} \cdots a_{nl} \quad (A.69)$$

In the above formula in each term the second subscripts $i, j, \cdots, l$ are permutations of numbers $1, 2, \cdots, n$. Additionally, terms whose subscripts are even permutations are given plus signs, and others with odd permutations are given minus signs.

2.2.1. Basic Properties of a Determinant
For a $n \times n$ matrix **A**

1. $\det \mathbf{I} = 1$
2. $\det(\mathbf{AB}) = (\det \mathbf{A})(\det \mathbf{B})$
3. $\det \mathbf{A}^T = \det \mathbf{A}$
4. $\det \alpha \mathbf{A} = \alpha^n \det \mathbf{A}$
5. The determinant of a diagonal or block-triangular matrix is equal to the product of the diagonal elements:

$$\det \begin{bmatrix} 2 & 0 \\ 0 & 3 \end{bmatrix} = 2 \cdot 3 = 6$$

$$\det \begin{bmatrix} 2 & 5 \\ 0 & 3 \end{bmatrix} = 2 \cdot 3 = 6$$

6. For a square matrix **A** such that some column is zero vector $A_i = 0$ for some i, a determinant is equal to zero. The same lemma holds for a matrix with at least one row equal to zero (having all elements equal to zero).
7. A determinant of a matrix having two identical rows is equal to zero.

Example:

$$\det \mathbf{A} = \det \begin{bmatrix} 0 & 5 & -2 \\ 0 & 3 & 8 \\ 0 & 6 & -1 \end{bmatrix} = 0$$

2.2.2. Rank of Matrix

Definition: The **rank** of matrix **A** is the **dimension** of the largest square matrix **A'** contained in **A** (obtained by deleting rows and columns in **A**) which has nonzero determinant.

We see that the nonsingular $n \times n$ matrix **A** has the rank n.

We also may say that the rank of a matrix is the maximum number of linearly independent column vectors of the matrix.

Example: The rank of matrix

$$\mathbf{A} = \begin{bmatrix} 1 & 0 & 0 & 0 \\ 0 & 0 & 0 & 0 \\ 0 & 1 & 0 & 0 \\ 0 & 0 & 0 & 0 \end{bmatrix}$$

is equal to 2 (*rank* $\mathbf{A} = 2$), since the columns from the set columns $\{[1, 0, 0, 0]^T, [0, 0, 1, 0]^T\}$ are linearly independent, whereas the columns from the set $\{[1, 0, 0, 0]^T, [0, 0, 1, 0]^T, [0, 0, 0, 0]^T\}$ are linearly dependent.

2.3. Transpose, Inverse, and Rank

2.3.1. Transpose of a Matrix

The transposition of matrix **A** denoted by $\mathbf{A}^T$ is defined as an operation of taking rows from matrix **A** and putting them as corresponding columns of $\mathbf{A}^T$ – the ith row of **A** becomes i^{th} column of $\mathbf{A}^T$, so for matrix **A** we have:

$$a_{ij}^T = a_{ji} \tag{A.70}$$

Example: If

$$\mathbf{A} = \begin{bmatrix} a_{11} & a_{12} \\ a_{21} & a_{22} \end{bmatrix}$$

then

$$\mathbf{A}^T = \begin{bmatrix} a_{11} & a_{21} \\ a_{12} & a_{22} \end{bmatrix}$$

For a rectangular matrix

$$\mathbf{A} = \begin{bmatrix} 2 & 0 \\ 1 & 0 \\ 4 & 5 \end{bmatrix}$$

we have

$$\mathbf{A}^T = \begin{bmatrix} 2 & 1 & 4 \\ 0 & 0 & 5 \end{bmatrix}$$

2.3.2. Properties of Matrix Transposition
1. $(\mathbf{AB})^T = \mathbf{B}^T \mathbf{A}^T$
2. $(\mathbf{A}^{-1})^T = (\mathbf{A}^T)^{-1}$
3. $(\mathbf{A}^{-1})^T \mathbf{A}^T = \mathbf{I}$
4. $\mathbf{A}^T (\mathbf{A}^{-1})^T = \mathbf{I}$
5. $(\mathbf{A} + \mathbf{B})^T = \mathbf{A}^T + \mathbf{B}^T$

2.3.3. Symmetric Matrix

A symmetric matrix is a matrix which equals its transpose

$$\mathbf{A}^T = \mathbf{A} \tag{A.71}$$

For example

$$\begin{bmatrix} 1 & 5 \\ 5 & 2 \end{bmatrix}^T = \begin{bmatrix} 1 & 5 \\ 5 & 2 \end{bmatrix}$$

2.3.4. Inverse of a Matrix

Consider square matrices. The inverse of a square matrix $\mathbf{A}$ is denoted by $\mathbf{A}^{-1}$ and defined such that

$$\mathbf{A}^{-1}\mathbf{A} = \mathbf{A}\mathbf{A}^{-1} = \mathbf{I}, \qquad (A.72)$$

where

$$\mathbf{A}^{-1} = \frac{1}{|A|} adj \mathbf{A} \qquad (A.73)$$

The $adj\ \mathbf{A}$ (the **adjoint of matrix A**) is the matrix formed by replacing each element with its cofactor and transposing the result.

The **cofactor** A_{ij} of matrix $\mathbf{A}$ is the the determinant of the submatrix formed by deleting from matrix $\mathbf{A}$ the ith row and j^{th} column, multiplied by $(-1)^{i+j}$.

The solution to the inverse of a matrix is time-consuming but algorithmically specified.

Algorithm: Inverse of matrix $\mathbf{A}$ with nonzero determinant.

1. Construct the matrix of cofactors of $\mathbf{A}$:

$$cof\ \mathbf{A} = \begin{bmatrix} A_{11} & A_{12} & \cdots & A_{1n} \\ \vdots & \vdots & \ddots & \vdots \\ A_{n1} & A_{n2} & \cdots & A_{nn} \end{bmatrix} \qquad (A.74)$$

2. Design the transposed matrix of cofactors matrix (**adjoint** of $\mathbf{A}$ matrix)

$$adj\ \mathbf{A} = (cof\ \mathbf{A})^T = \begin{bmatrix} A_{11} & A_{21} & \cdots & A_{n1} \\ \vdots & \vdots & \ddots & \vdots \\ A_{1n} & A_{2n} & \cdots & A_{nn} \end{bmatrix} \qquad (A.75)$$

3. Compute the inverse of $\mathbf{A}$ as

$$\mathbf{A}^{-1} = \frac{1}{\det \mathbf{A}} adj\ \mathbf{A} \qquad (A.76)$$

Example: For the square matrix $\mathbf{A}$

$$\mathbf{A} = \begin{bmatrix} a_{11} & a_{12} \\ a_{21} & a_{22} \end{bmatrix}$$

first replace each a_{ij} in $\mathbf{A}$ with the determinant of the 1×1 matrix formed by deleting the ith row and j^{th} column of $\mathbf{A}$, recall that the determinant of a matrix of one element is the element itself, we now have:

$$\begin{bmatrix} a_{22} & -a_{21} \\ -a_{12} & a_{11} \end{bmatrix}$$

next, the above matrix is transposed yielding:

$$\begin{bmatrix} a_{22} & -a_{12} \\ -a_{21} & a_{11} \end{bmatrix}$$

lastly, we multiply the above matrix by the determinant of the original matrix which gives:

$$\mathbf{A}^{-1} = \frac{1}{|\mathbf{A}|} adj\ \mathbf{A} = \frac{1}{a_{11}a_{22} - a_{12}a_{21}} \begin{bmatrix} a_{22} & -a_{21} \\ -a_{12} & a_{11} \end{bmatrix}$$

Another example is:

$$\mathbf{A} = \begin{bmatrix} 1 & -2 \\ 3 & 1 \end{bmatrix}$$

so, $\mathbf{A}^{-1}$ is:

$$\mathbf{A}^{-1} = \frac{1}{|\mathbf{A}|} adj\ \mathbf{A} = \frac{1}{1 \cdot 1 - (-2) \cdot 3} \begin{bmatrix} 1 & -3 \\ -(-2) & 1 \end{bmatrix}$$
$$= \frac{1}{7} \begin{bmatrix} 1 & -3 \\ 2 & 1 \end{bmatrix} = \begin{bmatrix} \frac{1}{7} & \frac{-3}{7} \\ \frac{2}{7} & \frac{1}{7} \end{bmatrix}$$

Not all square matrices have an inverse. **Nonsingular** matrices posses a nonzero determinant and can be inverted but **singular** matrixes have a zero determinant and the inverse for such a matrix does not exist. If the matrix is **nonsingular** then

$$(\mathbf{AB})^{-1} = \mathbf{B}^{-1}\mathbf{A}^{-1} \tag{A.77}$$

We see that the multiplication of a matrix by its inverse is commutative. Only square matrices can have inverses.

2.3.5. Application of an Inverse of Matrix

Matrix inverse is used in solving linear matrix equations (for given $\mathbf{A}$ and $\mathbf{y}$)

$$\mathbf{Ax} = \mathbf{y} \tag{A.78}$$

where $\mathbf{A}$ is a square $n \times n$ matrix and $\mathbf{x}$ and $\mathbf{y}$ are n-dimensional vectors. The solution is found by multiplying both sides of the matrix equation by $\mathbf{A}^{-1}$ as:

$$\mathbf{A}^{-1}\mathbf{Ax} = \mathbf{A}^{-1}\mathbf{y} \tag{A.79}$$

$$\mathbf{Ix} = \mathbf{A}^{-1}\mathbf{y} \tag{A.80}$$

$$\mathbf{x} = \mathbf{A}^{-1}\mathbf{y} \tag{A.81}$$

2.3.6. Matrix Pseudoinversion

Matrix inverse is defined only for square matrices. For a rectangular matrix with more rows then columns, the **pseudoinverse** is defined as:

$$\mathbf{A}^{\#} = (\mathbf{A}^T\mathbf{A})^{-1}\mathbf{A}^T \tag{A.82}$$

for nonsingular matrix $(\mathbf{A}^T\mathbf{A})$. The need for the pseudoinverse can be seen in solving the **overdetermined** linear equations (when there are more equations than unknowns)

$$\mathbf{Ax} = \mathbf{y} \tag{A.83}$$

with the solution

$$\mathbf{x} = \mathbf{A}^{\#}\mathbf{y} \tag{A.84}$$

If matrix $\mathbf{A}$ has more columns than rows the pseudoinverse if defined as

$$\mathbf{A}^{\#} = \mathbf{A}^T(\mathbf{A}\mathbf{A}^T)^{-1} \tag{A.85}$$

This case corresponds to matrix linear equations when there are fewer equations than unknowns.

2.4. Functions of Square Matrices

A matrix may be characterized by its **eigenvalues** and **eigenvectors**. These play a great role in linear algebra.

2.4.1. Eigenvalues and Eigenvectors

So far we have investigated determinant as a real scalar number characterizing a matrix. However, we will find other objects characterizing some properties of a square matrix. Consider an equation for square matrix $\mathbf{A}$ of dimension $n \times n$ and vector $\mathbf{x}$ of dimension n

$$\mathbf{Ax} = s\mathbf{x} \tag{A.86}$$

where s is a scalar. Any solution of this equation for $\mathbf{x}$ is called an **eigenvector** of matrix $\mathbf{A}$. The scalar s is called the eigenvalue of matrix $\mathbf{A}$. We see that multiplication of the eigenvector by the scalar s is equivalent to the matrix multiplication of the eigenvector. The **eigenvalue** s indicates how much $\mathbf{x}$ is shortened or lengthened after multiplication by the matrix $\mathbf{A}$ (without changing the vector "orientation"). The above equation may be rearranged into the form:

$$(\mathbf{A} - s\mathbf{I})\mathbf{x} = \mathbf{0} \tag{A.87}$$

where $\mathbf{I}$ is the identity matrix. It can be shown that a nontrivial (i.e., $\mathbf{x} \neq \mathbf{0}$) solution of this equation requires the matrix $(\mathbf{A} - s\mathbf{I})$ (or $s\mathbf{I} - \mathbf{A}$) to be singular, i.e.

$$\det(s\mathbf{I} - \mathbf{A}) = 0 \tag{A.88}$$

By expanding the exponents (in determinant formula) in the polynomial form we obtain:

$$s^n + a_1 s^{n-1} + \cdots + s a_{n-1} + a_n = 0 \tag{A.89}$$

where the s denotes a complex variable and the a_i denotes coefficients, **not** matrix elements. The last equation is called the **characteristic equation** for matrix $\mathbf{A}$.

Definition: For the square matrix $\mathbf{A}$ the equation

$$\det(s\mathbf{I} - \mathbf{A}) = s^n + a_1 s^{n-1} + a_{n-1} s + a_n = 0 \tag{A.90}$$

is called a **characteristic equation** for matrix $\mathbf{A}$, and $s^n + a_1 s^{n-1} + \cdots + s a_{n-1} + a_n = 0$ is called a **characteristic polynomial** of $\mathbf{A}$.

Definition: The **eigenvalues** $s_i (i = 1, 2, \cdots, n)$ of matrix $\mathbf{A}$ are the solutions (roots) of its characteristic equation

$$\begin{aligned}\det(s\mathbf{I} - \mathbf{A}) &= s^n + a_1 s^{n-1} + \ldots + a_{n-1} s + a_n \\ &= (s - s_1)(s - s_2) \cdots (s - s_n) = 0\end{aligned} \tag{A.91}$$

Some of the s_i may be identical (the same eigenvalues) and eigenvalues may be complex.

Definition: Each solution s_i of the characteristic equation (eigenvalue) has a corresponding **eigenvector** $\mathbf{x}_i$ which may be computed from the equation

$$(s_i \mathbf{I} - \mathbf{A}) \mathbf{x}_i = \mathbf{0}, \ or \ \mathbf{A}\mathbf{x}_i = s_i \mathbf{x}_i \tag{A.92}$$

We can find the general algebraic solution for the eignevalues of a square 2×2 matrix $\mathbf{A}$

$$\mathbf{A} = \begin{bmatrix} a_{11} & a_{12} \\ a_{21} & a_{22} \end{bmatrix} \tag{A.93}$$

First we compute the following characteristic equation of matrix $\mathbf{A}$

$$f(s) = \det(s\mathbf{I} - \mathbf{A}) = |s\mathbf{I} - \mathbf{A}| = \det \begin{bmatrix} s - a_{11} & -a_{12} \\ -a_{21} & s - a_{22} \end{bmatrix} = 0 \quad (A.94)$$

After expanding the determinant, the polynomial form of characteristic equation is

$$\begin{aligned} f(s) &= (s - a_{11})(s - a_{22}) - a_{12}a_{21} = s^2 - s(a_{11} + a_{22}) + (a_{11}a_{22} - a_{12}a_{21}) \\ &= (s - s_1)(s - s_2) = 0 \end{aligned} \quad (A.95)$$

From which we can easily find the equation for the two eigenvalues as

$$s_{1,2} = \frac{(a_{11} + a_{22}) \mp \sqrt{(a_{11} + a_{22})^2 - 4(a_{11}a_{22} - a_{12}a_{21})}}{2} \quad (A.96)$$

These eigenvalues may have different values (real or complex) depending of the values of the elements of matrix $\mathbf{A}$. For $(a_{11} + a_{22})^2 - 4(a_{11}a_{22} - a_{12}a_{21}) > 0$ the eigenvalues are real and for $(a_{11} + a_{22})^2 - 4(a_{11}a_{22} - a_{12}a_{21}) < 0$ are complex $s_i = p_i \mp jq_i; i = 1, 2$.

Example: Consider the eigenvalues of the diagonal square matrix $\mathbf{A}$

$$\mathbf{A} = \begin{bmatrix} 2 & 0 \\ 0 & 1 \end{bmatrix}$$

the characteristic equation is

$$\begin{aligned} f(s) = \det(s\mathbf{I} - \mathbf{A}) &= \det \begin{bmatrix} s - 2 & 0 \\ 0 & s - 1 \end{bmatrix} \\ &= (s - 2)(s - 1) = s^2 - 3s + 2 = 0 \end{aligned}$$

Thus matrix $\mathbf{A}$ has two distinct real eigenvalues

$$s_1 = 2, \quad s_2 = 1$$

We can conclude that for the diagonal matrix the eigenvalues are just the elements on the main diagonal. For the square matrix

$$\mathbf{B} = \begin{bmatrix} 2 & -1 \\ 2 & 1 \end{bmatrix}$$

the characteristic equation is

$$\begin{aligned} f(s) = \det(s\mathbf{I} - \mathbf{A}) &= \det \begin{bmatrix} s - 2 & 1 \\ -2 & s - 1 \end{bmatrix} \\ &= (s - 2)(s - 1) + 2 = s^2 - 3s + 4 = 0 \end{aligned}$$

Thus matrix $\mathbf{B}$ has two complex (here conjugate) eigenvalues

$$s_1 = \frac{3}{2} + j\frac{1}{2}\sqrt{7}; \quad s_2 = \frac{3}{2} - j\frac{1}{2}\sqrt{7}$$

since $(a_{11} + a_{22})^2 - 4(a_{11}a_{22} - a_{12}a_{21}) = -7 < 0$.

The eigenvectors corresponding to the eigenvalues may be found from the equation

$$(s\mathbf{I} - \mathbf{A})\mathbf{x} = \begin{bmatrix} s-2 & 0 \\ 0 & s-1 \end{bmatrix} \mathbf{x} = 0$$

The solution for the first eigenvalue $s_1 = 2$ is

$$\begin{bmatrix} 2-2 & 0 \\ 0 & 2-1 \end{bmatrix} \mathbf{x}_1 = 0$$

and gives eigenvector

$$\mathbf{x}_1 = \begin{bmatrix} c \\ 0 \end{bmatrix}$$

where c denotes any value. For the second eigenvalue $s_2 = 1$ we have

$$\begin{bmatrix} 1-2 & 0 \\ 0 & 1-1 \end{bmatrix} \mathbf{x}_2 = 0$$

and gives eigenvector

$$\mathbf{x}_2 = \begin{bmatrix} 0 \\ c \end{bmatrix}$$

where c denotes any value.

For square matrix

$$A = \begin{bmatrix} 4 & -5 \\ 2 & -3 \end{bmatrix}$$

the characteristic equation is

$$\det \begin{bmatrix} s-4 & 5 \\ -2 & s+3 \end{bmatrix} = (s-4)(s+3) + 10 = s^2 - s - 2$$

$$= (s-2)(s-(-1)) = 0$$

Thus matrix **A** has two distinct real eigenvalues

$$s_1 = 2, \quad s_2 = -1$$

since the solution of the characteristic equation is

$$s_{1,2} = \frac{1 \mp \sqrt{1 + 4 \times 2}}{2} = 2 \text{ or } -1$$

The eigenvectors may by found from the equation

$$\det \begin{bmatrix} s-4 & 5 \\ -2 & s+3 \end{bmatrix} \mathbf{x} = 0$$

For the eigenvalue $s_1 = 2$ we have the eigenvector being a multiple of

$$\mathbf{x}_1 = \begin{bmatrix} 5 \\ 2 \end{bmatrix}$$

and for the eigenvalue $s_2 = -1$ a multiple of

$$\mathbf{x}_2 = \begin{bmatrix} 1 \\ 1 \end{bmatrix}$$

We see that the eigenvectors for these distinct eigenvalues are linearly independent.

Theorem: The eigenvalues of **diagonal matrix** $\mathbf{A} = [a_{ii}]$ are equal to the elements on the main diagonal, since the characteristic equation is $(s - a_{11})(s - a_{22}) \cdots (s - a_{nn}) = 0$.

2.4.2. The Matrix Exponential e^A

An important role in linear algebra is played by **matrix exponential** for a square matrix which is expressed as:

$$e^A = I + A + \frac{1}{2!}A^2 + \frac{1}{3!}A^3 + \cdots \qquad (A.97)$$

and represents a square matrix. We have the following properties of the matrix exponent

1. $e^{A+B} = e^A e^B$, if $AB = BA$
2. $e^{ABA^{-1}} = A e^B A^{-1}$, for $|A| \neq 0$
3. $e^A = e^{traceA}$

For the diagonal square matrix A its exponent is

$$e^A = \begin{bmatrix} e^{a_{11}} & 0 & \cdots & 0 \\ 0 & e^{a_{22}} & \cdots & 0 \\ \vdots & \vdots & \ddots & \vdots \\ 0 & 0 & \cdots & e^{a_{nn}} \end{bmatrix} = \begin{bmatrix} e^{s_1} & 0 & \cdots & 0 \\ 0 & e^{s_2} & \cdots & 0 \\ \vdots & \vdots & \ddots & \vdots \\ 0 & 0 & \cdots & e^{s_n} \end{bmatrix} \qquad (A.98)$$

with elements $a_{ii}(i = 1, 2, \cdots, n)$ on the main diagonal equal to the system eigenvalues $a_{ii} = s_i (i = 1, 2, \cdots, n)$.

Example: For the matrix

$$A = \begin{bmatrix} 1 & 1 \\ 0 & 0 \end{bmatrix}$$

the exponent matrix can be found as

$$e^{At} = I + (At + \frac{A^2 t^2}{2!} + \cdots + \frac{A^k t^k}{k!} + \cdots) = I + A(t + \frac{t^2}{2!} + \cdots \frac{t^k}{k!})$$

$$= I + (e^t - 1)A = \begin{bmatrix} e^t & e^t - 1 \\ 0 & 1 \end{bmatrix}$$

since for above matrix we have $A^k = A$.

Other techniques of calculating the matrix exponents are available as well.

2.4.3. Kronecker Product of Two Matrices

The **Kronecker** poduct of two matrices $A \in \mathbb{R}^{n \times m}$ and $B \in \mathbb{R}^{p \times q}$ is the matrix of the dimension $np \times mq$

$$A \otimes B = [a_{ij} B] = [A b_{ij}] = \begin{bmatrix} a_{11} B & a_{12} B & \cdots & a_{1m} B \\ \vdots & \vdots & \ddots & \vdots \\ a_{n1} B & a_{n2} B & \cdots & a_{nm} B \end{bmatrix} \qquad (A.99)$$

2.4.4. Role of Eigenvalues and Eigenvectors

Eigenvalues and corresponding eigenvectors are a powerful tools in linear algebra and linear differential equations. For example, consider the linear mapping (transformation from x to y)

$$y = Ax \qquad (A.100)$$

where $\mathbf{x} \in \mathbb{R}^n$, $\mathbf{y} \in \mathbb{R}^n$ and $\mathbf{A} \in \mathbb{R}^{n \times n}$. Assume that $s_i (i = 1, 2, \cdots, n)$ are the distinct eigenvalues of matrix A, and $\mathbf{x}_i$ $(i = 1, 2, \cdots, n)$ are corresponding eigenvectors. Hence, since eigenvectors form the basis in $\mathbb{R}^n$ we can express the vector $\mathbf{x} \in \mathbb{R}^n$ as a linear combination of eigenvectors $\mathbf{x}_i$ $(i = 1, 2, \cdots, n)$

$$\mathbf{x} = c_1 \mathbf{x}_1 + c_2 \mathbf{x}_2 + \cdots + c_n \mathbf{x}_n \tag{A.101}$$

where c_i $(i = 1, 2, \cdots, n)$ are appropriately selected constants. We can rewrite the original equation as:

$$\mathbf{y} = \mathbf{A}\mathbf{x} = \mathbf{A}(c_1 \mathbf{x}_1 + c_2 \mathbf{x}_2 + \cdots + c_n \mathbf{x}_n) \tag{A.102}$$

Using linearity property we obtain:

$$\mathbf{y} = c_1 (\mathbf{A}\mathbf{x}_1) + c_2 (\mathbf{A}\mathbf{x}_2) + \cdots + c_n (\mathbf{A}\mathbf{x}_n) \tag{A.103}$$

Remembering that eigenvectors satisfy the equation:

$$\mathbf{A}\mathbf{x}_i = s_i \mathbf{x}_i, \quad (i = 1, 2, \cdots, n) \tag{A.104}$$

we finally obtain:

$$\mathbf{y} = c_1 s_1 \mathbf{x}_1 + c_2 s_2 \mathbf{x}_2 + \cdots + c_n s_n \mathbf{x}_n \tag{A.105}$$

Instead of matrix $\mathbf{A}$ we have used its eigenvalues and eigenvectors, with parameters $c_i(i = 1, 2, \cdots, n)$ selected for the given vector $\mathbf{x}$.

2.4.5. Solution of System of Linear Differential Equations

Let us consider, for example, the linear differential equation, where t represents time,

$$\dot{\mathbf{x}}(t) = \mathbf{A}\mathbf{x}(t) + \mathbf{B}\mathbf{u}(t) \tag{A.106}$$

$$\dot{\mathbf{x}}(t) = \mathbf{A}\mathbf{x}(t), \quad \mathbf{x}(0), \ t_0 = 0 \tag{A.107}$$

the time solution may be found in the form

$$x(t) = c_1 e^{s_1 t} \mathbf{d}_1 + c_2 e^{s_2 t} \mathbf{d}_2 + \cdots + c_n e^{s_n t} \mathbf{d}_n \tag{A.108}$$

where $s_i (i = 1, 2, \cdots, n)$ are the distinct eigenvalues of matrix $\mathbf{A}$, $\mathbf{d}_i$ $(i = 1, 2, \cdots, n)$ are corresponding eigenvectors, and coefficients $c_i (i = 1, 2, \cdots, n)$ are dependent on the initial state of the system.

2.4.6. Cayley-Hamilton Theorem

If the polynomial

$$f(s) = s^n + a_1 s^{n-1} + \cdots + a_n = 0 \tag{A.109}$$

(where the s terms are complex variables and the a_i terms are coefficients) is the characteristic equation of the square matrix $\mathbf{A}$, then matrix $\mathbf{A}$ satisfies the following equation

$$f(\mathbf{A}) = \mathbf{A}^n + a_1 \mathbf{A}^{n-1} + \cdots + a_n \mathbf{I} = 0 \tag{A.110}$$

where

$$\mathbf{A}^n = \mathbf{A}\mathbf{A} \cdots \mathbf{A} \ (n \text{ multiplications}) \tag{A.111}$$

That is, the matrix $\mathbf{A}$ satisfies its own characteristic equation.

2.5. Quadratic Forms

Let us consider the vector $\mathbf{x}$ of dimension n and square matrix Q of dimension $n \times n$. The scalar function of $\mathbf{x}$

$$J(\mathbf{x}) = \mathbf{x}^T \mathbf{Q} \mathbf{x} \tag{A.112}$$

or

$$J(\mathbf{x}) = q_{11}x_1^2 + q_{22}x_2^2 + \cdots + q_{nn}x_n^2 + 2(q_{12}x_1x_2 + q_{13}x_1x_3 + \cdots + q_{n-1,n}x_{n-1}x_n) \tag{A.113}$$

is called a **quadratic form**.

2.5.1. Definite Forms

The quadratic form is used to define properties of matrices. The square matrix $\mathbf{Q}$ is said to be:

1. **positive definite** ($\mathbf{Q} > 0$) if $\mathbf{x}^T \mathbf{Q} \mathbf{x} > 0$
2. **positive semidefinite** ($\mathbf{Q} \geq 0$) if $\mathbf{x}^T \mathbf{Q} \mathbf{x} \geq 0$
3. **negative semidefinite** ($\mathbf{Q} \leq 0$) if $\mathbf{x}^T \mathbf{Q} \mathbf{x} \leq 0$
4. **negative definite** ($\mathbf{Q} < 0$) if $\mathbf{x}^T \mathbf{Q} \mathbf{x} < 0$

for all real vectors $\mathbf{x}$.

We can test for definiteness of a square matrix $\mathbf{Q}$ independently of the vector $\mathbf{x}$. If matrix $\mathbf{Q}$ has the eigenvalues s_i, then

1. $\mathbf{Q} > 0$ if all $s_i > 0$ (positive definite)
2. $\mathbf{Q} \geq 0$ if all $s_i \geq 0$ (positive semidefinite)
3. $\mathbf{Q} \leq 0$ if all $s_i \leq 0$ (negative semidefinite)
4. $\mathbf{Q} < 0$ if all $s_i < 0$ (negative definite)

2.6. Matrix Calculus

Let $\mathbf{x}$ and $\mathbf{y}$ be n-dimensional vectors, g be a scalar function of vector $\mathbf{x}$, h be a scalar function of vectors $\mathbf{x}$ and $\mathbf{y}$, and

$$F(\mathbf{x}) = \begin{bmatrix} f_1(\mathbf{x}) & f_2(\mathbf{x}) & \cdots & f_m(\mathbf{x}) \end{bmatrix}^T \tag{A.114}$$

be an m-vector function of vector $\mathbf{x}$.

The differential of $\mathbf{x}$ is:

$$d\mathbf{x} = \begin{bmatrix} dx_1 \\ dx_2 \\ \vdots \\ dx_n \end{bmatrix} \tag{A.115}$$

Let us assume that vector $\mathbf{x}$ is a function of the independent scalar variable t,

$$\mathbf{x} = \mathbf{x}(t) = \begin{bmatrix} x_1(t)(t), & x_2(t), & \cdots, & x_n(t) \end{bmatrix}^T$$

If a vector $\mathbf{x}$ is a function of the independent variable t, then the derivative of $\mathbf{x}$ with respect to the variable t (for example time) is:

$$\frac{d\mathbf{x}}{dt} = \begin{bmatrix} \frac{dx_1}{dt} \\ \frac{dx_2}{dt} \\ \vdots \\ \frac{dx_n}{dt} \end{bmatrix} \tag{A.116}$$

2.6.1. Gradient Operations

The gradient of scalar function g with respect to $\mathbf{x}$ is the column vector

$$g_{\mathbf{x}} = \Delta g(\mathbf{x}) = \frac{\partial g}{\partial \mathbf{x}} = \begin{bmatrix} \frac{\partial g}{\partial x_1} \\ \frac{\partial g}{\partial x_2} \\ \vdots \\ \frac{\partial g}{\partial x_n} \end{bmatrix} \qquad (A.117)$$

The total differential of g is

$$dg = \left(\frac{\partial g}{\partial \mathbf{x}}\right)^T d\mathbf{x} = \sum_{i=1}^{n} \frac{\partial g}{\partial \mathbf{x}_i} d\mathbf{x}_i \qquad (A.118)$$

If $h(\mathbf{x}, \mathbf{y})$ is the scalar function of vectors $\mathbf{x}$ and $\mathbf{y}$, then

$$dh = \left(\frac{\partial h}{\partial \mathbf{x}}\right)^T d\mathbf{x} + \left(\frac{\partial h}{\partial \mathbf{y}}\right)^T d\mathbf{y} \qquad (A.119)$$

2.6.2. Hessian

The **Hessian** of scalar function g of $\mathbf{x}$ is the second derivative of g with respect to $\mathbf{x}$

$$g_{\mathbf{xx}} = \Delta_{xx} = \frac{\partial^2 g}{\partial \mathbf{x}^2} = \left[\frac{\partial^2 g}{\partial x_i \partial x_j}\right] \qquad (A.120)$$

Hessian is a symmetric $n \times n$ matrix.

2.6.3. Taylor Series Expansion

The Taylor series expansion of scalar function g of $\mathbf{x}$ around the vector value $\mathbf{x}_0$ (operational point) is defined as

$$g(\mathbf{x}) = g(\mathbf{x}_0) + \left(\frac{\partial g(\mathbf{x})}{\partial \mathbf{x}}\right)^T (\mathbf{x} - \mathbf{x}_0) + \frac{1}{2}(\mathbf{x} - \mathbf{x}_0)^T \left(\frac{\partial^2 g(\mathbf{x})}{\partial \mathbf{x}^2}\right)^T (\mathbf{x} - \mathbf{x}_0) + r \qquad (A.121)$$

where r represents a higher expansion term. The column vector $g_{\mathbf{x}} = \frac{\partial g(\mathbf{x})}{\partial \mathbf{x}}$ and the symmetric Hessian $n \times n$ matrix $g_{\mathbf{xx}} = \frac{\partial^2 g(\mathbf{x})}{\partial \mathbf{x}^2}$ are evaluated at the operational point $\mathbf{x}_0$.

2.6.4. Jacobian

For m dimensional function $F(\mathbf{x})$ a $m \times n$ dimensional Jacobian matrix (often called a Jacobian) with respect to vector $\mathbf{x}$ is defined as

$$F_{\mathbf{x}} = \frac{\partial F(\mathbf{x})}{\partial \mathbf{x}} = \begin{bmatrix} \frac{\partial f_1(\mathbf{x})}{\partial x_1} & \frac{\partial f_1(\mathbf{x})}{\partial x_2} & \cdots & \frac{\partial f_1(\mathbf{x})}{\partial x_n} \\ \frac{\partial f_2(\mathbf{x})}{\partial x_1} & \frac{\partial f_2(\mathbf{x})}{\partial x_2} & \cdots & \frac{\partial f_2(\mathbf{x})}{\partial x_n} \\ \vdots & \vdots & \ddots & \vdots \\ \frac{\partial f_m(\mathbf{x})}{\partial x_1} & \frac{\partial f_m(\mathbf{x})}{\partial x_2} & \cdots & \frac{\partial f_m(\mathbf{x})}{\partial x_n} \end{bmatrix} \qquad (A.122)$$

The total differential of $F(\mathbf{x})$ is

$$dF(\mathbf{x}) = \frac{\partial F(\mathbf{x})}{\partial \mathbf{x}} d\mathbf{x} = \sum_{i=1}^{n} \frac{\partial F(\mathbf{x})}{\partial \mathbf{x}_i} dx_i \qquad (A.123)$$

2.6.5. Useful Gradient Operations

Let us assume that the matrices $\mathbf{A}, \mathbf{B}, \mathbf{D}$ and $\mathbf{Q}$ as well as the vectors $\mathbf{x}$ an $\mathbf{y}$ have appropriate dimensions. We can find that:

$$\frac{d}{dt}(\mathbf{A}^{-1}) = -\mathbf{A}^{-1}\frac{d\mathbf{A}}{dt}\mathbf{A}^{-1} \tag{A.124}$$

The following properties of gradient operations can be derived:

1. $\frac{\partial}{\partial \mathbf{x}}(\mathbf{y}^T\mathbf{x}) = \frac{\partial}{\partial \mathbf{x}}(\mathbf{x}^T\mathbf{y}) = \mathbf{y}$
2. $\frac{\partial}{\partial \mathbf{x}}(\mathbf{y}^T\mathbf{A}\mathbf{x}) = \frac{\partial}{\partial \mathbf{x}}(\mathbf{x}^T\mathbf{A}\mathbf{y}) = \mathbf{A}\mathbf{y} = F_{\mathbf{x}}^T\mathbf{y}$
3. $\frac{\partial}{\partial \mathbf{x}}(\mathbf{x}^T\mathbf{A}x) = \mathbf{A}\mathbf{x} + \mathbf{A}^T\mathbf{x}$

However, for the symmetric matrix $\mathbf{Q}$ we have

$$\frac{\partial}{\partial \mathbf{x}}(\mathbf{x}^T\mathbf{Q}\mathbf{x}) = 2\mathbf{Q}\mathbf{x} \tag{A.125}$$

If $F(\mathbf{x})$ and $H(\mathbf{x})$ are vector functions, the chain rule of differentiation results in:

$$\frac{\partial}{\partial \mathbf{x}}(F^T(\mathbf{x})H(\mathbf{x})) = F_{\mathbf{x}}^T H(\mathbf{x}) + H_{\mathbf{x}}^T F(\mathbf{x}) \tag{A.126}$$

and Hessian property provides:

$$\frac{\partial^2 \mathbf{x}^T \mathbf{A}\mathbf{x}}{\partial \mathbf{x}^2} = \mathbf{A} + \mathbf{A}^T \tag{A.127}$$

For the symmetric matrix $\mathbf{Q}$ we have:

$$\frac{\partial^2}{\partial \mathbf{x}^2}\mathbf{x}^T \mathbf{A}\mathbf{x} = 2\mathbf{Q} \tag{A.128}$$

$$\frac{\partial^2}{\partial \mathbf{x}^2}(\mathbf{x}-\mathbf{y})^T \mathbf{Q}(\mathbf{x}-\mathbf{y}) = 2\mathbf{Q} \tag{A.129}$$

Furthermore, for the Jacobian we have:

$$\frac{\partial}{\partial \mathbf{x}}(\mathbf{A}\mathbf{x}) = \mathbf{A} \tag{A.130}$$

2.6.6. Some Properties of the Square Matrix

1. $\frac{\partial}{\partial \mathbf{A}} trace\,(\mathbf{A}) = \mathbf{I}$
2. $\frac{\partial}{\partial \mathbf{A}} trace\,(\mathbf{B}\mathbf{A}\mathbf{C}) = \mathbf{B}^T\mathbf{C}^T$
3. $\frac{\partial}{\partial \mathbf{A}} trace\,(\mathbf{A}\mathbf{B}\mathbf{A}^T) = 2\mathbf{A}\mathbf{B}$
4. $\frac{\partial}{\partial \mathbf{A}} trace\,(e^{\mathbf{A}}) = e^{\mathbf{A}}$
5. $\frac{\partial}{\partial \mathbf{A}} |\mathbf{B}\mathbf{A}\mathbf{D}| = |\mathbf{B}\mathbf{A}\mathbf{D}|A^{-T}$

where $\mathbf{A}^{-T} = (\mathbf{A}^{-1})^T$.

2.6.7. Derivative of Determinant

Let us consider square matrix $\mathbf{A} \in \mathbb{R}^{n \times n}$ which is a function $\mathbf{A} = \mathbf{A}(t)$ of the independent variable t. The derivative of the determinant is defined as:

$$\frac{d}{dt} \det \mathbf{A}(t) = \sum_{i=1}^{n} \sum_{j=1}^{n} \frac{\partial}{\partial a_{ij}} [\det \mathbf{A}(t)] \frac{da_{ij}}{dt} \qquad (A.131)$$

where a_{ij} denotes the ij^{th} element of $\mathbf{A}$.

We also find that:

$$\frac{d}{dt} \det \mathbf{A}(t) = \sum_{i=1}^{n} \sum_{j=1}^{n} \Delta_{ij} \frac{da_{ij}(t)}{dt} = \text{trace} \left[\frac{d\mathbf{A}(t)}{dt} adj \mathbf{A}(t) \right] \qquad (A.132)$$

where $adj\, \mathbf{A}(t)$ denotes the **adjacent** matrix:

$$adj\mathbf{A} = \begin{bmatrix} \Delta_{11} & \Delta_{21} & \cdots & \Delta_{n1} \\ \Delta_{21} & \Delta_{22} & \cdots & \Delta_{n2} \\ \vdots & \vdots & \ddots & \vdots \\ \Delta_{n1} & \Delta_{n2} & \cdots & \Delta_{nn} \end{bmatrix} \qquad (A.133)$$

with cofactors:

$$\Delta_{ij} = (-1)^{i+j} \det \mathbf{A}_{ij} \qquad (A.134)$$

and $\mathbf{A}_{ij}$ being a $(n-1) \times (n-1)$ matrix obtained by taking out the ith row and j^{th} column of $\mathbf{A}$.

Example: For matrix

$$\mathbf{A}(t) = \begin{bmatrix} 1 & t \\ 1-t & 2t \end{bmatrix}$$

we have

$$\det \mathbf{A} = \det \begin{bmatrix} 1 & t \\ 1-t & 2t \end{bmatrix} = t^2 + t$$

$$\frac{d\mathbf{A}(t)}{dt} = \frac{d}{dt} \begin{bmatrix} 1 & t \\ 1-t & 2t \end{bmatrix} = \begin{bmatrix} 0 & 1 \\ -1 & 2 \end{bmatrix}$$

$$adj \begin{bmatrix} 2t & -t \\ t-1 & 1 \end{bmatrix}$$

and eventually

$$\frac{d}{dt} \det \mathbf{A}(t) = \text{trace} \left[\frac{d\mathbf{A}(t)}{dt} adj\, \mathbf{A}(t) \right] = \text{trace} \begin{bmatrix} 0 & 1 \\ -1 & 2 \end{bmatrix} \begin{bmatrix} 2t & -t \\ t-1 & 1 \end{bmatrix} = 2t+1$$

2.7. Matrix Norms

It is important to find scalar measures of the matrix, which describes intrinsic matrix properties.

2.7.1. Trace of a Matrix

The **trace** of a square matrix is the scalar sum of its diagonal elements

$$\text{trace } \mathbf{A} = \sum_{i=1}^{n} a_{ii} \quad (A.135)$$

Example:

$$\text{trace } \mathbf{A} = \text{trace} \begin{bmatrix} 1 & 5 \\ 5 & 2 \end{bmatrix} = 1 + 2 = 3$$

For square matrices **A** and **B**

$$\text{trace } (\mathbf{AB}) = \text{trace } (\mathbf{BA})$$

2.7.2. Matrix Norms

The norm of a matrix is a matrix-associated quantity, like the length of vector is associated with a given vector. A norm of a matrix may be defined in many ways, but must satisfy the general requirements imposed on the norm in any set. For example, the **Euclidean matrix norm** is defined as the square root of the sum of squares of its elements

$$\|\mathbf{A}\|_E = \sqrt{\sum_{i=1}^{n} \sum_{j}^{n} a_{ij}^2} \quad (A.136)$$

Since *trace* **A** is defined as a sum of the matrix elements we can easily find the Euclidean norm of the matrix as:

$$\|\mathbf{A}\|_E = \sqrt{\text{trace } (\mathbf{A}^T \mathbf{A})} \quad (A.137)$$

Frequently, the matrix norm (spectral norm) is induced by vector Euclidean norm $\|\mathbf{x}\|$ in the following way.

Definition The matrix norm (or spectral norm) of square matrix **A** is:

$$\|\mathbf{A}\| = max_{\|\mathbf{x}\|=1} \frac{\|\mathbf{A}\mathbf{x}\|}{\|\mathbf{x}\|} = max_{\mathbf{x} \neq 0} \|\mathbf{A}\mathbf{x}\| \quad (A.138)$$

and hence

$$\|\mathbf{A}\mathbf{x}\| \leq \|\mathbf{A}\| \, \|\mathbf{x}\| \quad (A.139)$$

The norm $\|\mathbf{A}\|$ has the following properties:

1. $\|\mathbf{A}\| \geq 0$ and $\|\mathbf{A}\| = \mathbf{0}$ implies $\mathbf{A} = \mathbf{0}$
2. $\|\alpha \mathbf{A}\| = |\alpha| \|A\|$ for any scalar α
3. $\|\mathbf{A} + \mathbf{B}\| \leq \|\mathbf{A}\| + \|\mathbf{B}\|$ triangle inequality
4. $\|\mathbf{AB}\| \leq \|\mathbf{A}\| \|\mathbf{B}\|$

The first three properties are similar to the corresponding properties of a vector norm. We can also find that:

$$\|\mathbf{A}\mathbf{x}\|^2 = \mathbf{x}^T \mathbf{A}^T \mathbf{A} \mathbf{x} \leq s_{max}(\mathbf{A}^T \mathbf{A}) \|\mathbf{x}\|^2 \quad (A.140)$$

and equality holds when **x** is the eigenvector associated with maximum eigenvalue s_{max} of the matrix $\mathbf{A}^T\mathbf{A}$ and the resulting norm

$$\|\mathbf{A}\| = \sqrt{s_{max}} \tag{A.141}$$

We also have a bound of the spectral norm (with respect to the Euclidean norm):

$$\|\mathbf{A}\| \le \|\mathbf{A}\|_E \tag{A.142}$$

3. Linear Transformation

Recall that multiplication of an $m \times n$ dimensional rectangular matrix $\mathbf{A} \in \mathbb{R}^{m \times n}$ by an n-dimensional vector $\mathbf{x} \in \mathbb{R}^n$ produces a new vector $\mathbf{y} \in \mathbb{R}^m$:

$$\mathbf{y} = \mathbf{A}\mathbf{x} \tag{A.143}$$

This operation is called **linear transformation** from the space $\mathbb{R}^n$ (of starting vector $\mathbf{x} \in \mathbb{R}^n$) to the space $\mathbb{R}^m$ (of the resulting vector $\mathbf{y} \in \mathbb{R}^m$) and is denoted as $\mathbf{A}: \mathbb{R}^n \to \mathbb{R}^m$. Also, recall that

$$\mathbf{y} = \mathbf{A}\mathbf{x} = \begin{bmatrix} a_{11} & a_{12} & \cdots & a_{1n} \\ a_{21} & a_{22} & \cdots & a_{2n} \\ \vdots & \vdots & \ddots & \vdots \\ a_{m1} & a_{m2} & \cdots & a_{mn} \end{bmatrix} \begin{bmatrix} x_1 \\ x_2 \\ \vdots \\ x_n \end{bmatrix} = \begin{bmatrix} a_1^T \\ a_2^T \\ \vdots \\ a_m^T \end{bmatrix} \begin{bmatrix} x_1 \\ x_2 \\ \vdots \\ x_n \end{bmatrix} = \begin{bmatrix} a_1^T x \\ a_2^T x \\ \vdots \\ a_m^T x \end{bmatrix} \tag{A.144}$$

where $a_i^T (i = 1, 2, \cdots, m)$ denote the ith row in matrix $\mathbf{A}$.

It is convenient to visualize matrix by vector multiplication as a linear combination of columns from matrix $\mathbf{A}$:

$$\mathbf{y} = \mathbf{A}\mathbf{x} = \begin{bmatrix} a_1 & a_2 & \cdots & a_n \end{bmatrix} \begin{bmatrix} x_1 \\ x_2 \\ \vdots \\ x_n \end{bmatrix} = x_1 a_1 + x_2 a_2 + \cdots + x_n a_n \tag{A.145}$$

where $a_i (i = 1, 2, \cdots, n)$ are columns in a matrix $\mathbf{A}$. Matrix linear transformation satisfies the following:

1. It is impossible to transpose (i.e., move) the origin since $\mathbf{A} \cdot \mathbf{0} = \mathbf{0}$ for every matrix,
2. $\mathbf{A}(\alpha \mathbf{x}) = \alpha(\mathbf{A}\mathbf{x})$; for example if vector $\mathbf{x}$ is transformed (moved) to $\mathbf{x}'$ then $2\mathbf{x}$ is transforms to $2\mathbf{x}'$.
3. $\mathbf{A}(\mathbf{x} + \mathbf{y}) = \mathbf{A}\mathbf{x} + \mathbf{A}\mathbf{y}$; if vectors $\mathbf{x}$ and $\mathbf{y}$ go to $\mathbf{x}'$ and $\mathbf{y}'$, then $\mathbf{x} + \mathbf{y}$ goes to $\mathbf{x}' + \mathbf{y}'$.

We summarize these features as follows. For all scalars α and β matrix linear transformation satisfies the rule of linearity:

$$\mathbf{A}(\alpha \mathbf{x} + \beta \mathbf{y}) = \alpha(\mathbf{A}\mathbf{x}) + \beta(\mathbf{A}\mathbf{y}) \tag{A.146}$$

We see that for the transformations $F: \mathbb{R}^n \to \mathbb{R}^m$ the matrix transformation $\mathbf{y} = \mathbf{A}\mathbf{x}$ is linear, i.e., it preserves the vector space structure of $\mathbb{R}^n$ and $\mathbb{R}^m$ (preserves the notion of sum $F(\mathbf{x}+\mathbf{y}) = F(\mathbf{x}) + F(\mathbf{y})$ and the multiplication by scalar $F(\alpha \mathbf{x}) = \alpha F(\mathbf{x})$).

We find that the transpose of matrix A provides another transformation:

$$A^T : \mathbb{R}^m \to \mathbb{R}^n \tag{A.147}$$

$$\mathbf{x} = \mathbf{A}^T \mathbf{y} \tag{A.148}$$

We also notice that the $n \times n$ identity matrix **I** gives the **identity** transformation $I : \mathbb{R}^n \to \mathbb{R}^n$ which leaves **x** unchanged.

Example: The matrix

$$\mathbf{A} = \begin{bmatrix} 3 & 6 \\ 4 & 7 \\ 5 & 8 \end{bmatrix}$$

transfers vector

$$\mathbf{x}_1 = \begin{bmatrix} 1 \\ 0 \end{bmatrix} \text{ into } \mathbf{A}\mathbf{x}_1 = \begin{bmatrix} 3 \\ 4 \\ 5 \end{bmatrix}$$

and vector

$$x_2 = \begin{bmatrix} 0 \\ 1 \end{bmatrix} \text{ into } \mathbf{A}\mathbf{x}_2 = \begin{bmatrix} 6 \\ 7 \\ 8 \end{bmatrix}$$

Now recall the linear equation (transforming **x** into **y**):

$$\mathbf{y} = \mathbf{A}\mathbf{x} \tag{A.149}$$

where the square matrix $\mathbf{A}^{n \times n}$ and vector $\mathbf{y}^n$ are known. Vector **x**, the solution of this equation, is found as:

$$\mathbf{A}^{-1}\mathbf{y} = \mathbf{A}^{-1}\mathbf{A}\mathbf{x} \tag{A.150}$$

$$\mathbf{x} = \mathbf{A}^{-1}\mathbf{y} \tag{A.151}$$

3.1. Similarity Transformation

Definition Suppose that the matrices **A** and **T** are of dimension $n \times n$, as is $\mathbf{T}^{-1}$. Assume that matrix **T** is nonsingular (i.e., having nonzero determinant and possessing inverse matrix $\mathbf{T}^{-1}$). The $n \times n$ matrix

$$\mathbf{B} = \mathbf{T}^{-1}\mathbf{A}\mathbf{T} \tag{A.152}$$

is said to be **similar** to matrix **A**. The operation $\mathbf{T}^{-1}\mathbf{A}\mathbf{T}$ is called **similarity transformation**. The similar matrices **A** and $\mathbf{T}^{-1}\mathbf{A}\mathbf{T}$ share the same eigenvalues and an eigenvector **x** of matrix **A** corresponds to an eigenvector $\mathbf{T}^{-1}\mathbf{x}$ of matrix $\mathbf{T}^{-1}\mathbf{A}\mathbf{T}$. Similar matrices represent the same transformation with respect to different bases.

Theorem: The eigenvalues of the matrix are invariant under a similarity transformation, that is, the matrix **A** and $\mathbf{T}\mathbf{A}\mathbf{T}^{-1}$ have the same set of eigenvalues for any nonsingular transformation matrix **T**.

3.2. Change of Basis

Let us recall that a basis for a vector space is a set of linearly independent vectors that span that space. In spite of the tendency to think in the category of standard basis, there are situations that for convenience of analysis it is better to change the basis. The numbers representing the vector elements (coordinates) are relative to the choice of basis. When the basis is changed the coordinates change as well. To find new coordinates of vectors related to the new basis, we start from the original coordinates in the original basis. Consider the vector $\mathbf{x} = \begin{bmatrix} 2 \\ 1 \end{bmatrix}$ in the standard basis $\mathbf{e}_1 = \begin{bmatrix} 1 \\ 0 \end{bmatrix}$ and $\mathbf{e}_2 = \begin{bmatrix} 0 \\ 1 \end{bmatrix}$. We change the basis by choosing new basis vectors $\mathbf{d}_1 = \begin{bmatrix} 1 \\ -1 \end{bmatrix}$ and $\mathbf{d}_2 = \begin{bmatrix} 1 \\ 1 \end{bmatrix}$. Let $\mathbf{x}^*$ represent vector $\mathbf{x}$ in the new basis. We see that vector $\mathbf{x}$ expressed in terms of the new basis as $\mathbf{x}^*$ is obtained as a linear combination of d_1 and d_2:

$$\mathbf{x}^* = 0.5 \begin{bmatrix} 1 \\ -1 \end{bmatrix} + 1.5 \begin{bmatrix} 1 \\ 1 \end{bmatrix}$$

Thus, we have:

$$\mathbf{x}^* = \begin{bmatrix} 0.5 \\ 1.5 \end{bmatrix}$$

We can easily generalize this procedure. The coordinate elements x_i^*, $(i = 1, 2, \cdots, n)$ of the vector $\mathbf{x}^*$, obtained from the vector $\mathbf{x}$, are just coefficients in this equation:

$$\mathbf{x}^* = \alpha_1 \mathbf{d}_1 + \alpha_2 \mathbf{d}_2 + \cdots + \alpha_n \mathbf{d}_n \tag{A.153}$$

We express this procedure of basis change using matrix notation. For this purpose we form matrix $\mathbf{T}$ consisting of the rows being new basis vectors $\mathbf{d}_i$ $(i = 1, 2, \cdots, n)$:

$$\mathbf{T} = \begin{bmatrix} \mathbf{d}_1 & \mathbf{d}_2 & \cdots & \mathbf{d}_n \end{bmatrix} \tag{A.154}$$

Thus, the above coordinate change equation is written as:

$$\mathbf{x} = \mathbf{T}\mathbf{x}^* \tag{A.155}$$

where $\mathbf{x}^*$ is the unknown representation of vector $\mathbf{x}$ in the new basis. We find the solution to this equation as:

$$\mathbf{x}^* = \mathbf{T}^{-1}\mathbf{x} \tag{A.156}$$

For the above example we have

$$\mathbf{T} = \begin{bmatrix} 1 & -1 \\ 1 & 1 \end{bmatrix}$$

and

$$\mathbf{T}^{-1} = \begin{bmatrix} 0.5 & -0.5 \\ 0.5 & 0.5 \end{bmatrix}$$

We now easily find the new coordinates of vector $\mathbf{x} = \begin{bmatrix} 2 \\ 1 \end{bmatrix}$ in the new basis as

$$\mathbf{x}^* = \mathbf{T}^{-1}\mathbf{x} = \begin{bmatrix} 0.5 & -0.5 \\ 0.5 & 0.5 \end{bmatrix} \begin{bmatrix} 2 \\ 1 \end{bmatrix} = \begin{bmatrix} 0.5 \\ 1.5 \end{bmatrix}$$

We should remember that $\mathbf{x}$ and $\mathbf{x}^*$ are really the same vectors but described in terms of different bases.

Now we consider the situation for the transformation operation $\mathbf{y} = \mathbf{Ax}$ when we change the basis and express the vectors $\mathbf{x}$ and $\mathbf{y}$ in the new basis as $\mathbf{x}^*$ and $\mathbf{y}^*$. We look for the matrix $\mathbf{A}^*$ which provides, in the new basis, the same transformation as $\mathbf{A}$ does in the old basis, i.e.:

$$\mathbf{y}^* = \mathbf{A}^*\mathbf{x}^* \tag{A.157}$$

Using backward conversion from $\mathbf{x}^*$ to $\mathbf{x}$ we have

$$\mathbf{x} = \mathbf{T}\mathbf{x}^* \tag{A.158}$$

where matrix $\mathbf{T}$ consist of columns of the new basis vectors and:

$$\mathbf{y} = \mathbf{Ax} \tag{A.159}$$

$$\mathbf{y}^* = \mathbf{T}^{-1}\mathbf{y} \tag{A.160}$$

Joining the above equations we obtain:

$$\mathbf{y}^* = \mathbf{T}^{-1}\mathbf{y} = \mathbf{T}^{-1}\mathbf{Ax} = \mathbf{T}^{-1}\mathbf{A}\mathbf{T}\mathbf{x}^* \tag{A.161}$$

We obtain the matrix $\mathbf{A}^*$ as

$$\mathbf{A}^* = \mathbf{T}^{-1}\mathbf{A}\mathbf{T} \tag{A.162}$$

When the basis changes the vector coordinates change according to the equation $\mathbf{A}^* = \mathbf{T}^{-1}\mathbf{A}\mathbf{T}$. The matrices $\mathbf{A}^*$ and $\mathbf{T}^{-1}\mathbf{A}\mathbf{T}$ are similar.

3.3. Matrix Diagonalizing Transformation

It is difficult to handle matrix exponentiation and other complex matrix operations. Knowing that the operations on diagonal matrices are much simpler we look for a transformation of the original system with a nondiagonal matrix to the new coordinates with a diagonal matrix, which is easier to handle. We observe that a particular situation is created when, for equation $\mathbf{y} = \mathbf{Ax}$, we select a new basis $\mathbf{d}_i$ $(i = 1, 2, \cdots, n)$ being equal to the eigenvectors of the matrix $\mathbf{A}$. By definition, for every eigenvector $\mathbf{d}_i$ $(i = 1, 2, \cdots, n)$ of matrix $\mathbf{A}$ we have

$$\mathbf{A}\mathbf{d}_i = s_i \mathbf{d}_i, \quad (i = 1, 2, \cdots, n) \tag{A.163}$$

where s_i $(i = 1, 2, \cdots, n)$ are eigenvalues of matrix $\mathbf{A}$. If we select matrix $\mathbf{T}$ having columns that are the eigenvectors $\mathbf{d}_i$ $(i = 1, 2, \cdots, n)$ of matrix $\mathbf{A}$:

$$\mathbf{T} = \begin{bmatrix} \mathbf{d}_1 & \mathbf{d}_2 & \cdots & \mathbf{d}_n \end{bmatrix} \tag{A.164}$$

we can generalize the above equation to

$$\mathbf{AT} = \mathbf{T}\Lambda \tag{A.165}$$

where Λ is a diagonal matrix whose elements on the main diagonal are the eigenvalues s_i ($i = 1, 2, \cdots, n$) of matrix $\mathbf{A}$:

$$\Lambda = \begin{bmatrix} s_1 & . & . & . \\ . & s_2 & . & . \\ . & . & . & . \\ . & . & . & s_n \end{bmatrix} \tag{A.166}$$

Finally, multiplying both sides of the above equation by $\mathbf{T}^{-1}$ we conclude that:

$$\mathbf{A}^* = \mathbf{T}^{-1}\mathbf{A}\mathbf{T} = \Lambda \tag{A.167}$$

The conclusion is that, for the new basis, **being a set of eigenvectors of matrix A**, the transition matrix $\mathbf{A}^*$ in basis corresponding to the original basis matrix $\mathbf{A}$, is just the diagonal matrix consisting of the eigenvalues on the main diagonal. We know how to compute the eigenvector of matrix $\mathbf{A}$ as the (nontrivial) solutions of the equation:

$$(s\mathbf{I} - \mathbf{A})\mathbf{d}_i = 0, \quad (i = 1, 2, \cdots, n) \tag{A.168}$$

3.4. Jordan Canonical Form of a Matrix

From the previous section we know how to diagonalize a matrix by linear transformation. Now we consider a more general form of diagonal-like matrices called **Jordan canonical form**.

If $\mathbf{A}$ is $n \times n$ square matrix and if we let $\mathbf{A}_k(s)$ denote the $k \times k$ matrix of the form:

$$\mathbf{A}_k(s) = \begin{bmatrix} s & 1 & 0 & \cdots & 0 & 0 \\ 0 & s & 1 & \cdots & 0 & 0 \\ \vdots & \vdots & \vdots & \ddots & \vdots & \vdots \\ 0 & 0 & 0 & \cdots & s & 1 \\ 0 & 0 & 0 & \cdots & 0 & s \end{bmatrix} \tag{A.169}$$

then if the eigenvalues of the matrix $\mathbf{A}$ are real, then $\mathbf{A}$ is similar to the matrix of the form:

$$J(\mathbf{A}) = \begin{bmatrix} \mathbf{A}_{m_1}(s_1) & | & \mathbf{0} & | & \cdots & | & \mathbf{0} \\ \mathbf{0} & | & \mathbf{A}_{m_2}(s_2) & | & \cdots & | & \mathbf{0} \\ \vdots & | & \vdots & | & \ddots & | & \vdots \\ \mathbf{0} & | & \mathbf{0} & | & \cdots & | & \mathbf{A}_{m_p}(s_p) \end{bmatrix} \tag{A.170}$$

where $m_1 + m_2 + \cdots + m_p = n$ and $s_1, s_2, \cdots, s_p$ are the eigenvalues of the matrix $\mathbf{A}$ (not necessarily distinct). The matrix $J(\mathbf{A})$ is called the **Jordan canonical form** of the matrix $\mathbf{A}$. The Jordan canonical form of the matrix $\mathbf{A}$ has the properties that the elements on the main diagonal of the matrix are eigenvalues of $\mathbf{A}$ and the elements immediately above (or below) the main diagonal are either 1 or 0 and all other elements are zeros.

We notice that the diagonal matrix is a special case of the Jordan canonical form. The matrices $\mathbf{A}_{m_i}(s_i)$ are called m_i **order Jordan block**.

Example: Let us consider a 2×2 matrix $\mathbf{A}$ with a double eigenvalue s (i.e., roots of $\det(\mathbf{A} - s\mathbf{I}) = 0$). Then we may find that $\mathbf{A}$ is similar to either matrix of the Jordan canonical form:

$$J(\mathbf{A}) = \begin{bmatrix} s & 0 \\ 0 & s \end{bmatrix} \quad J(\mathbf{A}) = \begin{bmatrix} s & 1 \\ 0 & s \end{bmatrix}$$

Example: Let us consider a 3×3 matrix $\mathbf{A}$ with a triple s eigenvalue of s (i.e., roots of $\det(\mathbf{A} - s\mathbf{I}) = 0$). Then we find that $\mathbf{A}$ is similar to either matrix of the Jordan canonical form:

$$J(\mathbf{A}) = \begin{bmatrix} s & 1 & 0 \\ 0 & s & 1 \\ 0 & 0 & s \end{bmatrix} \quad J(\mathbf{A}) = \begin{bmatrix} s & 1 & | & 0 \\ 0 & s & | & 0 \\ \cdots & \cdots & | & \cdots \\ 0 & 0 & | & s \end{bmatrix} \quad J(\mathbf{A}) = \begin{bmatrix} s & | & 0 & 0 \\ \cdots & | & \cdots & \cdots \\ 0 & | & s & | & 0 \\ 0 & | & 0 & | & s \end{bmatrix}$$

Example: Let us consider the matrix **A** with double eigenvalue s_1 and one distinct eigenvalue s_2. The following Jordan forms of the matrix **A** are possible:

$$J(\mathbf{A}) = \begin{bmatrix} s_1 & 1 & 0 \\ 0 & s_1 & 0 \\ 0 & 0 & s_2 \end{bmatrix} \quad J(\mathbf{A}) = \begin{bmatrix} s_1 & 0 & 0 \\ 0 & s_1 & 0 \\ 0 & 0 & s_2 \end{bmatrix}$$

We see that all Jordan forms have the same characteristic equation $(s - s_1)(s - s_2)(s - s_3) = 0$.

3.5. Partitioned Matrices

If the symmetric matrix is composed of matrices on the main diagonal

$$\mathbf{B} = \begin{bmatrix} \mathbf{A}_{11} & 0 & \cdots & 0 \\ 0 & \mathbf{A}_{22} & \cdots & 0 \\ \vdots & \vdots & \ddots & \vdots \\ 0 & 0 & \cdots & \mathbf{A}_{nn} \end{bmatrix} \quad (A.171)$$

then **B** is called the **block diagonal matrix**.

If the composite matrices $\mathbf{A}_{ii}$ ($i = 1, 2, \cdots, n$) are square, then

$$\det \mathbf{B} = \det \mathbf{A}_{11} \det \mathbf{A}_{22} \cdots \det \mathbf{A}_{nn} \quad (A.172)$$

If $\det \mathbf{B} \neq 0$ then

$$\mathbf{B}^{-1} = \begin{bmatrix} \mathbf{A}_{11}^{-1} & 0 & \cdots & 0 \\ 0 & \mathbf{A}_{22}^{-1} & \cdots & 0 \\ \vdots & \vdots & \ddots & \vdots \\ 0 & 0 & \cdots & \mathbf{A}_{nn}^{-1} \end{bmatrix} \quad (A.173)$$

Appendix B

Probability

In this appendix we review basic concepts of probability theory.

1. Basic Concepts

Incomplete and imprecise knowledge mitigate against predicting an event with certainty, where **event** is defined as an outcome or collection of outcomes, and **outcome** is defined as an instance of single experiment result. Probability provides a mechanism for dealing with this uncertainty, and probability theory is the mathematical technique of describing uncertainty and randomness occurring in real-life situations. The probability concept of an uncertain, random event relates to measuring a chance, or a likelihood, that an event will occur, in a statistical experiment in which events are sampled from a defined population to make statistical inferences.

Conceptually, probability is defined by an experiment (process, observation) in which outcomes are considered to be both observed and justified. An experiment may be active or passive. In an active experiment, an investigator or researcher intentionally manipulates experimental variables and conditions according to established goals and objectives. For example, providing a researcher-controlled input voltage pulse to an electronic circuit and then measuring the output may be considered an active experiment. In a passive experiment, the role of the investigator or researcher is to observe and record experiment results without altering experiment conditions. For example, recording daily temperatures is an example of the passive experiment. In both active and passive experiments outcomes are also called "observed data."

Let us assume an experiment in which one possible outcome is an event denoted by A. The uncertain nature of the experiment is such that event A may – or may not – occur when the experiment is performed. We designate the probability that A will occur by $P(A)$. Let $n_{repetition}$ denote the number of times the experiment is performed and $n_{occurrence}$ denote the number of times A occurs during the $n_{repetition}$ executions of the experiment. With $n_{occurrence}$ and $n_{repetition}$ we can designate the likelihood that event A will occur by the following proportion:

$$\frac{n_{occurrence}}{n_{repetition}} \quad (B.1)$$

To interpret a probability, let us assume the experiments are repeated an infinite number of times. As the number of experiments increases, the ratio $\frac{n_{occurrence}}{n_{repetition}}$ converges to the probability of A. Hence,

$$P(A) = \lim_{n_{repetition} \to \infty} \frac{n_{occurrence}}{n_{repetition}} \quad (B.2)$$

Example: Let us consider an experiment that involves tossing a coin. The possible outcomes (results) of a coin-toss experiment are a **head** and a **tail**. Since an impartial observer cannot analyze or control the experiment conditions or predict the outcome, the results of tossing the coin are considered random. If we designate to A a **head** outcome for our coin-toss experiment, we assign to $n_{occurrence}$ the number of times A (a **head** outcome) occurs as we toss the coin. As the number of tosses approaches infinity, the proportion of **heads** to the number of repetitions, $\frac{n_{occurrence}}{n_{repetition}}$, approaches 0.5. Thus we conclude that the probability of a **head** occurring, $P(A)$, is 0.5. From Equation (B.2) we get:

$$P(A) = \lim_{n_{repetition} \to \infty} \frac{n_{occurrence}}{n_{repetition}} = 0.5$$

Since it is impossible to perform an infinity number of experiments, we must come up with an alternative approach that will approximate the same value. To do this we estimate the probability of an event occurring with a value equivalent to what we would obtain if we were able to perform the experiment an infinite number of times. Intuitively, then, we can consider **probability** as the relative frequency of a specific event occurring as an experiment is executed a large number of times.

2. Probability Laws

As we stated above, a single result of an experiment is called an **outcome**. A collection of all possible outcomes is called a **sample space**. Within the context of a sample space, a single outcome is generally referred to as an **element** or a **point** in that space, and because they are single elements or points, outcomes do not overlap in a sample space. The number of element or points in a sample space can be any number up to infinity, depending on the nature of the experiment, and are mutually exclusive (disjoint). Sample space is usually designated by $S = \{s_1, s_2, s_3, \ldots\}$. For example, in an experiment of tossing a fair coin, the outcomes (sample space) are \{**head, tail**\}. In an experiment determining whether a person has had the **rubella** disease, the outcomes (sample space) are \{**yes, no**\}.

Previously, we said that an **event** can be a specific outcome or a collection of outcomes. Events are usually denoted by capital letters – for example A, B. An event can be interpreted as a subset of an outcome set. For example, the event A may contain a collection of one or more elements in the sample space, S, which would form a subset of S. We designate the fact that event A occurs in the sample space, S, as:

$$A \subseteq S$$

It is important to understand that an event cannot be considered merely as an outcome. As event may contain zero, one or more outcomes (elements or points) in the sample space. Further, an event may occur as a consequence of more than one repetition of an experiment, or it may not occur at all. If the event contains zero outcomes it is referred to as an **empty set**, usually designated as $\emptyset$, and is an important concept for building probability theory.

Consider an experiment of tossing a fair, 6-sided die with a different number of spots, or "pips," on each face. The sample space (or outcome space) for this experiment is

$$S = \{1, 2, 3, 4, 5, 6\}$$

We can define an event, A, as "**tossing a number 4 in one toss**." Here, if A occurs, it contains only one outcome from the sample space. It is the subset $A = \{4\}$ of the outcome set, S.

$$A = \{4\} \subset \{1, 2, 3, 4, 5, 6\}$$

In another example, an event, B, can be defined as "**tossing an odd number in one toss**." This time, if B occurs, that is, an odd number is thrown, the subset of the outcome set, S, is $B = \{1, 3, 5\}S$ or

$$B = \{1, 3, 5\} \subset \{1, 2, 3, 4, 5, 6\}$$

Going back to our probability definition above, (B.2), for event A, if we want a single number, 4, to occur, we set $n_{occurrence} = 1$. Since there are 6 possible numbers that could come up, we set $n_{repetition} = 6$. From formula (B.2), we can see that

$$P(A) = \frac{1}{6}$$

3. Probability Axioms

The probability $P(A)$ of an event A to occur, as a measure of uncertainty, takes a real number value from the range $[0, 1]$.

The **system of probabilities** assigns probabilities $P(A)$ to events A based on three **probability theory axioms**.

Axiom 1. The probability is greater or equal to zero and less or equal to 1

$$0 \leq P(A) \leq 1 \quad for \quad each \; event \; A \tag{B.3}$$

The probability cannot be negative.

Axiom 2. An event $A = S$, which includes all outcomes from a sample space S, must have a probability equal to 1. An event $A = S$ is called a **certain event**. When $P(A) = 0$, an event cannot occur. $P(A) = 1$ means that an event will always occur.

Axiom 3. Let events $A, B, C, \cdots$ be mutually exclusive (disjoint). The probability of the event that "**A or B or C or** $\cdots$" will happen is

$$\begin{aligned} P(A \text{ or } B \text{ or } C \text{ or } \cdots) &= P(A \cup B \cup C \cup \cdots) \\ &= P(A) + P(B) + P(C) + \cdots \end{aligned} \tag{B.4}$$

The third axiom states that for non-overlapping events (disjoint) the probability of their union is equal to the sum of the probabilities of the individual events.

4. Defining Events With Set–Theoretic Operations

Once an event has been defined as a subset of a sample space (outcome set), we use set–theoretic operations to define events and their compositions.

Let us consider two events A and B from the sample space S ($A \subseteq S$, $B \subseteq S$). We define other events using set theoretic operators:

1 $\bar{A}$ (**not A, complement of A**) – an event consisting of all elements of a sample space not belonging to A.
2 $A \cup B$ (**A or B**) – an event consisting of all outcomes belonging to A, B, or A and B.
3 $A \cap B$ (**A and B**) – an event consisting of all outcomes belonging simultaneously to A and B.
4 $A \setminus B$ (**A but not B**) – an event consisting of all outcomes belonging to A, but not to B.
5 $A \triangle B$ (**A xor B but not both**) – an event consisting of all outcomes belonging to A or B, but not belonging simultaneously to A and B.

Complement $\bar{A}$ of a Given Event. The **complement** $\bar{A}$ of an event A can be understood as the nonoccurrence of event A. An event A and its complement $\bar{A}$ are mutually exclusive. Because an event A and $\bar{A}$ are all possible events that may happen in an experiment, thus the sum of their probabilities equals 1

$$P(\bar{A}) = 1 - P(A) \tag{B.5}$$

Example: A complement for an event "**having disease**" (denoted by A) is an event "**not having disease**" (denoted by $\bar{A}$). If the probability of an event "**having disease**" (A) is equal to 0.2, thus the probability of an event "**not having disease**" ($\bar{A}$) is $P(\bar{A}) = 1 - 0.2 = 0.8$

Union of Events $A \cup B$. An event "**A or B**," referred to as **union** $A \cup B$ of events A and B, defines the event "**either A or B will occur**."

For two mutually exclusive events (disjoint, non-overlapping) we have

$$P(A \text{ or } B) = P(A \cup B) = P(A) + P(B) \tag{B.6}$$

If events overlap, with a common region $A \cap B$, the resulting probability is defined as

$$P(A \text{ or } B) = P(A \cup B) = P(A) + P(B) - P(A \cap B) \tag{B.7}$$

In the above formula, the term $P(A \cap B)$ is extracted in order to avoid counting the sum $P(A)$ and $P(B)$ twice in the overlapping region.

Example: Let us consider an experiment of tossing a fair coin twice. Let an event A will be defined as "a head in the first toss" and event B as "a head in the second toss." Now we may define a new event C as "**at least one head in two tosses**" being union $A \cup B$ of events A and B. Since events A and B are not mutually exclusive, thus the probability of occurrence "**at least one head in two tosses**" (occurring of an event $C = A$ or $B = A \cup B$) is

$$P(A \cup B) = P(A) + P(B) - P(A \cap B) = 0.5 + 0.5 - 0.25 = 0.75$$

Example: As an example of mutually exclusive events (disjoint), let us define an event A – "**husband has had and wife has not had disease**," and an event B – "**husband has not had and wife has had disease**." We may define a new event C – "**husband has had and wife has not had disease**" or "**husband has not had and wife has had disease**," being union $A \cup B$. We may compute probability $P(C) = P(A \cup B)$ of an event $C = A \cup B$ knowing probabilities $P(A)$ and $P(B)$ for events A and B. If probabilities of events A and B are both equal to 0.2, then

$$P(A \cup B) = P(A) + P(B) = 0.16 + 0.16 = 0.32$$

Intersection of Events. Let us consider two events A and B with corresponding probabilities $P(A)$ and $P(B)$. We may define a new event "**both A and B occur**." From the set theory point of view this new event is a subset of a sample space, being a set intersection $A \cap B$ of sets representing events A and B. The probability of this event is denoted by

$$P(A \text{ and } B) = P(A \cap B) \tag{B.8}$$

For independent events A and B we have

$$P(A \cap B) = P(A)P(B) \tag{B.9}$$

For dependent events formula is more complicated and will be presented later.

Example: In the experiment of tossing a fair coin, A denotes an event "**head occurs in the first toss**" with the probability $P(A) = 0.5$, and an event B "**head occurs in the second toss**" with the probability $P(B) = 0.5$. Events A and B are independent, because result of the first experiment does not affect result of the second experiment. We may define a new event as intersection of events A and B, $A \cap B$ "**head occurs in the first toss**" **and** "**a head occurs in the second toss**":

$$P(A \cap B) = P(A)P(B) = 0.5 \cdot 0.5 = 0.25$$

While defining events and their combinations as a subset of a sample space, we must be sure that the events, sets are sufficiently complete. Operations on events must also yield properly defined events from a sample space. We must also provide a technique of assigning probability to defined events according to probability axioms.

Two events are **mutually exclusive** when they cannot occur simultaneously. In other words, the happening of one event excludes the happening of the other.

Two events A and B are considered to be **independent** if the probability of occurrence of one event is not affected by occurrence of the other.

Probability that two independent events A and B will occur simultaneously ($A \cap B$, **A and B**) is

$$P(A \cap B) = P(A \text{ and } B) = P(A)P(B) \tag{B.10}$$

$$P(A \cup B) = P(A \text{ or } B) = P(A) + P(B) \tag{B.11}$$

5. Conditional Probability

The fact that a particular event has already occurred may influence the occurrence of another event, and it influences value of this event's probability. A **conditional probability** is the probability of an event occurring given the knowledge that another event already occurred.

The conditional probability that an event A will happen given that an event B has already happened is denoted by $P(A|B)$.

Example: Consider an experiment of tossing a fair coin twice. Let us define the events A "**head in the first toss**" and B "**head in the second toss**." We may define the event C as "**two heads in the first two tosses**." Before any tossing the probability of tossing "**two heads in the first two tosses**" (probability of happening of the event C) is $P(\text{two heads in the first two tosses}) = 0.25$. If in the first toss the outcome was a head (the event A happened), then the conditional probability that "**two head in the first two tosses**" will happen (the event C will happen) is

$$P(\text{two heads in the first two tosses} \mid \text{a head in the firsttoss}) = 0.5$$

or

$$P(C|A) = 0.5$$

Knowledge of an outcome of the first toss being a head (happening of the event A), provides additional information that changes an estimation of the probability of tossing "**two heads in the first two tosses**" (happening of the event C).

In the extreme situation the knowledge of the first event may reveal that the second event never or always occurs.

For example, probability of an event C "**two heads in the first two tosses**" under condition that in the first toss a head has not occurr (the event A has not happened, which means that the event $\bar{A}$ happened) is equal to 0

$P(\text{two heads in the first two tosses} \mid \text{no head in the first toss}) = 0$

or

$$P(C|\bar{A}) = 0$$

The conditional probability allows us to use existing knowledge that concerns events that happened prior to a given event in order to improve a probability estimation of this event.

The probability $P(A)$ of an event A prior to any observation (what we know initially) is called **prior** probability. Subsequent observation can be used to improve this probability regarded now as a conditional probability which concerns increased knowledge of the situation.

A conditional probability, which is estimated after some measurements of prior event outcomes, is called **posterior** probability.

Definition: The conditional probability of event A given event B (probability of A conditioned on B) is defined as:

$$P(A|B) = \frac{P(A \text{ and } B)}{P(B)} \tag{B.12}$$

$$P(A|B) = \frac{P(A \cap B)}{P(B)} \tag{B.13}$$

with $P(B) \neq 0$.

If happening of event A has no influence on probability of event B, then these events are independent, i.e.,

$$P(A) = P(A|B) \text{ and } P(B) = A(B|A) \tag{B.14}$$

For instance, the happening of an event "**rain**" has no effect on outcomes of tossing twice a fair coin resulting in the happening of an event **two heads in the first two tosses**"

$P(\text{two heads in the first two tosses} \mid \text{rain}) = P(\text{ two heads in the first two tosses})$

When new information is available as a result of a prior event, this information can be used to improve the estimate of prior probability. This may result in three outcomes: new probability will increase, decrease or not change.

6. Multiplicative Rule of Probability

The definition of the conditional probability leads to **multiplicative rule** of probability

$$P(A \text{ and } B) = P(A \cap B) = P(A|B)P(B) \tag{B.15}$$

6.1. Independence

Two events A and B are independent if:

$$P(A|B) = P(A) \ P(B|A) = P(B) \tag{B.16}$$

and

$$P(A \cap B) = P(A)P(B)$$

6.2. Bayes Rule (Theorem)

Combining equations

$$P(A|B) = \frac{P(A \cap B)}{P(B)} \text{ and } P(B|A) = \frac{P(B \cap A)}{P(A)} \quad (B.17)$$

leads to the **Bayes rule (Bayes theorem)**

$$P(A|B) = \frac{P(B|A)P(A)}{P(B)} \quad (B.18)$$

The Bayes rule is useful to swap events in conditional probability evaluation. The conditional probability $P(A|B)$ can be expressed by the conditional probability $P(B|A)$, $P(A)$ and $P(B)$.

The Bayes rule can be extended to a collection of events $A_1, \cdots, A_n$ conditioned on the event B

$$P(A_i|B) = \frac{P(B|A_i)P(A_i)}{P(B)} = \frac{P(B|A_i)P(A_i)}{\sum_{i=1}^{n}[P(A_i)P(B|A_i)]} \quad (B.19)$$

where

$$P(B) = \sum_{i=1}^{n}[P(B|A_i)P(A_i)] \quad (B.20)$$

7. Random Variables

In some situations outcomes of experiments are non-numerical, like a tail in a fair coin tossing. However, statistical inferring is usually expressed in numerical terms. In physical or biological experiments the physical quantities represent outcomes of experiments. For example, we may measure pressure, velocity, and so forth. In these situations, happening or not of some events relates to numbers rather than to nominal values like head, tail, color, shape, and the like.

A random variable is related to random outcomes associated with real numbers. Let us assume that we have a set of all outcomes of a given experiment, which represents a sample space with associated probabilities defined in a sample space. A **random variable** is a function (a rule) that maps (assigns) outcomes of experiments to real numbers. A random variable assumes numerical values associated with each outcome from the original sample space. Assigning of outcomes to real numbers can be done in different ways, depending on the overall project goal.

Example: When tossing a die, we have 6 outcomes in a sample space. Each of them represents a pattern of dots (one dot, two dots, $\cdots$, six dots), which in fact are not numbers. In the random variable space, we can assign a real number for each outcome from a sample space in the following way:

```
one    dot   –  1
two    dots  –  2
three  dots  –  3
four   dots  –  4
five   dots  –  5
six    dots  –  6
```

The above assignment represents mapping from the original sample space into equivalent real number outcomes. This mapping is arbitrary and can be done in different ways, i.e., one dot could correspond to 6, two dots to 5, etc.

Events defined in the original sample space can be also defined in real number space according to provided mapping. If probability has been assigned (according to the probability axioms) for each event in the original sample space, then the probabilities can be assigned in an analogous way to corresponding events in the random variable space:

$P(\text{event in a random variable space}) = P(\text{corresponding event in an original sample space})$

Example: In the experiment of tossing a fair coin twice we can define in the original sample space events:

"**0 heads**," "**1 head**" and "**2 heads**"

with corresponding probabilities

$$\frac{1}{4}, \frac{2}{4} \text{ and } \frac{1}{4}.$$

We can define a random variable X that take real values 0, 1, 2 in the random variable space, with the following mapping of events from the original sample space:

Original sample space	Random variable sample space
0 heads	0
1 head	1
2 heads	2

Given probabilities that are assigned to the events in the original sample space, we can assign the same probabilities to corresponding events in a random variable space:

Original sample space	Random variable sample space	Probability
$P(0\ heads)$	$= P(0)$	$= \frac{1}{4}$
$P(1\ head)$	$= P(1)$	$= \frac{2}{4}$
$P(2\ heads)$	$= P(2)$	$= \frac{1}{4}$

7.1. Discrete and Continuous Random Variables

Random variables can be of two types, discrete and continuous, depending on a set of values taken.

A **discrete** random variable takes only a very small (countable) number of values. Its values may be isolated (separated) rather than continuously spaced.

Coin tossing outcomes are discrete. Consequently, the corresponding random variable obtained from a specific mapping of a coin face to real numbers is a discrete random variable.

A **continuous** random variable may take any real number value in a given interval of real numbers.

A process of measuring people's height results in a continuous outcome. Corresponding random variable obtained from a specific mapping of height to real numbers is a discrete random variable.

Value taken by a random variable as a result of an experiment is called a **realization** of the random variable.

8. Probability Distribution

Realizations of a random variable are governed by a random law. A realization of a random variable in a given experiment cannot by predicted with absolute certainty. Instead, we may predict a certain realization of a random variable by using a probability of occurrence of the given realization. To do that we define potential values of realizations of a random variable along with probabilities of likelihood of their occurrences. We have to assume that the likelihood of distinct realizations of a random variable may be different. This leads to probability distribution for a random variable, which is defined as a set of probabilities for all realizations of the variable.

Definition: A **probability distribution** of a random variable is a set of its possible values (realizations) with assigned probabilities as measures of likelihood that these values will occur. The probability distributions are introduced separately for discrete and continuous random variables.

8.1. Probability Distribution for Discrete Random Variables

A discrete **probability distribution** of a discrete random variable is a set of its possible values (realizations) with associated probabilities of their occurrences.

Let us consider an experiment of tossing a fair coin twice with two possible outcomes: a **head** and a **tail** in one toss.

For the defined (in the original sample space) events representing a number of heads obtained "**0 heads**," "**1 head**" and "**2 heads**" with the corresponding probabilities $\frac{1}{4}$, $\frac{2}{4}$ and $\frac{1}{4}$ we can define a discrete random variable, X, taking real values $0, 1, 2$ in the random variable space, that correspond to the events in original sample space. Probability distribution for the discrete random variable X is a set of probabilities for all realizations of X:

$$p(0) = \frac{1}{4}, \ p(1) = \frac{2}{4}, \ p(2) = \frac{1}{4}$$

or

$$p(X=0) = \frac{1}{4}, p(X=1) = \frac{2}{4}, p(X=2) = \frac{1}{4}$$

The sum of all probabilities must equal 1

$$\sum_{all x} p(x) = p(0) + p(1) + p(3) = \frac{1}{4} + \frac{2}{4} + \frac{1}{4} = 1$$

This probability distribution can be shown using a histogram with bar representing a probability value for each outcome.

A probability distribution can be presented as a probability distribution function of a random variable. Given that X denote a random variable taking possible values x_i from a discrete set $S = \{x_1, x_2, \cdots\}$, the probability distribution function $p(x)$ is defined as:

$$p(x) = P(X = x), \ for \ all \ x \ \{x_1, x_2, \ldots\} \tag{B.21}$$

$$\sum_{all \ x} p(x) = 1$$

where $P(X = x)$ denotes a probability that a random variable X takes a value (realization) x. In the same case, probability distribution function for a random variable can be expressed as an algebraic formula of x, for example

$$p(x) = \frac{x}{4}$$

A probability distribution function $p(x)$ gives probability of occurrence for all individual values (realizations) of a random variable X. In order to obtain a probability of occurrence for a specific subset of values of X, we need to simply add corresponding probabilities of values from this subset.

If S is a sample space for a random variable X, and A is a subset of a sample space $A \subseteq S$, then the probability that one value of a subset A will occur is

$$P(X \in A) = \sum_{x \in A} p(x) \tag{B.22}$$

Example: Let us consider a random variable X with corresponding realizations $S = \{x_1 = 1, x_2 = 2, x_3 = 3, x_4 = 4\} = \{1, 2, 3, 4\}$ and the associated probabilities for these realizations

$$\begin{aligned} p(1) &= \tfrac{1}{10} \\ p(2) &= \tfrac{2}{10} \\ p(3) &= \tfrac{3}{10} \\ p(4) &= \tfrac{4}{10} \end{aligned}$$

$$\sum_{1}^{4} p(x_i) = 1$$

The probability that X will take a value from the subset $A = \{2, 3\}$ ($A \subset S$ being a subset of S) is

$$P(X \in \{2, 3\}) = p(X = 2) + p(X = 3) = p(2) + p(3) = \frac{2}{10} + \frac{3}{10} = \frac{5}{10}$$

8.2. Cumulative Probability Function

Definition: A **cumulative probability** function $F(x)$ of a random variable X gives probability that X takes value x or smaller

$$F(x) = P(X \leq x) \tag{B.23}$$

For a discrete random variable with a discrete probability distribution we have

$$F(x) = \sum_{z \leq x} p(z) \tag{B.24}$$

It represents a sum of probabilities of all values of a random variable equal or less than x.

Example: For the previous example

$$F(2) = P(X \leq 2) = p(1) + p(2) = \frac{1}{10} + \frac{2}{10} = \frac{3}{10}$$

A cumulative probability function may be depicted as non-decreasing function graph.

8.3. Summarizing Measures for Probability Distribution

In order to represent a shape of probability distribution several characteristic measures are defined.

For a discrete probability distribution, a probability histogram depicts a shape of the distribution. The histogram corresponds to the relative frequency of occurrence of the experimental outcomes. However, it represents theoretical probabilities instead of experimental outcome frequencies.

Let us consider averaging of a distribution of a discrete random variable X. As in other numerical averaging processes, average for random variables is computed as the sum of variable values divided by their number. If we consider a finite number, N, of trials for a discrete random variable X, then for these samples, we can compute a sample **average** $\bar{X}$ (or a sample **mean**) as:

$$\bar{X} = \frac{X_1 + X_2 + \cdots + X_N}{N} \tag{B.25}$$

where $X_1, X_2, \cdots X_N$ are realizations from a sample obtained as outcomes of N trials. The above average is computed for a finite number of N realization of a random variable. If we extend computing this average for an infinite number of trials we can consider this as an average referred to as the **expected value** of a random variable X.

Assume that a discrete random variable X has n possible realizations $\{x_1, x_2, \cdots, x_n\}$ with the corresponding probabilities of occurrence $p(x_1), p(x_2), \cdots, p(x_n)$. For N trials (when N is large) we expect approximately $p(x_1) \cdot N$ occurrences of x_1, $p(x_2) \cdot N$ occurrences of x_2, and $p(x_i) \cdot N$ occurrences of x_i, where $i = 1, 2, \ldots, n$. By recalling that the probability $p(x_i)$ is defined as a limit of a ratio of a number of occurrences of x_i to a number of trials N ($\lim_{N \to \infty} \frac{n_{occurrences}}{N}$), a sample average for a discrete random variable X is given as:

$$\bar{X} = \frac{(p(x_1)N)x_1 + (p(x_2)N)x_2 + \cdots + (p(x_n)N)}{N}$$
$$= p(x_1)x_1 + p(x_2)x_2 + \cdots + p(x_n)x_n \tag{B.26}$$

This leads to the definition of an expected (mean) value of a random variable.

8.3.1. Expected value

The **expected value** (the **mean**) of a discrete random variable X with possible realizations $\{x_1, x_2, \cdots, x_n\}$ is defined as

$$Expected\ value\ of\ X = E[X] = \sum_{i=1}^{n} p(x_i) x_i \tag{B.27}$$

The expected value (also called *mean* of a discrete probability distribution) is denoted by $E[X]$ or μ. The expected value of a discrete probability distribution represents an expected value of a random variable.

We can also define the variance of a random variable as a measure of its spread over the mean.

8.3.2. Variance

The **variance** of a random variable X is defined as

$$Variance\ of\ X = VAR\ X = E[(X - E[X])^2] \tag{B.28}$$

We have

$$VAR\ X = \sum_{i=1}^{n} (x_i - E[X])^2 p(x_i) \tag{B.29}$$

The variance of X measures the dispersion of X over its mean. Since expectation of a sum is a sum of expectation, we have

$$VAR\ X = E[X^2 - 2XE[X] + E[X]^2] = E[X^2] - (E[X])^2 \tag{B.30}$$

The square root of of a variance $\sqrt{VAR}$, called the standard deviation, is also an important measure related to probability distribution.

8.3.3. Standard Deviation

The **standard deviation** of a random variable X, denoted as σ, is defined as

$$Standard\ deviation\ of\ X = \sigma = \sqrt{VAR X}. \tag{B.31}$$

The standard deviation is usually denoted by σ and the variance by σ^2 ($\sigma^2 = VAR\ X$).

The variance of a random variable is its mean square distance from the center of the probability distribution.

8.3.4. Moments

The expected value can be computed for a function $f(X)$ of a random variable X, as well as for X. The expected value for $f(X)$ is defined as

$$E[f(X)] = \sum_{i=1}^{n} f(x_i) p(x_i) \tag{B.32}$$

The expected value of a function $f(X) = X^k$ (power k of X) of a discrete random variable X is called the kth moment of X and it is defined as

$$E[X^k] = \sum_{i=1}^{n} x_i^k p(x_i) \tag{B.33}$$

We can see that the second order moment $E[X^2]$ of X

$$E[X^2] = \sum_{i=1}^{n} x_i^2 p(x_i) \tag{B.34}$$

is equal to the variance of a random variable X with the expected value equal to 0.

8.4. Probability Distribution of Discrete Random Variables

The widely used probability distributions for discrete variables are: uniform, binomial, geometric, and Poisson.

8.4.1. Uniform Distribution

A discrete random variable X has the uniform distribution when probabilities for its all realizations are equal and the sum of all probabilities equals 1

$$p(x_1) = p(x_2) = \cdots = p(x_n) \tag{B.35}$$

$$\sum_{i}^{n} p(x_i) = 1$$

Example: Let us consider a random variable X taking a finite number of n integer values as realizations $S = \{1, 2, \cdots, n\}$ with equal frequency of occurrence of each realization. The probability of obtaining in an experiment some realization from the set of possible realizations is given by the close algebraic formula

$$p(x_i) = \frac{1}{n}, \quad i = 1, 2, \ldots, n$$

$$\sum_{i}^{n} p(x_i) = 1$$

Having a close algebraic formula for a probability distribution we can compute summarizing measures for distribution like the mean or the standard deviation. For the uniform distribution of a random variable X taking integer values from the set $\{1, 2, \ldots n\}$ we can compute the mean as:

$$E[X] = \mu = (1)\frac{1}{n} + (2)\frac{1}{n} + \cdots + (n)\frac{1}{n} = \frac{n+1}{2}$$

$$\sigma^2 = \sum_{1}^{n}(x_i - \mu)^2 = \frac{n^2 - 1}{12}$$

Example: For tossing a fair die we can define a discrete random variable having a uniform probability distribution. Let us define a random variable X representing events being number of spots on a face of a die. The random variable X takes six discrete realizations x_i from the set $S_x = \{1, 2, 3, 4, 5, 6\}$. The probability of each realization is

$$p(x_i) = \frac{1}{6}, \quad x_i = 1, 2, 3, 4, 5, 6$$

$$\sum_{1}^{6} p(x_i) = \sum_{1}^{6} \frac{1}{6} = 1$$

The mean and the variance of X are

$$E[X] = \mu = (1)\frac{1}{6} + (2)\frac{1}{6} + \cdots + (6)\frac{1}{6} = \frac{6+1}{2} = 3.5$$

$$\sigma^2 = \frac{6^2 - 1}{12} = 2.917$$

We notice that neither realization of X is equal to a mean value.

8.4.2. Binomial Distribution

The probability distribution of a number of "positive" outcomes for an experiment with only two outcomes denoted by "positive" and "negative" is governed by the binomial distribution.

Let us consider a sequence of trials (experiments) with only two outcomes: "success" and "failure" in each trial. We also assume that probability of a "success" in one trial is constant and equal to p. For this sequence of trials we may define a discrete random variable X representing a number of "successes" obtained in n consecutive trials.

Definition: Let us assume that following conditions hold

1) The experiment consist of fixed number (sequence) of n trials.
2) There are only two outcomes of each trial: "success" and "failure."
3) The probability of "success" in one trial is constant and equal to p. This implies that the outcomes of trials are independent.

The **binomial** probability of the discrete random variable X, representing a number of "successes" in a sequence of n trials, is defined as:

$$p(x) = C_x^n p^x (1-p)^{n-x} = \frac{n!}{x!(n-x)!} p^x (1-p)^{n-x}, \text{ for } x = 0, 1, \ldots, n \quad \text{(B.36)}$$

$$0 < p < 1$$

The term

$$C_x^n = \binom{n}{x} = \frac{n!}{x!(n-x)!} \quad \text{(B.37)}$$

denotes the number of ways (combination) of labeling by "success" and "failure" in n trials sequence. The binomial distribution has only one parameter p representing the probability of "success" in one trial (with probability of a "failure" equal to $1-p$).

The $p(x)$ is the probability of getting exactly x successes in n trials. For n trials, we have $(n-x)$ failures and x successes. Since trials are independent, the probability of x "successes" is a product of probabilities of x individual "successes" $p^x = p \cdot p \cdot p \cdots p$, and $(1-p)^{n-x}$ is the probability of $n-x$ "failures" in the n trials with x successes. Therefore, the probability of x "successes" and $n-x$ "failures" in the n trial sequence is:

$$p^x(1-p)^{n-x} \tag{B.38}$$

The above probability was computed only for one sequence of n trials. The number of all possible combinations in which a sequence can be labeled by x "successes" and $n-x$ "failures" is given by:

$$C_x^n = \frac{n!}{x!(n-x)!} \tag{B.39}$$

Eventually, the probability of x "successes" and $n-x$ "failures" in n trials can be computed using the addition rule, which is given by multiplication of probability of a sequence by the number of all possible n trial sequences.

For the binomial distribution with the given probability p of "success" in one trail we find that:

$$\mu = np \tag{B.40}$$

$$\sigma^2 = np(1-p)$$

where

$$\sum_{x=0}^{n} p = 1$$

8.4.3. Geometric Distribution

Let us again consider a sequence of trials with two possible outcomes for each trial with fixed probability p of a "**success**" (and $1-p$ for a "**failure**").

We may define a discrete random variable X representing a "**number of trials required to first get a "success"**" outcome. The discrete probability of this variable is governed by the geometric distribution.

Definition: The **geometric distribution** is a discrete probability distribution governed by the following probability function:

$$p(x) = \begin{cases} p(1-p)^{x-1} & \text{for} \quad x = 1, 2, \cdots \\ 0 & \text{otherwise} \end{cases} \tag{B.41}$$

for $0 < p < 1$.

Given X with a realization x (x trials with the first "**success**" in x^{th} outcome), there must be initially $x-1$ consecutive "**failures**" in $x-1$ first trials, followed by the "**success**." Thus in the definition of the geometric distribution there is $x-1$ terms with probability $(1-p)$ (yielding a resulting probability as $(1-p)^{x-1}$), and one term p representing the first "**success**."

The name geometric distribution came from the fact that $p(1), p(2), p(3), \cdots$ represent terms in the infinite geometric series for $p\frac{1}{1-(1-p)}$.

For the geometric distribution we have:

$$E[X] = \mu = \sum_{i=1}^{\infty} x p(1-p)^{x-1} = \frac{1}{p} \tag{B.42}$$

$$VAR\ X = \sigma^2 = \sum_{i=1}^{\infty} (x-\mu)^2 p(1-p)^{x-1} = \frac{1}{p}$$

$$= \sum_{i=1}^{\infty} (x - \frac{1}{p})^2 p(1-p)^{x-1} = \frac{1-p}{p^2}$$

8.4.4. Poisson Distribution

In some applications we may be interested in probability distribution of events happening at random moments in time. For example, a bank customer may arrive at any time during working hours. We might be interested in probability distribution of a discrete random variable representing "**a count of number of occurrences of a random event across a specific time interval**." Similarly, we might be interested in probability distribution of a discrete random variable representing "**a count of number of occurrences of a random event across a specific fragment of a space**" (for example, area of a plane). These random events probabilities are governed by the Poisson probability distribution.

Definition: The **Poisson (λ) distribution** is a discrete probability distribution given by the formula

$$p(x) = \begin{cases} p(x) = \frac{\lambda^x e^{-\lambda}}{x!} & \text{for} \quad x = 0, 1, 2, \cdots \\ 0 & \text{otherwise} \end{cases} \tag{B.43}$$

where $\lambda > 0$ is a positive distribution parameter.

In order to better understand the role of the parameter λ let us assume that events happen at an average rate α per unit of a time. Then, the probability distribution of a discrete random variable X representing "**a count of number of occurrences of a random event across a specific time interval T**" is defined by the following Poisson distribution

$$p(x) = \begin{cases} p(x) = \frac{(\lambda T)^x e^{-\lambda T}}{x!} & \text{for} \quad x = 0, 1, 2, \cdots \\ 0 & \text{otherwise} \end{cases} \tag{B.44}$$

For the Poisson probability distribution with the the parameter λ we have:

$$E[X] = \mu = \sum_{x=0}^{\infty} x \frac{\lambda^x e^{-\lambda}}{x!} = \lambda \tag{B.45}$$

$$VAR\ X = \sigma^2 = \sum_{x=0}^{\infty} (x-\mu)^2 \frac{\lambda^x e^{-\lambda}}{x!} = \lambda$$

The shape of the Poisson distribution for values $x = 0, 2, 3, \cdots$ is right-skewed with probability histograms having the maximum near the parameter λ.

8.5. Continuous Probability Distribution

When results of experiments are measured by physical devices, continuous random variables well describe a reality.

A continuous random variable can take any value from a given range (interval). This means that a number of possible values that a continuous random variable can take is infinite (uncountable). Therefore, we must consider probability measures for a random variable taking an infinite number of realizations. This requires a different different conceptual approach when compared with measures defined for discrete random variables. This approach resembles the concept of a mass density in mechanics and leads to the idea of a probability density. In order to describe a continuous probability distribution, defined for infinite number of realizations of a continuous random variable, we need to find a way of describing "how high" (how intensive) the probability is for different sets of realizations. This measure is called a probability density function.

Definition: A **probability density function** for a continuous random variable X is a function $f(x)$ with the following characteristics:

1) $f(x)$ is non-negative function,
2) The total area under the curve representing a function plot and a horizontal axis of x must equal one

$$\int_{-\infty}^{\infty} f(x)\, dx = 1 \tag{B.46}$$

3) The probability that a random variable X will take a value within a given interval of realizations $[a, b]$ $(a \leq b)$ is

$$P[a \leq X \leq b] = \int_{a}^{b} f(x)\, dx \tag{B.47}$$

The probability value $P[a \leq X \leq b] = \int_{a}^{b} f(x)\, dx$ is equal to the area between a function is curve plot and horizontal axis for the interval $[a, b]$ of a random variable realization. This way of integrating the probability of having a realization of a continuous random variable within a given interval corresponds to adding probabilities for computing a probability of a discrete random variable.

The probability density functions typically have a symmetric centered shape or right or left skewed shape.

The idea of a probability density function can be better understood if we assume computing a probability of variable X realized over an infinitely small interval $[a, b]$ of the length $b - a$ denoted by Δx. The area between a curve $f(x)$ for Δx can be approximated by $f(a)\Delta x$. In other words, if $f(x)$ is a value of a probability density function around x, then for a small interval dx around x a probability of realization of X within dx can be the area $f(x)\, dx$. Hence, to obtain probability of X's realization from the interval $[a, b]$ we need to sum all values $f(x)\, dx$ within the interval ($\sum_{x=a}^{b} f(x)\, dx$) for which limit is $\int_{a}^{b} f(x)\, dx$.

The interpretation of a probability distribution $P(Y)$ for a discrete random variable Y is different than a probability density function $f(x)$ of a continuous random variable X. Actually $f(x)$ does not describe a probability $P[X = x]$ that a continuous random variable X takes a realization x like $P[y]$ does for a discrete variable Y. This is caused by the fact that X can take an infinite number of realizations, which means that it is impossible to assign a probability value for a given realization $X = x$. A probability density function may take any non-negative value (assuming that total area under a curve $f(x)$ equals 1). Like in continuous mass distribution in mechanics, where mass of the single point is equal to 0, a probability that X takes one given realization x is equal to 0

$$P[a \leq X \leq a] = P[x] = \int_{a}^{a} f(x)\, dx = 0 \tag{B.48}$$

This implies that the inequality $a \leq X \leq b$ in the probability $P[a \leq X \leq b]$ can be substituted by $a < X < b$

$$P[a \leq X \leq b] = P[a < X < b] \tag{B.49}$$

Similarly

$$P[a \leq X \leq b] = P[a \leq X < b] = P[a < X \leq b] \tag{B.50}$$

Therefore, a probability density function $f(x)$ is used to compute a probability $P[a \leq X \leq b] = \int_a^b f(x)\, dx$ that X will take a value from the interval $[a, b]$.

8.5.1. Cumulative Probability Function for a Continuous Random Variable

A cumulative probability function for a continuous random variable X is a function $F(x)$ that for a specific value x gives a probability that X will take that value or smaller

$$F(x) = P[X \leq x] \tag{B.51}$$

For a continuous random variable X a cumulative probability function can be expressed as an integral of its probability density function $f(x)$ for all arguments of $X \leq x$

$$F(x) = P[X < x] = \int_{-\infty}^{x} f(z)\, dz \tag{B.52}$$

From calculus we have that:

$$\frac{d}{dx}F(x) = f(x) \tag{B.53}$$

which shows that a probability density function $f(x)$ can be obtained by differentiation of a cumulative probability function $F(x)$.

A cumulative probability function for a continuous random variable X has to satisfy the following properties:

1) $F(x)$ is a non-decreasing function of x
2) $F(x) \to 0$ as $x \to -\infty$
3) $F(x) \to 1$ as $x \to \infty$

8.5.2. Expected Value and Variance of a Continuous Random Variable

Since shape of a probability density function plot reveals type of a probability distribution of a continuous random variable, two measures, namely the expected value and the variance, can be used to characterize it.

Definition: The **expected value** (the **mean**) of a continuous random variable X with a probability density function $f(x)$ is defined as

$$Expected\ value\ of\ X = E[X] = \mu = \int_{-\infty}^{\infty} x\, f(x)\, dx \tag{B.54}$$

The expected value for a continuous random variable corresponds to a mass center in mechanics.

By analogy to an inertia of a continuous mass distribution we define a variance for a continuous random variable.

Definition: The **variance** of a continuous random variable X with a probability density function $f(x)$ is defined as

$$Variance\ of\ X = VAR\ X = \sigma^2 = \int_{-\infty}^{\infty} (x - E[X])^2\, f(x)\, dx$$

$$= \int_{-\infty}^{\infty} (x - \mu)^2\, f(x)\, dx \tag{B.55}$$

We have that

$$\text{VAR } X = \sigma^2 = \int_{-\infty}^{\infty} x^2 f(x) \, dx - E[X]^2 \tag{B.56}$$

The **standard deviation** of a continuous random variable X is defined as a square root of the variance

$$\sigma = \sqrt{\text{VAR } X} = \sqrt{\sigma^2} \tag{B.57}$$

and it is denoted by σ^2.

8.5.3. Moments of Continuous Random Variables

Similarly as in the case of discrete random variables we define moments for continuous random variables as the expected values for a specific type of continuous random variables functions $f(X) = X^k$ with an integer k.

Definition: The kth **moment** of a continuous random variable X is defined as

$$E[X^k] = \int_{-\infty}^{\infty} x^k f(x) \, dx \tag{B.58}$$

The first moment ($k = 1$) is equal to the expected value μ of X.

The **second moment** ($k = 2$) of a continuous random variable X is defined as

$$E[X^2] = \int_{-\infty}^{\infty} x^2 f(x) \, dx \tag{B.59}$$

The second moment ($k = 2$) of the random variable X is equal to its variance σ^2 for the expected value μ equal to zero.

8.6. Normal or Gaussian Probability Distribution

The normal or Gaussian probability distribution for continuous random variables is one of the most commonly used in practice.

The normal probability distribution can be used to approximate probability distribution for outcomes of many natural phenomena.

8.6.1. Univariate Normal Distribution

Definition: A scalar continuous-valued random variable X has the **normal** or **Gaussian** probability distribution if its probability density function is defined as

$$f(x) = \frac{1}{\sqrt{2\pi\sigma^2}} e^{-\frac{(x-\mu)^2}{2\sigma^2}}, \quad -\infty < x < \infty \; (all \; x) \tag{B.60}$$

with $\sigma > 0$, and $e = 2.71828$ being the "Napier constant" (Napierian logarithm base), and $\pi \approx 3.14159$ being a ratio of circumference to a diameter of a circle. The μ is the **mean**, σ^2 is the **variance**, and σ (the square root of the variance) is the **standard deviation** of the normal probability density.

The Gaussian continuous random variable is a variable that has the normal probability density function and this probability density function plot is called the **normal curve**.

The area under the normal curve (between plot of a probability density function an the axis x) is equal to 1

$$\int_{-\infty}^{\infty} f(x) \, dx = \int_{-\infty}^{\infty} \frac{1}{\sqrt{2\pi\sigma^2}} e^{-\frac{(x-\mu)^2}{2\sigma^2}} \, dx = 1 \tag{B.61}$$

The $\frac{1}{\sqrt{2\pi\sigma^2}}$ ensures that the above integral equals to 1.

The normal distribution of a continuous random variable is specified by two parameters:

1) Expected value (mean) $E[X] = \mu$, and
2) Standard deviation $E[(X-\mu)^2] = \sqrt{VAR\ X} = \sigma^2$

For normal distribution:

$$E[X] = \int_{-\infty}^{\infty} x\ f(x)\ dx = \int_{-\infty}^{\infty} x \frac{1}{\sqrt{2\pi\sigma^2}} e^{-\frac{(x-\mu)^2}{2\sigma^2}} dx = \mu \tag{B.62}$$

$$E[(X - E[X])^2] = E[(X-\mu)^2] = \int_{-\infty}^{\infty} (x-\mu)^2\ f(x)\ dx$$

$$= \int_{-\infty}^{\infty} (x-\mu)^2 \frac{1}{\sqrt{2\pi\sigma^2}} e^{-\frac{(x-\mu)^2}{2\sigma^2}} dx = \sigma^2 \tag{B.63}$$

The normal density function (its shape) is computed for given values of parameters μ and σ^2. Often this density function is denoted by $N(\mu, \sigma^2)$. Random samples from normal distribution cluster around the mean and have spread proportional to the standard deviation σ^2.

A graph of the normal density function is bell-shaped with a peak at mean value and inflection points at $\mu - \sigma$ and $\mu + \sigma$.

The normal distribution and the normal curve representing the normal probability distribution are characterized by the following important properties:

1) They have simple analytical properties allowing many useful results in explicit analytical form. For example, any moment of normal distribution can be found as a function of μ and σ.
2) The normal curve is centered at the mean value μ.
3) The mean value μ corresponds to maximum value of the normal curve.
4) The normal curve is symmetric around the mean value μ.
5) Because the normal curve is a probability distribution, thus the total area under the curve is equal to 1.
6) The area under the normal curve for values above the mean and below the mean are equal and each equal to 0.5.
7) For given values of the mean μ and variance σ, the normal distribution has the maximum entropy, and thus minimal information of all distributions having the same mean and standard deviation.

The normal cumulative probability function for the normal random variable is defined as:

$$P(X \leq x) = \int_{-\infty}^{x} \frac{1}{\sqrt{2\pi\sigma^2}} e^{-\frac{(z-\mu)^2}{2\sigma^2}} dz \tag{B.64}$$

This cumulative probability function for the normal distribution has no close algebraic form and must be computed numerically. For $x = \mu$ we have that $P(\mu) = 0.5$.

8.6.2. Standard Normal Distribution

A special case of the normal distribution with $\mu = 0$ and $\sigma = 1$ is called the **standard normal distribution**.

Definition: The normal distribution of a random variable with $\mu = 0$ and $\sigma = 1$ is called the **standard normal distribution** and is denoted by:

$$\phi(z) = \frac{1}{\sqrt{2\pi}} e^{-\frac{z^2}{2}} \tag{B.65}$$

The cumulative probability for the standard normal distribution is:

$$\Phi(z) = \frac{1}{\sqrt{2\pi}} \int_{-\infty}^{z} e^{-\frac{y^2}{2}} dy \tag{B.66}$$

Since there is no close form for computing probabilities for the normal distribution (for example a cumulative distribution), we use probability tables, which are defined for the standard normal distribution. The probabilities $P(X \leq x)$ and $P(a \leq X \leq b)$ for the normal distribution with any μ and σ can be computed by using transformation of variables to a standard distribution and computing probability for a transformed variable from the standard normal distribution.

Let us assume that we have a random variable X normally distributed with a given mean μ and standard deviation σ (generally with values different that 0 and 1). If we wish to compute a cumulative probability $F(x) = P(X \leq x)$ for the value x of the variable X we first transform the variable X into the standard normal distribution with $\mu = 0$ and $\sigma = 1$ using the following general formula:

$$Z = \frac{X - \mu}{\sigma} \tag{B.67}$$

and for a given value x as:

$$z = \frac{x - \mu}{\sigma} \tag{B.68}$$

Next, we compute the cumulative probability value $F(z)$ for z using the table for the standard normal distribution. This probability equals $F(x)$ for the realization x of the variable X. This is by virtue of the fact that an area under the normal curve from $-\infty$ to x equals the area under a standard normal curve from $-\infty$ to $z = \frac{x-\mu}{\sigma}$. Similarly we can compute the area under the normal curve for all realizations (values) of the variable X from the interval $[a, b]$:

$$P(a \leq X \leq b) = \frac{1}{\sqrt{2\pi\sigma^2}} \int_a^b e^{-\frac{(y-\mu)^2}{2\sigma^2}} dy = \frac{1}{\sqrt{2\pi}} \int_{\frac{a-\mu}{\sigma}}^{\frac{b-\mu}{\sigma}} e^{-\frac{z^2}{2}} dz$$

$$= \Phi(\frac{b-\mu}{\sigma}) - \Phi(\frac{a-\mu}{\sigma}) \tag{B.69}$$

The algorithm for computing the cumulative probability $F(x) = P(X \leq X)$ for the the random variable X the normal distribution is as follows.

Algorithm: Given a variable X normally distributed with parameters μ and σ, and a value x of a variable X.

1) Compute a transformed value $z = \frac{x-\mu}{\sigma}$.
2) Find a value $F(z) = \Phi(z)$ for z from a table for the standard normal distribution.
3) The demanded value $F(x)$ for the cumulative probability for the normally distribute variable equals $F(z)$.

8.6.3. Percentages of Probabilities under the Normal Curve

Here we evaluate percentages under the normal curve for three characteristic ranges of realizations $[\mu - \sigma, \mu + \sigma]$, $[\mu - 2\sigma, \mu + 2\sigma]$ and $[\mu - 3\sigma, \mu + 3\sigma]$ of the normally distributed random variable with parameters μ and σ^2.

1) $P(\mu - \sigma \leq x \leq \mu + \sigma) = 0.6826$
   This tells that 68.26% of realizations of the normally distributed variable will fall in the region $[\mu - \sigma, \mu + \sigma]$.
2) $P(\mu - 2\sigma \leq x \leq \mu + 2\sigma) = 0.9544$
   This tells that 95.44% of realizations of the normally distributed variable will fall in the region $[\mu - 2\sigma, \mu + 2\sigma]$.
3) $P(\mu - 3\sigma \leq x \leq \mu + 3\sigma) = 0.9975$
   This tells that 99.75% of realizations of the normally distributed variable will fall in the region $[\mu - 3\sigma, \mu + 3\sigma]$.

Appendix C

Lines and Planes in Space

In this appendix we cover the basic operations of lines on plane.

1. Lines on Plane

Definition: The **point-slope equation** of a line through a point (x_1, y_1) with a slope m is

$$y - y_1 = m(x - x_1) \text{ or } \frac{y - y_1}{x - x_1} = m, \tag{C.1}$$

where $m = \tan \phi$ with ϕ being an angle between the x-axis and the line (see Figure C.1).

Example: The line passing the point $P = (2, 1)$ ($x_1 = 2$, and $y_1 = 1$) with the slope $m = 2$ has the equation

$$y - y_1 = m(x - x_1)$$
$$y - 1 = 2(x - 2)$$
$$y = 2x - 3$$

Definition: The x-coordinate of the intersection point of a line with the x-axis is called the **x-intercept**. To find this point we must put in a line equation $y = 0$ and solve resulting equation for x. The y-coordinate of the intersection point of a line with y-axis is called the **y-intercept**. To find this point we must put in a line equation $x = 0$ and solve resulting equation for y.

Definition: The **slope-intercept equation** of a line with slope m and a y-intercept b is (see Figure C.2)

$$y = mx + b \tag{C.2}$$

Example: The line equation with the slope $m = 2$ and the intercept $b = -2$

$$y = 2x - 2$$

The x-intercept may be found by setting $x = 1$ in the line equation $y = 0$. The horizontal and vertical lines may be written as equations

$$y = a$$
$$x = c$$

567

1. Lines on Plane

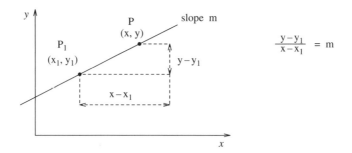

Figure C.1. A line on the plane as a point-slope equation.

Definition: The line equation in the term of nonzero *x*-**intercept** and nonzero *y*-**intercept** is

$$\frac{x}{a} + \frac{y}{b} = 1 \tag{C.3}$$

Example: The line $y = -2x + 4$ with x and y intercepts equal to 2 and 4, respectively, may be written as

$$\frac{x}{2} + \frac{y}{4} = 1$$

Definition: Let us assume that d denotes the perpendicular distance from the origin **0** to a line, and α be an angle between the perpendicular (from the origin to a line) with positive *x*-axis. The **normal** for the equation of a line is given as the equation (see Figure C.3)

$$x \cos \alpha + y \sin \alpha = p \tag{C.4}$$

Definition: Every line may always be described by the **general linear equation**

$$Ax + By + C = 0 \tag{C.5}$$

where A and B are not both zero.

Example: The line $y = 2x - 2$ may be written as a general line equation

$$2x - 1 \cdot y - 2 = 0$$

We remember that the distance between two points $P_1 = (x_1, y_1)$ and $P_2 = (x_2, y_2)$ is defined as

$$d = \sqrt{(x_1 - x_2)^2 + (y_1 - y_2)^2} \tag{C.6}$$

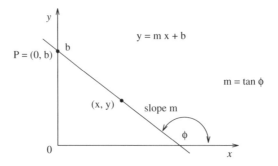

Figure C.2. A line on the plane for slope-intercept equation.

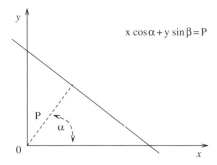

Figure C.3. Normal equation of a line.

Definition: The **distance from a point** $P = (x_1, y_1)$ to a line $Ax + By + C = 0$ is

$$d = \frac{Ax_1 + By_1 + C}{\pm\sqrt{A^2 + B^2}} \qquad (C.7)$$

where the sign is chosen so that the distance is positive.

Example: The distance from the point $P = (1, 2)$ to the line $y = x + 2$ can be found by first converting the line equation to the general line form

$$x - y + 2 = 0, \ A = 1, \ B = -1, \ C = 2$$

and then using the formula for the distance of point to a line form

$$d = \frac{1 \cdot 1 - 2 \cdot 1 + 2}{\sqrt{1^2 + (-1)^2}} = \frac{1}{\sqrt{2}}$$

Definition: The **angle ψ between two lines** with slopes m_1 and m_2 is

$$\tan \psi = \frac{m_2 - m_1}{1 + m_1 m_2} \qquad (C.8)$$

For parallel lines, $m_1 = m_2$ (and $\psi = 0$). For perpendicular lines, $m_1 = -\frac{1}{m_2}$.

Definition: The **area of a triangle on a plane** with the vertices $P_1(x_1, y_1)$, $P(x_2, y_2)$, and $P(x_3, y_3)$ is given by

$$area = \pm \begin{vmatrix} x_1 & y_1 & 1 \\ x_2 & y_2 & 1 \\ x_3 & y_3 & 1 \end{vmatrix} = \pm (x_1 y_2 + y_1 x_3 - y_2 x_2 - y_1 x_2 - x_1 y_3) \qquad (C.9)$$

where the sign must be chosen so that the area is nonnegative.

2. Lines and Planes in a Space

Lines and planes are special kind of sets in 3D space with the Cartesian coordinate system (see Figure C.4).

We will first recall that the distance between two points $P_1 = (x_1, y_1, z_1)$ and $P_2 = (x_2, y_2, z_2)$ is defined as

$$d = \sqrt{(x_1 - x_2)^2 + (y_1 - y_2)^2 + (z_1 - z_2)^2} \qquad (C.10)$$

570   2. Lines and Planes in a Space

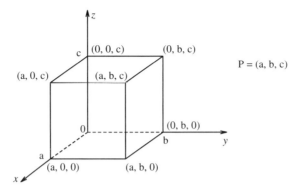

Figure C.4. Cartesian coordinate system in 3D.

Definition: The **direction cosines** of a line joining points $P_1(x_1, y_1, z_1)$ and $P_2(x_2, y_2, z_2)$ in 3D space are defined as

$$l = \cos \alpha = \frac{x_2 - x_1}{d}, \quad m = \cos \beta = \frac{y_2 - y_1}{d}, \quad n = \cos \gamma = \frac{z_2 - z_1}{d} \tag{C.11}$$

where d is the distance between two points P_1 and P_2. The angles α, β, and γ are the angles which the line passing through the points P_1 and P_2 makes with the positive axes x, y, and z (see Figure C.5).

The following relations hold for direction cosines

$$\cos^2 \alpha + \cos^2 \beta + \cos^2 \gamma = 1 \quad or \quad l^2 + m^2 + n^2 = 1 \tag{C.12}$$

Definition: The **standard line equation** passing through points $P_1(x_1, y_1, z_1)$ and $P_2(x_2, y_2, z_2)$ is

$$\frac{x - x_1}{x_2 - x_1} = \frac{y - y_1}{y_2 - y_1} = \frac{z - z_1}{z_2 - z_1} \quad or \quad \frac{x - x_1}{l} = \frac{y - y_1}{m} = \frac{z - z_1}{n} \tag{C.13}$$

where l, m, and n are direction cosines for a line. The direction cosines l, m, and n may be substituted in the above line equation by any numbers L, M, and N proportional to them. L, M and N are called **direction numbers**.

Definition: The **angle between two lines** with direction cosines l_1, m_1, n_1, and l_2, m_2, n_2 is

$$\cos \phi = l_1 l_2 + m_1 m_2 + n_1 n_2 \tag{C.14}$$

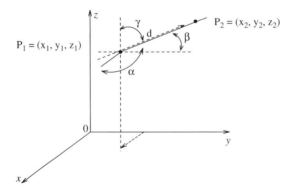

Figure C.5. Line in 3-D scope with direction cosines.

Definition: The **parametric line equation** passing through points $P_1(x_1, y_1, z_1)$ and $P_2(x_2, y_2, z_2)$ is

$$x = x_1 + l\,t, \quad y = y_1 + m\,t, \quad z = z_1 + n\,t \qquad \text{(C.15)}$$

where one value of the parameter t produces one point on line. The set of these equations covers a line completely as t value runs from $-\infty$ to ∞.

Again, the direction cosines l, m, and n may be substituted in the above line equation by any numbers L, M, and N proportional to them.

The parametric equation of a **line segment** joining two points P_1 and P_2 may be found by first writing the equation of the line passing through the points and then by finding the closed segment of the parameter t values $[t_1, t_2]$ with t_1 corresponding to a point P_1 and t_2 corresponding to a point P_2. The points belonging to a segment joining points P_1 and P_2 are points on the lines for the parameter $t \in [t_1, t_2]$. The line equation with the restriction $t \in [t_1, t_2]$ is called the **line segment equation** joining points P_1 and P_2.

Example: The line passing through the points $P_1(2, -3, -3)$ and $P_2(-1, 1, 4)$, with distance between them $d = \sqrt{(2+1)^2 + (-3-1)^2 + (-3-4)^2} = \sqrt{74}$, may be written in the standard form as

$$\frac{x-2}{-1-2} = \frac{y-(-3)}{1-(-3)} = \frac{z-(-3)}{4-(-3))}$$

The parameterized equation for this line may be written, if we first compute the direction cosines $l = \frac{3}{\sqrt{74}}$, $m = \frac{4}{\sqrt{74}}$, and $n = \frac{7}{\sqrt{74}}$,

$$x = 2 + l\,t, \quad y = -3 + m\,t, \quad z = -3 + n\,t$$

Definition: Let us assume that we have a vector in a space $\mathbf{v} = A\mathbf{i} + B\mathbf{j} + C\mathbf{k}$, (which also can be written in vector space notation as $\mathbf{v} = \begin{bmatrix} A \\ B \\ C \end{bmatrix}$), with the basis vectors $\mathbf{i}, \mathbf{j}$, and $\mathbf{k}$ (base) being the vectors from the origin $\mathbf{0} = (0, 0, 0)$ to the points $(1, 0, 0)$, $(0, 1, 0)$, and $(0, 0, 1)$, respectively.

The **line equation** passing through a point $P_0(x_0, y_0, z_0)$ and parallel to a vector $\mathbf{v} = A\mathbf{i} + B\mathbf{j} + C\mathbf{k}$ can be written as

$$x = x_0 + t\,A, \quad y = y_0 + t\,B, \quad z = z_0 + t\,C \qquad \text{(C.16)}$$

where t is the parameter.

For a given value of the parameter t, the above equation describes one point $P(x, y, z)$ on a line. As the parameter t increases from $-\infty$ to ∞, the point $P(x, y, z)$ traces the entire line exactly one time. When the parameter t traverses the closed interval $a \leq t \leq b$, the point $P(x, y, z)$ corresponding to the t traces the line segment from the point P_a where $a = b$ to the point P_b where $t = b$.

Example: The line passing through the point $P_0(2, 2, 2)$ and parallel to the vector $\mathbf{v} = 2\mathbf{i} + 3\mathbf{j} - 2\mathbf{k}$ (or just written as $\mathbf{v} = \begin{bmatrix} 2 \\ 3 \\ -2 \end{bmatrix}$) is

$$x = 2 + 2t, \quad y = 2 + 3t, \quad z = 2 - 2t$$

For $t = 1$ we have the point $P_1(4, 5, 0)$, i.e. the point where the line crosses the xy-plane (with $z = 0$) (see Figure C.6).

3. Planes

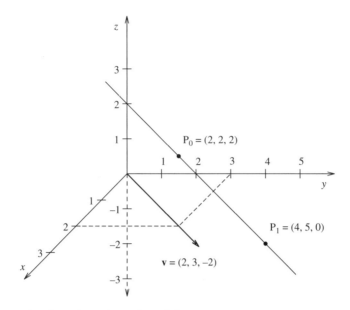

Figure C.6. Line passing through a point and parallel to a vector.

3. Planes

Loosely speaking, a plane in 3-dimensional space is the analog of a line on the plane.
Definition: The **general equation of a plane** in $3D$ space is written as

$$Ax + By + Cz + D = 0 \qquad (C.17)$$

where A, B, C, and D are constants which determine the position of the plane in the space in the Cartesian coordinate system.

More generally, using set notation, we can write the following definition of a plane being a specific set of points

$$L(\mathbf{v}) = \{\mathbf{v} : \mathbf{v} \in \mathbb{R}^3, \quad Ax + By + Cz + D = 0\} \qquad (C.18)$$

where a vector $\mathbf{v}$ is defined as $\mathbf{v} = \begin{bmatrix} x \\ y \\ z \end{bmatrix}$.

Example: The equation

$$x + 2y + 3z + 4 = 0$$

is an example of the plane in the space. The point $P_1(1, 2, -3)$ belongs to the plane, while the point $P_2(1, 2, -4)$ does not.

Definition: The **equation of a plane passing through three points** $P_1(x_1, y_1, z_1)$, $P_2(x_2, y_2, z_2)$, and $P_3(x_3, y_3, z_3)$ is written as

$$\det \begin{bmatrix} x - x_1 & y - y_1 & z - z_1 \\ x_2 - x_1 & y_2 - y_1 & z_2 - z_1 \\ x_3 - x_1 & y_3 - y_1 & z_3 - z_1 \end{bmatrix} = \begin{vmatrix} x - x_1 & y - y_1 & z - z_1 \\ x_2 - x_1 & y_2 - y_1 & z_2 - z_1 \\ x_3 - x_1 & y_3 - y_1 & z_3 - z_1 \end{vmatrix} = 0 \qquad (C.19)$$

where det denotes the determinant of the square matrix. Expanding in part the determinant around the first row we can also find

$$\begin{vmatrix} y_2 - y_1 & z_2 - z_1 \\ y_3 - y_1 & z_3 - z_1 \end{vmatrix}(x - x_1) + \begin{vmatrix} z_2 - z_1 & x_2 - x_1 \\ z_3 - z_1 & x_3 - x_1 \end{vmatrix}(y - y_1)$$

$$+ \begin{vmatrix} x_2 - x_1 & y_2 - y_1 \\ x_3 - x_1 & y_3 - y_1 \end{vmatrix}(z - z_1) = 0 \qquad (C.20)$$

Example: The plane passing through the three points $P_1(2, 0, 0)$, $P_2(0, 1, 0)$ and $P_3(0, 0, 3)$ may be be written as

$$\begin{vmatrix} 1 & 0 \\ 0 & 3 \end{vmatrix}(x - 2) + \begin{vmatrix} 0 & -2 \\ 3 & -2 \end{vmatrix}(y - 0) + \begin{vmatrix} -2 & 1 \\ -2 & 0 \end{vmatrix}(z - 0) = 0$$

which leads to the general equation of this plane

$$3x + 6y + 2z - 6 = 0$$

We can easily test to see that the coordinates of the points P_1, P_2 and P_3 satisfy the general plane equation.

Definition: The plane may be described by the **equation in the intercept form**

$$\frac{x}{a} + \frac{y}{b} + \frac{z}{c} = 1 \qquad (C.21)$$

where a, b, and c are the intercepts (coordinates of cutting a plane by axes) of the x, y, and z axes respectively (Figure C.7).

Definition: The **equation of plane which passes through a point** $P_0(x_0, y_0, z_0)$ and is **perpendicular to the nonzero vector** $\mathbf{n} = A\mathbf{i} + B\mathbf{j} + C\mathbf{k}$ can be written as (see Figure C.8)

$$Ax + By + Cz = D, \quad \text{where} \quad D = Ax_0 + By_0 + Cz_0 \qquad (C.22)$$

and the plane coefficients of x, y, and z are coordinates A, B, and C of the vector $\mathbf{n}$.

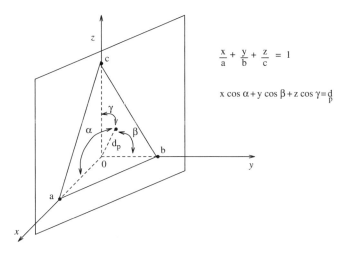

Figure C.7. Intercepts of a plane and the Normal Form of a plane equation.

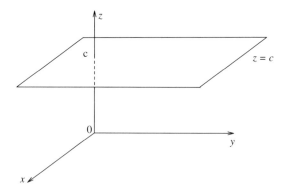

Figure C.8. Plane perpendicular to the z-axis and parallel to the x and y axes.

Example: The plane equation passing through the point $P_0(-3, 0, 4)$ and perpendicular to the vector $\mathbf{n} = 5\mathbf{i} + 2\mathbf{j} - 1\mathbf{k}$ may be described by

$$5x + 2y - z = -19$$

Definition: The **line which passes a point** $P_0(x_0, y_0, z_0)$ and is **perpendicular to a plane** $Ax + By + Cz + D = 0$ can be described by the equation

$$\frac{x - x_0}{A} = \frac{y - y_0}{B} = \frac{z - z_0}{C} \tag{C.23}$$

or

$$x = x_0 + A\,t, \quad y = y_0 + B\,t, \quad z = z_0 + C\,t \tag{C.24}$$

We note that the direction numbers for the line perpendicular to a plane $Ax + By + Cz + D = 0$ are equal to the plane coordinate coefficients A, B, and C.

The vector **normal** to a plane $Ax + By + Cz + D = 0$ is defined as

$$\mathbf{N} = A\mathbf{i} + B\mathbf{j} + C\mathbf{k} \tag{C.25}$$

with the length $d = ||\mathbf{N}|| = \sqrt{A^2 + B^2 + C^2}$.

Definition: The **distance from a point** $P_0(x_0, y_0, z_0)$ **to a plane** $Ax + By + Cz + D = 0$ can be found as

$$d = \frac{Ax_0 + By_0 + Cz_0 + D}{\pm \sqrt{A^2 + B^2 + C^2}} \tag{C.26}$$

where the sign must be chosen so that the distance is nonnegative.

Example: The distance of the point $P_0(1, 2, 4)$ from the plane $2x + 3y + 6z - 6 = 0$ is $d = \frac{26}{7}$.

Definition: If d_p denotes the perpendicular distance from the plane point to P the origin $\mathbf{0}$ then the **normal form of a plane equation** is

$$x \cos \alpha + y \cos \beta + z \cos \gamma = d_p \tag{C.27}$$

where α, β, and γ are angles between the line passing through the points $\mathbf{0}$ and P (normal to a plane) and the positive axes x, y, and z (see Figure C.7).

Definition: For two planes $A_1 x + B_1 y + C_1 z + D_1 = 0$ and $A_2 x + B_2 y + C_2 z + D_2 = 0$ the **angle between them** is defined as

$$\cos\theta = \frac{\mathbf{N}_1^T \mathbf{N}_2}{\|\mathbf{N}_1\|\|\mathbf{N}_2\|}, \quad \theta = \cos^{-1}\frac{\mathbf{N}_1^T \mathbf{N}_2}{\|\mathbf{N}_1\|\|\mathbf{N}_2\|} \tag{C.28}$$

where $\mathbf{N}_1^T \mathbf{N}_2$ denotes the dot product of the vectors

$$\mathbf{N}_1 = A_1 \mathbf{i} + B_1 \mathbf{j} + C_1 \mathbf{k}, \quad \mathbf{N}_2 = A_2 \mathbf{i} + B_2 \mathbf{j} + C_2 \mathbf{k} \tag{C.29}$$

normal to the first and second plane correspondingly, and $\|\mathbf{N}_1\|$ and $\|\mathbf{N}_2\|$ denote the lengths of the normal vectors. We see that the angle between two planes is equal to the angle between the normal vectors to these planes.

Example: The angle between two planes $3x - 2y - 6z - 12 = 0$ and $x + 2y - 2z - 6 = 0$ is equal to

$$\cos\theta = \frac{\mathbf{N}_1^T \mathbf{N}_2}{\|\mathbf{N}_1\|\|\mathbf{N}_2\|} = \frac{11}{7 \cdot 3}, \quad \theta = \cos^{-1}\frac{11}{21}$$

because the normal vectors to the planes are

$$\mathbf{N}_1 = 3\mathbf{i} - 2\mathbf{j} - 6\mathbf{k}, \quad \mathbf{N}_2 = 1\mathbf{i} + 2\mathbf{j} - 2\mathbf{k}$$

4. Hyperplanes

Loosely speaking, if we consider the *n*-**dimensional vector space** $\mathbb{R}^n$, a hyperplane is analog of the plane in 3-**dimensional** space. From another point of view, a hyperplane in $\mathbb{R}^n$ may be considered as a translation of $\mathbb{R}^{n-1}$ subspace of $\mathbb{R}^n$.

Definition: Let $L(\mathbf{x})$ be the real-value function on the space $\mathbb{R}^n$ defined by

$$L(\mathbf{x}) = \mathbf{a}^T \mathbf{x} - b = [a_1, a_2, \cdots, a_n] \begin{bmatrix} x_1 \\ x_2 \\ \vdots \\ x_n \end{bmatrix} - b = \sum_{i=1}^{n} a_i x_i - b \tag{C.30}$$

where $\mathbf{a}$ is nonzero vector of $\mathbb{R}^n$ (a point in $\mathbb{R}^n$) and b is a given real number, then the subset $L \subseteq \mathbb{R}^n$, defined by

$$L = \{\mathbf{x}: \ \mathbf{x} \in \mathbb{R}^n, \quad L(\mathbf{x}) = \sum_{i=1}^{n} a_i x_i - b = 0\}, \tag{C.31}$$

is called a **hyperplane in** $\mathbb{R}^n$.

The equation

$$a_n x_n + a_{n-1} x_{n-1} + \cdots + a_1 x_1 - b = 0 \tag{C.32}$$

is called a **linear equation** in $\mathbb{R}^n$.

Example: The set L being a subset of $\mathbb{R}^3$ and defined as

$$L = \{\mathbf{x} = \begin{bmatrix} x_1 \\ x_2 \\ x_3 \end{bmatrix} : \ 1 \cdot x_1 + 1 \cdot x_2 - 2 \cdot x_3 - 4 = 0\}$$

is a hyperplane in $\mathbb{R}^3$. Of course we can recall that it is just the plane

$$1 \cdot x + 1 \cdot y - 2 \cdot z - 4 = 0$$

if we change the names of the coordinates from x_1, x_2, and x_3 to x, y, and z.

We see that a plane divides the space on two subspaces.

Definition: The sets $L^+ \in \mathbb{R}^n$ and $L^- \in \mathbb{R}^n$ defined as

$$L^+ = \{\mathbf{x} : \mathbf{x} \in \mathbb{R}^n, L(\mathbf{x}) = \sum_{i=1}^{n} a_i x_i - b > 0\} \tag{C.33}$$

$$L^- = \{\mathbf{x} : \mathbf{x} \in \mathbb{R}^n, L(\mathbf{x}) = \sum_{i=1}^{n} a_i x_i - b < 0\} \tag{C.34}$$

are called the **open half-spaces** in $\mathbb{R}^n$ (see Figure C.9).

Definition: The set $L^+ \cap L \in \mathbb{R}^n$ and $L^- \cap L \in \mathbb{R}^n$ are called **closed half-spaces** in $\mathbb{R}^n$ determined by the hyperplane L. We also see that the set $L^+ \cap L$ is the closure of L^+ and that $L^- \cap L$ is the closure of L^-, and that the hyperplane L is boundary of both sets L^+ and L^-. Of course $L^+ \cup L^- \cup L = \mathbb{R}^n$.

Definition: If A and B are two sets in $\mathbb{R}^n$, then we say that a hyperplane L **separates** sets A and B if A is contained in one closed half-space determined by a hyperplane L and B in the other, that is

$$A \subseteq L^+ \cup L \quad \text{and} \quad B \subseteq L^- \cup L \tag{C.35}$$

or

$$A \subseteq L^- \cup L \quad \text{and} \quad B \subseteq L^+ \cup L \tag{C.36}$$

A hyperplane **separates strictly** two sets A and B if A is contained in one open half-space determined by a hyperplane L and B in the other open half-space.

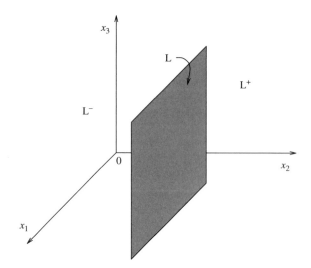

Figure C.9. Half-space in $\mathbb{R}^3$.

Example: The hyperplane $x_1 - 1 = 0$ in $\mathbb{R}^3$ separates the set

$$\overline{S}\left(\begin{bmatrix}0\\0\\0\end{bmatrix}, 1\right)$$

being the closed sphere at the origin with the radius $\rho = 1$, and

$$\overline{S}\left(\begin{bmatrix}2\\0\\0\end{bmatrix}, 1\right)$$

being the closed sphere at the point $\mathbf{x}_0 = \begin{bmatrix}2\\0\\0\end{bmatrix}$ with the radius $\rho = 1$.

We observe the a hyperplane divides all points in $\mathbb{R}^n$ into two half-spaces. This fact is important in the pattern recognition when one tries to classify points belonging to two separate classes. The hyperplane equation

$$a_n x_n + a_{n-1} x_{n-1} + \cdots + a_1 x_1 + b = 0 \tag{C.37}$$

is called the **classifying hyperplane**. For points "above" a hyperplane we have

$$a_n x_n + a_{n-1} x_{n-1} + \cdots + a_1 x_1 + b > 0 \tag{C.38}$$

and for points "below" a hyperplane

$$a_n x_n + a_{n-1} x_{n-1} + \cdots + a_1 x_1 + b < 0 \tag{C.39}$$

We will revise the definition of line segment and we will introduce the concept of half-rays and cones.

Definition: A **line segment** joining two vectors $\mathbf{y}$ and $\mathbf{z}$ (points) in $\mathbb{R}^n$ is the following subset of $\mathbb{R}^n$

$$\{\mathbf{x} : \mathbf{x} \in \mathbb{R}^n, \ \mathbf{x} = r\mathbf{y} + s\mathbf{z}, \ r \geq 0, \ s \geq 0, \ r + s = 1\} \tag{C.40}$$

and **half-ray** joining $\mathbf{y}$ and $\mathbf{z}$ emanating from $\mathbf{y}$ is the subset of $\mathbb{R}^n$ defined as

$$\{\mathbf{x} : \mathbf{x} \in \mathbb{R}^n, \ \mathbf{x} = \mathbf{y} + r(\mathbf{z} - \mathbf{y}), \ r \leq 0\} \tag{C.41}$$

Note that half-ray joining $\mathbf{y}$ and $\mathbf{z}$ end emanating from the point $\mathbf{y}$ is not the same as the half-ray joining $\mathbf{y}$ and $\mathbf{z}$ end emanating from the point $\mathbf{z}$ (see Figure C.10).

Definition: A subset K of $\mathbb{R}^n$ is called a **cone** with vertex $\mathbf{x}_0$ if, for any given point in K with $\mathbf{x} \neq \mathbf{x}_0$, all points on the half-ray joining $\mathbf{x}$ and $\mathbf{x}$, emanating from $\mathbf{x}_0$, belong to a cone K (see Figure C.11).

Definition: A subset $A \in \mathbb{R}^n$ is **convex** if, for any point $\mathbf{x}$ and $\mathbf{y}$ in a set A, and for $r, s \in \mathbb{R}$ with $r \leq 0$, $s \leq 0$ and $r + s = 1$, the point $r\mathbf{x} + s\mathbf{y}$ is in a set A.

Loosely speaking, a subset of $\mathbb{R}^n$ is convex if, for any two points in the set, the line segment between two points is contained within a set.

578   4. Hyperplanes

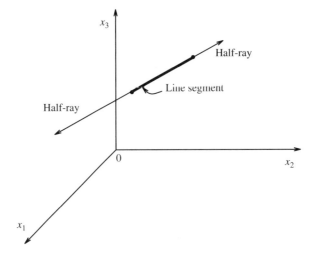

Figure C.10. Line segment and half-rays.

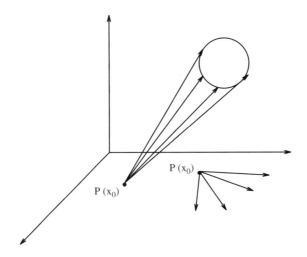

Figure C.11. Cones.

Example: The set in $\mathbb{R}^3$

$$A = \left\{ \begin{bmatrix} x_1 \\ x_2 \\ x_3 \end{bmatrix} \in \mathbb{R}^3 : |x_1| \leq 1, \ |x_2| \leq 1, \ |x_3| \leq 1 \right\}$$

which is a cube, is convex. The set in $\mathbb{R}^3$

$$A = \left\{ \begin{bmatrix} x_1 \\ x_2 \\ x_3 \end{bmatrix} \in \mathbb{R}^3 : x_1 + 2x_2 + x_3 - 2 = 0 \right\}$$

which is a plane, is convex.

Definition: A cone K in $\mathbb{R}^n$ which is also a convex set is called a **convex cone**.

Appendix D

Sets

In this appendix we review the basic concepts of sets.

1. Set Definition and Notations

Mathematics deals with statements and operations on objects. Objects may be grouped in different collections. The criteria for the selection of objects for a given collection may vary and yet are normally oriented toward some specific purpose. For example, we may create a collection of objects having some common feature important for further storing or processing of information. For example, we may collect names of all students in a class for storage and further processing purposes. The common feature of the objects from this collection is that they belong to the one class.

The area of mathematics that deals with the collections of objects, their properties and operations is called **set theory**.

Definition: A collection (or group) of objects (or items) is called a **set**. The objects belonging to a set are called **members** or **elements** of the set. They may be numbers, people, names, or other objects that related by the fact that they belong to the set.

Sets are normally denoted by capital letters and the members of a set by lowercase letters. A set of members given explicitly is denoted by an unordered list of objects enclosed in { } braces.

For example

$$A = \{a, b, c, d\} \tag{D.1}$$

denotes the set A with four members a, b, c, d.

The fact that an element a belongs to a set A is denoted by

$$a \in A \tag{D.2}$$

and the notation

$$a \notin B \tag{D.3}$$

denotes that the element a is not a member of the set B. The notation

$$a, b \in C \tag{D.4}$$

denotes that both elements *a* and *b* belong to the set *C*. Note that the order of elements inside the set denoted by braces is not important, thus

$$A = \{a, b, c, d\} \text{ and } A = \{c, d, a, b\} \tag{D.5}$$

represent the same set.

We may also say that a set may consist of elements which have common properties that distinguish them from other elements.

A set must be defined so that it is possible to determine whether a given element is or is not a member of a set.

Example: The set

$$X = \{1, 2, 3, 4\}$$

consists of natural numbers less than 5. We write $2 \in X$ to denote the fact that the number 2 is a member of the set *X*. The common feature of all members of the set *X* is that they are natural numbers less than 5. Because the set *X* is well-defined, we have no difficulty deciding that the number 7 is not a member of this set (i.e., $7 \notin X$).

1.1. Ways of Defining a Set

The most natural way of defining a set is to just provide a list of all elements of a set. Such a list is called a **roster**. The members of a set are enclosed in the curly braces { }.

Example:

$$A = \{Ania, Mary, Halina\}$$

denotes the set *A* having all members *Ania, Mary*, and *Halina* listed explicitly.

For large sets we can use an abbreviation to define the set, particularly when the members may be put in some well-defined sequence.

Example: The set

$$M = \{1, 2, 3, 4, \cdots, 99, 100\}$$

denotes all natural numbers less than or equal to 100.

A set may be defined in a written description, expressing the properties of the set members.

Example: The set $S = \{\text{even integers between 0 and 100}\}$ denotes all even integer numbers between 0 and 100. Because the set *S* is well defined, we may decide that $16 \in S$.

The common way of denoting a set in mathematical analysis is to define **a variable** to denote arbitrary elements of the set and then make a statement about the variable which specifies the property or properties of the members of the set. The statement about the variable may be descriptive or given as a mathematical formula (for example, inequality).

Example: The set

$$B = \{x : x \text{ is an odd integer and } 3 \leq x < 9\}$$

is equal to the set $B = \{3, 5, 7\}$ which denotes all odd integers equal to or greater than 3 and less than 9. Here "*x*" is a variable denoting an element and "*x* is an odd integer and $3 \leq x < 9$" is the statement.

The set

$$C = \{x : x \in R \text{ and } |x| \leq 1\} \tag{D.6}$$

denotes all real numbers whose absolute value is less than or equal to 1.

A set may consist of any collection of well-defined objects. For example, a set may consist of pairs as members

$$D = \{(a, b), (c, d), (a, d)\}. \tag{D.7}$$

Thus, we may say that the pair (a, b) is an element of the set D, i.e., $(a, b) \in D$.

2. Types of Sets

Mathematics defines some basic types of sets depending on their contents and properties.

Definition: The set having no members is called the **empty set** (or **null set**) and is denoted by $\emptyset$ or $\{\ \}$.

Example:

$$A = \{integers\ between\ 4\ and\ 5\} = \emptyset$$
$$B = \{people\ living\ on\ Mars\} = \{\ \}$$

Definition: A set consisting of a single element is called a **singleton set**.

Example:

$$A = \{blue\}$$

Definition: Two sets are the **same** (**equal**) if and only if they have the same elements.

Example: The sets

$$A = \{\alpha, \beta, \gamma\}, B = \{\beta, \gamma, \alpha\}$$

are equal because they have the same members.

Definition: A set with finitely many members or elements is called a **finite set**.

Example: The set

$$F = \{10, 11, 12, \cdots, 98, 99, 100\}$$

consisting of 91 elements is a finite set.

Definition: A set with infinitely many members or elements (whose elements may be counted) is called an **infinite set**.

Example: The set of all natural numbers

$$F = \{1, 2, 3, \cdots\}$$

is infinite, having infinitely many members.

2. Types of Sets

Definition: A set whose elements may be enumerated in a sequence $a_1, a_2, \cdots$ which may or may not terminate is called a **countable** or **denumerable set**.

In a countable set A we can find a one-to-one correspondence between the infinite set of all natural numbers $\mathbb{N}$ and the members of the set A.

Every finite set is a countable set. However, not every countable set has a finite number of elements.

Example: The set

$$E = \{2, 4, 6, 8, 10, \cdots\} \tag{D.8}$$

that is, the set of even numbers, is countable since the function $f(n) = \frac{n}{2}$ maps one-to-one onto the set E.

The set

$$A = \{1, 2, 3, 4\}$$

is finite and therefore is a countable set. On the contrary, the set

$$B = \{1, \frac{1}{2}, \frac{1}{3}, \cdots, \frac{1}{n}, \cdots\}$$

is a countable set which is not finite. The set

$$C = \{x : |x| \leq 1\}$$

is not countable, because we cannot enumerate the members in a defined sequence.

The set of all real numbers is not countable.

Definition: A set U containing (but not limited to) all possible elements being considered in a given problem is called a **universal set**.

Example: The set

$$U = \{all\ students\ in\ an\ elementary\ school\}$$

can be considered as a universal set from the point of view of given school members.

The set $\mathbb{R}$ of all real numbers is also an example of universal set.

Definition: For the finite set A, **cardinality**, denoted by $|A|$, is the number of set elements.

Two sets A and B have the same cardinality if there is a one-to-one correspondence between them (see the definition of one-to-one correspondence).

Example: The cardinality of the set

$$A = \{a, b, c, d, g\}$$

is $|A| = 5$.

Definition: The infinite set of all **natural numbers**

$$\mathbb{N} = \{1, 2, 3, \cdots\} \tag{D.9}$$

is denoted by $\mathbb{N}$ or N.

The infinite set of all **integer numbers**

$$\mathbb{Z} = \{\cdots, -2, -1, 0, 1, 2, 3, \cdots\} \tag{D.10}$$

Appendix D: Sets

is denoted by $\mathbb{Z}$ or Z.

The infinite set of all **rational numbers** is denoted by $\mathbb{Q}$ or Q.

The infinite set of all **real numbers** is denoted by $\mathbb{R}$ (or R) and contains: all natural numbers

$$\mathbb{N} = \{1, 2, 3, \cdots, \} \tag{D.11}$$

zero and all negative integers

$$\{0, -1, -2, -3, \cdots, \} \tag{D.12}$$

the rational numbers

$$\{\frac{1}{2}, -\frac{3}{5}, \frac{61}{32}, \cdots\} \tag{D.13}$$

and the irrational numbers

$$\{\sqrt{2}, \pi, \cdots\}. \tag{D.14}$$

The infinite set of all **complex numbers** is denoted by $\mathbb{C}$ or C.

Definition: The **infinite set of all points** on the real line is defined as

$$\{x : x \in \mathbb{R}, -\infty < x < \infty\} \tag{D.15}$$

The infinite set of all points on a real plane, denoted by a pair of coordinates (x, y) (or just a 2-dimensional vector $\mathbf{a} = \begin{bmatrix} x \\ y \end{bmatrix} \in \mathbb{R}^2$), is defined as

$$\{(x, y) : x, y \in \mathbb{R}, -\infty < x < \infty \text{ and} - \infty < y < \infty\} \tag{D.16}$$

And the infinite set of all points on a real three dimensional space, denoted by coordinates (x, y, z) (or just a 3-dimensional vector $\mathbf{a} = \begin{bmatrix} x \\ y \\ z \end{bmatrix} \in \mathbb{R}^3$), is defined as

$$\{(x, y, z) : x, y, z \in \mathbb{R}, -\infty < x < \infty \text{ and} - \infty < y < \infty \text{ and } \infty < z < \infty\} \tag{D.17}$$

Note: These notations are equivalent: $(x, y, z \in \mathbb{R})$ and $(x \in \mathbb{R}, y \in \mathbb{R}, z \in \mathbb{R})$.

Definition: **Line segments**, **regions** on real plane and spheres with their interiors on the 3-dimensional planes may be expressed as sets of points.

Example The set of points on the real line (Figure D.1) is defined as the set

$$A = \{x : x \in \mathbb{R}, -2 \le x \le 4, x = 6.2\}$$

In this example, the set A contains the line segment which is all the points lying between and including -2 and 4, plus the single point 6.2.

Figure D.1. The segment of line $[-2, 4]$ plus the single point 6.2

2. Types of Sets

Definition: An **interval** is a subset of the set of all real numbers (the real line).
The **closed interval** denoted by $[a, b]$ is the set

$$[a, b] = \{x : x \in \mathbb{R}, a \leq x \leq b\} \tag{D.18}$$

The **half-open interval** denoted by $[a, b)$ is the set

$$[a, b) = \{x : x \in \mathbb{R}, a \leq x < b\} \tag{D.19}$$

The **half-open interval** denoted by $(a, b]$ is the set

$$(a, b] = \{x : x \in \mathbb{R}, a < x \leq b\} \tag{D.20}$$

Definition: The **regions** above the line on the real plane

$$y = ax + b \tag{D.21}$$

may be defined as a set of points

$$S = \{(x, y) : x \in \mathbb{R}, y \in \mathbb{R}, y - ax - b > 0\} \tag{D.22}$$

and below and on line as

$$T = \{(x, y) : x \in \mathbb{R}, y \in \mathbb{R}, y - ax - b \leq 0\} \tag{D.23}$$

Example: The region between the lines

$$y = x + 1$$

and

$$y = 4x - 1$$

may be defined as the set of points

$$W = \{(x, y) : x \in \mathbb{R}, y \in \mathbb{R}, y - x - 1 < 0 \text{ and } y - 4x + 1 > 0\}$$

The set of points (x, y) on the unit circle and within the interior of the unit circle on the real plane may be denoted as (Figure D.2)

$$V = \{(x, y) : x \in \mathbb{R}, y \in \mathbb{R}, x^2 + y^2 \leq 1\}$$

The **interior** of a ellipsoid in 3-dimensional real space with points given by coordinates (x, y, z) can be described as a set

$$E = \{(x, y, z) : x, y, z \in \mathbb{R}, \frac{x^2}{a^2} + \frac{y^2}{b^2} + \frac{z^2}{c^2} \leq 1 \text{ and } |x| < a, |y| < b, |z| < c\}$$

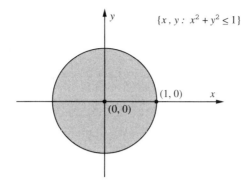

Figure D.2. The set of points on and within the interior of the unit circle

3. Set Relations

Definition: The illustrative description of sets and their relation may be provided by so called **Venn diagrams** (named for John Venn, an English mathematician). Venn diagrams are simply figures (usually circles) on a plane (see Figure D.3).

Despite the fact that Venn diagrams are planar figures, they are used to illustrate membership, relations and operations among different sets. Frequently, a Venn diagram will illustrate a set or sets being part of some well-defined universe (see Figure D.4).

Definition: As we mentioned, two or more sets are **equal** ($=$) if they have exactly the same elements. **Nonequal** sets are denoted as $A \neq B$.

Example: The following sets are equal ($=$)

$$\{a, b, c\} = \{b, c, a\}$$

Definition: Two or more sets that each have the same number of elements are called **equivalent sets** (denoted by $\leftrightarrow$).

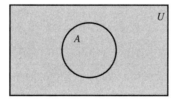

Figure D.3. A Venn diagram

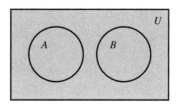

Figure D.4. A Venn diagram showing two sets as part of a well-defined universe

3. Set Relations

Example: The sets

$$\{a, b, c\} \leftrightarrow \{1, 4, 6\}$$

are equivalent because every element of one set may be paired with exactly one element in the other set, and vice versa.

Definition: A set A is a **subset** ($\subseteq$ or $\subset$) of set B

$$A \subseteq B \tag{D.24}$$

if every element of A is also contained in B.

Example: The set $A = \{1, 2, 3\}$ is a subset of the set $B = \{1, 2, 3, 4, 5\}$

$$A \subseteq B; \; \{1, 2, 3\} \subseteq \{1, 2, 3, 4, 5\}$$

because all members 1, 2, 3 of the set A are also members of the set B (see Figure D.5).

If the set A is not a subset ($\not\subseteq$) of the set B we denote this fact by

$$A \not\subseteq B \tag{D.25}$$

Example: The set $C = \{0, 2, 3\}$ is not a subset of the set $B = \{1, 2, 3, 4, 5\}$

$$A \not\subseteq B; \quad \{0, 2, 3\} \not\subseteq \{1, 2, 3, 4, 5\}$$

Definition: The set A is a **proper subset** ($\subset$) of the set B if A is a subset of B containing at least one element and if A is not equal to B.

The set B must include at least one element which does not belong to A.

Example: The set $A = \{a, b\}$ is a proper subset of the set $B = \{a, b, c\}$, but set $C = \{a, b, c\}$ is not a proper subset of the set A, even though it is a subset of A

$$A \subset B; \quad \{a, b\} \subset \{a, b, c\},$$
$$C \not\subset B; \quad \{a, b, c\} \not\subset \{a, b, c\}.$$

Example: The empty set $\emptyset$ is subset of every set because it contains no members that are not contained in other sets

$$\{\,\} \subseteq \{a, b, c\}$$

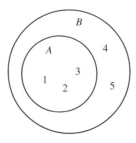

Figure D.5. The set A is a subset of set B, $A \subseteq B$

Definition: Sets A and B are disjoint if they have no elements in common.

Example: The sets $\{a, b, c\}$ and $\{d, e, f, g, h\}$ are disjoint.

Definition: The **power set** of a given set (denoted by 2^A) is the collection of all the possible subsets of the given set A.

Example: The power set of the set $A = \{1, 2, 3\}$ is

$$2^A = 2^{\{1,2,3\}} = \{\{1, 2, 3\}, \{1, 2\}, \{1, 3\}, \{2, 3\}, \{1\}, \{2\}, \{3\}, \{\ \}\}$$

The **number of subsets** in the power set of a given finite set is equal to 2^n, where n is the number of elements in the given set.

Example: The power set 2^A of the set $A = \{1, 2, 3\}$ (having $n = 3$ members) consists of $2^3 = 8$ subsets.

3.1. Properties of Subset Operation

The subset operation $\subseteq$ is in fact the relation called the **inclusion** operation.

The inclusion operation has the following properties

1) $A \subseteq A$ (reflexivity); set is a subset of itself
2) $A \subseteq B$ and $B \subseteq C$ implies $A \subseteq C$ (transitivity)
3) $A \subseteq B$ and $B \subseteq A$ if and only if $A = B$ (equality)
4) $\emptyset \subseteq A$ for all sets A (empty set)

Two sets can be combined to form a third set by various set operations.

4. Set Operations

Definition: The **intersection** of two sets A and B, denoted by $A \bigcap B$, is the set C consisting of elements common to both A and B (see Figure D.6)

$$C = A \bigcap B = \{c : c \in A \text{ and } c \in B\} \tag{D.26}$$

Example: The intersection of sets

$$A = \{1, 2, 3, 7\} \text{ and } B = \{0, 1, 2, 3, 4, 8\}$$

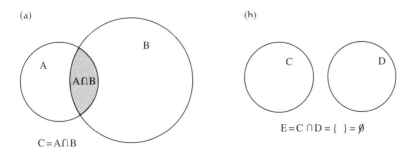

Figure D.6. (a) Intersection of two sets where the shaded region shows the region in the intersection, (b) The intersection of disjoint sets is the empty set

is the set

$$C = A \cap B = \{1, 2, 3\}$$

More than two sets may be involved in some set operations.

Example:

$$D = A \cap B \cap C$$

denotes intersection of three sets, which is the set consisting elements common to A, B and C.

The intersection obeys the conditions

1) $A \cap A = A$ independence
2) $A \cap B = B \cap A$ commutativity
3) $A \cap \emptyset = \emptyset$ empty set property
4) $(A \cap B) \cap C = A \cap (B \cap C)$ associativity
5) $A \cap B = A$ if and only if $A \subseteq B$
6) $A \cap B \subseteq A$ and $A \cap B \subseteq B$

The intersection of two disjoint sets is the empty set.

Definition: The **union** of two sets A and B, denoted by $A \cup B$, is the set C of all elements contained either in A or in B or in both.

$$C = A \cup B = \{c : c \in A \text{ or } c \in B\} \tag{D.27}$$

The number of elements in the union of two finite sets A and B is equal to the sum of number of elements of A and B, minus the number of elements in $A \cap B$.

Example: The union of two sets

$$A = \{1, 2, 3, 7\} \text{ and } B = \{0, 1, 2, 3, 4, 8\}$$

is the set

$$C = A \cup B = \{0, 1, 2, 3, 4, 7, 8\}.$$

Set union satisfies the following conditions

1) $A \cup A = A$ independence
2) $A \cup B = B \cup A$ commutativity
3) $A \cup \emptyset = A$ empty set property
4) $(A \cup B) \cup C = A \cup (B \cup C)$ associativity
5) $A \cup A \subseteq B$ if and only if $A \subseteq B$ inclusion
6) $A \subseteq A \cup B$, $B \subseteq A \cup B$ ordering

Definition: The **difference** of two sets A and B, denoted by $A - B$, is the set C of all elements of A which are not in B (see Figure D.7)

$$A - B = \{c : c \in A \text{ and } c \notin B\} \tag{D.28}$$

Example: The difference $A - B$ of the two sets

$$A = \{blue, green, black, orange, grey\} \text{ and } B = \{blue, grey, red, purple\}$$

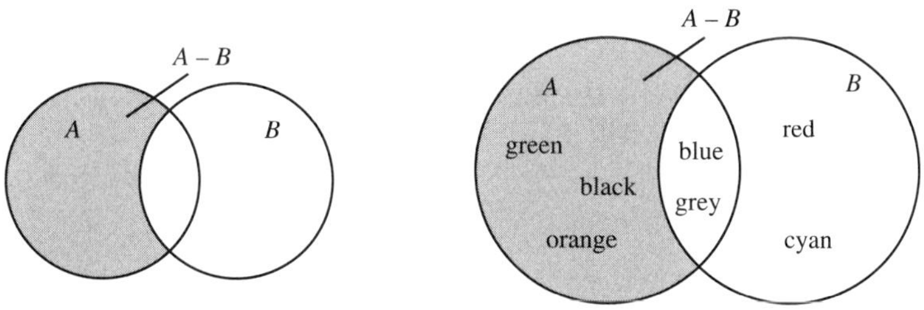

Figure D.7. Set difference

is the set

$$C = A - B = \{green, black, orange\}$$

but

$$D = B - A = \{red, purple\}.$$

Definition: If the set A is a subset of B, then the **complement** of A in B, denoted by $\overline{A}$, is the set of elements contained in B but not in A (see Figure D.8)

$$\overline{A} = \{c: \ c \in B \text{ and } c \notin A\} \tag{D.29}$$

and, of course, $\overline{A} = B - A$.

Example: Let us consider two sets

$$A = \{1, 3, 4\} \text{ and } B = \{1, 2, 3, 4, 5, 6, 7\}$$

We see that $A \subseteq B$, and the complement $\overline{A}$ of the set A in B is

$$\overline{A} = B - A = \{2, 5, 6, 7\}.$$

We use also use the phrase "complement of A" alone, without specifying "complement of A in B," understanding the complement of A in some well-defined universal set U

$$\overline{A} = \{c: \ c \in U \text{ and } c \notin A\}. \tag{D.30}$$

Of course $\overline{A} = U - A$.

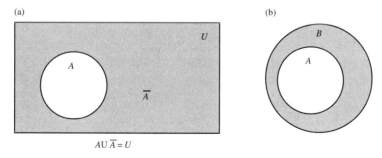

Figure D.8. Set complement (a) $A \bigcup \overline{B} = U$, (b) $\overline{A}$ where $A \subseteq B$

5. Set Algebra

Algebra for sets provides some properties and theorems related to set operations. Certain properties of the sets follow easily from their definitions.

5.1. Properties of the Set Complement

Let us assume that U is the well-defined universal set. The following properties hold for the set complement:

1) $\overline{\emptyset} = U$
2) $\overline{U} = \emptyset$
3) $\overline{\overline{A}} = A$ the complement of the complement is the set itself
4) $A \cup \overline{A} = U$
5) $A \cap \overline{A} = \emptyset$

5.2. Basic Set Algebra Laws

A fundamental role in set algebra is played by DeMorgan's theorem (named for the English logician, Augustus DeMorgan), relating the basic set operations: union, intersection and complement.

DeMorgan's Theorem: The complement of the union of two sets is the intersection of these sets' complements (Figure D.9)

$$\overline{A \cup B} = \overline{A} \cap \overline{B}. \tag{D.31}$$

The complement of the intersection of two sets A and B is the union of these sets' complements (Figure D.10)

$$\overline{A \cap A} = \overline{A} \cup \overline{B}. \tag{D.32}$$

If A, B, and C are sets, the following laws hold:

1) $A \cup A = A$ idempotency
2) $A \cap A = A$ idempotency
3) $A \cup B = B \cup A$ commutativity
4) $A \cap B = B \cap A$ commutativity
5) $(A \cup B) \cup C = A \cup (B \cup C)$ associativity
6) $(A \cap B) \cap C = A \cap (B \cap C)$ associativity

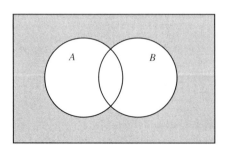

Figure D.9. $\overline{A \cup B} = \overline{A} \cap \overline{B}$

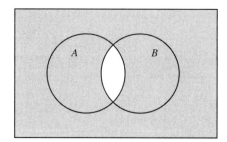

Figure D.10. $\overline{A \cap B} = \overline{A} \cup \overline{B}$

7) $A \cup (B \cap C) = (A \cup B) \cap (A \cup C)$ distributivity
8) $A \cap (B \cup C) = (A \cap B) \cup (A \cap C)$ distributivity
9) $A \cap (A \cup B) = A$ absorption
10) $A \cup (A \cap B) = A$ absorption

5.3. Union of Finite Sequence of Sets

Definition: Let $A_1, A_2, \cdots, A_n$ be a finite sequence of n sets. The union of n sets is defined as

$$B = \bigcup_{i=1}^{n} A_i = \left(\bigcup_{i=1}^{n-1} A_i \right) \cup A_n \tag{D.33}$$

The above definition may be extended to the infinite sequence of sets.

$$B = \bigcup_{i=1}^{\infty} A_i = \{b: \text{ there is an } i_0 \text{ with } b \in A_{i_0}\} \tag{D.34}$$

Example: The union of three sets

$$A = \{3, 2, 1\}, B = \{1, 2, 5, 4, 5\}, C = \{8, 9\}$$

is the set

$$D = A \cup B \cup C = \{1, 2, 3, 4, 5, 8, 9\}.$$

5.4. Intersection of Finite Sequence of Sets

Definition: Let $A_1, A_2, \cdots, A_n$ be a finite sequence of n sets. The intersection of n sets is defined as

$$B = \bigcap_{i=1}^{n} A_i = \left(\bigcap_{i=1}^{n-1} A_i \right) \cap A_n \tag{D.35}$$

The above definition may be extended to the infinite sequence of sets.

$$B = \bigcap_{i=1}^{\infty} A_i = \{b: \ b \in A_i \text{ for every } i\} \tag{D.36}$$

Example: The intersection of three sets

$$A = \{3, 2, 1\}, B = \{1, 2, 5, 4, 5\}, C = \{2, 8, 9\}$$

is the set

$$D = A \cap B \cap C = \{2\}.$$

If $A_i (i = 1, 2, \cdots)$, are subsets of a set A, then

1) $A - \bigcup_{i=1}^{\infty} A_i = \bigcap_{i=1}^{\infty} (A - A_i)$
2) $A - \bigcap_{i=1}^{\infty} A_i = \bigcup_{i=1}^{\infty} (A - A_i)$

6. Cartesian Product of Sets

6.1. Ordered Pair and n-tuple

First we will define the **ordered pair** of two objects a and b denoted by (a, b). The objects a and b are called the components of the ordered pair (a, b). The ordered pair (a, a) is a valid pair, since the pair elements need not be distinct. In the ordered pair (a, b), the element a is first and element b second. The pair (a, b) is different than (b, a). The ordered **pair** (a, b) is not the same as the **set** $\{a, b\}$, since $\{a, b\} = \{b, a\}$ but $(a, b) \neq (b, a)$. Two ordered pairs (a, b) and (c, d) are equal only when $a = c$ and $b = d$.

We can generalize the concept of ordered pairs to many ordered objects.

Definition: Let n be the natural number. If $a_1, a_2, \cdots, a_n$ are any n objects, not necessary distinct, then $(a_1, a_2, \cdots, a_n)$ is ordered n-**tuple**.

For each $i = 1, 2, \cdots, n$, a_i is the ith **component** of n-tuple $(a_1, a_2, \cdots, a_n)$. The element a_{i-1} is placed before the element a_i in n-tuple.

The ordered n-tuples are also denoted as n-dimensional vectors

$$\mathbf{a} = \begin{bmatrix} a_1 \\ a_2 \\ \vdots \\ a_n \end{bmatrix} \qquad (D.37)$$

An ordered tuple m-tuple $(a_1, a_2, \cdots, a_m)$ is the same as n-tuple $(a_1, a_2, \cdots, a_m)$ if and only if $m = n$ and $a_i = b_i$ for $i = 1, 2, \cdots, n$.

The ordered 2-tuple is the same as the ordered pair.

6.2. Cartesian Product

Definition: Let A be the set with the elements $\{a_1, a_2, \cdots\}$, and B the set with the elements $\{b_1, b_2, \cdots\}$. The **Cartesian** product of two sets A and B, denoted by $A \times B$, is the set of all ordered pairs (a, b) of elements $a \in A$ of the set A and elements $b \in B$ of the set B.

$$A \times B = \{(a, b) : a \in A, b \in B\} \qquad (D.38)$$

In notation of the pair (a, b) we understand that the first element a comes from the set A and the second element b from the set B.

The Cartesian product $A \times B$ is not equal to the $B \times A$, since the order of the elements in the pairs is important

Example: The Cartesian product $A \times B$ of two sets

$$A = \{1, 2\} \text{ and } B = \{3, 5\}$$

is the set

$$A \times B = \{(1, 3), (1, 5), (2, 3), (2, 5)\}.$$

The Cartesian product $B \times A$ of sets

$$A = \{1, 2\} \text{ and } B = \{3, 5\}$$

is the set

$$B \times A = \{(3, 1), (3, 2), (5, 1), (5, 2)\}.$$

The Cartesian product $A \times B$ of two sets

$$A = \{blue, red\} \text{ and } B = \{grey, brown\}$$

is the set

$$A \times B = \{(blue, grey), (blue, brown), (red, grey), (red, brown)\}.$$

The Cartesian product may be generalized to more than two sets.

Definition: The Cartesian product of n sets $A_1, A_2, \cdots, A_n$, denoted by $\prod_{i=1}^{n} A_i$, is defined as a set of n-tuples $(a_1, a_2, \cdots, a_n)$

$$\prod_{i=1}^{n} A_i = \{(a_1, a_2, \cdots, a_n) : a_i \in A_i \text{ for } i = 1, 2, \cdots, n\}. \tag{D.39}$$

6.3. Cartesian Product of Sets of Real Numbers

If $\mathbb{R}$ is the set of all real numbers (representing the real line), then the Cartesian product $\mathbb{R} \times \mathbb{R}$, denoted by $\mathbb{R}^2$ (called the real plane), is the set of all ordered pairs (x, y) of real numbers. The set $\mathbb{R}^2 = \mathbb{R} \times \mathbb{R}$ is the set of all Cartesian coordinates of points in the plane.

Example: Let $\mathbb{R}^2 = \mathbb{R} \times \mathbb{R}$. We may define a region (subset of points in $\mathbb{R}^2$) in $\mathbb{R}^2$ by defining the Cartesian product.

For example, let the set

$$A = \{x : x \in \mathbb{R}, \ 0 \le x \le 1\}$$

represent the line segment $[0, 1]$ (horizontal axis), and the set

$$B = \{y : y \in \mathbb{R}, \ 1 < y < 2\}$$

the line segment $[1, 2]$.

The set of points on the square and in the interior of the square adjacent to the vertical axis $x = 0$ in $\mathbb{R}^2$ (on the real plane) may be defined as the Cartesian product (Figure D.11)

$$A \times B = \{(x, y) : x \in \mathbb{R}, y \in \mathbb{R}, \ 0 \le x \le 1, 1 \le y \le 2\}.$$

The square adjacent to the horizontal axis $y = 0$ in $\mathbb{R}^2$ (on the real plane) may be defined as the Cartesian product (see Figure D.12)

$$B \times A = \{(y, x) : y \in \mathbb{R}, x \in \mathbb{R}, \ 1 \le y \le 2, 0 \le x \le 1\}.$$

594   6. Cartesian Product of Sets

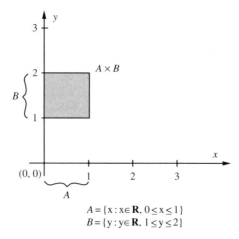

$A = \{x : x \in \mathbb{R}, 0 \leq x \leq 1\}$
$B = \{y : y \in \mathbb{R}, 1 \leq y \leq 2\}$

Figure D.11. $A \times B = \{(x, y) : x \in \mathbb{R}, y \in \mathbb{R}, 0 \leq x \leq 1, 1 \leq y \leq 2\}$

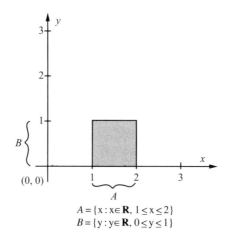

$A = \{x : x \in \mathbb{R}, 1 \leq x \leq 2\}$
$B = \{y : y \in \mathbb{R}, 0 \leq y \leq 1\}$

Figure D.12. $A \times B = \{(x, y) : x \in \mathbb{R}, y \in \mathbb{R}, 1 \leq x \leq 2, 0 \leq y \leq 1\}$

The region in the interior of a unit square centered at the origin $(0, 0)$ and on the circle on the real plane may be defined as a Cartesian product (see Figure D.13)

$$A \times B = \{(x, y) : x \in \mathbb{R}, y \in \mathbb{R}, -1 \leq x \leq 1, -1 \leq y \leq 1\}.$$

If $\mathbb{R}$ is the set of all real numbers (representing the real line), then the Cartesian product $\mathbb{R} \times \mathbb{R} \times \mathbb{R}$, denoted by $\mathbb{R}^3$ (called the real 3-dimensional space), is the set of all ordered triplets (x, y, z) of real numbers. The set $\mathbb{R}^3 = \mathbb{R} \times \mathbb{R} \times \mathbb{R}$ is the set of all Cartesian coordinates of points in the real 3-dimensional space.

Example: Let $X = \{x : x \in \mathbb{R}, 0 \leq x \leq 1\}$, $Y = \{y : y \in \mathbb{R}, 0 \leq y \leq 1\}$, and $Z = \{z : z \in \mathbb{R}, 0 \leq z \leq 1\}$. Thus in the 3-dimensional real Cartesian space we can define the cube region by providing the Cartesian product

$$X \times Y \times Z = \{(x, y, z) : x \in \mathbb{R}, y \in \mathbb{R}, z \in \mathbb{R}, 0 \leq x \leq 1, 0 \leq y \leq 1, 0 \leq z \leq 1\}.$$

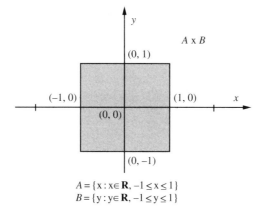

Figure D.13. $A \times B = \{(x, y): x \in \mathbb{R}, y \in \mathbb{R}, -1 \leq x \leq 1, -1 \leq y \leq 1\}$

We may generalize the real n-dimensional space. If $\mathbb{R}$ is the set of all real numbers (representing the real line), then the Cartesian product $\prod_{i=1}^{n} \mathbb{R}$, denoted by $\mathbb{R}^n$ (called the real n-space), is the set of all ordered n-tuples $(x_1, x_2, \cdots, x_n)$ of real numbers.

Since an n-tuple may be viewed as an n-dimensional vector, we say that the n-dimensional vector $\mathbf{x}$ is defined in $\mathbb{R}^n$ if

$$\mathbf{x} = \begin{bmatrix} x_1 \\ x_2 \\ \vdots \\ x_n \end{bmatrix} \in \mathbb{R}^n \tag{D.40}$$

Example: The vector $\mathbf{x}$ is defined in $\mathbb{R}^4$

$$\mathbf{x} = \begin{bmatrix} 1.5 \\ -6 \\ 6 \\ 0 \end{bmatrix} \in \mathbb{R}^4$$

6.4. Laws of Cartesian Product

If A, B, C, and D are sets, then

1) $A \times B = \emptyset$ if and only if $A = \emptyset$ or $B = \emptyset$
2) $(A \times B) \bigcup (C \times B) = (A \bigcup C) \times B$
3) $(A \times B) \bigcap (C \times D) = (A \bigcap C) \times (B \bigcap D)$

6.5. Cartesian Product of Binary Numbers

Let us consider the set B of two distinct objects denoted by 0 and 1

$$B = \{0, 1\}. \tag{D.41}$$

The objects may represent **binary numbers** or two constant elements ("0" and "1") in the Boolean algebra.

The Cartesian product of sets B and B consist of all $2^2 = 4$ possible pairs of two binary digits 0 and 1.

$$B \times B = \{0, 1\} \times \{0, 1\} = \{(0, 0), (0, 1), (1, 0), (1, 1)\}$$

Similarly, the binary (Boolean) n-tuple $(b_1, b_2, \cdots, b_n)$ (or n-dimensional **binary vector b**) consists of ordered elements $b_i \in B$ ($i = 1, 2, \cdots, n$).

The Cartesian product $\Pi_{i=1}^{n} B_i$ of n sets $B_i = B$ ($i = 1, 2, \cdots, n$) is defined as

$$\Pi_{i=1}^{n} = \{(b_1, b_2, \cdots, b_n) : b_i \in \{0, 1\} \; for \; i = 1, 2, \cdots, n\} \tag{D.42}$$

and has 2^n binary n-tuples $(b_1, b_2, \cdots, b_n)$.

It is easy to observe that we have 2^n possible instances of binary n-tuple.

Example: For example to define all possible instances of the binary 3-tuple (b_1, b_2, b_3), we compute the Cartesian product

$$\begin{aligned} B \times B \times B &= \{0, 1\} \times \{0, 1\} \times \{0, 1\} \\ &= \{(0, 0, 0), (0, 0, 1), (0, 1, 0), (0, 1, 1), \\ &\quad (1, 0, 0), (1, 0, 1), (1, 1, 0), (1, 1, 1)\} \end{aligned}$$

Of course this Cartesian product consists of $2^3 = 8$ binary triplets (equal to the number of all possible binary triplet instances).

7. Partition of a Nonempty Set

Definition: A **partition** of a nonempty set A is a subset Π of the power set 2^A of a set A, such that $\emptyset$ is not an element in Π and such that each element of A is in one and only one set in Π.

In other words, the set Π constitutes a partition of a set A if a set Π is a set of subsets of A such that

1) each element of the set Π is nonempty
2) distinct members of Π are disjoint
3) the union of all members (which are sets) of the set Π is equal to a set A ($\bigcup \Pi = A$)

Example: Let us consider the set $A = \{a, b, c, d\}$. The $2^4 = 16$ element power set of A is

$$\begin{aligned} 2^A = &\{\{a\}, \{b\}, \{c\}, \{d\}, \{a, b\{a, c\}, \{a, d\{b, c\}, \{b, d\}, \{c, d\}, \\ &\{a, b, c\}, \{a, b, d\}, \{a, c, d\}, \{b, c, d\}, \{a, b, c, d\}, \{\ \}\} \end{aligned}$$

The set $\Pi_1 = \{\{a, b\}, \{c\}, \{d\}\}$ is a partition of the set A. However, the set $\Pi_2 = \{\{a, b, c\}, \{c, d\}\}$ is not a partition, because its members are not disjoint.

Index

Absolute refractory period, 424
Abstraction, 87, 257, 490
Accuracy, 478
Addition, vector, 505
Adjacent matrix, 538
Adjoint, 528
Agglomerative approach, 260
Aggregate functions, 101
Akaike information criterion (AIC), 476
AIC methods, 485
Algorithm, speed up, 39
    processing, 127, 129
Anand and Buchner's eight-step model, 21
Ancestor, 303
Angle, 512
    between two lines, 569
Antimonotone property, 296
Approximate search, 455
Approximation-generalization dilemma, 65
Apriori
    algorithm, 296, 304
    property, 295
Area
    under curve (AUC), 483
    of triangle on plane, 569
Artificial neural networks, 419
AS SELECT query, 102
Assessments of data quality, 45
Association rule, 290, 304
    confidence, 290
    hierarchy, 291
    mining, 304
        algorithms, 304
    multidimensional, 291
    multilevel, 291
    single-dimensional, 291
    single-level, 291
    support, 290
Attributes, 31
Augmented objective function, 495
Autocorrelation, 360
Axis, vertical, 517
Axon, 421

Backward selection, 226
Bag-of-words model, 464
    inverted index, 456
    search database, 456
Basis, 516
    function, 434
    vectors, 506
Bayes' decision, 311
    for multiclass multifeature objects, 313
Bayesian classification rule, *see* Bayes' decision
Bayesian information criterion (BIC), 476, 477
    method, 485
Bayesian methods, 307
Bayes risk, 315
Bayes' theorem, 310, 313, 553
Between-class scatter matrix, 148, 149, 217
Bias, 471
Biased estimators, 472
Bias-variance
    dilemma, 209, 471
    tradeoff, 209
Binary features
    nominal, 28
    ordinal, 28
Binary numbers, 595
Binary vector, 596
Biological neuron, 449
Biomimetics, 419
Bitmap
    indexing, 124
    join index method, 129
Black box models, 444
Blind source separation (BSS), 157, 228
Bootstrap, 474
    samples, 474
Bottom-up mode, 260
Boundary point (BP), 246
Branches, 388

Cabena's five-step model, 21
Cardinality, 582
Cartesian product, 592
Case-based learning, 384
Centroid, 169
Certainty factors, 443
Characteristic equation for matix, 530
Characteristic polynomial, 530
Children nodes, 388

598 Index

Cios's six-step model, 21
Class, likelihood of, 310
Class-attribute
   interdependence redundancy, 239
   interdependence uncertainty, 239
   interdependency maximization algorithm, 240
   mutual information, 239
Class conditional probability density function, 309, 313
Classification, 49, 65
   boundary, 61
   error, 61
Classifiers, 470
   feedback methods, *see* closed loop method
   two-class and many-class problems, 55
Classifying hyperplane, 577
Class interference algorithm, 439
CLIP4 algorithm, 400
Closed half-spaces, 576
Closed interval, 584
Closed loop method, 211
Cluster, 36, 171
   validity, 281
Clustering, 164
   mechanisms, 50
Cocktail-party problem, 155
Codebook, 164, 165, 167, 279
Codevectors, 164
Coefficient
   of determination, 361
   of multiple determination, 371
Cofactor, 528
Collaborative clustering, 491, 492–494
Column vector, 506, 508
Compactness, 281
Competitive learning, 173
Complement, 550
   set, 589
Complete link method, 261
Complex numbers, 583
Component density, 340
Composite join indices, 125
Computer eye, 274
Concept
   hierarchy, 112, 129
   learning, 388
Conceptual clustering, 384
Conditional entropy, 214
Conditional risk, 315
Cone, 577
Confidence measure, 443
Conformable matrices, 520
Confusion matrix, 457, 477
Consensus, clustering, 491, 498
Context-oriented clustering, 448
Contingency matrix (table), 300, 477
Continuous quantitative data, 69
Continuous random variable, 554
Convex, 577
Convolution, 188
Correlation coefficient, 300, 360

Cosine measure, 461, 464
Countable (denumerable) set, 582
Covariance matrix, 134, 136
Cover, 382
CREATE VIEW, 102
CRISP-DM model, 21
Criterion variable, 348
Cross product, 510
Cross-validation, 473, 485
   10-fold, 473
   n-fold, 473, 474
Cryptographic methods, 489, 499
Cumulative probability, 556
CURE, 272

Data
   amount and quality of, 27
   cleaning, 45
   cube, 33, 112, 129
   design matrix, 369
   distortion, 489, 499
   indexing, 124, 129
   items, 289
   large quantities of, 44
   mart, 108, 129
   matrix, 156
   perturbation, *see* data distortion
   preparation step, 21
   privacy, 499
   quality problems
      imprecision, 44
      incompleteness, 44
      noise, 44
      redundancy missing values, 44
   randomization, *see* data distortion
   retrieval, 95, 454
   reuse measures, 471, 485
   sanitation, 489, 499
   security, 499
   sets, 27, 29, 44
      partitioning, 297, 304
   similarity or distance between, 50
   storage techniques, 27
   types of data, 27
      binary (dichotomous), 44
      categorical (discrete), 44
      continuous, 44
      nominal (polytomous), 44
      numerical, 44
      ordinal, 44
      symbolic, 44
   vector, 368
   visualization, 349
Database, 27, 31, 44, 128, 454
   populating, 99
Database management systems (DBMS), 31, 95, 128
Data control language (DCL), 98, 129
Data definition language (DDL), 98, 129
Data manipulation language (DML), 98, 129
Data mining (DM), 9, 10, 16, 127, 129

algorithms
  incremental, 40
  nonincremental, 40
  definition, 3
  domain, 5
  privacy in, 489–490
  RDBMS, 99
  scalability, 44
Data warehouses (DW), 27, 32, 44, 95, 128, 129
Data warehousing, 106
Davies-Bouldin index, 281
Decision
  attribute, 382
  boundaries, 318
  layer, 342
  regions, 318
  surfaces, 318
  trees, 59, 74, 250, 388, 416
Declarative language, 98
Decoding, 166, 280
Deduction, 384
Default hypothesis, 397
Degree of freedom, 367, 470
  of subspace, 516
DELETE, 104, 105
Demixing (separation) matrix B, 160
DeMorgan's theorem, 590
Dendrites, 421
Dendrogram, 260
Density-based clustering, 271
Dependent variable, 348
Determinant, 523, 526
Diagonal matrix, 521, 532
Diameter of cluster, 282
DICE, 116, 118
Dichotomizer, 321
Difference
  distortion measures, 168
  of two sets, 588
Differential entropy, 161
Dilations, 194
Dimension, 526
  of subspace, 516
  tables, 109
Dimensionality reduction, 146
Direction
  numbers, 570
  of vector, 508
Discrete
  random variable, 554
  value, 236
    nominal, 236
    numerical, 236
  wavelets, 195
    transformation, 197
Discrete Fourier transform (DTF), 181, 183, 229
  inverse Fourier transform, 181
Discretization, 28, 235

algorithim, 28
  supervised, 235, 236, 237
  unsupervised, 235, 236, 237
Discriminant functions, 319
Distance, 42, 485
  from point, 569
  from a point to a plane, 574
Distortion measures, 164
Distributed clustering, 500
Distributed databases, 493
Distributed dot product, 490
Divide-and-conquer
  method, 299
  strategy, 388
Divisive approach, 260
Document, 454, 464
Domain knowledge, 469
Dot product, 509
DRILL DOWN, 116, 117–118
Dunn separation index, 282
Dynamic
  attribute discretization, 235
  data, 44
  discretization algorithm, 253
  model, 347

Eigenvalues, 134, 155, 529, 530
  problem, 137
Eigenvectors, 134, 155, 529, 530
Eight-step model, Anand and Buchner, 21
Elements, 548, 579
Empty set (null set), 548, 581
Encoding, 166, 280
  regions, 165
Energy, 187
Enterprise warehouse, 108, 129
Entity-relational (ER) data model, 31
Entropy, 161, 214, 341
Equal sets, 585
Equal within-class covariance matrices, 323
Equation in intercept form, 573
Equation of plane, 572, 573
Estimator, 470, 472
Euclidean distance, 168
Euclidean matrix norm, 539
Event, 547
Excitation matrix, 370
Expected value, 563
  of random variable, 557
Explanation-based learning, 384
Explanatory variable, 348
Extension, principle, 79
Extraction method, 133

Fact
  constellation, 110
  table, 109
"Failure," 560
FASMI test, 129

Fast analysis of shared multidimensional information (FASMI), 126
Fast Fourier transform (FFT), 182–183, 229
Father wavelet, 194
Fayyad's nine-step model, 21
Feature
    extraction, 133
    goodness criterion, 207
    scaling, 228
    selection, 133, 207
    selection criterion, 212
    space, 136
    subset, 207
    vectors, 134
Feature/attribute, 44
    value of, 27
Filter, *see* open loop methods
Finite set, 581
First-generation systems, 21
Fisher F-ratio, 148, 218
Fisher's linear discriminant method, 146, 228
Five-step model, Cabena, 21
Flat files, 29
Flat (rectangular) files, 27, 44
F-measure, 480
10-fold cross-validation, 473
Folding frequency, 182
Formal measures, 485
Forward selection, 224
Fourier spectrum, 179, 186
Fourier transform, 178, 229
    and wavelets, 133
Frequency
    domain, 177, 229
    increment, 181
    spectrum, 184
    variable, 178
Frequent itemset, 294
Frequent-pattern
    growth (FP growth), 299
    tree (FP-tree), 299
Front end methods, *see* open loop methods
Full cube materialization, 124
Full stepwise search, 227
Fuzzy
    C-Means clustering, 269
    context, 447
    numbers, 79
    production rules, 445
    relations, 80
    sets, 78

Gain ratio, 253, 388
Galaxy schema, 110, 111, 129
Gaussian basis function, 439
Gaussian probability distribution, 564
General equation of plane, 572
Generalization, 63, 386
    error, 470
    threshold, 397

General linear equation, 201, 204, 229, 568
    magnitude of, 192
General moments, 203
General quadratic distortion, 168
Geometric distribution, 560
Gini
    coefficient, 484
    index, 389
Global, 236
Goodness of prediction, 470
Granular computing, 76, 89
Granular information in rule-based computing, 88
Graph structure, 89
Grid-based clustering, 272
Group average link, 261

Haar transform, 198
Half-open interval, 584
Half-ray, 577
Hashing, 297, 304
Head, 548
Hebb's rule, 430
Hessian of scalar function, 536
Heterogeneous databases, 37
Heuristics, 471
    measures, 485
Hierarchical clustering, 50, 258
Hierarchical topology, 265
Hierarchy, 69
Horizontal clustering, 492
Horizontal modes, 492
Horizontal plane, vectors, 517
Hot deck imputation, 42
Hotelling, 139
Human-centricity, 76
Hybrid(s), 416
    dimension association rule, 303
    inductive machine learning algorithms, 399
    models, 493
    OLAP (HOLAP), 129
Hyperboxes, 272
Hypercube-type Parzen window, 335
Hyperlinks, 37
Hyperplanes, 367, 516, 575
    classifiers, 437
Hypertext, 34, 44

Identity matrix, 522
Ill-posed problems, 431
Image
    normalization, 202
    translation, 206
Imprecise data objects, 40
Inclusion operation, 587
Incomplete data, 40
Inconsistency rate, 216
Incremental data mining methods, 44
Independent components, 154, 228
    vectors, 160

Independent component analysis (ICA), 133, 154, 228
Independent variable, 348
Indices, 98, 524
  terms, 459
Induced partition matrices, 494
Induction, 384
Inductive bias, 389
Inductive machine learning, 384, 416
  process, 382, 383
Inductive *versus* deductive, 383
Infinite set, 581, 583
Information
  gain, 253, 388
  granularity, 491, 499
  granulation, 29, 89
  granules, 71, 76, 89
  processing, 127, 129
  retrieval (IR), 453, 454–462, 464
    approximate search, 454
    core IR vocabulary, 454
    inner multiplication, 511
    relevance, notion of, 454
    semistructured, 464
    unstructured, 464
  theoretic approach, 253
Inner product, 509, 511, 513
  space, 510
Input
  layer, 342
  variable, 348
INSERT, 105
Integer
  numbers, 582
  programming (IP), 401
Integrate-and-fire model, 420, 421
Interclass separability, 216
Interdimension association rule, 303
Interestingness measures, 300, 469, 471, 484, 485
Interior, 584
Internal state, 347
Intersection, 587
Interval, 584
Inverse
  continuous wavelet transform, 200
  discrete Fourier transform (DTF), 181
  discrete wavelet reconstruction, 201
  document frequency, 460
  Fourier transform, 178
  transform, 207
Inverted index, 464
Irrelevant data, 41
IR systems, 455
  approximate search, 455
Items, 35
Itemset, 293
  mining, 304

Join indexing, 124, 125, 129
Join operation, 111
Jordan

block, order, 544
canonical form, 544

Karhunen-Loéve (KLT), 139
KDP, *see* knowledge discovery processes (KDP)
Kernel
  -based method, 334
  classifiers, 437
  function, 266
  (window) function, 335
Key, 31, 96
K-fold, 473
K-itemset, 293
K-means clustering, 264
K-medoids clustering algorithm, 266
K-nearest neighbors, 336
  classification rule, 337
  method, 336
Knowledge
  conclusion, 71
  condition, 71
  discovery in databases archive, 45
  extraction, 10
  representation, 89
    graphs, 89
    networks, 89
    rules, 89
  representation schemes, 71
  retrieval, 454
Knowledge-based clustering, 500
Knowledge discovery processes (KDP), 9, 10, 20
  data preparation step, 21
  iterative, 20
  multiple steps, 20
  sequence, 20
  standardized process model, 9–10
Kullback-Leibler distance, 341
Kurtosis of random variable, 161

Latent semantic analysis, 454, 507
Latent semantic indexing, 462–463, 464
Latent variables, 155
Lattice of cuboids, 115
LBG centroid algorithm, 170
Learner, 382
Learning, 49
  algorithms, 61
  by analogy, 384
  from examples, 383
  with knowledge-based hints, 54
  phase, 382
  rule, 275, 419, 420, 428, 449
  vector quantization, 229
Learning vector quantization (LVQ) algorithm, 173
Least squares error (LSE), 354
Leave-one-out, *see* cross-validation, n-fold
Leaves, 388
Level-by-level independent method, 302
Level-cross-filtering

by k-itemset method, 302
by single item method, 302
Level of collaboration, 497
Likelihood, 331
Line, passes a point, 574
Linear classifier, 59
Linear combination, 514
   non-trivial, 514
Linear dependent vector, 514
Linear discriminant function, 323
Linear equation, 575
Linearly independent, 514
Linear manifolds, 516
Linear transformation, 540
Line equation, 571
Line segment, 571, 577, 583
   equation, 571
Linguistic preprocessing, 453, 456, 464
Local frequent itemsets, 297
Logic operators, 79
Loss matrix, 314
LVQ1, 174

Machine learning, 381
   concept, 382
   hypothesis, 382
   ineductive *versus* deductive, 383
   repository, 45
Mahalanobis distortion, 168
Main diagonal, matrix, 521
Manual inspection, 43
Market-basket analysis, 289
Materialization of cuboids, 124, 129
Materialized cuboids, selection of, 129
Matrix, 519
   block diagonal, 545
   characteristic equation, 530
   determinant, 523
   eigenvectors, 544
   exponential, 533
   main diagonal, 521
Maximal class separability, 208
Maximum likelihood
   classification rule, 317
   estimation, 270, 331
M-dimensional vectors, 522
Mean imputation, 42
Median, 266
Membership functions, 79
Memorization, 64
Messages, 35
Metadata repository, 109, 124
Metrics, 434
Minimal representation, 208
Minimum
   concept description paradigm, 214
   construction paradigms, 208
   description length, 208
   distance classifier, 329
   Euclidean distance classifier, 329

Mahalanobis distance classifier, 327
   message length, 208
   support count, 294
Minimum description length (MDL)
   principle, 475
Mining itemsets, 304
Minkowski
   distance, 51
   norm, 168
Minor, 524
Misclassification error, 389
Misclassification matrix, 477
Missing values, 41
Mixing
   matrix, 154, 228
   parameter, 340
Mixture models, 340
Model, 470, 472
   assessment, 469, 485
   based algorithms, 269
   based clustering, 258
   degree of freedom, 470
   error, 470
   in multiple linear regression, 353
   selection, 470, 485
   structure, 352
Moment of continuous random
   variable, 564
Moments, concept of, 201
Monotonic function, 439
Monte Carlo techniques, 227
Mother
   function, 194
   Haar wavelets, 198
Multidimensional association rules, 303, 304
Multidimensional data model, 112, 129
Multidimensional OLAP (MOLAP), 129
Multilevel association rules, 302, 304
Multimedia
   data, 34, 44
   databases, 36
Multiple correlation, 372
Multiple imputations method, 45
Multiplicative rule of probability, 552
Mutual information (MI), 214, 215
Mutually exclusive event, 551

Natural logarithmic, 321
Natural numbers, 582
N-dimensional vector space, 575
Nearest neighbor
   approximation, 204
   classification, 337
   classification rule, 337
   classifier, 58
   selection rule, 169
Negentropy, 161
Neighbor function, 275
Networks, 75
Neural network

algorithms, 449
topologies, 431, 449
    feedforward, 431
    recurrent, 431
Neuron(s), 419, 421
    at data point (NADP) method, 436
    radii, 438
N-fold cross-validation, 474
Nine-step model, Fayyad, 21
No cube materialization, 124
Noise, 42
    threshold, 411
Nominal qualitative data, 69
Nonequal sets, 585
Nonincremental learning, 383
Nonlinear projection, 276
Nonparametric methods, 330, 333
Nonsingular matixes, 529
Non-trivial, 514
Normal curve, 564
Normalized images, 206
N terms, 505
N-tuples, 505
Nucleus, 421
Number of inversions, permutation, 524
Number of subsets, 587
N-vector, 505
Nyquist frequency, 182

Object feature variable of, 309
Objective function, 263, 269
Object-oriented databases, 35
Object-relational databases, 35
Occam's razor, 208, 474, 485
One-dimensional Fourier transform, 229
One-rule discretizer, 250
On-line analytical processing (OLAP), 34, 95, 107,
        116, 128
    commands, 116, 129
    queries, 129
On-line transaction processing (OLTP),
        107, 129
Ontogenic neural networks, 420, 431
Open half-spaces, 576
Open loop methods, 211
Optimal feature selection, 208
Optimal parameters, 357, 369
    values, in minimum least squares sense, 360
Order, of model, 368
Ordered pair, 592
Ordinal qualitative data, 69
Orthogonal, 513, 516
Orthogonal basis, 518
Orthogonal complement, 517
Orthogonal least squares (OLS)
    algorithm, 436
Orthogonal projection, 517
Orthogonal vectors, 509, 513
Orthonormal, 518
Outer product, 510, 511

Output layer, 342
Overfitting, 470

Parametric line equation, 571
Parametric methods, 330
Parsimonious, 476
Parsimonious model, 470
Partial cube materialization, 124, 129
Partial supervision, 54
Partition, 257, 388
    coefficient, 283
    data set, 40
    entropy, 283
    matrices, 264, 490
Partitioning, 304
Part-of-speech tagging, 456
Parzen window, 334
Pattern, 133
    dimensionality reduction, 207
    layer, 342
    processing, 344
    recognition, 65
Performance
    bias, see closed loop method
    criterion, 353
    index, 263
Periodicity and conjugate symmetry, 187
Permutation, 524
Perpendicular to a plane, 574
Phase
    angle, 179
    spectrum, 186
PIVOT, 116, 118–119
Pixels, 202
Plane equation, normal form of, 574
P-nearest neighbors method, 438
Point-slope equation, 567
Poisson ($\wp$)distribution, 561
Polysemy problem, 454
Position vector, 507
Posteriori (posterior) probability, 310, 313, 552
Postsynaptic neuron, 421
Postsynaptic potential (PSP), 421
    modification, 429
        excitatory (EPSP), 429
        inhibitory (IPSP), 429
Power set, 587
Power spectrum, 179, 186
Precision, 455, 457, 464
Prediction accuracy, 210
Predictor variable, 348
Preset bias, see open loop methods
Presynaptic neuron, 421
Principal components, 140
Principal component analysis (PCA), 133, 134, 228
Principal eigenvectors, 140
Priori probability (prior), 308, 313
Privacy issues, 500
Privileges, 99
Probabilistic neural network (PNN), 342, 344

normalized patterns, 345
radial Gaussian Kernel, 344
radial Gaussian normal kernel, 345
Probability, 547–548
   distribution, 555
      of discrete random variable, 561
   theory axioms, 549
Probability density function, 160, 562
Procedural language, 98
Process model, 9, 21
   independence, 21
Projection operation, 111
Proper subset, 586
Prototypes, 263, 490
Proximity hints, 54
Proximity matrix, 499
Pruning techniques
   postpruning, 391
   prepruning, 391
Pruning threshold, 397, 411
Pseudoinverse, 529
Pseudo-inverse of matrix, 370

Quadratic discriminant, 322
Quadratic form, 535
Qualitative data, 69
Quality, linear regression model and linear correlation analysis, 360
Quanta matrix, 238, 253
Quantitative association rules, 303, 304
Queries, 96, 129, 454, 464
   optimization, 105, 129
   processor, 97, 128

Radial basis function (RBF)
   networks, 431, 434, 449
Random variable, 553
Rank of matrix, 526
Raster format, 36
Rational numbers, 583
Realization, 554
Real numbers, 505, 583
Real scalar, 507
Recall, 455, 457, 458, 464
Receiver operating characteristics (ROC), 481
Reduced support, 302
   based methods, 304
Reduction of dimensionality, 133–134
Redundant data, 41
Regions, 584
Regression, 49, 57, 67
   analysis, 347
   equation (regression model), 348
   errors, 353, 367
   line, 348
   model
      computing optimal values of, 356
      sum of squared variations, 355

Regularization theory, 431
Reinforcement learning, 53
Rejection, 312
Relational database, 31
   management system (RDBMS), 96, 109, 125
Relative refractory period, 424
Relevance, 209, 210, 454, 455, 478
   documents, 457
   feedback, 454, 463, 464
      mechanism, 464
Removal of transactions, 304
Repositories, 45
Resampling, 485
Retrieved documents, 457
Robustness of clustering, 267
ROC curves, 485
ROLL UP, 116–117
Roster, 580
Rotation, 202
Rotational invariance, 187, 201, 229
Rote learning, 384
Rough sets, 84
Row vectors, 508
Rubella disease, 548
Rule, 382
   algorithms, 393, 416
   learners, 393
   refinement, 484

Sammon's projection method, 278
Sample space, 548
Sampling, 284, 297, 304
   frequency, 180, 181
   period, 180
   rate, 180
   time, 180
Scalability, 289
Scalable algorithms, 38, 45
Scalar
   Hessian, 536
   multiplication, 505
   trivial combinations of, 514
   vector, 509
Scale-invariance, 201, 203
Scale normalization, 206, 207
Scaling, 187, 202
   function, 196
Scatter
   matrices, 146
   plot, 349
Schema, 31, 97, 98, 109
Search
   database, 464
   methods, 133
   procedure, 212
Second-generation systems, 22
Second type fuzzy sets, 81
SELECT, 99, 100

Index

Selection
  criteria, 133
  operation, 111
Self-organizing feature maps, 274
Semiparametric methods, 330, 338
Semi-structured data, 453
Sensitivity, 478, 485
Separability, 187, 281
Sets
  complement, 589
  covering, 401
  difference, 588
  power, 587
  proper subset, 586
  subset, 586
  of transactions, 292
  union of two, 588
Set theory, 579
Shadowed sets, 82
Shannon's entropy, 239, 247, 388
Signal-to-distortion ratio (SDR), 171
Similarity
  concept of, 258
  measure, 280
Similarity transformation, 541
Simple linear regression analysis, 351, 356
Single link method, 261
Singleton set, 581
Singular matrixes, 529
Singular value decomposition (SVD), 133, 153, 228, 462
Six-step model, Cios, 21
Skewed basis, 518
Skewed datasets, 397
SLICE, 116, 118
Smoothing parameter, 336
Snowflake schema, 110, 111, 129
Spanning subspace, 515
Spatial data, 34, 44
Spatial databases, 36
Specialization, 386
Specificity, 478
  analyses, 485
Specificity/generality, 87
Spectral coefficients, 178
Spectral density, 186
Spectrogram, 183
Spectrum, 177
Sphering, 158
Spiking neuron model, 421, 449
Split information, 389
Splitting algorithm, 171
SQL, *see* Structured Query Language (SQL)
Squared errors, sum of, 353
Standard basis, 518
Standard deviation, 558, 564
Standardized process model, 9–10
Standard line equation, 570
Star schema, 109, 111, 129
Static attribute discretization, 235
Static model, 347

StatLib repository, 45
Stemming, 456, 464
Stopping criterion, 269
Stop threshold, 411
Stop words, 464
  removal of, 456
    conjunction, 456
    determiner, 456
    preposition, 456
Storage manager, 97, 128
Strong association rules, 293
Strong rules, 386
Structured algorithms, 384
Structured data, 69
Structured Query Language (SQL), 31, 95, 98, 128, 129
  commands
    from, 99
Structured Query Language (SQL) (*Continued*)
    select, 99
    where, 99
    DML, 99, 105
Subgaussian, 161
Subjective evaluation of association rules, 300
Subset, 586
Subspace, 515
  degree of freedom, 516
  dimension, 516
"Success," 560
Summation layer, 342
Supergaussian, 161
Supervised dynamic discretization, 251
Supervised Fisher's linear discriminant analysis, 133
Supervised learning, 52, 65, 146, 173
Supervised ML algorithms, 383
Support count, 293
Synapses, 421
Synaptic activity plasticity rule (SAPR), 429
Synaptic time-delayed plasticity (STDP), 429
Synonyms, 457, 464
System of fuzzy rules, 449

Tagging, part-of-speech, 456
Tail, 548
Targets, 209
Teacher/oracle, 382
Temporal data, 34, 44
Temporal databases, 36
Temporal spiking, 420
Term, 454
  frequency of, 460
Term-document matrix, *see* term-frequency matrix
Term frequency matrix, 453, 459, 464
Testing, 64
  phase, 382
Text databases, 36, 453
  unstructured, 36
Text mining, 453, 464

Text similarity measures, 454, 461
  cosine measure, 461, 464
Tf-idf
  measure, 460
  weighting, 453
Third-generation systems, 22
Three-dimensional scatter plot, 366
Three-valued logic, 83
Time-series, 133
Top-down approach, 260
Topological properties, 274
Topology, 420
Total data mean, 147, 149, 217
Total scatter matrix, 147, 217
Trace square matrix, 539
Training, 473
  data, 63
  data set, 383
Train of spikes, 421
Transaction, 35, 96, 129, 289
  manager, 97, 128
  removal, 297, 304
Transactional data, 34, 44
Transactional databases, 35
Translation, 194, 202, 207
  invariance, 201, 203
  and phase, 187
Transposition operation, 508
Tree-projection algorithm, 299
Trial and error method, 436
Triangle inequality, 512
Triangular matrix, 521
  lower triangular matrix, 521
  upper triangular, 521
Tuples, 31, 505, 592
Two-dimensional continuous Fourier transform (2DFT), 184, 185
  inverse of, 185
Two-dimensional continuous wavelet
  expansion, 200
Two-dimensional data, 346
Two-dimensional Fourier transform, 229
Two-dimensional wavelet transform, 200

Unbiased estimators, 472
Unconditional probability density
  function, 310, 313
Uncorrelatedness, 161
Underfitting, 470
Uniform support, 302
  based method, 304
Union of two sets, 588
Unit vector, 508
Universal approximators, 420
Universal set, 582
Unstructured algorithms, 384
Unsupervised data mining, 304
  method, 289
Unsupervised learning, 65, 135, 257, 384
  techniques, 164

Unsupervised ML algorithms, 383
UPDATE, 105

Validation set, 63
Values
  features, 27
    discrete (categorical) or continuous, 27
  numerical, 27
  symbolic, 27
Variables, 35
Variance, 472, 557
Vector, 505, 506
  addition, 507
  angle, 512
  component of, 506
  cross product, 510
  format, 36
  inner product, 509, 511, 513
  norm, 509, 511
  normal, 574
  outer product, 510, 511
  quantization (VQ), 164, 229, 279
  space, 505, 507
    model, 453, 459, 464
Vector's dimension, 506
Venn diagrams, 585
Vertical clustering, 493
Vertical modes, 492
Very simple spiking neuron (VSSN) model, 421, 424
Virtual data warehouse, 107
Vocabulary, 459
Voronoi
  cell, 167
  quantizer, 165
  tessellation, 165

Wavelets, 193, 229
  analysis, 229
  patterns, 201
  transform, 193
Weight, 421, 459
Weighted-squares distortion, 168
WHERE, 104
Whitening matrix, 158
Wilks' lambda, 218
Windowing, 392
Within-class scatter matrix, 147, 149, 217
Wrapper method, *see* closed loop method
WWW, 34, 37, 44

Xie-Benie index, 282
X-intercept, 567, 568

Y-intercept, 567, 568

Zero vector, 508

Printed in the United States of America.